Ordering in Two Dimensions

Ordering in Two Dimensions

Proceedings of an International Conference held
at Lake Geneva, Wisconsin, U.S.A., May 28–30, 1980

Editor:

SUNIL K. SINHA

Solid State Science Division
Argonne National Laboratory
Argonne, Illinois, U.S.A.

NORTH HOLLAND
New York • Amsterdam • Oxford

Published by:

Elsevier North Holland, Inc.
52 Vanderbilt Avenue, New York, New York 10017

Distributed by:

North-Holland Publishing Company
P.O. Box 104
Amsterdam, The Netherlands

Library of Congress Cataloging In Publication Data

Main entry under title:

Ordering in two dimensions.

Bibliography p.
Includes indexes.
1. Surface (Physics)—Congress. 2. Order—disorder models—Congress. I. Sinha, Sunil K.
QC173. 4. S94072 530.4'1 80-26060
ISBN 0-444-005811

Manufactured in the United States of America.

CONTENTS

SECTION II

A. PHYSISORBED AND CHEMISORBED SYSTEMS

C. THEORY OF 2-D PHASE TRANSITIONS

D. INTERCALATED MATERIALS

E. 2-D MAGNETISM

F. MOLECULAR MONOLAYERS, BILAYERS, MEMBRANES AND LIQUID CRYSTALS

G. 2-D SUPERCONDUCTIVITY AND TRANSPORT

PREFACE

Various aspects of ordering in two dimensions have become areas of intense theoretical and experimental interest over the last few years. For instance, the field of monolayer and submonolayer molecular films physisorbed on solid surfaces, particularly graphite surfaces, has been growing steadily, with a variety of experimental and theoretical techniques being brought to bear on the problem. It has been the topic of several conferences, symposia and summer schools in the last three years. Similarly, ordering in chemisorbed monolayers has been an increasingly important topic at meetings devoted to Surface Physics. There have also been several recent conferences devoted to intercalated compounds, inhomogeneous superconductors, molecular monolayers and liquid crystals where aspects of two-dimensional ordering have played an important role. In view of the growing recognition of the universal aspects of many of these phenomena and their connection with the dimensionality of the studied systems, it was felt that the time was ripe for an International Conference specifically devoted to the topic of ordering in two dimensions, regardless of the particular systems involved. It was our hope that such a conference would lead to increased contact and cross-stimulation between experimental physicists, chemists, biologists and theorists working in many of the above areas. It was also hoped that an increased appreciation of certain general unifying principles would emerge, or at least that the major issue of whether such unifying principles existed or not would be faced. It appears that this hope was indeed realized at this Conference. Two hundred and ten scientists from the U.S. and abroad participated in the Conference, and twenty-eight invited papers and seventy-eight contributed papers were presented. Many of these papers contained important new results which are presented for the first time in these proceedings. We hope that these proceedings remain useful to workers in the area for a long time as an outpouring of vigorous activity in a field which has reached maturity.

I wish to acknowledge with thanks the excellent work done by the other members of the Organizing Committee, consisting of B. M. Abraham, M. Bretz, M. B. Brodsky, M. W. Cole, S. Doniach, P. M. Horn, J. B. Ketterson, U. Landman, J. P. McTague, L. Passell, M. Schick and P. D. Vashishta, in helping to organize the scientific program, and the useful input from the members of the International Program Advisory Committee. Special thanks are due to P. D. Vashishta for the steadfast support received from the Solid State Science Division of Argonne National Laboratory.

The Conference undoubtedly owes much of its success to the organizational abilities of Miriam Holden and Stephen C. Smith, who helped take care of every little detail. Special mention should be made of Nora Meneghetti who worked perhaps the longest and hardest on this Conference since its conception. Valuable and experienced help was also provided by members of the Local Committee; George Crabtree, Lennox Iton, Torben Brun, Ken Miyano, Debra McCann, and Carole Kasper. I am also very grateful to our contingent of tireless student volunteers; Parul Vora, Peter Bancel, Cynthia McIntyre, Jagmohen Bajaj, Hugh Goodman, Doug Keszler and Andrew Kueny, who worked so hard during the meeting; to Ron Koopman for his excellent work with the audiovisual equipment and to the staff of the Abbey Hotel, Fontana, Wisconsin for their excellent cooperation. Special thanks are due to Jonathan Rotholz for his professional help in editing these proceedings, and to Bonnie Russell, Karen Beres and Delores Shutak for excellent secretarial help in typing some of the material.

Finally, I wish to acknowledge with gratitude the generous support provided by the agencies which sponsored this Conference: the U.S. Department of Energy and Argonne National Laboratory, the Argonne Universities Association, the Office of Naval Research, the Advanced Research Projects Agency, and the National Science Foundation.

Sunil K. Sinha
Argonne National Laboratory
June 1980

LIST OF PARTICIPANTS

B. M. Abraham
Argonne National Laboratory
Argonne, Illinois

F. F. Abraham
IBM Research Laboratory
San Jose, California

D. W. Allender
Kent State University
Kent, Ohio

A. A. Antoniou
National Research Council
Ottawa, Ontario, Canada

R. E. Avila
Case Western Reserve University
Cleveland, Ohio

J. Bajaj
Northwestern University
Evanston, Illinois

P. Bak
H. C. Orsted Institute
Copenhagen, Denmark

P. A. Bancel
Argonne National Laboratory
Argonne, Illinois

J. A. Barker
IBM Research Laboratory
San Jose, California

M. T. Béal-Monod
Université Paris-Sud
Orsay, France

M. Bender
Fairleigh Dickinson University
Teaneck, New Jersey

A. N. Berker
Massachusetts Institute of Technology
Cambridge, Massachusetts

S. Bhattacharya
University of Chicago
Chicago, Illinois

J. P. Biberian
University of Marseille
Marseille, France

R. J. Birgeneau
Massachusetts Institute of Technology
Cambridge, Massachusetts

A. R. Bishop
Los Alamos Scientific Laboratory
Los Alamos, New Mexico

D. J. Bishop
Bell Laboratories
Murray Hill, New Jersey

D. N. Bittner
University of Michigan
Ann Arbor, Michigan

E. B. Bradley
University of Kentucky
Lexington, Kentucky

M. Bretz
University of Michigan
Ann Arbor, Michigan

F. Bridges
University of California
Santa Cruz, California

W. Brinkman
Bell Laboratories
Murray Hill, New Jersey

M. Brodsky
Argonne National Laboratory
Argonne, Illinois

J. Broughton
Bell Laboratories
Murray Hill, New Jersey

M. Brown
Massachusetts Institute of Technology
Cambridge, Massachusetts

L. W. Bruch
University of Wisconsin
Madison, Wisconsin

T. O. Brun
Argonne National Laboratory
Argonne, Illinois

D. M. Butler
Carnegie-Mellon University
Pittsburgh, Pennsylvania

D. A. Cadenhead
State University of New York
Buffalo, New York

D. J. Callaway
University of Washington
Seattle, Washington

C. E. Campbell
University of Minnesota
Minneapolis, Minnesota

J. H. Campbell
University of Michigan
Ann Arbor, Michigan

F. X. Canning
University of Massachusetts
Amherst, Massachusetts

Y. J. Chabal
Bell Laboratories
Murray Hill, New Jersey

M. W. Chan
Pennsylvania State University
University Park, Pennsylvania

P. Choquard
Laboratoire de Physique Theorique-EPFL
Lausanne, Switzerland

D. D. L. Chung
Carnegie-Mellon University
Pittsburgh, Pennsylvania

R. Clarke
University of Michigan
Ann Arbor, Michigan

C. L. Cleveland
Georgia Institute of Technology
Atlanta, Georgia

M. W. Cole
University of Pennsylvania
University Park, Pennsylvania

C. A. Condat
University of Massachusetts
Amherst, Massachusetts

G. W. Crabtree
Argonne National Laboratory
Argonne, Illinois

S. B. Crary
Amherst College
Amherst, Massachusetts

J. G. Dash
University of Washington
Seattle, Washington

D. R. Day
Massachusetts Institute of Technology
Cambridge, Massachusetts

M. P. den Nijs
University of Chicago
Chicago, Illinois

R. D. Diehl
University of Washington
Seattle, Washington

M. J. DiPirro
National Bureau of Standards
Gaithersburg, Maryland

S. Doniach
Stanford University
Stanford, California

M. Dritschel
Northwestern University
Evanston, Illinois

P. Dutta
Argonne National Laboratory
Argonne, Illinois

R. E. Ecke
University of Washington
Seattle, Washington

P. M. Eisenberger
Bell Laboratories
Murray Hill, New Jersey

P. J. Estrup
Brown University
Providence, Rhode Island

S. C. Fain, Jr.
University of Washington
Seattle, Washington

A. J. Fedro
Northern Illinois University
De Kalb, Illinois

G. P. Felcher
Argonne National Laboratory
Argonne, Illinois

A. T. Fiory
Bell Laboratories
Murray Hill, New Jersey

D. S. Fisher
Bell Laboratories
Murray Hill, New Jersey

R. M. Fleming
Bell Laboratories
Murray Hill, New Jersey

S. E. Friberg
University of Missouri
Rolla, Missouri

F. M. Gasparini
State University of New York at Buffalo
Amhurst, New York

A. Georgallas
St. Francis Xavier University
Antigonish, Nova Scotia, Canada

R. E. Glover, III
University of Maryland
College Park, Maryland

R. Gomer
University of Chicago
Chicago, Illinois

W. H. Goodman
Northwestern University
Evanston, Illinois

J. F. Gouyet
Ecole Polytechnique
Palaiseau Cedex, France

K. E. Gray
Argonne National Laboratory
Argonne, Illinois

S. Gregory
Cornell University
Ithaca, New York

J. M. Greif
California Institute of Technology
Pasadena, California

A. Griffin
University of Toronto
Toronto, Ontario, Canada

R. B. Griffiths
Carnegie-Mellon University
Pittsburgh, Pennsylvania

C. C. Grimes
Bell Laboratories
Murray Hill, New Jersey

D. U. Gubser
Naval Research Laboratory
Washington, District of Columbia

R. B. Hallock
University of Massachusetts
Amherst, Massachusetts

B. I. Halperin
Harvard University
Cambridge, Massachusetts

M. J. Harrison
Michigan State University
East Lansing, Michigan

J. B. Hastings
Brookhaven National Laboratory
Upton, New York

A. F. Hebard
Bell Laboratories
Murray Hill, New Jersey

M. Hértier
Université Paris-Sud
Orsay, France

G. B. Hess
University of Virginia
Charlottesville, Virginia

A. Holz
Universität des Saarlandes
Saarbrücken, W. Germany

P. M. Horn
IBM - T. J. Watson Research Lab
Yorktown Heights, New York

D. G. Howard
Portland State University
Portland, Oregon

C. I. Huber
Brown University
Providence, Rhode Island

D. L. Huber
University of Wisconsin
Madison, Wisconsin

B. A. Huberman
Xerox Palo Alto Research Center
Palo Alto, California

D. L. Hunter
St. Francis Xavier University
Antigonish, Nova Scotia, Canada

A. Isihara
State University of New York at Buffalo
Amherst, New York

L. E. Iton
Argonne National Laboratory
Argonne, Illinois

L. P. Kadanoff
University of Chciago
Chicago, Illinois

G. Kaindl
Freie Universität Berlin
Berlin, W. Germany

R. Kalia
Argonne National Laboratory
Argonne, Illinois

W. A. Kamitakahara
Iowa State University
Ames, Iowa

C. R. Kannewurf
Northwestern University
Evanston, Illinois

D. Keszler
Northwestern University
Evanston, Illinois

J. B. Ketterson
Northwestern University
Evanston, Illinois

P. H. Kleban
University of Maine
Orono, Maine

M. A. Klenin
North Carolina State University
Raleigh, North Carolina

H. Krakauer
Northwestern University
Evanston, Illinois

A. W. Kueny
Northwestern University
Evanston, Illinois

M. G. Lagally
University of Wisconsin
Madison, Wisconsin

J. R. Lalanne
Domaine Université
Talence, France

D. P. Landau
University of Georgia
Athens, Georgia

U. Landman
Georgia Technical Institute
Atlanta, Georgia

J. Z. Larese
Wesleyan University
Middletown, Connecticut

Y. Larher
DPC/SCM CEN
Saclay, France

H. J. Lauter
Institute Max Von Laue
Grenoble, France

E. Lerner
Universidade Federal Do Rio De Janeiro
Rio de Janeiro, Brasil

Z. Lihua
University of Wisconsin
Madison, Wisconsin

M. Listvan
University of Chicago
Chicago, Illinois

J. E. Litzinger
University of Pittsburgh
Pittsburgh, Pennsylvania

T. M. Lu
University of Wisconsin
Madison, Wisconsin

J. W. Lyding
Northwestern University
Evanston, Illinois

J. L. McCauley, Jr.
University of Houston
Houston, Texas

C. R. McIntyre
Argonne National Laboratory
Argonne, Illinois

J. P. McTague
University of California
Los Angeles, California

D. B. McWhan
Bell Laboratories
Murray Hill, New Jersey

T. D. Mai
University of Minnesota
Minneapolis, Minnesota

R. S. Markiewicz
General Electric Research and Development
Schenectady, New York

L. G. Marland
University of Guelph
Guelph, Ontario, Canada

J. D. Maynard
Pennsylvania State University
University Park, Pennsylvania

D. Mitchell
National Science Foundation
Washington, District of Columbia

K. Miyano
Argonne National Laboratory
Argonne, Illinois

D. E. Moncton
Bell Laboratories
Murray Hill, New Jersey

M. Monkenbusch
Kernforschungsanlage Jülich
Jülich, W. Germany

R. C. Morris
Florida State University
Tallahassee, Florida

S. C. Moss
University of Houston
Houston, Texas

W. J. Mullin
University of Massachusetts
Amherst, Massachusetts

G. E. Murch
Argonne National Laboratory
Argonne, Illinois

D. R. Nelson
Harvard University
Cambridge, Massachusetts

M. Nielsen
Risø National Laboratory
Roskilde, Denmark

A. Novaco
Lafayette College
Easton, Pennsylvania

S. S. Ostlund
Harvard University
Cambridge, Massachusetts

R. L. Park
University of Maryland
College Park, Maryland

L. Passell
Brookhaven National Laboratory
Upton, New York

R. A. Pelcovits
Brookhaven National Laboratory
Upton, New York

K. A. Penson
Université Paris-Sud
Orsay, France

T. A. Perry
University of Michigan
Ann Arbor, Michigan

P. Pershan
Harvard University
Cambridge, Massachusetts

R. Pindak
Bell Laboratories
Murray Hill, New Jersey

D. A. Pink
St. Francis Xavier University
Antigonish, Nova Scotia, Canada

M. Pomerantz
IBM Research and Development
Yorktown Heights, New York

J. H. Quateman
University of Michigan
Ann Arbor, Michigan

A. Rahman
Argonne National Laboratory
Argonne, Illinois

T. V. Ramakrishnan
Princeton University
Princeton, New Jersey

M. T. Ratajack
Northwestern University
Evanston, Illinois

H. A. Resing
Naval Research Laboratory
Washington, District of Columbia

E. K. Riedel
University of Washington
Seattle, Washington

R. J. Rollefson
Wesleyan University
Middletown, Connecticut

J. E. Rowe
Bell Laboratories
Murray Hill, New Jersey

L. Sander
University of Michigan
Ann Arbor, Michigan

R. Savit
University of Michigan
Ann Arbor, Michigan

M. Schick
University of Washington
Seattle, Washington

R. R. Sharma
University of Illinois
Chicago, Illinois

H. Shechter
Technion-Israel Institute of Technology
Haifa, Israel

G. Shirane
Brookhaven National Laboratory
Upton, New York

S. K. Sinha
Argonne National Laboratory
Argonne, Illinois

A. Sjolander
Argonne National Laboratory
Argonne, Illinois

S. C. Smith
Argonne National Laboratory
Argonne, Illinois

J. B. Sokoloff
Northeastern University
Boston, Massachusetts

B. W. Southern
University of Manitoba
Winnipeg, Manitoba, Canada

M. O. Steinitz
St. Francis Xavier University
Antigonish, Nova Scotia, Canada

G. A. Stewart
University of Pittsburgh
Pittsburgh, Pennsylvania

R. M. Suter
IBM - T. J. Watson Research Lab
Yorktown Heights, New York

R. H. Swendsen
IBM Zurich Research Laboratory
Rüeschlikon, Switzerland

Y. Takano
Argonne National Laboratory
Argonne, Illinois

H. Taub
University of Missouri
Columbia, Missouri

E. D. Thompson
Case Western Reserve University
Cleveland, Ohio

D. J. Thouless
Yale University
New Haven, Connecticut

J. Tobochnik
Cornell University
Ithaca, New York

T. Tokuhiro
Argonne National Laboratory
Argonne, Illinois

M. F. Toney
University of Washington
Seattle, Washington

G. J. Trott
University of Missouri
Columbia, Missouri

D. C. Tsui
Bell Laboratories
Murray Hill, New Jersey

P. Ukleja
Southeastern Massachusetts University
North Dartmouth, Massachusetts

M. Úlehla
Oak Ridge National Laboratory
Oak Ridge, Tennessee

J. Unguris
University of Wisconsin
Madison, Wisconsin

V. Urumov
University of Florida
Gainesville, Florida

D. Van Vechten
National Bureau of Standards
Washington, District of Columbia

P. D. Vashishta
Argonne National Laboratory
Argonne, Illinois

O. E. Vilches
University of Washington
Seattle, Washington

J. Villain
Centre d'Etudes Nucleaires de Grenoble
Grenoble, France

P. Vora
Argonne National Laboratory
Argonne, Illinois

N. Wada
University of Chicago
Chicago, Illinois

R. W. Wang
University of Missouri
Columbia, Missouri

M. B. Webb
University of Wisconsin
Madison, Wisconsin

J. D. Weeks
Bell Laboratories
Murray Hill, New Jersey

M. Weissmann
Universidad Simon Bolivard
Caracas, Venezuela

C. T. White
Naval Research Laboratory
Washington, District of Columbia

F. I. B. Williams
Commissariat Energie Atomique
Gif-Sur-Yvette, France

G. A. Williams
University of California
Los Angeles, California

M. Wortis
University of Illinois
Urbana, Illinois

A. Y. S. Yang
Merck Sharp & Dohme Research Lab
West Point, Pennsylvania

S. C. Ying
Brown University
Providence, Rhode Island

H. Zabel
University of Illinois
Urbana, Illinois

A. Zippelius
Harvard University
Cambridge, Massachusetts

A. B. Zisook
University of Chicago
Chicago, Illinois

G. Zografi
University of Wisconsin
Madison, Wisconsin

J. A. Zollweg
University of Maine
Orono, Maine

SECTION I

INVITED PAPERS

Sinha, ed. Ordering in Two Dimensions

PHYSICS IN d-1 DIMENSIONS: SURFACES AND INTERFACES

MICHAEL WORTIS

Physics Department, University of Illinois, Urbana, IL 61801

ABSTRACT

Surfaces and interfaces of bulk samples are (d-1)-dimensional thermodynamic systems. These systems coexist with the bulk but have their own phase diagrams, phase transitions, and thermodynamic functions. We give a brief introduction to and review of what is known about such systems from a statistical mechanical point of view.

I. INTRODUCTION

At a conference on two-dimensional ordering it is perhaps not out of place to remind ourselves that all "real" two-dimensional systems couple to some extent into a three-dimensional "bulk." For uniformly layered systems this gives the ordering ultimately a three-dimensional character and two-dimensionality is seen only by virtue of the existence of a regime where the in-plane correlation length is large while the out-of-plane correlation length remains microscopic. By contrast, films, surfaces, and interfaces are genuinely two-dimensional [1]. Indeed, there is a smooth interpolation connecting an ideal film supported by a perfectly inert rigid substrate and the "surface phase" of an otherwise homogeneous, bulk sample with an exposed face. Only for the ideal film-on-substrate is there a clean separation of surface and bulk degrees of freedom. More generally, both surfaces and interfaces possess well-defined (d-1)-dimensional thermodynamic properties which, however, must be separated from the coexisting d-dimensional bulk properties by an appropriate thermodynamic limiting procedure [2].

It is the purpose of this presentation to review the separation of d- and (d-1)-dimensional thermodynamic properties and then to illustrate by some simple examples the richness of the (d-1)-dimensional physics. It will turn out that a number of diverse physical situations can be simulated by simple Ising models. The model and its various representations are introduced in Sec. II. Section III then sketches the phenomenology of some applications involving semi-infinite samples with a single free surface. Section IV is devoted to the phenomenology of a simple interface between two coexisting bulk phases. The work reviewed is not new. Selected references to the recent and/or principal literature are given for the interested reader.

II. (d-1)-DIMENSIONAL THERMODYNAMICS OF THE ISING MODEL

We consider for simplicity a d-dimensional hypercubical lattice of $N=L^d$ sites, of which $N_s=2dL^{d-1}$ are surface sites. Each site i is occupied by a spin $\sigma(i) = \pm 1$ with nearest-neighbor coupling only, so the Hamiltonian of the system is

$$-\beta\mathcal{H}_{N,N_s} = \sum_i h(i)\sigma(i) + \sum_{\substack{<ij>\\ nn}} K(ij)\sigma(i)\sigma(j) . \quad (1)$$

In a homogeneous bulk system the fields and couplings are uniform,

$$h(i) = h, \quad K(ij) = K. \quad (2)$$

However, in a sample with free surfaces it is reasonable to expect that physical parameters change near the surface, so magnetic fields and couplings may be modified in the first few atomic layers. This can be modelled by taking surface fields and surface couplings which differ from their bulk values,

$$h(i) = h+h_n \text{ for i in the } n^{th} \text{ layer,}$$

$$K(ij) = K_n \text{ for i and j in the } n^{th} \text{ layer, etc.} \quad (3)$$

The bulk free energy per site is defined by the usual thermodynamic limit [3],

$$f_b(h,K) \equiv \lim_{N\to\infty} F_{N,N_s}/N \text{ with}$$

$$F_{N,N_s}(\{h(i),K(ij)\}) \equiv \ln \mathrm{Tr}\, e^{-\beta\mathcal{H}_{N,N_s}} . \quad (4)$$

$f_b(h,K)$ is independent of the surface fields and surface couplings. It contains a complete thermodynamic description of the d-dimensional bulk, for example, $\partial f_b/\partial h = m_b$, the bulk magnetization, etc. In particular, the loci of singularities of f_b as a function of the bulk variables h and K define the bulk phase boundaries. The behavior of f_b near second-order phase boundaries defines the usual bulk critical exponents.

To extract the surface thermodynamics, it is necessary to subtract off the bulk contributions and take the limit [2],

$$f_s(h,K,h_1,h_2,\ldots K_1,K_2,\ldots) \equiv \lim_{N,N_s\to\infty} \frac{1}{N_s}(F_{N,N_s} - Nf_b). \quad (5)$$

f_s is the surface free energy per surface site. It depends on both bulk and surface variables and contains a complete thermodynamic description of the (d-1)-dimensional surface, for example [4], $\partial f_s/\partial h_n = m_n$, the nth layer magnetization, etc. The loci of singularities of f_s as a function of its variables describe the surface-phase phase boundaries in precise analogy to the bulk. The presence of surface variables in addition to bulk variables makes the surface phase diagram in general more complicated than that of the bulk and can require new, surface critical exponents to describe behavior near second-order transitions.

When bulk variables lie on a first-order phase boundary, then two bulk phases can coexist. By appropriate manipulation of boundary conditions [5], the interface between the coexisting phases can be made planar (on the average). Local properties like layer magnetizations vary spatially in directions perpendicular to the interface and are associated with a (d-1)-dimensional "interfacial" contribution to the free energy [6],

$$f_i(h,K) \equiv \lim_{N,N_s,N_i\to\infty} \frac{1}{N_i}(F_{N,N_s} - Nf_b - N_s f_s), \quad (6)$$

where N_i measures the area of the undistorted, flat interface, e.g. $N_i = L^{d-1}$ for a (1,0,0) interface. The interfacial free energy f_i depends only on bulk variables as restricted to the bulk first-order phase boundary, e.g. for the ferromagnet h=0, $K_c(d) \leq K < \infty$. It contains a complete thermodynamic description of the (d-1)-dimensional interface.

The system defined by Eqs. (1)-(3) evidently applies to the surface of a uniaxial magnet and to the interface between coexisting "up" and "down" magnetic regions, when they are present. In particular, the situation $K = K_2 = K_3 = \ldots = 0$ but $K_1 > 0$ models an ideal ferromagnetic monolayer, which will order (provided d>2) for $K_1 > K_c(d-1)$. If, instead, $K \neq 0$ with $0 < K << K_1$, then the bulk is weakly magnetic and we expect a regime $K_1 \gtrsim K_c(d-1)$, $K < K_c(d)$ where there is no bulk order but surface magnetism is enhanced by coupling into the paramagnetic bulk.

Another set of phenomena is accessible to our model via replacement of the magnetic variables by AB binary-alloy variables, according to [7]

$$n_A = \frac{1}{2}(1+\sigma), \quad n_B = \frac{1}{2}(1-\sigma). \quad (7)$$

Dominant A-A and B-B attraction promotes phase separation at low temperatures and corresponds in magnetic language to ferromagnetic ordering (K>0). The development of a surface-layer magnetization $m_n > m_b$ is referred to in metallurgical terms as "surface enrichment," while the occurrence of a surface magnetization in the presence of bulk paramagnetism is called "surface segregation." f_i gives the free energy of an AB interface. On the other hand, dominant A-B attraction promotes an AB ordered phase at low temperatures and corresponds in magnetic language to antiferromagnetic ordering (K<0).

Finally, the lattice-gas transcription [7],

$$n = \frac{1}{2}(1+\sigma), \quad (8)$$

in which a filled (unfilled) site corresponds to spin up (down), allows the model to describe (a) a liquid-vapor interface or (in the surface geometry) (b) condensation of a vapor on an inert substrate (Sec. IIB) and even (c) surface reconstruction (Sec. IIA).

III. SURFACES

Magnetic surfaces have been studied by mean-field [4,8,9,10], series [4], Monte Carlo [11], ε-expansion [10,12], and position-space renormalization-group [13] techniques. In addition certain exact results are available [14] for d=2. We organize discussion around representative phase diagrams in two special cases.

A. Zero magnetic fields ($h = h_1 = h_2 = \ldots = 0$)

Consider ferromagnetic interactions $0 < K = K_2 = \ldots$ with variable first-layer coupling K_1. The surface phase diagram in the variables K, K_1 is sketched in Fig. 1, as first proposed by Binder and Hohenberg [8]. Note that the surface free energy $f_s(K,K_1)$ is always singular along $K = K_c(d)$, where the bulk orders and the surface is, therefore, subject to a singular change in its environment. The additional phase boundaries correspond to surface ordering in regions where the bulk is analytic and show up only for d>2 (surface dimensionality greater than unity), where the surface can support independent order. The phase denoted SF(BP) exhibits surface ferromagnetism in the absence of bulk order and corresponds to surface segregation in binary-alloy language, as remarked in Sec. II. The analogous antiferromagnetic surface phase SA(BP) describes a binary ordering of the surface before bulk phase separation. For $K > K_c(d)$ surface phases coexist with bulk ferromagnetism. For SF(BF) the surface is simply ferromagnetic; but, for SF+A(BF) the surface has antiferro-

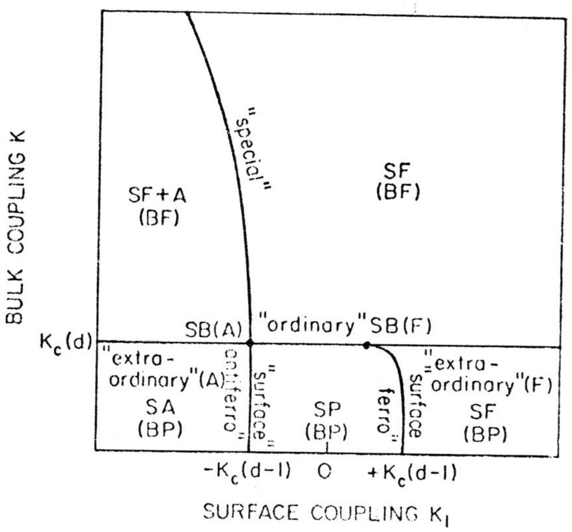

Fig. 1 Surface phase diagram for d>2. Phases of the surface S (with coexisting bulk B) are denoted paramagnetic (P), ferromagnetic (F), and antiferromagnetic (A). The phase boundaries are all believed second order; however, the different transitions are characterized by different critical behavior. The intersections SB, at which surface and bulk are separately and simultaneously critical, are multicritical points.

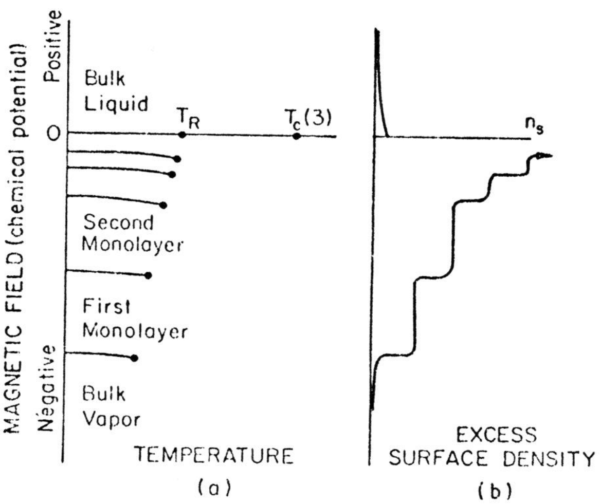

Fig. 2(a) Phase diagram for the d=3 adsorption problem. Each line of first-order transitions terminates in a critical point.
(b) Excess surface density as a function of chemical potential at a fixed (low) temperature. Each step represents the condensation on the substrate of an additional monolayer.

magnetic order in addition to ferromagnetic order, which in the lattice-gas transcription models surface reconstruction. Each second-order phase boundary is characterized by its own critical behavior. For example, at the "ordinary" transition (Fig. 1) h_1 is relevant but K_1 is not, so

$$f_s(h,K,h_1,K_1) \sim t^{2-\alpha-\nu} F(h/t^{\Delta}, h_1/t^{\Delta_s^{(0)}}) , \qquad (9)$$

involving a surface gap exponent $\Delta_s^{(0)}$.

B. Nonzero magnetic fields, adsorption on a substrate

The van der Waals attraction between vapor atoms and an inert planar substrate can be modelled in the lattice-gas language (8) by taking [15] the semi-infinite geometry with layer fields $h_n \sim 1/n^3$. The vapor density is controlled by a chemical potential which is essentially the bulk field h. The phase diagram for $K=K_1=K_2 = \ldots > 0$ is shown [15] in Fig. 2(a). At sufficiently large negative chemical potential the density of the tenuous vapor is weakly enhanced near the substrate surface because of the attraction represented by the positive layer fields h_n. If the bulk density is now increased by increasing h at a fixed low T, there is a first-order surface phase transition at which the density of atoms in the first atomic layer increases discontinuously, corresponding to monolayer condensation on the substrate. Subsequent increase in density condenses successive layers, as shown in Fig. 2(b). When h passes zero, the bulk vapor condenses to liquid and the excess surface density falls off again to a low value. Each layer transition is first-order but terminates in a critical point at a temperature T_n. It is conjectured [15] for d=3 that, as $n\to\infty$ (many adsorbed layers), $T_n \to T_R$, the "roughening temperature" ($T_R < T_c(3)$) at which the distinction between successive layers is lost (see Sec. IV).

IV. INTERFACES

Magnetic interfaces have been studied by mean-field theory [16], series [17], Monte Carlo [18], ε-expansion [19], and position-space renormalization groups [20]. Exact results [21] are available only for d=2. For a ferromagnetic bulk coupling (K>0) coexistence occurs only for h=0 and $0 \leq T \leq T_c(d)$ (i.e., $K_c(d) \leq K < \infty$), so the interfacial phase diagram, shown in Fig. 3, is one-dimensional. Note the dependence on the bulk dimensionality d. At T=0 a (1,0,0) interface is always perfectly flat. Perpendicular deviations z of the interface from this refer-

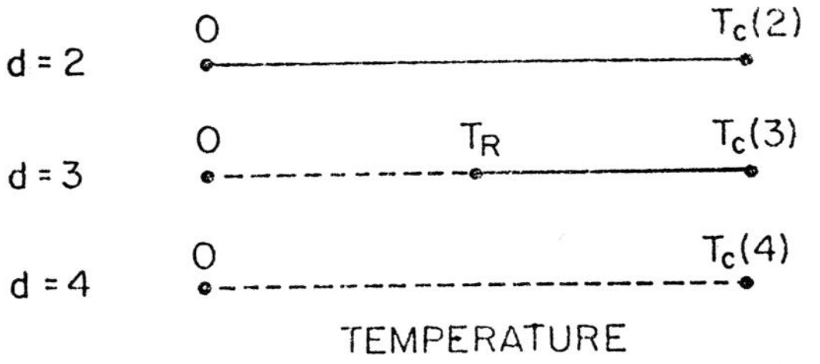

Fig. 3 Phase diagram of an Ising magnetic interface for various bulk dimensionalities d. A solid line represents the rough phase; a dotted line, the smooth phase. In d=3 the two phases occur separated by a roughening transition at the "roughening temperature" T_R.

ence plane occur as a result of thermal fluctuations when T>0. A "smooth" phase is characterized by finite fluctuations $\langle z^2 \rangle < \infty$. When $\langle z^2 \rangle$ diverges in the thermodynamic limit, the interface is "rough." The sharp "roughening transition" which separates smooth and rough phases in d=3 was originally noted in the context of crystal growth [22].

The information summarized in Fig. 3 can be understood qualitatively by the following two non-rigorous [23] but appealing physical arguments: First, consider a continuum model [24] in which the interface behaves like a stretched drumhead of dimensionality (d-1). Because the surface tension is positive, it costs energy to increase the area of the drumhead, so "capillary-wave" excitations of small wavenumber k have energy $\varepsilon_k \sim Ck^2$, where C is proportional to the surface tension. We may expect $\langle z^2 \rangle$ to be proportional to the equilibrium number of such excitations, so, if the excitations are taken to be independent and commutative,

$$\langle z^2 \rangle \sim \int (d^{d-1}k) n_k \xrightarrow[k\to 0]{} \int \frac{dk\, k^{d-2}}{Ck^2} , \qquad (10)$$

which converges (smooth phase) for d>3 but diverges [25] (rough phase) for d≤3. The fact that this argument misses the discreteness of the Ising lattice makes it suspect, particularly at low temperature, where proper representation of the lowest-energy excitations is crucial. Here, however, there is a complementary argument, based on minimization of the free energy F=E-TS. At T=0, entropy is irrelevant and the planar interface minimizes E. At T>0 there is always some local fuzziness to the interface due to short wavelength thermal excitations; however, if excitations with wavelengths comparable to the sample size are favorable, then it is reasonable to suppose that the interface becomes rough in the thermodynamic limit. These long wavelength excitations have perimeter of order L^{d-2}, so $E \sim L^{d-2}$ while $S \sim \ln L^{d-1} \sim \ln L$. Thus, for d=2 the entropy term dominates as $L\to\infty$ for any T>0, long wavelength excitations are favorable, and the interface is rough. Similarly, for d>2 the energy term wins, long wavelength excitations are suppressed, and the interface is smooth at sufficiently low T. In d=3 entropy considerations force a smooth interface at low T; however, at higher T the discreteness of the lattice no longer plays a crucial role and the capillary-wave analysis applies.

The roughening transition of the d=3 interface is plausibly believed to be in the universality class of the d=2 XY model [26]. This can be understood as follows: Suppose in (1) that all couplings $K_\perp$ in the direction perpendicular to the interface are allowed to go to infinity, while the couplings $K_\parallel$ in parallel directions remain finite. This limit, which we assume does not change the character of the transition, suppresses all "wrong-phase" fluctuations and reduces the Ising model to a (d-1=2)-dimensional system depending only on the L variables z(i) which describe the wandering of the interface from its reference plane. This model, which has the Hamiltonian

$$-\beta\mathcal{H} = K_\parallel \sum_{\substack{<ij>\\ nn}} |z(i)-z(j)|^p , \quad z=0,\pm1,\pm2,\ldots , \qquad (11)$$

with p=1, is called the solid-on-solid (SOS) model [22,27]. The SOS model is probably in the same universality class as the discrete Gaussian model [28], described by Eq. (11) with p=2, which in turn is equivalent to the d=2 Coulomb gas [29] and via duality [30] to the XY model.

REFERENCES

1. Real samples are finite. Strictly speaking, dimensionality only appears when sample dimensions $L \to \infty$.
2. M. E. Fisher and G. Caginalp, Commun. Math. Phys. 56, 11 (1977); G. Caginalp and M. E. Fisher, ibid. 65, 247 (1979).
3. R. B. Griffiths, in Phase Transitions and Critical Phenomena, edited by C. Domb and M. S. Green (Academic, London, 1972), Vol. 1.
4. K. Binder and P. C. Hohenberg, Phys. Rev. B 6, 3461 (1972).
5. One method is appropriate pinning of surface spins; another is application of an inhomogeneous magnetic field $h_n = gn$, $g\to 0$.
6. G. Gallavotti, A. Martin-Löf, S. Miracle-Sole, in Mathematical Methods in Statistical Mechanics, edited by A. Lenard (Springer, Berlin, 1973).
7. K. Huang, Statistical Mechanics (Wiley, New York, 1963), Chapter 16.

8. K. Binder and P. C. Hohenberg, Phys. Rev. B 9, 2194 (1974).
9. T. C. Lubensky and M. H. Rubin, Phys. Rev. B 12, 3885 (1975).
10. A. J. Bray and M. A. Moore, J. Phys. A 10, 1927 (1977).
11. K. Binder, in Phase Transitions and Critical Phenomena, edited by C. Domb and M. S. Green (Academic, New York, 1976), Vol. 5B.
12. T. C. Lubensky and M. H. Rubin, Phys. Rev. Lett. 31, 1469 (1973); Phys. Rev. B 11, 4533 (1975).
13. N. M. Švrakić and M. Wortis, Phys. Rev. B 15, 396 (1977); Th. Burkhardt and E. Eisenriegler, Phys. Rev. B 16, 3213 (1977) and ibid. B 17, 318 (1978); N. M. Švrakić, R. Pandit, and M. Wortis, Phys. Rev. B (in press).
14. M. E. Fisher and A. E. Ferdinand, Phys. Rev. Lett. 19, 169 (1967); B. M. McCoy and T. T. Wu, Phys. Rev. 162, 436 (1967).
15. M. J. de Oliveira and R. B. Griffiths, Surf. Sci. 71, 687 (1978).
16. J. W. Cahn and J. E. Hilliard, J. Chem. Phys. 28, 258 (1958); B. Widom, in Phase Transitions and Critical Phenomena, edited by C. Domb and M. S. Green (Academic, New York, 1972), Vol. 2.
17. J. D. Weeks, G. H. Gilmer, and H. J. Leamy, Phys. Rev. Lett. 31, 549 (1973).
18. H. J. Leamy and G. H. Gilmer, J. Cryst. Growth 24/25, 499 (1974).
19. T. Ohta and K. Kawasaki, Prog. Theor. Phys. Jpn. 58, 467 (1977); J. Rudnick and D. Jasnow, Phys. Rev. B 17, 1351 (1978).
20. M. J. de Oliveira, D. J. Furman, and R. B. Griffiths, Phys. Rev. Lett. 40, 977 (1978).
21. L. Onsager, Phys. Rev. 65, 117 (1944); M. E. Fisher and A. E. Ferdinand, Phys. Rev. Lett. 19, 169 (1967); D. B. Abraham, G. Gallavotti, and A. Martin-Löf, Physica (Utrecht) 65, 73 (1973).
22. W. K. Burton, N. Cabrera, Disc. Faraday Soc. 5, 33, 40 (1949); W. K. Burton, N. Cabrera, and F. C. Frank, Phil. Trans. Roy. Soc. A 243, 299 (1951).
23. It is rigorously proven that the interface is rough for d=2 (G. Gallavotti, Commun. Math. Phys. 27, 103 (1972)) and that it is smooth for d=3 at sufficiently low T (H. van Beijeren, Commun. Math. Phys. 40, 1 (1975)).
24. F. P. Buff, R. A. Lovett, and F. H. Stillinger, Jr., Phys. Rev. Lett. 15, 621 (1965).
25. In practice the divergence is often cut off by a "gravitational" field ($\varepsilon_k \sim g + Ck^2$) or by finite sample size $1/L < k < \infty$.
26. $T > T_R$ in the roughening transition corresponds to $T < T_c$ in the XY model.
27. D. E. Temkin, in Growth of Crystals, ed. by N. Sheftal, 5A, 71 (1968).
28. V. J. Emery and R. H. Swendsen, Phys. Rev. Lett. 39, 1414 (1977).
29. S. T. Chui and J. D. Weeks, Phys. Rev. B 14, 4978 (1976).
30. J. V. José, L. P. Kadanoff, S. Kirkpatrick, and D. R. Nelson, Phys. Rev. B 16, 1217 (1977); R. Savit, Phys. Rev. Lett. 39, 55 (1977); H.J.F. Knops, Phys. Rev. Lett. 39, 766 (1977).

DISCUSSION

M. Schick:

In the phase marked SF + A you stated that the surface is "mixed ferro and antiferromagnetic." What do you mean by this term?

M. Wortis:

In this phase the surface exhibits both ferromagnetic order (uniform magnetization) and antiferromagnetic order (staggered magnetization) simultaneously. The situation is analogous to an antiferromagnet in an external magnetic field. Here, the bulk ferromagnetism acts as a unifrom field on the antiferromagnetically ordered surface.

R. R. Sharma:

Could there be a "disordered" phase at the interface? In the limiting cases the "rough" phase may go into the "disordered" phase. It appears that the inclusion of a "disordered" phase would make the treatment and discussion more complete.

M. Wortis:

I am not aware of the "disordered" phase at the interface.

COMMENSURATE ORDER, MULTICRITICAL POINTS, AND FINITE SIZES OF ADSORBED SYSTEMS

A. N. BERKER

Department of Physics, M.I.T., Cambridge, MA 02139

ABSTRACT

Multicritical phase diagrams of commensurate ordering in adsorbed systems are discussed, in terms of qualitative expectations and quantitative calculations for actual experimental systems. The usage of prefacing and renormalization transformations is outlined. A phenomenological scaling argument is given in support of a cusped coexistence boundary due to finite sizes and heterogeneous boundaries of substrate crystallites. In this respect, analogies are made to ^3He-^4He superfluidity, in bulk and films. Microscopic calculations of finite systems are reported, showing double-peaked specific heat versus temperature curves.

When a periodic array of substrate sites is exposed to adsorbing atoms oversized with respect to the spacing between sites, ordered superlattice arrangements can result at low temperatures and appropriate coverages. Although mean-field theories dictate first-order phase transitions for the disordering of most superlattice arrangements, experiments [1], and theories [2-5] including fluctuations (also rigorous work [6] on models related by universality [7,8]), indicate higher-order phase transitions for a range of coverages. First-order transitions do occur at low coverages and temperatures, as a condensation of vacancies. If the attractive part of the inter-adsorbate potential is strong, the multicritical point separating the first and higher-order transitions is experimentally accessible. At this point, an interesting "real-system" phenomenon occurs [4]: The combined effects of the finite sizes and heterogeneous boundaries of the substrate crystallites create a cusped coexistence boundary. This illustrates the need for particulate theories which can incorporate real-system conditions.

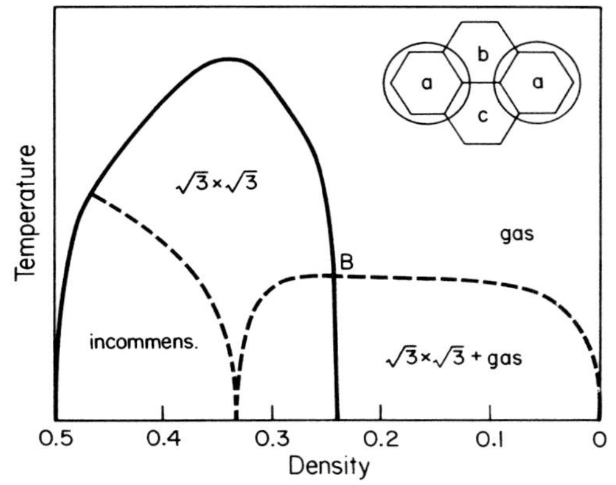

Fig.1 In dark, calculated [2] phase diagram of helium on graphite. Dashed, a hypothetical multicritical structure. Inset shows adsorption of oversized atoms on basal graphite.

A PHASE DIAGRAM

The graphite basal surface presents a triangular array of adsorption sites (Fig.1). A helium atom, for instance, is slightly oversized, so that simultaneous adsorption on two nearest-neighbor sites is energetically unfavorable. Accordingly, when one third of all sites is to be occupied, the minimum energy configuration has one of three sublattices (e.g., sublattice a in Fig.1) completely occupied, and the other two sublattices (b and c) completely empty. Indeed, at low temperatures, an ordered phase is achieved, characterized by the preferential occupation of one sublattice. Upon increasing temperature, a phase transition occurs into a disordered phase with, on the average, equivalent occupation of the sublattices. The ordered phase has a three-fold, permutation-symmetric degeneracy (either a, or b, or c is spontaneously preferred), so that its transition should be [7] in the universality class of the three-state Potts model. This model, in two dimensions, was rigorously shown to have a higher-order phase transition [6], in spite of a third-order term in the Landau free energy expansion. Evidently, this two-dimensional phase transition is strongly fluctuation dominated, which invalidates classical concepts. Upon changing coverage from the optimum value n=1/3, less energy is minimized by the (zero-temperature) ordered configuration, due to the vacancies or interstitials. Therefore, the system disorders at a lower temperature. This suggests, qualitatively, the full-line phase boundary in Fig.1. The ordered arrangement just discussed is denoted by $(\sqrt{3}\times\sqrt{3})R30°$, since its two lattice vectors are $\sqrt{3}$ times those of the graphite surface and they are rotated by 30° from these graphite vectors.

Consider the dilution limit, say n=0.1. According to the full boundary of Fig.1 (calculated [2] with nearest-neighbor repulsion only, which is sensible for helium on graphite), the adsorption system must be in a two-dimensional gas phase, dilute and disordered. However, in addition to near-neighbor repulsion, physical systems can have important further-neighbor attractions which, for example, could favor second-neighbor pairs such as in the Fig.1 inset (i.e., assign a negative energy compared to infinite separation). At temperatures low enough for entropy to be unimportant, free energy is minimized by formation of two types of domains, one of near-perfect $\sqrt{3}\times\sqrt{3}$ phase, the other of near-empty gas. Thus, in a schematic representation in Fig.1, a region of coexistence of $\sqrt{3}\times\sqrt{3}$ and gas phases (i.e., a first-order phase transition) is drawn with dashes. A multicritical point B is implied, its temperature reflecting the strength of the attractive potential.

The dense region is not expected to mirror-image the dilute side, due to the asymmetry of the inter-adsorbate force, being increasinly repulsive at small separation, but vanishing at large separation. It is believed that the excess density organizes itself into domain walls, thereby destroying commensuration (registry) to the substrate [9]. Thus, a commensurate-incommensurate phase transition line is drawn on the dense side of Fig.1, under assumption that the $\sqrt{3}\times\sqrt{3}$ phase accomodates more isolated interstitials (as well as vacancies) at higher temperatures, but none at zero temperature. This type of phase diagram was obtained in a heat capacity study of krypton on graphite [10].

CALCULATIONS

The renormalization-group method is an iterative solution of a statistical mechanics problem [11]. By a partial calculation of the partition sum, a given system is mapped onto another system with same structure, but increased length scale. This

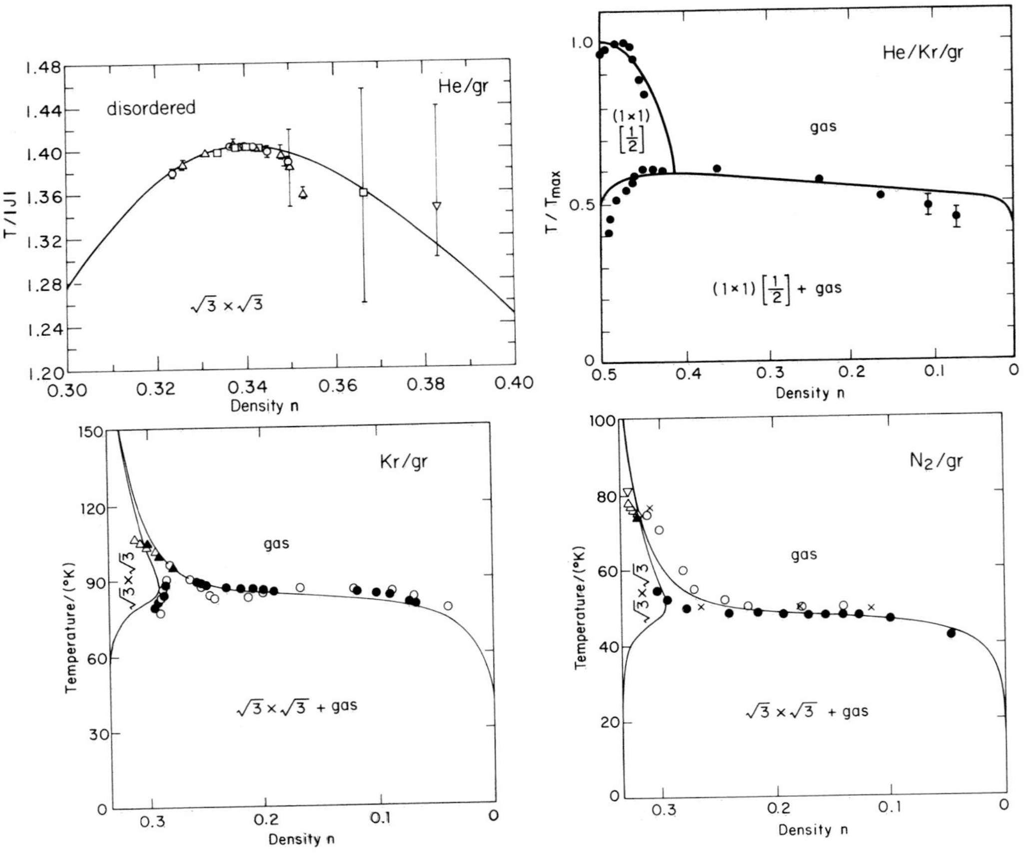

Fig. 2 Calculated phase diagrams for helium [2], krypton [3], and nitrogen [3] on graphite, and helium on krypton-plated graphite [4]. Experimental points are also shown.

means that a fraction of the degrees of freedom has been summed out of the problem. This is compensated by new, effective values for thermodynamic fields such as temperature and chemical potential. Since the structure of the problem is conserved, this transformation can be recycled, either algebraically or numerically, and an infinite repetition completely solves the problem. In practice, for a non-trivial problem, an approximate transformation is used, since an attempt at the exact procedure generates many new types of thermodynamic fields. One version of this approach, which manipulates the position-space representation of the partition sum, is well developed for two-dimensional systems.

A realistic microscopic description of an adsorption system, such as adsorbates moving across the potential wells of the substrates and interacting with each other over distances of several substrate lattice constants, may be cumbersome to renormalize directly. This situation is dealt with by a "prefacing" transformation [12], in which the original system is quantitatively mapped onto another one with different structure more amenable to renormalization. This transformation is performed once, followed by repeated renormalization transformations. The prefacing (restructuring) transformation is inspired by and shares characteristics with the renormalization (rescaling) transformation: (i) Both transformations rely on *a priori* physical intuition. One has to decide what the important degrees of freedom are, keep those intact, and approximately sum out the unimportant ones. (ii) Both transformations are effected by a partial trace of the partition function. Thus, if the commensurate multicritical point is studied, prefacing involves integration over the continuously varying positions of the adsorbate [4], yielding a discrete Hamiltonian. On the other hand, if superfluidity is studied, the continuous degrees of freedom are dealt with in the renormalization transformations [13].

Quantitative phase diagrams (Fig.2) have been obtained for the physisorption systems of helium [2], krypton [3,4] and nitrogen [4] on basal graphite, and helium on krypton-plated basal graphite [5], showing satisfactory agreement with experiments. Figs. 3 display a sequence of phase diagrams obtained [14] by prefacing/renormalization, for a Lennard-Jones system on a triangular lattice, as the ratio of core diameter σ to lattice constant a is varied. The krypton-on-graphite type phase diagram (a) evolves to a triple-point phase diagram (c). The liquid phase is identified as a melted (2×2)R0°. In fact, this ordering appears in the next phase diagram (d), which, on the dilute side, makes contact with experiments on methane physisorbed on basal graphite [15]. One important correction is necessary, though. At an inter-adsorbate dilation corresponding to the 2×2 structure, a methane

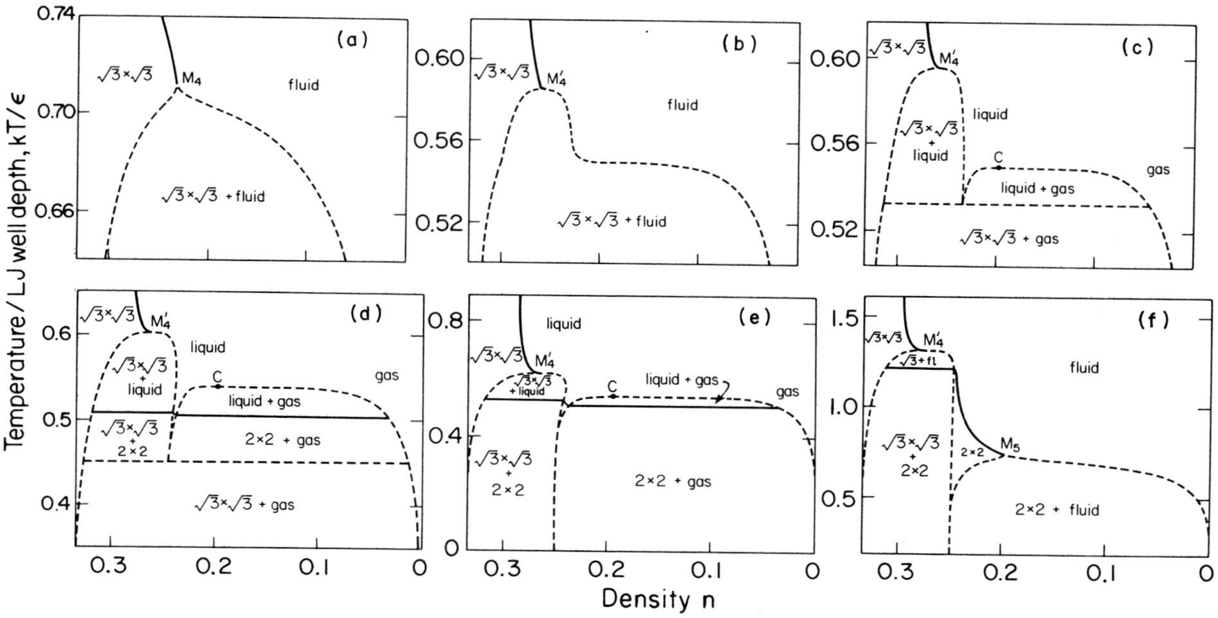

Fig. 3 Calculated [14] phase diagrams for Lennard-Jones system on a triangular lattice. The σ/a values are: (a) 1.63, (b) 1.6455, (c) 1.646, (d) 1.6465, (e) 1.648, and (f) 1.7.

molecule can gain more entropy, as compared with the same situation in the $\sqrt{3}\times\sqrt{3}$ structure, by traveling across substrate potential barriers and fully occupying its Wigner-Seitz cell. When substrate barriers are ignored, commensuration is destroyed. One could imagine realizing a 2×2 physisorbed ordering by increasing the core diameter, i.e. using xenon, but when σ is much larger than a, the adsorbate atom does not fit into the substrate well. This qualitatively explains experimental results on methane [15] and xenon [16] on graphite. Thus, in physisorbed systems, the 2×2 phase is preempted by a dilated incommensurate phase, not projected in the prefacing leading to Figs.3. This does not occur in chemisorbed systems, where covalent bonding gives a variety of commensurate structures. In fact, the phase diagram 3(f) is very similar to that observed [17] for oxygen on nickel (111). In this physical situation, a slightly more energetic triangular lattice of sites is present in addition to the one on which the $\sqrt{3}\times\sqrt{3}$ and 2×2 ground states occur. The possible excitations are modified, requiring an extension of the calculation leading to Fig.3(f). Another renormalization treatment of 2×2 ordering in is [18].

In addition to phase diagrams, critical exponents seem to justify the approach described for commensurate problems, which eventually involves the renormalization-group solution of a discrete Hamiltonian. The specific heat exponent measured [19] for helium on graphite is $\alpha = 0.36\pm0.02$, that for helium on krypton-plated graphite [20] is $\alpha \simeq 0$, and the order parameter exponent measured [21] at the multicritical point of krypton on graphite is $\beta \simeq 0.08$. The corresponding exponents from Ising and Potts models [22,14] have the values 1/3, 0, and 7/96, respectively. Finally, these methods can be fully extended [13], beyond prefacing, to continuous degrees of freedom.

FINITE-SIZE INDUCED COEXISTENCE

This section discusses the suggestion [4] that finite sizes and heterogeneous boundaries of substrate crystallites induce something very much resembling phase coexistence, and consequently a cusped phase boundary, near the commensurate multicritical point. Consider a graphite substrate crystallite bounded by the bottom of a step on one side and the top of a step on the other side [23] (Fig.4 is a side view). Due to increased substrate proximity and covalent bonding to dangling bonds, site S at the bottom of the step is always occupied. For adsorbates with nearest-neighbor exclusion, such as krypton or nitrogen, maximum commensurate density is achieved along the bottom of the step, with one sublattice entirely full and the two other sublattices entirely empty. This will be called a strong boundary. Conversely, position W at the top of the step is always empty, due to the lack of substrate and to the adatom being pulled down to the bottom of the step. This will be called a weak boundary. Since the inter-adsorbate potential is attrac-

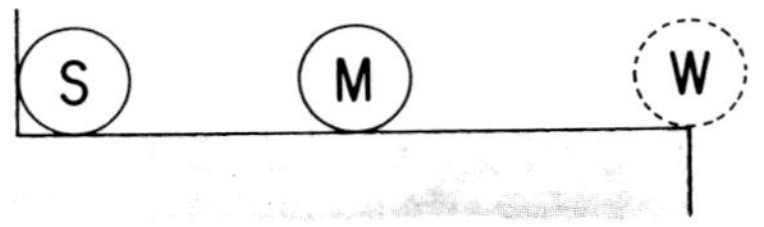

Fig.4 Substrate steps.

tive beyond nearest-neighbor exclusion, strong or weak boundaries locally cause an increase or decrease in density, respectively. Furthermore, the strong boundary is a nucleation line for the $\sqrt{3}\times\sqrt{3}$ phase.

The boundary effects penetrate the interior over the distance of the correlation length ξ. When ξ is small, the more dense or dilute regions at each boundary are unimportant. However, as a higher-order phase transition is approached $\xi \sim |T-T_c|^{-\nu}$ increases. Consider a region R of size ξ next to either boundary. Within R, a sum rule gives

$$\frac{dn}{d\mu} \sim \xi^{(2/\nu)-d} \qquad (1)$$

where n is density, μ is chemical potential, and d=2 is dimensionality. The effect of the boundary is equivalent to a chemical potential change [24],

$$\delta\mu \sim \pm\xi^{-1} \qquad (2)$$

where plus and minus signs apply to strong and weak boundaries. Combining gives

$$\delta n = \frac{dn}{d\mu}\,\delta\mu \sim \pm\xi^{(2/\nu)-3} \quad . \qquad (3)$$

Using conjectured [22] exact values, $(2/\nu)-3$ is 3/7 at the multicritical point B, and -3/5 at the higher-order boundary away from B, with an effective crossover exponent smoothly interpolating in between. Consider the neighborhood of B. Since n is bounded, Eq.(3) is not literally fulfilled, but the positive exponent indicates a discrete deviation from the density of the interior. Thus when $\xi \sim L$, the crystallite size, the finite system should be composed of two regions with distinct densities. Moving away from B along the higher-order line, the difference between the two densities diminishes, as the exponent $(2/\nu)-3$ crosses over to its negative value. However, since from the latter $\delta n \sim \pm L^{-3/5}$, and L is finite, it does not completely disappear until monolayer saturation leaves no room for density variation. The resulting cusped coexistence boundary is depicted in Fig.5, and in krypton and nitrogen diagrams of Fig.2.

The above appears to be a reenactment of observations on the tricritical phase diagram of bulk ^3He-^4He mixtures, where gravity causes a chemical potential gradient across the sample. Thus, a small cusplike shape occurs on top of the i-

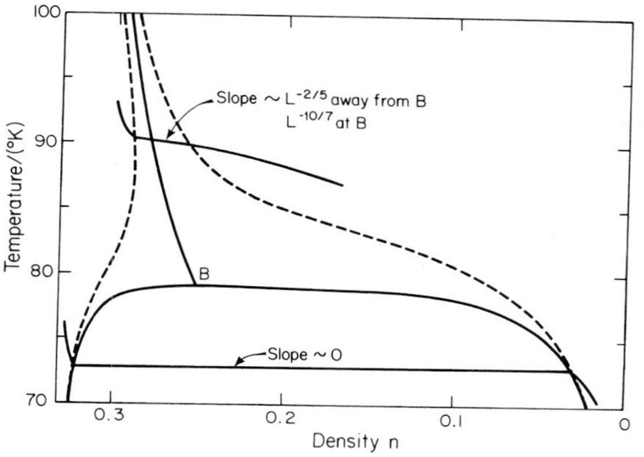

Fig.5 Ideal system phase boundaries (full lines) and finite-size induced cusp (dashed) from [3]. Two isobaric scans are dash-dotted. Isothermal vapor pressure curves should have the same *L* dependences as the scans here.

deal system coexistence boundary, Fig.2 in [25], and induced coexistence has been discussed [26]. In physisorbed systems, heterogeneous boundary conditions on a finite system with large correlation length could be thought of as a "lateral gravity field". Another connection may exist, with films of ^3He-^4He mixtures. Because of zero-point motion, ^3He is on the average further away from the substrate. Thus, the ^4He concentration is larger close to the bottom of a substrate step, but smaller close to the top of a step. Indeed, at the onset of two-dimensional superfluidity in the mixtures, an interval of coexistence has been observed [27], similar to the above phenomena, but in disagreement with calculations for infinite systems [12]. This was observed using mylar as substrate, but is not reported for quartz substrate [28]. The hills and valleys of the mylar surface should occur in a complicated, interpenetrating, and multiply connected pattern, causing even a small superfluid fraction to be connected across the sample. On the other hand, the steps of the quartz surface probably occur in discrete locations, so that locally superfluid islands do not percolate and cannot give superflow response.

SPECIFIC HEAT MAXIMA

Two microscopic calculations for finite adsorption systems, using approximate renormalization methods, have been completed [23,29]. In this section, preliminary results of another such calculation are reported. They apply to a sample composed of triangular-shaped crystallites of side *L*. The crystallites are completely surrounded by either strong, or weak boundaries, with equal probability. A triangular lattice, Potts model representation of adsorption [23] and a Migdal-Kadanoff renormalization procedure [30] were used, with special attention to the different recursions at the boundary and in the interior. In the previous finite-size calculations, it has been useful to reproduce some known property in the infinite system limit, by adjusting a parameter in the treatment. In the present calculation, as in [23], the number of Potts states q was adjusted to yield a reasonable value, $\alpha \sim 1/3$, for the specific heat exponent of the infinite system.

The recursion Hamiltonians for this problem are

$$-\mathcal{H}/kT = \sum_{<ij>} J\,\delta_{s_i s_j} \qquad (6)$$

for the interior bonds, and

$$-\mathcal{H}/kT = \sum_{<ij>} \left\{K\delta_{s_i s_j} + L\delta_{s_i a}\delta_{s_j a} + H(\delta_{s_i a} + \delta_{s_j a})\right\} \qquad (7)$$

for the boundary bonds, where the variable at each site i, $s_i = a, b, \ldots$, can be in one of q states, $\delta_{s_i s_j} = 1\ (0)$ for $s_i = s_j$ $(s_i \neq s_j)$, and $<ij>$ indicates summation over nearest-neighbor pairs of sites. The initial conditions are $K^{(o)} = J^{(o)}$, $L^{(o)} = 0$, and $H^{(o)} = 0$ $(H^{(o)} = J^{(o)})$ for weak (strong) boundaries (also, $H^{(o)} = 2J^{(o)}$ for corner sites in strong boundaries). The renormalization-group transformation [30], with length rescaling factor of 2, is carried out in two steps. First, a bond-moving approximation gives

$$\tilde{J} = 2J, \qquad\qquad (8)$$
$$\tilde{K} = K + J/2,\ \tilde{L} = L,\ \tilde{H} = H$$

This is followed by decimation along the boundary and in the interior. After several such normalizations, a remaining small system is solved exactly.

Two resulting specific heat curves are in Fig.6, for crystallites of side length L=16 and 32 sites corresponding to 68 and 136 A on graphite. The double-peaked feature is noteworthy. In this calculation, only the two extreme types of boundaries were included, namely either all strong or all weak. A distribution of partly strong and partly weak boundaries could very possibly fill the valley between the two peaks, linearly interpolating between them. This would be the equivalent signal of crossing a coexistence region. Conversely, if there were any reason to suspect a bimodal distribution of graphite crystallite features in an actual experimental sample, two cearly resolved specific heat maxima would not be surprising, based on Fig.6. (The important distinction of strong and weak boundaries occurs only with nearest-neighbor exclusion, such as in krypton or nitrogen. For adsorbed helium, one the other hand, a step bottom and top have essentially equivalent effects.)

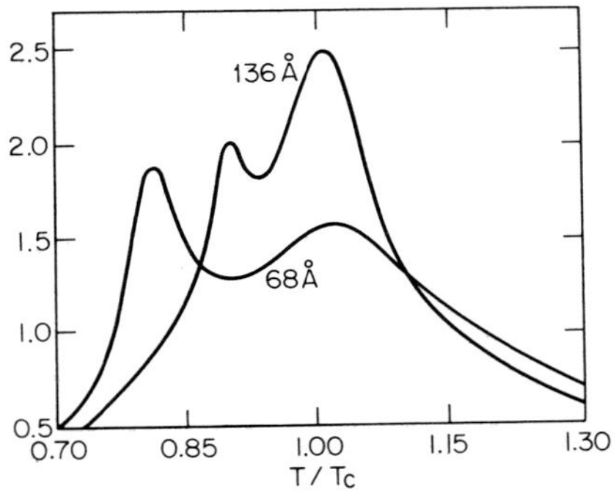

Fig.6 Finite-system specific heat per nearest-neighbor pair (Sec. V).

Finally, a word of caution. It could well be that this double-peak phenomenon is a figment of the Migdal-Kadanoff *approximation*. A phenomenological argument, similar to the one in Sec. IV and relying on a strongly positive specific heat exponent, indicates that with strong (weak) boundaries the critical region propagates from the periphery to the interior (vice-versa) and supports the present results. However, more calculation and thinking are needed to strengthen (or destroy) these preliminary findings.

I would like to thank Professors R.J. Birgeneau, M.E. Fisher, R.B. Griffiths, and Dr. S. Ostlund for useful discussions. This work is supported by NSF Grant No. DMR 78-24185.

REFERENCES

1. A. Thomy and X. Duval, J. Chim. Phys. 66, 1966 (1969); M. Bretz and J.G. Dash, Phys. Rev. Lett. 27, 647 (1971); D.M. Butler, G.B. Huff, R.W. Toth, and G.A. Stewart, ibid. 35, 1719 (1975).
2. M. Schick, J.S. Walker, and M. Wortis, Phys. Lett. A58, 479 (1976); Phys. Rev. B16, 2205 (1977).
3. A.N. Berker, S. Ostlund, and F.A. Putnam, Phys. Rev. B17, 3650 (1978).
4. S. Ostlund and A.N. Berker, Phys. Rev. Lett. 42, 843 (1979).
5. W. Kinzel, M. Schick, and A.N. Berker, in these precedings.
6. R.J. Baxter, J. Phys. C6, L445 (1973).
7. S. Alexander, Phys. Lett. A54, 353 (1975).
8. E. Domany, M. Schick, and J.S. Walker, Phys. Rev. Lett. 38, 1148 (1977); E. Domany, M. Schick, J.S. Walker, and R.B. Griffiths, Phys. Rev. B18, 2209 (1978); E. Domany and E.K. Riedel, Phys. Rev. Lett. 40, 561 (1978).
9. J. Villain, in these proceedings.
10. D.M. Butler, J.A. Litzinger, G.A. Stewart, and R.B. Griffiths, Phys. Rev. Lett. 42, 1289 (1979).
11. C. Domb and M.S. Green, eds., Phase Transitions and Critical Phenomena (Academic, 1977), Vol. 6.
12. A.N. Berker, Phys. Rev. B12, 2752 (1975).
13. J.L. Cardy and D.J. Scalapino, Phys. Rev. B19, 1428 (1979); A.N. Berker and D.R. Nelson, ibid. 19, 2488 (1979).
14. S. Ostlund and A.N. Berker, Phys. Rev. B21 (1 June 1980).
15. J.P. Coulomb, M. Bienfait, and P. Thorel, Phys. Rev. Lett. 42, 733 (1979); P. Vora, S.K. Sinha, and R.K. Crawford, ibid. 43, 704 (1979).
16. E.M. Hammonds, P. Heiney, P.W. Stephens, R.J. Birgeneau, and P. Horn, J. Phys. C13, L301 (1980).
17. A.R. Kortan, P.I. Cohen, and R.L. Park, J. Vac. Sci. Technol. 16, 541 (1979); L.D. Roelofs, T.L. Einstein, P.E. Hunter, A.R. Kortan, R.L. Park, and R.M. Roberts, ibid. 17, 231 (1980).
18. J.S. Walker and M. Schick, Phys. Rev. B20, 2088 (1979).
19. M. Bretz, Phys. Rev. Lett. 38, 501 (1977).
20. M.J. Tejwani, O. Ferreira, and O.E. Vilches Phys. Rev. Lett. 44, 152 (1980).
21. D.E. Moncton *et al.*, preliminary result.
22. M.P.M. den Nijs, J. Phys. A12, 1857 (1979); B. Nienhuis, A.N. Berker, E.K. Riedel, and M. Schick, Phys. Rev. Lett. 43, 737 (1979).
23. A.N. Berker and S. Ostlund, J. Phys. C12, 4961 (1979).
24. M.E. Fisher and A.E. Ferdinand, Phys. Rev. Lett. 19, 169 (1967).
25. E.H. Graf, D.M. Lee, and J.D. Reppy, Phys. Rev. Lett. 19, 417 (1967).
26. G. Ahlers, in Physics of Liquid and Solid Helium, Part I, eds. K.H. Benneman and J.B. Ketterson (Wiley, 1976), p. 101.
27. D.J. Bishop and J.D. Reppy, preprint.
28. E. Webster, G. Webster, and M. Chester, Phys. Rev. Lett. 42, 243 (1979).
29. D.J.E. Callaway and M. Schick, in these proceedings.
30. A.A. Migdal, Zh. Eksp. Teor. Fiz. 69, 1945 (1975) [Sov. Phys.JETP 42, 753 (1976)]; L.P. Kadanoff, Ann. Phys. (N.Y.) 100, 359 (1976).

DISCUSSION

P. Bak:

The phase diagrams that you have shown include commensurate phases, and regions with coexistence of commensurate phases and liquid phases only. On the other hand, real physical systems often exhibit incommensurate phases. Do you believe that it is an artifact of your renormalization group method that you do not obtain incommensurate phases, and is there any hope of generalizing your method to include incommensurate phases?

A. N. Berker:

Yes, it is an artifact of the limited prefacing that incommensurate phases are not obtained.

A calculation of the commensurate-incommensurate transition is in principle possible, by treating continuously varying degrees of freedom in the renormalization. Such calculations have been done for XY models and superfluidity in films.

J. A. Barker:

The "prefacing" transformation seems to be the point at which the physics goes in, and therefore "critical." Could the "prefacing" required vary from point to point in the phase diagram--for example in "commensurate" solid phases the periodic substrate potential is important, but in the transition to liquid or gaseous phase it probably becomes unimportant, so that the Potts model probably doesn't describe the entropy correctly. Is this a problem?

A. N. Berker:

The prefacing transformation is indeed a crucial step of the treatment, where one has to decide (more or less a priori) which degrees of freedom are important, to be treated as a cooperative problem, and which are unimportant, to be immediately summed out. This is also a feature of renormalization-group transformations. If different degrees of freedom are important at different regions of the phase diagram, a different prefacing can be used. For a unified treatment, both sets of degrees of freedom should be projected in the prefacing, as, for example, the Potts-lattice-gas model does for the order-disorder transition on the one hand, and the liquid-gas transition on the other hand.

S. Doniach:

I would like to bring up another variable, in addition to the commensurable/incommensurable variable, that of zero point quantum fluctuations, of particular importance near to the 50% occupancy leading to the highly degenerate antiferromagnetic Ising state (as opposed to 1/3 occupation for 3-state Potts). Have you seen evidence of the importance of zero point motion in connection with fitting classical Potts models in data for He on grafoil?

A. N. Berker:

This should be non-negligible at the dense coverages of light adsorbates. Indeed, the recent position-space renormalization-group calculation of Kinzel et al. (these proceedings) suggests that zero-point motion does affect the effective classical potential between helium atoms on krypton-plated graphite.

J. P. McTague:

With regard to John Barker's question concerning the effect of treating the fluid phase as a disordered lattice gas, our molecular dynamics calculations of Kr on graphite at 0.7 registered monolayer [F. Hanson and J. P. McTague, J. Chem. Phys., June 15, (1980)] show a first order melting at a reduced temperature of $T_m = 0.43$, while the lattice model of Ostlund and Berker [Phys. Rev. Lett. 42, 843 (1979)] has $T_m = 0.6$. The lower entropy of the lattice gas appears to raise the transition temperature, but not to qualitatively change the nature of the transition.

Sinha, ed. Ordering in Two Dimensions

MULTI-CRITICAL PHASE DIAGRAM OF A CHEMISORBED LATTICE GAS SYSTEM-O/Ni(111)*

Robert L. Park†, T. L. Einstein‡, A. R. Kortan†, and L. D. Roelofs‡

Department of Physics and Astronomy, University of Maryland, College Park, Md. 20742

Chemisorbed overlayers on single-crystal metal surfaces are realizations of truly two-dimensional lattice gas systems since the adsorption is strongly site specific. Such overlayers therefore offer experimental tests of many lattice gas model predictions. Using LEED, we have determined the phase diagram for O adsorbed on Ni(111) for coverages $\theta \leq .4$ monolayers. Three distinct ordered phases are observed: p(2x2) below $\theta = .27$, $(\sqrt{3}x\sqrt{3})R30°$ above $\theta = .29$, and a complex intermediate phase involving short range 2x2 coordination and two different three-fold binding sites, with with dense antiphase domains. The intermediate structure is suggestive of those predicted by recent theories of melting in 2-D. A variety of cooperative phenomena is observed including first and second order phase transitions and a tricritical point. In particular, the p(2x2) phase undergoes a continuous order-disorder transition predicted to be in the 4-state Potts universality class. To determine critical exponents for this transition, we measured the adlayer-induced (1/2,0) beam profile at closely spaced temperature intervals. After deconvoluting the instrument response from these profiles and subtracting background terms, the nearly Lorentzian short-range-order contribution can be separated from the long-range order Bragg part. The Bragg intensity is then fitted to the function $(T_c-T)^{2\beta}$ to obtain T_c and β. The exponents γ and ν can be determined from the amplitude and the width, respectively, of the short range order part. Our measured critical exponents will be contrasted to the predictions of various theories.

INTRODUCTION

For nearly two decades low energy electron diffraction (LEED) has been largely preoccupied with the problem of determining the two-dimensional crystallography of static surface structures. The possibility of using LEED to follow the details of phase transitions in two dimensions, however, seems to have been recognized by Davisson and Germer [1] in the first full-length paper describing the diffraction of electrons from crystal surfaces. In that 1927 paper they observed that "when the target is not free of gas atoms, anomalous diffraction beams occur in each of the principal azimuths. These constitute a family attaining greatest intensity when the quantity of adsorbed gas has a certain critical value. These anomalous beams could be accounted for as radiation scattered by a layer of gas atoms of the same structure and orientation as the nickel atoms but of twice the scale factor." The structure they inferred for the adsorbed layer is shown in Fig. 1, taken from their original paper.

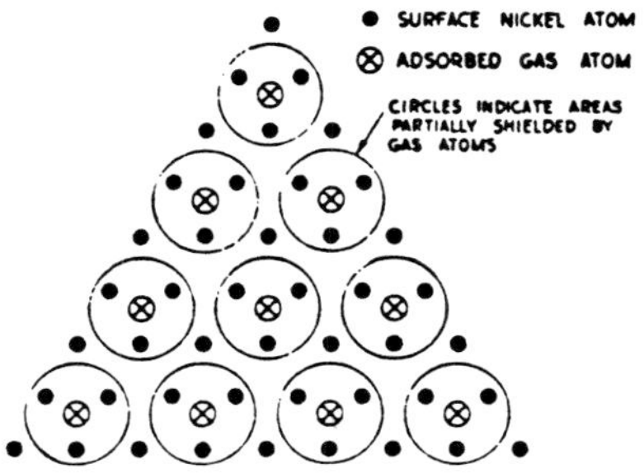

Fig. 20. Arrangement of gas atoms on the surface, and the topmost layer of nickel atoms.

Fig. 1: Fig 20 of Davisson and Germer's classic paper (1), proposing the p(2x2) superlattice arrangement of O atoms on a Ni(111) surface (and conjecturing that the observed new LEED beams arise from shielding of 3 out of 4 surface Ni atoms).

They went on to report that "we have further observed that these beams cannot be made to appear when the temperature of the target is somewhat above that of the room, although under these conditions oxygen still collects on the target. The explanation of this behavior may be that the melting point of the two-dimensional crystal is not far above room temperature. We have not yet observed whether the beams disappear sharply at a critical temperature." We can find no evidence that Davisson and Germer ever made that final observation but it certainly seemed, more than half a century later, that somebody should.

We have accordingly obtained the complete phase diagram for the system consisting of oxygen atoms on the (111) face of nickel. We have also carried out a Monte Carlo simulation of the phase diagram in order to determine the adatom interactions. In addition to LEED, we used Auger electron spectroscopy to determine the relative coverage of the surface, as well as work function measurements to verify that the chemical state is not a function of temperature or coverage.

Since oxygen atoms are strongly bound to specific sites on the nickel surface, the adlayer provides an excellent realization of a two-dimensional lattice gas system and displays a rich variety of critical phenomena. In particular, the continuous order-disorder transition that Davisson and Germer reported for the p(2x2) structure is predicted to lie in the 4-state Potts universality class [2]. We have attempted to test this prediction by extracting critical exponents from the diffraction beam profiles.

Experimental

Experiments [3,4] were carried out in an ultra-high-vacuum stainless steel bell-jar system with a total base pressure below 10^{-10} Torr. A quadrupole mass filter was used to monitor the composition of gases in the system. All measurements were made with the standard four-grid LEED optics, which was operated as a retarding potential analyzer to obtain the Auger spectrum. Work function changes were determined from the low-energy cutoff of the secondary electron spectrum.

Experiments were carried out on five different samples, all of which were nominally 99.995% pure. Initially the samples exhibited mosaic spreads of 30' to 60'. Two of the samples, however, were zone-refined to eliminate the mosaic spread and oriented to the (111) direction to better than 20'. The results presented here, however, were found not to be measurably dependent on the mosaic spread.

Prior to any measurements the samples were cleaned by argon-ion sputtering and annealing at 1000 K. Oxygen exposures of the cleaned sample were in the range of 10^{-9} - 10^{-8} Torr and were all made at 273 K sample temperature. The temperature was measured by a calibrated chromel-alumel thermocouple. The output of the thermocouple was also used to provide feedback for a circuit that controlled the current to the specimen. A liquid nitrogen cooling system allowed measurements to be made at temperatures as low as 100 K.

An optical system mounted external to the vacuum was used to project the real image of a given diffraction beam onto the aperture of a photon counter. The angular profile of the beam was obtained by varying the incident beam energy to move the beam across the aperture. The equivalent angle of acceptance of the aperture was 1°.

Results

At room temperature and low coverages oxygen adsorption results in no new diffraction features. As the coverage approaches a quarter of a monolayer, however, the oxygen orders in p(2x2) phase. The linear increase of the work function with coverage suggests that at all coverages the oxygen atoms occupy the same sites. This lattice gas behavior is a natural consequence of a strong and site-localized adatom-substrate interaction. At coverages above about 0.4 of a monolayer, however, nickel oxide in the sodium chloride structure begins to form [5], and at temperatures above about 500 K the oxygen diffuses into the bulk. Since our interest concerns 2-d cooperative behavior, we have stayed below these limits in our study of the overlayer.

To obtain the phase diagram, the surface was exposed to oxygen at 273 K to reach a given relative coverage, as indicated by the Auger electron spectrum. With the coverage held fixed at this value, the temperature was varied from about 100 K to 500 K, and the changes in the LEED pattern with changing temperature monitored.

Varying the temperature between these limits at fixed coverages produced no change in either the work function or the Auger electron signal from the oxygen. The only changes were in the LEED. An ordered p(2x2) structure could be obtained at oxygen coverages well below a quarter of a monolayer by lowering the temperature sufficiently. The integral-order diffraction beams corresponding to the substrate spacings were unaffected by temperature changes, except for the Debye-Waller effect. In contrast, the half-order diffraction beams, produced by the double spacing of the overlayer, broadened abruptly and disappeared above a critical temperature. It is thus evident that at low temperatures the oxygen condenses into ordered islands. The formation of these islands in the p(2x2) structure demonstrates that the third-nearest-neighbor interaction is attractive. Similar island formation and dissolution has been studied by Lagally and co-workers for O/W(110) [6]. As the oxygen coverage approaches a quarter of a monolayer the extinction of the half-order beams is smooth and very reversible. This behavior is characteristic of a single-phase region with a continuous order-disorder transition. We will discuss this order-disorder transition in greater detail in the next section.

When the oxygen coverage is increased above about .27, the half-order LEED beams abruptly broaden. With increasing coverage the (1/2,1) beams split into two components along the (0,1) directions, while the (1/2,0) beams remain as single broad beams, as depicted in fig. 2. The splitting of the (1/2,1) beams increases continuously with coverage.

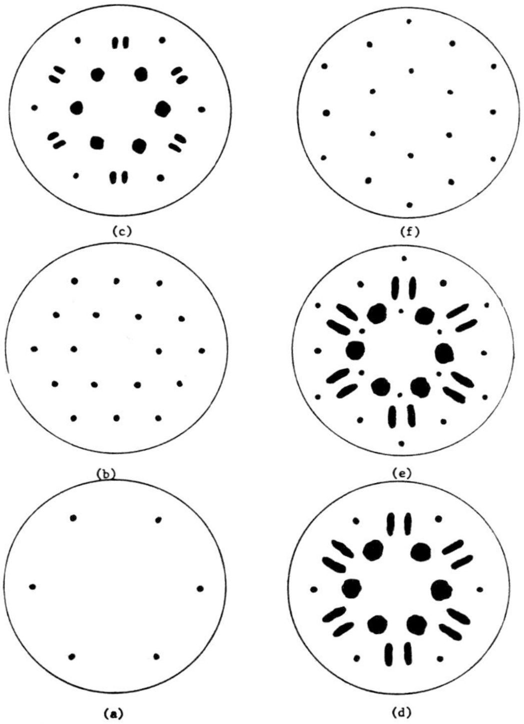

Fig. 2: Schematic of the LEED beam pattern for the intermediate structure found at coverages between .27 and .29. The (1/2,0) beams broaden circularly, while the (1/2,1) beams split and elongate.

Broadening of a diffraction beam, of course, represents a loss of long-range order. However, in the case of a simply related surface structure, (whose unit mesh dimensions are simple multiples of the substrate lattice dimensions) [7], this broadening does not affect all diffraction beams. The fact that no broadening is observed around the integral order positions is evidence that the adsorbed atoms occupy positions of high symmetry, presumably three-fold hollows, as Marcus et al. [8] concluded on the basis of dynamic LEED calculations. There are, however, two types of these sites: fcc and hcp, and the LEED calculations were unable to distinguish between them.

The loss of long-range order in a simply related surface structure requires the formation of anti-phase boundaries between ordered domains [9]. The splitting of diffraction beams, as opposed to simple broadening, requires that the anti-phase boundaries have a relatively well-defined separation. However, the fact that the (1/2,1) beams exhibit splitting while the (1/2,0) beams do not provides further information on the nature of the anti-phase boundaries. The key point is that these beams differ by substrate reciprocal lattice vectors from unsplit half-order beams. The result is that the anti-phase boundaries must separate regions in which only fcc sites are occupied from regions in which only hcp sites are occupied. The density of atoms along such a boundary is slightly greater than the density of atoms in the p(2x2) structure. If the coverage is increased to .29, however, very sharp 1/3 order diffraction beams, corresponding to a ($\sqrt{3}$x$\sqrt{3}$)R30° structure, abruptly appear super-imposed on the pattern of the intermediate structure. At slightly higher coverages the diffraction pattern of the intermediate structure is extinguished, leaving only diffraction beams from a well-ordered ($\sqrt{3}$x$\sqrt{3}$)R30° structure.

There has been some dispute in the literature regarding the existence of a ($\sqrt{3}$x$\sqrt{3}$)R30° phase. The reason for this controversy was evident when the sample temperature was increased. At about 300 K the 1/3 order beams are abruptly extinguished. The phase transition takes place over a small temperature interval with no beam broadening but considerable hysteresis, and is thus first-order. All this information is summarized in the phase diagram shown in fig. 3.

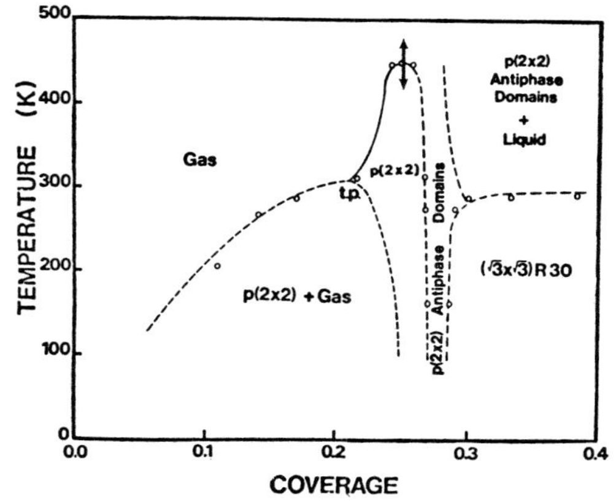

Fig. 3: Phase diagram for O atoms on Ni(111). Solid lines indicate continuous phase transitions, while dashed lines are first order. A tricritical point is denoted by "t.p.".

First-order phase boundaries are represented by dashed lines, continuous phase boundaries by solid lines. The sharp rise in the transition temperature near θ = .21 strongly suggests the presence of a tricritical point, where three phases meet. The phase boundary separating the co-existence and long-range ordered p(2x2) regions is conjectured rather than observed experimentally. We do not have measurements below 100 K, since the equilibration times for the overlayer at such low temperatures exceed practical experimental time-scales.

The p(2x2) Order-Disorder Transition

As pointed out in the previous section, near a coverage of .25 the p(2x2) structure disorders in a reversible second-order transition. According to Domany et al. [2], this continuous transition should lie in the 4-state Potts universality class, in which case the critical exponents are theoretically well characterized. To test this prediction, diffraction beam profiles were measured at closely spaced temperature intervals extending through the transition temperature. Ideally the diffraction beam profiles should consist of the sum of a delta function contributed by the long-range order and a nearly Lorentzian contribution resulting from scattering from short-range order. This sum is then broadened by convolution with an instrument response function. The broadening effect of the LEED instrument limits the range on the surface over which correlations are detected. The instrument response function was taken to be the measured beam profile at very low temperatures. It was deconvoluted from each profile measured at higher temperatures by two different two-dimensional methods. To reduce truncation effects in the deconvolution, each profile was analytically extended and a uniform background was subtracted. Additionally, the Debye-Waller dependence of the intensity was removed.

In fact, of course, the extent of long-range order in the p(2x2) structure was limited at all temperatures by the degree of perfection of the substrate. Steps are thought to be the dominant contribution limiting the extent of long-range order in the measurements reported here.

An additional important factor which must be considered in the analysis of LEED beam profiles is multiple scattering. Indeed, in the present case the intensities of the half-order diffraction beams are found to be strongly modulated in energy, an indication of the significance of multiple scattering. However, the strong peaking of the atomic differential scattering cross-section in the forward and back directions [10] insures that multiple scattering within the oxygen overlayer is unimportant relative to multiple scattering between adatom and substrate layers [11]. Therefore LEED still measures the pair correlation function of the overlayer; higher-order adatom correlation functions are not significantly included in the measurement. Variation in scattered intensity as a function of diffraction angle does, however, influence the diffracted beam profiles. To test the significance of this effect, we assumed that the plot of beam intensity as a function of electron energy represented an approximate multiple scattering envelope. The effect on the diffraction beam profiles was found to be negligible in comparison to other uncertainties.

To compare the observed order-disorder transition with the predictions of theoretical models, three critical exponents were determined from the variation of the diffraction beam profiles in the vicinity of T_c. The exponent β, which describes the variation of the order parameter, was obtained from the amplitude of the long-range order component of the intensity. The exponent ν, which gives the behavior of the correlation length, was obtained from the width of the short-range order part of the intensity. The exponent γ, which describes the divergence of critical scattering and is analogous to the susceptibility in a magnetic phase transition, was obtained from the amplitude of the short-range order term.

The individual diffraction beam profiles were analyzed in two different ways. The first method consisted of Fourier space deconvolution of the two-dimensional profiles. The Bragg (long-range order) intensity was separated from the critical (short-range order) scattering by fitting to the wings of the deconvoluted Fourier transformed diffraction beam. In the second method, the sum of a delta function and a Lorentzian was convoluted with the (measured) instrument response function. The width of the Lorentzian and the two heights were adjusted to optimize the fit of the profiles. The results for the Bragg intensity ($T \leq T_c$) agreed with those obtained by the first method. For $T \geq T_c$ the second method produced better results since less noise was added to the data. By assuming that the critical scattering contribution can be represented by a Lorentian, this second method ignores line shape corrections described by the exponent η. We found, however, that our results were insensitive to the use of a function with nonzero η. Our results for the exponents ρ, ν and γ are listed in table 1, which also displays the values appropriate to various relevant model systems.

Table 1

EXP.	MEASURABLE	ISING	POTTS (4-State)	POTTS (Fisher Ren.)	p(2x2)O/ Ni(111)
α	Specific Heat	0	2/3	2	
β	Magnetization	1/8	1/12	1/4	0.2±.05
γ	Susceptibility	7/4	7/6	7/2	1.6±0.2
δ	Coexistence	15	15	45	
ν	Correl. length	1	2/3	2	0.9±0.1

A sample fit of the long-range order part of the intensity (for $T \leq T_c$) is shown in figure 4.

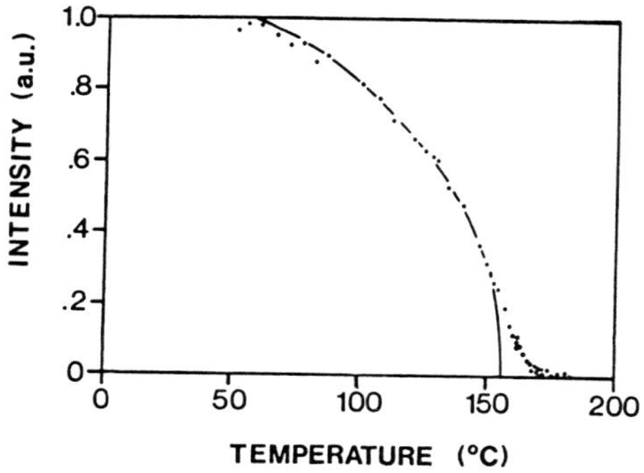

Fig. 4: Sample plot of extracted Bragg intensity of the (1/2,0) beam in arbitrary units, versus temperature. The rounding near T_c comes from finite-size effects. The line, based on a least squares fit, has the form $I \propto (T_c-T)^{2\beta}$, with T_c = 157.0 C and β = 0.2.

The first method described above was used to obtain the data points, which were then fit to a power law by non-linear curve-fitting techniques [12]. Since the rounding left in the data was insensitive to increases in the 'strength' of the deconvolution, it must come from the size limitations imposed by substrate imperfections. This characteristic size is of order 300 Å. Because of the rounding, our determination of β has rather large error bars.

Figures 5a and b show sample fits for the exponents ν and γ, respectively.

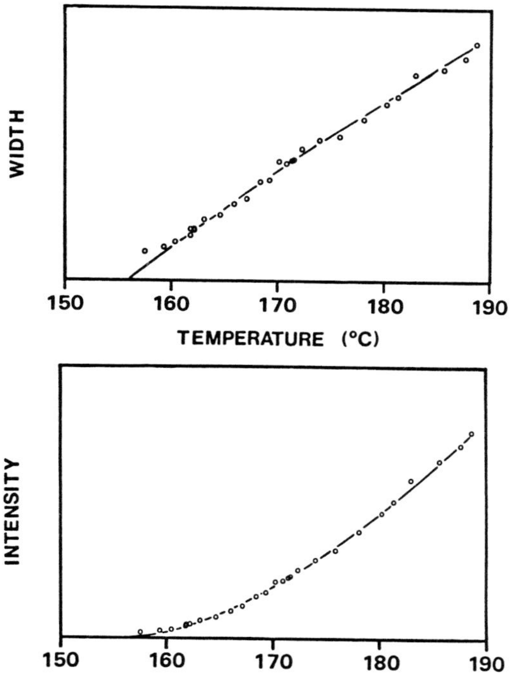

Fig. 5: Plots of the (a) width and (b) height of the scattering intensity for $T > T_c$ when no Bragg part need be separated. The lines are the results of nonlinear fits with respect to T_c and the critical exponent ν and γ, respectively, (as well as the relevant overall amplitude).

Here the second method was used to obtain the data points from the measured profiles. The data for ν and for γ are consistent in that nonlinear fits to the two curves give nearly identical transition temperatures. The error bars for these values quoted in table 1 were obtained by examining the variations of the deduced exponents as data points were removed from the analysis.

For a chemisorbed adlayer, the coverage (rather than its conjugate field, the chemical potential) is the experimentally accessible variable. Since our scan of the order-disorder transition is done at fixed coverage, one might expect that Fisher renormalization [13] applies. However, the region over which this correction is appropriate shrinks as one approaches an extremum in the phase boundary. For several reasons [14] we found it desirable to make the measurements as close as possible to the maximum in the p(2x2) phase boundary at θ = .25 (cf. double arrow in figure 3). In fact our values for ν and γ do not agree with renormalized values of exponents for any plausible lattice model in which the renormalization is non-trivial (i.e. $\alpha \neq 0$). Since the specific heat exponent α is not experimentally accessible in these systems, we cannot determine directly whether renormalization would have any noticeable effect.

Even with their large error bars, the exponent values we have determined are not consistent with those of the predicted universality class (4-state Potts model). The finite size limitations restrict the analysis to reduced temperatures $t \equiv (T - T_c)/T_c$ such that $|t| \geq 0.01$. Thus it is possible that the critical region is extremely small and thus that Potts-like exponents only obtain for $|t| < 0.01$. There is however another possible explanation. Domany and Schick [2] have pointed out that a definitive 4-state Potts classification is possible only if there is a non-zero binding energy difference, h_s (i.e. a staggered field), between sites on the two triangular sub-lattices which form the hexagonal net. For h_s=0 the continuous-transition possibilities are Heisenberg, or being followed by 4-state Potts (as T increases). The previously discussed analysis of the complex intermediate structure between the p(2x2) and $(\sqrt{3}\times\sqrt{3})R30°$ structures demonstrates that both kinds of sites can be occupied; thus while crystal field effects dictate that h_s be finite, it is likely to be quite small. While our values for ν and γ (see table 1) are suggestive of Ising results, our β is strikingly not. A conceivable resolution of this unclear situation might come from including logarithmic corrections to the singular part of the free energy [15].

Interactions between Adatoms

Monte Carlo simulation methods provide a means of determining the values of the adatom-adatom interactions which underpin the phase diagram. We have developed the first such simulation for the case in which the net of potential occupation sites has hexagonal symmetry. More accurately, the hexagonal net of sites is actually a triangular lattice with a basis of two sites, which have different binding energies. This staggered field is a parameter in the simulation along with six adatom-adatom interaction energies between distinct pairs. This number of parameters is far greater than in any previous simulation and reflects the richness of the system under investigation.

We have carried out the simulation in two modes. In the first, in close correspondence to the experimental situation, the coverage is fixed, and adatoms hop from site to site. From a computational standpoint the fixed chemical potential mode, in which atoms hop onto and off the lattice, is often preferable. It has been used by most others in previous simulations. We find, however, that the relative economy of the two schemes depends drastically on which region in the phase diagram is being investigated; both are crucial for a complete treatment of the phase diagram.

A determination of adatom interactions proceeds iteratively by assumption of a set of interactions, generation of the resulting phase diagram, comparison to experimental phase diagrams and adjustment of the assumed interaction strengths. Our initial simulation shown in fig. 6 can be compared to the experimental phase diagram in fig. 3.

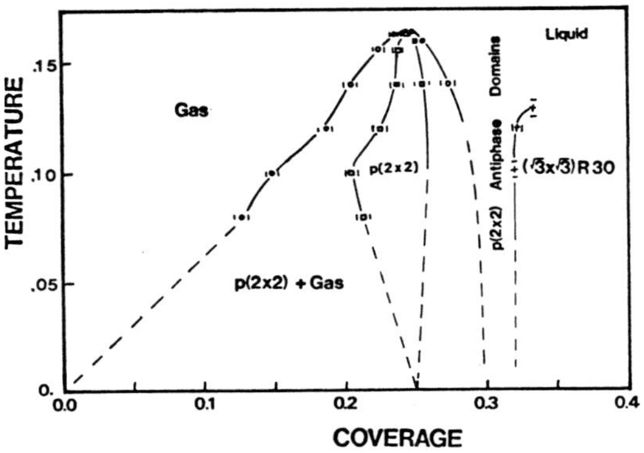

Fig. 6: Phase diagram obtained by Monte Carlo simulation using initial set of parameters, to be compared with Fig. 3. Subsequent versions more accurately reproduced the tricritical point region and coverages below .25 in general.

Our results for the adatom interactions support the indirect interaction picture for this case. The results are not consistent with a van der Waals-like interaction, as has been proposed by Gallagher and Haydock [16] for O/Ni(100), since all but one of the lateral interactions are repulsive, many weakly so. Because of the expense of the Monte Carlo calculations, we also developed a relaxation mean-field method for this system. We had hoped that the meanfield method might be reasonably accurate near some of the first-order transitions which occur in this system. Unfortunately it generally overestimates transition temperatures by a factor of 2 or 3 [17] and cannot reproduce the intermediate-coverage, dense-antiphase-boundary regime. We expect this problem to occur for most two-dimensional systems due to the importance of configurational entropy even near first-order transitions.

Discussion

We have seen that chemisorbed overlayers provide stimulating realizations of two-dimensional lattice gas models. They thus offer an area in which experimentalist and theoretician, statistical mechanic and solid state phyisicist can fruitfully collaborate. We have shown how, using ideas from theory, critical exponents and lateral interaction energies can be obtained from experimental results. The chief limitation imposed by experimental apparatus is the restricted coherence of the LEED instrument; improvements in this regard will permit useful information to be obtained closer to T_c. Less subject to human effort is the paucity of ideal closed systems, in which the substrate geometry is rigid over the phase diagram and the adatoms are completely confined to the surface. A fairly complete catalogue [18] has been made for relatively simple systems. It must still be ascertained how widespread is the substrate reconstruction found on the (100) faces of some bcc metals [19]. The major immediate theoretical problem is to understand the numbers we extract for critical exponents. More generally, we do not know over what range it is valid to invoke universality between models having the same Landau-Ginzburg-Wilson free energy expansion but different microscopic Hamiltonians and over what temperature range criticality holds. To round out understanding, it will be important to compute quantitatively rather than qualitatively the lateral interaction energies between adatoms that constitute the parameters of the model Hamiltonians.

Acknowledgements

We are grateful for helpful discussions with Profs. A. N. Berker, M. G. Lagally, and M. Schick.

REFERENCES AND FOOTNOTES

*Some of this work is from a dissertation to be submitted to the Graduate School, University of Maryland, by Lyle D. Roelofs in partial fulfillment of the requirements for the Ph.D. degree in Physics.

†Supported by the National Science Foundation under grant DMR 7900323.

‡Supported by the Dept. of Energy under grant DE AS05-79ER-10427. Computer time and facilities supplied by the U. of Maryland Computer Science Center.

1. C. Davisson and L. H. Germer, Phys. Rev. 30 (1927) 705.

2. E. Domany and M. Schick, Phys. Rev. B20 (1979) 3828. See also E. Domany, M. Schick and J. S. Walker, Phys. Rev. Lett. 38 (1977) 1148; E. Domany, M. Schick, J. S. Walker and R. B. Griffiths, Phys. Rev. B18 (1978) 2209.

3. A. R. Kortan, P. I. Cohen and R. L. Park, J. Vac. Sci. Tech. 16 (1979) 541.

4. A. R. Kortan, Ph.D. thesis, U. of Maryland (1980).

5. P. H. Holloway and J. B. Hudson, Surf. Sci. 43 (1974) 141.

6. T. M. Lu, G. C. Wang and M. G. Lagally, Phys. Rev. Lett. 39 (1977) 411; G.-C. Wang, T. M. Lu, and M. G. Lagally, J. Chem. Phys. 69 (1978) 479.

7. R. L. Park and H. H. Madden, Surf. Sci. 11 (1968) 188.

8. P. M. Marcus, J. E. Demuth and D. W. Jepsen, Surf. Sci. 53 (1975) 501.

9. R. L. Park and J. E. Houston, Surf. Sci. 18 (1969) 213; and J. E. Houston and R. L. Park Surf. Sci. 21 (1970) 209.

10. M. Fink and J. Ingram, At. Data 4 (1972) 1.

11. L. D. Roelofs, Ph.D. thesis, U. of Maryland (1980).

12. P. R. Bevington, Data Reduction and Error Analysis for the Physical Sciences, McGraw-Hill, New York, 1969.

13. M. E. Fisher, Phys. Rev. 176 (1968) 257.

14. At the maximum the transition temperature is most reproducible, the mobility is highest, and the critical region is largest in terms of real (and reduced) temperature.

15. M. Nauenberg and D. J. Scalapino, Phys. Rev. Lett. 44 (1980) 837; Alan C. Brown, unpublished.

16. J. M. Gallagher and R. Haydock, Surf. Sci. 83 (1979) 117; J. M. Gallagher, R. Haydock and V. Heine, J. Phys. C12 (1979) L13.

17. K. Binder and D. P. Landau, Phys. Rev. B21 (1980) 1941, have noted similar difficulties in simpler 2-d lattice-gas systems.

18. Eg., G. A. Somorjai and L. L. Kesmodel, in MTP Review on Surface Science, ed. M. Kerker, Butterworth, London, 1974.

19. M. K. Debe and D. A. King, Phys. Rev. Lett. 39 (1977) 708; Surf. Sci. 81 (1979) 193. R. A. Barker and P. J. Estrup, Phys. Rev. Lett. 41 (1978) 1307.

DISCUSSION

S. Fain:

What types of problems will limit the applicability of simple theoretical models for critical behavior to LEED experiments on real chemisorption systems? For example, will the different types of sites be a general feature? Will long-range adatom-adatom forces be important?

R. L. Park:

The problem of different sites is not necessarily confined to hexagonal surfaces, but that is where the problem is likely to be worst, since the energy difference between hcp and fcc sites is expected always to be small, but never zero.

Forces beyond third nearest neighbor are probably less important than trio interactions.

It would probably not help much to build a higher resolution LEED system unless we can find a way to produce more perfect surfaces.

M. Shick:

A few comments. First, even if there are two kinds of sites on the hexagonal lattice, which is surely the case on the actual physical surface, the transition to the p (2x2) state is still expected to be in the class of the four-state Potts model. This is discussed explicitly by Domany and me.

Second, if your measurements were carried out at the maximum of the phase diagram, i.e., T_c is not changing with coverage, there is no Fisher renormalization, a point discussed by Fisher in his paper.

Third, the four-state Potts model has logarithmic corrections to its power law singularities. These have made difficult the extraction of exponents from theoretical "experiments" (i.e., series) and may be the source of difficulty in your work. Such logarithmic terms should be included in your fit.

R. L. Park:

We took our order-disorder measurements at an extremun of the phase diagram to minimize the region over which Fisher renormalization would be important. We agree that it is not likely to a factor.

We will attempt to include the logarithmic corrections to the power law singularities.

S. C. Moss:

Did you follow the main beam intensity through the phase transition and was there any change through T_c (or at T_c) other than the normal Debye-Waller effect?

R. L. Park:

Yes, we followed the integral order beams. The intensity-energy profiles do not change as the sample is taken through the order-disorder transition.

A. N. Berker:

At a second-order phase transition, the LEED spot intensities from the substrate should also have a singularity, with critical exponent $1-\alpha$, reflecting the behavior of overall density. However, this may be hard to extract, due to background.

R. L. Park:

The background is indeed changing rapidly near T_c.

Sinha, ed. Ordering in Two Dimensions

HEAT CAPACITY STUDIES WITH MONOLAYER FILMS

O. E. VILCHES

Department of Physics
University of Washington
Seattle, WA 98195

Heat capacity measurements on a variety of monolayer films adsorbed on solid and liquid substrates have been reported in the literature. Results have been used to identify phases, to establish phase boundaries, to find the singular behavior at order-disorder transitions, to calculate ground state energies and to study the influence of different substrates on the monolayer. Particular results are shown as they apply to some of these specific problems.

INTRODUCTION

Heat capacity measurements with physisorbed films have been used to identify phases and phase transitions in monolayers of atoms and molecules. With the development of highly sophisticated elastic and inelastic scattering techniques the characterization of solid phases is now made by their use rather than calorimetry. Nevertheless, thermal measurements complement other methods, and in some cases are unique in the information they provide, like in the study of fluid phases and of order-disorder transitions where results can be compared to particular theoretical models. Furthermore, when combined with vapor pressure measurements, a rather accurate description of the thermodynamic properties of the system under study can be obtained. In this article we review some of the specific problems that have been addressed with calorimetric measurements. Very recent developments can be found in other contributions to these Proceedings.

EXPERIMENTAL TECHNIQUE

All the heat capacity results reported in the physisorption literature have been obtained by adiabatic calorimetry. The substrate is in the form of a large specific area powder or porous solid, to which the desired monolayer can be adsorbed [1]. The vast majority of measurements have been done on various forms of expanded graphite (Grafoil, graphite foam, ZYX [2]) but very interesting results have been reported for ^{3}He films adsorbed on ^{4}He films [3]. All the data are extracted by a substraction technique where the film heat capacity is the difference between the adsorber plus adsorbate and the adsorber alone heat capacities. This presents no serious problems for measurements at helium temperature on helium monolayers on Grafoil since the desired signal is a substantial fraction of the total heat capacity, but it becomes a larger problem for monolayers of other films at higher temperatures or when relatively small specific area substrates are used at any temperature.

RESULTS

A few specific examples of heat capacity measurements and their use to understand various phases or phase transitions is given below.

I. - The Gas Phase. A two dimensional (2-D) monoatomic gas should give a specific heat signal that at high temperatures asymptotes to $C/Nk = 1$. In addition, one should expect, as in the case of helium, that at low enough temperature quantum effects would become important. It is not surprising then that the best studied gas phases are the ones of ^{4}He, ^{3}He, [1] and ^{3}He-^{4}He mixtures [4]. Partial results also exist for H_2 adsorbed on graphite [5].

The 2-D gas-like behavior of the helium isotopes and their mixtures when adsorbed on graphite can be almost exactly accounted for by considering the effect of interactions on the ideal system. The quantum mechanical second virial coefficient, and the heat capacity contribution from its second derivative, were calculated by Siddon and Schick [6] who found excellent agreement between their calculations and the experimental results. A detailed review of the measurements and calculations has been given by Dash and Schick [7]. The effect of the solid substrate on the gas-like behavior of helium has been probed in measurements using rare gas plated graphite as adsorber [8 - 10]. Coating the graphite with one compressed (incommensurate) layer of Ar [8,9] produces an increase of approximately a factor of two on the temperature at which the rise of the ^{4}He specific heat above $C/Nk = 1$ occurs. A similar effect is observed after a one layer plating of Kr [11]. On the other hand, Ne plating [12] or one solid layer of ^{4}He [13] produces a 40% decrease in that temperature. Similar substrate effects are seen on ^{3}He adsorbed on all the plated substrates; the observed property is the occurrence of a heat capacity peak at about 0.8 to 0.9 K for the Ar/graphite [8] or Kr/graphite [11] substrates, while Ne/graphite [12], and ^{3}He/graphite [14] substrates results in no ^{3}He peak at all.

The second virial contribution to the heat capacity of ^{4}He and ^{3}He on the plated substrates has been studied by Rehr and Tejwani [15,16], who assume that the helium atoms form a 2-D gas that is partially localized, interacting mainly at the distances determined by the substrate sites. Their theoretical results are in reasonable agreement with the experimental results.

The free surface of bulk ^{4}He at low enough temperature is another system where ^{3}He can be adsorbed to study its 2-D Fermi-system behavior [17]. Heat capacity results for ^{3}He adsorbed on relatively thick (d > 1.8 nm) ^{4}He films, themselves formed on Nuclepore filters [18], in the 0.05 to 0.3 K region show in fact a 2-D gas like behavior [3]. Depending on the ^{4}He film thickness and the temperature, evaporation of ^{3}He from the surface into the ^{4}He film can also be observed.

II. The Commensurate Phase and Order-Disorder Transition in Helium. It is well known that many atoms and molecules will form commensurate phases when adsorbed on a variety of substrates. While the actual structure and commensurability can only be found through scattering experiments, heat capacity measurements permit the study of the order-disorder transition and the determination of the phase boundary and the critical exponent α associated with those measurements. In a rather careful experiment Bretz [19] studied the commensurate-disorder transition of helium adsorbed on graphite. In the ordered state there is a helium atom occupying one of every three sites of the underlying graphite lattice. In the disordered state the helium atoms form a dense 2-D gas. While earlier experiments on the same system [1] had been compared to the heat capacity result of the Ising model ($\alpha = 0$), Alexander [20] pointed out that the He/graphite system should be compared instead to the 3-state Potts model (for which α has been proved recently [21] to be 1/3). Bretz's experiment gave approximately this value ($\alpha = 0.36$).

In a series of papers Domany et al. [22-25] studied a variety of hypothetical commensurate systems on triangular, rectangular, and honeycomb lattices and predicted the universality classes to which their order-disorder transition should belong. One of the systems was a honeycomb lattice where half the sites are occupied. The order-disorder transition should belong to the same universality class of the Ising model, thus have a logarithmic singularity ($\alpha = 0$) in the specific heat at the transition. This particular system can be realized by adsorbing He on a close-packed substrate like Kr, Ar, or Xe. Recent results for ^{4}He/Kr/graphite [10] show in fact that $\alpha = 0$, while removal of the Kr plating changes α to about 0.28. The difference between this result and that of Bretz [19] for the ^{4}He/graphite system is not understood. The rounding of the peaks at rather large values of $(T - T_c)/T_c$ is due to finite size effects (see article by Callaway and Schick in these Proceedings).

III. - Melting of Helium Adsorbed on Graphite. For surface densities larger than 0.078 atoms per square angstrom both ^{4}He and ^{3}He form an incommensurate 2-D solid when adsorbed on graphite. Of particular interest is the melting of these solids, specifically whether it occurs via a first order or a continuous transition. When melting occurs via a first order transition there should be a latent heat associated with it and a fluid-solid coexistence region. The constant area specific heat should show discontinuities associated with entering the two phase from the pure solid region, and leaving the two phase and entering the pure liquid region, with the excess heat capacity determined by the latent heat of transformation and the rate of conversion of solid to fluid.

The experimental results so far have not shown this type of melting, but they have characteristics not entirely understood [26-28]. In particular one should compare ^{3}He and ^{4}He at the lowest solid densities, mainly those that show melting below 4 K [28], since at these coverages layer promotion is not a factor. While the ^{4}He melting has "rounded" anomalies, ^{3}He shows well defined peaks.

Comparisons have been made between the ^{3}He melting and theoretical results of Halperin and Nelson [29] for 2-D melting on a smooth substrate. The theory develops a model for 2-D melting advanced earlier independently by Feynmann (see Elgin and Goodstein [27]) and Kosterlitz and Thouless [30]. In particular experimental results using nuclear magnetic resonance [31,32] appear to indicate the presence of two transitions, one occurring at a lower temperature than the specific heat peak (as theoretically expected for the melting temperature), the other one essentially in agreement with it. New experimental results obtained with elastic neutron scattering measurements by Lauter et al. (reported elsewhere in these Proceedings) indicate a complex melting mechanism. The situation is not yet resolved.

IV. - Melting at an Apparent Triple Line. In three dimensional (3-D) systems (except helium) there is a triple line (in the P, V, T space) where the solid, liquid and vapor phases coexist. When heat capacity measurements are made and this line is crossed melting occurs into the two fluid phases (liquid and vapor) at constant P with a latent heat that produces an "infinite" singularity in C/Nk.

Using adsorption isotherms one can construct phase diagrams for adsorbed atoms and molecules where such triple lines exist. In other systems where isotherms are not complete such lines are expected from combination of structural and vapor pressure measurements. One should realize that crucial to the occurrence of a triple line is the existence of two fluid phases.

Prime candidates for these studies appear to be Xe, CH_4, and O_2 [33], where isotherms and scattering measurements show evidence for a low temperature solid and a large coexistence region between the expected triple point temperature and critical point (T_c = 60 K, T_t = 26K for O_2; T_c = 117 K, T_t = 100 K for Xe; T_c = 72 K T_t = 56 K for CH_4).

Evidence for a large specific heat peak at what appears to be the triple point has been observed on Ne [34], O_2 [35,36], N_2 [37], Xe [38], and Kr [39] on graphite, but whether they really correspond to a triple point or not depends on identifying the liquid and vapor phases and their coexistence region. Here is where isotherm and heat capacity measurements should complement each other, but it is sometimes very difficult, if not impossible, to do heat capacity measurements on the same high quality substrates (like exfoliated graphite) used in the isotherm measurements. Thus, although isotherms show liquid-vapor coexistence, the only system where heat capacity measurements appear to indicate its existence and a critical point is Ne/graphite [40].

V. - The Ground State of Helium Adsorbed on Graphite. Elgin and Goodstein [27] and Elgin, Greif, and Goodstein [41] have combined their own high temperature (T > 2 K) vapor pressure and heat capacity measurements with lower temperature heat capacity measurements of other authors [1,28] to calculate the chemical potential of the ^{4}He or ^{3}He/graphite systems at T = 0 K. This is equivalent to calculating the ground state energy. For low densities, and after considerable numerical work, these authors found $E_b = -143 \pm 2$ K for ^{4}He and $E_b = -136 \pm 2$ K for ^{3}He. These results are in remarkable agreement with values deduced from atomic beam scattering experiments.

ACKNOWLEDGEMENTS

The experimental studies done by this author on physisorbed systems are supported by the NSF, Grant DMR 78-22697.

REFERENCES

1. M. Bretz, J. G. Dash, D. C. Hickernell, E. O. McLean, and O. E. Vilches, Phys. Rev. A8, 1589 (1973).
2. Grafoil, ZYX, and graphite foam are products marketed by Union Carbide Corp., Carbon Products Division.
3. M. J. di Pirro, and F. M. Gasparini, Phys. Rev. Lett. 44, 269 (1980).
4. D. C. Hickernell, E. O. McLean, and O. E. Vilches, J. Low Temp. Phys. 23, 143 (1976).
5. M. Bretz, and T. T. Chung, J. Low Temp. Phys. 17, 479 (1974).
6. R. L. Siddon and M. Schick, Phys. Rev. A9, 907 (1974), and A9, 1753 (1974).
7. M. Schick and J. G. Dash in "The Physics of Liquid and Solid Helium", Part II, pp. 497-571, K. H. Bennemann and J. B. Ketterson, eds., Wiley (New York) (1978).
8. S. B. Crary and O. E. Vilches, Phys. Rev. Lett. 38, 973 (1977).
9. C. M. Koutsogeorgis and J. G. Daunt, J. de Physique (Paris) Colloque 39C6, 308 (1978).
10. M. J. Tejwani, O. Ferreira, and O. E. Vilches, Phys. Rev. Lett. 44, 152 (1980).
11. O. Ferreira, Doctoral Dissertation, UNICAMP, Campinas, Brazil (1978) (unpublished).
12. S. B. Crary, Ph.D. Thesis, U. of Washington (1978) (unpublished).
13. S. E. Polanco and M. Bretz, Phys. Rev. B17, 151 (1978).
14. S. W. van Sciver and O. E. Vilches, Phys. Rev. B18, 285 (1978).
15. J. J. Rehr and M. J. Tejwani, Phys. Rev. B19, 345 (1979).
16. M. J. Tejwani, Ph.D. Thesis, U. of Washington (1979) (unpublished).
17. D. O. Edwards and W. F. Saam, in Prog. Low Temp. Phys., Volume VII A, 283 (1978).
18. Nuclepore filters are polymer sheets with approximately cylindrical holes produced by irradiation and etching, Nuclepore Corp., California.
19. M. Bretz, Phys. Rev. Lett. 38, 501 (1977).
20. S. Alexander, Phys. Lett. 54A, 353 (1975).
21. B. Nienhuis, A. N. Berker, E. K. Riedel, and M. Schick, Phys. Rev. Lett. 43, 737 (1979).
22. E. Domany, M. Schick, and J. S. Walker, Phys. Rev. Lett. 38, 1148 (1977).
23. E. Domany, M. Schick, J. S. Walker, and R. B. Griffiths, Phys. Rev. B18, 2209 (1978).
24. E. Domany and M. Schick, Phys. Rev. B20, 3828 (1979).
25. See article by M. Schick in "Phase Transitions in Surface Films", pp 65 - 113, J. G. Dash and J. Ruvals, eds. Plenum (New York, London) (1980).
26. M. Bretz, G. B. Huff, and J. G. Dash, Phys. Rev. Lett. 28, 729 (1974).
27. R. L. Elgin and D. L. Goodstein, Phys. Rev. A9, 2657 (1974).
28. S. V. Hering, S. W. van Sciver, and O. E. Vilches, J. Low Temp. Phys. 25, 793 (1976).
29. B. I. Halperin and D. R. Nelson, Phys. Rev. Lett. 41, 121 (1978); Phys. Rev. B19, 2457

(1979).
30. J. M. Kosterlitz and D. J. Thouless, J. Phys. C6, 1181 (1973).
31. A. Widom, J. R. Owers-Bradley, and M. G. Richards, Phys. Rev. Lett. 43, 1340 (1979).
32. See article by M. Richards in Ref. 25, in particular Figure 16 of his article.
33. See articles by M. Bienfait, pp 29-64, and M. Nielsen, J. P. McTague, and L. Passell, pp. 127-163 in Ref. 25.
34. G. B. Huff, and J. G. Dash, J. Low Temp. Phys. 24, 155 (1976).
35. R. Marx and R. Braun, Solid State Comm., 33, 229 (1980).
36. J. Stoltenberg and O. E. Vilches, to be published in Phys. Rev. B 15.
37. T. T. Chung and J. G. Dash, Surf. Sci. 66, 559 (1977).
38. J. A. Litzinger and G. A. Stewart, private communication and this Proceedings.
39. D. M. Butler, J. A. Litzinger, G. A. Stewart, and R. B. Griffiths, Phys. Rev. Lett. 42, 1289 (1979).
40. E. Lerner, private communication.
41. R. L. Elgin, J. M. Greif, and D. L. Goodstein, Phys. Rev. Lett. 41, 1723 (1978).

DISCUSSION

M. J. Harrison:

For very low coverage of the surface by adsorbed atoms are there any experimental results or outcomes which would reveal any details of the forces between adatoms at large distances; for example such as possible effects of retarded Van der Waals interactions between appropriately separated adatoms?

O. E. Vilches:

In all the calorimetric studies of helium or graphite or rare gas plated graphite it is impossible to get results at very low densities that are not influenced by the substrate. One may do numerical fittings to the data to extract the very low density behavior (like done by Elgin and Goodstein) but I don't believe this is accurate enough to extract retarded effects.

M. W. Cole:

In the paper by W. E. Carlos et al. (these proceedings), we show how both band structure effects and screening of the adatom-adatom interaction are manifested by He on graphite.

Published 1980 by Elsevier North Holland, Inc.
Sinha, ed. Ordering in Two Dimensions

STRUCTURE AND TRANSITIONS OF MONOLAYER KRYPTON AND XENON ON GRAPHITE

R. J. BIRGENEAU, E. M. HAMMONS, P. HEINEY, and P. W. STEPHENS

Department of Physics, Massachusetts Institute of Technology,
Cambridge, MA 02139, USA

P. M. HORN

IBM T. J. Watson Research Center,
Yorktown Heights, NY 10598, USA

ABSTRACT

We have performed a series of x-ray diffuse scattering measurements on monolayers of krypton and xenon physiadsorbed onto the (0001) surfaces of ZYX exfoliated graphite. In the submonolayer regime, krypton is commensurate, with the $\sqrt{3}x\sqrt{3}R30$ structure; the commensurate melting transition is well-described by a Potts lattice-gas model provided that one explicitly includes finite-size effects. Above one monolayer krypton undergoes a hexagonal commensurate to hexagonal incommensurate transition; this transition is at least nearly second order with the simple universal behavior $\varepsilon \sim (\mu-\mu_c)^{0.29\pm0.04}$, where ε is the fractional misfit and μ is the chemical potential. Xenon is incommensurate at all coverages for all temperatures above $\sim$ 70 K. This system exhibits a 2D triple point, first order and possibly second order melting; unusual lineshape and intensity behavior is observed for the solid in the gas-solid co-existence region possibly due to interface roughening.

I. INTRODUCTION

As first demonstrated in the pioneering vapor-pressure isotherm measurements of A. Thomy and X Duval [1], the systems krypton on graphite and xenon on graphite have rich and interesting phase diagrams in the monolayer regime. These systems present examples of two dimensional (2D) commensurate melting, commensurate-incommensurate transitions (C-IT), continuous symmetry 2D melting, a 2D gas-liquid-solid triple point, and novel coexistence phenomena. At the same time the important interactions are simple and well known so that there is the possibility of obtaining a very detailed understanding both theoretically and experimentally. Both monolayer krypton and monolayer xenon on graphite have been extensively investigated with the above-mentioned vapor-pressure measurements [1], electron diffraction [2,3], specific heat [4], Auger spectroscopy [5], and, as we shall discuss extensively in this paper, x-ray diffuse scattering [6-9]. Indeed, x-ray techniques are very well matched to these problems and it is somewhat puzzling that only recently has x-ray diffraction been used as a probe of monolayer physiadsorbed structures and transitions.

In this paper we shall give a brief survey of experiments which we have carried out over the past two years on the krypton and xenon systems. Certain of these results have already appeared in the literature [7-9], albeit in a very concise form. We shall give special emphasis to krypton commensurate melting and to the general phase diagram of xenon in the incommensurate regime. We shall also mention very briefly some recent results obtained at the Stanford Synchrotron Radiation Laboratory by David Moncton, George Brown and three of us (R.J.B., P.M.H. and P.W.S.) [10]. These initial measurements suggest that synchrotron radiation experiments may have a revolutionary impact on this field.

The format of this paper is as follows. In Section II we describe the experimental techniques. Section III contains the results for krypton on graphite while Section IV gives our results for xenon on graphite. Conclusions are given in Section V.

II. EXPERIMENTAL TECHNIQUE

Our original work in this area was inspired by the pioneering experiments of L. Passell and his coworkers at Brookhaven using neutron scattering techniques [11].

Indeed, we first realized that X-ray techniques might be advantageous for certain physiadsorbed systems by a simple scaling of probe particle flux, scattering particle density and scattering cross sections between neutrons and X-rays. We shall discuss the detailed experimental arrangement below. Here we note simply that, using our 12 kW Rigaku rotating anode generator in a spectrometer configuration that yields a longitudinal resolution width of 0.008 $Å^{-1}$ half-width-at-half-maximum (HWHM) with Cu Kα X-rays (λ = 1.542 Å), the photon flux incident on the sample is about 10^{10} photons/cm^2 sec. This is about five orders of magnitude larger than the corresponding flux of neutrons at the same resolution. This flux difference is, however, partially cancelled by the very large illuminated volumes possible in neutron experiments (~ 10 cm^3) compared with X-rays (~ 0.02 cm^3). An important difference occurs in the scattering cross sections; for krypton the X-ray scattering length is ~ $Z(e^2/m\ c^2)$ = 10.2×10^{-12} cm as compared with a neutron scattering length of 0.74×10^{-12} cm. For neutrons, the graphite substrate phonon background may be largely eliminated by energy analysis of the diffracted beam [11]. For X-rays, the graphite thermal diffuse scattering is always visible; however, its relative intensity may be diminished either by studying high Z adsorbates or by going to very high resolution with a long coherence-length substrate.

Fig. 1 gives a schematic representation of the experimental arrangement. The x-ray beam is collimated by successive slits S_1, S_2, and S_3 before the sample and by Soller slits after the sample. We use a 2 cm high vertically-bent pyrolytic graphite monochromator both to monochromatize the beam and to focus it in the vertical plane onto the sample. The graphite sample is contained in a thin square-shaped sample holder which in turn is placed in a Displex closed cycle refrigerator; the temperature control and stability are 5×10^{-3} K between 11 K and room temperature. The sample is composed of a stack of exfoliated ZYX graphite slices $24 \times 24 \times 3mm^3$ in

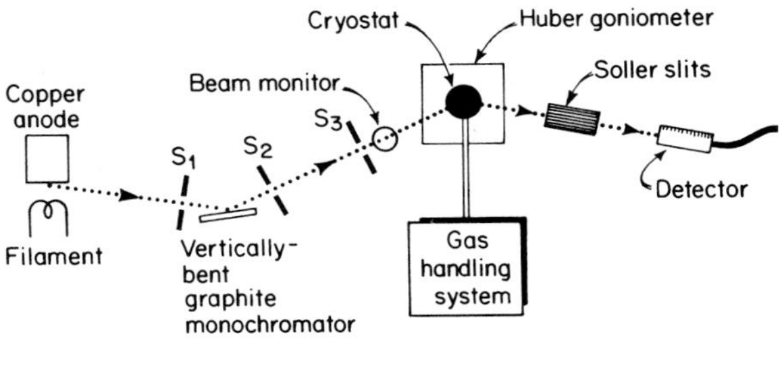

Fig. 1: Schematic diagram of the experimental configuration. S_1, S_2 and S_3 are single slits which collimate the X-ray beam. The vertically-bent monochromator is positioned so that the beam is focussed onto the sample in the vertical plane. The beam monitor measures the flux just before the sample.

volume with the average C-axis vertical. Scattering experiments are carried out in transmission through the 3 mm direction using Cu-Kα radiation (λ = 1.5418 Å) from a Rigaku 12 kW rotating anode x-ray source operating typically at 8 kW. A beam monitor system is used so that count rates are per unit photon flux rather than per unit time. In most of the experiments discussed here, the slits S_1, S_2, S_3, the Soller slits, and the distances are chosen so that the in-plane longitudinal resolution is 0.008 $Å^{-1}$ HWHM while the vertical resolution is 0.05 $Å^{-1}$ HWHM.

The amount of gas adsorbed and its equilibrium vapor pressure are measured using a Baratron pressure gauge with a 10 Torr maximum pressure and a full scale precision of about 1 part in 10^4. The amount of gas adsorbed can be measured to an accuracy of about 0.002 monolayers.

III. KRYPTON ON GRAPHITE

(i) Phase Diagram, Theory, and Basic Diffraction Results

As we noted in the Introduction, monolayer krypton on graphite has already been the subject of extensive experimental and theoretical investigation. Here we shall give only a brief discussion of the salient features. The Lennard-Jones separation of a pair of krypton atoms is 4.041 Å compared with the $\sqrt{3}x\sqrt{3}R30°$ graphite superlattice separation of 4.26 Å. Thus, in order to form a commensurate phase, the krypton mean separation need only be expanded by 5.4%. The relevant energies are a substrate binding energy of ~ 1400 K, a Kr-Kr Lennard-Jones well depth of ε/k = 145 K and a substrate anisotropy of 25 to 50 K [12,13]. As is by now well known experimentally, this anisotropy is sufficient to stabilize the commensurate $\sqrt{3}x\sqrt{3}R30$ phase for certain values of coverage and temperature. In this phase the krypton atoms rest in the center of every third graphite hexagon.

We show in Fig. 2 a schematic representation of the empirical krypton phase diagram in the 0 to 1.6 monolayer region as determined by a variety of measurements including our own [1,4]. A detailed lattice-gas theory for the sub-monolayer regime has been proposed by Berker, Ostlund and Putnam (B.O.P.)[13]. This theory rests on the observation by Alexander [14], that the $\sqrt{3}x\sqrt{3}R30$ phase has the symmetry of a three component Potts model; B.O.P. then develop a detailed lattice gas theory for the three component Potts model with vacancies. This theory predicts a line of second order "melting" transitions terminating at a novel multicritical point at a coverage of 0.87; below 0.87 monolayers they predict a first order fluid-solid transition. We shall discuss this model in more detail in the context of our experiments.

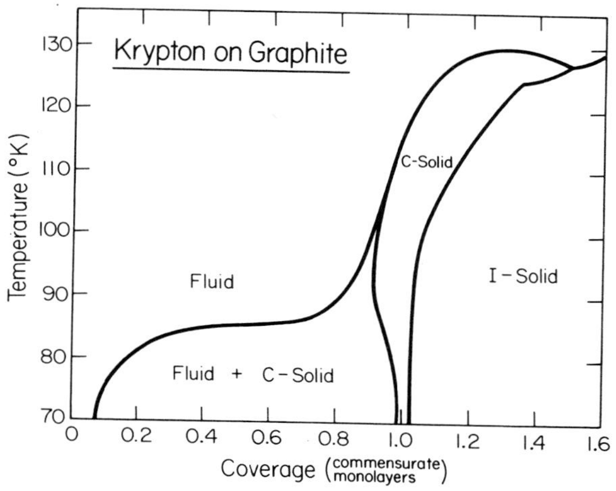

Fig. 2: Phase diagram of Krypton on graphite. The C-solid has the $\sqrt{3}x\sqrt{3}R30$ commensurate structure. The phase boundaries come primarily from the specific heat studies of Butler et al. [4].

As expected from simple lattice constant and energy considerations, above one commensurate monolayer the krypton undergoes a commensurate-incommensurate transition [1,2,8,15]. The actual shape of the phase boundaries, as shown in Fig. 2, is however, rather surprising; there is currently no realistic theory for this region of the phase diagram. There have, of course, been a remarkably large number of attempts at a theoretical description of the commensurate-incommensurate transition [16]; from our point of view none of them have represented a fundamental advance over the early work of Frank and Van der Merwe [17] although the most recent work by Villain [16] and coworkers [18] does hold some promise. We shall discuss both our experimental results and the essential theoretical ideas which emerge from these measurements in subsection (iii).

(ii) Measurements in the Commensurate Phase

Our experiments in the commensurate phase have addressed three issues: (a) the nature of the X-ray scattering from monolayer of krypton on ZYX graphite, (b) the general shape of the phase diagram, (c) the detailed behavior of the solid and liquid scattering near the predicted multicritical point. We show in the top panel of Fig. 3 scattering from a monolayer krypton at T = 89.3 K with a density near the gas-solid phase boundary and away from the C-IT. The solid line is the best fit to a finite-size broadened 2D Gaussian profile as discussed in detail by Stephens et al. [8]. As a side comment we should note that the computation time involved in a proper diffraction lineshape theory is equivalent to that required for the Kjems et al. [11] adaptation of the Warren lineshape and, as is well known, the latter may lead to serious errors for a variety of real situations. We note from Fig. 3 that the Gaussian lineshape describes the measured profile remarkably well. This in turn implies that the overlayer thermal diffuse scattering (TDS) is relatively flat as expected for a commensurate system since the "acoustic" phonon spectrum has a gap at q = 0. The crystallite size, as determined from the width of the leading edge, is 500 ± 50 Å while the distribution of tilts (vertical mosaicity) of the krypton is ~ 14 degrees HWHM.

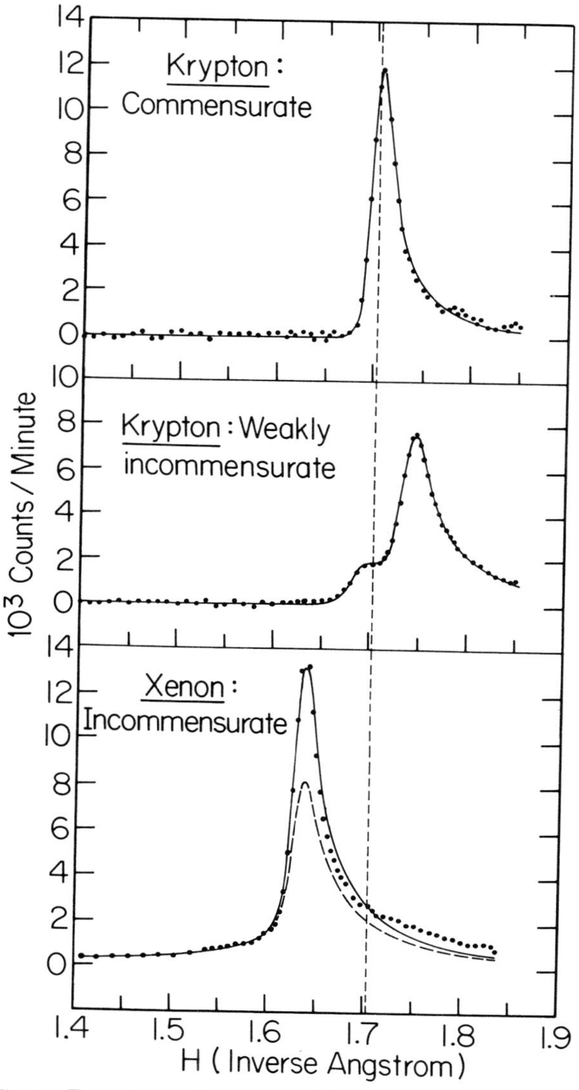

Fig. 3: Experimentally measured (1,0) diffraction profiles with the graphite background subtrated upper pannel: krypton at T = 89.3 K, coverage = 0.98 monolayers; the solid line is the best fit of a Gaussian profile; middle pannel: Krypton at T = 89.3 K, coverage ≃ 1.08 monolayers; the solid line represents the sum of two Gaussians; lower panel: xenon at T = 129.02 K, coverage = 1.2 monolayers; the solid line represents the sum of a Gaussian plus a Lorentzian as described in the text.

Solid-liquid phase boundaries may be mapped out using x-rays simply by observing the coverage-temperature points at which the solid Bragg scattering vanishes. We have so far carried out such studies only in a coarse fashion; our results are, in general, consistent with the phase boundaries shown in Fig. 2. We plan a much more detailed study in the near future.

We have, however, performed a very careful set of measurements in the neighborhood of the predicted multicritical point. In order to characterize the melting transition we have carried out measurements of the peak intensities at (1,0) and (2,0) as a function of temperature. We have also monitored the amplitude of the liquid scattering by intensity measurements at (0.97, 0), that is, $Q = 1.65\ \text{Å}^{-1}$; as is evident from Fig. 3 this is just below the Bragg diffraction region. The experiments are carried out in the closed cell configuration so that the coverage changes slightly with increasing temperature, albeit in a known fashion. It should be noted that these measurements were carried out in a small-volume cell configuration [7] so that the absolute coverages are only known approximately; the temperatures are, however, quite precise. We show in Fig. 4 the measured (1,0), (2,0) and (0.97,0) intensities at a coverage of ~ 0.9 monolayers, just below the predicted multicritical point. It is evident that over the range of 90 K to 100 K both the (1,0) and (2,0) Bragg intensities go linearly to zero while the liquid scattering as estimated from the (0.97,0) intensity increases linearly. This, of course, is just the behavior expected as one traverses the coexistence region in a first order melting transition.

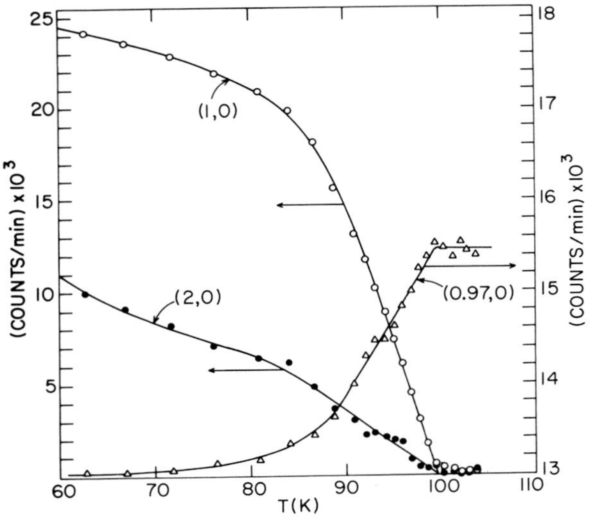

Fig. 4: Peak intensities for krypton on graphite at a coverage of ~ 0.9 monolayers versus temperature; (1,0) and (2,0) are bragg peak intensities while the point (0.97,0) represents the graphite background plus the liquid amplitude.

We have also carried out a series of scans in the liquid phase above the melting temperature of 99.7 K. We show two representative profiles, at 101 K and 124 K just above the phase boundary and deep into the fluid regime, respectively. The solid lines represent the results of fits to an azimuthally powder-averaged Lorentzian lineshape. It is evident that at 101 K the liquid is poorly correlated, $\xi \sim 12$ Å, but that nevertheless, the peak in the structure factor is centered about the commensurate position; thus even in the first order melting regime, the behavior near the phase boundary is consistent with the lattice gas description anticipated by B.O.P. [13]. At higher temperature, the peak in S(Q) moves off the commensurate position.

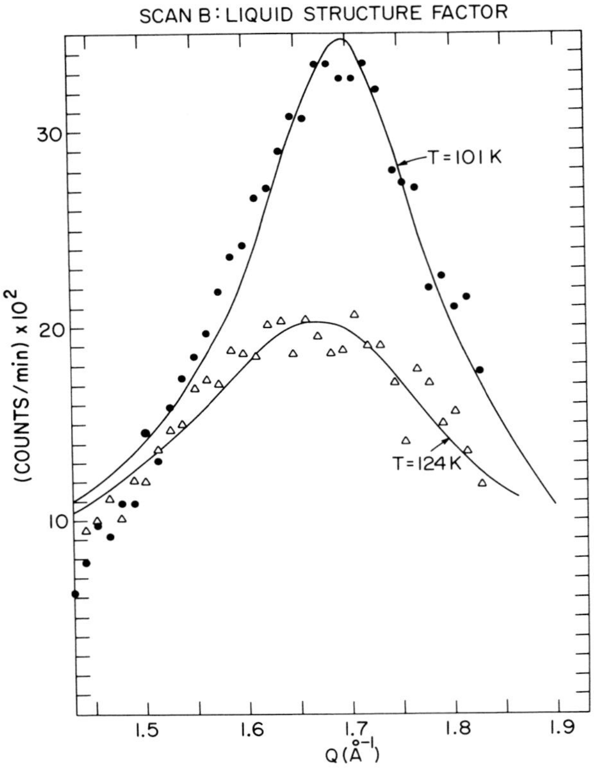

Fig. 5: Liquid krypton structure factor at 0.9 monolayers coverage at 101 K and 124 K; the solid lines represent the results of fits to an azimuthally powder averaged Lorentzian profile.

From the relative intensities of the (1,0) and (2,0) peaks one may deduce the Debye-Waller factor and thence the mean square atomic displacement $\langle u^2 \rangle$. At the melting temperature for the above data one deduces $\langle u^2 \rangle = 0.17\ \text{Å}^2$; it is amusing to note that the Lindemann criterion applied to the system predicts $\langle u^2 \rangle \sim 0.18\ \text{Å}^2$ at melting.

When the coverage is increased above approximately 0.92 monolayers, there is a drastic change in the behavior near melting. Fig. 6 gives the (1,0) intensity data near the melting transition for coverages such that the melting temperatures are 99.7 K (the coverage shown in Figs. 3, 4, 5) and 105.87 K. It is evident that with a very small change in coverage (~ 3%) one has passed from one melting regime to another, that is, from first order to at least nearly second order melting. As we shall discuss below this behavior was explicitly predicted by B.O.P. in advance of our experiments. According to theory, in this coverage range one should pass over from a first order to a second order lattice-gas transition; furthermore, the lattice-gas melting belongs in the universality class of the three component Potts model [14]. It is clear, however, that in order to analyze the data quantitatively one must explicitly incorporate finite-size effects as previously noted by Bretz [19]. As a first approximation we have modelled the finite size effects via a distribution of melting temperatures of width σ. From simple scaling ideas we expect σ/T_c to be of the order of $(L/a)^{-1/\nu} \simeq 0.003$ for our system, where ν is the correlation length critical exponent. The solid line in Fig. 6 represents the best fit of a smeared power law to the melting data. This particular fit gives $\beta = 0.08 \pm 0.01$ and $\sigma/T_c = 0.004$; as an average we find $\beta = 0.09 \pm 0.03$. At the multicritical point B.O.P. predict $\beta = 0.072$ whereas along the lambda line β becomes 0.105; we note, however, that the crossover lines [20] are such that in the region of our experiment the measured intensities will be dominated by the multicritical point value $\beta = 0.072$ and one should cross over to the 3-component Potts value only very close to T_c. Our experimental results are clearly consistent with this predicted behavior. Since we first reported these experiments, Ostlund and Berker [13] have explicitly incorporated finite size effects into the theory; the general agreement with our measurements and with the known phase diagram is excellent.

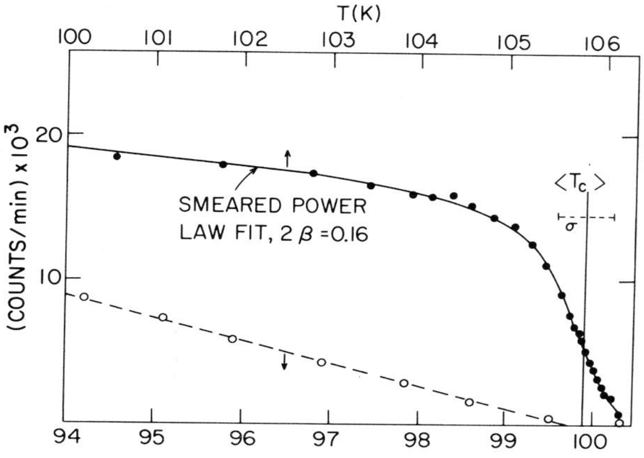

Fig. 6: (1,0) peak intensity vs. temperature for krypton scans just below and just above the multicritical point; the dashed line is a guide to the eye while the solid line is the result of a least-squares fit of a smeared power law to the data shown.

It is clear that finite size effects represent a fundamental limitation in this work. Very recently, David Moncton, George Brown and three of us (R.J.B., P.M.H., and P.W.S.) [10] have repeated these melting experiments using high resolution synchrotron techniques together with specially-handled ZYX graphite. These experiments yield a surface coherence length of many thousands of Angstroms. From our vantage point, at least, this is a very important experimental development; this work will be published separately. We note here only that these new experiments support the above description in detail, including the finite-size scaling.

(iii) Commensurate-Incommensurate Transition

In Ref. 8 we have given a rather full description of our C-IT work. Accordingly, we shall make only a few comments here, primarily on recent developments. It is believed [16] that the C-IT occurs by the generation of domain walls, or solitons, [17] at which the krypton atoms shift from one sublattice to another. The walls may form either a one-dimensional array, the so-called stripe-domain (SD) phase, or a two-dimensional hexagonal incommensurate (HI) array. At T = 0 the domain walls will form a lattice. Bak et al. [18] give a T = 0 mean field theory which predicts, for a positive domain wall crossing energy, a second order hexagonal commensurate (HC) to the SD incommensurate structure followed by a first order transition to the HI phase. More recent work by Villain [16] has suggested that at finite temperatures the domain walls may be disordered so that there may be a disordered near-hexagonal incommensurate (DHI) phase intermediate between the SD and HI phases. Pokrovsky and Talapov [21] predict that at finite temperatures for a one-dimensionally modulated system the HC-SD transition will be second order with an incommensurability $\varepsilon \sim |\mu-\mu_c|^{1/2}$ where μ_c is the chemical potential at the transition.

In the sharp-domain wall region, the diffraction pattern will be markedly altered. In particular, the individual commensurate peaks should split into a series of superlattice peaks, the explicit pattern and separation being determined by the geometry and separation of the domain walls. There is a wide variety of possible configurations; we discuss here explicitly only the two most probable cases. As discussed by Venables and Schabes-Retchkiman [12] the HI phase may be generated simply by uniformly contracting the atoms in the HC phase and adjusting the atom positions to the nearest graphite hexagonal centers, this gives a sublattice array of the form A$\overset{C}{Y}$B with the principal axes along the $\sqrt{3}x\sqrt{3}R30$ axes. From simple diffraction theory, it is straightforward to show that this leads to strong peaks in a powder pattern near the original (1,0) commensurate position at $Q/Q_o = 1 - \varepsilon/2$, $1 + \varepsilon$ and $1 + 5\varepsilon/2$. For large ε the intensity should be concentrated in the $Q/Q_o = 1 + \varepsilon$ peak. For an SD phase with the domain walls along the same directions as the HI phase and a sequence A-B-C-A one predicts peaks at $Q/Q_o = 1 - \varepsilon/2$, $1 + \varepsilon/4$, $1 + \varepsilon$ etc., with the $1 + \varepsilon/4$ and $1 + \varepsilon$ peaks being prominent and with an intensity ratio of ~ 2:1. All of this, of course, assumes that the domain walls are perfectly ordered.

The krypton C-IT was first studied by Chinn and Fain [2] using LEED techniques, on single crystal graphite surfaces; in the temperature region 50 K to 57 K they find a nearly second order HC to HI transition. We have carried out similar measurements at 80 K and 89.3 K. We show a typical x-ray profile in the weakly incommensurate region in Fig. 3. The solid line is the standard diffraction theory albeit with peaks only at $1 - \varepsilon/2$ and $1 + \varepsilon$, at least partially consistent with our expectations for the HI phase. We note several features of the spectrum: (a) the integrated intensity does not change appreciably in going from the HC to HI phases - this contradicts the prediction of Ref. [12], (b) the peaks are much broader and there is considerable extra scattering on the high Q-side, (c) the Gaussian theoretical lineshape is no longer adequate - presumably because of the crossover from discrete to continuous symmetry. In this region, the sample is clearly in the HI phase although the apparent absence of a sharp peak at $1 + 5/2\ \varepsilon$ is somewhat puzzling; we may nevertheless clearly exclude the SD phase. Indeed, we never observe any evidence for the SD phase although a rigorous differentiation becomes difficult very near the transition. The recent high resolution synchrotron measurements by Moncton et al. [10] give a much richer picture of the details of the C-IT. In particular, they show clearly that the broadening which is barely evident in Fig. 3, is a central feature of the C-IT and indeed near the transition the domain walls are almost certainly disordered.

In our original publication [8], we suggested that the extra high-Q scattering reflected primarily the effects of interstitial krypton atoms. Here we should like to propose an alternative explanation. As noted above, in the HI phase there should be a peak at $1 + 5/2\ \varepsilon$. In model calculations for a variety of domain wall geometries we find that the $5/2\ \varepsilon$ peak should be at least 0.5 times as intense as the $-\ \varepsilon/2$ peak. Thus, the $5/2\ \varepsilon$ peak should be observable, provided that the peak widths are all comparable. The synchrotron experiments show that near the C-IT the peaks are severely broadened. In this situation we expect some dependence on the harmonic number. For example, if the broadening reflects a distribution of effective incommensurabilities, ε, due to domain wall disorder, then the width would scale linearly with the index. It is evident from Fig. 1 of Ref. 8, that the extra scattering on the high-Q side of the $1 + \varepsilon$ peak could be readily explained by assigning it to the predicted $5/2\ \varepsilon$-peak albeit with this peak being 5/2 times broader than that at ε. This also would explain why the extra high-Q intensity disappears at the same rate as the $-\ \varepsilon/2$ peak. We thank David Moncton for first drawing our attention to this possibility.

We show in Fig. 7 the incommensurability versus chemical potential difference for both the x-ray and the LEED data. As first noticed by Fain et al., these form one universal curve [22]. Further, this curve is described to within the experimental error by a simple power law $\varepsilon \sim (\mu-\mu_c)^\beta$ with $\beta = 0.29 \pm .04$. No current theory is able to predict this universal behavior. Clearly, therefore, in contrast to the commensurate melting problem where theory has led recent experiment, there is a pressing need for an improvement in the theories for the C-IT.

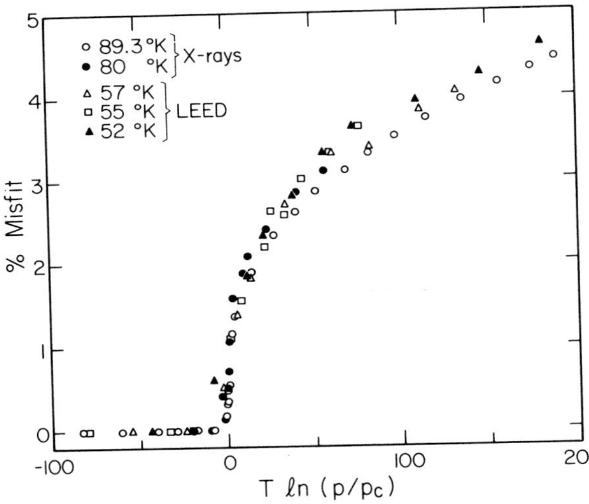

Fig. 7: Percent incommensurate versus chemical potential difference $\mu - \mu_c$ at a series of temperatures for krypton on graphite. The squares and triangles are LEED data from Ref. 2 whereas the circles are x-ray data from Ref. 8.

IV. XENON ON GRAPHITE

At the present time somewhat more limited information is available on the xenon system. We show in Fig. 8 the phase diagram of xenon in the monolayer region as deduced by Thomy and Duval [1] from their vapor-pressure isotherm measurements. From the total amounts absorbed, Regnier, Thomy and Duval [1] suggest that at least above 100 K, solid monolayer xenon is incommensurate with respect to the graphite. We note that the xenon system in the sub-monolayer regime exhibits a 2D gas-liquid critical point, a gas-liquid-solid triple point and concomitant first order transitions. Hence, monolayer xenon seems to be a model case for studies of 2D continuous symmetry melting [23] and 2D co-existence phenomena.

Our measurements of the xenon system to date have been directed largely towards a basic characterization of the x-ray diffraction profiles. We have also carried out a very careful study of the melting transition above one monolayer. As our analysis of that data is not yet complete we shall not discuss those results in detail here. We might note, however, that for this simple Lennard-Jones-like system, the nature of the 2D melting process depends in detail on the thermodynamic conditions and

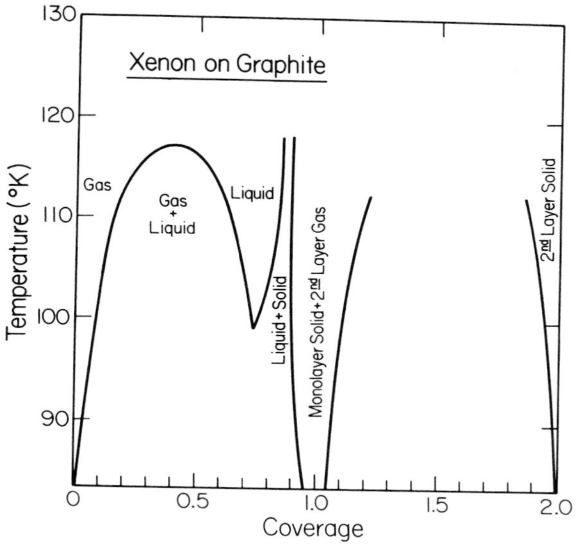

Fig. 8: Phase diagram of monolayer and bilayer xenon on graphite; the sub-monolayer region is taken from Thomy and Duval (1970). The one-to-two layer region is conjectured; the coverage scale is given by Thomy and Duval.

there is no straightforward universal behavior. In our view, this vitiates much of the controversy which has recently arisen in molecular dynamic studies of 2D melting [23]. A typical x-ray diffraction profile for a coverage of 1.2 monolayers at T = 129 K is shown in the bottom panel of Fig. 3. The graphite background, which is about 2 × 10^3 counts per minute has been subtracted off; we note that the background is first determined from an empty cell scan and then scaled uniformly from data around Q ≃ 1.2 $Å^{-1}$ to correct for absorption of the x-rays by the xenon overlayer, etc. It is evident from Fig. 3 that the xenon lattice is indeed incommensurate; measurements of higher order Bragg peaks confirm that the xenon has a simple triangular structure. The lattice constant is 4.42 Å; this is identical to that calculated for the triangular lattice from the Lennard-Jones parameter.

It is clear from Fig. 3 that the xenon line profile differs significantly from that observed for commensurate krypton; this difference originates from the continuous versus the discrete symmetry. For an infinite continuous symmetry 2D solid one anticipates $q^{-2+\eta_G}$ Landau-Peierls singularities rather than true Bragg peaks [24,25]. For a finite size overlayer crystal with L ~ 120 atoms on a side it is probably more appropriate to describe the diffraction as reflecting a finite-size-limited Bragg peak plus strong $1/q^2$ TDS extending in to $|Q-Q_o| \sim \pi/L$. We have fitted the diffraction lineshapes to a variety of models. We find, in fact, that the profiles are adequately described by a single sharp Lorentzian convoluted with the Gaussian resolution function. However, equally good fits may be obtained from the sum of Lorentzian TDS plus a Gaussian Bragg component of comparable amplitude. The solid line in Fig. 3 is the result of such a fit. Thus, in our view, it is not possible to prove unambiguously the existence of algebraic decay of positional correlations from the (1,0) Bragg peaks in these surface overlayer systems. Similar experiments in 3D smectic liquid crystals [26] do, however, explicitly demonstrate the Landau-Peierls effect.

We show in Fig. 9 scans at a series of coverages at T = 112 K. The profiles are all consistent with the identifications suggested by Thomy and Duvai [1]. The solid line through the liquid profile represents an azimuthally powder-averaged 2D Lorentzian with correlation length ξ = 54 Å; thus, even though the melting transition at this temperature is first order, the liquid nevertheless is very well correlated.

We have also carried out scans as a function of temperature for coverages ranging from 0.36 to 0.89 monolayers; in each case there is an abrupt transition from a liquid to a solid lineshape at 99 K thus verifying the existence of the triple point at 99 K. We do, however, observe one very unusual feature. In a simple coexistence picture the solid intensity should scale linearly

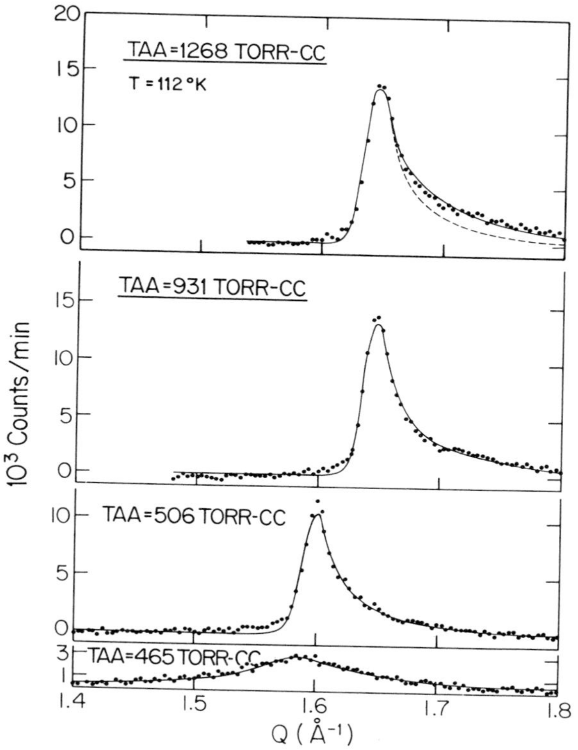

Fig. 9: X-ray scans of the xenon (1,0) peak at four coverages at T = 112 K. The coverages are 1.91, 1.33, 0.90, and 0.83 after correction for alternate-site adsorption [9].

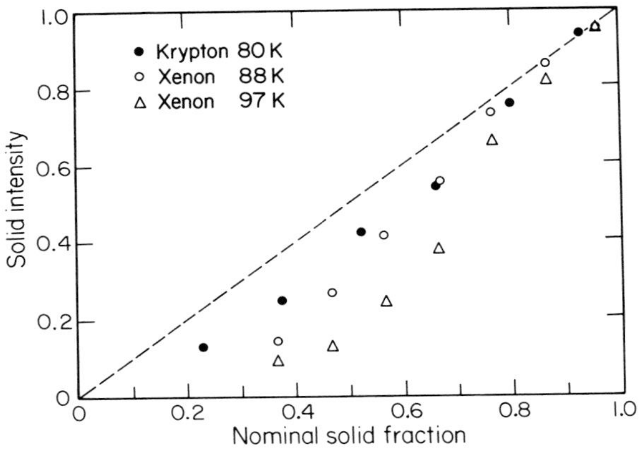

Fig. 10: Relative solid intensity versus the nominal solid fraction across the gas-solid coexistence region.

as one traverses the co-existence region. Fig. 10 gives the normalized integrated solid intensity versus the nominal solid fraction across the gas-solid coexistence region for krypton at 80 K and for xenon at 97 K and 88 K. It is evident that the usual "law-of-levers", is only approximately obeyed for krypton while for xenon at 97 K it is seriously violated. Indeed, below the 50% density point the solid signal is barely observable. In this range the profile has particularly large wings. We find, further, that the peak intensity varies considerably in intensity on cycling through the triple point temperature. It is interesting to speculate that this destruction of the solid may be connected with interface roughening [27]; in an infinite system, it is impossible to define an interface in two dimensions. Further, this effect should be more important for a system with continuous symmetry such as xenon than for a discrete symmetry system such as commensurate krypton. A sufficiently diffuse solid-gas interface would give rise to a correspondingly broad diffraction signal. Clearly, more work should be done on this aspect of the problem. High resolution synchrotron studies on a long coherence length substrate would be particularly valuable.

We should like to conclude our discussion of monolayer xenon with a brief comment on the C-IT. In the 60-70 K region for coverages above one monolayer xenon undergoes a C-IT to the $\sqrt{3}x\sqrt{3}R30$ structure. This was originally demonstrated in THEED measurements by Venables et al. [3]; our initial x-ray results are general consistent with their measurements. It appears, however, that the xenon C-IT is much more complicated than the corresponding transition for krypton. In general, we observe broad, complicated diffraction profiles with large hysteresis effects indicative of a first order transition. We plan a much more detailed study of the xenon C-IT in the near future.

V. CONCLUSIONS

We began this article by noting that it was somewhat puzzling to us that x-rays have barely been used for surface overlayer studies. We hope that it is clear from this paper that in spite of their lack of surface selectivity, x-rays represent a powerful probe of surface structures and transitions, especially for high-Z elements adsorbed on light surfaces. We believe that the extension to single crystal surfaces will be straightforward once appropriate surface cleanliness techniques are incorporated in the experiments. It is also clear that synchrotron-based diffuse scattering studies will have an enormous impact on this field.

In terms of the explicit systems studied, our work on krypton and xenon has been primarily exploratory in character. A large number of questions remain to be addressed in detail. Our results to-date have given strong support to the B.O.P. lattice gas model of submonolayer krypton on graphite. We also have obtained rather complete information on the krypton C-IT which lacks a quantitative theoretical explanation. However, we still need thorough studies of the krypton one-to-two layer region, the xenon C-IT and the xenon melting, especially the transition from first to the (apparently) second order region. It is clear that there is still a large amount of new physics to be derived from these systems.

This work was supported by the Joint Services Electronics Program under Contract No. DAAG-29-78-C-0020.

REFERENCES

(1) A. Thomy, and X. Duval, J. Chim. Phys. **66**, 1966 (1969), **67**, 286 (1970), **67**, 1101 (1970). J. Regnier, A. Thomy and X. Duval, ibid **74**, 926 (1977).

(2) M. D. Chinn and S. C. Fain, Jr., Phys. Rev. Lett. **39**, 146 (1977).

(3) J. A. Venables, H. M. Kramer, and G. L. Price, Surf. Sci. **55**, 373 (1976); **57**, 782 (1976).

(4) D. M. Butler, J. A. Litzinger, and G. A. Stewart, Phys. Rev. Lett. **44**, 466 (1980).

(5) J. Suzanne, J. P. Coulomb and M. Bienfait, Surf. Sci. **44**, 141 (1974).

(6) T. Ceva and C. J. Marti, J. de Physique **39**, L-221 (1978).

(7) P. M. Horn, R. J. Birgeneau, P. Heiney, and E. M. Hammonds, Phys. Rev. Lett. **41**, 961 (1978).

(8) P. W. Stephens, P. Heiney, R. J. Birgeneau and P. M. Horn, Phys. Rev. Lett. **43**, 47 (1979).

(9) E. M. Hammonds, P. Heiney, P. W. Stephens, R. J. Birgeneau and P. M. Horn, J. Phys. C **13**, L301 (1980).

(10) D. E. Moncton, P. W. Stephens, R. J. Birgeneau, P. M. Horn and G. E. Brown, (unpublished work).

(11) See for example J. K. Kjems, L. Passell, H. Taub, J. G. Dash and A. D. Novaco, Phys. Rev. **B13**, 1446 (1976).
(12) J. A. Venables and P. S. Schabes-Retchkiman, J. de Physique **38**, C4-105 (1977).
(13) A. N. Berker, S. Ostlund and F. A. Putnam, Phys. Rev. **B17**, 3650 (1978); S. Ostlund and A. N. Berker, Phys. Rev. Lett. **42**, 843 (1979); J. Phys. C **12**, 4961 (1979).
(14) S. Alexander, Phys. Lett. **A54**, 353 (1975).
(15) Y. Larher, J. Chem. Phys. **68**, 2257 (1978).
(16) For a recent review see the lecture notes by J. Villain in "Order in Strongly Fluctuating Condensed Matter Systems", p. 221, Ed. T. Riste, Plenum Press, New York (1980).
(17) F. C. Frank and J. H. van der Merwe, Proc. Roy. Soc., London, Ser **A198**, 205, 216 (1949).
(18) P. Bak, D. Mukamel, J. Villain, and K. Wentowska, Phys. Rev. **B19**, 1610 (1979).
(19) M. Bretz, Phys. Rev. Lett. **38**, 501 (1977).
(20) A. N. Berker, (private communication).
(21) V. L. Pokrovsky and A. L. Talapov, Phys. Rev. Lett. **42**, 65 (1979).
(22) S. C. Fain, Jr., M. D. Chinn and R. D. Diehl, Phys. Rev. **B21**, 4170 (1980).
(23) D. Frenkel and J. P. McTague, Phys. Rev. Lett. **42**, 1632 (1979); F. F. Abraham, Phys. Rev. Lett. **44**, 463 (1980); S. Toxvaerd, Phys. Rev. Lett. **44**, 1002 (1980).
(24) J. M. Kosterlitz and D. G. Thouless, J. Phys. C **6**, 118 (1973); B. I. Halperin and D. R. Nelson, Phys. Rev. **B19**, 2457 (1979); A. P. Young, Phys. Rev. **B19**, 1855 (1979).
(25) B. Jancovici, Phys. Rev. Lett. **19**, 20 (1967); Y. Imry and L. Gunther, Phys. Rev. **B3**, 3939 (1971).
(26) J. Als-Nielsen, R. J. Birgeneau, M. Kaplan, J. D. Litster, and C. R. Safinya, Phys. Rev. Lett. **39**, 1668 (1977) and Phys. Rev. **B21**, June 1 (1980).
(27) S. T. Chui and J. D. Weeks, Phys. Rev. **B14**, 4978 (1976); Phys. Rev. Lett. **40**, 733 (1978).

DISCUSSION

R. B. Griffiths:

It appears to me that there is a discrepancy between your results and the theory which Dr. Berker presented to us this morning. That is, Berker expects a multicritical point at about 80 K, and thus that is the highest temperature for solid-fluid coexistence. On the other hand, you apparently observe the coexistence of these two phases up to about 100K. Could you comment on this?

R. J. Birgeneau:

The agreement with the original Berker, Ostlund, Putnam calculations is clearly only approximate. Since their experiments were just performed, Ostlund and Berker have explicitly included finite size effects in their theory and they find that in finite systems the two phase coexistence region appears to extend up to higher temperatures. This is a very interesting idea and it may well be correct. I believe that using our new ZYX graphite, which has a much longer coherence length, together with a synchotron source we will be able to explore the role of finite size effects (or, hopefully, their absence in this new graphite) in detail; thence we should be able to give a more complete empirical description of the behavior of the real system. As expressed in our paper, my basic view is that the B.O.P. theory contains most of the essential physics of the ordering of krypton on graphite in the sub-monolayer regions. However, the quantitative agreement they seem to have obtained in their most recent papers may be partially accidental.

M. Nielsen

In the commensurate-incommensurate transition in the region where the domain-wall superstructure is not "melted" in your model you have a scattering profile with a low-Q satellite and a higher-Q main peak (+ higher Q satellites). Assuming a domain wall thickness you can estimate the relative satellite to main peak intensity ratio as a function of coverage and thus obtain a check on the analysis. Have you tried that for your new data?

R. J. Birgeneau:

We have not done this for our new data but Peter Stephens of our group has carried out very detailed calculations of the satellite intensities for various domain wall shapes and has made detailed comparisons with our M.I.T. rotating anode data. In general, he obtains quite reasonable results.

D. E. Moncton:

In our incommensurate data, the lineshape is never as sharp as the commensurate peak. I don't believe that the hexagonal domain lattice has substrate-limited long-range order at any value of epsilon in this regime of the phase diagram. Domain wall solidification may occur at lower temperature. A quick check of incommensurate Xenon at the synchrotron suggested substrate-limited peaks.

S. C. Fain:

In your new synchrotron data for the commensurate-incommensurate transitron, what is the misfit range ($\frac{\varepsilon-\varepsilon_0}{\varepsilon_0}$) over which you observe the "disordered domain wall" hexagonal incommensurate phase (characterized by the absence of a satellite peak and the large breadth of the main peak)? Also, what if the misfit range over which you find a power law behavior $(T-T_c)^{0.30}$ for the new data?

R. J. Birgeneau:

The peak appears to be rather broad up to an incommensurability of about 2%. The data appear to follow the simple $\Delta T^{0.29}$ power law over a range of misfits from 0.3% to 4%.

J. Sokoloff:

Why are coherent regions in the graphite work now $\approx$ 5000 Å?

R. J. Birgeneau:

We do not fully understand why our most recent ZYX graphite gives such long surface coherence lengths although, of course, we are extremely pleased. There were some differences in our handling of this piece of ZYX; a systematic study may show up what aspects are truly important. On the other hand, it may turn out to be just Divine Providence.

A. N. Berker:

I would like to note that G. Mazenko (University of Chicago) and co-workers have recently developed the position-space renormalization-group method to calculate S(Q).

Sinha, ed. Ordering in Two Dimensions

SUBSTRATE REARRANGEMENT AND 2D ORDERING: HYDROGEN ADSORPTION ON TUNGSTEN(001) AND MOLYBDENUM(001)

PEDER J. ESTRUP R.A. BARKER*

Department of Physics and Department of Chemistry
Brown University
Providence, R.I. 02912

A variety of ordered structures are produced by chemisorption of hydrogen and other adsorbates on the (001) faces of tungsten and molybdenum. The phase diagram in the temperature-coverage plane have been studied and they show examples of order-order, order-disorder, and commensurate-incommensurate transitions. These structural changes are found to involve displacive transformations ("reconstruction") of the metal surface and conventional models which treat the overlayer as a lattice gas on a rigid substrate therefore fail for these systems.

1. INTRODUCTION

The chemisorption of hydrogen on the (001) faces of tungsten and molybdenum is known to produce many interesting surface phenomena, including the formation of a series of two-dimensional (2D) ordered structures [1]. Since these surfaces can be quite accurately controlled, they have been chosen frequently as model systems in adsorption studies and have been investigated by almost all of the surface diffraction and spectroscopy techniques now available. It seems a good guess that more experimental data have been obtained for H/W(001) than for any other adsorbate-substrate combination. This has not yet led to a full understanding of the physics and chemistry of the surface; however, it has permitted rather detailed tests of models for the adsorbate-substrate (AS) and adsorbate-adsorbate (AA) interactions. These tests indicate that it is necessary to change the conventional approach to the problem of 2D ordering and phase formation, at least for this surface. In particular it is found that the substrate plays a more active role than previously suspected; in the presence of the adsorbate the metal surface undergoes a displacive transformation, a "reconstruction" [2].

2. RIGID LATTICE GAS MODEL

The conventional model for the behavior of chemisorbed layers is based on the assumption that the substrate surface can be treated as a rigid matrix of discrete sites for the adsorbed species (the "adatoms"). When many adatoms are on the surface, the occupation of the sites is non-random because of lateral AA interactions. For chemisorbed species on a metal substrate this interaction may oscillate and have a relatively long range [3] so that ordered 2D structures are expected even at low coverage. Most results concerning the geometry of chemisorbed layers are interpreted in this manner. As an illustration, consider the (001) surface of a cubic crystal shown schematically in Fig. 1. The formation of a (2 x 2) structure (Fig.1A) at a fractional coverage of $\theta = \frac{1}{4}$ monolayer is explained in the lattice gas treatment by the existence of AA interactions which give a repulsion between atoms on nearest-neighbor sites ($\epsilon_1 > 0$) and on next-nearest

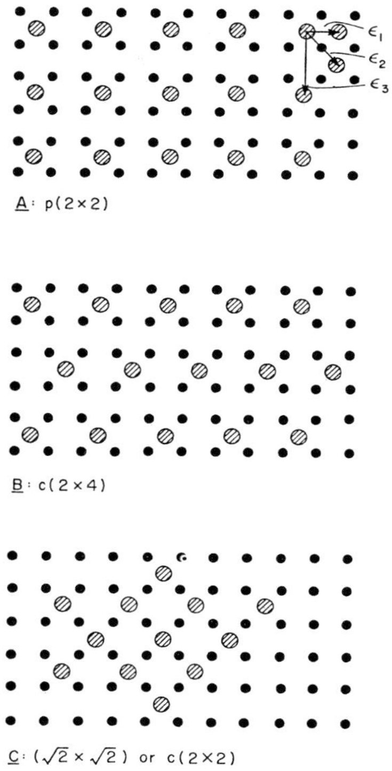

Fig. 1. Examples of 2D structures according to lattice gas model. The large circles represent adatoms. A) p(2 × 2) structure; B) c(2 × 4) structure; C) Island of $\sqrt{2} \times \sqrt{2}$ -- or c(2 × 2) -- structure.

sites ($\varepsilon_2 > 0$), combined with an attraction at greater distances. If instead the repulsion extended beyond next-nearest neighbors, a c(2 x 4) arrangement (Fig. 1B) would be favored at this coverage. If the AA repulsion were short-ranged and only $\varepsilon_1 > 0$, a ($\sqrt{2} \times \sqrt{2}$) structure - also known as c(2 x 2) - would be expected; if in addition $\varepsilon_2 < 0$, islands of this geometry would form at low coverage (Fig. 1C). In principle the magnitude of $\varepsilon_1, \varepsilon_2, \ldots$ can be found by calculating the phase diagram for different sets of these parameters until a fit with the experimental diagram is achieved. Such calculations have been done for O/Ni(111) as discussed at this Conference by Park et al. [4], and for O/W(110) as discussed by Lagally et al. [5].

3. PHASE DIAGRAM OF H/W(001)

In the case of H/W(001) the lattice gas model fails to explain the observed 2D periodicity. The first pattern seen by LEED during hydrogen adsorption at room temperature [2,6-10] corresponds to a c(2 x 2) structure, indicating that the dominating AA interaction is a nearest-neighbor repulsion perhaps combined with a longer-range attraction. The system should then be equivalent to a 2D Ising antiferromagnet and should have the phase diagram shown in Fig. 2. As the temperature T is raised the c(2 x 2) structure will undergo a continuous transition to a disordered phase, and the maximum stability will be at $\theta = 0.5$. The critical temperature at this coverage is $T_c = \varepsilon_1/1.76k_B$ where k_B is Boltzmann's constant [11]. At higher and lower θ the transition temperature is known from the Monte-Carlo calculations by Binder and Landau [12] who also investigated the formation of c(2 x 2) islands which occurs when $\varepsilon_2 < 0$. Assuming the validity of the lattice gas model, an independent estimate of ε_1 can be obtained by measuring the desorption kinetics [13]. Due to the AA repulsion the desorption energy E_d is highest when θ is small and decreases as the coverage goes up. For H/W(001) E_d decreases by $\approx$ 6 kcal/mol [8] giving an apparent value of the pair repulsion $\varepsilon_1 \approx 0.033$ eV, and hence $T_c \approx 220$K at $\theta = 0.5$.

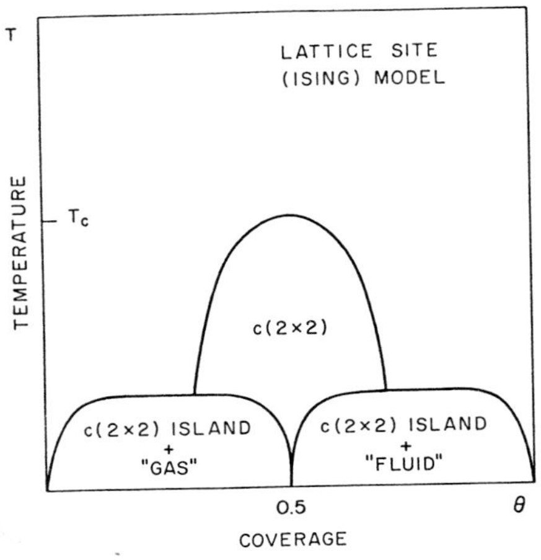

Fig. 2. Schematic phase diagram in the temperature-coverage plane for a lattice gas on a square net with nearest neighbor repulsion and a second-neighbor attraction (Ref. 12).

These predictions turn out to be incorrect as shown by a comparison with the experimental phase diagram [10] in Fig. 3. Some of the qualitative features are as expected; a c(2 x 2) structure is prominent in this diagram and an apparently continuous disordering occurs as T is raised [10]. However, the maximum stability is not at $\theta = 0.5$ and $T \approx 220$k but at $\theta \approx 0.3$ and $T \approx 400$k. Furthermore the phase diagram contains a series of incommensurate structures which are not predicted by the lattice gas model.

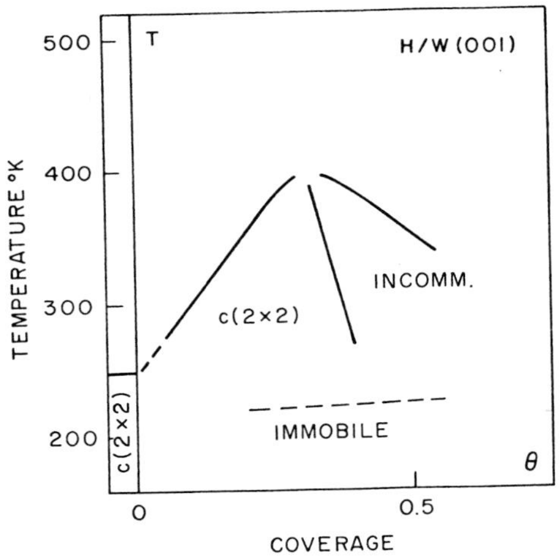

Fig. 3. Experimental phase diagram for the H/W(001) surface (Ref. 10) showing the approximate regions where the c(2 × 2) and the incommensurate structures are stable. The region to the left of $\theta = 0$ refers to clean W(001). No ordered structures appear for adsorption in the region labeled "immobile."

Some of the observed LEED patterns [6-10] are sketched in Fig. 4. At high T and low θ only the "normal" spots from a (1 x 1) structure are seen (Fig. 4A). Adsorption at $\gtrsim$ 300K produces a pattern with extra ½ ½ spots (Fig. 4B) which weaken and broaden if T is raised [10]. Additional adsorption induces a commensurate-incommensurate transition and each ½ ½ spot

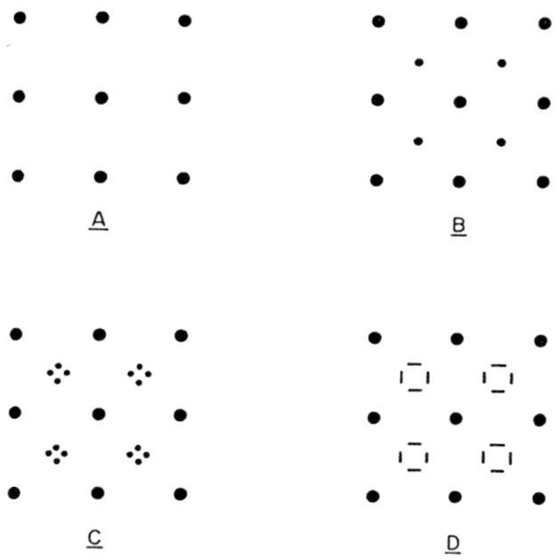

Fig. 4. Sketch of LEED patterns observed during hydrogen adsorption on W(001) at room temperature. A) (1 × 1) structure on clean W(001); B) c(2 × 2) pattern with extra $\frac{1}{2}\frac{1}{2}$ spots; C) Incommensurate structure showing split $\frac{1}{2}\frac{1}{2}$ spots; D) Partially disordered structure giving streaky diffraction features.

splits into four new spots (Fig. 4C) which gradually move apart and eventually become streaky (Fig. 4D), indicating a loss of long-range order. (At higher coverages other poorly developed patterns appear [6,9], but they will not be discussed here.) At room temperature the saturation coverage approaches $\theta = 2$, i.e. almost 2H atoms per W atom. The various LEED patterns are not observed in adsorption below ≈ 200K and this region is labeled "immobile" in Fig. 3. If the temperature exceeds 500-600K, rapid desorption of H_2 takes place.

The commensurate-incommensurate transition can be induced either by raising the temperature (i.e. by moving vertically in the T-θ diagram) or by increasing the coverage (i.e. by moving horizontally in the T-θ diagram). The "incommensurability", Δ, i.e. the separation of two split spots, is shown in Fig. 5 as a function of temperature and coverage. In both cases the experimental data [10] are well represented by a straight line over a fairly large range.

The phase diagram for H/Mo(001)[2] is even more complex. A c(2 x 2) structure is among the observed phases also for this system but it appears only at low temperature and in a small coverage range below $\theta = 0.5$; an increase in temperature causes a reversible transformation to a (4 x 2) structure [2]. Incommensurate structures are seen at both higher and lower coverages and are more prevalent in this system than in H/W(001). Detailed studies of the various phase transitions have not yet been completed.

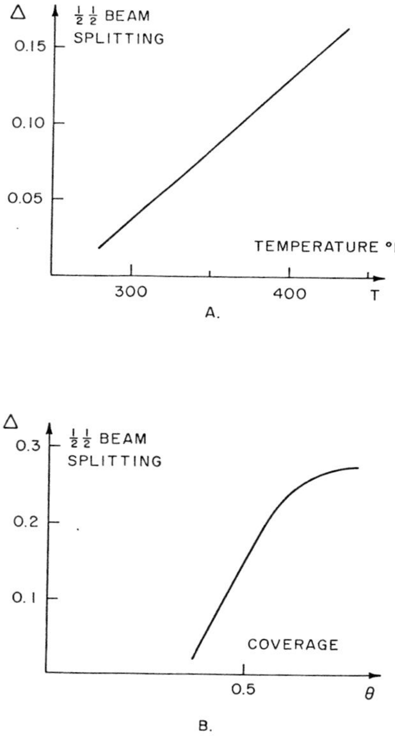

Fig. 5. Degree of spot splitting for the incommensurate H/W(001) structure. A) Δ as function of T at $\theta \simeq 0.25$; B) Δ as a function of θ at room temperature. The maximum value is ≈ 0.3 corresponding to ≈ 1/3 order positions of the split spots (Ref. 10).

4. SUBSTRATE REARRANGEMENT

The principal reason for the differences between the expected and observed phase diagrams is the occurrence of a rearrangement of the substrate surface atoms. Strong evidence of this phenomenon is obtained in studies of the clean (001) surfaces of W [14,15] and Mo [14]. Well above room temperature the W(001) surface has the expected (1 x 1) periodicity but upon cooling the structure changes gradually to c(2 x 2), as indicated on the left ($\theta = 0$) in the diagram of Fig. 3. The phase change is rapid, completely reversible and occurs without hysteresis; it must therefore be a displacive rather than a reconstructive transformation. Subsequent studies of H/W(001) have shown, furthermore, that the structures formed in the presence of hydrogen also involve displacements of the W atoms. For example, the LEED intensities are very similar for the temperature-induced and the adsorbate-induced structures [2,16,17], and calculations indicate that the H atoms themselves contribute negligibly to the LEED intensity. Confirmation of the hydrogen-induced displacive transformation has been obtained by ion-scattering measurements [18] and by electron energy loss spectroscopy, ELS [19]. All three methods suggest that in the c(2 x 2) structures the W atom displacements have a component in the surface plane of ≥ 0.2Å. The direction of this component, $\mathbf{d}_s$, can be inferred from the symmetry of the LEED patterns. In most experiments the pattern appears to have four-fold symmetry about the origin, as indicated in Fig. 4, but this is because the pattern is a sum of intensities from equivalent surface domains rotated 90° with respect to each other. Under certain circumstances this rotational degeneracy is lifted, for example, if the surface contains narrow terraces [10]. Each ½ ½ beam is then seen to split into only two spots, and not four as in Fig. 4C. It turns out that the incommensurate periodicity develops in the direction parallel to the terrace edge but not in the direction perpendicular to the edge [10]. Even before the splitting, the pattern has only two-fold symmetry and the intensity variation across the pattern reveals the direction of $\underset{\sim}{d}_s$. The kinematical diffraction amplitude is

$$A \propto \sum_j e^{i\underset{\sim}{k} \cdot \underset{\sim}{r}_j}$$

where $\underset{\sim}{k} = \underset{\sim}{K} - \underset{\sim}{K}_o$ is the difference between the wave vectors of the diffracted and incident beam. The atoms are displaced in a periodic lattice distortion (PLD):

$$\underset{\sim}{r}_j = \underset{\sim}{r}_{jo} + \underset{\sim}{d} \sin(\underset{\sim}{q} \cdot \underset{\sim}{r}_{jo}) + \ldots$$

The leading term for the diffracted intensity of the extra spots, at normal incidence, becomes

$$I(\underset{\sim}{k}) = A(\underset{\sim}{k})A^*(\underset{\sim}{k}) \propto (\underset{\sim}{k} \cdot \underset{\sim}{d}_s)^2$$

Weak or absent spots identify the condition $\underset{\sim}{k} \perp \underset{\sim}{d}_s$. For H/W(001) $\underset{\sim}{d}_s$ is found to be along <01> [10]. The low-temperature c(2 x 2) structure on clean W(001) also has lower than four-fold symmetry, as reported by Debe and King [20], but the intensity variation shows that in this surface $\underset{\sim}{d}_s$ must be along <11> . Fig. 6 depicts models of the two c(2 x 2) structures. In the clean surface (Fig. 6A) the W atoms in the top layer form zig-zag chains, but hydrogen adsorption causes a switch in the direction of $\underset{\sim}{d}_s$ producing the W "dimers", shown in Fig. 6B. The H atoms are not shown in the figure; however, although they cannot be located by LEED, ELS data [19] indicate that the favored site is in a bridged position on top of a dimer.

The same type of transformations occur on Mo(001). The clean cold surface has a $(\sqrt{2} \times \sim(8 \text{ to } 9)\ \sqrt{2})$ structure [14] and appears to be incommensurate. The displacement $\underset{\sim}{d}_s$ is along <11> but switches direction to <1$\tilde{0}$> when hydrogen adsorbs [21] to give an arrangement similar to that in Fig. 6B [2].

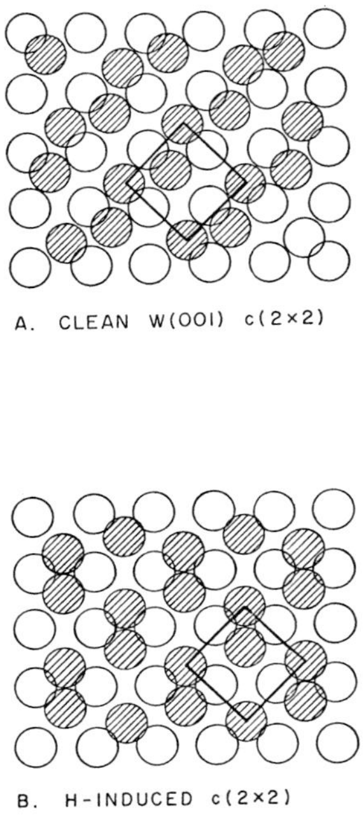

Fig. 6. Models of the c(2 × 2) structures on W(001) illustrating the displacements in the plane of the surface. The shaded circles represent the W atoms in the top layer, the open circles represent the second layer. A) The clean W(001) surface at low temperature; B) The hydrogen-induced structure. The H atoms (not shown) are located on top of the dimers.

Models of the hydrogen-induced incommensurate structures can be constructed by introducing a modulation of the c(2 x 2) structure [2]. As described above this modulation depends on both the temperature and the coverage.

The picture of the surface resulting from these considerations is obviously quite different from the lattice gas model. Evidently, it is not a good approximation to treat the substrate as a rigid assembly of adsorption sites. As a consequence of the lattice distortion, the adatom-substrate interaction will vary across the surface and, as illustrated in Fig. 7, the probability that a site is occupied will oscillate with the same period as the PLD. The modulation need not be purely sinusoidal as in Fig 7A but may have any number of higher harmonics; the LEED data give only the periodicity (and the direction) of the displacement and, for example, an antiphase domain model, which corresponds to a square wave, is equally acceptable (Fig. 7B) [22].

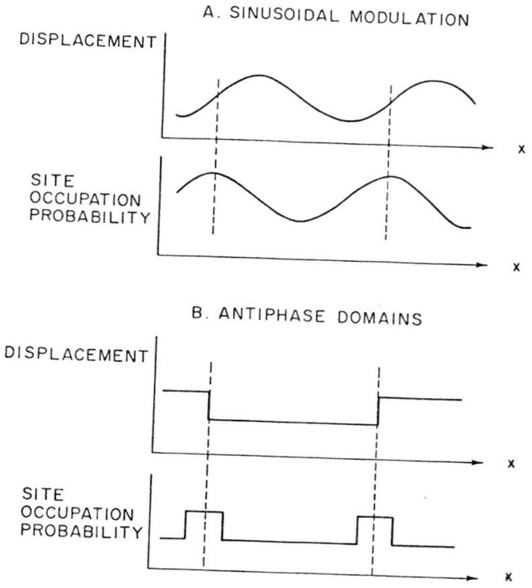

Fig. 7. One-dimensional illustration of the rearranged surface. A) Sinusoidal PLD. The probability that a site is occupied varies with the same period as the lattice distortion. B) Antiphase domain model. The modulation has the same period as in the case of the sinusoidal PLD (Ref. 22).

5. PHASE TRANSITIONS

The experiments have not yet provided a satisfactory explanation of the microscopic driving mechanism of the phase transitions, and the statistical thermodynamics of these systems has not been fully developed.

The most likely driving mechanism is an electronic instability such as a Peierls transition or a Jahn-Teller transition. As discussed by Tossatti [23], if the electrons in the surface states behave as a 2D gas, a peak is expected in the electronic susceptibility at $q = 2k_F$ where k_F is a nesting Fermi wave vector. The electronic energy may be lowered by the formation of a charge density wave, CDW, and a PLD is produced simultaneously, most probably having the form of a frozen-in phonon [24]. The 2D Fermi surface geometry required to produce a c(2 x 2) structure, i.e., $2k_F = q = \left(\frac{\pi}{a}, \frac{\pi}{a}\right)$, is sketched in Fig. 8A. Calculations by Krakauer et al. [25] of the electronic density of states of clean W(001) support this mechanism. The interpretation of calculations for clean Mo(001) seems less clear [26] but the unusual periodicity $q \approx 0.9 \left(\frac{\pi}{a}, \frac{\pi}{a}\right)$ of the cold surface strongly suggests a Fermi surface effect. It is less likely, however, that the periodicity of the hydrogen-induced structures

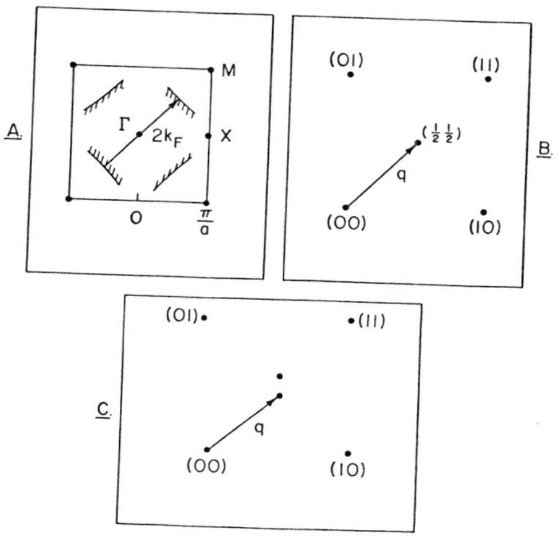

Fig. 8. Schematic illustration of the 2D fermi surface geometry A) required to give $q = 2k_F$ when the reciprocal lattice B) is that of the c(2 × 2) structure. In the case of the incommensurate structures C), $q = \left((1 - \Delta)\ \frac{\pi}{a}, \frac{\pi}{a}\right)$.

can be explained entirely by a $2k_F$-instability. For example, in the pattern from the incommensurate H/W(001) structures, the split spots always move parallel to [10] (or[01]), i.e., $q = \left((1 - \Delta)\ \frac{\pi}{a}, \frac{\pi}{a}\right)$ as shown in Fig. 8C. An evolution of the 2D Fermi surface to give precisely these nesting vectors (as T and θ are varied) would be a remarkable coincidence. It is easier to rationalize the direction of the splitting by assuming that the structures reflect the local geometry of the adsorbate-substrate complex [10].

A promising approach to a general understanding of the hydrogen-induced effects is the theoretical model by Lau and Ying [27] which includes the lattice distortion, as well as the AS and AA interactions. This model can account for the stabilization of the W c(2 x 2) structure by the adsorbate (Fig. 3), the switch in the direction of the displacement (Fig. 6), and the qualitative effect of coverage on the commensurate-incommensurate transition (Fig 5B).

6. CONCLUDING REMARKS

Many details of the H/W(001) and H/Mo(001) systems remain to be understood but it is clear that the 2D ordering can be treated only by taking substrate rearrangement into account. It is possible that such a "reconstruction" will turn out to be important also in many other chemisorption systems.

This work was supported by the Brown University Materials Sciences Program, funded by the National Science Foundation.

*Present address: Bell Laboratories, Murray Hill New Jersey 07974.

REFERENCES

1. See, for example, L. D. Schmidt, in Interactions on Metal Surfaces (R. Gomer, ed.) Springer, New York, 1975, Page 63; and E. W. Plummer, ibid., Page 143.
2. P. J. Estrup, J. Vac. Sci. Technol. 16, 635 (1979).
3. T. L. Einstein, CRC Crit. Rev. Solid State & Mat. Sci. 7, 260 (1978).
4. R. L. Park, T. L. Einstein, A. R. Kortan and L. D. Roelofs, in Proceedings of this Conference.
5. M. G. Lagally, T. M. Lu, and G. C. Wang, in Proceedings of this Conference.
6. P. J. Estrup and J. Anderson, J. Chem. Phys. 45, 2254 (1966).
7. K. Yonehara and L. D. Schmidt, Surface Sci. 25, 238 (1971).
8. R. A. Barker and P. J. Estrup, Phys. Rev. Lett. 41, 3107 (1978).
9. D. A. King and G. Thomas, Surface Sci. 92, 201 (1980).
10. R. A. Barker and P. J. Estrup, to be published.
11. See, for example, K. Huang, Statistical Mechanics (Wiley, New York 1963) Chap. 16.
12. K. Binder and D. P. Landau, Surf. Sci. 61, 577 (1976).
13. D. L. Adams, Surf. Sci. 42, 12 (1974).
14. T. E. Felter, R. A. Barker and P. J. Estrup, Phys. Rev. Lett. 38, 1138 (1977).
15. M. K. Debe and D. King, J. Phys. C 10, L303 (1977).
16. R. A. Barker, P. J. Estrup, F. Jona and P. Marcus, Solid State Commun. 25, 379 (1978).
17. M. K. Debe and D. King, Surface Sci. 81, 193 (1979).
18. I. Stensgaard and L. C. Feldman, Phys. Rev. Lett. 42, 247 (1979); L. C. Feldman, P. Silverman and I. Stensgaard, Surf. Sci. 87, 410 (1979).
19. M. R. Barnes and R. F. Willis, Phys. Rev. Lett. 41, 1729 (1978).
20. M. K. Debe and D. King, Phys. Rev. Lett. 39, 708 (1977).
21. R. A. Barker, S. Semancik and P. J. Estrup, Surf. Sci. 94, L162 (1980).
22. S. Semancik and P. J. Estrup, to be published.
23. E. Tosatti, Solid State Commun. 25, 637 (1978).
24. A. Fasolini, G. Santoro and E. Tosatti, to be published.
25. H. Krakauer, in Proceedings of this Conference; H. Krakauer, M. Posternak and A. J. Freeman, Phys. Rev. Lett. 43, 1885 (1979).
26. J. E. Inglesfield, J. Phys. C 12, 149 (1979); K. Terakura, I, Terakura and Y. Teraoka, Surf. Sci. 86, 535 (1979).
27. K. H. Lau and S. C. Ying, in Proceedings of this Conference; Phys. Rev. Lett. 44, 1222 (1980).

DISCUSSION

D. E. Moncton:

How do you know that multiple displacement waves do not occur in one coherent surface domain?

P. J. Estrup:

The LEED patterns from the incommensurate structures usually show a splitting into four spots because of domain degeneracy. When we remove this degeneracy, as we have done for W(001) using a "terraced" sample, then the patterns show a splitting into only two spots. This is symptomatic of a single displacement wave. In the case of clean Mo(001) we do not yet have data from terraced surfaces, but the intensity variation in the low-temperature pattern suggest domains with a single displacement wave. Some details are published in Surface Sci. 94, L162 (1980).

B. I. Halperin:

What is the density of steps that you had in your experiments on a terraced substrate?

P. J. Estrup:

In these experiments we have used a (001) sample prepared in such a way that it has long narrow terraces at the edges. The advantage is that the center of the sample gives the ordinary, domain-averaged behavior, while domains of a particular orientation may exist at the edges. We estimate, from beam broadening, that the terrace width must be 100 A to remove the domain degeneracy. However, another factor is the direction of the terrace edge; for example, if the edge is along <10> we can select the domain orientation of the hydrogen-induced structures but not of the temperature-induced (clean) structures.

T. V. Ramakrishnan:

(1.) Is the 'immobile' phase characterizable as a 2-d glass-like phase?

(2.) What is the nature of the transition between the immobile phase and the ordered C and IC phases?

(3.) Up to what coverage does the immobile phase persist?

P. J. Estrup:

The surface layer produced by low-temperature adsorption on W gives a (1x1) pattern, i.e., the clean (2 x 2) structure is destroyed and no new superstructure is induced. We don't know the position of the H-atoms; we suppose they are arranged randomly and that they have low mobility. The H-atom density is rather low, the coverage range over which we have studied the phase diagram is 0.6. The "immobile phase" is not an equilibrium state: if we stop the adsorption and then raise the temperature to 200K, the expected surface structure, commensurate or incommensurate appears and persists if we again cool the sample. We have not been able to vary the temperature rapidly enough to study the kinetics of the change.

GENERALIZED SUSCEPTIBILITIES AND ELECTRONICALLY DRIVEN SURFACE PHASE TRANSITIONS*

H. KRAKAUER, M. POSTERNAK[+], and A. J. FREEMAN

Northwestern University, Evanston, Illinois 60201, USA

ABSTRACT

The temperature dependent reconstruction of the clean W(001) surface to the c(2x2) structure is one of the best known examples of a surface phase transition, although the precise nature and origin have not been unambiguously determined. Calculations for the surface generalized susceptibility reveal a prominent peak in $\chi(\vec{q})$ at $\overline{M}$ [$\vec{q} = (1,1)\pi/a$] when matrix elements and local-field corrections are incorporated, providing theoretical evidence that the transition may be electronically driven via the onset of a CDW screened by an accompanying surface phonon softening. In order to elucidate possible effects of a surface interlayer contraction, results obtained for a 6% contracted surface are also presented.

In this paper, the possibility that the reconstruction of some transition metal surfaces may be explained in terms of electronically driven phase transitions is examined. Focusing on the reconstructions of the W(001) surface and the related Mo(001) surface reconstruction, we show that the existence of electronic instabilities, which can lead to the onset of a surface charge density wave (CDW), are related to Fermi surface nesting of electronic surface resonance states (states localized near the surface).

The reconstructions of the clean W and Mo(001) surfaces were first observed in low-energy-electron-diffraction (LEED) experiments [1-3], which showed a reversible termperature-dependent phase transition on cooling below about 300 K. Below this temperature, extra LEED spots at the half-order positions appear producing a $(\sqrt{2}\times\sqrt{2})R45^\circ$ LEED pattern on the W(001) surface. On Mo(001), a quartet of extra LEED spots appears centered on the $(\frac{1}{2},\frac{1}{2})$ positions, and this has been related [2] to the occurrence of an incommensurate reconstructed phase on the Mo surface, in contrast to the commensurate structure appearing on the W surface. The transformation seems to be of second order and is reversible on varying only the temperature, and the variation in LEED intensity of the extra spots versus temperature is similar for both the W and Mo surfaces. Recent investigations [2,3] conclude that no chemisorbed impurity (including hydrogen) need be present on the surface when the low temperature phase is observed, showing that the transition is characteristic of the clean surface.

The interpretation of this phase transition has led to controversy about its origin and has centered about the possible formation and role of a surface CDW as in the layered transition metal dichalcogenides. Experimental LEED symmetry analysis has led to the suggestion of displacive phase transitions involving parallel shifts of surface atoms on both the W and Mo surfaces [4-6]. This is supported by a theoretical LEED intensity analysis [7] which favors parallel shifts of the atoms on the W surface with the shifts being about 0.15-.30Å. The Debe and King [4] parallel shift model for W involves alternating atomic shifts in a [110] direction, which results in a loss of four-fold symmetry in the reconstructed phase. This model can be described in terms of a displacement wave with vector, $\vec{q}=(1,1)\pi/a$. A similar model for the Mo(001) surface [6] also involves parallel shifts along the [110] direction. In this case, however, the wave vector of the displacement wave is incommensurate with the two-dimensional periodicity of the substrate atoms. Two such rotationally equivalent domains result in the quartet of extra LEED spots which appear centered at the $(\frac{1}{2},\frac{1}{2})$ positions on cooling below about 300 K. Noting the chemical similarity of W and Mo, the fact that commensurate structures appear on the W surface while incommensurate structures appear for Mo, and the continuous nature of the transition on both W and Mo, various authors [2-9] have proposed a CDW mechanism for the reconstructions as in the layered transition metal dichalcogenides.

Surface reconstructions were first discussed in terms of CDW's by Tosatti and Anderson [10]. They showed that if the two-dimensional Fermi surface of a surface-state band had large, nearly parallel sections spanned by a two-dimensional nesting vector, $q=2k_F$, this could lead to the occurrence of an electronic instability at this wave vector driving the phase transition via surface-phonon softening and gapping of the two-dimensional Fermi surface accompanying the onset of a CDW. The associated periodic-lattice-distortion (PLD) leads to the reconstructed phase and the observed extra LEED spots.

As noted before, some conflicting structural evidence has been presented for the W(001) reconstruction. High-energy ion scattering results [11] obtained at room temperature indicate that only about 67% of the surface atoms are displaced on the W(100) surface. Results obtained with a field ion microscope [12] at 21 K show no loss of four-fold symmetry, as predicted by the parallel-shift model [4] for W. Finally, results obtained after field evaporation [13] show evidence for substantial vertical components of the displacements, and, in addition, find evidence for the reconstruction over the entire temperature range studied, 15-460 K. Thus, the precise nature and origin of the reconstruction have not yet been determined.

In order to help resolve these problems, we have theoretically investigated the possible existence of electronic instabilities on the W(001) surface by calculating the surface generalized susceptibility, $\chi(\vec{q})$, including matrix elements and local-field effects [14]. These calculations were based on previously obtained self-consistent thin-film energy band results for the unreconstructed W(001) surface [15]. These self-consistent results (for a seven-layer film) provide excellent agreement with the experimentally determined work function and spectroscopic features of the various surface states and surface resonances on W(001) and, in agreement with results obtained for the Mo(001) surface [16], it was found that self-consistency was crucial in obtaining reliable results for the surface states on W(001). We briefly summarize those results which are pertinent to the present discussion of the reconstruction.

A good overview of the surface electronic structure is provided by the layer projected density of states (DOS) pictured in Fig. 1. The central layer DOS of the seven-layer W(001) film is essentially identical to the bulk DOS which is characterized by a broad peak below E_F, representing the occupied bonding t_{2g} d-orbitals, and an unoccupied peak above E_F, due to anti-bonding e_g d-orbitals. The bonding and anti-bonding orbitals are separated by what is almost a gap in the DOS centered at E_F where the DOS is very small. These general features persist up to the sub-surface layer (S-1). There are pronounced differences, however, in the surface layer DOS. The most significant difference is the large peak occurring in what is essentially the "gap" region of the bulk DOS This large peak is due to a high density of surface states and surface resonance states which are quite localized in the surface layer. This already suggests that the reconstruction of the surface might lower this high peak via gapping of the two-dimensional Fermi surface, thus favoring the reconstructed phase.

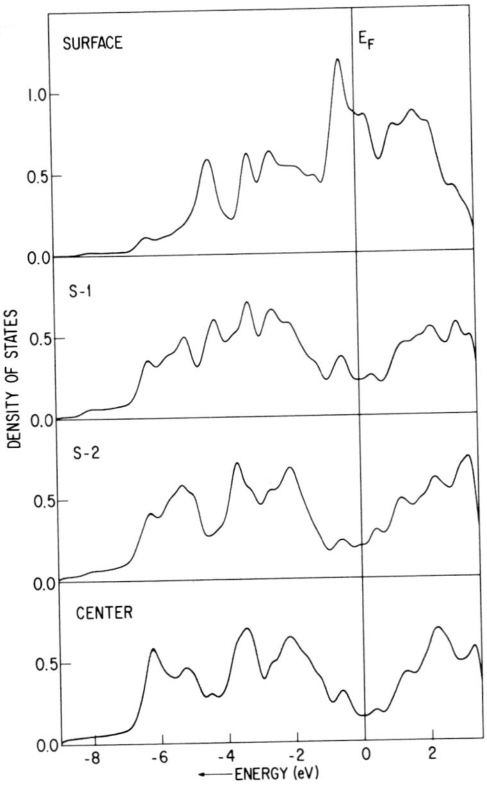

Fig. 1. Layer projected density of states for the seven-layer film. S-1 is the first layer in from the surface.

Figure 2 shows the total valence charge density, $\rho(\vec{r})$, in a (110) plane for the upper half of the seven-layer W(001) film. The bonding characteristics discussed above in connection with Fig. 1 are also evident in Fig. 2. In particular, it is seen that the bonding xy, xz, yz (t_{2g}) d-orbitals form fairly localized lobes pointing along the body diagonals to the nearest-neighbor atoms. In addition, there is a rather uniform metallic bonding charge density in the interior interstitial regions. These features persist up to the second layer from the surface, and then there are marked changes in the bonding character. Compared to an interior atom, substantial weight has been removed from the lobes pointing towards the missing nearest neighbor above the film, although there still remains a localized "dangling" d-bond pointing out into vacuum. Similar unsatisfied dangling bonds on semiconductor surfaces have been associated with a tendency for the surface to reconstruct [10], and it is possible that the reconstruction of the W(001) surface would also reduce the energy associated with this dangling bond. Finally, there is a rapid variation in $\rho(\vec{r})$ in the surface layer interstitial region going out into the vacuum. The

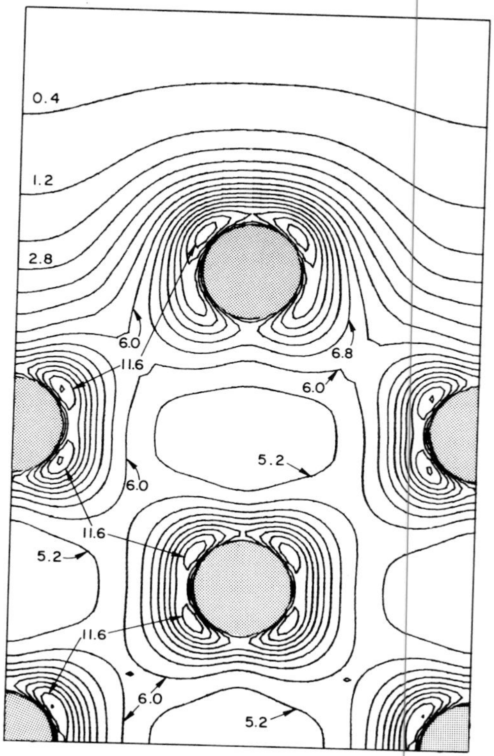

Fig. 2. Contour plot of the total valence charge density. Successive contours are separated by 0.8 in units of electrons per bulk unit cell.

sizeable redistribution of charge near the surface is associated with the formation of the dipole layer which sensitively determines the work function. Self-consistency is found to be crucial in obtaining an accurate value of the work function, and our theoretical value, 4.5 eV, is in excellent agreement with experiment [17].

The surface states and surface resonance states for the seven-layer film are displayed in Fig. 3. For ease of reference, the bottom the conduction band is shown along the lower portion of the figure (the bulk-like states have been suppressed for greater clarity of presentation). These results provide very good agreement with detailed angular resolved photo-emission measurements [18], both with respect to energy position and the symmetry of the various bands. The surface states which are relevant to the present discussion are the $\bar{\Sigma}_1$ and $\bar{\Sigma}_2$ pair of surface states which cross E_F about midway between $\bar{\Gamma}$ and $\bar{M}$. The nesting vector which spans the two-dimensional Fermi surface of this pair of states is $\vec{q}=(1,1)\pi/a$, and this is also the distance between the $\bar{\Gamma}$ point and the $\bar{M}$ point. This is precisely the wave vector of the displacement wave used to describe the Debe and King [4] parallel shift model of the reconstructed W(001) surface; and immediately suggests the possible role of these states leading to the onset of a CDW. Contour charge density plots for these states are shown in Figs. 4 and 5. The $\bar{\Sigma}_1$ state (Fig. 4) has xy and (xz+yz) orbital character, while the $\bar{\Sigma}_2$ state has (x^2-y^2) and (xz-yz) character. Both of these states are quite localized in the surface layer and have substantial bonding charge density directed towards the nearest

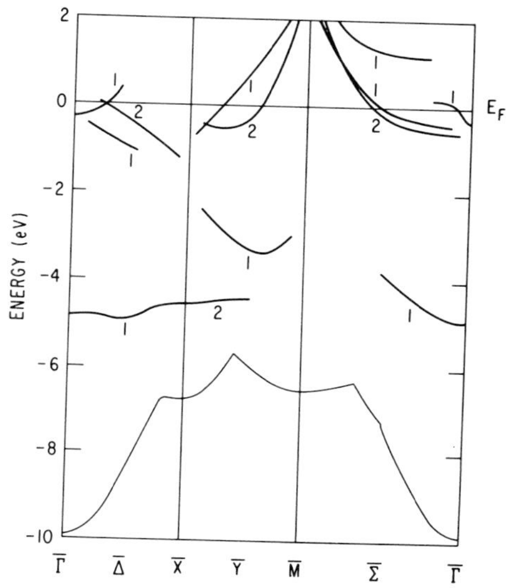

Fig. 3. Surface states and resonances. The bottom of the conduction band is outlined along the lower portion of the figure.

neighbor atoms in the second layer. In addition, they both have unsatisfied "dangling" bonds pointing out towards the vacuum, and one can speculate that the reconstruction would reduce the energy associated with these dangling bonds.

Using the self-consistent energies and wavefunctions obtained for the seven-layer film, one can investigate the possible existence of electronic instabilities which might drive the reconstruction, by calculating the generalized susceptibility, χ, and looking for peaks in χ throughout the two-dimensional Brillouin zone. The expression for χ is given by

$$\chi_{layer}(\vec{q}+\vec{G}, \vec{q}+\vec{G}') = \sum_{\substack{k,k' \\ n,n'}} \frac{f_k^n(1-f_{k'}^{n'})}{E_k^n - E_{k'}^{n'}} \times M_{kk'}^{nn'}(\vec{q}+\vec{G}) M_{kk'}^{nn'*}(\vec{q}+\vec{G}'). \quad (1)$$

Here $M_{kk'}^{nn'} = \langle \psi_k^n | e^{-i(\vec{q}+\vec{G})\vec{r}} | \psi_{k'}^{n'} \rangle_{layer}$, $\vec{q} = (q_x, q_y, 0)$, $\vec{k}$ is the 2D Bloch momentum, $E_{\vec{k}}^n$

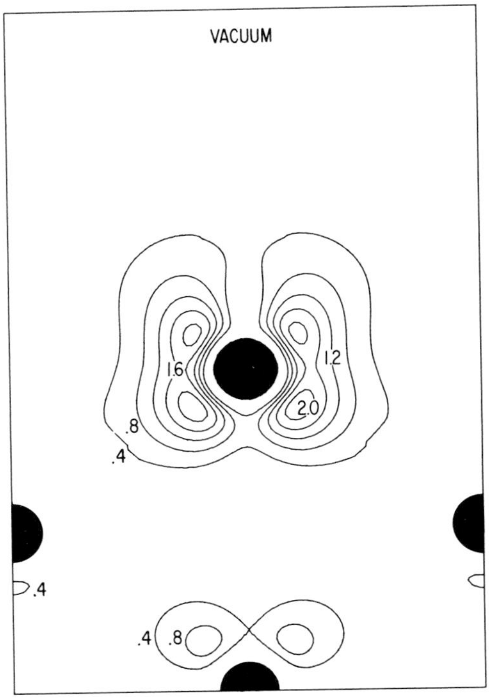

Fig. 4. Charge density for the $\overline{\Sigma}_1$ surface resonance near E_F at $k_{||}$ = (0.5, 0.5) π/a. Successive contours are separated by 0.4 in unites of electrons per bulk unit cell.

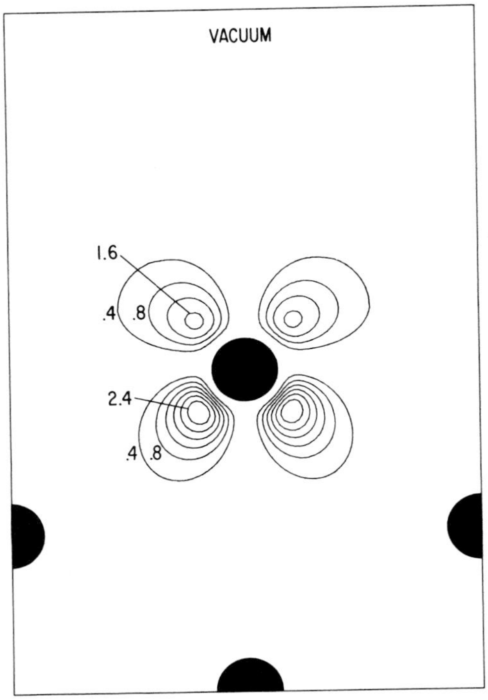

Fig. 5. Charge density for the $\overline{\Sigma}_2$ surface resonance near E_F at $k_{||}$ =(0.5, 0.5)π/a. Successive contours are separated by 0.4 in units of electrons per bulk unit cell.

is the energy of the 2D Bloch wave function Ψ_k^n with band index n and occupation number $f_{\vec{k}}^n$, and $\vec{G}$ and $\vec{G}'$ are 2D reciprocal-lattice vectors. The matrix element M imposes the restriction $\vec{k}'=\vec{k}+\vec{q}+\vec{K}$, where K is a 2D reciprocal-lattice vector which reduces $\vec{k}'-\vec{k}$ to the first 2D Brillouin zone. It should be emphasized that the calculations described below include all of the film states: surface states as well as bulk states.

Surface sensitivity is achieved by decomposing χ into layer by layer contributions through restricting the volume of integration in the matrix element, M, to be over the surface layer only. The details of the calculation have been described elsewhere [14], and we focus here on the salient features of χ resulting from different approximations for the matrix elements $M_{kk'}^{nn'}$.

Curve (a) in Fig. 6 corresponds roughly to the constant matrix element approximation [for the intraband (n=n') part of Eq. (1)] which is almost exclusively employed in bulk calculations of χ. In the present case, this consists of replacing the product of the matrix elements in Eq. (1) by the product of the layer weights of the occupied and unoccupied states $W_k^n W_{k+q}^n$, where $W_k^n = \langle \Psi_k^n | \Psi_k^n \rangle_{layer}$. There is a definite peak in curve a at the M point, which corresponds to $\vec{q}$ =(1,1)π/a. This peak is due to two-dimensional Fermi surface nesting of the surface states midway between $\overline{\Gamma}$ and $\overline{M}$ in Fig. 3 as discussed above, and supports the CDW interpretation for the phase transition.

The effect of properly calculating the matrix elements in Eq. (1) is known to be important in bulk calculations of χ, and similar behavior is found in curve b in Fig. 6 which pictures the total (intraband plus the interband) χ(q). The most dramatic effect of properly including the matrix element is to remove completely the peak at $\overline{M}$. The overall behavior in going from curve a to b in Fig. 6 is similar to that found previously for bulk Cr [19] and Sc [20]. In these cases, absolute peaks were found at physically significant nesting vectors when the constant-matrix element approximation was used for χ. In both cases, however, including matrix elements completely removed the peak at the nesting vector. This behavior is related to the fact that in a transition metal, there is a

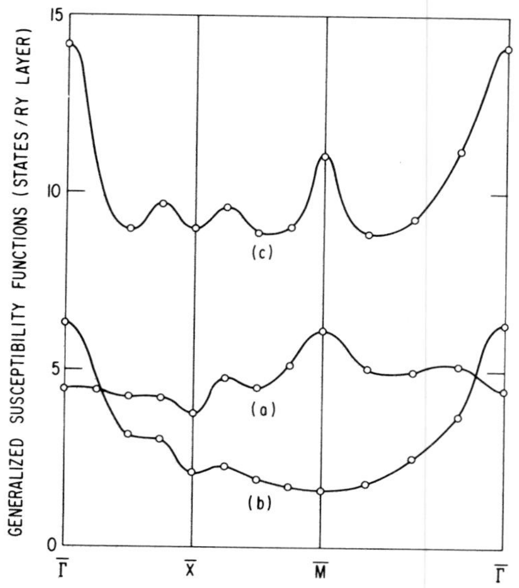

Fig. 6. Susceptibility functions for the surface layer. Curves a, b, and c are described in the text.

rapid spatial variation of fields, and one can't neglect the short wavelength components (so-called local field corrections) in the susceptibility. These short wavelength terms enter through the occurrence of the two-dimensional reciprocal lattice vectors, $\vec{G}$ and $\vec{G}'$, in Eq. (7) in the matrix elements, M. The complete treatment of these terms is extremely difficult, so we follow Gupta and Sinha [19] in estimating these corrections by evaluating the function $F(\vec{q}) = \sum_{G} \chi(\vec{q}+\vec{G}, \vec{q}+\vec{G})$. The local field corrections are thus approximately included by summing over higher reciprocal lattice vectors in the function $F(\vec{q})$. This function is plotted as curve c in Fig. 6, which reveals a prominent peak restored at the nesting vector $\vec{q}=(1,1)\pi/a$ ($\bar{M}$). As in the calculations for bulk Cr [19] and Sc [20], including local field corrections restores the peak. We conclude, therefore, that this is strong evidence that the reconstruction of the W(001) is indeed electronically driven.

While the W(001) surface is in the unreconstructed phase above room temperature, there is evidence for a contraction of the topmost surface interlayer spacing. LEED intensity analysis predicts contractions in the range 4-11% [21], while high energy ion-scattering results set an upper limit of 6% [11]. In order to elucidate the possible effects of a surface interlayer contraction on our conclusions, similar self-consistent results were obtained for 6% contracted surface [22]. Detailed comparison of surface states and other electronic properties such as work functions and surface core level shifts, show that these are relatively insensitive to a contraction of this magnitude. For example, the energy position of the various surface states changes by less than ~3mRy. Thus our earlier conclusion that the temperature dependent reconstruction is electronically driven remains valid.

An instability with respect to CDW formation may also explain why hydrogen could act as an impurity stabilizer in the formation of W(001)c (2x2)-H at room temperature and above, where it has been suggested that the superstructure involves reconstruction of the W substrate [5,6]. Current models [5,6] suggest that the substrate reconstruction can be described in terms of a displacement wave which is similar but not the same as that occurring for the clean W surface. For very low H coverage (less than a few percent of a monolayer coverage) the W atoms are thought to be displaced along the [010] direction (unlike the [110] shifts on the clean surface), but the wave vector of the displacement wave is still given by $\vec{q}=(1,1)\pi/a$. Fasolino [9] et al. have shown that the CDW couples to a $\bar{M}_5$ soft phonon which is a two-fold degenerate vibrational mode. One of these modes leads directly to the parallel shift model observed on the clean W surface, and the other mode leads to the suggested structure for the H-induced reconstructed phase. The H-induced structure is, furthermore, consistent with the fact that H always seems to bond at the bridge site on the W surface [23]. Thus the presence of small amounts of H atoms randomly located on the W surface seems to preferentially select one CDW mode over the other.

Finally, we note that similar conclusions are probably valid for the origin of the reconstruction on the Mo(001) surface, given the similarity of the surface electronic structure of these metals.

*Supported by NSF-DMR, the AFOSR and in part by the Swiss NSF.

+Permanent address: EPFL, Laboratoire de Physique Appliquée, Lausanne, Switzerland.

REFERENCES

1. K. Yonehara and L. D. Schmidt, Surface Science 25, 238 (1971).
2. T. E. Felter, R. A. Barker and P. J. Estrup, Phys. Rev. Lett. 38, 1138 (1977).
3. M. K. Debe and D. A. King, J. Phys. C10 L303 (1977).
4. M. K. Debe and D. A. King, Phys. Rev. Lett. 39, 708 (1977).
5. D. A. King and G. Thomas, Surface Science 92, 201 (1980).
6. R. A. Barker, S. Semancik and P. J. Estrup, to be published.
7. R. A. Barker, P. J. Estrup, F. Jona, and P. M. Marcus, Solid State Commun. 25, 375 (1978).

8. E. Tosatti, Solid State Commun. 25, 637 (1978).
9. A. Fasolino, G. Santoro, and E. Tosatti, to be published.
10. E. Tosatti and P. W. Anderson, Solid State Commun. 14, 773 (1974).
11. I. Stensgaard, L. C. Feldman, and P. J. Silverman, Phys. Rev. Lett. 42, 247 (1979). Also see L. C. Feldman, R. L. Kaufman, P. J. Silverman, and R. A. Zuhr, Phys. Rev. Lett. 39, 38 (1977).
12. T. T. Tsong and J. Sweeney, Solid State Commun. 30, 767 (1979).
13. A. J. Melmed, R. T. Tung, W. R. Graham, and G. D. W. Smith, Phys. Rev. Lett. 43, 1521 (1979).
14. H. Krakauer, M. Posternak, and A. J. Freeman, Phys. Rev. Lett. 43, 1885 (1979).
15. M. Posternak, H. Krakauer, A. J. Freeman and D. D. Koelling, Phys. Rev. B, to appear.
16. G. P. Kerker, Y. M. Ho, and M. L. Cohen, Phys. Rev. Lett. 40, 1593 (1978).
17. R. L. Billington and T. N. Rhodin, Phys. Rev. Lett. 41, 1602 (1978).
18. Shang-lin Weng, E. W. Plummer, and T. Gustafsson, Phys. Rev. B18, 1718 (1978).
19. R. P. Gupta and S. K. Sinha, Phys. Rev. B3, 2401 (1971).
20. R. P. Gupta and A. J. Freeman, Phys. Rev. B13, 4376 (1976).
21. M. N. Reed and G. J. Russell, Surface Science 88, 95 (1979).
22. M. Posternak, H. Krakauer, and A. J. Freeman, Bull. Am. Phys. Soc. 25, 235 (1980).
23. R. F. Willis, Surface Science 89, 457 (1979).

DISCUSSION

R. L. Park:

Your band of surface states in the gap at the Fermi level strongly resembles that calculated by Feibeman et al. for the basal plane of Ti.

Has the surface state been studied as a function of reconstruction?

H. Krakauer:

I do not know of any studies in which these surface states have been monitored as a function of temperature.

P. Kleban:

Can you make any comments on how specific this mechanism is to this particular surface?

H. Krakauer:

The mechanism may be quite general. Tosatti and Anderson, for example, suggested that the reconstruction of Si(111) may be explained in terms of a CDW instability.

Sinha, ed. Ordering in Two Dimensions

STAGING AND ORDERING IN ALKALI-METAL GRAPHITE INTERCALATION COMPOUNDS

ROY CLARKE

Physics Department, The University of Michigan, Ann Arbor, Michigan 48109

ABSTRACT

Ordering processes in the intercalant substructures of various alkali-metal graphite intercalation compounds are reviewed. X-ray diffuse scattering studies of the most extensively studied Cs compounds are described. The saturated Stage 1 compound CsC_8 is fully three-dimensional and exhibits a discontinuous order-disorder transition. The correlation length within unregistered Cs layers of the more dilute Stage 2 structure CsC_{24} grows continuously towards low temperatures whereas transverse to the layers it is limited to ∿40Å. These layers never achieve true long-range order and show unusual, possibly 2-d, intralayer X-ray scattering. No significant change in the behavior is observed on going to the still more dilute Stage 3 structure.

INTRODUCTION

Lamellar intercalation compounds are formed by the diffusion of atoms or molecules into the Van der Waals gap between the weakly bonded crystallographic planes of a layered material. In the graphite intercalation compounds (GIC's) considered in the present study, layers of alkali-metal ions are interspersed with the planes of carbon atoms from the graphite structure.

A key feature of GIC's is the formation of *regular* sequences or 'Stages', of carbon and intercalant layers [2,3] and a sequence in which there are n layers of carbon atoms interposed between every pair of intercalant layers is referred to as a Stage n compound (see Fig. 1). A further important aspect of GIC's is that the intercalation reaction involves *charge exchange* between the intercalant and host layers. In the alkali metal case, the intercalant atoms donate an average 0.5-0.8 electrons to the GIC conduction band [4]. Acceptor compounds, on the other hand, may be formed by intercalating with strongly acidic molecules. Both donor and acceptor compounds exhibit metallic conductivities within the intercalant layers [5].

During intercalation the carbon layers retain their microscopic integrity with <1% change in the intralayer C-C bond length [6] a consequence of the strong covalent planar bonding of the carbon layer. The carbon layers bounding each intercalant layer provide a rather smooth lamellar containment for the intercalant ions which can thus be quite mobile at modest temperatures. These remarkable characteristics of the GIC's promote an unusually rich variety of order-disorder phenomena associated with the intercalant substructure [7-11].

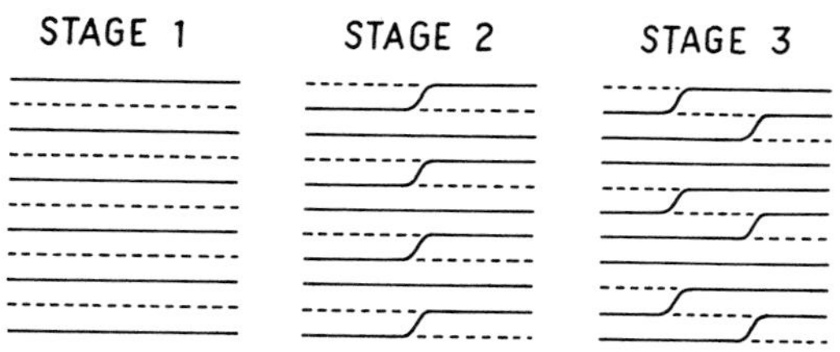

Fig. 1: Staging in GIC's according to Daumas and Hérold [1]. Solid lines represent carbon layers and broken lines, intercalant layers. Domains in Stages $n > 1$ are macroscopic ($\gtrsim$ 400Å).

Graphite intercalation compounds present us with several novel experimental situations. Among these are: quasi two-dimensional (2-d) metals in various ordered and disordered forms; physical realizations of several lattice gas models; and a unique configuration in which disordered layers are embedded in an otherwise ordered 3-d matrix, providing additional types of disorder to those found in liquid crystal systems. A further point of interest in the GIC's is that the behavior of the intercalant layer is, to some degree, controlled by the interaction of the intercalant ions with the graphite bounding layers. In this respect there is an interesting connection to be made with the physics of monolayers adsorbed on graphite substrates.

There are two questions of primary concern in

the present study of GIC's, other than the general aim to gain some understanding of the physical properties of this fascinating class of materials. Firstly, how does the variable interlayer interaction, afforded by the staging property, affect ordering processes within intercalant layers? And secondly, to what extent can the intercalant substructures be considered as two-dimensional entities? In an attempt to provide answers to these questions, a systematic X-ray diffuse scattering study has been made of the temperature dependent inter- and intralayer correlations of the first three Stages of the cesium GIC's. These have chemical compositions [12] CsC_8 (Stage 1), CsC_{24} (Stage 2), and CsC_{36} (Stage 3) The choice of cesium as the intercalant in our studies is based partly on its large X-ray cross-section and also because the alkali metal GIC's in general are the best characterized systems.

EXPERIMENTAL

X-ray scattering experiments have been performed on two different types of intercalated samples, namely those prepared from highly oriented pyrolytic graphite (HOPG) and from single crystal forms of graphite. In the former, polycrystalline samples, the c-axes (basal plane normals) of individual grains are aligned with each other to within 2° but the a-axes are randomized. The advantage of the HOPG samples is that they can be made in larger volume than available single crystals. However, because of the powder averaging in the basal plane, they provide information principally concerning the *positional* correlations of the intercalant atoms. Definitive details of the in-plane intercalant-intercalant *bond orientations* relative to the graphite host are only obtained from single crystal samples [10].

Samples were intercalated using the standard two-bulb technique of Hérold [13]. The intercalated samples are highly reactive to water vapor and oxygen and must be carefully protected against contamination. To this end they were sealed-off in evacuated Pyrex containers. Finally, their stacking sequences were confirmed from 00ℓ X-ray diffraction profiles. The quality of our single crystal samples, prepared by intercalating natural Madagascar graphite crystals (0.5 mm × 0.5 mm ×0.05 mm), is reflected by their a-axis mosaic spread of only 0.4° after intercalation. The c-axis spread was approximately 1.3° compared to the values of ∿2° usually found in HOPG samples.

X-ray intensities were measured using a Picker four-circle diffractometer fitted with either a closed-cycle helium refrigerator or resistive heating elements. The X-ray source was a graphite-monochromated MoKα beam from a 12 kW rotating anode generator. The instrumental resolution in the horizontal plane was approximately $0.015Å^{-1}$.

RESULTS AND DISCUSSION: STAGE 1, CsC_8

At 300K CsC_8 has a three-dimensionally ordered structure in which the Cs ions form a commensurate $2a_o \times 2a_o \times c_o$ superlattice, where a_o and c_o are respectively the intralayer and interlayer repeat distances of the carbon planes. A projection of this structure is shown in Fig. 2 (insert). Each successive layer of Cs ions occupies one of the symmetry equivalent sublattices, α, β, or γ, so that the room-temperature stacking sequence in CsC_8 may be described as AαAβAγAαA... where A refers to the stacking of carbon layers [14] whereas in KC_8 and RbC_8 the full four-layer repeat is observed [15].

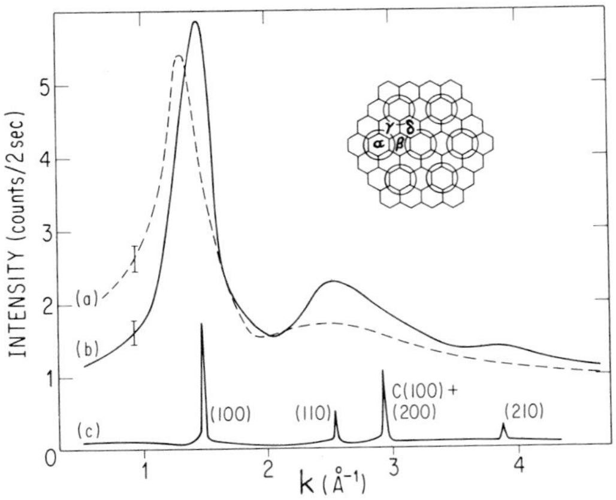

Fig. 2: Scattered X-ray intensity from:(a) the 'liquid' phase of CsC_8 (b) the disordered phase of $Cs_{1-x}C_8$ ($x \approx 0.2$) (c) CsC_8 at 300K, Bragg peaks indexed on 2×2 structure shown in insert.

At a temperature of approximately 650K the Cs superlattice structure is observed [8] to undergo a first-order transition to a liquid-like phase in which the average Cs-Cs intralayer separation (5.78Å) does not correspond to a distance between preferred carbon hexagon sites [8,16]. In this sense the structure of the metal layer is dominated more by Cs-Cs interactions than by interactions with the bounding layers and is akin to the bulk liquid metal configuration. High temperature diffractometer scans [16] of CsC_8 reveal short-range Cs-Cs interlayer correlations showing that three-dimensional interactions are still important in the liquid phase.

At high temperatures ($T \gtrsim 550K$) there is a tendency for CsC_8 to de-intercalate causing vacancies in the Cs occupation. This is the reason for the shift in the position of the peaks of the diffuse scattering profiles shown in Fig. 2. A detailed study of these decomposition effects has been made by Caswell [16]. In the partially depleted state ($Cs_{1-x}C_8$) a type of disorder is possible that is distinctly different than the liquid case described above: one in which Cs ions are statistically distributed over preferred lattice sites (centers of carbon hexagons) and hopping can now take place to vacant lattice sites [8]. Note that the large ionic size of Cs forbids occupation of nearest (a_0) and next-nearest ($\sqrt{3}\ a_0$) lattice sites. This is the origin of the short range order at $2a_0 \times 2a_0$ positions reflected in the diffuse scattering profile of Fig. 2(b). Such 'lattice gas' configurations in the Stage 1 GIC's have recently been analyzed in a Landau-Ginzburg theory by Bak et al. [17,18] . CsC_8 was shown to belong to the universality class of the three-dimensional, 6-component, Potts model corresponding to the six equivalent wavevectors of the mass density waves which describe the 3-d Cs substructure. The Landau expansion of the free energy in terms of these mass density waves contains a third-order invariant which precludes a continuous transition in this theory. In practice the order-disorder transition in $Cs_{1-x}C_8$ is found to be of first order, but weakly so [8]. It is interesting to note that in certain truly 3-d structures, namely the rare earth GIC's (e.g. EuC_6) *continuous* melting of the intercalant superlattice may be possible [18]. To date no experiments have been performed to confirm this.

RESULTS AND DISCUSSION: STAGE 2, CsC_{24}

The higher-stage alkali metal intercalants present a qualitatively different behavior than the saturated Stage 1 compounds: at no temperature do they take on an ordered structure in the way that Stage 1 compounds exhibit simple 3-d superlattices. They do, however, show an interesting variety of ordering phenomena some of which will be seen to arise from the reduced dimensionality of their highly anisotropic structures.

The separation of Cs layers in the Stage 1 CsC_8 structure depicted in Fig. 1 is 5.95Å. This is close to the Cs-Cs distance of 5.61Å found in bulk liquid Cs metal. It is not surprising therefore that essentially 3-d Cs-Cs correlations are observed in the Stage 1 case, even in the unregistered liquid phase of CsC_8. However, the separation of Cs layers in Stage 2 increases to 9.30Å and, in Stage 3 to 12.65Å. We may therefore expect much more anisotropic behavior of the Cs substructure in the higher Stages. Intralayer correlations. In order to study the structure of the Cs layers in CsC_{24} a 2-d radial distribution function (RDF) analysis has been made [10]. Fig. 3(a) shows the room-temperature structure factor S(k) ($k \equiv 4\pi\sin\theta/\lambda$) for the Cs layer in CsC_{24}, where the diffraction vector $\vec{k}$ is confined to the Cs plane. In Fig. 3(b) is plotted the 2-d RDF obtained from the Fourier-Bessel inversion of S(k) [10]. Here it is assumed that interlayer positional correlations of Cs ions are effectively absent. We shall see later that this is a valid assumption and that, indeed, a 2-d RDF analysis is appropriate.

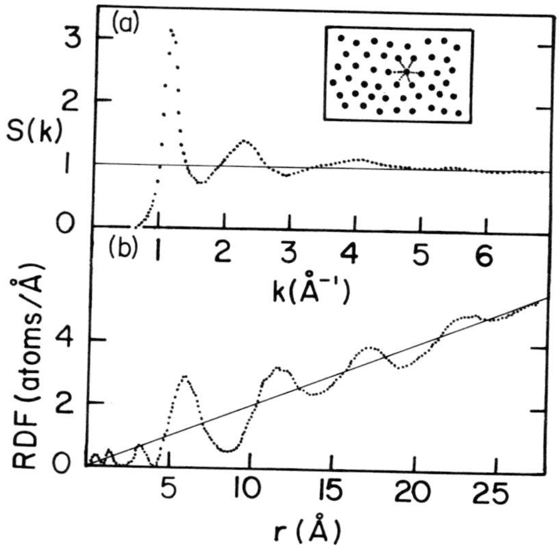

Fig. 3: (a) Structure factor S(k) for the Cs layer in HOPG CsC_{24} at 300K (b) Radial distribution function (RDF) from S(k) in (a). Insert: short-range order of Cs layer.

The results shown in Fig. 3 indicate that the Cs layer is disordered with a simple close-packed structure (see insert of Fig. 3). The average nearest neighbor Cs-Cs distance of 5.95±0.1Å, given by the position of the first peak in the RDF, is close to the value of 6.02Å one would expect given the supposed *layer* stoichiometry of CsC_{12} and assuming a uniform close-packed structure. The distance 5.95Å is not related to a simple multiple of the carbon layer translation vector and so it is concluded that, on average, the Cs ions are not in registry with carbon lattice positions. This is a similar situation to the high-temperature liquid-like phase of CsC_8 described above, except that now the areal density of Cs is somewhat smaller. The Cs substructure in CsC_{24} is therefore pictured as a stack of uncorrelated layers within which the Cs ions have a simple close-packed disordered triangular structure.

In Fig. 4 the in-plane X-ray scattering profiles, in the region of the first peak in S(k) are compared at several temperatures. It is seen that as the temperature is decreased this diffuse peak continuously becomes narrower and its peak height increases down to the lowest temperature achieved (10K), indicating a continuous growth of positional correlations within the Cs layer as the temperature is reduced. The wavevector of the primary Cs diffraction peak remains constant with temperature at $1.158 Å^{-1}$ confirming that the spacing of the unregistered triangular structure does not change with temperature to within experimental resolution.

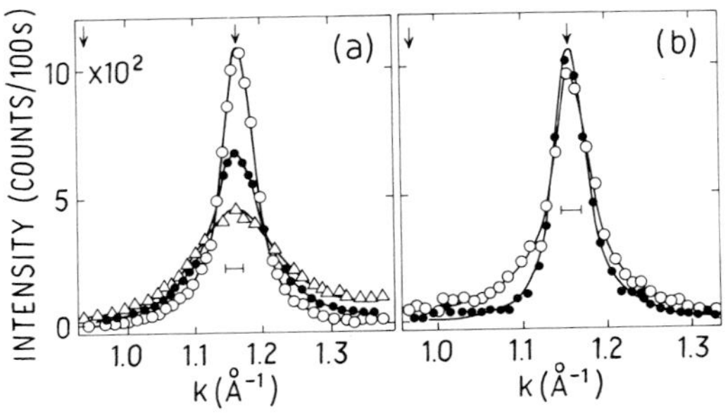

Fig. 4: Temperature dependence of X-ray scattering in vicinity of first peak in S(k)
(a) Δ 236K, ● 170K, o 157K
(b) o 90K, ● 10K
The solid lines between the arrows are fits to Lorentzians (see text). The asymmetry in the peaks is due to the presence of other features at $k > 1.3 Å^{-1}$. Bars represent instrumental resolution.

The temperature dependent lineshapes shown in Fig. 4 may be fit with a Lorentzian form $I(\varepsilon) \sim [1 + \xi^2(T)\varepsilon^2]^{-1}$ where $\varepsilon = q_o - k$, $q_o = 1.158 Å^{-1}$ and $\xi(T)$ is an in-plane correlation length. Given the relatively poor resolution of the experiment, we would not claim to be able to distinguish between a Lorentzian lineshape and the power-law cross section, $I(\varepsilon) \sim \varepsilon^{-\alpha(T)}$ expected for a finite 2-d crystal [19]. However, we can say that the X-ray scattering from this unregistered Cs layer never develops into true Bragg scattering even though at low temperatures the observed correlation lengths are at least an order of magnitude larger than those in a common liquid or amorphous material. ξ, corrected for the instrumental resolution, is plotted in Fig. 5. Interlayer correlations. In an early study [7] of CsC_{24} it had been reported that the Cs layer underwent an order-disorder transition at a temperature $T_L \approx 165K$. Fig. 5 shows no evidence for this although ξ is certainly starting to increase rapidly at about this temperature. One might expect that, with a growing value of ξ, neighboring layers of Cs ions would eventually begin to become three-dimensionally correlated. Indeed, if we probe the interlayer Cs-Cs correlations by measuring the diffraction profile along the (10ℓ) direction of reciprocal space (Fig. 6) it can be seen that at approximately 170K, 3-d structure does begin to appear. The flat profile at temperatures above 170K confirms the lack of

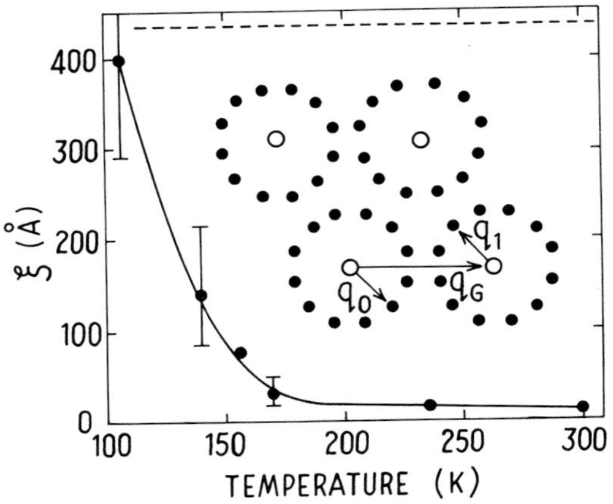

Fig. 5: In-plane correlation length, ξ, from Lorentzian fits. Values have been corrected for instrumental resolution. Broken line is resolution limit. Insert: Schematic of in-plane reciprocal lattice of CsC_{24} at 150K. $\vec{q}_o$ and $\vec{q}_G$ are the wavevectors of the primary Cs and graphite reflections respectively. $q_1 = q_G - q_o$ is a satellite reflection.

interlayer correlations supposed above. By 150K the 3-d structure has fully developed but the interlayer correlation length, as determined from the width of the profile along c* (the reciprocal c-axis), never exceeds 40Å [10].

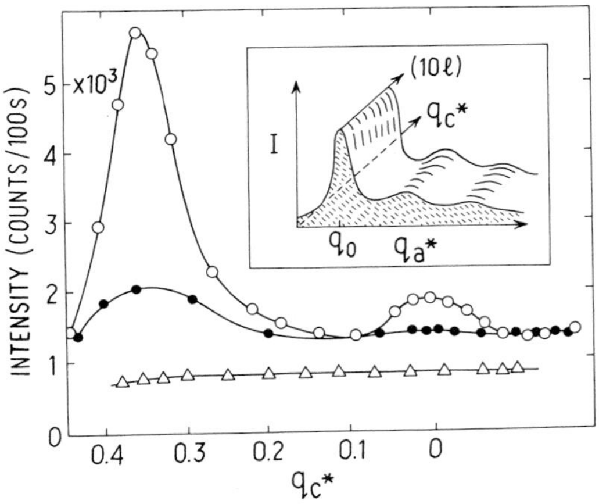

Fig. 6: Diffractometer scans along (10ℓ) direction of CsC_{24} shown in insert. q_{c*} (in $Å^{-1}$) = $2\pi\ell/c$ where c is the Stage 2 c-axis repeat distance. The peak at $q_{c*} \approx 0.34 Å^{-1}$ (ℓ=3) gives c = 55.4 Å.
Δ 300K; ● 170K; o 130K

This important phenomenon was first observed in KC_{24} by Hastings et al. [9] and was attributed to the existence of stacking faults in the carbon host structure. The temperature T_L, therefore can be identified not with a simple order-disorder transition (as in the stage 1 compounds) but rather with a crossover in the *dimensionality* of the positional correlations from 2-d (for $T > T_L$) to 3d (for $T < T_L$). However, the 3-d interlayer correlations are not long range but are limited to only 4-5 Cs layers.

Further effects of host structure on ordering. In our picture of the disordered triangular layer in which Cs ions are not registered with carbon layer sites, we have largely been able to ignore the effects of the carbon layer potential on the ordering of the Cs ions. At lower temperatures, when thermal energies become much smaller than the binding energy to preferred carbon hexagon sites it may be energetically favorable for the uniform unregistered structure to break up into a non-uniform registered structure. In practice this is observed as the appearance of X-ray scattering from macroscopic ($\gtrsim$400Å) regions with the 2×2 superlattice structure, coexisting along with regions of the unregistered phase. This begins to occur at approximately 140K and continues down to at least 10K, the area of registered regions growing at the expense of the unregistered phase. An in-plane diffraction pattern from HOPG CsC_{24}, Fig. 7, clearly shows this coexistence. It is interesting to compare the lineshapes of the diffraction from each region and these are shown in more detail in the insert of Fig. 7.

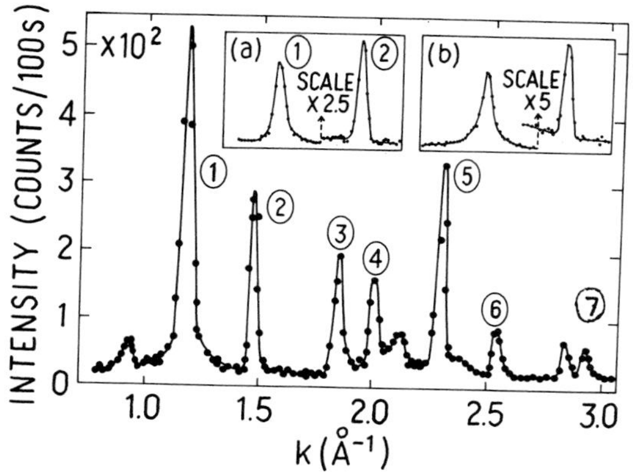

Fig. 7: In-plane diffraction profile of HOPG CsC_{24} at 10K. Peaks 1, 4 and 5 are from the unregistered Cs layer and 2 and 6 are from the registered 2×2 regions discussed in the text. Peak 3 is a q_1 satellite (see Fig. 5, insert) and Peak 7 a graphite/2×2 Cs reflection. Insert: (a) Peaks 1 and 2 at 10K; (b) at 90K.

The peaks from the unregistered structure clearly have the long-tail character of Lorentzian (or power law) scattering while the diffraction from the registered regions has a temperature independent resolution-limited lineshape imposed by the long range 3-d nature of the graphite host layers. One may speculate whether this is also the origin of the mixed lineshapes observed in KC_{24} [11]. The registered regions in this case however would have the $\sqrt{3} \times \sqrt{3}$ superlattice rather than the 2×2 superlattice of the larger Cs ions. At 50K there is a discontinuous transition to a different, but as yet unidentified, stacking sequence in CsC_{24}. More details are given in Ref. [10].

A further manifestation of the effects of the graphite host on the Cs layer structure appears most clearly in single crystal samples. A schematic of the in-plane diffraction [10] from a single crystal of CsC_{24} at 150K is shown in the insert of Fig. 5. All the main features of this pattern can be explained on the basis of a uniform unregistered Cs layer with triangular symmetry and fundamental wavevector q_o=1.158Å^{-1}. The 12 primary spots at this wavevector are interpreted [10] as arising from a *rotation* of the Cs layer relative to the graphite symmetry axes. Six of the spots arise from domains which are rotated in a positive sense relative to the graphite <100> direction and six from domains rotated in the opposite sense. The rotation angle is not related to a symmetry angle of the carbon layer. In the simpler case of an incommensurate rare-gas monolayer physisorbed onto a graphite substrate, Novaco et al. [20] have shown that transverse forces arising from the incommensurability of the overlayer lead to a rotation, ϕ, or 'orientational epitaxy' of the overlayer. The pattern shown in Fig. 5 is very similar to that observed by Shaw et al. [21] for an incommensurate monolayer of argon on graphite. In this case the satellite reflections around each main carbon reflection (see Fig. 5) are assumed to arise from the modulation of the overlayer by the substrate potential. In the intercalate case, where there is reason to believe that the intercalant ions cause elastic deformations [22,23] of the carbon layers, it may be more reasonable to suppose that the Cs layer modulates the graphite periodicity [24]. At temperatures above ~150K, as ξ decreases and the Cs layer becomes more disordered, the Cs-Cs bond directions become more random. The behavior of the average bond direction ϕ_p relative to <100> is shown in Fig. 8 and suggests an orientational phase transition of the Cs layer at ~228K where the Cs-Cs bond directions become equally distributed over two counter-rotated Ising-like states $\pm\phi_i$. With this analogy ϕ_p would be the order parameter for the transition. An exponent, $\beta = 0.47$ is measured [10] and is thought to reflect the 3-d nature of the orientational order of the graphite host.

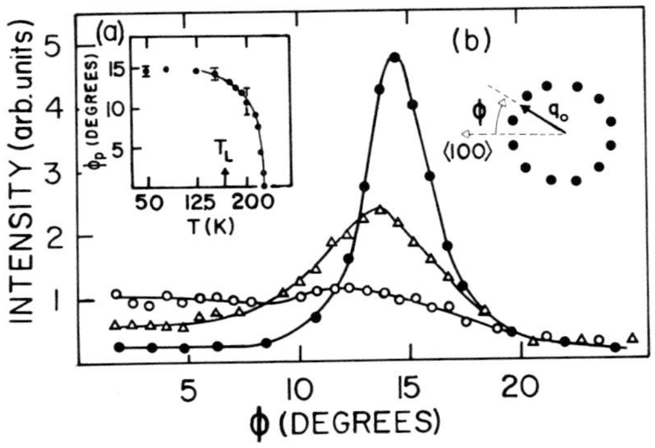

Fig. 8: Detailed temperature dependence of one of the 12 primary q_o reflections in single crystal CsC_{24}.
● 148K; Δ 169K; o 185K. Insert:
(a) Temperature dependence of angular position of reflection relative to a graphite ⟨100⟩ direction. (b) Geometry of χ-scans used to obtain data in this figure.

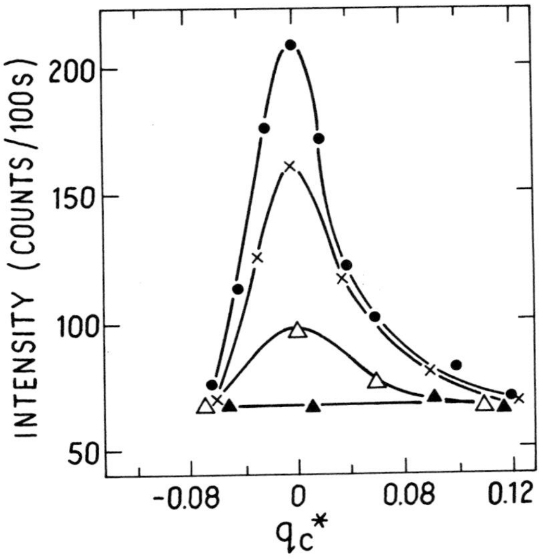

Fig. 9: (10ℓ) scan for CsC_{36}. q_{c^*}(in $Å^{-1}$) $= 2\pi\ell/c'$ where c' is the Stage 3 c-axis repeat distance. ● 130K; × 140K; Δ 170K; ▲ 180K.
(Peak asymmetry due to absorption of sample container.)

RESULTS AND DISCUSSION: STAGE 3, CsC_{36}

The behavior of the Cs layer in Stage 3 samples is essentially identical to that of CsC_{24}. The peaks in S(k) occur at the same positions as in CsC_{24}, in particular the peak at $1.158Å^{-1}$. This latter observation is expected, since the areal density of Cs is supposedly [12] the same in both Stage 2 as Stage 3 compounds and it is the areal density which determines the peak position in the case of the postulated uniform, unregistered, triangular structure. This situation may be contrasted with the case of KC_{24} where the peaks in S(k) appear not to be at the same positions in Stage 2 and Stages 3 and 4 [25]. It is not yet understood why this occurs. Also not resolved is the general question of why the stoichiometry should be precisely MC_{12n} in these compounds, as is generally accepted. It is easy to see how the ordered 2×2 superlattice of the Stage 1 compounds imposes the formula MC_8 but it is not at all clear what mechanism would give rise to a fixed stoichiometry in these disordered compounds.

The development of 3-d correlations in CsC_{36} occurs over the same temperature range as in CsC_{24} (see Fig. 9). Low temperature single crystal in-plane diffraction patterns also appear to be very similar to those of CsC_{24} with minor variations in the saturation value of ϕ (14.5° in CsC_{24} and 15.5° in CsC_{36}).

CONCLUSIONS

The Stage 1 $Cs_{1-x}C_8$ structure displays fully 3-d order-disorder transitions which are of first order. The ordering of the intercalant layers in Stage 2 CsC_{24} proceeds in a qualitatively different fashion. In particular, the Cs layer, which is unregistered with the carbon layers, becomes more and more correlated as the temperature is decreased but never achieves true long-range order. At low temperatures, Cs-Cs positional correlations are found to extend over hundreds of Ångstroms within the Cs layer but over only ∿40Å in the transverse direction. This may be the origin of the unusual form of the X-ray scattering from the Cs layers. According to the results presented above, the Cs layers already have the character of isolated quasi 2-d entities in the Stage 2 structure since further decreasing the dilution to Stage 3 has an insignificant effect on their behavior. These experiments suggest that CsC_{24} may be a useful system for the study of ordering in two dimensions and high resolution X-ray scattering experiments are now required to characterize the lineshapes more accurately. Further work along these lines in a wider range of GIC's is suggested. If the anisotropy of the electrical conductivity, σ, is taken as a measure of the relative strengths of the intralayer and interlayer interactions, the acceptor compounds (with ratios $\sigma_a/\sigma_c \sim 10^5$ [5]) are even more likely to show 2-d effects than the donor compounds.

ACKNOWLEDGEMENTS

It is a pleasure to acknowledge stimulating and fruitful collaborations with S. A. Solin, N. Caswell, and P. M. Horn during the course of the work described above. Thanks are also due to N. Wada, S. R. Nagel, T. Witten, S. C. Moss, and H. Zabel for useful discussions. This work was supported by the NSF-MRL program of the University of Chicago and by the Office of Energy Research, University of Michigan.

REFERENCES

1. N. Daumas and A. Hérold, C. R. Acad. Sci. 268, 373 (1969).
2. W. Rüdorff and E. Schulze, Z. Anorg. Chem. 277, 156 (1954).
3. W. Metz and D. Holwein, Carbon 13, 87 (1975).
4. I. Spain and D. J. Nagel, Mater. Sci. Eng. 31, 183 (1977).
5. For a review of electronic properties, see: J. E. Fischer, Physica 99B, 383 (1980).
6. D. E. Nixon and G. s. Parry, J. Phys. C 2, 1732 (1969).
7. G. S. Parry, Mater. Sci. Eng. 31, 99 (1977); G. S. Parry. D. E. Nixon, K. M. Lester and B. C. Levene, J. Phys. C 2, 2156 (1969).
8. Roy Clarke, N. Caswell and S. A. Solin, Phys. Rev. Lett. 42, 61 (1979).
9. J. B. Hastings, W. D. Ellenson and J. E. Fischer, Phys. Rev. Lett. 42, 1552 (1979).
10. Roy Clarke, N. Caswell, S. A. Solin and P. M. Horn, Phys. Rev. Lett. 43, 2018 (1979); same authors in Physica 99B, 457 (1980).
11. H. Zabel, S. C. Moss, N. Caswell and S. A. Solin, Phys. Rev. Lett. 43, 2022 (1979).
12. F. J. Salzano and S. Aronson, J. Chem. Phys. 47, 2978 (1967).
13. A. Hérold, Bull. Soc. Chim. Fr. 187, 999 (1955).
14. D. Guérard, P. Lagrange, M. El Makrini and A. Hérold (preprint).
15. W. D. Ellenson, D. Semmingsen, D. Guérard, D. G. Onn and J. E. Fischer, Mater. Sci. Eng. 31, 137 (1977).
16. N. Caswell (preprint).
17. Per Bak and Eytan Domany, Phys. Rev. B20, 2818 (1979).
18. Per Bak, Phys. Rev. Lett. 44, 889 (1980).
19. Y. Imry and L. Gunther, Phys. Rev. B3, 3939 (1971).
20. A. D. Novaco and J. P. McTague, Phys. Rev. Lett. 38, 1286 (1977).
21. C. G. Shaw, S. C. Fain, Jr., and M. C. Chinn, Phys. Rev. Lett. 41, 995 (1978).
22. N. Caswell, S. A. Solin, T. M. Hayes and S. J. Hunter, Physica 99B, 463 (1980).
23. S. A. Safran and D. R. Hamann, Phys. Rev. Lett. 42, 1410 (1979).
24. S. C. Moss and H. Zabel (private communication) - see also H. Zabel, this conference.
25. H. Zabel, Y. M. Jan and S. C. Moss, Physica 99B, 453 (1980).

DISCUSSION

D. D. L. Chung:

In contrast to your report of a first order transition for the in-plane melting in C_8Cs, we observed a higher order transition for the in-plane melting in high stage graphite-Br_2. On the other hand, we have observed with calorimetry in C_8Cs a large enthalpy change associated with this phase transition, indicating that it is first order. Thus, the order of this phase transition probably depends on the stage and the intercalate species. It is also interesting to note that graphite-halogens of stage ≥ 3 exhibit so little interlayer correlation that the compounds are almost ideally two-dimensional.

R. Clarke:

It is notoriously difficult to establish the order of a phase transition using calorimetry alone, especially with a technique such as DTA. What you are probably seeing in CsC_8 is the expulsion of Cs atoms from the sample upon melting, which we have previously discussed.

D. Nelson:

You observe two-phase coexistence between a commensurate solid and an incommensurate structure with large in-plane translational correlations. Could this be interpreted as evidence for a first order commensurate-incommensurate transition?

R. Clarke:

The mixed phase can be described as macroscopic regions with registered 2x2 superlattice structure coexisting with an unregistered fluid with very long range, temperature dependent, in-plane correlations. The very large coexistence region of 140K, coupled with the unusual nature of the unregistered structure make the situation distinct from both previously observed commensurate-incommensurate transitions and normal liquid-solid transitions.

ORDER AND DISORDER IN GRAPHITE-POTASSIUM INTERCALATION COMPOUNDS

H. ZABEL

Department of Physics
University of Illinois at Urbana-Champaign
Urbana, IL 61801

Potassium intercalated between graphite basal planes undergoes order-disorder phase transitions. The ordered and disordered structures, as well as the transition temperatures depend on the stage of the material. In all cases, the disordered state shows true two-dimensional character, whereas three-dimensional coupling takes place on ordering. X-ray scattering experiments on HOPG and graphite single crystals permit the study of commensurate, incommensurate and modulated structures of the intercalation compounds. Recent X-ray experiments will be reviewed and discussed in the context of elastic interaction and the distortion modes of the graphite planar structure of the higher stage compounds ($n \geq 2$) induced by the incommensurate potassium intercalant.

I. INTRODUCTION

Graphite is well known as a favorable substrate for studying phase transitions of adsorbed rare gas and hydrocarbon monolayers [1]. It is also becoming clear that graphite serves as well as an ideal host lattice for the study of quasi two-dimensional ordered and disordered structures. In particular the graphite lamellar structures $C_{12n}M$($n \geq 2$, M = K,Rb,Cs) have received considerable experimental interest [2-6] especially in light of recent theoretical progress in describing two-dimensional [2-D] phase transitions [7].

Pure graphite is characterized by a layered structure, originating from a strong sp^2 bonding of the carbon atoms in the plane and weak Van de Waals coupling between the planes. The in-plane C-C bond length is 1.42Å, and the separation is 3.35Å. The planes are stacked together in the sequence ABA and are related to each other by the translation vector $\vec{g} = (2\vec{a} - \vec{b})/3$, where $\vec{a}$ and $\vec{b}$ are lattice vectors of the graphite unit cell (fig. 1).

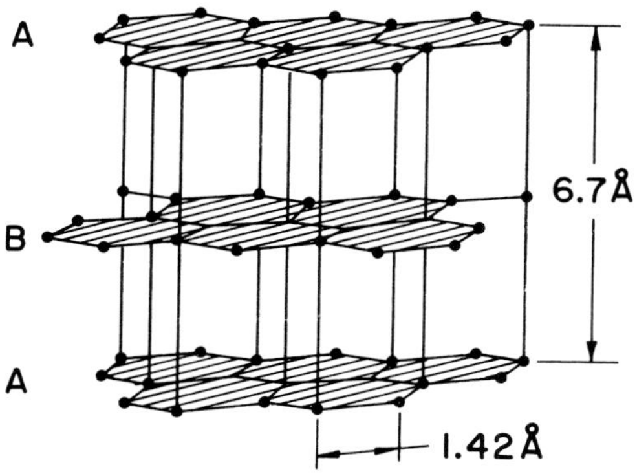

Fig. 1. Structure of graphite.

Under appropriate conditions of graphite temperature and alkali vapor pressure, intercalation of alkali atoms between graphite basal planes takes place [8]. The dissolved atoms cause three main structural changes of the graphite host lattice: 1. the alkali atoms intercalate in a regular fashion so that an intercalated layer is followed by a constant number of graphite planes (n), where n defines the stage of the compound, 2. alkali layers are always flanked by equivalent carbon planes, thereby causing a rearrangement of the sequential order of the graphite planes, 3. the distance between adjacent graphite planes increases considerably (from 3.35Å to 5.4Å for K) upon intercalation, while the in-plane structure of the graphite remains almost uneffected (fig. 2). It has been

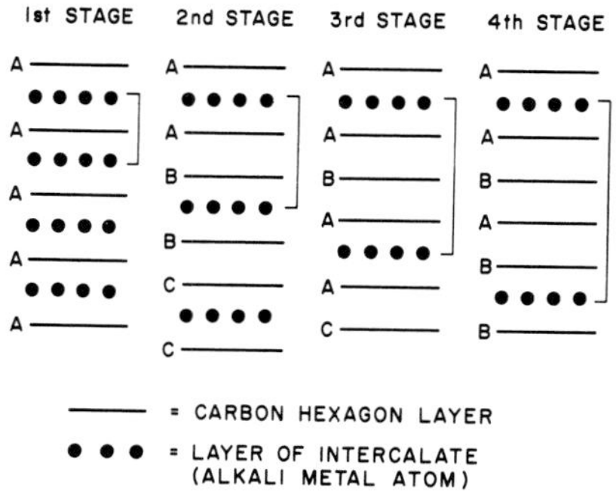

Fig. 2. Stacking sequence of carbon and alkali layers in stages 1 to 4 compounds.

commonly accepted that all compounds have a constant stochiometric in-plane alkali concentration of 1/8[M/C] for stage 1 and 1/12[M/C] for stages $n \geq 2$ (M,C number of metal and carbon atoms, resp.). During intercalation approximately one electron per alkali atom is transferred to the graphite. Recent experiments indicate that these extra electrons not only compensate for the charge of the remaining ionic alkali layers but also screen them [9]. Therefore, long range Coulomb interaction between essentially neutral CMC units is negligible and one expects 2-D alkali properties with increasing separation of these units in higher stage compounds. In the following we will concentrate on graphite-potassium compounds. A review of the experimental results on structures and phase transitions of different stage compounds will be given in part III, and discussed in the framework of the elastic interaction and graphite with modulation structures in part III.

II. STRUCTURE AND PHASE TRANSITIONS

Most of the structural work has been derived so far from highly oriented pyrolytic graphite (HOPG). This material is characterized by a well defined $\vec{c}$-axis but cylindrical averaged in-plane $\vec{a}$ and $\vec{b}$ directions. In diffraction experiments, normally, the $\vec{c}$-axis is kept in the scattering plane which allows a study of in-plane structures and inter-plane correlations. However, because of the random orientation of the $\vec{a}$ and $\vec{b}$ axis, the relative orientation of intercalates to the graphite host cannot be resolved and questions about the commensurability of the alkali layers are more difficult to answer. It is only recently that some single crystal work has become available which, of course, greatly enhances our understanding of structures and phase transitions in these compounds.

1. ORDER AND DISORDER IN C_8K

In stage 1 compounds C_8K, the potassium ions are centered over carbon hexagons and form a 2×2 commensurate in-plane structure (fig. 3) [10,11]. Four equivalent interstitial sites for the alkali ions α,β,γ and δ are available, which alternate with the carbon basal planes in an ordered sequence $A\alpha A\beta A\gamma A\delta A$...(capital roman letters refer to graphite layers, greek letters to alkali layers) to form a 3-D superlattice. In contrast to C_8Rb [12] and C_8Cs [13] order-disorder transitions of potassium in C_8K compounds have not been investigated as yet.

2. ORDER AND DISORDER IN $C_{24}K$

In $C_{24}K$ compounds (stage 2) the potassium layers undergo two structural phase transitions: an upper transition at $T_U = 123K$, followed by a transition at $T_L = 98K$. Onn et al. [14] observed these two transitions by anomalies in the basal

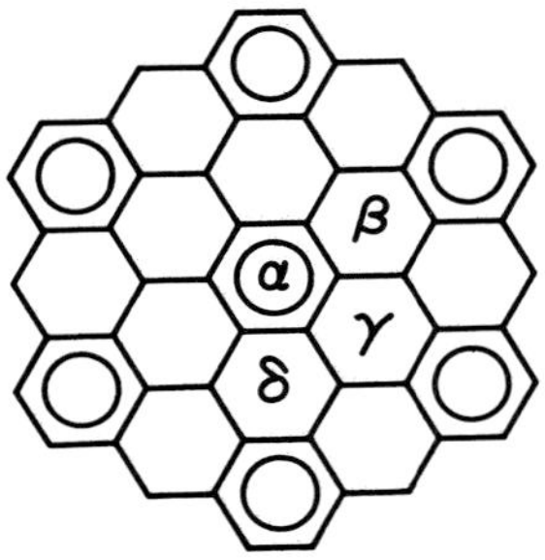

Fig. 3 In plane 2×2 structure of potassium in C_8K. $\alpha,\beta,\gamma,\delta$ denotes the position of alkali atoms in subsequent layers.

plane conductivity. X-ray scattering experiments [15,2,4] led to the same results. Above T_U the potassium layers are disordered. Short range order correlation exists only in the plane but not between different potassium planes. In X-ray studies Bragg-reflections from the graphite sublattice are accompanied by diffuse scattering in the (hk0) planes (fig. 4a), which originates from the 2-D potassium correlation. This diffuse scattering shows no structure normal to the (hk0) planes [15,4]. Computer calculations of the structure factor showed that the diffuse scattering cannot be explained with a model of

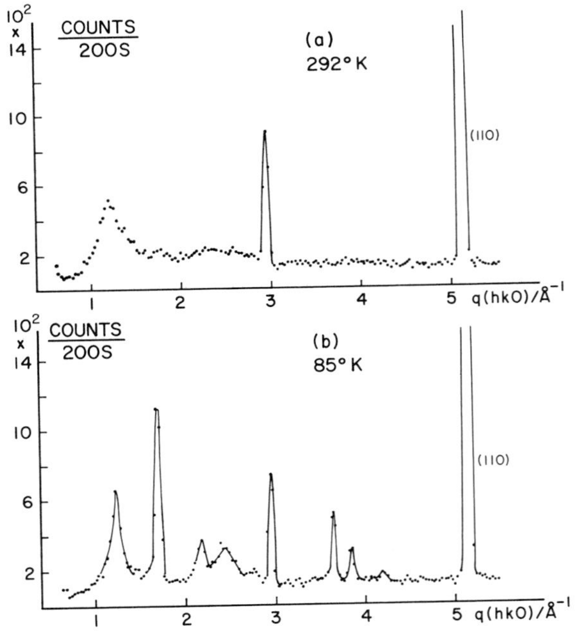

Fig. 4 X-ray diffraction pattern of the powder-averaged (hk0) plane: (a) disordered phase at 292°K; (b) ordered phase at 85°K. The curves are drawn as an aid to the eye [4].

a planar lattice gas, where the potassium ions are restricted to occupy interstitial sites over carbon hexagon centers [16]. In the light of new results on the low temperature structure [17], it is reasonable to describe the high temperature disorder by a closed packed, liquid-like alkali structure, which is incommensurate to the graphite host planes.

At T_U the potassium-graphite lattice undergoes an order-disorder transition, first observed by Nixon and Parry [15]. The transition appears to be of second order with a critical exponent $\beta = 0.18$ [2,4]. At the transition a 3-D coupling of the alkali planes takes place. Crossover effects from 2-D to 3-D structure were not observed. Between T_U and T_L the alkali layers are stacked together in the sequence $\alpha\beta\gamma$. The lower transition at T_L is reported to be first order and relates to an alkali superstacking $\alpha\beta\gamma\alpha'\beta'\gamma'$ [15,2].

Below T_U a whole set of superlattice reflections appear in the cylindrical averaged (hk0) plane (fig. 4b). It has been a long standing question to which alkali planar order these reflections are related. Since only structural information on HOPG material was available so far, a commensurate $2\sqrt{3} \times 2\sqrt{3}$ structure, as proposed by Rüdorff and Schulze [10], could not be definitely excluded. Very recently, Mori et al. [17] have performed X-ray experiments on single crystal $C_{24}K$, which reveal for the first time the low temperature (hk0) plane within the {110} graphite reflections. A schematic representation of their X-ray photograph is shown in figure 5. From the diffraction pattern three basic observations have been made [17]: 1. the potassium ordered structure has hexagonal symmetry with a fundamental reciprocal lattice vector $|\vec{G}^K_{100}| = 1.26\text{Å}^{-1}$, which is incommensurate with the graphite structure with $|\vec{G}^G_{100}| = 2.94\text{Å}^{-1}$, 2. the alkali layer is rotated by $\pm|\theta| = 7.5°$ with respect to the graphite [H00] direction, 3. each graphite reflection is surrounded by a hexagonal arrangement of spots with distance $|G| = 1.26\text{Å}^{-1}$ from the center spot. These reflections can be interpreted as side bands of the graphite Bragg-reflections, due to a strain induced modulation of the graphite planes by the alkali superstructure. This will be discussed in more detail in part III.

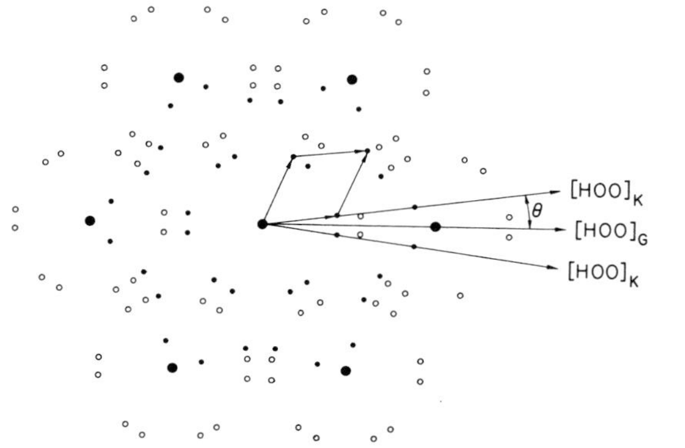

Fig. 5 Schematic representation of the (hk0) plane of $C_{24}K$, taken from an X-ray single crystal photograph (while the original photo is richer in higher order modulation effects, the principle aspects of intensity and position are reproduced here) [17]. Large solid ● denote graphite fundamental reflections, small solid • potassium fundamental reflections. Modulation peaks are indicated by O. For explanation, see text.

The interpretation of the HOPG results now follows immediately by a rotation of the single crystal pattern around (000). In particular, by measuring the graphite, alkali and modulation reciprocal lattice vectors $|\vec{G}^G|$, $|\vec{G}^A|$ and $|\vec{G}^M|$ resp., the rotation angle θ can be deduced directly from the HOPG diffraction pattern, as outlined in fig. 6.

3. ORDER AND DISORDER IN $C_{36}K$ AND $C_{48}K$

Similar to stage 2, the alkali layers in stage 3 ($C_{36}K$) and 4 ($C_{48}K$) are disordered at room temperature. Measurements of the basal plane conductivity by D. G. Onn et al. [14] again yield two anomalies, which indicate order-disorder transitions of the alkali structure at $T_U \sim 250$K and $T_L = 95$K in stage 3, and $T_U = 250$K and $T_L = 92$K in stage 4. To date, the potassium structure factor of both stage compounds has only been reported for the high temperature disordered state [16]. In agreement with stage 2, the experiments exhibit a 2-D potassium short range order correlation. But a remarkable difference occurs in the in-plane diffraction pattern. The main diffuse peak centered around

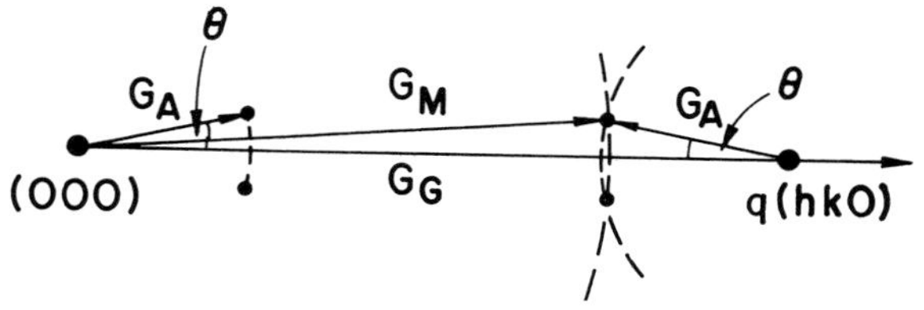

Fig. 6. Projection of modulation reflection in the cylindrical averaged q(hk0) direction of HOPG crystals. $\vec{G}_G$, $\vec{G}_A$ and $\vec{G}_M$ denote graphite, alkali and modulation reciprocal lattice vectors, resp. The rotation angle θ between the alkali and graphite structure follows from $G_M^2 = G_G^2 + G_A^2 - 2G_G G_A \cos\theta$.

$K = 1.26\text{Å}^{-1}$ in stage 2 is shifted to $K = 1.16\text{Å}^{-1}$ in stage 3 and 4. This difference certainly reflects different effective ion sizes or, equivalently, different inter- and intraplane interactions and one, in principle, may expect a different sequence of phase transitions and low temperature potassium structures in stage 3 and 4, compared to those in stage 2.

III. DISCUSSION

One of the main features of order disorder transformations in graphite-alkali compounds is the appearance of modulation effects in the low temperature ordered state [2,17,18]. The structural modulation via elastic distortion modes is directly connected to the incommensurability of the intercalates with the graphite basal plane. For rare gas monolayers adsorbed on graphite the formalism was developed by McTague and Novaco [19]. Unlike the gas monolayers on graphite, in the case of alkali layers in graphite there exist in principle two possibilities for the modulation: either graphite modulates the intercalates, or the intercalates modulate the graphite planes. Which case actually dominates, can be decided from the X-ray diffraction pattern, and may be seen in the following way: The structure factor of an ordered plane is given by

$$S(\vec{K}) = \sum_m f_m(|\vec{K}|)\exp[i\vec{K}(\vec{R}_m + \vec{U}_m)], \qquad (1)$$

where $f_m(|\vec{K}|)$ is the atomic form factor and $\vec{K}$ the scattering vector. The ideal location of atom m in the plane is given by $\vec{R}_m$, and $\vec{U}_m$ describes deviations from $\vec{R}_m$. Equation (1) can be written

$$S(\vec{K}) = \sum_m f_m(|\vec{K}|)\exp i\vec{K}\vec{R}_m[\exp(i\vec{K}\vec{U}_m) - 1] + \sum_m f_m(|\vec{K}|) \exp i\vec{K}\vec{R}_m \quad . \qquad (2)$$

For small deviations $\vec{U}_m (\vec{K}\vec{U}_m \ll 1)$ one may expand the exponential to give for the structure factor:

$$S(\vec{K}) = \sum_m f_m(|\vec{K}|)\exp(i\vec{K}\vec{R}_m) + i\, \vec{U}(\vec{K})\cdot\vec{K}, \qquad (3)$$

where

$$\vec{U}(\vec{K}) = \sum_m f_m(|\vec{K}|)\vec{U}_m \exp(i\vec{K}\vec{R}_m) \qquad (4)$$

is the Fourier transform of the distortion field $\vec{U}_m$. If the distortion field is periodic, we may write

$$\vec{U}_m = \sum_\tau \vec{U}_\tau \exp(i\vec{\tau}\vec{R}_m) \quad , \qquad (5)$$

and it follows

$$S(\vec{K}) = \sum_m f_m(|\vec{K}|) \exp(i\vec{K}\vec{R}_m) + i\sum_\tau (\vec{U}_\tau\vec{K}) \sum_m f_m(|\vec{K}|)\exp[i(\vec{K}+\vec{\tau})\vec{R}_m]. \qquad (6)$$

From (6) follows the intensity

$$S(\vec{K})^2 = A(\sum_G \delta(\vec{K}-\vec{G}) + \sum_G \sum_\tau (\vec{U}_\tau\vec{K})^2 \delta(\vec{K}-(\vec{G}\pm\vec{\tau}))) \qquad (7)$$

where the reciprocal lattice vector $\vec{G}$ is defined by $\vec{G}\vec{R}_m = 2\pi n$ and A is a proportionality factor. The first part of eq. (7) yields Bragg-reflections from the average lattice at positions $\vec{K} = \vec{G}$ in reciprocal space, whereas the second part is due to the periodic modulation with wave vector $\vec{\tau}$, and creates side bands at positions $\vec{K} = \vec{G} \pm \vec{\tau}$. The intensity of the side bands is proportional to $(\vec{U}_\tau\vec{K})^2$[20], which is characteristic for scattering at lattice distortions [21,22]. Therefore the intensity increases with $\vec{K}^2$, but decreases for higher Fourier components $\vec{U}_\tau$ and depends on the direction cosine between $\vec{U}_\tau$ and $\vec{K}$.

In order to demonstrate the possible modulation effects in graphite intercalation compounds, we show the reciprocal (hk0) plane of graphite modulated by the wave vector $\vec{\tau} = \vec{G}^A_{100}$ of the alkali structure (fig. 7a), and (hk0) plane of the alkali layer which is modulated by the graphite wave vector $\vec{\tau} = \vec{G}^G_{100}$ (fig. 7b). It is assumed that $\vec{U}_\tau \parallel \vec{K}$ and that only the first Fourier component $\vec{U}_\tau$ significantly contributes to the intensity of the side bands. In the second case of the alkali planes modulated by the graphite structure, the modulation wavevector $\vec{\tau} = \vec{G}^G_{100}$ lies outside of the first Brillouin zone of the alkali reciprocal lattice and therefore must be folded back, which in turn generates the pattern shown in fig. 7b.

In both examples the side bands appear on circles around their parent fundamental reciprocal lattice spot of the modulated structure. Thus, only by investigation of the centering of the first order side bands one already comes to a decision which lattice dominately is modulated. Neglecting the splitting of the reflections in fig. 5, a comparison between the X-ray pattern from $C_{24}K$ and fig. 7 strongly suggests a modulation of the graphite planes by the alkali incommensurate low temperature structure. Refined conclusion can be drawn out of X-ray intensity measurements of the side bands and comparison with the predictions of the second part of eq. (7). This is discussed in detail in ref. [17].

The modulation of the graphite planes can be understood as elastic distortion modes, generated by the alkali layers upon ordering. As pointed out by Safran and Hamann [23], the

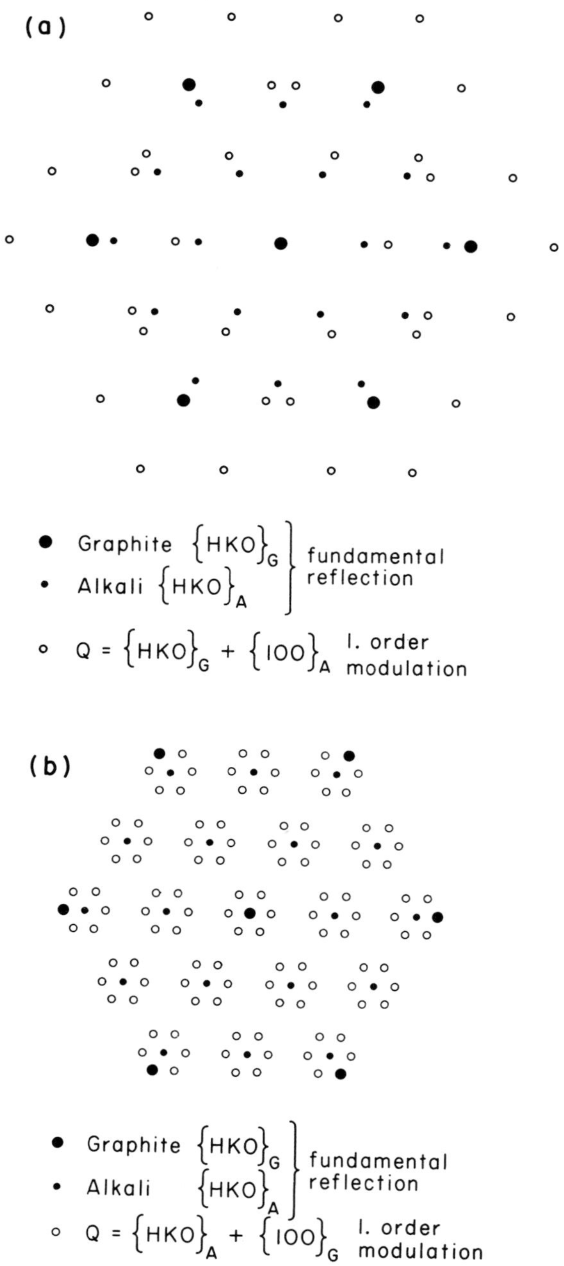

Fig. 7 Schematic representation of modulation effects in the (hk0) plane of graphite and alkali layers, assuming same orientation of both layers: (a) diffraction pattern, expected for modulation of graphite by the alkali structure, (b) diffraction pattern, expected for modulation of the alkali layer by the graphite planar structure.

alkali atoms represent elastic dipoles, which set up long range strain fields in the graphite host lattice. The deformation of the graphite planes, in turn, leads to an attractive elastic interaction between the intercalates. With increasing distortion amplitude $\vec{U}_\tau$, the elastic energy decreases [24]. In the case of an incommensurate alkali layer, $\vec{U}_\tau$ may increase by rotation of the alkali plane with respect to the adjacent graphite planes, to take advantage of transverse displacements as well as longitudinal displacements (for commensurate interstitial atoms, i.e. hydrogen in metals, only the longitudinal part enters in the dipole-dipole term of the elastic interaction [24]). The elastic energy can be minimized with respect to the rotation angle θ, where θ depends in principal on the elastic coefficients C_{11} and C_{66} of the graphite planes as well as on the ratio of the reciprocal lattice vectors of the two structures [19]. This rotation in turn causes the splitting of the alkali (hk0) reflections and graphite modulation peaks, i.e. in fig. 5. Fig. 8 contains all data available so far on the rotation angle θ and the ratio $|\vec{G}^A_{100}|/|\vec{G}^G_{100}|$ of low temperature modulated graphite-alkali structures. Obviously there exists a functional relationship between θ and $|\vec{G}^A_{100}|/|\vec{G}^G_{100}|$ for all alkalis studied. This observation supports the interpretation of the modulation effect as induced through elastic distortion modes by the incommensurate alkali structure. In conclusion: unlike gas monolayers on graphite, in graphite

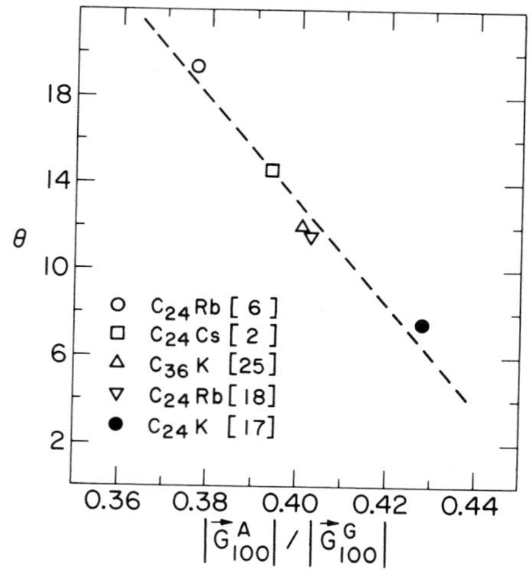

Fig. 8 Rotation angle θ between $\vec{G}^A_{100}$ and $\vec{G}^G_{100}$ reciprocal lattice vectors in dependence of the ratio $|\vec{G}^A_{100}|/|\vec{G}^G_{100}|$ for different graphite-alkali compounds. The dashed line is drawn as an aid to the eye.

intercalation compounds of stages $n \geq 2$, the almost rigid alkali planar structure rotates by a certain degree θ against the graphite planes, thereby inducing elastic distortion modes in the graphite host lattice.

The author thanks S. C. Moss for many helpful discussions.

REFERENCES

1. See i.e. contributions of R. J. Birgenau, H. Taub and P. Dutta in this volume.
2. J. B. Hastings, W. D. Ellenson, J. E. Fischer, Phys. Rev. Lett. 42, 1552 (1979).
3. R. Clarke, N. Caswell, S. A. Solin, and P. M. Horn, Phys. Rev. Lett. 43, 2018 (1979).
4. H. Zabel, S. C. Moss, N. Caswell, and S. A. Solin, Phys. Rev. Lett. 43, 2022 (1979).
5. M. Suzuki, H. Ikeda, H. Suematsu, Y. Endok, H. Shiba, and M. T. Hutchings, to be published.
6. N. Kambe, G. Dresselhaus and M. S. Dresselhaus, to be published.
7. D. R. Nelson and B. I. Halperin, Phys. Rev. B19, 2457 (1979).
8. A. Hérold, Bull. Soc. Chim. Fr. 187, 999 (1955).
9. J. E. Fischer, Physica 99B, 383 (1980).
10. W. Rüdorff and E. Schulze, Z. Anorg. und Allg. Chemie. 277, 156 (1954).
11. D. E. Nixon and G. S. Parry, J. Phys. D1, 291 (1968).
12. W. D. Ellenson, D. Semmingsen, D. Guérard, D. G. Onn and J. E. Fisher, Mater. Sci. Eng. 31, 137 (1977).
13. R. Clarke, N. Caswell, and S. A. Solin, Phys. Rev. Lett. 42, 61 (1979).
14. D. G. Onn, G. M. T. Foley and J. E. Fischer, Phys. Rev. B19, 6474 (1979).
15. G. S. Parry and D. E. Nixon, Nature 216, 909 (1967).
16. H. Zabel, Y. M. Jan and S. C. Moss, Physica 99B, 453 (1980).
17. M. Mori, Y. M. Jan, S. C. Moss and H. Zabel, to be published.
18. Y. Yamada, p. communication.
19. J. P. McTague and A. D. Novaco, Phys. Rev. B19, 5299 (1979).
20. An exact treatment, without the assumption $\vec{K}\vec{U}<<1$, leads to the factor $J_1(\vec{K}\vec{U}_T)$ instead of $(\vec{K}\vec{U}_T)^2$ for the intensity of the modulation peaks, where J_1 is the first order Bessel function, compare R. W. James, The Optical Properties of the Diffraction of X-rays, G. Bell and Sons LTD, London 1958.
21. P. H. Dederichs, J. Phys. F: Metal Phys. 3, 471 (1973).
22. H. Peisl, J. of Appl. Crystall. 8, 143 (1976).
23. S. A. Safran and D. R. Hamann, Phys. Rev. Lett. 42, 1410 (1979).
24. H. Wagner and H. Horner, Adv. Phys. 23, 587 (1974).
25. S. C. Moss, p. communication.

DISCUSSION

S. K. Sinha:

Don't you have to rely on more than geometrical arguments to explain whether K is modulating graphite or vice-versa--after all one has the same set of combinations of reciprocal lattice vectors in either case--one must thus rely on some intensity calculations to decide.

H. Zabel:

My argument was that taken into account only first order modulation then the geometry of the modulation effects are different. If one considers also higher order modulation effects, the geometry is certainly the same and intensity calculations have to decide which lattice is dominantly modulated.

A. N. Berker:

Electron diffraction on single crystallites of stage 2 graphite-potassium has revealed a variety of behavior. At the intermediate temperature range, 86 K < T < 130K, incommensurately ordered islands coexist in a disordered background and are rotationally unlocked with graphite. At 130K, the larger of the islands coalesce and rotationally lock to the graphite, as indicated by real-space images and diffraction spots superimposed on rings.

R. Clarke:

a) Is it possible that in your calculation of the lattice gas S(q) the inclusion of next nearest neighbor occupation would improve agreement with experiment?

b) Would a truly random decoration of the graphite matrix achieve the required density of KC_{12}?

H. Zabel:

a) The first maximum of the lattice gas $S(|\vec{K}|)$ lies at $K = 1.36 Å^{-1}$ compared to the maximum at $K = 1.26 Å^{-1}$ observed in the scattering experiment at $C_{24}K$. Inclusion of next nearest neighbors in the lattice gas structure factor calculation would shift the first maximum up and make the disagreement worse.

b) In a truly random decoration of graphite matrix the averaged distance between two alkali ions is about 3-4 a_G (a_G = graphite lattice constant) which yields a too small in-plane density. In our case of a nucleation type generation of the lattice gas the averaged distance lies between 2-3 a_G and we achieve the required density of 1/12.

Sinha, ed. Ordering in Two Dimensions

MELTING IN PHOSPHOLIPIDS AND IN SMECTICS - A COMPARISON

S. DONIACH*

Dept. of Applied Physics, Stanford University, Stanford, California 94305 USA

and Groupe de Physique des Solids, Univ. de Paris VII, Pl. Jussieu, Paris 5ème, France

ABSTRACT

Phase transitions in multilamellar phospholipid-water systems and in smectic liquid crystals are contrasted. A model of the first-order lipid melting transition is reviewed. In contrast to the smectic melting transition (B → A), the lipid transition is driven by intramolecular excitations of polymeric character. The intermediate, rippled phase, of phospholipids is discussed briefly. The effects of interlayer coupling on intermolecular ordering are shown to be strongly relevant in the case of smectics but to play at most a minor role in the lipid case.

1. The phase diagram of pure hydrated phospholipids in multilamellar form

Phospholipids are essential structural components of biological membranes. They consist of a hydrophylic head group (dipolar or acidic) attached to two polyethelyne chains $(CH_2)_nCH_3$ forming a hydrophobic region. In molecules found in biological membranes, $12 \lesssim n \lesssim 20$, and one of the chains always has one unsaturated C = C bond. In the presence of water in excess of about 3% by weight, the lipid molecules spontaneously arrange themselves, at normal temperatures, into multilamellar structures consisting of a periodic array of bilayers with the hydrophobic tail regions in the interior, coated on each side with H_2O[1]. Typical bilayer dimensions are on the order of 50 Å in thickness with 25 Å of H_2O separating the bilayers under saturation conditions (of order 30% H_2O). For water concentrations below saturation, the multilamellar structure appears to form a single thermodynamic phase; above saturation, phase separation of some H_2O occurs. Under sonication in excess H_2O, vesicles may be formed consisting of a single bilayer closed in a spherical form containing (and swimming in) H_2O. These may typically be of order a few hundred Å in diameter. The periodic arrangement of the hydrated bilayers in the multilamellar phase suggests a comparison of the structural and phase transition properties of this phase with those of liquid crystals in the smectic phases. The aim of this article is to provide this comparison by contrasting simplified theoretical models of the phase transitions in these systems. The literature dealing with the physical properties of phospholipids is quite extensive and no attempt will be made to survey it here. The reader is referred to a recent review by Nagle for such a survey [2].

Biological membranes are chemically very complex systems containing a mixture of phospholipids of various head group compositions and varying chain lengths, in addition to many other components (cholesterol, proteins, etc.). However physical studies may be made on single component phospholipids of which dipalmitoyl phosphatydil choline (DPPC), for which n = 16, is an extensively studied example [3]. A sketch of its phase diagram is shown in Fig. 1. (For a more complete version see ref. [4].)

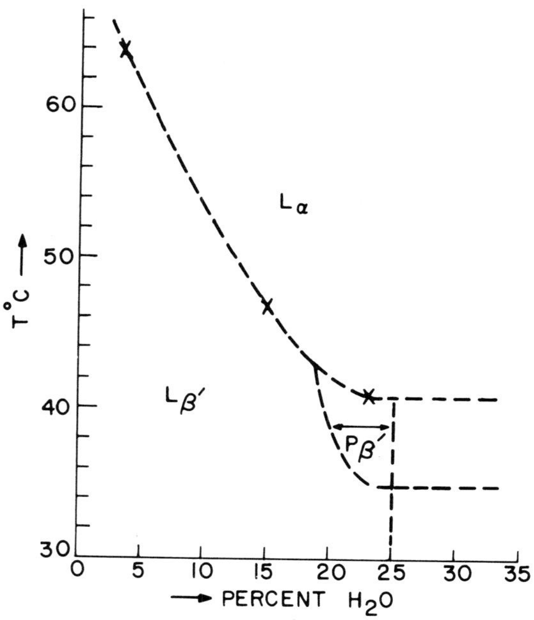

Fig. 1. Sketch of the phase diagram for the dipalmitoyl-phosphatidyl choline-water system.

In the high temperature or L_α phase X-ray studies show that the polyethylene chains have liquid-like, or short-range order, while preserving the multilamellar arrangement. They are therefore somewhat analogous to the smectic A phase of liquid crystals, although the hydrophylic character of the head group surface guarantees that no intermingling occurs of molecules of one bilayer with those of adjacent bilayers, so that there is no analogy possible with the gradual disappearance of smectic ordering associated with a smectic-nematic transition.

In the low temperature, or L_β phase, the chains are found to be closed-packed in a hexagonal 2-D solid array. Possible ordering of the head groups, linking hexagonal sites in pairs, has not been detected to date. In some cases (such as DPPC), denoted by $L_{\beta'}$, the chains are oriented at a tilt angle of about 30° to the bilayer plane. Thus the L_β ($L_{\beta'}$) phase is analogous to smectic-B (smectic C) phases. However, the latter have recently been confirmed to display 3-D order, i.e., registry of molecular positions from one phase to the next. [5]. Evidence for similar interlayer correlations in lipids, leading to a three-dimensional x-ray diffraction pattern, is so far lacking. (Owing to difficulties of oriented sample preparation, lipid diffraction patterns tend to be averaged over orientations leading to powder patterns, thus making structural determinations more difficult than in the liquid crystal case where external magnetic fields may often be used to orient the sample.)

In the intermediate, or $P_{\beta'}$, phase, the chain arrangement is also hexagonally close packed, thus very similar to the L_β phase. The structure of this phase will be discussed further in Section 3.

2. The main melting transition of phospholipids - physical origin of the latent heat

The phase transition on raising the temperature of the $L_{\beta'}$ phase (below 17% H_2O in Fig. 1) or of the $P_{\beta'}$ phrase (above 17% H_2O) is found experimentally to be quite strongly first order in the sense that it appears to have a latent heat of order 9 kCal/mole [6]. This is to be contrasted with the melting of the smectic B phase of butyloxybenzylidine-octylaniline (40.8) which has a latent heat of about 0.5 kCal/mole at 49°C [6]. At the same time, the bilayer thickness undergoes a contraction of 8% with corresponding increase in area occupied per head group (on going to the liquid phase). This large change of total entropy corresponding to 14.6 K_B per molecule (compared to 0.7 K_B/molecule for smectic melting) is clearly dominated by intramolecular excitation in the lipid case. The character of this excitation, principally involving polymer-like kinking and flexing of the polyethylene chains, has been discussed by Marcelja in terms of a molecular field picture [7]. A very simplified picture of the resulting phase transition, based on a mapping onto a 2-state, or Ising-like model was proposed by Caillé et al [8] and elaborated by the present author [9]. The basic idea behind this picture is that the internal free energy of a single hydrocarbon chain changes from a function $f^\alpha_{chain}(T)$ in the low temperature close-packed state (denoted by α) to a function $f^\beta_{chain}(T)$ in the high temperature or fluid state (denoted by β). The partition function of a single chain is then approximated by

$$Z_{chain} = e^{-f^\alpha/K_BT} \sum_{\sigma=\pm1} \exp\{-(1+\sigma)\Delta f/2K_BT\} \qquad (1)$$

where $\Delta f = f^\beta - f^\alpha$ and σ is an Ising variable taking on values -1 (α state) and +1 (β state). The simplest contribution to the internal excitation of a chain is a kink. If only zero or single kink excitations are considered, f^α(0-kink)=0 and f^β (1 kink)=$E_{kink}-K_BT\log m$, where m is the number of ways a kink can be excited, 2(n-2) for a chain with n CH_2 groups. (In general, account must be taken of many possible excitation modes [7].) In spin language, Eq. (1) represents the partition function of a spin in an external magnetic field, the "entropy driving field," $h(T)=\Delta f(T)/2$, whose temperature dependence, arising from the internal entropy of the chain, favors the excited, or β, state at high temperatures. The cooperative nature of the phase transition now arises from the dependence of the interchain forces - van-der Waals and dipolar-head attractions together with short range steric hindrance - on the mean interchain distance, which is lower in the close-packed state, than in the chain-disordered, fluid state. The change of attractive interchain energy on going from the α to β state may be simulated by an interaction energy of the form

$$H_{int} = -\frac{W}{8} \sum_{<ij>} (1-\sigma_i)(1-\sigma_j) \qquad (2)$$

while the steric hindrance term may be simulated by an "internal pressure" term, following Marcelja, with an energy of the form

$$H_\pi = \sum_i \pi\Delta A(1+\sigma_i)/2 \qquad (3)$$

where ΔA is the change in area occupied by the chain ongoing from α to β state. Putting these terms together, the resulting partition function for the bilayer, in which the effects of spatial disordering of the hexagonal chain lattice have been neglected, is that of an Ising model for a set of spins in a temperature dependent effective magnetic field

$$H = \sum_i h_{eff}(T)\sigma_i - \frac{W}{8}\sum_{<ij>} \sigma_i\sigma_j \qquad (4)$$

where

$$h_{eff}(T) = \frac{1}{2}(\Delta E_{chain} + \pi\Delta A + 3W - S_{eff}T) \qquad (5)$$

where Δf_{chain} has been expanded to first order in T: $f_{chain} = E_{chain} - S_{eff}T$ and S_{eff} is a resulting "internal chain entropy."

As discussed in ref.[9], the resulting thermodynamical properties of this bilayer model follow directly from those of the 2-D Ising model. In particular, a first order transition occurs at a temperature T_M given by

$$h_{eff}(T_M) = 0 \qquad (6)$$

provided the resulting T_M is less than the critical temperature T_c of the Ising model. The resulting dependence of $T_M = (\Delta E(\text{chain}) + \pi\Delta A + 3W)/S_{eff}$ on chain length is found to be in semiquantitative agreement with experiment [9]. The latent heat of the transition is given by $\Delta H = \frac{1}{2} T_M S_{eff}$ $(\langle\sigma\rangle_+ - \langle\sigma\rangle_-)$ and comparison with experiment leads to a value of $S_{eff} \cong 6.4 K_B$ per chain consistent with an increase in the chain excitations at the melting transition of about 4 gauche bonds observed by Raman scattering [10]. Thus it may be seen that a physical picture of the lipid melting transition as a cooperative transition principally driven by the entropy due to "chain melting" accounts quite well for the observed change in intra-chain conformation and latent heat.

How important is the neglect of entropy due to interchain disordering on going from hexagonal to fluid arrangements as the bilayer melts? A rough estimate may be made by applying the theory of two dimensional melting due to Kosterlitz and Thouless [11]. Although the unbinding of dislocation pairs leads to no specific heat singularity at the Kosterlitz-Thouless phase transtion T_{2D}, an estimate of the disordering entropy, which comes about as more pairs dissociate above T_{2D}, has been made by Berker and Nelson for the X-Y model analog [12]. They do not quote a number, but a rough estimate of the total entropy change due to pair unbinding taken from the plot in their Fig. 5 gives $\Delta S \lesssim 2K_B$ per molecule. So, in contrast to smectic melting, where molecular disorder presumably dominates the thermodynamics of the transition [13] (see also Section 4 below), it only plays a relatively minor role in lipid melting.

3. The ripple phase of phospholipids - an analogy with helical magnetism

In the intermediate temperature phase of multilamellar phospholipids, denoted $P_{\beta'}$ in Fig. 1, the basic chain configuration is hexagonal close packed, but the bilayer orientation appears to undergo a 1-dimensional wavelike modulation, or ripple, of period about 150 Å, reminiscent of incommensurate density modulation observed in various simple crystalline systems. It is not yet clear whether the form of the modulation consists of alternate compressions and expansions of the bilayer thickness [14] or of a bilayer whose thickness is constant but in which the relative angles of the chains to the local surface tangent is an oscillating function of positions; however a recent x-ray study appears to favor the latter picture [4]. A Ginzburg-Landau-type model describing the latter structural deformation has recently been proposed by the author [15]. In this model the order parameter is taken to be the angle $\theta(x)$ between the chain orientation and the local tangent to the bilayer surface. The tendency to local curvature is described by a Ginzburg-Landau free energy functional containing terms which lower the free energy when the local curvature of the membrane increases. The curvature is compensated by a potential energy term measuring the tendency of adjacent chains to be in positions of well-defined relative registry (i.e., position of good "fit" for adjacent zig-zag $(CH_2)_n$ chains). Since in the $L_{\beta'}$ phase, the chains are oriented at an angle of about 30° to the bilayer normal, it is assumed for simplicity that there are two such minima of potential energy, tending to lead to chain angles $\theta = \pm 30°$. In the continuum limit, the resulting free energy functional has the form (in suitable units)

$$F = \int dx \left\{ \tfrac{1}{3}\left[\left(\frac{d\theta}{dx}\right)^2 - 3\theta_0'^2\right]^2 + (\theta^2 - \theta_0^2)^2 \right. \qquad (7)$$

where θ_0' is a constant measuring the tendency to spontaneous curvature and θ_0 is of order 30°. On minimizing Eq. (7), it is found that two different types of wave form result depending on the magnitude of the parmeter $\eta = \theta_0'^2/\theta_0^2$. For $\eta < 1$ an asymetric solution is found in which $\theta(x)$ oscillates, as a function of x, about a nonzero mean value, θ_{eff} (which may be positive or negative) while for $\eta > 1$ a symmetric profile is found in which the mean value of θ is zero. The point $\eta=1$ is a singular one in which the ripple period diverges logarithmically. For values of η slightly greater than 1 (say 1.1) the period is found to be considerably longer than the period of solutions for η a little less than 1 (say 0.9). The coexistence of rippling with two different periods (differing by almost a factor of 2) in samples containing excess water, observed by Luna and McConnell [16] may be evidence for coexistence of two different phases with slightly different values of η on each side of unity.

The nature of the rippling may be understood, via the model of Eq.(7) as a phase in which a local order ($\theta \cong \pm\theta_0$) is perturbed by a coupling tending to distort this ordering, due to the dependence of F on $(d\theta/dx)^2$. In the discrete version of the model, $d\theta/dx \sim (\theta_{i+1} - \theta_i)$ may be seen to depend on a second neighbor interaction when expressed as a function of local chain displacements. In this sense the rippling phenomenon may be thought of as analogous to the incommensurate phases of helical ferromagnets in which a second neighbor antiferromagnetic coupling leads to a slow twisting of the local ferromagnetic ordering direction induced by a nearest neighbor ferromagnetic coupling.

4. The effects of interlayer coupling - a theory of smectic melting

As emphasized above, the main melting transition of phospholipids is undoubtedly driven by intramolecular excitation. There remains, however, the question of the effects of interlayer coupling on intermolecular disorder, i.e., on changes in entropy on going from the hexagonal close-packed to liquid-like phase. In this section a theory of smectic melting is presented. It will be shown that interlayer coupling plays an important role in this phase transition, but is unlikely to play a similar role in the corresponding lipid melting transition.

The nature of the smectic melting (B→A) transition was discussed by DeGennes and Sarma [17] in terms of a harmonic model, and by Huberman, Lublin and Doniach [13] (referred to as HLD) in terms of the Kosterlitz-Thouless picture of 2-D melting. More recently, Birgeneau and Litster [18] have pointed out that interlayer coupling, even if weak, is likely to have a strongly relevant effect on two-dimensional melting. The following is an estimate of the way in which interlayer coupling modifies the HLD picture of smectic melting.

An approximate representation of the interaction of pairs of dislocations in a 2-D solid is to consider only the logarithmic term in the interaction

$$U(r_i - r_j) = +q^2 \log(|r_i - r_j|/\xi_0) \qquad (8)$$

As shown by Nelson [19], this is a reasonable approximation for square lattices where x-directed dislocations are disconnected from y-directed dislocations, but would require modification for triangular lattices. Eq. (8) results from an x-y model approach to two-dimensional melting when the local phase, $\phi(x)$, represents the atomic misfit to the ideal lattice in a given lattice direction. The effect of interlayer coupling may now be represented by a weak commensuration potential tending to align the phase in adjacent planes:

$$V = -g \sum_j \int \left(\cos \phi_j(x) - \phi_{j+1}(x)\right) d^2x \qquad (9)$$

where j is a plane index. If it is now supposed that 3-D long-range order is established in the smectic-B phase, then the inplane order parameter $\psi(x) = \psi_0 e^{i\phi(x)}$ has a nonzero thermal average $\langle\psi\rangle$. On replacing the effect of coupling in Eq. (9) by a thermal average over the phase in adjacent planes, the equation for minimizing the local free energy in a given plane in the presence of a vortex becomes of sine Gordon form

$$\nabla^2\phi + 2g\langle\psi\rangle \sin\phi = \delta(x) \qquad (10)$$

Following Venables and Schabes-Retchkiman [20] it may now be seen that the presence of interlayer coupling, however weak, leads to the formation of a line of discommensuration or "string" between a vortex and an antivortex. For x very close to the vortex core, the solution of (10) has the form

$$\phi(r,\sigma) \cong \theta - \frac{2}{3} r^2 g\langle\psi\rangle \sin\theta \qquad (11)$$

while for x in between the vortex and antivortex (assumed far apart), Eq. (10) reduces to the 1-D pendulum equation as a function of the distance, y, measured normal to the string direction. From Eq. (11) it may be seen that the phase field due to a vortex reduces to that for the pure 2-D system for $r \ll r_g$, where $r_g = 1/\sqrt{g\langle\psi\rangle/3}$. Hence at short distances compared to r_g the vortex-antivortex interaction is logarithmic. For $r \gg r_g$, the vortex-antivortex pair will be joined by a string of width of order r_g; hence the interaction grows linearly with pair separation. It is straightforward to show that the string tension (energy per unit length) is of order $[g\langle\psi\rangle]^{1/2}$.

Summarizing these results, the effect of interplanar coupling in the ordered state, treated in the above "mean field" type of approximation is to replace Eq. (8) by

$$\left.\begin{aligned} U(r_{ij}) &\cong q^2 \log(|r_{ij}|/\xi_0) & r \ll r_g \\ &\cong \sqrt{g\langle\psi\rangle}\,|r_{ij}|/\xi_0 & r \gg r_g \end{aligned}\right\} \qquad (12)$$

From the form of Eq. (12), it may be seen that the effect of interlayer coupling on the Kosterlitz-Thouless transition is to strongly inhibit the formation of dislocation pairs with spacing $\gtrsim r_g$. It may further be argued that the effect of vortex pairs on the order parameter correlation function will be to diminish it for $|r| < r_g$ and will have little effect for $|r| > r_g$:

$$\begin{aligned} g(r) = \langle\psi(r)\psi^*(o)\rangle &\cong 1/|r|^{\eta(T)} & r \ll r_g \\ &\cong 1/|r_g|^{\eta(T)} & r \gg r_g \end{aligned} \qquad (13)$$

Hence it may be seen that the mean order parameter $\langle\psi\rangle = \{\lim_{r\to\infty} g(r)\}^{1/2}$ has a value of order $1/|r_g|^{\eta/2}$ in the presence of interlayer coupling. On the other hand, if it is assumed from the start that $\langle\psi\rangle = 0$, then within the above mean field-type approach, the interlayer coupling (9) averages to zero and the strings do not appear.

The above arguments suggest a new model for the smectic phases: in the Smectic A phase, positional order decays exponentially (for $T > T_{2D}$); in the Smectic B phase positional order is 3-dimensional, but $\langle\psi\rangle$ is reduced by the thermal excitation of dislocation pairs of limited range. The transition between these phases will now be first order in character and will be determined by the temperature at which the difference in their opinion and other non-dislocation contributions to the free energy resulting from a difference in their density (Smectic A being of lower density than Smectic B) compensates the lowering of the free energy of the Smectic-A phase due to the existence of dislocation pairs with $|r_{ij}| > r_c$. In the limit that the interplanar coupling, $g \to 0$, these differences will become very small and $T_{AB} \to T_{2D}$. Thus T_{2D} may provide a reasonable estimate of the smectic melting temperature, in agreement with the empirical finding of HLD.

The above arguments lead to the conclusion that interlayer commensuration coupling can have very little effect on the melting transition in the multilamellar lipid system, since the dislocation unbinding mechanism, as argued in Section 2, plays at most a minor role in the lipid melting process. On the other hand, at low water concentrations, the presence of "chemically bound water" bridging the phospholipid head groups may perturb the intermolecular interaction energies, and hence alter the melting temperature through the mechanism of Section 2.

Acknowledgement: Part of the work reported here was done at the Aspen Center for Physics. I would also like to thank members of the "Physique des solides" laboratories at Université of Paris VII and Orsay centers for their hospitality.

REFERENCES

1. A. Tardieu, V. Luzzati, and F.C. Reman, J. Mol. Biol 75, 711 (1973).
2. J. Nagle, Annual Reviews of Physical Chemistry (1980).
3. M.J. Janiak, D.M. Small and G.G. Shipley, Biochemistry 15, 4575 (1976).
4. M.J. Janiak, D.M. Small, and G.G. Shipley, J. Biol. Chem. 254, 6068 (1978).
5. D.E. Moncton and R. Pindak, Phy. Rev. Letts. 43, 701 (1979).
6. H.J. Hinz and J.M. Sturtevant, J. Biol. Chem. 19, 6071 (1972).
7. S. Marcelja, Biochim. Biophys. Acta, 367, 165 (1974).
8. A. Caille´, A. Rapini, M.J. Zuckerman, A.Gros, and S. Doniach, Can. J. Phy. 56, 348 (1978).
9. S. Doniach, J. Chem. Phys. 68, 4912 (1978).
10. N. Yellin and I.W. Levin, Bichem. 16, 642 (1977).
11. M. Kosterlitz and D.J. Thouless, J. Phys. C., 6, 1181 (1973).
12. A.N. Berker and D.R. Nelson, Phys. Rev. 19, 2488 (1979).
13. B.A. Huberman, D. Lublin, and S. Doniach, Solid-State Communs. 17, 485 (1975).
14. C. Gebhardt, H. Gruler, and E. Sackmann, Z. Naturforsch. C32, 581 (1977).
15. S. Doniach, J. Chem. Phys. 70, 4587 (1979).
16. E.J. Luna and H.M. McConnel, Biochim. Biophys. Acta, 470, 303 (1977).
17. P. deGennes and G. Sarma, Phys. Lett. A38, 219 (1972).
18. R. Birgeneau and D. Litster, J. de Physique Lettres 39, 399 (1978).
19. D. Nelson, Phys. Rev. B18, 2318 (1978).
20. J.A. Venables and P.S. Schabes-Retchkiman, J. Phys. C. 11, L913 (1978).

*Research partially supported by NIH Grant #GM25217-02 and by the Centre National de la Recherche Scientifique.

DISCUSSION

D. Nelson:

Do you think it might be possible to modify or inhibit this chain-driven first order melting transition by, say, using polyunsaturated polyethylene hydrocarbon chains?

S. Doniach:

Yes, the effect of unsaturated C=C bonds is to freeze in "gauche" excitations, thus reducing the intrachain entropy and altering the melting temperature. Another somewhat analagous effect occurs on the incorportion of cholesterol in the bilayer. This gets in the way of chain-chain interaction and hence produces a local reduction of the cooperative character of the phase transition. In fact, cholesterol acts as a sort of "antifreeze" and results in a much broadened melting transition.

D. A. Cadenhead:

Recently Sturtevant (Biochem 1978) has gone to extraordinary measures to purify his DPPC (zone-melting of palmiticacid) in order to reduce the finite width of the main (41-42°C) transition. The transition, however, still has a finite width and he ascribes this as possibly due to domain-type transitions. How do you explain the finite width of this first order transition?

S. Doniach:

Data on ion diffusion across bilayers had been used to suggest that in fact the main melting transition is quite close to a critical point (Doniach, J. Chem. Phys. 1978). This would give pretransitional contributions to the specific heat, in addition to the first order delta function spike of the first order transition, but would not broaden the transition itself. It should be noted that the main transition is relatively sharp compared to the premelting transition ($L_{\beta'} \to P_{\beta'}$) which seems broadened well beyond resolution of the measurements.

M. Schick:

Your Ising model of the melting of the bilipid would predict the existence of a critical point. Is that ever observed?

S. Doniach:

Yes, for lipid films on water, measured by the Langmuir balance technique, it is possible to take the system up to what looks like a critical point. However, for the bulk multilamellar materials the intermolecular tension is determined mainly by the chemistry and cannot be altered enough to bring the system to the critical point.

A. Isihara:

a) I wonder whether the distances between the neighboring phospholipid molecules increase as the Gauche structures start appearing?

b) Then, the Ising model might not be applicable.

S. Doniach:

a) The Raman scattering technique does allow a direct observation of the degree of intrachain excitation, since gauche bonds lead to different C-C stretching frequencies from trans bonds. In this way it is possible to correlate mean intermolecular distances directly with the state of chain excitation.

b) The model does include a surface pressure term taking account of changes in intermolecular distances (following Marcelja). However, it does not include the effects of disorder in the interchain arrangement and is in this sense approximate.

B. I. Halperin:

In principle, if you have a stack of two dimensional solids, the stack will form a three dimensional solid, for any non-vanishing coupling between the layers. If this does not occur in the stacked liquid structures, there may be several explanations. (1) The forces between layers may be sufficiently weak and the relaxation time sufficiently long, that equilibrium is not achieved. (2) There may be frozen-in defects or arrays of impurities, within the layers, resulting in a finite correlation length for translational order in the layer.

S. Doniach:

I agree, this is precisely what my analysis shows. The point I perhaps should have made clearer for the lipid case is that as regards the mechanism of the melting transition itself, the effects of interlayer coupling on intermolecular role can play, at most, a very minor role.

P. Pershan:

a) Dr.S. Asher, in our group at Harvard, has grown 3D-crystals of DPPL at 25°C containing 2% water. We have also observed that these crystals do not change when the chemical potential for water in equilibrium is raised up to 70% of the saturated water vapor pressure. Thus, I believe that interlayer coupling will produce full 3D order in an equilibrium situation.

b) In response to Dr. Cadenhead's comment on the width of the "main transition;" I believe the width of the transition might be due to the fact that the system is a two component system (water and lipid) and the transition is <u>not</u> observed at fixed chemical potential, but rather at fixed total water/lipid ratio.

S. Doniach:

It is important to note that at low water concentrations NMR studies have led to the idea that the water is chemically bound to the head groups, hence could form a rigid 3-dimensional structure (e.g., by bridging between layers). Only at higher water concentrations does ordinary "liquid" water appear between the layers. Presumably this would lead to a strong decrease in the interlayer coupling. In principle, I agree that the solid phase could exhibit long range 3-dimensional order. My main point was that interlayer commensuration is unlikely to contribute significantly to the melting mechanism, in contrast to the smectic melting case where it is important.

E. K. Riedel:

What is the reason for the form of the interlayer-coupling term in your Sine-Gordon model? [i.e., $\cos(\phi^a(r) - \phi^{a+1}(r))$]

S. Doniach:

This term is put in to approximate the tendency of adjacent layers to become registered and, hence, commensurate. It would correspond to the first Fourier component of the actual interlayer coupling resulting from the intermolecular force laws.

Published 1980 by Elsevier North Holland, Inc.
Sinha, ed. Ordering in Two Dimensions

PHASE TRANSITIONS IN INSOLUBLE ONE AND TWO-COMPONENT FILMS AT THE AIR/WATER INTERFACE

D. A. CADENHEAD and F. MÜLLER-LANDAU
Department of Chemistry, State University of New York at Buffalo
Buffalo, New York 14214

B. M. J. KELLNER
IBM-East Fishkill, Hopewell Junction, New York 12533

The physical states in monomolecular films of insoluble amphipathic lipids are discussed. Transitions between differing physical states are described with particular attention to that between the liquid expanded (LE) and the liquid condensed (LC) states. Isotherms of dihexadecyl phosphatidylcholine are cited as examples. The LE/LC transition is thought to be first order within a microscopic domain, but diffuse first order on a macroscopic level. The LE state is shown to consist of two states differing in their degree of anisotropy. The addition of a second component such as cholesterol can affect and eliminate the LE/LC transition to produce an intermediate state. In contrast, epicholesterol condenses the LE state less and perturbs the LC state more.

INTRODUCTION

In this paper we will examine phase changes occurring in insoluble films of amphipathic molecules at the air/water interface and the effect on such phase changes of the addition of a second component. We will concentrate particularly on the liquid expanded/liquid condensed (LE/LC) phase change since this appears the most interesting, controversial and useful. There is a need to describe such transitions in detail from an experimental viewpoint since many theories ignore or contravene obvious experimental facts. Paying attention to such data allows us to select those theories most likely to provide a better understanding of two-dimensional phase transitions.

With two-component films we will restrict ourselves to those systems where the addition of a second component produces a significant reduction in the mean molecular area when the first component exhibits an expanded state in a pure film. In order to do this use will be made of previously published data [1,2,3] for the systems dipalmitoyl phosphatidylcholine (DPPC)/cholesterol and myristic acid/cholesterol. The model developed, however, will be broadly applicable to other two-component films.

PHASE TRANSITIONS IN SINGLE COMPONENT FILMS

Gaines [4] in his well known text, describes the various physical states that have been defined in films of insoluble amphipathic molecules. Here we will use the terminology of N. K. Adam, a pioneer in this field [5] without accepting his interpretation of their nature or of transitions between them. At sufficiently high areas/molecule or low surface pressure (see Figure 1), all films exist in a gaseous state (G). Gershfeld [6] has shown that such films generally behave as van der Waals two-dimensional gases, however in some instances their behavior approaches that of an ideal two-dimensional gas [7]. Films with low intermolecular forces or at higher temperatures will remain in the gaseous state as compression occurs. Otherwise condensation can occur, directly or indirectly, to a liquid or a liquid-crystalline-like state (liquid expanded, LE), to a gel-like state (liquid condensed, LC), or to a solid state or states (SC).

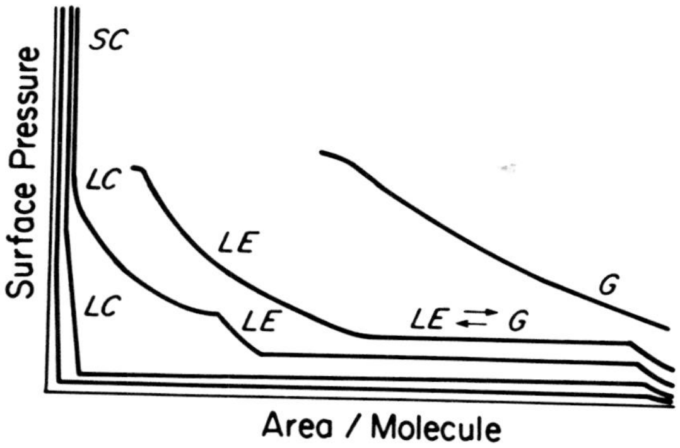

Figure 1. A schematic representation of the various physical states of insoluble monomolecular films: SC, solid condensed; LC, liquid condensed; LE, liquid expanded; G, gaseous.

The G/LE transition is clearly a first order phase transition with a pronounced break in the π/A plot. The transition only takes place if the film is below a van der Waals type critical temperature. The presence of the two different phases was clearly demonstrated by Adam and Jessop [8] through electrode scanning of a myristic acid film. In contrast the LE/LC phase change has been described as diffuse first order or second order or a combination of both. It is usually observed that the LE/LC transition (Figure 2) occurs as a single slope change in the π/A plot with the increase in compressibility occuring at lower areas/molecule. At low temperatures the slope of the isotherm beyond the transition may approach, but never attains, a completely horizontal slope (1st order).

At higher temperatures the slope increases and the width of the transition decreases until another critical temperature is observed. It has generally been considered that this critical temperature is also of the van der Waals type, but recently this has been contested by Albrecht et al. [10] who have described the critical temperature as tricritical (i.e. the transition is diffuse first order below T_c and second order above). The data they have to substantiate this claim is, however, very limited. Compressional hysteresis for this phase change has been reported by ourselves [11] and by Albrecht et al. [10]. At low compressional speeds, the magnitude of this hysteresis does not vanish and only attains a minimum value. Following transition initiation a linear variation of π is usually observed with decreasing area/molecule. This leads without interuption into a non-linear increasing π region until the LE/LC transition is complete.

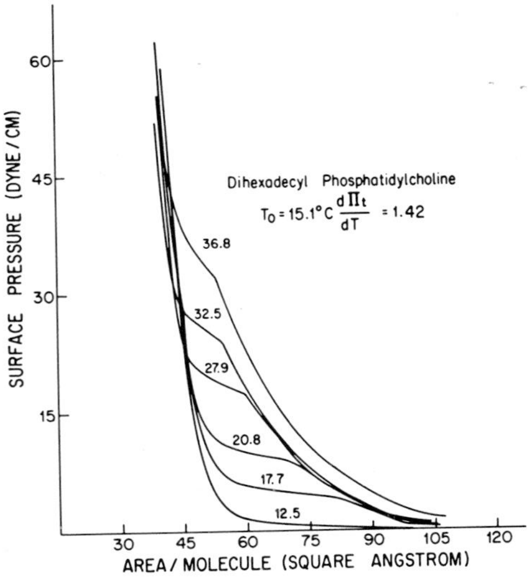

Figure 2. The temperature dependence of films of dihexadecyl phosphatidylcholine. The temperatures are as indicated on the individual isotherms. T_o is the lowest temperature at which a liquid expanded phase can exist and $d\pi_t/dT$ gives the linear shift of the onset of the LE/LC phase change with respect to temperature [9].

The LC/SC transition almost always appears second order showing a break in the slope of the π/A plot. However, once again, at lower temperatures there is evidence that the transition is diffuse first order [10,12,13]. It may be that this transition similarly represents a transition with a tricritical point, however it seems unlikely, with the very small area changes involved, that accurate experimental evidence will be easy to come by. It should also be pointed out that LC/SC diffuse first order transitions have been observed for fatty acid, fatty ester and diglyceride films at low temperatures (unpublished work).

THEORIES OF TRANSITIONS IN MONOMOLECULAR FILMS

(For a review of the various theories, see reference 14). Monolayer transition theories generally concentrate on the LE/LC phase change, partially because of the lack of reliable data on low area (LC/SC) and low pressure (G/LE, G/LC) phase changes. In addition, most theories have little problem reproducing some kind of 1st order transition, as is exhibited by G/LE or G/LC transitions. Theories dealing with the LE/LC phase change fall into two categories: order/disorder and orientational.

Order/disorder theories usually allow for a limited number of gauche transformations in otherwise all-trans chains. [A kink may be produced in a straight hydrocarbon chain (all trans) by rotating about a C-C bond by an angle of 120° while rotating either of the neighboring C-C bonds by -120°. By doing this, two trans configurations are transformed into gauche configurations.] One of the more successful attempts [15] involves two C_{14} chains in a hexagonal cage of eight other molecules with an intramolecular energy describing the energy difference between rotational isomers and an intermolecular energy based on a Lennard-Jones potential. An important restriction of this model was that the two chains must have the same conformation simultaneously. This suggests that cooperativity, at least on a microscale, is essential. While the theory reproduced several experimental features, it still results in two slope discontinuities and a flat portion. The difficulty may be overcome if it is assumed that cooperativity is complete only within microscopic domains [10,11] (non-horizontal slope) and that the status of a domain in an all or nothing transition is not independent of that of neighboring domains (only one slope discontinuity). That cooperativity is indeed important, is emphasized by the low temperature flat portions of the isotherms of similar films at the oil/water interface [16], that micro-domains play a role is supported by the observed compressional/decompressional hysteresis [11]. Such an approach is attractive in that while the transition may be regarded as first order (all or nothing) within a domain, it can be treated as a diffuse first order (the average change in orientation is gradual) on a macroscopic level (Figure 3).

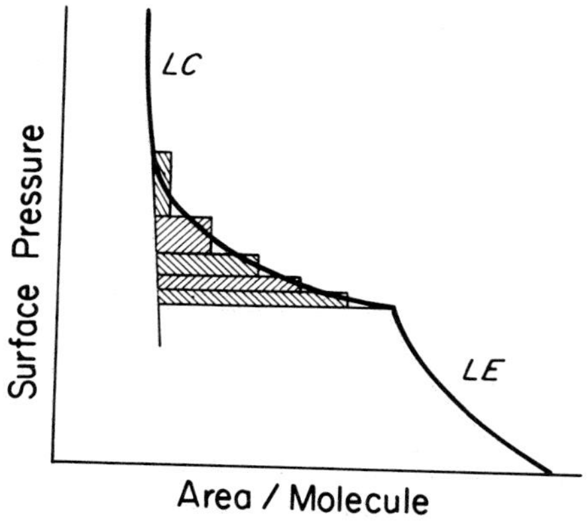

Figure 3. Theoretical Interpretation of the LE/LC phase change: The transition may be considered as first order within domains of limited size (∿ 100 molecules/domain [10]). The transition within a domain is cooperative and limited interaction between domains is possible. On a macromolecular scale the transition is diffuse first order.

Orientation models can also reproduce apparent second order phase transitions. Kaye and Burley [17] have demonstrated that with isotropic point molecules only first neighbor exclusion is required. Bell et al. [18] use two molecular orientations with two molecules in a similar orientation having the lower net energy. Thus cooperative interactions are emphasized and the transition has the correct shape. Firpo et al. [19] have a rectangular molecule in one or other perpendicular orientation. In addition allowance is made for holes. The model is capable of exhibiting a low pressure first order gas/liquid and a liquid/solid transition which changes from first order at lower pressure to second order at higher pressure. A tricritical point can thus be produced in agreement with the predictions of Albrecht et al. [10]. Orientational models can be interpreted in terms of order/disorder of flexible chain molecules, but this is not necessarily the case. At this point it is important to remember that no rigid amphipathic molecule has ever been demonstrated to exhibit a LE/LC type transition [20] whereas if appropriate conditions can be attained, all flexible chain amphipaths can show such a transition. Until such transitions have been demonstrated for rigid amphipathic molecules the picture presented in Figure 3, based on order/disorder transitions, would seem to be the most reasonable.

Particularly at lower temperatures, the LE/LC transition approaches the appearance of a first order transition. If we treat the transition as diffuse first order we can evaluate either the heat or entropy of the transition using a two dimensional Clapeyron equation:

$$\frac{d\pi_t}{dT} = \frac{\Delta S_t}{\Delta A} = \frac{Q_t}{T\Delta A} \qquad (1)$$

where $d\pi_t/dT$ is the shift of the transitions with temperature, ΔS_t and Q_t are the respective entropy and enthalpy changes, ΔA the change in area/molecule and T the temperature of the transition.

Heats of the LE/LC transition may be evaluated from either isotherms [21] or isobars [10]. From an isotherm ΔA may be taken as the difference between the area of the discontinuity π_t (see Figure 3), and that where the extrapolated LC state intercepts the horizontal at pressure π_t (ΔA_1). Alternatively ΔA may be taken as corresponding to only the linear portion of the isotherm after compression to π_t (ΔA_2). From an isobar ΔA may be taken either as the total area change between an expanded and condensed state (see Figure 4, ΔA_3), or as the area shift corresponding to the linear portion only (Figure 4 arrows, ΔA_4). For isobars it is logical to take the temperature as that corresponding to the mid-point of ΔA_3 or ΔA_4. Since the slope of π_t versus T is positive and fixed (i.e. not a function of temperature), $d\pi_t/dT$ may be taken from values previously tabulated by us when the same or a similar substrate is selected [22]. These procedures are illustrated using the data of M. Cowden [23] for DPPC with the isobar shown in Figure 4. The heats Q_1, Q_2, Q_3 and Q_4 are plotted as a function of temperature in Figures 5 and 6. From equation (1), Q and ΔA will be directly proportional, hence $Q_1 < Q_2$ and $Q_3 < Q_4$. The Q values all extrapolate to zero at temperatures within experimental error of the critical temperature T_c. Similar observations can be made with dimyristoyl phosphatidylcholine, myristic acid, pentadecanoic acid, 3-hydroxy palmitic (hexadecanoic) acid (3HHA) and presumably are general for all films exhibiting an LE/LC phase change. Values for the various T_c values, including those estimated directly from the isotherms (T_c^m), and the main bilayer transition (T_c^b), where the substance forms a bilayer, are given in Table 1.

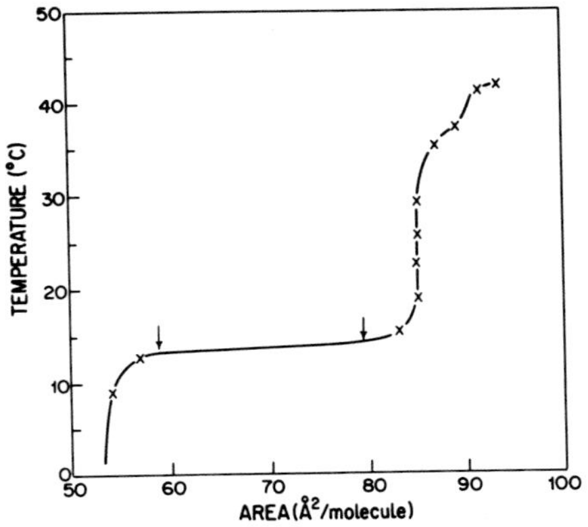

Figure 4. Isobar (π = 1.3 dynes/cm for L-α-dipalmitoyl phosphatidylcholine. Data taken from isotherms in reference 23. The arrows indicate ΔA_4. ΔA_3 includes the curved regions at either end of the linear section ΔA_4 (from 53 Å^2 to 86Å^2).

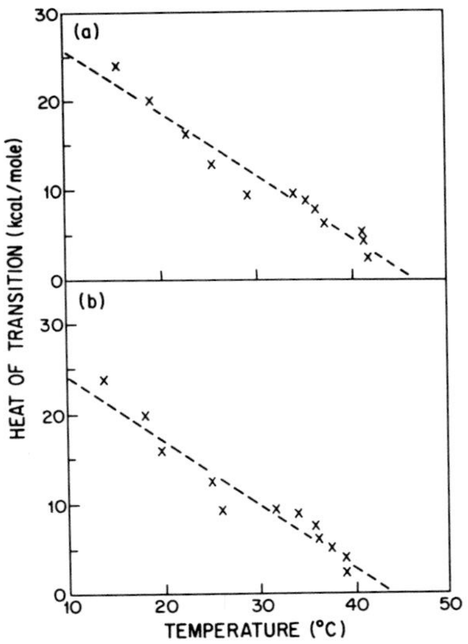

Figure 5. Heat of transition (k cal/mole) Vs. Temperature (°C) for the LE/LC phase transition in L-α-dipalmitoyl phosphatidylcholine. Date taken from reference 23.
(a) Q_1 Vs. Temperature.
(b) Q_3 Vs. Temperature.

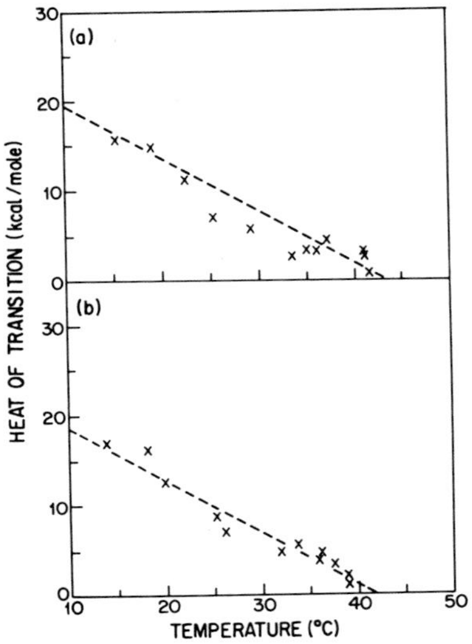

Figure 6. Heat of transition (k cal/mole) Vs. Temperature (°C) for the LE/LC phase transition in L-α-dipalmitoyl phosphatidylcholine. Data taken from reference 23.
(a) Q_2 Vs. Temperature.
(b) Q_4 Vs. Temperature.

Table I
MONOLAYER/BILAYER T_c VALUES

Substance	T_c^m	T_c^b	T_c^1	T_c^2	T_c^3	T_c^4	Ref.
DPPC	∿42*	41**	---	---	---	43.8	10
			46.4	43.3	44.1	41.9	23
			---	43.9	---	40.2	24
DMPC	∿24*	23**	---	25.4	---	24.4	10
C_{14} acid	27.5***	---	---	30.6	---	28.8	8
C_{15} acid	---	---	---	61.5	---	60.5	25
3HHA	∿45	---	---	45.6	---	46.4	26

T_c^m The critical temperature of the monolayer, isotherm determined

T_c^b The main transition temperature of the lecithin bilayer

T_c^1-T_c^4 Critical temperature of the monolayer determined by extrapolation of Q_1-Q_4 to zero.

Albrecht et al. [10] prefer T_c^4 values. We feel that T_c^3 may be a better choice.

* Hui et al. [28]

** D. Chapman, Quart. Rev. Biophys. 8, 125 (1975).

*** P. Joos, Bull. Soc. Chim. Belges. 79, 291 (1970).

Two important points can be made. Firstly there is a higher order phase change apparent in Figure 4 which is barely detectable as a slope change in the original isotherms. Similar pre-LE/LC transitions were found with the other film data listed in Table 1. Albrecht et al. [10] through compressibility studies have shown that they are second order and have attributed it to an increase in the chain anisotropy prior to the main transition. What we can add is that this pre-LE/LC transition would appear to be quite general and independent of the polar group and will have to be taken into account in any theory of two-dimensional film transitions.

Secondly the extrapolation of the various ΔQ values to zero at or about the T_c^m of the monolayer and the main phosphatidylcholine transion in the bilayer, suggests a correspondence between them. This was first proposed by Phillips and Chapman [21] and supported by Nagle [27] and Hui et al. [28]. However Nagle [27] and Hui et al. [28] have compared the physical state of the monolayer at its critical temperature to the bilayer at its main transition. Albrecht et al. [10] have disagreed with this, instead they equate the entropy and area changes in the monolayer and bilayer. When this is done they conclude that the monolayer at about 10-15 dynes/cm and 25-30°C is equivalent to the bilayer at 42°C and relate the differences in the two temperatures to coupling of monolayers in the bilayer. All of this leaves unexplained why the T_c^m values in a monolayer correspond so well with the main bilayer transition T_c^b as is indicated in Table 1.

PHASE TRANSITIONS IN TWO-COMPONENT LIPID/STEROL FILMS

When cholesterol is added to lipid monomolecular films in an expanded state a well documented condensation takes place. In bilayer vesicles, the addition of cholesterol to a lipid in a liquid crystalline state reduces the vesicle permeability to ions and small molecules. Less well known is that when added to a SC or LC monolayer state or a gel-like bilayer state, cholesterol will fluidize the lipid. The action of cholesterol is to move a condensed or expanded state to one of intermediate fluidity. Current thinking ascribes the bulk of this effect to hydrophobic interactions with a possible minor role to polar group interactions [29].

Recently [1,2,3] we published results which confirmed the qualitative statements listed above but put the picture on a quantitative molecular footing. Condensation of expanded lipids (myristic acid and DPPC) was accompanied by a delay in the onset of the LE/LC transition i.e. a delay in the formation of the LC state.

Our findings for both systems are summarized in Figure 7. The LE/LC phase change shifts in a linear fashion until a cholesterol concentration of 4.5 mole % (with myristic acid) or 8.7 mole % (with DPPC) is attained. At this point both systems have 21 acyl chains/cholesterol and in Figure 7 it is proposed that 7 of these chains constitute an inner ring of significantly affected chains while the remaining 14 constitute an outer partly affected ring. Cholesterol is proposed to affect its nearest neighbors significantly, its next-nearest neighbors partially, while acyl chains twice removed are not affected by the cholesterol. It should be kept in mind, however, that fluidity studies show that all three types of acyl chain are still part of a fluid film and should therefore be exchangeable.

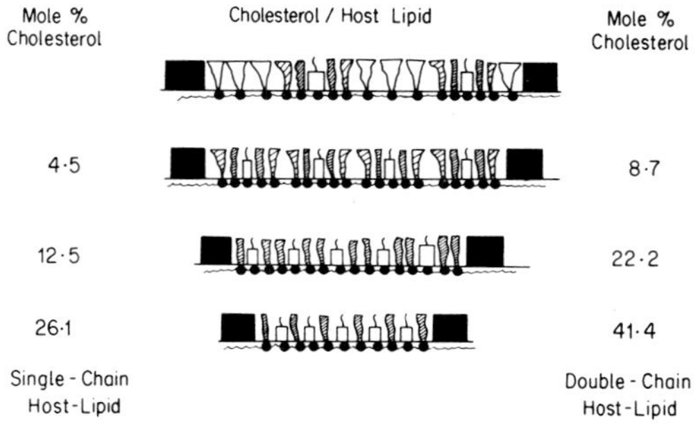

Figure 7. Schematic Representation (sideview) of the proposed changes taking place in a two-component mixed film where cholesterol ( ) is one component and a flexible chain(s) host-lipid ( ) is the other. The numbers indicate (in mole % cholesterol) the compositions at which the various arrangements are presumed to occur. For further details see the text.

At 12.5 mole % (myristic acid/cholesterol) or 22 mole % (DPPC/Cholesterol) the phase change is eliminated, the molecular ratio of acyl chains to cholesterol is once again the same for both systems: 7:1. The disappearance of the phase change is consistent with the model presented above, since now only an inner ring of seven acyl chains remains. Finally at 26 mole % (myristic acid/cholesterol) or at 42 mole % (DPPC/cholesterol) i.e. at about 3 acyl chains/cholesterol, each cholesterol can preserve a ring of seven acyl chains only by sharing. Cholesterol/acyl chain contacts are maximized, as is the condensation.

In that the behavior of these two mixed systems were identical on a unit chain basis, a direct role for the polar groups was discounted. Thus hydrophobic interactions were postulated to give rise to essentially all the effects described and the only role seen for the polar groups would be when they, in turn, affect the hydrophobic interactions.

We now turn our attention to an improved understanding of how cholesterol affects existing hydrophobic interactions. In the liquid expanded state fluidity increases as we proceed from the polar head-group to the chain-terminating methyl groups. Spin-labelling studies suggest a conical shape for flexible lipids, however deuterium labelling NMR studies indicate that significant "fanning out" only begins about carbon atom C_9 of the acyl chains [30]. In either case it is possible to incorporate rigid cholesterol molecules within the film such that geometric accommodation or "space-filling" can reduce the overall area of the mixed film, a suggestion first put forward by Shah and Schulman [31]. Their suggestion, however, was a purely entropic one, and it is clear from CPK models that any such accommodation will result in reduced distances between adjacent hydrophobic segments, in enhanced dispersive interactions and in an enthalpic contribution to the condensation. Thus, as we had previously pointed out [32], the effect of cholesterol will have both an entropic and an enthalpic contribution.

In the condensed state the host lipid molecules approach their limiting areas/molecule, about 20Å^2/alkane chain. For DPPC the limiting area is about 42Å^2/molecule and the molecule no longer resembles a cone but has a somewhat cylindrical shape. Cholesterol accommodation is no longer possible and the areas of the two components are essentially additive, however CPK models again indicate that cholesterol would only reach as far as C_{12} to C_{14} on the C_{16} acyl chains. This means that, following cholesterol introduction in an LC phase, intermolecular forces will be slightly reduced. As a result, cholesterol condenses an expanded film, delays the formation of a condensed phase, and favors an intermediate state.

Sterols like coprostanol, which has a bent or partially tilted shape, and epicholesterol, whose polar group forces the sterol to tilt even in a mixed film [3], cannot be accommodated in the way that cholesterol can. Previously [3], we suggested that epicholesterol because of a tilted conformation might try to line up. Initial results show that, on the first compression, a condensation, only about half of that of cholesterol occurs. Recompression shows segregation at about 25 mole % epicholesterol indicating that partial segregation, due to alignment, or complete segregation at higher concentrations results in a reduced epicholesterol condensation efficiency. In addition the shift on initial addition of epicholesterol of the LE/LC phase change to higher pressures is much greater than with cholesterol, indicating that the tilted conformation produces considerable disruption of a close-packed condensed state. Careful observation of the behavior of phase changes in pure and mixed films can provide considerable insight into the molecular events taking place.

ACKNOWLEDGEMENTS

We would like to acknowledge the financial assistance of the National Heart, Lung and Blood Institute through Grant No. HL-24535 in the completion of this work. We also wish to thank Dr. Helmut Hauser for providing the original sample of dihexadecyl phosphatidylcholine.

REFERENCES

1. F. Müller-Landau and D. A. Cadenhead, Chem. Phys. Lipids 25, 299 (1979).
2. F. Müller-Landau and D. A. Cadenhead, Chem. Phys. Lipids 25, 315 (1979).
3. D. A. Cadenhead and F. Müller-Landau, Chem. Phys. Lipids 25, 329 (1979).
4. G. L. Gaines Jr., Insoluble Monolayers at Liquid/Gas Interfaces Wiley-Interscience, New York, N.Y. (1966).
5. N. K. Adam, The Physics and Chemistry of Surfaces, 3rd Edition, Oxford University Press, Oxford, England (1941).
6. N. L. Gershfeld and R. E. Pagano, J. Am. Chem. Soc. 76, 1231 (1972).
7. D. A. Cadenhead and M. C. Phillips, J. Colloid Interface Sci. 24, 491 (1967).
8. N. K. Adam and G. Jessop, Proc. Roy. Soc. Ser. A 110, 423 (1926).
9. B. M. J. Kellner, F. Müller-Landau and D. A. Cadenhead, J. Colloid Interface Sci. 66, 597 (1978).
10. O. Albrecht, H. Gruler and E. Sackmann, J. Physique 39, 301 (1978).
11. D. A. Cadenhead in Structure of Biological Membranes, Plenum Publishing Corp., New York, N.Y. p. 63 (1976).
12. D. A. Cadenhead and R. J. Demchak, J. Colloid Interface Sci. 24, 484 (1967).
13. D. A. Cadenhead, J. Colloid Interface Sci. 24, 544 (1967).
14. J. F. Baret, Prog. Surface and Membrane Sci. 14, in press (1980).
15. P. Bothorel, J. Belle and B. Lamaire, Chem. Phys. Lipids 12, 96 (1974).
16. B. A. Pethica, J. Mingins and J. A. G. Taylor, J. Colloid Interface Sci. 55, 2 (1976).

17. R. D. Kaye and D. M. Burley, J. Phys. A: Math. Nucl. Gen. 7, 1303 (1974).
18. G. M. Bell, J. Mingins and J. A. G. Taylor, J.C.S. Faraday I 74, 223 (1978).
19. J. L. Firpo, J. J. Dupin, G. Albinet, A. Bois, L. Casalta and J. F. Baret, J. Chem. Phys. 68, 1369 (1978).
20. D. A. Cadenhead and R. J. Demchak, J. Chem. Phys. 49, 1372 (1968).
21. M. C. Phillips and D. Chapman, Biochim. Biophys. Acta 163, 301,(1968).
22. B. M. J. Kellner, F. Muller-Landau and D. A. Cadenhead, J. Colloid Interface Sci. 66, 597 (1978).
23. M. Cowden, Ph.D. Thesis, Biochemistry, State University of New York at Buffalo, (1975).
24. H. Trauble, H. Eibl and H. Sawada, Naturwissenschaften 61, 344 (1974).
25. W. D. Harkins, T. F. Young and E. Boyd, J. Chem. Phys. 8, 954 (1940).
26. B. M. J. Kellner and D. A. Cadenhead, J. Colloid Interface Sci. 63, 452 (1978).
27. J. F. Nagle, J. Membr. Biol. 27, 233 (1976).
28. S. W. Hui, M. Cowden, D. Papahadjopoulos and D. F. Parsons, Biochim. Biophys. Acta 382, 265 (1975).
29. R. A. Demel and B. de Kruyff, Biochim. Biophys. Acta 457, 109 (1976).
30. J. Seelig and W. Niederberger, Biochem. 13, 1585 (1974).
31. D. O. Shah and J. Schulman, Adv. Chem. Ser. 84, 189 (1968).
32. D. A. Cadenhead, Recent Progress in Surface Sci. 3, 169 (1970).

DISCUSSION

R. B. Griffiths:

a) Is a diffuse first-order transition the same as a second-order transition?

b) If the transition is second order, what is the order parameter?

c) Is it possible that the order parameter is some sort of spatial ordering, e.g., in a hexagonal array, in a manner which would be analagous to, say, gases adsorbed on graphite?

D. A. Cadenhead:

The liquid expanded/liquid condensed (LE/LC) phase transition has the appearance of a second order, not a first order transition. There are some who think that it has the appearance of a second order phase transition because that is what it is [J. F. Baret, Prog. Surf. and Membrane Sci. 14, in press (1980). We have noted (D. A. Cadenhead in The Structure of Biological Membranes, Plenum Press, N.Y., p.63(1976)] that there is a compressional/decompressonal hysteresis associated with this transition which justifies the term "diffuse first order" since this hysteresis has a minimum value which is independent of the kinetics of compression/decompression. The diffuse nature of the transition arises, we think, through the limited size of the regions for a cooperative phase change (domains). There is some evidence that above the critical point the transition is second order and that the critical point is tricritical and not a van der Waals critical point [Albrecht et al., J. Physique 39, 301 (1979)] however, the evidence at this point is very limited. Data above the critical point are very difficult to get since film collapse frequently intervenes. The only attempt to treat transitions in insoluble monomolecular films in terms of an ordering parameter is that of Albrecht et al. (1979). They use the stretching vector J first suggested by de Gennes [Phys. Lett. 47A, 123)1974)]. This parameter is not associated with any particular molecular model, but is a function of the density of monomeric units of chain-like molecules. The SC phase(s) are thought to consist of triangular lattices (tilted and non-tilted) rather than hexagonal. However, other solid phases may be possible and a hexagonal perched phase cannot be ruled out.

B. M. Abraham:

My colleagues, J. Ketterson, K. Miyano, and I have a presentation (these proceedings) on the question of the presumed phases displayed in the pressure-area diagram of a film. We showed that the shear modulus as the signature for a solid contradicts the assignment of solid to a low temperature stearyl alcohol film. Perhaps some of the concepts need to be re-examined.

D. A. Cadenhead:

While I have not made a study of the temperature dependency of stearyl alcohol films, a comparison with stearic acid, which at room temperature shows a liquid condensed state at low pressures and a solid condensed state at higher pressures, leads me to expect that stearyl alcohol would be liquid condensed. The lowering in temperature (not defined) below room temperature could well be more than offset by the change from a carboxyl to a hydroxyl polar group. Intemolecular film-molecule hydrogen-bonding would be significantly reduced and a liquid condensed state should persist. The liquid condensed state is gel-like in nature and is not a true 2-D solid. That this state retains significant fluidity may be very simply demonstrated by applying talc to the film and gently blowing on it with a stream of nitrogen. This would be consistent with your shear modulus data.

DISCUSSION

D. Nelson:

I was interested in your observation that no phase transitions have been observed in rigid chain systems. There are subtle phase transitions possible, which would not show up in either pressure-area isotherms or specific heat curves. Is there any hope of scattering experiments which prove the microscopic configurational order?

D. A. Cadenhead:

We have specifically looked for rigid amphipathic insoluble lipid molecules which exhibit a LE/LC phase transition [apparent second order, probably diffuse first order] and have never found any examples [see reference 20]. Among others we tried 4 amino-p-terphenyl, which approximates to a rigid rotor in that the three phenyl groups are almost in the same plane, permitting a significant contribution to the molecular wave function by a fully conjugated molecule, as confirmed by the exceptionally high surface potential (over 1 volt).Kirkwood predicted [J. G. Kirkwood in Surface Chemistry, American Ass. for Adv. Sci., Washington, D.C., p. 157 (1943)] that such a molecule would produce a second-order phase transition when it moved from free rotation to hindered rotation on close-packing. We believe that the failure of a slightly expanded film of 4-amino-p-terphenyl to exhibit a second order phase transition was because the molecules were already undergoing hindered rotation before achieving a close-packed state due to significant intermolecular forces. A film of -estradiol diacetate [see reference 7] with very weak intermolecular forces, however, also failed to show a LE/LC phase change. [It did however show a transition from a bipolar to a pseudo-monopolar film on compression.] Flexible amphipaths undergo expansion through chain order/disorder transitions with increasing temperature and all show an LE/LC phase change if physical conditions permit. Rigid amphipaths undergo significant expansion of asymetric [reference [20] also D. A. Cadenhead and R. J. Demchak, J. Colloid Interface Sci. 35, 154 (1971)] but so far have not been seen to undergo LE/LC type transitions. It is interesting that cholesterol with a rigid sterol portion and a hindered short alkane chain undergoes slight expansion with increasing temperature [0.5 A /10 shift] whereas the completely rigid and fairly symmetrical molecule 3-doxyl-17-hydroxyl-androstane shows essentially no expansion over 26 [reference 1]. The temperature dependency of cholesterol would therefore seem to be due to enhanced movement of the short alkane chain and not increased rotation of the steriod moeity. As such it could be that cholesterol would exhibit an LE/LC phase change at higher temperatures. However the film becomes unstable over 60°C.

Subtle phase changes can be detected in several ways: isobars (see this paper), compressibility studies, surface potential and surface viscosity studies have all been used. Scattering studies, though difficult to set up and having a low signal/noise ratio in monolayers, have been successfully carried out primarily by a group of Swedish workers [L. Lundström in Structure of Biological Molecules, Plenum Press, New York and London, p. 293 (1977)].

J. G. Dash:

I have some questions or suggestions related to the non ideality of the first order transitions, i.e. the appreciable variation of π within the presumably two-phase coexistence regions. First, do you think it possible that substrate contamination, and changes in contamination with film density, can be a factor here? Dissolved gases, specifically CO_2, have been shown to play important roles. Perhaps contamination from the walls of the trough can also affect the films. The overall purity of the water must be extremely high, particularly since there could be strong segregation of some impurities within the top few layers of the surface. As far as gases are concerned, should it not be advisable to operate in controlled inert atmospheres? Second, do you consider that the experimental films are completely equilibrated? Have there been any attempts to anneal the phases, and to search for long term relaxation effects? Finally, can you say whether the films are indeed two dimensional objects, or on the contrary whether the surface-normal dimension is perhaps not fully quenched? Have current theories taken the third dimension as fully quenched?

D. A. Cadenhead:

The presence of film impurities can affect the Π/A slope of a typical LE/LC Phase transition. Generally the slope is increased and the abrupt change in slope at Π_t rounded off. In the isotherms given as example in Figure 2, the dihexadecyl phosphatidylcholine (DHPC) was received as 99 + % pure from Dr. H. Hauser of the ETH in Zurich. The material showed only single spots using excess concentrations with thin layer chromatographic plates, but did show very slightly sharper phase transitions following several recrystallizations from hexane/ethanol (9/1 volume). We would estimate the purity as not less than 99.5 + % at this point. The isotherms were obtained immediately following the recrystallizations the films being spread from a hexane/ethanol solvent. Use of up to ten times the normal amount of solvent without DHPC produced no detectable change in the surface pressure during complete compression.

Substrate effects can similarly affect the film behavior [D. A. Cadenhead and K. E. Bean, Biochem. Biophys. Acta 290, 43 (1972)]. Typical effects of dissolved surface active materials are to expand the film unless the surface tension of the substrate solution is lowered even more [D. A. Cadenhead and J. E. Csonka J. Colloid Interface Sci. 33, 188 (1979)]. This illustrates the competitive nature of chain-chain and chain-substrate interactions in deciding the physical state of the film. The data in Figure 2 of this paper were obtained on a pure water substrate quadruply distilled [once from alkaline permanganate, once from a slightly acid H_2SO_4 solution and twice from quartz. With neutral films it is more important to avoid trace organic impurities (usually surface active) than to eliminate the last μ-mole of ionic content. As with the spreading solvent, the critical test is that Π remain essentially constant, in the absence of a film, during complete compression.

Generally the nature of the gaseous phase is of less concern unless the film is susceptible to some form of chemical attack. Typically unsaturated molecules eg cholesterol [D. A. Cadenhead, B. M. J. Kellner and M. C. Phillips, J. Colloid Interface Sci. 57, 224 (1976)] will oxidize at the air/water interface in less than thirty minutes and extended studies must be carried out in an inert atmosphere. As the question implies the substrate contains dissolved gases and we find that several hours of purging through and over the substrate are essential to remove dissolved oxygen. Oxidized impurities usually expand films and with flexible amphipaths produce Π/A slope changes. Because of dissolved CO_2 aqueous substrates are usually at pH 6. If this is delaterious, either buffering (possible contamination) or gaseous flushing (eg with N_2) can solve the problem. It should be remembered, however, that some gases may have trace organic impurities in them and this must be taken care of first. With regard to the gaseous phase, it should be noted that failure to allow water vapor equilibrium can result in the slope change at Π_t being less well defined and may also affect the film temperature.

Not all experimental films are in complete equilibrium. Thus at 25°C the DPPC isotherm must be regarded as metastable over almost its entire pressure range with respect to the three-dimensional crystalline monohydrate state [F. Muller-Landau, D. A. Cadenhead and B. M. J. Kellner, J. Colloid Interface Sci. 73, 264 (1980)] because of the low equilibrium spreading pressure of the latter state. The compressional/decompressional hysteresis indicates that either the compressional path or the decompressional path, or both, mixing is enforced by the use of a spreading solvent technique and because of the kinetics of segregation significant metastability of the mixed state may exist [D. A. Cadenhead, B. M. J. Kellner and M. C. Phillips J. Colloid Interface Sci. 57, 224 (1976)].

Your final question concerns the dimensionality of such films. Obviously lipid films bridge the gap between two and three dimensions. If it were not for the fact that only point molecules are truly 2-D, I might hesitate to introduce my work at this conference. We are really talking about pseudo-2-D films and although point molecules can reproduce some features of lipid films [see reference 17] the better solutions involve three dimensional character [see reference 14].

In all of this we must emphasize that we find the non-zero slope of the Π/A plots at areas/molecule less than Π_t during an LE/LC phase change to be real and to depend on the temperature (see Figure 2). We support the idea of a domain dominated transition that is diffuse first order for the LE/LC change and note that even with extreme care in purification Sturtevant [N. Albon and J. M. Sturtevant Biochem. 75, 2258 (1978)] still concludes the main bilayer transition of DPPC (which we can show corresponds to the T_c in the monolayer) is diffuse first order.

Published 1980 by Elsevier North Holland, Inc.
Sinha, ed. Ordering in Two Dimensions

X-RAY STUDIES OF SMECTIC-B LIQUID-CRYSTAL FILMS

D. E. Moncton and R. Pindak
Bell Laboratories
Murray Hill, New Jersey 07974

We present the results of extensive x-ray experiments aimed at understanding the degree of in-plane lattice order in smectic-B liquid crystals, particularly in regard to recent theories of two-dimensional melting and the hexatic phase. Using the freely-suspended film technique, it is possible to study both the 3D bulk phases by preparing thick films and 2D behavior by preparing films as thin as two molecular layers. We demonstrate that there are two distinctly different 3D smectic-B phases. The first is an apparently more common phase in which the lattice positional correlations are fully three dimensional (i.e., crystalline) and extend to macroscopic distances. We have also studied four-layer-thick films and observed the power-law lineshape due to the rigorous absence of true long-range order in a 2D solid. The second smectic-B phase is a newly discovered phase with exponential decay of in-plane positional correlations but long-range in-plane bond-orientational correlation. This new phase is the first example of a system in which bond orientation is the relevant order parameter which distinguishes the phase from a liquid (in this case the smectic A phase). This phase may be related to models which involve the stacking of interacting hexatic layers.

I. Introduction

Our x-ray scattering studies[1] of freely-suspended smectic liquid-crystal films have a twofold objective. First we are interested in establishing the order-parameters which distinguish each of the higher-order smectic phases and studying the development of order-parameter correlations at phase transitions. More generally, however, the present work is motivated by the second objective: to exploit the unusual opportunity presented by the thin-film technique to study ordering as a function of sample thickness, thus changing the dimensionality of the system.

This paper is organized as follows. First we will briefly review the relevant theoretical work on the structure of 2D solids and the melting transition, particularly as applicable to the smectic-B phase of liquid crystals. In the experimental section we will review the liquid-crystal film technique which has so dramatically improved the quality of samples and enabled the study of both 3D and 2D structures. The improvement in sample quality permits us to extract meaningful data cleanly from conventional rotating-anode x-ray sources and to realize the high- resolution potential of synchrotron-based x-ray scattering experiments, which have recently begun.[2] Following this discussion, we will present our main results. These are based on the discovery and study of two microscopically distinct smectic- B phases. The first is more commonly seen and has conventional 3D long-range order. Although previous x-ray data[3,4] claimed that lattice correlations were only short-ranged, there are now at least two independent investigations[5,6] in addition to our own thick-film study which show 3D crystalline order. By reducing the film thickness, we have observed the evolution from 3D order to 2D. In samples which are only four layers thick, we observe the power-law structure factor expected for a 2D solid. These are the first measurements of this structure factor in a bona fide 2D system. We will describe these results and present the temperature dependent values for the exponent η.

The second smectic-B phase has been discovered[7] in a new material which will be described shortly. This phase is considerably more intriguing since it does not have 3D long-range order. We have found that the salient structural feature which distinguishes this phase from the well-known smectic-A phase is the presence of six-fold periodic modulation of the ring of scattering due to the short-range in-plane order. We believe that this structure may be related to the presence of bond-orientational order anticipated to occur in bulk phases by Birgeneau and Litster.[8] They suggested that the smectic-B phase could result from the stacking of 2D layers which are in the hexatic phase[9] discussed below.

II. Theoretical Background

Since a review of the application of the theory of 2D melting and its extensions to the physics of liquid crystals is being presented by B. I. Halperin at this conference, we will only briefly review the results of immediate experimental relevance. It has been known for over a decade[10] that a 2D solid possesses both conventional long-range bond-orientational order and power-law decay of positional correlations. The delta-function Bragg scattering of a 3D solid is replaced by the structure factor

$$S\,(\vec{q}) \propto (\vec{q}-\vec{G})^{-2+\eta_{\vec{G}}(T)} \tag{1}$$

where $\vec{G}$ is a 2D reciprocal-lattice vector and $\eta_{\vec{G}}(T)$ is a wave-vector and temperature dependent function involving the 2D elastic constants. Although this lineshape is based on sound theoretical arguments, there has been no verification of it prior to the present work.

During our experiments Halperin and Nelson,[9] building upon ideas of Kosterlitz and Thouless,[11] showed that melting of the 2D solid could occur in two steps. With increasing temperature, one would first observe a transition to an intermediate phase they called hexatic which resulted from the unbinding of dislocation pairs. This new phase would have power-law decay of lattice-orientational order and short-range (exponentially decaying) positional correlations. Then in an additional transition at higher temperatures, this phase would transform to a 2D liquid by the unbinding of disclination pairs.

The development of the relationship between the theory of 2D lattices and the structure of the 3D smectic-B phase is quite interesting. Prior to 1970 the B phase was thought to involve the development of substantial in-plane positional order of the molecules as shown schematically in Fig. 1(a). The relevant question was obviously to what extent was the order local or long-range in nature and how did this phase differ from the higher-temperature smectic-A phase and the lower-temperature "crystalline" phases.

In 1972 a suggestion was made by DeGennes and Sarma[12] that the physics of the smectic-B was dominated by the fact that a 2D solid cannot have rigorous long-range order. They suggested that, in a system of stacked layers, the interactions might not be relevant when the power-law singularities of Eq. (1) become sufficiently weak [$\eta_{\vec{G}} \gtrsim 2$ for $\vec{G} = (1,0)$]. Subsequently however the Kosterlitz-Thouless theory established the fact that η_G could not have such a large value since a 2D solid would be unstable against dislocation unbinding when $\eta \geq 1/3$. In 1975 Huberman et al.[13] suggested that some of these ideas of dislocation-mediated melting might be relevant to the smectic-B phase. However, it was not until the Halperin-Nelson theory of 2D melting was available that a consistent picture of the smectic B phase was suggested by Birgeneau and Litster[8] in 1978. They proposed that a 3D system of stacked hexatic layers could have 3D long-range bond-orientational order but exponential decay of all in-plane positional correlations, as well as exponential decay of transverse interlayer correlations. Bruinsma and Nelson[14] have calculated the structure factor for such a phase and obtained the general form

$$S(\vec{q}) = \sum_{n=0}^{\infty} C_n(\vec{q},T) \cos(6n\theta_{\hat{q}}) \langle \psi_n(\vec{r}) \rangle \qquad (2)$$

where $\theta_{\hat{q}}$ is the angle the $\vec{q}$-vector makes with an axis chosen along a hexagonal lattice vector of the unit cell at the origin. In this expression $\psi_n(\vec{r}) = e^{i6n\theta(\vec{r})}$ is the orientational order parameter. If $\langle \psi_1(\vec{r}) \rangle$ is finite then the ring of scattering characteristic of the 2D liquid layers of the smectic A phase is modulated into six maxima by the function $\cos(6\theta_{\hat{q}})$. As orientational order develops in this phase, higher order parameters become increasingly important and sharp peaks will presumably develop. The in-plane positional correlations are short-ranged and this behavior is manifested in $S(\vec{q})$ in the coefficients $C_n(\vec{q},T)$. We expect that the simplest case where only $\psi_1(\vec{r})$ is important, the dependence of $C_n(\vec{q},T)$ on $|\vec{q}|$ will be Lorentzian and provide a direct measure of the in-plane correlation length. It is, therefore, straightforward to identify this phase by x-ray scattering since it differs from a liquid in having a modulated ring of scattering and it differs from a solid in having only short-range positional order.

III. Freely-Suspended Film Technique

To prepare a freely-suspended film,[15] the edges of a 6x6 mm^2 hole in a glass cover slide are wet with the liquid crystal sample heated to its smectic-A phase. Using a glass spreader, films are drawn across the hole and the thickness measured by optical reflectivity. Uniform, stable films can be prepared from two to hundreds of molecular layers thick. Once prepared, the films can be cooled into the smectic-B phase. The thin films are the first truly two-dimensional systems which allow the study of structural correlations. Since they are freely-suspended there are no substrate interactions. Furthermore, by increasing the number of layers, the evolution to three dimensions can also be studied. As 3D systems the films offer several advantages over other preparations. Since there is no layer pinning at sample-cell walls, the films have negligible mosaic spread in their layers ($<.01°$). In the smectic-B phase, the films also have large in-plane domains ($1 mm^2$). This alignment is achieved without the use of extraneous forces due to alignment fields. For x-ray studies the films are drawn inside an oven with 0.01°C temperature control, evacuated to 500 mTorr to minimize background scattering. It is crucial to have no sample-cell windows which produce background scattering because of the small scattering signal from 100-Å thick film of hydrocarbons.

IV. X-ray Apparatus

Experiments have been performed on x-ray spectrometers based on both a 50 KW rotating Cu anode generator and the Stanford Synchrotron. For the rotating-anode experiment, a vertically bent pyrolytic graphite(002) crystal focused CuK_α x-rays to $1x3 - mm^2$ spot on the film and scattered radiation was analyzed with a flat pyrolytic graphite(002) crystal and scintillation detector. The resolution function is Gaussian with full-widths-at-half-maxima given by

$$\Delta q_x = 0.046\cos\theta \text{ Å},^{-1}$$

$$\Delta q_y = 0.046\sin\theta \text{ Å},^{-1}$$

$$\Delta q_z \approx 0.1\text{Å},^{-1}$$

where $\hat{x}$, $\hat{y}$ are in the scattering plane, parallel and perpendicular, respectively, to the nominal momentum transfer $\vec{Q}$, and $\hat{z}$ is perpendicular to the scattering plane.

The synchrotron experiments were carried out on a beamline which incorporates a doubly bent mirror and a parallel double-crystal Ge(111) monochromator and single Ge(111) analyzer operating at an incident wave vector of 3.69 Å^{-1}. For this system the resolution is $\Delta q_x = 6x10^{-4}\text{Å}^{-1}$ and $\Delta q_z = 4x10^{-2}\text{Å}^{-1}$ and the spot size is $3x3\ mm^2$. Initial measurements indicate that, with the synchrotron running at 3.3 GeV and 100 ma, the flux available is approximately the same as the rotating-anode experiments. Thus, the intrinsic collimation of synchrotron radiation enables a factor of ~ 100 improvement in Δq_x with similar signal rates.

V. Crystalline Smectic-B Phases in 3D

Although liquid crystal chemists have been very successful at correlating the different observed macroscopic textures with microscopically distinct phases,[16] there is no guarantee that this procedure is unambiguous--that different phases always have different textures. In particular a long debate has surrounded the extent to which long-range lattice order exists in the smectic-B phase; there is of course no better way to answer this question than direct x-ray measurement of the mass-density correlation functions. Although we are principally concerned with the physics of two-dimensional systems, there are obvious reasons for our interest in 3D phases. First, it is natural to want as much information as possible on the 3D phase of a material so that one has a starting point for examining the phase diagram as a function of thickness. Secondly, the phases in 3D may be examples of previously unknown states of order. In particular, these states may have important connections with the physics of 2D as in the models of the 3D smectic-B described in section II.

The material chosen for the bulk of this work was butyloxybenzylidene octylaniline (4O.8):

$$C_4H_9O - \bigcirc - CH = N - \bigcirc - C_8H_{17}$$

X → B ↔ A ↔ N ↔ I
33°C   49°C   63°C   79°C

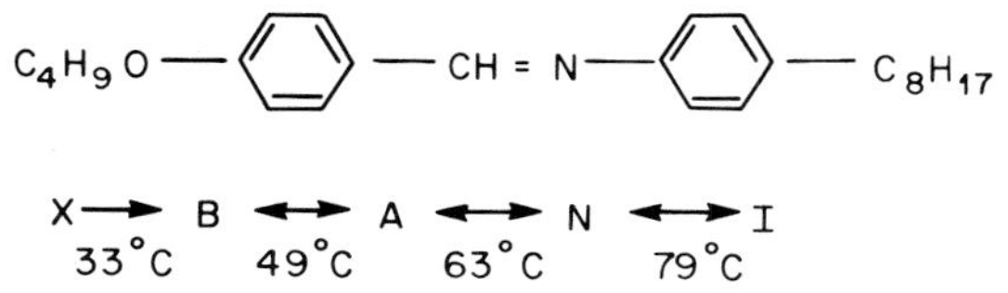

Although in 3D it was once thought to have only short-range interlayer correlations[3] (this was one of the important reasons for chosing it for our studies), we now know that this is not the case. Nevertheless it has become one of the most thoroughly studied smectic-B materials. Fig. 1 shows a simplified representation of the structure and reciprocal lattice which we have found appropriate to the smectic-B phase of 4O.8. Molecules are packed in hexagonal sheets and stacked in an ABAB sequence.

The majority of the data from which we draw our structural conclusions consists of scans taken in the vicinity of the (100) Bragg peak scanning momentum transfer $\vec{Q} = \vec{k}_f - \vec{k}_i$ (see insert to Fig. 5) in directions labeled $Q_{\parallel}, Q_{\perp}$, and χ. Each of these scans measures correlations which are important to the physics of the smectic-B. First, $Q_{\parallel}$ scans measure the in-plane longitudinal positional correlations. Next, a scan in the χ angle measures the extent to which the system has macroscopic orientational order. Finally, if a system has in-plane positional order, this order may be correlated from plane to plane. A scan in $Q_{\perp}$ then measures the extent of such interlayer correlation. Sharp, delta-function Bragg peaks, which are shown as black dots in Fig. 1(b) result if the system has crystalline long-range order. In a real solid there will also be temperature-dependent thermal diffuse scattering which is generally broad in $\vec{Q}$-space. Particularly in the present materials with small elastic constants we expect large diffuse signals.

Let us now look at how the quality of data obtained on 4O.8 depends on the method of preparation of the specimen and the resolution of x-ray scattering technique employed. This comparison, shown in Fig. 2, demonstrates how the separation between thermal diffuse scattering and the Bragg scattering becomes increasingly clear with technical improvements. The scans shown are along $Q_{\perp}$ and, therefore, measure the interlayer correlations. In (a) the data were obtained by Leadbetter et al.[5] on a magnetic-field-cooled sample using x-ray photographic techniques. In (b) a triple-axis x-ray spectrometer was used in studies by Pershan et al.[6] also with a field-cooled sample. The data in scan (c) were obtained using the triple-axis spectrometer described above in conjunction with samples prepared by the film technique. The fact that a substantial improvement in sample crystallinity occurs with the film samples is shown in scan (d) where a sample similar to that used in (c) is studied using the ultimate resolution available in a synchrotron-based triple-axis spectrometer. In cases (a) and (b), the sample probably has reduced interlayer correlations due to structural imperfections. The widths of the peaks are due to the spread in nominal layer orientation throughout the sample. In (c) the widths of the peaks are purely instrumental and good definition of the Bragg peaks is seen relative to the diffuse scattering. In (d) the volume of the synchrotron resolution function is ~10⁴ times smaller than for the instrument used in (c), so the diffuse scattering is correspondingly weaker. Although it is easily measured, it is not significant on the scale chosen. In order to actually see the width of the synchrotron/film data we show it on an expanded scale in Fig. 3. These data are not resolution limited, but rather they measure a combination of finite sample thickness (~100 *layers*) and warping of the layers (~0.01°).

It is clear, therefore, that when prepared by the film technique and studied with a high-resolution instrument, the Bragg structure characteristic of 3D long-range order is seen in the smectic-B phase of 4O.8. Scans along $Q_{\parallel}$ and χ are also consistent with this conclusion. The ABAB stacking arrangement is deduced from measured intensity ratios observed for $(10\ l)$ and $(11\ l)$ peaks in agreement with other workers.

Before going on to describe our studies on thin films, we make two more important points. The first concerns the observed diffuse scattering. As can most easily be seen in Fig. 2(c), the diffuse scattering in the range $-0.5 \leq \bar{Q}_{\perp} \leq 0.5$ is essentially flat unlike that expected for phonons with a conventional dispersion relation. Pershan et al.[6] maintain that, this scattering cannot be reconciled with a phonon dispersion relation which fits the diffuse scattering very close to the Bragg peaks. We believe that in this system with extremely weak interlayer interactions, the lattice dynamics is dominated by dispersionless independent-layer excitations. The second point demonstrates the weakness of the interlayer interactions. In studies where we cooled samples to monitor the transition to the triclinic crystalline state, the samples supercooled to 5°C. However a series of three new states were observed in transitions at 21°C, 12°C and 8°C. All of these transitions involve different reordering of the layer stacking. In order for such transitions to proceed, the interlayer potential energy must be a very weak function of the relative transverse displacement of near-neighbor layers.

VI. Crystalline Smectic-B Phase in 2D

We begin our discussion of the behavior of thin films by showing scans along $Q_{\perp}$ as a function of layer thickness in Fig. 4. The sharp Bragg structure characteristic of thick films (N>100) is destroyed and a broad interlayer interference function evolves. In principle these data could be used as a meas-

(a)

(b) **RECIPROCAL SPACE**

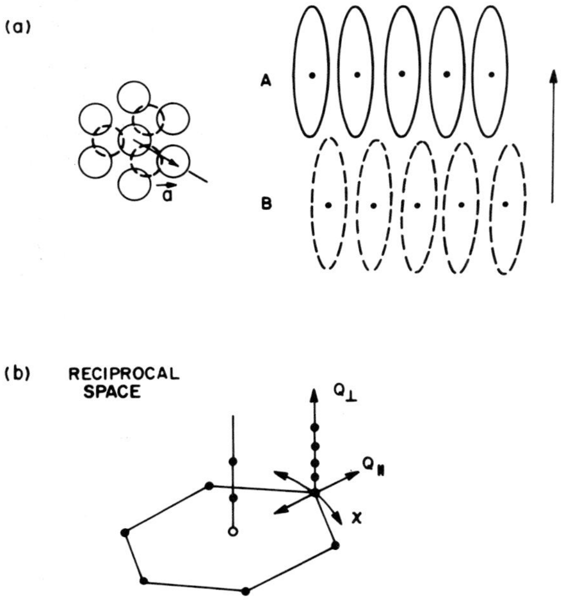

Fig. 1. (a) A schematic representation of the structure of 4O.8 in its smectic-B phase, and (b) the reciprocal lattice showing directions of the principal scans taken.

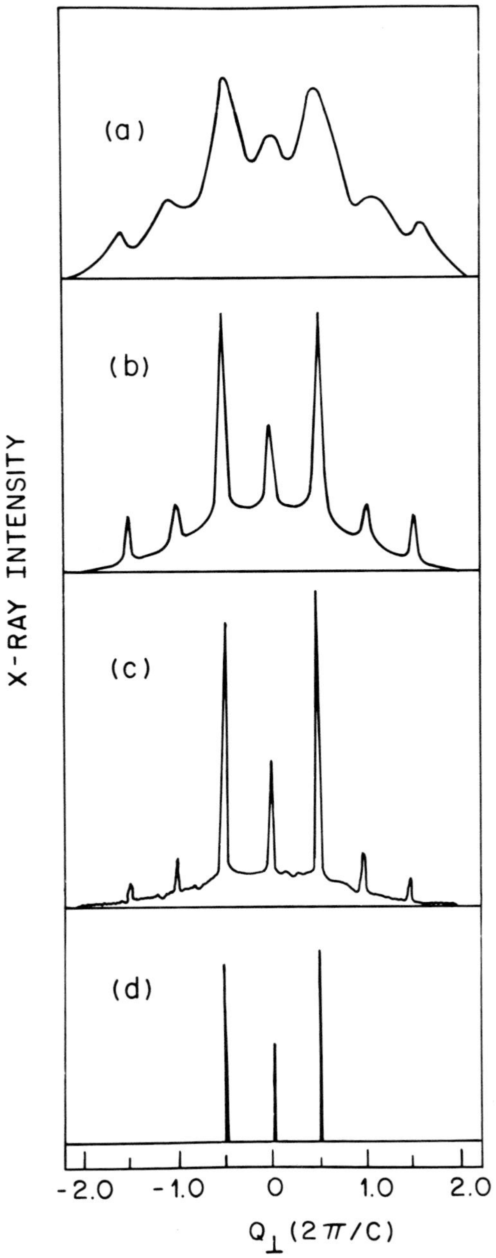

Fig. 2. Transverse scans taken at $Q = (1,Q_\perp)$ for various experimental conditions: (a) after Leadbetter et al. (ref. 5), using photographic x-ray techniques and field-cooled sample; (b) after Pershan et al. (Ref. 6), using a triple-axis spectrometer and field-cooled sample; (c) after Moncton et al. (Ref. 1), using a triple-axis spectrometer and liquid crystal films; and (d) after Moncton et al. (Ref. 2), using a synchrotron -based triple-axis spectrometer and liquid crystal films.

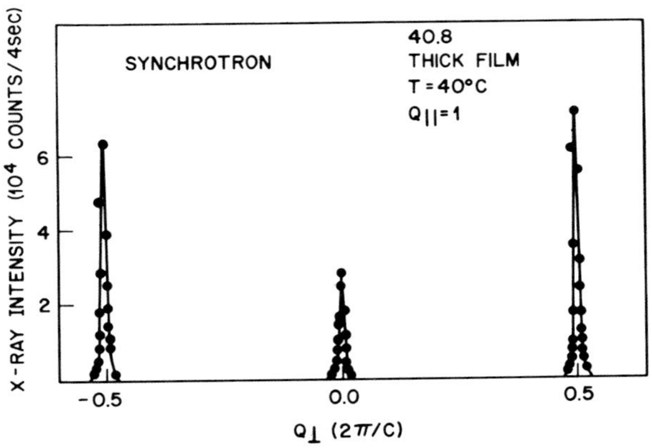

Fig. 3. The data of Fig. 2(d) is shown on an expanded scale. The width in these scans is due to a combination of sample warp (.01°) and finite sample thickness. The synchrotron resolution width is negligibly small.

ure of sample thickness, but optical reflectivity is a considerably more efficient technique for that purpose.

To establish the nature of in-plane correlations, data are obtained as a function of $Q_{\parallel}$ as shown in Fig. 5 for an N=4 film. The Gaussian resolution function is shown by the dashed line and this is what one would see if the structure factor $S(\vec{q})$ were a delta-function. This is clearly not the case here and, although the peak is nearly as sharp as the resolution, it has large wings. The theory of 2D solids to which we referred in Section II maintains that there is no Bragg scattering in a rigorous sense. Long-wavelength fluctuations cause the Debye-Waller exponent to diverge. The scattering which may be considered to come from phonons reduces to a simple function of the 2D elastic constants as given in Eq. (1). The data in Fig. 5 have been least-squares fitted using a routine that convolutes $S(\vec{q})$ with the instrumental lineshape. It also takes into account the fact that the data are obtained as azimuthal powder averages over 60° χ segments, a procedure which is necessary because slow random orientational motion of the 2D lattice occurs with time. As can be seen, the fit is within the statistical errors associated with the data points. This scan took 24 hours to obtain and radiation damage concerns prohibit longer exposures. Therefore, it is not possible to easily obtain better statistics. In fact since the total scattering cross-section of a 100Å thick hydrocarbon film is so small, obtaining data of this quality is a substantial achievement. The key experimental accomplishment is the low background--less than 0.025 counts/sec. with the beam ($\sim 10^{11}$ photons/sec.) irradiating the sample. Both the absence of sample-cell windows and the triple-axis spectrometer technique are crucial in this regard.

The important question now is how to interpret the fit. First it is important to address the possibility that this state is not a 2D solid but rather has exponentially-decaying positional correlations with a long correlation length. Since we observe clear orientational order at this temperature, the system cannot be liquid and the question is whether hexatic order may be relevant. Fits of the data in Fig. 5 to a Lorentzian can be made but they require a length ~800Å. If this state were hexatic it would be extremely close to the hexatic-solid transition and increasing temperatures should rapidly increase the population of free dislocations and broaden the peak. Since this does not occur, the width must either be extrinsic (due to defects and/or

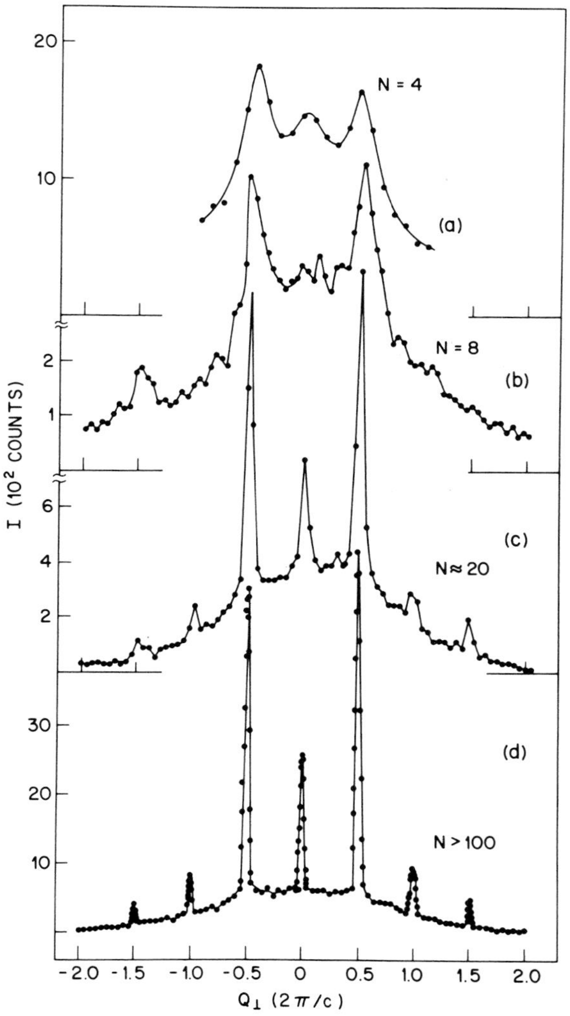

Fig. 4. Scans taken at $\vec{Q} = (1,Q_\perp)$ for four different film thicknesses show the development of interlayer correlations. The data for the N = 4 film was integrated over χ.

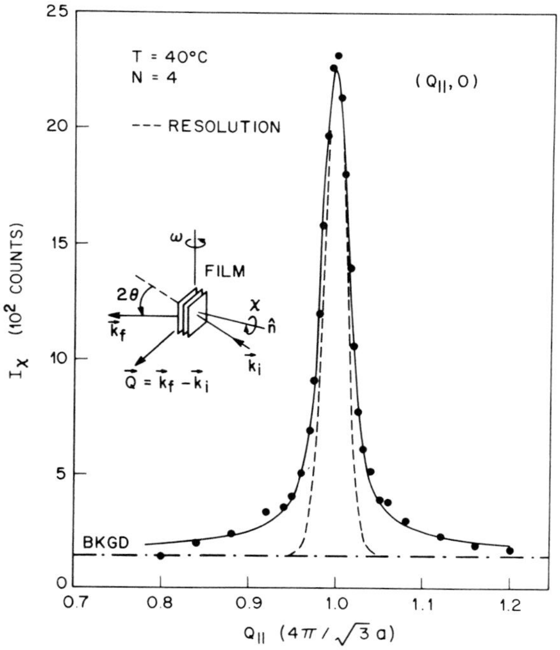

Fig. 5. Intensity integrated over a 60° χ segment for a four-layer film, versus the in-plane momentum transfer $Q_\parallel$. The solid line represents the structure factor described in the text. The inset describes the experimental geometry, with $\hat{n}$ being the film normal.

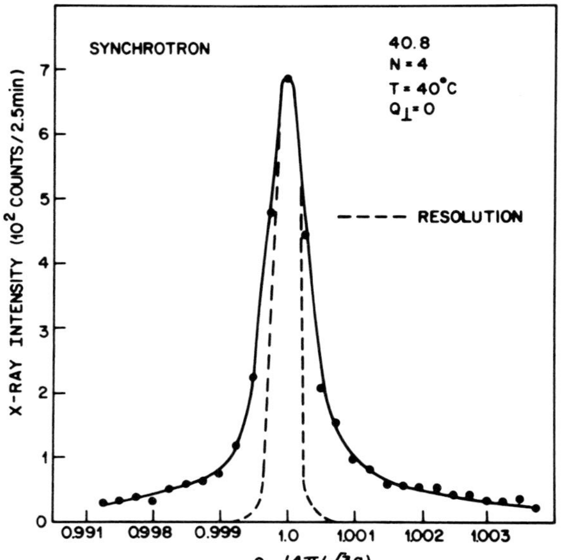

Fig. 6. Synchrotron data obtained on a four-layer film similar to that of Fig. 5, except that it had been damaged due to exposure to a high-humidity environment. This scan demonstrates that severe degradation results in only a small increase in width of the scan at this temperature.

damage) or the lineshape must not be Lorentzian. In Fig. 6 we show a scan taken at the synchrotron with a factor of 80 better $Q_\parallel$ resolution. The sample used here had been degraded by exposure to a high humidity environment and yet the excess width is much smaller than the Lorentzian width required to fit the data of Fig. 5. We conclude, therefore, that the power-law, not the Lorentzian, must be the relevant lineshape. The question now concerns the accuracy of the values of η which are extracted. From studies of many such fits we are cautious. In general, when data are available over less than a decade in $\Delta Q_\parallel$ the exponent η and the background are strongly correlated in a least-squares fit. Independent measurements can be made of all sources of background except diffuse scattering from the film which arises due to intramolecular vibrations and rotations. We can therefore set rather tight limits on the background: it is not greater than the scattering observed away from the peak (at $Q_\parallel$

= 0.7 or 1.3) for example and it is not less than the background measured with no film in the sample oven. All the data have been fit assuming the background to be the latter minimum value and η is displayed in Fig. 7. The single data point shown with the schematic dashed line is a fit to a scan taken over a large enough Q range to use the values in the wings as a maximum background. This reduced the value of η for that temperature and would have a similar effect on other data. We believe that this is the level of accuracy that can be confidently placed in our measurements. The values of η extracted are certainly consistent with predictions. A value of $\eta = 1/3$ cannot be exceeded according to the 2D melting theory. Perhaps independent measurements of the elastic constants and a theory of finite thickness samples will produce estimates for η which could be compared to these values.

From the combined experience of a number of research groups studying samples of 4O.8 in various forms, it is now clear that only the film technique produces single-crystal quality samples necessary to clearly establish the existence of long-range order by high-resolution x-ray scattering. We have, therefore, given high priority to examining the materials which various workers have suggested exhibit short-range correlations in the B phase. For this reason we have briefly studied three other materials. One of these is also from the same series, (4O.4), but with four fewer carbon atoms. It exhibits a bulk B phase structurally similar to 4O.8 although it has an ABC stacking. A second material TCOB[17] was examined[18] and it was also found to have a 3D crystalline B phase this time with a single-layer stacking arrangement. The third material examined does not exhibit a 3D crystalline B phase and we discuss it in more detail in the next section.

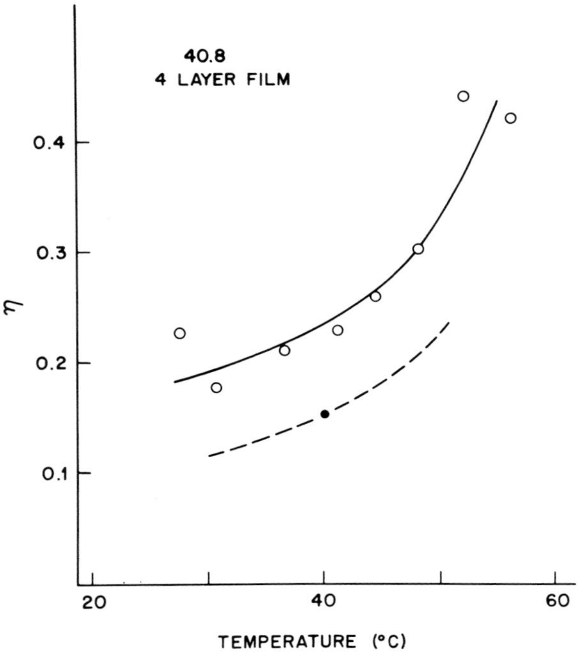

Fig. 7. The values of the exponent η versus temperature extracted from the fits of power-law lineshapes described in the text to a series of data similar to that in Fig. 5.

VII. Orientationally Ordered Smectic-B Phase

The material of interest is this section, referred to as 65OBC, is shown below:

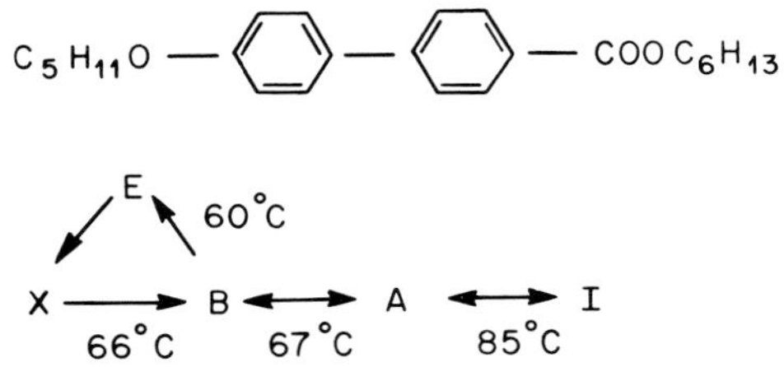

This compound was first prepared by J. W. Goodby[7] following a report by Leadbetter et al.[19] that related materials[20] did not have long-range interlayer correlations. Our work on this new phase has been done very recently and it is premature to discuss it here in detail. However we will present the two important observations we have made concerning the structure of its smectic-B phase. First, as shown in Fig. 8. the scattering is not in the form of Bragg peaks. A scan as a function of $Q_{\parallel}$ shows a Lorentzian lineshape with a correlation length $\xi = 150$Å. This length is longer than for a smectic-A (typically $\xi = 20$Å) but is nevertheless finite. This indicates of course that the in-plane positional correlations decay exponentially in space. Also the scattering is a very broad function of $Q_{\perp}$ indicating that the correlations between the layers extend only about one layer. Secondly, as shown in Fig. 9, the scattering is distributed over χ angles. In the A phase this scattering is a constant independent of χ, but in the B phase it develops substantial modulation. This structure can only originate from bond-orientational order which is correlated over the area illuminated by the x-ray beam (approximately $2mm^2$). Since the development of this structure in a χ-scan is associated with the development of long-range correlation in the bond orientation, we identify it as the order parameter for the B phase of this material.

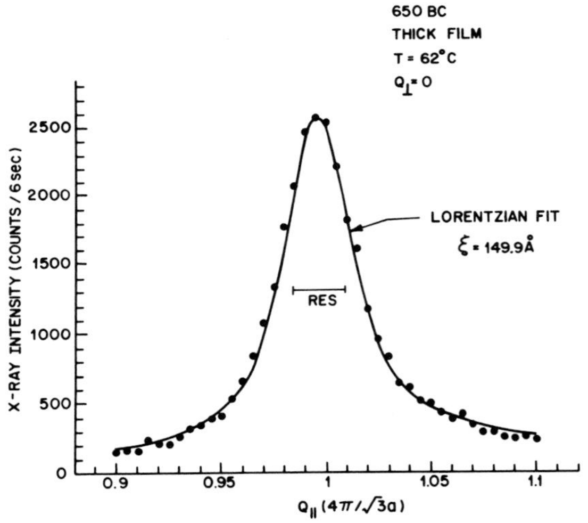

Fig. 8. Scattering as a function of $Q_{\parallel}$ in the smectic-B phase of 65OBC demonstrates that this phase does not have long-range positional correlations. Rather they decay exponentially with a length of $\xi = 150$Å at this temperature.

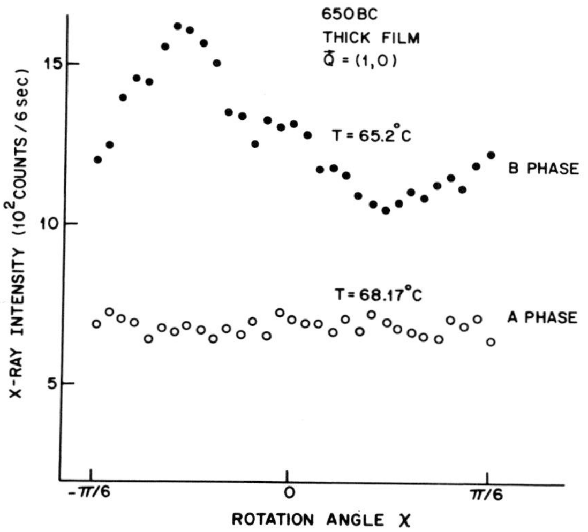

Fig. 9. Scans as a function of the angle χ in 65OBC. These data show the evolution from the liquid-like scattering of the smectic-A to the modulated ring of scattering indicating the presence of long-range lattice-orientational order in the smectic-B phase.

The mechanism for the development of bond-orientational order in a 3D smectic-B has been worked out in a picture based on the stacking of 2D hexatic layers as described in section II. The structure factor for such a system is a ring of scattering modulated into six maxima with the general form given by Eq. 2.[14] At the $B \rightarrow A$ transition the orientational order parameter goes to zero as measured by the amplitude of the modulation. Near the transition the six-fold modulation would presumably be dominated by the first harmonic, with the higher-order harmonics growing as the sample temperature is lowered. This behavior is observed qualitatively, and a direct measurement of the temperature-dependence of the orientational order parameter may be possible, although it has not yet been accomplished. One problem in this regard concerns the effect of having more than one orientational domain within the scattering volume. If each domain produces cosine-modulated scattering with a different χ phase shift, the resulting scattering will still be a single cosine wave. Under these circumstances it is not possible to unambiguously interpret the amplitude of the modulation as the orientational order parameter. One needs to probe the film over a large area with a small beam and demonstrate that multiple domains do not exist. So far attempts to do this have revealed that a few domains are in fact present.

At this point we should remark that we have evidence for substantial local correlation of the orientation of the molecules with respect to rotation about their long axis. If this molecular orientation field developed long-range order it could drive lattice order simultaneously. However, the molecular order would have a two-fold rotational symmetry and the films would be optical anisotropic. This anisotropy should show up readily in examining the films between crossed polarizers. Films have been examined specifically for this effect, and no evidence of optical anisotropy has been found. Nevertheless, more sensitive ellipsometer experiments are in progress. Certainly further thought needs to be given to the possible role of anisotropic forces in the development of bond-orientational order.

VIII. Conclusions

We hope that we have presented a convincing case for the use of the liquid-crystal film technique in combination with x-ray scattering to obtain high-quality data on ordering in smectic liquid crystals. Even the simplest problem of determining the existence of 3D crystalline order in the prototypical smectic-B compounds, has proved difficult historically. This is a result of the fact that these systems have weak interlayer forces which result in (1) a strong tendency to form highly defected structures in conventional sample preparations and (2) large diffuse scattering signals. The single-crystal quality samples which naturally result from the film preparation provide the sharp diffraction structure necessary to distinguish Bragg scattering from the diffuse background.

Obviously this sample quality is necessary to reveal the 3D ordered nature of the crystalline smectic-B phase. But it is also exceedingly important in the study of 2D films and the 3D stacked hexatic phase to know that the lack of long-range order in these systems is not some antifact of sample preparation. These systems exhibit new and exciting structural features which are characteristics of thermal equilibrium and not induced by sample-cell walls, applied fields or preparation history. In the case of the 2D films, we have demonstrated that rigorous long-range order is absent as expected, and that the Bragg scattering is reduced to a power law singularity. For the case of 65OBC we have shown that the smectic-B state is most likely composed of 2D hexatic layers. When stacked together such a system develops long-range lattice-orientational order which manifests itself clearly in our diffraction experiment. It is of interest to explore the behavior of this material into the 2D thin-film limit. At present it is the most likely candidate for the hexatic phase. This and other problems, particularly in tilted phases, will be pursued in the future as the full potential of synchrotron x-ray scattering studies of liquid crystal films is realized.

References

(1) D. E. Moncton and R. Pindak, Phys. Rev. Lett. *43*, 701 (1979).

(2) D. E. Moncton, R. Pindak and G. S. Brown, to be published.

(3) A. -M. Levelut, J. Doucet, and M. Lambert, J. Phys. (Paris) *35*, 773 (1974).

(4) A. DeVries, A. Ekachai, and N. Spielberg, J. Phys. (Paris), Colloq. *40*, C3-147 (1979); A. J. Leadbetter, J. Frost, J. P. Gaughan, and M. A. Mazid, J. Phys. (Paris), Colloq. *40*, C3-185 (1979); J. Doucet, A.-M. Levelut, and M. Lambert, Ann. Phys. (N.Y.) *3*, 157 (1978).

(5) A. J. Leadbetter, M. A. Mazid, B. A. Kelley, J. Goodby and G. W. Gray, Phys. Rev. Lett. *43*, 630 (1979).

(6) P. S. Pershan, G. Aeppli, J. D. Litster and R. J. Birgeneau, to be published.

(7) R. Pindak, D. E. Moncton, J. W. Goodby and S. C. Davey, to be published.

(8) R. J. Birgeneau and J. D. Litster, J. Phys. Lett. (Paris) *39*, 399 (1978).

(9) B. I. Halperin and D. R. Nelson, Phys. Rev. Lett. *41*, 121 (1978); D. R. Nelson and B. I. Halperin, Phys. Rev. B *19*, 2457 (1979).

(10) B. Jancovici, Phys. Rev. Lett. *19*, 20 (1967); N. D. Mermin, Phys. Rev. *176*, 250 (1968).

(11) J. M. Kosterlitz and D. J. Thouless, J. Phys. C *6,* 1181 (1973).

(12) P.G. DeGennes and G. Sarma, Phys. Lett. *A38,* 219 (1972).

(13) B. A. Huberman, D. M. Lublin and S. Doniach, Solid State Commun. *17,* 485 (1975).

(14) R. F. Bruinsman and D. R. Nelson, submitted to Phys. Rev.

(15) C. Y. Young, R. Pindak, N. A. Clark, and R. B. Meyer, Phys. Rev. Lett. *40,* 773 (1978); C. Rosenblatt, R. Pindak, N. A. Clark, and R. B. Meyer, Phys. Rev. Lett. *42,* 1220 (1979).

(16) D. Coates and G. W. Gray, The Microscope *24,* 117 (1976); D. Demus and L. Richter, "Textures of Liquid Crystals", (Verlag Chemie-Weinheim, New York, 1978).

(17) TCOB refers to Trans-l,4-cyclohexane-di-n-octyloxybenzoate. See A. DeVries, A. Ekachis and N.Spielberg, J. de Physique *40,* C3-147 (1979).

(18) R. Pindak, D. E. Moncton, M. E. Neubert and M.E. Stahl, unpublished.

(19) A. J. Leadbetter, J. C. Frost and M.A. Mazid, J. de Physique Letts. *40,* L-325 (1979).

(20) J. W. Goodby and G. W. Gray, J. de Physique *40,* 363 (1979); G. W. Gray and J. W. Goodby, Molecular Crystals Liquid Crystals *37,* 157 (1976).

Sinha, ed. Ordering in Two Dimensions

STUDY OF ORIENTATIONAL ORDER IN HYDROCARBON FILMS ADSORBED ON GRAPHITE

H. TAUB, G.J. TROTT, F.Y. HANSEN, and H.R. DANNER

University of Missouri-Columbia, Columbia, Missouri 65211, U.S.A.

J.P. COULOMB, J.P. BIBERIAN, J. SUZANNE, and A. THOMY

Croissance Cristalline, Centre Universitaire de Luminy

13288 Marseille, France

ABSTRACT

Deuterated paraffin films are exceptionally favorable for study by elastic neutron diffraction because of the large number of Bragg reflections observable at monolayer coverages. In this paper we review recent progress in analyzing the diffraction patterns of these films to obtain the orientation of the adsorbed molecule. Emphasis is placed on the system of ethane (C_2D_6) physisorbed on exfoliated graphite for which three different solid structures depending on coverage have been observed at low temperatures. The three Euler angles defining the molecular orientation have been determined for two of these structures and the transitions between the phases studied as the coverage is varied.

INTRODUCTION

Elastic neutron diffraction has developed rapidly in the past few years as a method of studying the structure of gases physisorbed on high-surface-area substrates. As discussed in several review articles [1-3], the technique has been used primarily to investigate the translational order of adsorbed monolayers and to elucidate the phase diagrams of simple gases physisorbed on various exfoliated graphite substrates. Only very recently has neutron diffraction begun to be applied to the study of orientational order in molecular monolayers. It is our purpose here to review some of this work on deuterated hydrocarbon films adsorbed on graphite. These films are probably the most favorable which have yet been found for study of orientational order by elastic neutron diffraction.

Conventional diffraction techniques for bulk structure determination involve comparison of the relative intensities of observed Bragg reflections with those calculated from a model structure factor. Such an approach is difficult to apply to low-energy electron diffraction (LEED), one of the most commonly used techniques for surface structure determination, because of the multiple scattering effects associated with the strongly interacting charged probe. Neutron diffraction has the advantage of such a weak interaction of the probe with the film and substrate that multiple scattering can usually be neglected. Indeed, the problem is more often one of obtaining sufficient scattered intensity from the overlayer. In the majority of films studied thus far by neutron diffraction, only one or two Bragg reflections are intense enough to be observed. Consequently, a conventional structure analysis using the relative intensity of a large number of Bragg reflections is again difficult.

Neutron diffraction can sometimes be used to infer molecular orientational order if it results in a structural phase transition which alters the translational symmetry of the film. The orientationally ordered phase may then exhibit Bragg reflections which are not present in the orientationally disordered state. Experiments of this kind have inferred orientational order in monolayers of oxygen [4] and nitrogen [5] adsorbed on exfoliated graphite. However, these experiments did not observe a sufficient number of Bragg reflections to determine the molecular orientation in the ordered state.

This situation has now improved considerably with the discovery of adsorbed films whose neutron diffraction patterns contain a larger number of Bragg reflections. Nitric oxide adsorbed on exfoliated graphite was one of the first of these systems to be studied. Suzanne et al [6] were able to use the relative intensity of several reflections to infer one of the Euler angles defining the orientation of the NO molecule.

Recently, deuterated hydrocarbon films have been found which yield very rich diffraction patterns. Bragg reflections out to third order have been observed from monolayers of neopentane (C_5D_{12})

adsorbed on TiO_2 [7], and ethane (C_2D_6) [8] and butane (C_4D_{10}) [9] adsorbed on exfoliated graphite. In a paper of this length it is impossible to summarize all the work which has been done on these systems. Instead, we shall concentrate on the ethane films for which the most effort has been made to characterize the orientational ordering.

In the next section we motivate the selection of paraffin molecules for studies of orientational ordering and discuss the importance of deuteration. A brief description of the neutron spectrometers and the sample cells is given. The third section describes the diffraction patterns and their analysis for the three solid phases observed in ethane monolayers on graphite at low temperature. The paper concludes with a short summary and a discussion of future prospects.

EXPERIMENTAL CONSIDERATIONS

The paraffin molecules consist of a carbon chain to which hydrogen atoms are tetrahedrally bonded to form methyl (CH_3) and methylene (CH_2) groups. Ethane (CH_3CH_3) as shown in Fig. 2 consists of only two methyl groups joined by a C-C bond while butane ($CH_3(CH_2)_2CH_3$) contains two interior methylene groups. At the outset, these molecules may seem unnecessarily complex for studies of orientational order. However, calculations with empirical potentials [10] indicate only small static distortions of the molecules when adsorbed on a graphite basal plane so that the intramolecular degrees of freedom are thought to be unimportant in determining the film structure. Thus the paraffins provide a homologous series of nearly rigid rod-shaped molecules for studies of orientational order on graphite. The rods have a width of ~4 Å and increase in length from ~5 Å (ethane) in increments of 1.25 Å for each additional CH_2 group. It should be noted, though, that the rods do not possess cylindrical symmetry about their long axis.

Deuteration was used both to enhance the coherent neutron cross-section of the films and, more importantly, to reduce the incoherent cross-section. The extremely large incoherent cross-section of hydrogen would otherwise result in the domination of the elastic diffraction pattern by an intense isotropic component. However, even with deuteration, the exceptional features of the paraffin films for neutron diffraction experiments were not apparent before these studies began. Since the coherent scattering length of deuterium is nearly equal to that of carbon, the deuterated hydrocarbon film has the same coherent cross-section per atom as the graphite substrate but comprises only ~1% of the atoms in the sample cell. In comparison, ^{36}Ar scatters much more strongly, having a coherent cross-section about 13 times larger than carbon. Nevertheless, previous experiments with ^{36}Ar monolayers adsorbed on Grafoil [11] had observed only three Bragg reflections, although these were sufficient to confirm an incommensurate triangular structure.

Rather than an exceptionally favorable ratio of film-to-substrate scattering, the large number of Bragg reflections observed from the deuterated hydrocarbon monolayers results from a combination of several factors. Because the 2D unit cells are generally larger and of lower symmetry that the commensurate $\sqrt{3} \times \sqrt{3}$ structure on the graphite basal plane, there are more distinct Bragg reflections at wavevector transfers $\lesssim 4$ Å^{-1}. Moreover, the structure factors are sufficiently large and the Debye-Waller factors small enough that Bragg peaks are observable over this entire Q-range. Since the deuterium is as effective as the carbon in scattering the neutrons, the relative intensity of these Bragg reflections is sensitive to the deuterium atom positions. This situation should be contrasted with that for LEED and x-ray studies of adsorbed hydrocarbons. Here scattering would be dominated by the carbon atoms which have the larger electron concentration.

The neutron scattering experiments described below were conducted at the University of Missouri Research Reactor Facility (MURR) and at the Institut Laue-Langevin (ILL) in Grenoble. The MURR experiments were performed on the two-axis D-port spectrometer with an incident neutron wavelength of 1.29 Å and a Grafoil substrate. The ILL measurements used the D1B spectrometer equipped with a multidetector and an incident wavelength of 2.52 Å. The Papyex substrate [6] in the ILL cell had about the same specific area, ~20 m^2/g, as the MURR Grafoil cell. In both experiments the scattering vector $\vec{Q}$ was aligned parallel to the foil planes and normal to the preferred c-axis direction.

RESULTS

Figure 1(a) contains the elastic neutron diffraction pattern obtained with 1.29 Å neutrons incident upon a 0.8-layer deuterated ethane film adsorbed on Grafoil at a temperature of 8.6 K [8]. The background scattering from the substrate has been subtracted. Seven peaks can be distinguished in this difference spectrum, the most intense of which clearly exhibit the asymmetric shape characteristic of diffraction from a 2D polycrystal [12]. In order to establish the reproducibility of some of the weaker peaks, measurements were repeated with the same sample and wavelength and also with the Papyex cell at λ = 2.52 Å. Except near Q = 1.9 Å^{-1} and 3.1 Å^{-1} where the graphite scattering is particularly strong, the diffraction pattern reproduced well.

The analysis of the diffraction pattern proceeded in two steps: 1) determination of a unit cell which would index the observed peaks; and 2) determination of the molecular orientation by fitting the relative intensity of the peaks to a model structure factor. The van der Waals dimensions of the C_2D_6 molecule together with the assumption of close-packing suggested an oblique cell containing one molecule as shown in Fig. 2. The cell of minimum area found to index the diffraction peaks has primitive vectors $|\vec{a}| = 4.91$ Å, $|\vec{b}| = 3.89$ Å, and the included angle $\theta = 86.4°$. We define unity coverage to be a complete monolayer having the density of this cell, i.e., 19.1 Å^2 per molecule.

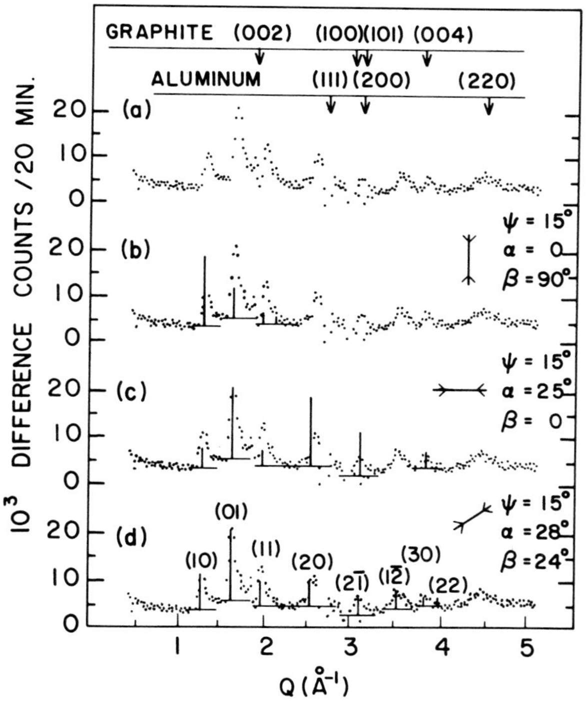

Fig. 1. (a) Diffraction pattern of the submonolayer S_1 phase of ethane (C_2D_6) adsorbed on Grafoil at a coverage of 0.8 layers and a temperature of 8.6 K [Ref. 8]. Arrows at the top mark angles at which Bragg reflections from the graphite and the aluminum sample holder occur. The observed pattern is compared with calculated spectra for different angles of the C-C bond with respect to the surface: (b) 90°, (c) 0°, and (d) 24°. The spectra are normalized so that the strongest peak of the calculated and observed spectrum, respectively, have equal intensity. The tilted configuration in (d) gives the best agreement with the observed spectrum.

The intensity of the hk Bragg reflection from a 2D polycrystal is given by the theory of Warren [12] appropriately modified for neutron scattering [13]:

$$I_{hk} \propto \frac{m_{hk}|S_{hk}|^2 e^{-2W}}{(\sin\theta)^{3/2}} \left(\frac{L}{\sqrt{\pi}\,\lambda}\right) F(a) \qquad (1)$$

where 2θ is the scattering angle, m_{hk} is the multiplicity of the reflection, S_{hk} is the geometrical structure factor, and e^{-2W} is the Debye-Waller factor. The other quantities in Eq. (1) are the wavelength λ, the characteristic dimension L of the 2D crystallites, and the function

$$F(a) = \int_0^\infty e^{-(x^2-a)^2} dx$$

where $a = (2\sqrt{\pi}\, L/\lambda)(\sin\theta - \sin\theta_{hk})$ and θ_{hk} is the Bragg angle. It is this function which gives the reflection the characteristic sawtooth shape.

To calculate the diffraction pattern for the ethane submonolayer, the geometrical structure factor S_{hk} was expressed in terms of the Euler angles describing the orientation of the single rigid molecule in the unit cell. The three Euler angles α, β, and ψ are defined in Fig.

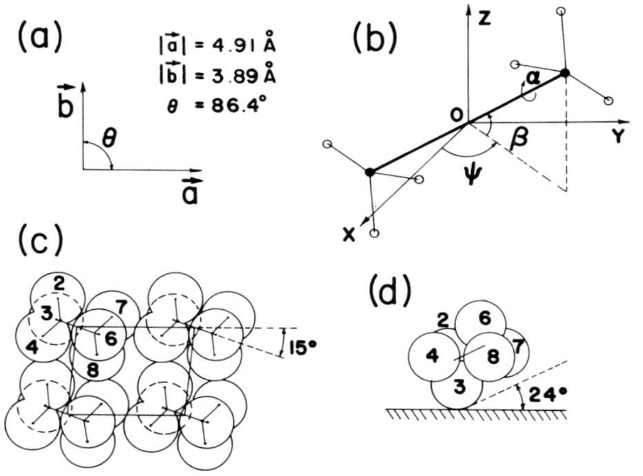

Fig. 2. Structure of the S_1 phase [Ref. 8]. (a) The 2D unit cell. (b) Definition of the Euler angles α, β, and ψ. For clarity in labeling, the molecule is not drawn to scale. The zero of α corresponds to the configuration in Fig. 4 of Ref. 10. (c) Projection (to scale) of the ethane molecules on the 2D unit cell. Circles represent approximate van der Waals radii of the D atoms. (d) Illustration of the tilting of the molecules with respect to the surface. Atoms numbered 2,4,7 and 8 are nearly coplanar.

2(b). The Debye-Waller factor in Eq. (1) was set equal to unity in order to limit the number of free parameters in the model. To simplify the analysis further, only the relative integrated intensity of the reflections was calculated as shown by the vertical bars in Fig. 1. Because of the asymmetric shape of the Bragg peaks, considerable overlapping of the reflections occurs making it difficult to estimate the integrated peak intensities from the diffraction pattern. For this reason, the integrated intensity of a reflection was assumed proportional to the peak height. The peak overlap also necessitated taking a separate background level for each reflection as shown by the horizontal lines in Fig. 1.

The Euler angles of the molecule were systematically varied to find the best agreement between the calculated and observed peak intensities. Fig. 1(b) contains the best fit which could be obtained with the C-C bond constrained perpendicular to the surface ($\beta = 90°$, α varied) while Fig. 1(c) shows the best fit for the C-C bond parallel to the surface ($\beta = 0$, α and ψ varied). The sensitivity of the calculated intensities to the molecular orientation is easily understood. The molecule is sufficiently large with respect to the 2D unit cell that even a small rotation can move an atom through a significant fraction of a lattice spacing.

We found the best fit to the observed spectrum to be obtained with the C-C bond tilted at an angle $\beta = 24°$ to the surface. The error in the Euler angles is estimated to be ±5° [8]. Calculated intensities for this orientation are shown in Fig. 1(d) and the corresponding crystal structure is illustrated in Figs. 2(c) and 2(d). We shall refer to this submonolayer structure of ethane on graphite as the S_1 phase.

A series of measurements [8] have shown the S_1 phase to be stable between coverages of 0.4 and 0.8 layers and at temperatures up to 60 K. However, as the coverage is increased above 0.8 layers for temperatures ≲40 K, a first order transition is observed to a second solid phase which we denote S_2. At a coverage of 0.92 layers, the neutron diffraction pattern clearly shows the coexistence of the S_1 and S_2 phases [14], establishing the first order character of the transition. The pure S_2 phase is identified with the diffraction pattern in Fig. 3(a) at 0.97 layers and a temperature of 9 K. The spectrum differs remarkably from that of S_1 with peaks occurring at several new scattering angles and having a qualitatively different intensity pattern. There is also a slowly increasing component to the scattering below the first Bragg peak which was not evident in the S_1 pattern.

The analysis of the S_2 diffraction pattern has proved to be more difficult than for the S_1. In order to index the Bragg peaks and also obtain a reasonable density for the film it was necessary to assume a considerably larger unit cell containing two molecules. The unit cell shown in the lower portion of Fig. 3 has primitive vectors $|\vec{a}| = 4.30$ Å, $|\vec{b}| = 7.45$ Å, and the included angle $\theta = 82°$. The area per molecule is 15.9 Å^2 which is a 17% increase in density over S_1.

With two molecules per cell, a total of six Euler angles must be determined from the analy-

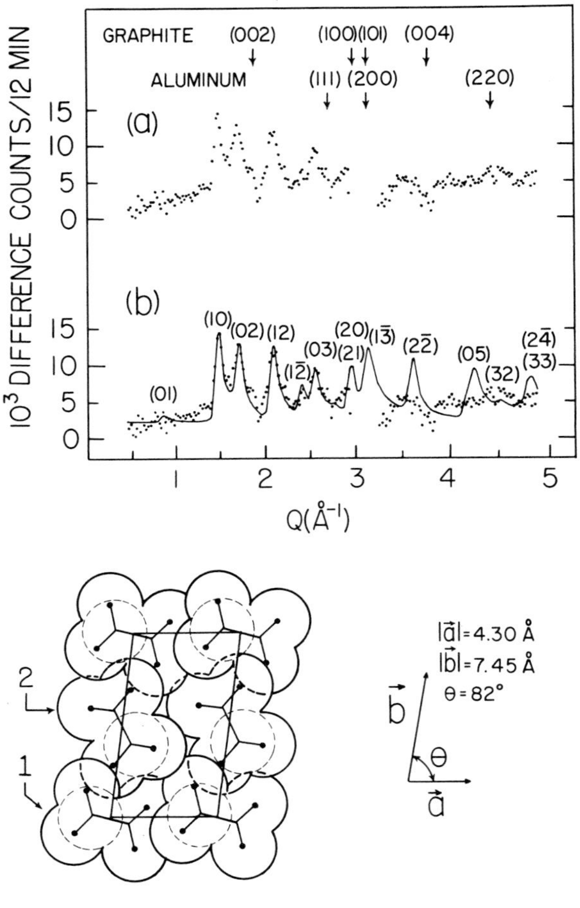

Fig. 3. (a) Diffraction pattern ($\lambda = 1.29$ Å) of the S_2 phase of ethane adsorbed on Grafoil at T = 9 K. Coverage is 0.97 layers. (b) Profile analysis of the diffraction pattern. Solid curve is the composite spectrum calculated for the crystal structure shown in the lower panel. Molecule numbered 1 has the same orientation as in S_1 (cf. Fig. 2) except for a smaller tilt angle $\beta_1 = 18°$ of the C-C bond to the surface. Molecule 2 has been rotated from the S_1 orientation so that its C-C bond makes an angle of 33° with $\vec{b}$ and is tilted at an angle $\beta_2 = 15°$ to the surface. The molecule is also rotated -10° about the C-C bond relative to the S_1 orientation.

sis of the relative peak intensities. There are, however, constraints which provide some direction to the fitting procedure. The first Bragg reflection corresponding to a d-spacing of 7.38 Å ($Q = 0.85$ Å^{-1}) must be very weak or completely extinguished. Also, it seemed reasonable to begin with molecular orientations which were not too dissimilar from that in the S_1 phase. Due to the higher density of the unit cell, it was found that the structure factor depended even more sensitively on the molecular orientations than in the S_1 phase.

Since the problem of peak overlap is even more severe in the S_2 diffraction pattern, we were motivated to develop a profile analysis technique to facilitate comparison of a model structure factor with the observed diffraction pattern. The technique is similar to that introduced by Rietveld [15] for the analysis of bulk powder diffraction patterns. The Warren profile of Eq. (1) is computed for each Bragg reflection and the intensities summed to form a composite spectrum. As described in Ref. 13, the shape function F(a) was altered to account for the preferred orientation of crystallites in the Grafoil substrate. Also, because the atoms in the molecules are situated at different heights above the surface, the profile must be corrected for multilayer effects which modulate the intensity along a Bragg rod [13]. In addition to the Euler angles of the molecules, two other parameters are introduced. One is a scale factor which sets the integrated intensity of the composite spectrum, and the other is an additive constant to simulate an isotropic component to the scattering. In this way, the assumption of a separate background level for each reflection is eliminated. In the last step of the calculation, the instrumental resolution function is folded with the composite spectrum.

A preliminary profile analysis of the S_2 diffraction pattern and the corresponding crystal structure are shown in Fig. 3. The model correctly predicts a very weak (01) reflection and reproduces the relative intensity of the next five peaks very well. Presumably, the strong peaks predicted near $Q = 3.1$ Å^{-1} are obscured by intense graphite scattering which causes the difference spectrum to go negative. At larger Q, the calculated peak intensities may also be affected by both substrate and aluminum cell reflections. The effect of a Debye-Waller factor on the high-angle reflections is now being investigated.

The model structure of the S_2 phase in Fig. 3 bears an interesting relationship to that of S_1. One of the molecules in the cell has the same orientation as in S_1 except for a slightly smaller tilt angle of the C-C bond to the surface ($\beta = 18°$). The second molecule has been rotated around so that it is wedged in the cleavage between methyl groups of neighboring molecules. The Euler angles α and β are close to the values for the first molecule. Since a change in the height of a molecule above the surface only alters the shape of the reflections and not their integrated intensity, it is difficult to determine whether some displacement of the molecules normal to the surface accompanies the transition to the S_2 phase.

Experiments at higher fillings and low temperature indicate that the S_2 phase is stable over an extremely narrow range of coverage. A small change in the relative peak intensities can be observed after only a 4% increase in coverage. As the coverage is increased further, the peak at $Q = 1.7$ Å^{-1} grows markedly [14] until at about 1.5 layers only this peak remains for $Q < 3$ Å^{-1}. This single-peak diffraction pattern shown in Fig. 4 is identified with a third structure S_3. The coverages between 0.97 and ~1.5 layers ($T < 65$ K) are interpreted as a region of coexistence between the S_2 and S_3 phases.

The peak in Fig. 4 can be indexed as the (10) reflection of a commensurate $\sqrt{3} \times \sqrt{3}$ structure on the graphite basal plane. We suggest that the ethane molecule may be oriented with the C-C bond perpendicular to the surface in this phase. The molecule would then be resting on a methyl hydrogen tripod as has been proposed [16] for methane (CH_4) in its $\sqrt{3} \times \sqrt{3}$ registered phase on graphite. In an effort to confirm this orientation, a search for higher order reflections in the S_3 phase is planned.

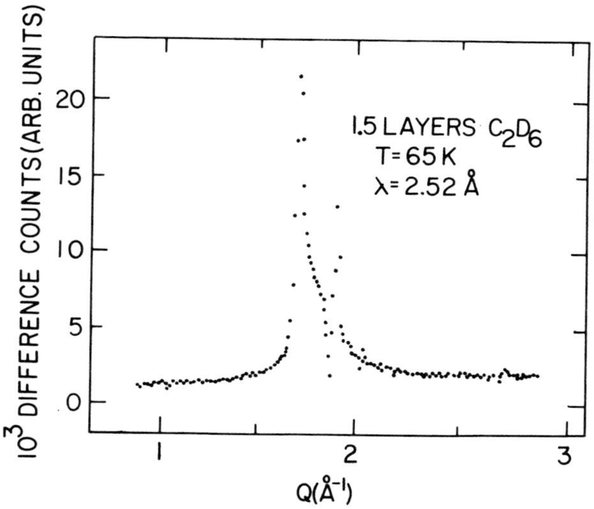

Fig. 4. Diffraction pattern ($\lambda = 2.52$ Å) of the S_3 phase of ethane adsorbed on Papyex at T = 65 K. Coverage is 1.5 layers.

SUMMARY AND CONCLUSIONS

Three distinct solid phases have been observed for ethane adsorbed on graphite at low temperature and coverages below 1.5 layers. The S_2 structure with two molecules per cell appears to be stable in an extremely narrow range of coverage near monolayer completion and coexists with both the S_1 structure (one molecule per cell) at lower coverage and the S_3 structure at higher coverage. The diffraction patterns of the S_1 and S_2 phases have been analyzed to yield the three Euler angles defining the molecular orientation. In both phases the ethane C-C bond is found to be tilted at a small angle $\lesssim 25°$ to the surface.

Experiments are now in progress to investigate the melting of these three solid phases of ethane on graphite. Some results have already been reported for the submonolayer phase S_1 [8]. A first order transition to an "intermediate" phase of unknown structure was observed to precede a continuous transition to the disordered state. It will be of interest to determine whether the S_2 and S_3 structures melt similarly via an intermediate phase and whether some orientational disorder occurs in the solid phase prior to melting.

Although the technique is still in an early stage of development, we believe that the results already obtained demonstrate elastic neutron diffraction to be a powerful method for studying the orientational order of adsorbed hydrocarbons. In addition to ethane, the profile analysis technique described here has been used to solve the structure of submonolayer butane adsorbed on graphite [9]. It is straightforward to treat larger molecules if the number of reflections observed is comparable to that for ethane, and the molecule does not distort in some unknown manner upon adsorption.

An attractive feature of the hydrocarbon-on-graphite systems is that empirical atom-atom potentials are available for calculating the intermolecular and molecular-substrate interactions [10]. Equilibrium structures can then be predicted for comparison with those determined experimentally. Such work is already in progress for butane monolayers [9,10]. In the case of ethane, calculations with empirical potentials may contribute to an understanding of why the molecules tilt when adsorbed on the graphite basal plane.

ACKNOWLEDGEMENT

We are grateful to P. Thorel for numerous discussions and the D1B staff for assistance in the neutron measurements at Grenoble. We acknowledge partial support of this research by the National Science Foundation under Grant No. DMR-7905958, the Petroleum Research Fund of the American Chemical Society, a Dow Chemical Company Grant of the Research Corporation, the NATO Research Grants Program, the University of Missouri Research Council, and the University of Missouri Research Reactor. One of us (G.J.T.) is in receipt of an American Chemical Society - Petroleum Research Fund Fellowship.

Permanent addresses: F.Y. Hansen - Fysisk-Kemisk Institut, The Technical University of Denmark, DK-2800 Lyngby, Denmark. A. Thomy - Laboratoire M. Letort, 54600 Villers-les-Nancy, France.

REFERENCES

1. J.W. White, R.K. Thomas, I. Trewern, I. Marlow, and G. Bomchil, Surf. Sci. 76, 13 (1978).
2. P.G. Hall and C.J. Wright in Chemical Physics of Solids and Their Surfaces, M.W. Robert and J.M. Thomas, eds. (Chemical Society, London, 1978), Vol. 7, p. 89.
3. J.P. McTague, M. Nielsen, and L. Passell, Crit. Rev. Solid St. Sci. 8, 135 (1979).
4. M. Nielsen and J.P. McTague, Phys. Rev. B 19, 3096 (1979).
5. J. Eckert, W.D. Ellenson, J.B. Hastings, and L. Passell, Phys. Rev. Letters 43, 1329 (1979).
6. J. Suzanne, J.P. Coulomb, M. Bienfait, J. Matecki, A. Thomy, B. Croset, and C. Marti, Phys. Rev. Letters 41, 760 (1978).
7. I. Marlow, Part II thesis, Oxford University, 1977 as quoted in Ref. 1.
8. J.P. Coulomb, J.P. Biberian, J. Suzanne, A. Thomy, G.J. Trott, H. Taub, H.R. Danner, and F.Y. Hansen, Phys. Rev. Letters 43, 1878 (1979).
9. G.J. Trott, H. Taub, F.Y. Hansen, and H.R. Danner, to be published.
10. F.Y. Hansen and H. Taub, Phys. Rev. B 19, 6542 (1979).
11. H. Taub, K. Carneiro, J.K. Kjems, L. Passell, and J.P. McTague, Phys. Rev. B 39, 215 (1977).
12. B.E. Warren, Phys. Rev. 59, 693 (1941).
13. J.K. Kjems, L. Passell, H. Taub, J.G. Dash, and A.D. Novaco, Phys. Rev. B 13, 1446 (1976).
14. J.P. Coulomb, J.P. Biberian, J. Suzanne, A. Thomy, G.J. Trott, H. Taub, H.R. Danner, and F.Y. Hansen, to be published.
15. H.M. Rietveld, J. Appl. Cryst. 2, 65 (1969).
16. M.W. Newbery, T. Rayment, M.V. Smalley, R.K. Thomas, and J.W. White, Chem. Phys. Letters 59, 461 (1978).

DISCUSSION

M. Bender:

Could you comment on the degree of tilt vs. degree of completion of the layer and also going to more than one layer--i.e., influence of packing on degree of tilt.

H. Taub:

Our preliminary analysis of the low-temperature ethane diffraction patterns at 0.4 and 0.8 layers does not establish a change in the tilt angle between these two coverages. However, we intend to perform a more rigorous profile analysis of the diffraction patterns.

R. B. Griffiths:

What is the nature of the evidence for continuous melting in contrast to, say, coexistence of two phases?

H. Taub:

The evidence is not yet compelling. The single-peak diffraction pattern associated with the "intermediate" phase at a coverage of 0.8 layers appears to weaken and broaden in a continuous manner above 65K. There is no qualitative evidence of a two-component structure of the peak (e.g., a sharp peak superimposed on a broader line) to suggest liquid-solid coexistence. We have not yet performed a quantitative line-shape analysis.

J. G. Dash:

You have shown results that seem to indicate three phase coexistence extending over a finite temperature range. That is, in a 0.4 coverage ethane film you see coexistence of two dense structures near 62 K. Since the coverage is relatively low there is also a low coverage vapor phase. The phase rule tells us that three phases can coexist only at a discrete temperature. It is a surface triple point, and there are no remaining degrees of freedom in a uniform surface system. Since you do observe coexistence over a finite temperature range, I can only conclude that (1) the system is not uniform, or (2) the temperature of the experiment varied, so that the system explored both sides of the phase boundary, or (3) the two dense phases are not in fact the same over the range, but are actually continuously varying. This last possibility seems to me quite improbable, since it requires that the two different continuously varying phases have identical free energies over a finite range of T. Still another possibility (4), exists which a few people have suggested to me in the few hours between the session and this writing. It may be that the two dense phases are in the form of a fine dispersion or mixture on the surface. Then if the boundary energy is appreciable, and if changes in the sizes of the islands would change the total free energy continuously with temperature, this mechanism seems to me to be quite plausible. Of course the phase rule would not apply in this case, since the system of small islands would not be in true thermodynamic equilibrium.

H. Taub:

I would only add to your analysis of the situation that we observe the "intermediate" phase in a range of coverage from 0.4 to 0.8 layers and in a temperature range from 60 K to 65 K.

LAYER DEPENDENT CORE-LEVEL SHIFTS FOR RARE-GAS ADSORBATES ON METALS

G. KAINDL+, T.-C. CHIANG, D.E. EASTMAN, AND F.J. HIMPSEL

IBM Thomas J. Watson Research Center
Yorktown Heights, New York 10598

Photo- and Auger-electrons from rare-gas atoms (Xe, Kr) physisorbed in form of monolayers, bilayers, and multilayers on metal surfaces (Pd(001)) exhibit well-resolved increases in kinetic energy with decreasing distance from the surface. The observed shifts are well described as extra-atomic hole-relaxation effects, using a point-charge image-potential model. They allow a direct labeling of the first few layers of an adsorbed multilayer configuration and are used to study the thermally activated inversion of a Kr/Xe bilayer on Pd(001).

It is well established that core-level ionization potentials and Auger-electron kinetic energies for adsorbed atoms and molecules are shifted relative to their gas-phase values by both initial-state chemical effects and by final-state hole-relaxation effects. The effective binding energy (ionization potential) of core electrons decreases by final-state hole-relaxation shifts. Because of the importance of final-state hole-relaxation effects for a quantitative analysis of chemisorption shifts of core-hole ionization potentials, several theoretical workers have dealt with the problem [1-4]. As a result, it is suggested that the metallic screening of a positive hole of charge q in an adsorbed atom at a distance x_1 from a metal surface can be described by a semiclassical image potential of the form $-q^2/4(x_1-x_o)$, if the image plane is moved out of the classical surface of the metal by x_o to the center of gravity of the induced surface-charge density [2]. Despite extensive experimental work on chemisorption shifts of ionization potentials of adsorbed atoms on metal substrates [5], no clear-cut tests of the validity of these theoretical models describing final-state hole-relaxation effects have been performed, mainly because of the difficulty of separating the initial-state and the final-state effects.

In order to achieve this goal, we have studied photoemission (PE) and Auger lines from rare-gas atoms adsorbed in monolayer, bilayer, and multilayer configurations on Pd(001) [6]. Due to the weak physisorption bonds initial-state chemical effects are expected to be rather small in these cases, allowing a detailed study of final-state hole-relaxation shifts.

The experiments were performed at the Synchrotron Radiation Center of the University of Wisconsin-Madison, using a double-cylindrical mirror analyzer system behind a toroidal-grating monochromator. The Pd(001) crystal was spot-welded to two 10-mil tungsten wires in thermal contact with the coldfinger of a closed-cycle helium refrigerator. In this way the temperature of the crystal, measured by a Chromel/Konstantan thermocouple spotwelded to it, could be continuously varied in the range from about 40 K to 1200 K. The Pd(001) surface was cleaned in the usual way by repeated Ar-ion sputtering and annealing. Rare gases were adsorbed at a substrate temperature of about 40 K, and close-packed monolayers and bilayers of both Xe and Kr were prepared from slightly thicker layers by controlled annealing at temperatures below the desorption temperatures of monolayers and bilayers, respectively. The appearance of rare-gas atoms in the second or third layer could be quantitatively checked using the layer-dependent shifts to be reported. Thicker multilayer configurations were produced by controlled rare-gas exposures after the exposure corresponding to a full monolayer had been calibrated.

+IBM Visiting Scientist 1979-1980. Permanent address: Institut für Atom- und Festkörperphysik, Freie Universität Berlin, Boltzmannstr. 20, D-1000 Berlin 33.

For three different Xe configurations physisorbed on Pd(001) the resulting Xe-4d and Xe-5p PE spectra as well as the Xe-NOO Auger spectra are shown in Fig. 1 and Fig. 2, respectively. The influence of final-state hole relaxation effects on the position of PE lines is most dramatically seen in the bilayer Xe-4d spectrum (Fig. 1b). The monolayer Xe-4d spectrum exhibits a spin-orbit-split doublet in agreement with the splitting observed in free atoms, but shifted by 2.14 eV to smaller effective binding energies (referred to the vacuum level). For a bilayer Xe-coverage, however, two well-resolved sets of Xe-4d peaks split by 0.72 ± 0.03 eV are observed. The less intense lines originate from the underlayer (first layer), while the more intense ones originate from the overlayer (second layer). In case of a 4-layer thick Xe film (Fig. 1c) the Xe-4d PE spectrum is further shifted to higher effective binding energies and exhibits contributions from all the different layers, with relative intensities given by the attenuation of the PE lines through the overlayers. Essentially the same observations are made with the Xe-5p PE spectra, but the observed shifts (relative to free atoms) and the bilayer splitting (0.59 eV) are approximately 30% smaller than in the Xe-4d case. Detailed numbers are given in Tabel 1, where the experimental results are summarized. The Xe-5p spectrum, especially the Xe-$5p_{3/2}$ line shows appreciable broadening due to band-dispersion effects [7].

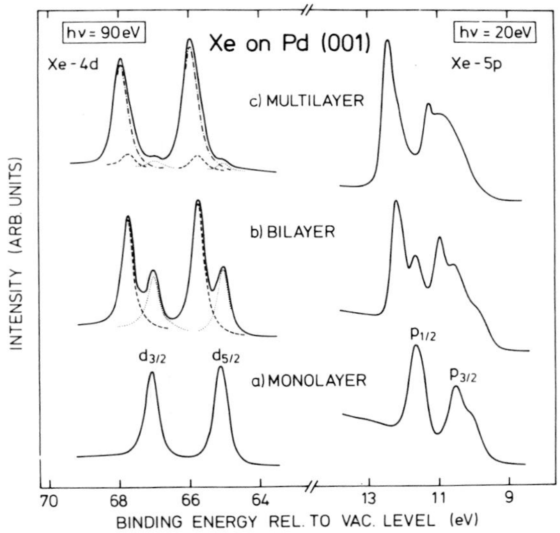

Fig. 1. Xe-4d and Xe-5p PE spectra for (a) a monolayer, (b) a bilayer, and (c) a multilayer (4 layers) of Xe physisorbed on Pd(001).

The Xe-NOO Auger spectra (Fig. 2) exhibit very similar, but by a factor of approximately 3 larger layer-dependent shifts. The monolayer spectrum is shifted by 6.57 eV to higher kinetic energies, while the bilayer spectrum is a superposition of two single-layer Auger spectra, shifted against each other by 2.01 eV. The deconvolution is indicated in Fig. 2b: the dotted (dashed) lines represent the contribution from the first (second) Xe layer. It is obvious that the monolayer spectrum (Fig. 1a) is broadened as compared to the second-layer contribution,

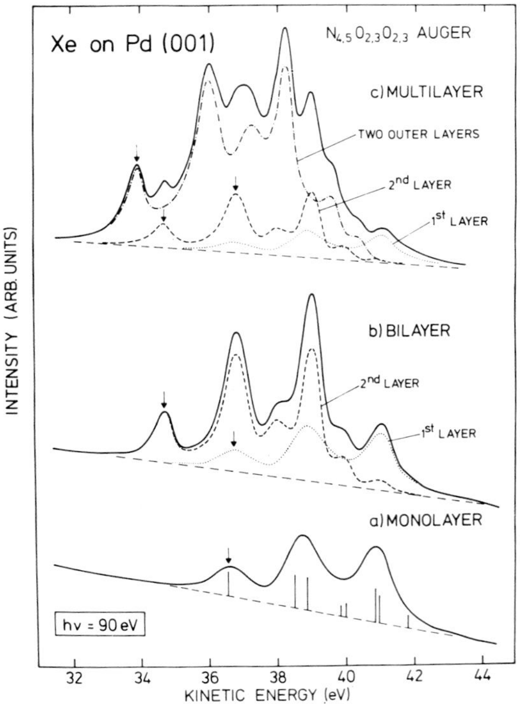

Fig. 2. Xe-NOO Auger spectra for the same three physisorbed Xe-configurations as in Fig. 1. In the monolayer spectrum the relative positions and intensities of the individual NOO components from gas-phase Xe data [9] are represented by vertical bars. The splitting of the lowest-kinetic energy NOO-Auger component is indicated by arrows.

which may be caused by inhomogeneous effects of the Pd(001) surface on the incommensurate Xe overlayer [8]. The Xe-NOO Auger spectrum of a 4-layer thick Xe film (Fig. 2c) is deconvoluted in contributions from the first Xe layer (dotted line), the second Xe layer (dashed line), and the two unresolved outer layers (dashed-dotted line).

The results of our analysis for the three Xe-layer configurations studied are summarized in Table 1. As stated above, the Auger-kinetic energy shifts are by approximately a factor of three larger than the Xe-4d binding energy shifts, while the Xe-5p shifts are about 30% smaller than the latter. The first-layer contributions of both the Xe-4d PE and the Xe-NOO Auger spectra of a Xe bilayer are further shifted to slightly larger kinetic energies as compared to the respective monolayer spectra. These additional shifts are caused by the screening of a core-hole in the first layer by the Xe-overlayer.

After the effect had been established, we have also studied distance-dependent relaxation shifts for Xe atoms physisorbed on Kr-plated Pd(001), using Xe core holes as probes for studying the potential on a metal surface as a function of distance (Fig. 3). In this way, single-layer PE and Auger spectra are observed, which are shifted according to the number of Kr layers used as dielectric spacer. The shifts are indicated by dashed lines for one Auger component of the NOO Auger spectra and for the Xe-$4d_{3/2}$ PE line. Again, the Auger-line shifts are about 3 times as large as the 4d-PE line shifts.

The observed distance-dependent shifts for localized hole excitations (Xe-4d and Xe-NOO Auger) can be described in terms of final-state hole-relaxation

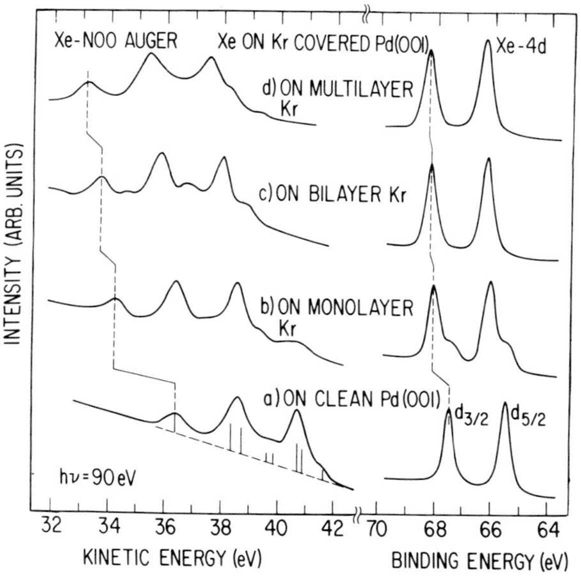

Fig. 3. Xe-NOO Auger and Xe-4d PE spectra of a submonolayer of Xe (about one-half of a monolayer) adsorbed on Kr films of various thicknesses on Pd(001).

TABLE I

Xe-5p and Xe-4d binding energy shifts ΔE_B and Xe-NOO Auger kinetic-energy shifts ΔE_K relative to free atoms for Xe overlayers on Pd(001). All energies are given in eV. The error bars for the quoted numbers are estimated to be ± 0.05 eV.

Xe configuration	Layer	ΔE_B(5p)	ΔE_B(4d)	ΔE_K
monolayer	1st	-1.71	-2.14	6.57
bilayer	1st	-1.69	-2.21	6.70
	2nd	-1.10	-1.49	4.69
multilayer	1st		-2.24	6.79
	2nd		-1.54	4.75
	outer	-0.90[a]	-1.28[b]	4.89

[a] 5 layers of Xe; [b] 4 layers of Xe.

effects for adsorbed atoms on a metal surface, using a semiclassical point-charge image-potential model. In this model, we describe the interaction of a core hole of charge q at a distance x_1 from the metal surface by an image-charge potential of the form $-q^2/[4(x_1-x_0)]$ as suggested by local-density functional theory for a jellium surface [1-3]. The metal surface is identical with the "jellium" edge, which is defined to be one-half of an interplanar distance outside of the last plane of ions. For a jellium surface with the electron density of Pd($r_s \simeq 2$) a value of x_0=0.85 Å results for the position of the image plane (center of mass of the induced surface charge distribution) outside of the metal surface [2]. In our model the dielectric effects of rare-gas underlayers on the relaxation shifts are described by a homogeneous dielectric slab with the dielectric constant ε(ε=2.25 for Xe and 1.90 for Kr) which extends from the image plane to a distance (x_1-x_d) from the metal surface. An important parameter in the model is the distance x_1 from the metal surface. Since no experimental values are yet available for the Xe-on-Pd(001) system, a value of x_1=2.38 Å was calculated as the mean distance of an incommensurate close-packed Xe monolayer, using a hard-sphere model with radii taken from the lattice constants of rare-gas solids and bulk Pd metal [10]. For thicker layers, the distances x_1 were calculated assuming close-packed ordered-layer structures.

Fig. 4 shows the results of our model analysis in a $1/x_1$ plot for both Xe multilayers and for Xe on Kr-plated Pd(001). The dotted line results from our model calculation with x_0=0.85 Å, as given by local-density functional theory (r_s=2) [2], and by fitting x_d, the boundary of the dielectric slab, to the experimental shifts of the outermost layers. This results in x_d=1.29 Å (1.18 Å) for the case of Fig. 4a (Fig. 4b), respectively. Such a fit is justified because initial-state chemical effects are expected to be negligible for outer layers. This one-parameter model describes the observed shifts rather well except for the monolayer point. It is possible that this deviation is caused by a small initial-state chemical effect for the first-layer Xe atoms directly adsorbed on Pd(001). The observed decrease of the Pd(001) workfunction with Xe coverage ($\Delta\phi \simeq -0.65$ eV) suggests such an initial-state chemical effect of the order of 0.2 eV, which would essentially eliminate the small discrepancy for the monolayer point. If we neglect such an initial-state effect and fit our model to the first-layer shifts, a value of x_0=0.68 Å results, which may therefore be considered as a lower limit for x_0 (dashed line in Fig. 4). The dashed-dotted line results from a classical image-charge potential with x_0=0, which clearly does not describe the experimental results well.

The fact that the Xe-NOO Auger shifts are 3 times as large as the Xe-4d PE shifts can be readily understood in terms of our model if full metallic screening of the initial core hole and of the final two-hole Auger state is assumed. In this case the hole-relaxation shift of a two-hole state would be 4 times as large as the one of a one-hole state, giving rise to the observed net factor of 3. Since the factor of 3 is also observed for first-layer Xe atoms (Fig. 4) our assumption of relatively small initial-state chemical effects (which should be approximately the same for Xe-4d PE and Xe-NOO Auger lines) even in this case is further supported.

We have also studied these distance-de-

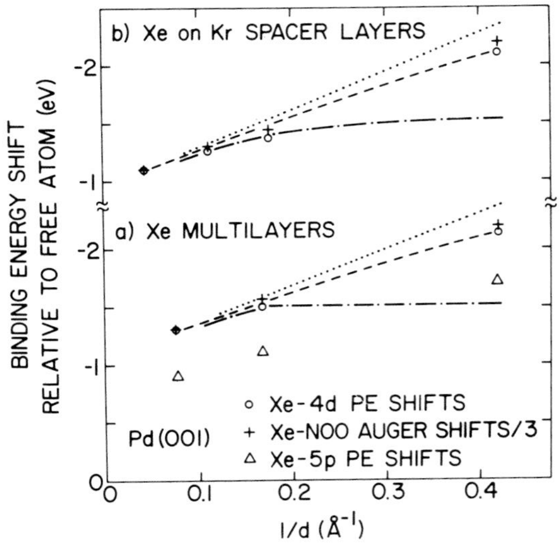

Fig. 4. Plot of Xe-4d and Xe-5p PE shifts and Xe-NOO Auger shifts (devided by a factor 3) as a function of the reciprocal distance $1/x_1$ from the Pd(001) surface: (a) for Xe multilayers and (b) for submonolayer Xe on top of Kr spacer layers (see text).

pendent shifts for Ar and Kr layers on Pd(001) as well as for Ar, Kr, and Xe layers on Pd(111) and on a widely different Al(111) substrate (more jellium like), with very consistent results [11]. Kr-4p PE spectra for four different Kr configurations on Pd(001) are shown in Fig. 5. As in the Xe case, band-dispersion effects are especially visible in the Kr-$4p_{3/2}$ line. Again, the Kr-$4p_{1/2}$ line shifts to larger effective binding energies with increasing distance from the metal surface, as indicated by the dashed line.

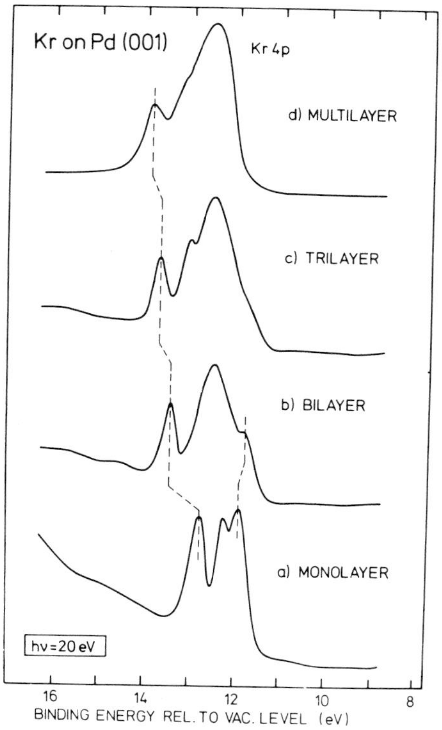

Fig. 5. Kr-4p PE spectra for (a) a monolayer, (b) a bilayer, (c) a trilayer, and (d) a 8-layer thick multilayer of Kr on Pd(001).

The observed layer-dependent shifts provide a very direct tool for distinguishing between the first few layers of an adsorbed rare-gas multilayer. This labeling feature allows several applications of these shifts to e.g. the study of layer-dependent properties [12] as well as to the study of interlayer diffusion processes in an adsorbed multilayer [11].

As a demonstration, we will discuss in the following the thermally activated inversion of a physisorbed Kr/Xe bilayer sandwich on Pd(001). The observation is clearly demonstrated by the data of Fig. 6. First, a monolayer of Kr is prepared on Pd(001) as described above (Fig. 6a). Then, at a substrate temperature of 50 K, 0.8 of a monolayer of Xe is deposited on top of the Kr monolayer. This results in a Kr/Xe bilayer sandwich (Fig. 6b), where most of the Kr atoms reside in the first layer on Pd(001), while most of the Xe atoms are staying in the second layer. The spectra of Fig. 6b, however, show quantitatively, that about 20% of the deposited Xe atoms reach right after deposition the Pd(001) surface, pushing an equivalent number of Kr atoms from the first into the second layer. After the sandwich was annealed at 58 K for 60 seconds, the sequence of rare-gas layers is inverted (Fig. 6c). Now, almost all of the Xe atoms are residing in the first layer, with an equivalent number of Kr atoms pushed into the second-layer position. The driving force for this inver-

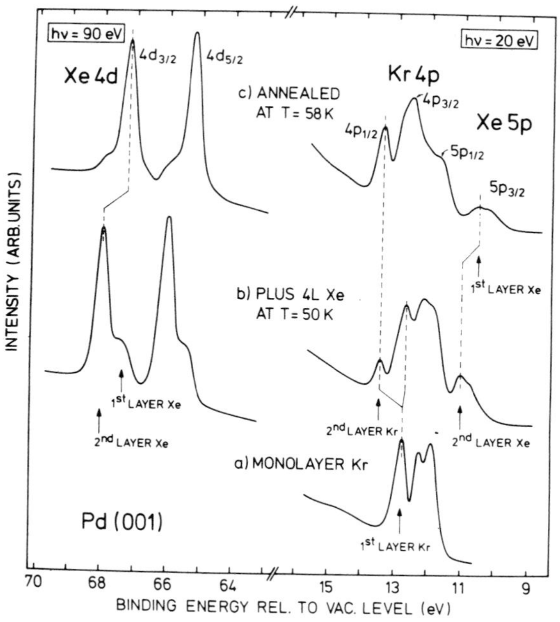

Fig. 6. Thermally activated inversion of a Kr/Xe bilayer sandwich on Pd(001). Both the Xe-4d and the Xe-5p/Kr-4p spectra are shown.

sion is the difference in the heat of adsorption for Xe and Kr, respectively, on Pd(001).

We have studied this thermally activated layer inversion more carefully by step-annealing the Pd(001) sample at increasing temperatures for periods of 60 sec. each. The results are shown in Fig. 7, by observing both the Xe-4d and the Xe-5p/Kr-4p PE spectra after each annealing step and after the sample had cooled down to 50 K. As can be seen especially clearly from an inspection of the Xe-4d spectra, the fraction of Xe atoms residing in the first layer grows with increasing annealing temperature. Accordingly, an increasing fraction of Kr atoms is pushed up into the second layer, as

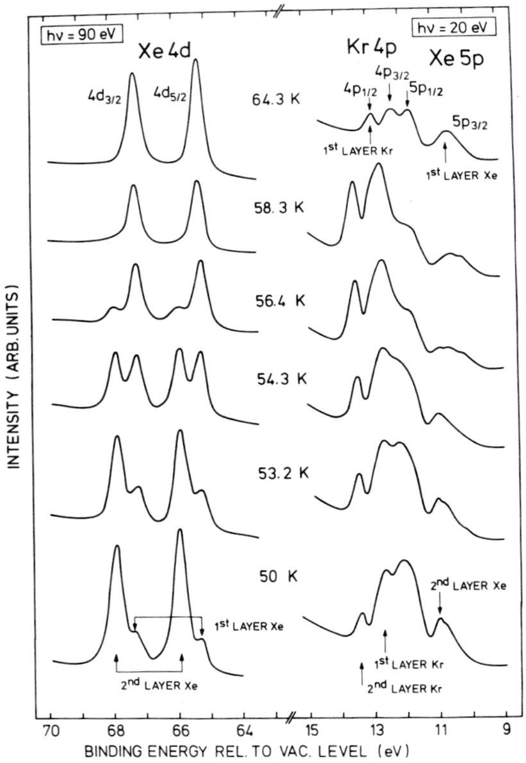

Fig. 7. Xe-4d and Xe-5p/Kr-4p PE spectra taken from a Kr/Xe bilayer sandwich on Pd(001) at T=50 K after the indicated annealing steps (60 sec. at each temperature). The error bars for the temperature differences are ± 0.1 K. The absolute temperatures, however, are only accurate to ± 5 K.

shown by the Kr-4p spectra. After the annealing step at 58.3 K, the inversion process has been completed. Furthermore, all second-layer Kr atoms have desorbed after the annealing step at 64.3 K.

We can derive an activation energy E_a for this thermally activated process from the data, if we assume that the rate of inversion is proportional to the fraction of the Kr atoms in the first layer times $\exp(-E_a/k_B \cdot T)$. The rate of inversion as a function of temperature can be obtained directly from the ratio of the intensity of Xe-4d lines originating from the first layer to those from the second layer. To this purpose, a correction for attenuation of the PE signal from the first layer by Xe and Kr atoms in the overlayer has to be applied. A mean attenuation factor was calculated for each inversion step from the measured attenuation factors for the Xe-4d lines at a photon energy of 90 eV for a full overlayer of Xe as well as for a full overlayer of Kr. This analysis results in a value of E_a=0.18±0.04 eV for the activation energy. A more detailed report of the analysis will be presented elsewhere.

The authors acknowledge constant technical support by J.J. Donelon, A. Marx, and the staff of the Synchrotron Radiation Center of the University of Wisconsin-Madison. The work was supported in part by the Air-Force Office of Scientific Research.

REFERENCES

1. J. A. Appelbaum and D. R. Hamann, Phys. Rev. B6, 1122 (1972).
2. N. D. Lang and W. Kohn, Phys. Rev. B7, 3541 (1973).
3. N. D. Lang and A. R. Williams, Phys. Rev. B16, 2408 (1977).
4. N. D. Lang and A. R. Williams, Phys. Rev. B20, 1369 (1979).
5. See e.g. Interactions on Metal Surfaces, R. Gomer Ed. (Springer-Verlag, 1975).
6. G. Kaindl, T.-C. Chiang, D.E. Eastman, and F.J. Himpsel, submitted to Phys. Rev. Lett. (1980).
7. K. Horn, M. Scheffler, and A. M. Bradshaw, Phys. Rev. Lett. 41, 822 (1978).
8. P. W. Palmberg, Surf. Sci. 25, 598 (1971).

9. L.O. Werme, T. Bergmark, and K. Siegbahn, Physica Scripta 6, 141 (1972).
10. This procedure reproduces within 0.05 A the distance derived by LEED analysis for the equivalent case of Xe on Ag(111): N. Stoner, M.A. van Hove, S.Y. Tong, and M.B. Webb, Phys. Rev. Lett. 40, 243 (1978).
11. G. Kaindl, T.-C. Chiang, D.E. Eastman, and F. J. Himpsel, unpublished results (1980).
12. T.-C. Chiang, G. Kaindl, and D.E. Eastman, submitted to Sol.-State Comm. (1980).

DISCUSSION

S. Fain:

1) Do you have information on the Auger peak shift at submonolayer coverages due to lateral interactions?

2) Do you have results for Xe on graphite? Such results might be very useful for understanding one detail of the extensive Auger isotherm measurements of Suzanne for Xe on graphite: the derivative Auger peak used for monolayer condensation measurements actually dropped for second layer condensation, perhaps due to a superposition of shifted Auger peaks similar to what you have shown for palladium.

J. Suzanne, J. P. Coulomb, and M. Bienfait, Surface Science 44, 141 (1974) and Thesis, L'Universite d'Aix-Marseilles II (1974, unpublished).

3) It would be very useful for the calculations of L. Bruch (Univ. Wisconsin) for you to measure the work function change on adsorption for rare gases on graphite.

G. Kaindl:

1) For Xe/Pd(001) the Auger peaks shift only very slightly with increasing submonolayer coverage to higher kinetic energies. The difference between 0.5 monolayer and a full monolayer of Xe is of the order of 0.2 eV.

2) We have not yet studied the Xe/Graphite system, but it is one of our next systems to look at. I certainly expect layer-dependent Auger-peak shifts in this case.

3) We have done careful work function measurements as a function of rare-gas adsorptions on Al(111) (G. Kaindl, F. C. Chiang, D. E. Eastman, to be published). Similar measurements are planned for the rare-gas/Graphite systems.

U. Landman:

You have two unknown distance parameters, X_0 and X_1. For X_1 you have adopted a "sum of hard-sphere radii" prescription and then chose X_0 to fit the data for shifts vs. $\frac{1}{X_1}$. Certain theoretical work and atom scattering data indicate that the above prescription for X_1 might not be accurate, i.e., distances X_1 larger than the sum of radii may be more accurate. How would a different assignment of values for X_1 influence the values of X_0 determined by best-fit to the data?

G. Kaindl:

We believe that the procedure used to calculate X_1 is accurate to about 0.2 Å in the present case. At least, in the equivalent Xe/Ag(111) case it reproduces the X_1 result of a LEED analysis within 0.05 Å (see Ref. 10). In the Xe/Pd(001) case, recent LEED measurements seem to indicate a slightly larger value for X_1 than used here (B. Webb, University of Wisconsin-Madison, private communication). An increase of X_1 by 0.1 Å, for example, would lead to a decrease in the X_0 value by approximately the same amount, in our data-fit procedure.

Published 1980 by Elsevier North Holland, Inc.
Sinha, ed. Ordering in Two Dimensions

MAGNETIC STATE AT THE SURFACE OF A FERROMAGNET BY POLARIZED ELECTRON DIFFRACTION*

G. P. FELCHER and S. D. BADER
Argonne National Laboratory, Argonne, Illinois 60439

and

R. J. CELOTTA, D. T. PIERCE, and G. C. WANG
National Bureau of Standards, Washington, D.C. 20234

The results of the first experiment by polarized electron diffraction (PLEED) on a (110) surface of magnetized nickel are presented and discussed. The conclusion is that nickel is magnetically active at the surface; and further, that this new technique might become a powerful analytical tool for the study of surface magnetism.

INTRODUCTION

The problem of the magnetization at the surface of a ferromagnet is of great theoretical interest. The presence of the boundary with the vacuum alters deeply the magnetic properties of the bulk, in a way that is still to be understood. For instance, the magnetic moment for nickel at the surface is considered to be different from that of the bulk, although a general consensus is emerging, negating the existence of magnetically "dead layers" at the surface of nickel. Even if the surface is magnetically active at zero Kelvin, its temperature dependence is expected to be significantly different from that of the bulk. In the low-temperature limit the presence of surface spin waves should give a linear temperature dependence of the magnetization, based on a two-dimensional modification of the famous Bloch $T^{3/2}$ law for the bulk. Close to the ordering temperature it should be possible to extract a critical exponent β_s that characterizes the surface magnetization. Existing theoretical treatments suggest that $\beta_s > \beta_{bulk}$, although surface exponents specifically for the case of present concern, that of itinerant ferromagnets such as Ni, have not been treated adequately to date. [1]

The magnetic properties at the surface of a ferromagnet are difficult to detect experimentally, since few techniques are sufficiently surface- and spin-sensitive to yield direct information. Perhaps the most novel of the new techniques is polarized low energy electron diffraction (PLEED) in which the polarization dependence of the diffracted intensity is related to the magnetization of the surface. The present paper is meant to give an outline of the current status of this technique and of the experiments so far collected.

In order to use the new technique of polarized LEED to probe the top atomic layers of ferromagnets and extract out physically relevant quantities, it is first necessary to gain a firm understanding of the detailed process by which the magnetization contributes to the spin-dependent scattering. We begin by considering strong elastic scattering due to Coulomb and exchange potentials around the atoms close to the surface and, further, we derive expressions (valid within the single-scattering approximation) for the diffracted intensities. These we compare with the first PLEED experiments on Ni(110). Finally we progress to a brief discussion of the more complete theory of multiple scattering within the top layers of the solid, and present the outlook that the interplay between experiment and calculation should indeed furnish a detailed knowledge of the relevant magnetic quantities at the surface.

The interaction between an incident electron of low energy (∿100 eV) and a localized potential (as that of an atom close to the surface) can be described by the Hamiltonian [2]

$$H_{int} = V(\vec{r} - \vec{r}_A) + J(\vec{r}-\vec{r}_A)\, \vec{s} \cdot \vec{S}_A + H_{S-O}, \quad (1)$$

where V and J are the Coulomb and the exchange potentials respectively, $\vec{r}$ and $\vec{r}_A$ are the positions of the electron and the scattering atom, and $\vec{s}$ and $\vec{S}_A$ are their spins.

The second term in Eq. (1) is the only portion of the interaction that depends explicitly on the spin of the incident electron and of the atom. The H_{S-O} term, due to the spin-orbit interaction is not of primary interest here and will be discussed later. The scattering of the incident electron from such a potential can be described by a scattering factor which depends on the scattering angle Θ and energy E of the incoming electron. If the quantization axis of $\vec{s}$ is alongside that of the atomic spin, the scattering factor f± can be written

$$f^{\pm}(\Theta,E) = V(\Theta,E) \pm J(\Theta,E) \cdot M, \qquad (2)$$

where V and J are the transforms of the real-space quantities of Eq. (1) - usually expressed in terms of partial wave phase shifts, and M is the atomic magnetization. The plus superscript refers to $\vec{s}$ parallel to $\vec{S}$, and minus to antiparallel alignment.

In conventional LEED experiments the intensities of the beams back-diffracted from the outermost layers of a crystal are measured. The electron beam is scattered by the two-dimensional net formed by the first layer of atoms, such that the parallel component $\vec{K}_{||}$ of the incident momentum $\vec{K}$ is conserved to within a reciprocal lattice vector $\vec{G}$. Scattering occurs also by the subsequent layers of the crystal down to the limited penetration of the beam (of the order of two or three layers); as a result the total back-diffracted beam intensities are modulated as a function of the incoming electron energy, and in particular they tend to be reinforced when the Bragg conditions for the three-dimensional lattice are satisfied. In this simplified picture the intensity of the back reflection of a ferromagnet is

$$I^{\pm}_{HK} \propto | \Sigma_{\ell}\, F^{\pm}_{\ell} \cdot \phi_{\ell} |^{2} \qquad (3)$$

where F_{ℓ} is the sum of the atomic scattering factor of Eq. (2) over all atoms of the ℓth layer, and ϕ_{ℓ} is the phase of that layer. The damping of the electron beam in its penetration of the crystal is included in Φ_{ℓ}. When the diffraction is reinforced, as when the conditions for the three-dimensional Bragg reflections are satisfied, the phase factors are unity - except for the damping factors. In these circumstances, the diffracted intensities become

$$I^{\pm}_{HKL} \propto |\bar{V}(\theta,E) + \bar{J}(\theta,E) \cdot \bar{M}|^{2} \qquad (4)$$

where the bars indicate layer-weighted averages of the quantities.

Equation (4) allows us to give a simple microscopic interpretation of the quantities that are obtained experimentally, when a polarized electron beam is scattered from a magnetized target. These quantities are the difference function:

$$\Delta I = I^{+} - I^{-} \propto - 2\bar{V}\ (\bar{J}\cdot\bar{M}) \qquad (5)$$

and the normalized difference function:

$$S = \frac{I^{+} - I^{-}}{I^{+} + I^{-}} \propto - \frac{\bar{J}}{\bar{V}} \cdot \bar{M} \qquad (6)$$

Note that the function S has a wider range of validity than the straight diffracted intensity. In the single scattering approximation, if all the atoms have the same scattering amplitude, the phases ϕ_{ℓ} are eliminated in the ratio defining the S-function. Hence this is valid even for energies not corresponding to the Bragg reflections. By the same token, the only temperature dependence of S is from the magnetization M, since Debye-Waller factors in the numerator and the denominator tend to cancel.

THE EXPERIMENT

The choice of nickel for the experiment was dictated by the ease in which magnetic saturation can be reached, as well as the amount of information available on the band structure and surface properties of this d-band metal. The experimental configuration of the single crystal sample is presented in Fig. 1. In the vacuum chamber of a LEED spectrometer the crystal is made an integral part of a magnetic circuit, forming the yoke of a horseshoe-shaped electromagnet. The [111] crystal axis, corresponding to the easy magnetization direction, is oriented along the external magnetic field. As determined by the magneto-optic Kerr effect, the crystal can be magnetized to saturation in either direction with only a minimum leakage field in the electron path. Heating is accomplished by passing a current through the thin crystal, which is spot welded to tantalum posts; the temperature can be monitored with a calibrated thermocouple attached to one of the posts.

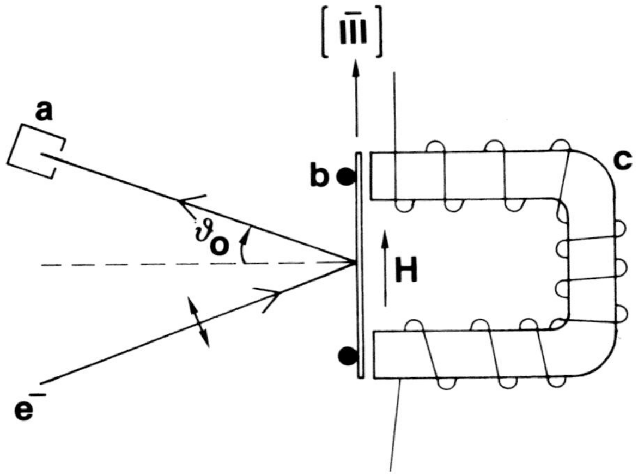

Fig. 1 Experimental arrangement for the PLEED experiment on Ni(110). e⁻ : incoming electron beam, polarized in the diffraction plane. a: Faraday cup detector b: Nickel crystal, anchored to tantalum posts c: electromagnet

The polarized electron beam is photoemitted by a GaAs source, treated with Cs and oxygen to produce a negative electron affinity surface [3]. Electron optics transport, collimate and accelerate the beam from the source chamber to a conventional LEED chamber, where the experiments are performed. The circularly polarized radiation incident on the GaAs source is modulated by a rotating quarter-wave plate to produce spin reversals in the polarized electron beam at a 30 Hz rate. It is important to note that the electron beam current incident on the target crystal is constant; any modulation of the scattered beam current is due to spin-dependent effects during scattering. Operationally, the difference function ΔI is measured by means of a Faraday cup collector and a.c. lock-in amplification techniques, while the denominator of the normalized difference function is the d.c. signal measured with an electrometer, as in conventional LEED experiments. The direction of polarization is normal to the incident beam, and almost parallel to the direction of magnetization of the crystal for the chosen angle of incidence $\theta = 12^{\circ}$.

The first measurements made [4] were of S and the total intensity corresponding to specular diffraction. Data were taken as a function of energy over the range of ∿ 20-150 eV. The S function for a magnetized target exhibited maxima with both positive and negative values less than 0.05 over this range of energies. The dependence of the features upon the magnetic ordering of the substrate was tested by reversing the magnetic field direction. A typical hysteresis curve is presented in Fig. 2 for an electron energy of 125 eV. The saturating field is similar to that obtained for the bulk by the magneto-optic measurements. The result contained in Fig. 2 clearly indicates the presence of exchange scattering in the surface region. (The electron mean free path is ∿5Å at this energy).

It is of interest to briefly contrast how polarization effects due to the spin-orbit interaction can be discriminated against. Its spin dependence [5] arises due to the coupling of the incident electron's spin to its own orbital angular momentum as it scatters from the atom core; hence, it is maximal when the electron spin is polarized normal to the scattering plane. In the present geometry the polarization vector of the incident electron is kept within the scattering plane. In the single-scattering approximation, the experimental polarization effect, due to this spin-orbit scattering, would be identically zero. Even if multiple scattering events are present, since the spin-orbit interaction depends on the spin state of the incident electron and not on the magnetic state of the atom, it would not be affected by externally applied magnetic fields. Hence, in the hysteresis curve shown in Fig. 2 H_{SO} could at most provide a minor zero offset of the polarization.

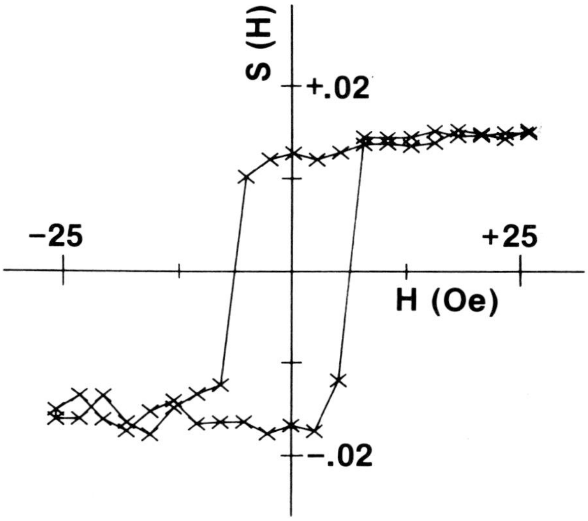

Fig. 2 Hysteresis loop for a specularly reflected beam of electrons at 125 eV. The coercive field is comparable to that of bulk nickel, as determined by magneto-optic measurements.

Hysteresis loops were obtained also for electron energies of 20 eV and 90 eV, corresponding approximately to the diffraction maxima of the (220) and (440) reflections. The results are presented in Fig. 3, together with a theoretical estimate derived along the lines discussed in the preceding section. The spin-polarized potential given by Wakoh [6] for nickel in the bulk was used as a local potential. The continuous line was obtained from scattering amplitudes including the first ten partial waves. The excellent agreement with the experimental data has to be interpreted only in the sense that the basic assumptions made here are qualitatively correct; hence metallic nickel at the (110) surface is ferromagnetically ordered.

The temperature dependence of S at fixed energy (125 eV, at $\theta = 12$) was measured in the temperature interval from 0.5 to 0.8 T_{Curie}. In this temperature interval the polarization signal decreased roughly linearly by a factor of three from $S(0.5\ T_c) = 0.015$ to $S(0.8\ T_c) = 0.005$. This is in sharp contrast to the functional form of the temperature dependence of the bulk magnetization in this same interval, where a decrease of only ∿30% occurs. In this first experiment we were not working in a temperature region that can rigorously be considered low or high, hence formal contact with existing theories must await a redesign of the sample environment which is now under way. (See Fig. 4).

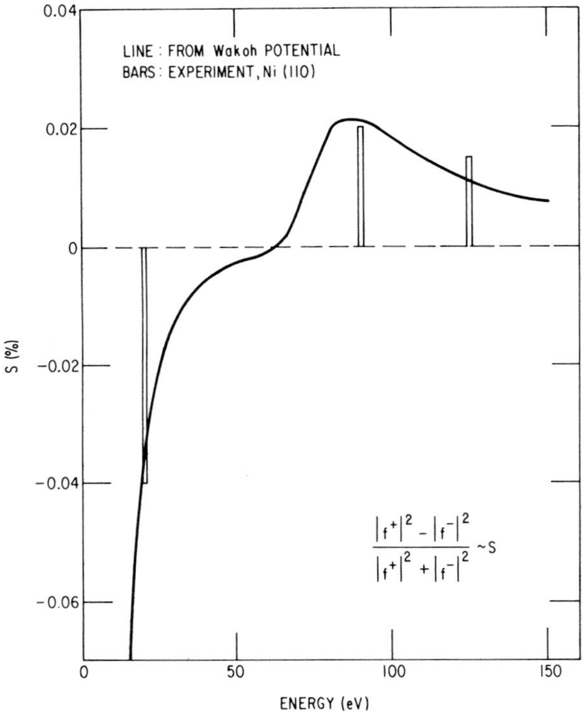

Fig. 3 Calculated and measured S-function for Ni(110). data were obtained at room temperation.

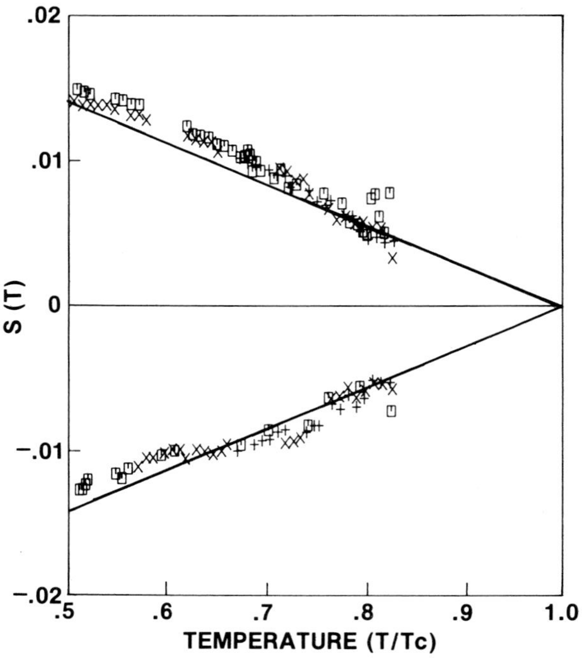

Fig. 4 Temperature dependence of S(T) for Ni (110), as measured on the specular reflection for electrons of 125 eV. The upper and lower sets of points correspond to the extremes of the hysteresis loop. The straight lines are for guidance.

DISCUSSION

The ultimate goal of a PLEED experiment is the determination of the magnetic moment of a ferromagnet, layer by layer starting from the surface. This moment is expected to be layer dependent, as for example indicated by the result of a self-consistent calculation on the (100) surface of nickel [6]. The question is, can such a goal be reached, and can the PLEED technique become a quantitative tool of analysis of the magnetism at the surface?

In the presentation above, multiple-scattering effects were not considered explicitly. However, it has been amply demonstrated that dynamical scattering theory, in general, is necessary to rigorously describe unpolarized LEED intensities. In the input to such LEED computations inelastic scattering is as important as elastic scattering in limiting the mean free path of low energy electrons within the solid. Conventionally, the electrons travelling in the crystal from the vacuum are considered as in an optical potential, whose imaginary part represents the inelastic scattering losses. In these loss processes an important role is played by the formation of plasmons and electron-hole (e-h) pairs. [8]

Feder [9] has suggested that of the principal loss mechanisms e-h excitations may produce significant polarization effects. In Feder's ansatz the imaginary potential V_{eh} for one spin state is inversely proportional to the density of states of the electrons in that spin state. Feder has performed a full multiple-scattering calculation (including the effect of this interaction) of the S-function for Fe, and very recently Harold Davis [10] did likewise for Ni(110). It was found in both cases that the resultant spin dependence rivaled the importance of that due to the direct effect of the exchange interaction on the elastic scattering.

In these circumstances, the retrieval of detailed physical information (as the layer-by-layer magnetization) from a comparison of experiments and calculations seems an arduous task. However, hope stems from the fact that the exchange terms are much weaker than the Coulomb interaction. (In our experiment the S-function never exceeded 5%). Of great importance in this regard are the multiple-scattering calculations of Davis on nickel. He showed that the polarization effects due on one hand to the direct exchange scattering and on the other hand to the electron-hole losses are completely additive. Hence, the exchange terms give rise to single scattering, while the Coulomb interaction gives rise to multiple scattering effects. This observation raises the possibility that a renormalization scheme can be devised that would allow a much more direct calculation of the S-function from a given set of layered magnetic moments, thus enabling the extraction of truly quantitative magnetic surface properties from PLEED experiments.

ACKNOWLEDGMENTS

We would like to thank D. Koelling for giving us assistance in generating the phase shifts of nickel, and H. Davis for enlightening discussions and for providing us with the results of his calculations prior to publication.

BIBLIOGRAPHY

[1] Symposium on Surfaces and Magnetism: Recent Developments and Future Opportunities, A. J. Freeman, Chairman. In: Proceedings of the International Conference on Magnetism, München 1979, published by Journal of Magnetism and Magnetic Materials, 15 - 18, 1070 (1980)

[2] J. B. Pendry, Low Energy Electron Diffraction, Academic Press, London, (1974)

[3] D. T. Pierce and Felix Meier, Physical Review B, 13, 5484 (1976)

[4] R. J. Celotta, D. T. Pierce, G.-C. Wang, S. D. Bader and G. P. Felcher, Phys. Rev. Letters 43, 728 (1979)

[5] J. Kessler, Polarized Electrons, Springer-Verlag, Berlin (1976)

[6] S. Wakoh, J. Phys. Soc. Jap. 20, 1984 (1965)

[7] C. S. Wang and A. J. Freeman, J. Magnetism and Magnetic Mat., 15 - 18, 869 (1980). See also O. Jepsen, J. Madsen and O. K. Anderson, J. Magnetism and Magnetic Mat., 15 - 18, 867 (1980)

[8] Leonard Kleinman, Phys. Rev. B, 3, 2982 (1971)

[9] R. Feder, Solid State Communications, 31 821 (1979)

[10] Harold Davis, private communication

DISCUSSION

M. Pomerantz:

What is the source of the J used in your calculation?

Are theories of J accurate, or can experimental values be used?

G. Felcher:

Calculations were made by Harold Davis using both Wakoh's and Moruzzi's potentials. The results were virtually identical. There is not experimental proof yet of how accurate these potentials are.

M. Wortis:

The crudely linear behavior of the surface magnetization which you measure is fully consistent with what would be expected from what is known theoretically about surface magnetization: We expect $M_{bulk} \sim t^{\beta}$, $M_{surface} \sim t^{\beta_s}$ with $\beta_s = \beta + \nu$, ie $\beta_s > \beta$, so the surface magnetization falls off near t_c much more slowly than the bulk magnetization. Would you care to comment on this?

G. Felcher:

I would rather wait until the next round of experiments - with better temperature control close to T_c - is completed.

U. Landman:

In order to get a better estimate about the adequacy of the kinematic, single-scattering approximation versus a full multiple-scattering description, it may be useful to perform both experiments and calculations (multiple scattering to compare with kinematic) for various angles of incidence. Since the degree of multiple-scattering depends upon the angle of incidence, you may be able to find a range of angles in which the kinematical approximation is adequate. Comparison of interlayer and intralayer contributions to the multiple-scattering may also be useful.

G. Felcher:

Thank you for your comment. We shall try. However, Lagally's technique of averaging cannot be fully exploited here, for the insertion of the magnetic circuitry limits the experimental range.

Sinha, ed. Ordering in Two Dimensions

OBSERVATIONS OF ISLAND FORMATION AND DISSOLUTION IN OXYGEN ON W(110) BY LOW-ENERGY ELECTRON DIFFRACTION

M.G. LAGALLY*, T.-M. LU, and G.-C. WANG**

Department of Metallurgical and Mineral Engineering and Materials Science Center
University of Wisconsin—Madison
Madison, Wisconsin 53706

Some thermodynamic properties related to the coexistence region of ordered phase and lattice vapor in chemisorbed overlayers at low coverages are discussed, with examples for O/W(110). Using low-energy electron diffraction, parts of the phase diagram are established. A phase boundary is calculated for the coexistence region. Finite-size effects due to substrate heterogeneity are described. A comparison of measured island sizes of the ordered p(2×1) overlayer with the calculated phase boundary gives an indication of a boundary energy contribution.

1. INTRODUCTION

Any material in which the net interactions between the particles are attractive can exhibit two-phase coexistence between a dense or condensed phase and a dilute or vapor phase. The analogue for multicomponent systems occurs when the net interactions between like particles are more attractive than between differing particles. For a binary system of A and B atoms, for example, this leads to phase coexistence between an α phase rich in A and a β phase rich in B. The boundary in a temperature-composition phase diagram separating the two-phase coexistence region from the one-phase α or β regions give the equilibrium compositions of the two phases as a function of temperature. Equivalently, the phase boundary represents the maximum solubility of one component in the other as a function of temperature. The amounts of each phase present for any average composition in the two-phase region are given by the lever rule [1], i.e., once the phase boundary is known, both the amounts and the compositions of the phases in equilibrium at any temperature can be readily calculated from the phase boundaries.

Because the phase boundary is just the locus of compositions and temperatures that define the maximum solubility of one component in the other, it is possible to extract from it the partial molal enthalpy of mixing of one component into the other and thus, in principle, the interaction energies. If the system in question can be modeled adequately with a simple interaction potential, as, for example, in the quasichemical model [1], it is possible to determine the interaction energies quantitatively.

An analogous situation exists in two-dimensional systems. If there is a net attractive interaction energy between atoms deposited on a surface, these atoms will attempt to cluster into regions of "condensed" phase and a dilute phase that is essentially "empty" substrate surface. Numerous examples exist of such two-phase regions in a variety of two-dimensional systems [2]. It is particularly simple to discuss two-phase coexistence in terms of a lattice gas in which the number of particles is fixed. Such systems are realized by chemisorption systems that are "closed," i.e., not in equilibrium with either the three-dimensional vapor or the solid. This occurs when the heats of adsorption and of solution into the bulk for the adsorbate atoms are high relative to energetic barriers to diffusion or ordering in the surface. The most thoroughly studied chemisorbed layers that can be considered "closed" are O/W(110) [3], O/Ni(111) [4], and H/W(100) [5], although phase diagrams or partial phase diagrams for a number of other chemisorption systems have appeared in the literature [6]. Studies of two-phase coexistence in closed systems should be especially important in understanding epitaxial growth [7].

The ordering in such systems can be simply illustrated by means of Fig. 1. At any coverage less than saturation coverage for a particular overlayer structure (represented in the figure by $\theta/\theta_s = 1$), an ordered phase (lattice solid) and a disordered phase (lattice vapor) must coexist below some finite temperature if a net attractive adatom interaction exists. The densities of the two phases at any

*H.I. Romnes Fellow.

**Present address: Oak Ridge National Laboratory, Oak Ridge, TN.

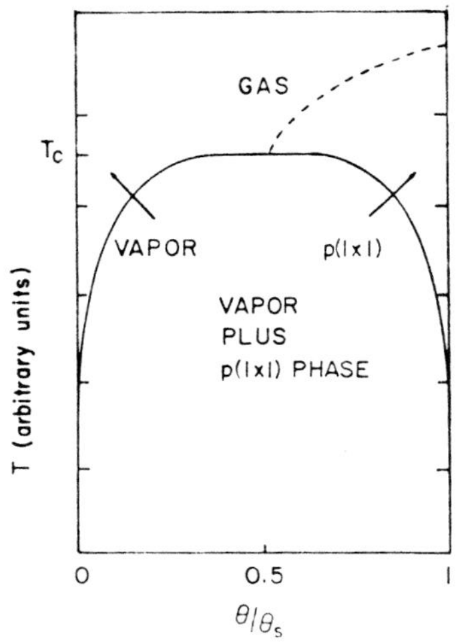

Fig. 1. Schematic phase diagram of the ordering of a two-dimensional system with net attractive interactions, showing a two-phase region. $\theta/\theta_s = 1$ corresponds to saturation coverage for a given overlayer structure. Because the ordered phase may itself undergo a phase transition to a disordered phase, there is generally another branch in the phase diagram, shown by the dashed line, that reaches a maximum temperature at $\theta/\theta_s = 1$.

temperature are given by the phase boundaries. The low-coverage phase boundary represents (in analogy with the three-dimensional binary alloy) the solubility of adatoms in the two-dimensional lattice-vapor phase, or equivalently, the phase boundary is proportional to the "lateral vapor pressure" of the ordered lattice-solid phase at any temperature. Similarly, the high-coverage phase boundary represents the solubility of vacancies in the lattice-solid phase. At coverages or temperatures outside the two-phase region, only one phase exists, respectively, a lattice vapor at low coverages or a lattice solid covering the whole surface at high coverages, but not completely dense at $\theta/\theta_s < 1$. Above the critical temperature, T_c, the thermal energy exceeds the interaction energies and only a disordered phase is stable. Because the ordered phase, even for saturation coverage, may itself undergo a phase transition, such as melting, order-disorder, or order-order, there is in general an additional branch in the phase diagram connecting the (multi-) critical point with a point on the saturation coverage isostere that represents the disordering temperature for the dense lattice solid. This is shown as the dashed curve in Fig. 1.

The partial phase diagram of Fig. 1 should be part of the total phase diagram of any adsorbed layer in which the adatoms have a net attractive interaction. Details of the diagram, such as its shape, position, and, in fact, the possible presence of additional phase regions, depend on the nature of the interactions and, more generally, on any contributions to the free energy of the ordered phase. Thus, in particular, defects and finite-size effects may change the phase boundaries in significant ways [6,8].

The purpose of this paper is to discuss phenomena in the two-phase region at low coverages for a particular system, a chemisorbed partial monolayer of O on W(110). This system is well described by a lattice gas model. In the next section, we discuss the types of experiments, using a particular technique, low-energy electron diffraction (LEED), that can be done to determine the phase diagram and to study overlayer island size distributions and substrate surface defects. In Sec. 3, we present a partial phase diagram for O/W(110), and calculate an effective cohesive energy for the ordered structure observed at low coverages. In Sec. 4, we consider the effect of a lack of long-range order in the substrate on overlayer ordering, extract sizes for the ordered regions that form, and discuss the equilibrium size of the islands in terms of a possible boundary energy contribution to their stability. In Sec. 5, we discuss the possible effects of strong adsorption sites and other defects on the ordering, as well as the uncertainties in this simple treatment.

2. LOW-ENERGY ELECTRON DIFFRACTION

The most straightforward and readily interpretable measurements of changes in the structural order of a two-dimensional system are made, as for three dimensions, by a diffraction technique. Low-energy electron diffraction (LEED) is ideal for this purpose because the large inelastic and elastic cross sections for scattering of electrons of 10 to 1000 eV limit the mean free path of these electrons to ~10Å. General aspects of LEED, and especially its application to the determination of equilibrium positions of atoms on a surface, have been amply reviewed [9,10]. Its use in the study of the thermodynamics and kinetics of ordering on surfaces is much simpler. Although qualitative measurements of this nature have been made since the discovery of electron diffraction [11], the number of quantitative studies is still quite limited.

In LEED, a monoenergetic beam of electrons impinges on a crystal surface. The resulting diffraction pattern is simply the Fourier transform of the autocorrelation function of the surface structure. If an adsorbed layer has a unit mesh larger than the substrate, extra (superlattice) reflections appear in the diffraction pattern. An analysis of the intensity distribution in these superlattice reflections as a function of adsorbate coverage, temperature, and time gives the thermodynamics and kinetics of ordering of the superlattice. Two kinds of measurement are important. These can be explained with reference to Fig. 2. In the simplest case of true order-

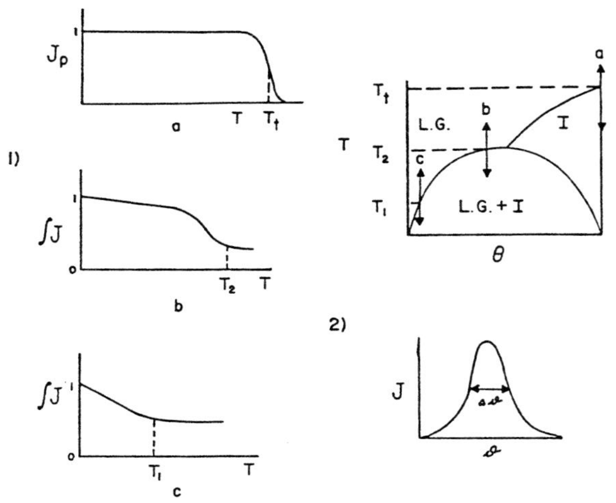

Fig. 2. LEED measurements in overlayer phase transition studies. 1) Intensity vs. temperature: a) the decay with temperature of the peak intensity of superlattice reflections for order-disorder transitions is proportional to the order parameter and critical exponents can be extracted from this decay; b) and c) the integrated intensity in first-order transitions out of a coexistence region measures the decrease in the quantity of ordered phase, and mirrors the shape of the phase boundary. (2) Intensity vs. angle: the angular distribution of intensity in superlattice reflections is related to the size of ordered absorbate islands on the surface.

disorder transitions the intensity decay with rise in temperature, J vs. T, at a fixed coverage gives the change of the order parameter with temperature. Phase changes of this nature occur, for example, at or near saturation coverage for a particular overlayer structure, where the transition is from a one-phase region to another one-phase region. LEED measurements of critical exponents for such a case, a saturated p(2×2) layer of O on Ni(111), are reported elsewhere in this volume [4]. A discussion of the theoretical interpretation of the LEED intensity of any superlattice beam for this type of transition also appears in this volume [12]. The shape of the J vs. T curve and the position of the inflection point will be a function of coverage. For example, at somewhat less than saturation coverage the overlayer may still exist as a single phase covering the whole surface, as discussed earlier, but it will contain vacancies. The structure should therefore be less stable and should be easier to disorder. Hence the transition temperature will drop. The shape of the curve may also change because of changes in the critical exponents, which are unknown for such structures.

For transitions from a two-phase region to a one-phase region, a measurement of the temperature decay of the LEED intensity has an entirely different meaning. Since the diffracted superlattice beam intensity reflects only the ordered phase, a measurement of the integrated intensity in a diffracted beam vs. temperature is equivalent to a disappearing-phase measurement; the integrated intensity decreases as the ordered phase "dissolves" or "evaporates" into the lattice gas. Thus, the decay of the integrated intensity with temperature mirrors the shape of the phase boundary; in coverage and temperature regions where the solubility increases approximately linearly with temperature, the integrated intensity decays linearly, and where the solubility changes rapidly with temperature, the integrated intensity will also. In any case, the position of the phase boundary in the T-θ plane is determined from the temperature where the integrated intensity goes to zero, because at that T, all of the ordered phase has evaporated. In actuality, the interpretation of the integrated intensity versus temperature is not quite this simple, because the density of the ordered phase also changes in accord with the high-coverage phase boundary.

Because the size of the ordered regions shrinks as more phase evaporates into the 2-D lattice gas, the diffraction peaks broaden with rising temperature. This implies that the peak intensity decays generally in a different manner than the integrated intensity, although both should be zero at the same temperature. Thus, although the peak intensity can also be used to determine the phase boundary, the shape of the peak intensity decay with temperature does not in general mirror the shape of the phase boundary. The peak intensity will fall more rapidly than the integrated intensity.

The second major type of measurement, the intensity as a function of angle, follows directly from the above. The angular distribution of intensity mirrors the size of the coherently scattering domains. The smaller these domains are, the broader will be the diffracted beam [13]. In fact, the behavior of the angular distribution as diffraction parameters are changed can be used to separate different limitations on the size of coherently scattering domains [14-18]. Thus, for example, finite-size islands sitting on a flat coherent substrate can be distinguished from antiphase domains in a dense phase [16,17]. Substrate defects such as steps [18], mosaic structure [14,15], and misorientation [15] can also be distinguished.

The ability to interpret diffraction spots in terms of island sizes is important here for two reasons. One, as the amount of ordered phase decreases, the size of the ordered region will shrink. Conversely, if the temperature is lowered from the vapor phase into the two-phase region, small islands will nucleate and grow, and this growth can be observed in the sharpening of the diffracted beams with time at a fixed temperature, from which it should be possible to extract activation energies for "recrystallization" or island growth [19]. Second, if the substrate contains defects that limit the obtainable order in the overlayer, the diffraction beam profile can be used to quantify the extent of the ordered regions and the overlayer island sizes. Limitations to overlayer ordering are, in fact, observed, and an illustration of the overlayer island size dependence on temperature and coverage in O/W(110) represents the main thesis of this paper.

In order to interpret quantitatively the angular profiles of diffraction beams, as well as the temperature dependence of the intensities, the resolving power of the LEED instrument [20] must be known. The resolving power is limited by measurement accuracy and by a series of contributions [21] that cause a spread of the Fourier components around a nominal set of diffraction conditions. These contributions are collectively referred to as the instrument response function [21]. An accurate measurement of the response function can be obtained [21-23] by considering the major factors contributing to it, i.e., the incident electron beam divergence, the beam diameter, the detector diameter, and the energy spread of the beam. Because the finite resolving power causes a spread in the angular profile of a beam, it is obvious that it will affect the quantitative determination of finite-size effects. It is less obvious that it will also affect the intensity decay with temperature [24] and hence determinations of phase boundaries and of critical exponents in order-disorder transitions [4,25].

3. THE O/W(110) PHASE DIAGRAM

The adsorption of O on W(110) represents one of the most interesting chemisorption systems that have been investigated. Three different ordered overlayers form at different coverages, a p(2×1) structure with saturation coverage $\theta_s = 1/2$, a p(2×2) structure with $\theta_s = 3/4$, and a p(1×1) structure with $\theta_s = 1$. It is noteworthy that no p(2×2) structure forms at $\theta = 1/4$. Figure 3 illustrates the adsorbate unit mesh for the p(2×1) structure. The periodicity is twice the substrate primitive unit mesh distance in one direction and the same as the substrate in the other. The p(2×2) structure occurs with additional adsorption of O into every other site in the empty rows of the p(2×1) structure. It can be visualized by viewing the symmetry of the remaining vacancies and invoking Babinet's principle [26]. For both structures, two orientationally equivalent domains exist because of the symmetry of the substrate.

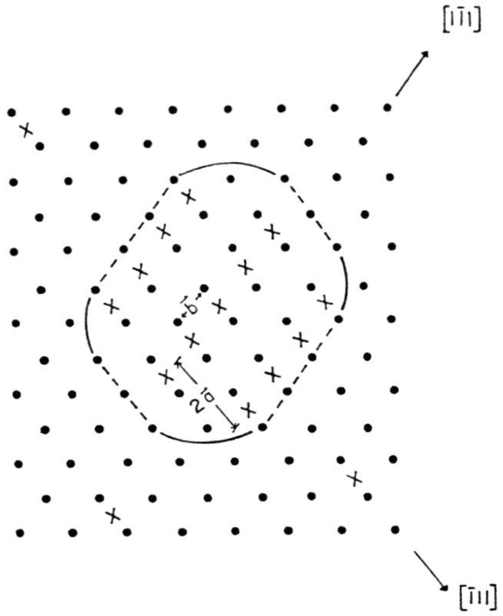

Fig. 3. Symmetry of a p(2×1) island on a W(110) substrate. The X's are overlayer atoms. Their lateral position is arbitrary. X's not attached to the island represent the two-dimensional vapor pressure of the p(2×1) phase at finite temperatures. The p(2×2) structure is formed by filling every other site in the empty rows.

This chemisorption system is "closed" i.e., not in equilibrium with the 3-D vapor or with the bulk. Desorption temperatures are about 2000°K, the solubility of O in W is very small, and measurable oxide formation does not take place below ~1300°K and coverages below $\theta = 1/2$. At the temperatures of interest in ordering of the overlayer, $T < 800°K$, the coverage can thus be considered constant. The system can be modeled by a lattice gas because the diffusional barriers between adsorbate sites are of the order of 1 eV.

Because the O layer forms a superlattice, the diffraction pattern contains extra reflections. The measurements described in the last section have been made on some of these superlattice reflections. From the temperature dependence of the intensity, a possible phase diagram for the O overlayer can be constructed. This is shown in Fig. 4. The data points are from the (1/2 1/2) superlattice beam at various diffraction conditions. The phase boundaries are schematic only. For a variety of reasons, notably the absence of a p(2×2) phase at low coverages [27] and the change in diffracted beam width as the temperature is raised at low coverages [24], there must be a coexistence of p(2×1) phase and vapor at low temperatures and coverages. As shown in Fig. 1, a branch must then connect this region with the phase boundary point for the order-disorder transition

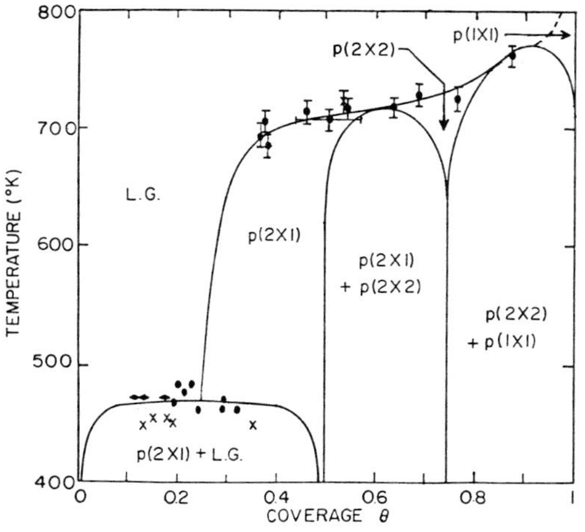

Fig. 4. Possible phase diagram for O/W(110). The data points are measured for the (1/2 1/2) superlattice reflection at various incident-beam energies. ●:E = 115eV, X:E = 80eV, ◆:E = 52eV. Other phase boundaries are schematic only. L.G.:lattice gas.

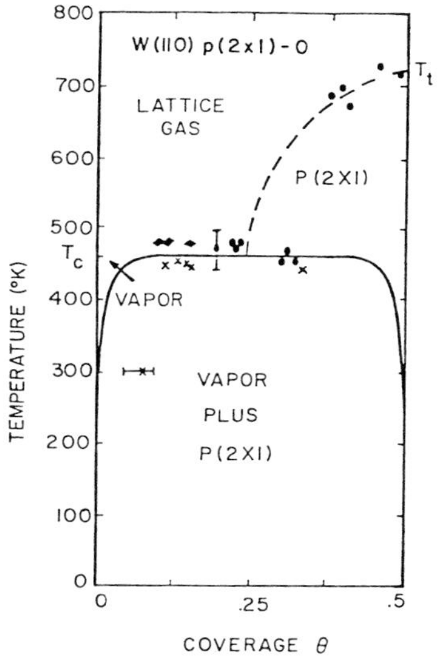

Fig. 5. Ising model fit to the coexistence region for O/W(110) at low coverages, assuming a single attractive interaction, ε = 0.07eV/atom. See Ref. 27 for details. The star at 300°K represents the position of the phase boundary calculated from the island size dependence on coverage shown in Fig. 8.

of the p(2×1) phase at its saturation coverage, θ_s = 0.5. The same features are likely to occur for the p(2×2) and p(1×1) phases. The phase boundaries are drawn so as not to violate the Gibbs phase rule [1]; there is at present no evidence to support the positions of the phase boundaries above θ = 0.5, except for the data points above which no ordered phase exists.

The continuous order-disorder phase transition of a p(2×1) overlayer on a bcc(110) substrate belongs to the universality class called the XY model with cubic anisotropy [28]. The p(2×2) structure also belongs to this class. To our knowledge, O/W(110) is so far the only experimental example of this model. This model does not obey universality and thus unique critical exponents cannot be calculated. However, the anisotropy term changes between the p(2×1) and p(2×2) phases. A comparison of transitions for these two structures, which provides an interesting test of the model, will be presented elsewhere [29].

Since the phase boundary separating the low-coverage two-phase region from the vapor phase is proportional to the solubility or lateral vapor pressure of the ordered p(2×1) phase, it should be possible to extract a net cohesive energy for the ordered phase by fitting the phase boundary. This has been done [24] with a very simple model for the interaction energies, using an Ising model and an infinite system. The results, shown in Fig. 5, yield a net attractive interaction energy of 0.07 eV/atom. The phase boundaries obtained with this simple model are very steep at low temperatures, and rather suddenly turn over to close the two-phase region at a multicritical point. The implication is that the vapor pressure for the p(2×1) phase is quite low (and conversely, that the solubility of vacancies in the ordered phase is quite low) until temperatures are reached that are very near the maximum temperature for phase coexistence.

4. FINITE-SIZE EFFECTS

For a perfectly ordered substrate, the equilibrium condition at any coverage or temperature in the two-phase region should be that all the ordered phase be condensed into one large island rather than to be dispersed into a number of small islands. For any coverage that we are able to observe, this provides many more atoms than are required to make an island much larger than is resolvable with a LEED instrument [17,20]. As a result, the diffracted beam profiles should at all coverages be as sharp as the instrument is able to measure. This is not the case, as is demonstrated in Fig. 6, which shows the minimum full widths at half-maximum of the angular profile of a superlattice reflection as a function of coverage that we are able to measure. We propose that these correspond to the equilibrium average size of the ordered regions. The broadness of the profiles implies that the ordered phase is broken up into small islands. Although kinetic limitations to the

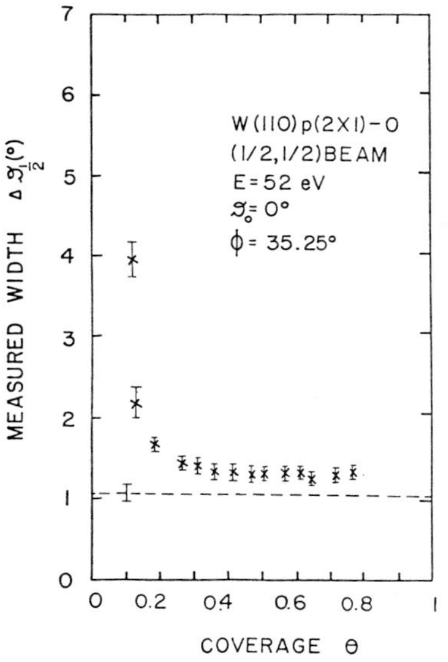

Fig. 6. Full width at half-maximum of the (1/2 1/2) superlattice reflection as a function of coverage at equilibrium. Adsorption at 300°K, anneal for up to four hours at ~425°K, measurement at 300°K. The dashed curve is the instrument response function width.

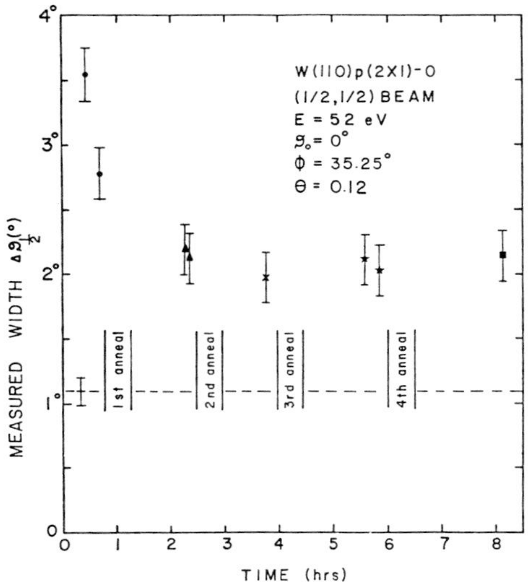

Fig. 7. Time dependence of the full width at half-maximum of the (1/2 1/2) superlattice reflection at a coverage of θ = 0.12. Each box marked "anneal" refers to a half-hour anneal at ~425°K. The overlayer was then allowed to cool slowly and all measurements were made at room temperature. After 3-4 hours the superlattice beams achieved their minimum widths. The dashed curve is the instrument response function width.

growth of large islands at the expense of small islands cannot be totally excluded, Fig. 7 indicates that rapid island growth takes place at short times, and that further annealing (to the limits of time available before contamination from CO becomes a problem) has little effect.

An alternative explanation of the finite size of islands is substrate heterogeneity, in particular steps. From independent measurements [22], it has been possible to determine an average size on the W(110) substrate between 80 and 150Å (depending on the details of the model [30,31] used to interpret the measurements). If the edges of steps, both up and down, can be considered as significant diffusion barriers (for a discussion of this point, see Ref. 6), then each terrace can be considered to act as an independent thermodynamic system, i.e., atoms once adsorbed on a terrace are with high probability constrained to stay there at the temperatures of the phase transitions encountered here, and there is no communication between atoms on different terraces.

The angular distribution of the diffracted intensity in a superlattice beam for ordered regions adsorbed on a stepped surface [17] can be interpreted in terms of a simple island-size broadening model [16,17]. Depending on the size distribution function assumed, one can extract the average size of islands. Figure 8 shows island sizes as a function of coverage at 300°K derived from the data in Fig. 6, using the model that gives the largest possible island sizes [17]. At ~425°K, the island sizes are smaller. They are also shown in Fig. 8. The maximum size island is reached at saturation coverage, and corresponds roughly to a diameter of 25 atomic rows. The spacing between adsorbate atoms in the direction in which the profiles of Fig. 6 were measured is 5.46Å. Hence the maximum island size is roughly 130Å, consistent with the terrace size determined independently from the substrate [22,30,31].

5. DISCUSSION

If the observed small islands in fact represent an equilibrium size distribution on substrate lattices of limited extent, as we suggest, one can compare these sizes with those expected on the basis of the phase diagram for the infinite system shown in Fig. 5. The calculation to fit the data in Fig. 5 involves no contribution to the free energy of the island other than the areal term, i.e., it neglects boundary energy and contributions from strong adsorption sites that may act as centers where islands nucleate. Using as a normalization the island size observed at saturation coverage, where presumably all the terraces are filled with adsorbed phase, one can calculate from Fig. 5 the size of

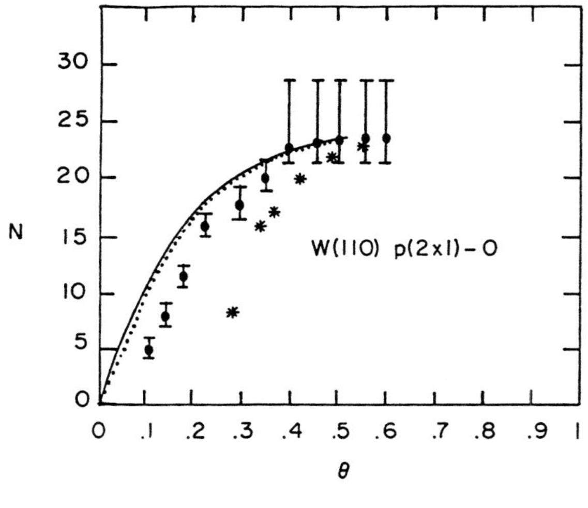

Fig. 8. Average island size versus coverage at two temperatures. The diameter is obtained by multiplying N, the average number of atoms in a row, by 5.46Å, the overlayer lattice constant in this direction. ●: Equilibrium size at 300°K; *: Equilibrium size at 425°K. The solid curve is the island size as a function of coverage at T = 0°K calculated from the phase diagram in Fig. 5, with the size normalized to the experimental values at θ = 0.5, the saturation coverage for the p(2×1) structure. The dotted curve is the same calculation at 300°K. The difference in the dotted curve and the experimental values may be an indication of a boundary energy contribution for small islands.

islands at any temperature and coverage in the two-phase region from the lever rule. At absolute zero there will be no adsorbate atoms in the gas phase, hence all the atoms in each terrace must be condensed into one island and the island size goes as the square root of the coverage. This is shown as the solid line in Fig. 8. At 300°K, the calculated phase boundary has not deviated significantly from its position at absolute zero, hence still very few atoms are in the gas phase (the system has a low 2-D vapor pressure). The calculated island size is shown as the dotted line. Comparison of these with the data shows that the measured island size is smaller than one would expect from the phase boundary calculated for an infinite system, and therefore more atoms must be in the vapor phase. It thus appears that there is a solubility of the ordered p(2×1) phase consisting of small islands that is greater than expected if the phase formed an infinite structure, a result that is well known in three dimensions and is described by the Thompson-Freundlich equation [1]. The increased solubility is caused by lesser stability of small particles because of a surface energy contribution to the total free energy. In two dimensions this is an indication of a boundary energy contribution to the stability of islands. The island size should increasingly deviate from the calculated values for lower coverages, as observed.

If one uses a simple bond-breaking model, the boundary energy for a 35Å diameter island is 10% of the areal energy.

Using the measured island size to calculate the solubility of islands at 300°K gives one point of the "actual" phase boundary for finite-size islands. This is shown as the star on Fig. 5.

A number of factors have been neglected in this very simple, and perhaps simplistic, view of finite-size effects. For one, there may be other contributions to the free energy of the island, such as strong adsorption sites. All of these that we can think of will make the island more stable and hence its size larger than that calculated without consideration of any but the areal energy. If such sites are postulated, it would imply a greater boundary energy contribution. Additionally, the model and the interactions used to calculate the phase boundary in Fig. 5 may be too simple to describe the physical system adequately. A more realistic calculation might move the phase boundary inward sufficiently to give agreement with the island size without the need to invoke a boundary energy contribution. The island size determination itself may be inadequate. The island on each terrace may not be perfect but may have antiphase boundaries in it, which would broaden the reflection farther. It would also add to the boundary energy in some unknown way. Islands may not be on the average be round.

Despite these limitations to the interpretation, it is clear that it is possible to do quantitative experiments of a nature that should be highly useful in exploring initial stages of epitaxy and growth, in investigating the various contributions to the cohesive energy of islands growing on a crystalline substrate, and generally in studying the nature of two-phase regions in overlayers. O/W(110) happens to be a convenient and well-behaved system for such studies, but it is by no means the only one. The major application for these types of studies would seem to be in processes involving epitaxial growth, e.g., for semiconductor materials.

6. ACKNOWLEDGEMENTS

This research was supported by ONR. We have benefited from discussions with D.L. Huber, A.N. Berker, J.G. Dash, and M.B. Webb.

REFERENCES

1. See, for example, R.A. Swalin, Thermodynamics of Solids, Wiley, New York (1972).
2. A representative sample of the many types of systems that exhibit two-phase coexistence is given in various chapters of this volume.
3. J.C. Buchholz and M.G. Lagally, Phys. Rev. Letters 35, 442 (1975); T.-M. Lu, G.-C. Wang, and M.G. Lagally, Phys. Rev. Letters 39, 411 (1977); T. Engel, H. Niehus, and E. Bauer, Surface Sci. 52, 237 (1975).
4. R.L. Park, A.R. Kortan, L.D. Roelofs, and T.L. Einstein, this volume; L.D. Roelofs, T.L. Einstein, P.E. Hunter, A.R. Kortan, R.L. Park, and R.M. Roberts, J. Vac. Sci. Technol. 17, 231 (1980); A.U. McRae, Surface Sci. 1, 319 (1964).
5. P.J. Estrup, this volume; M.K. Debe and D.A. King, Surface Sci. 81, 193 (1979).
6. For a list of references, see M.G. Lagally, T.-M. Lu, and D.G. Welkie, J. Vac. Sci. Technol. 17, 223 (1980).
7. See, for example, R.J.H. Vorhoeve, in Treatise in Solid State Chemistry, ed. N. B. Hannay, Plenum, New York, Vol. 6A (1976).
8. A.N. Berker, this volume, A.N. Berker and S. Ostlund, J. Phys. C12, 4961 (1979).
9. M.B. Webb and M.G. Lagally, Solid State Phys. 28, 301 (1973).
10. J.B. Pendry, Low-Energy Electron Diffraction, Academic, New York (1974).
11. Prof. R.L. Park has brought to our attention that Davisson and Germer observed phase transitions in overlayers in their original work on electron diffraction; C. Davisson and L.H. Germer, Phys. Rev. 30, 705 (1927).
12. T.-M. Lu, this volume.
13. B.E. Warren, X-Ray Diffraction, Addison-Wesley, Reading, MA (1969).
14. R.L. Park, J. Appl. Phys. 37, 295 (1966).
15. D.G. Welkie, M.G. Lagally, and R.L. Palmer, J. Vac. Sci. Technol. 17, 453 (1980); M.G. Lagally and D.G. Welkie, in Advanced Techniques for the Characterization of Microstructures, ed. F.W. Wiffen, TMS/AIME (1980) (to be published).
16. J.C. Tracy, Ph.D. Dissertation, Cornell University, unpublished (1969); J.C. Tracy and J.M. Blakely, in Structure and Chemistry of Solid Surfaces, ed. G.A. Somorjai, Wiley, New York (1969).
17. M.G. Lagally, G.-C. Wang, and T.-M. Lu, in Chemistry and Physics of Solid Surfaces, Vol. II, ed. R. Vanselow, CRC Press, Boca Raton, FL (1979): T.-M. Lu, G.-C. Wang, and M.G. Lagally, submitted to Surface Sci.
18. For a thorough review of surface steps, see M. Henzler, in Electron Spectroscopy for Surface Analysis , ed. H. Ibach, Springer, Berlin (1977).
19. G.-C. Wang, Ph.D. Dissertation, University of Wisconsin-Madison, unpublished (1978), and in preparation; J. Perepezko, J.T. McKinney, and M.G. Lagally, to be published.
20. T.-M. Lu and M.G. Lagally, Surface Sci., to appear.
21. R.L. Park, J.E. Houston, and D.G. Schreiner, Rev. Sci. Instrum. 42, 60 (1971).
22. G.-C. Wang and M.G. Lagally, Surface Sci. 81, 69 (1979).
23. D.G. Welkie and M.G. Lagally, Appl. Surface Sci. 3, 272 (1979).
24. G.-C. Wang, T.-M. Lu, and M.G. Lagally, J. Chem. Phys. 69, 479 (1978).
25. T.-M. Lu and M.G. Lagally, in preparation.
26. G.S. Monk, Light: Principles and Experiments, Dover, New York (1963).
27. T.-M. Lu, G.-C. Wang, and M.G. Lagally, Surface Sci. 92, 133 (1980).
28. E. Domany, M. Schick, and J.S. Walker, Phys. Rev. Letters 38, 1148 (1977); E. Domany, M. Schick, J.S. Walker, and R.B. Griffiths, Phys. Rev. B18, 2209 (1978).
29. T.-M. Lu, D. Teske, M.G. Lagally, and G.-C. Wang, in preparation.
30. M. Henzler, Surface Sci. 73, 240 (1978).
31. T.-M. Lu, S.R. Anderson, M.G. Lagally, and G.-C. Wang, J. Vac. Sci. Technol. 17, 207 (1980), and to be published.

DISCUSSION

A. Berker:

I would think that, for the occurrence of many small domains rather than few large ones, a network of substrate crystallite boundaries is sufficient. As long as these boundaries interrupt the ordered structure of the overlayer, it is not necessary to invoke a diffusion mechanism.

M. Lagally:

I would agree that this is another possibility for interrupting the long-range periodicity of a dense overlayer, in effect introducing domain boundaries. I am not as certain how such energetic barriers can cause small islands to exist that are widely separated. Without a mechanism to limit diffusion it would seem that atoms would still attempt to aggregate in large ordered regions that may be traversed by domain boundaries. As a result, the island, or more properly domain, size would be independent of coverage.

P. Horn:

We have observed similar effects for both Kr and Xe adsorbed on exfoliated graphite. Our observations cannot be easily related to diffusional barriers because of the ability for mass transport through the three-dimensional vapor.

Have you considered other possible explanations for the formation of small islands, like for example, a mechanism based on interface roughening?

M. Lagally:

It would certainly be true that a diffusional barrier is not realistic if the system is in equilibrium with the three-dimensional gas phase. Energetically unfavorable boundaries, as suggested by A. N. Berker in the previous question can certainly be invoked if there is no coverage dependence of the island size. This would also be true for an island boundary roughening mechanism. If there are no limitations to the growth of large ordered regions, they should form up to the limits of domain walls. Any roughening at that interface will be coverage-independent.

B. Halperin:

Presumably the presence of a step will strongly modify the local attraction to the adatoms and will either tend to nucleate, or tend to repel a droplet, depending on the sign. Why can this be neglected in your theory?

M. Lagally:

Clearly any defect can modify the local attraction and thus increase the stability of islands. It should be taken into account in any quantitative treatment of the contributions to the free energy of the ordered phase. We have no way to estimate the strength of attraction of a step except possibly through band counting, but there is no evidence that it is more important than boundary energy effects. In any case, its effect is opposite to the boundary energy contribution, and if we postulate a step attraction, the boundary energy contribution will have to be even larger in our very qualitative model.

Defects that have a repulsive interaction with the adsorbed phase should not be important at low coverages where the boundary energy might be significant, because the island would form outside the range of influence of such defects. At or near saturation coverage, such defects would, of course, reduce the stability of the ordered phase.

U. Landman:

Could not there be a sufficient concentration of point defects on the terraces of the steps which would nucleate small islands around them? A point defect or a planar partial dislocation (a small one) could have a rather long-range influence, and thus influence the kinetics of growth and island size distribution.

M. Lagally:

It is certainly true that point defects can influence the growth and island size distribution. This reduces to a quantitative question of their range of influence. If this range is small, one might expect to have many small (e.g. 1 or 2 lattice constant size) islands decorating the defects, with the rest of the atoms in one large island. If the number of point defects is very large, the measured island size should be independent of coverage and should reflect the range of influence of the point defect.

If the range of influence of the defect is large, one will obtain the observed coverage dependence, except that the islands should be more stable. This was addressed in the previous question.

S. K. Sinha:

Did you check the variation of the intensity of your peaks with θ (coverage)? If your picture of a number of fixed cells dividing the surface is correct, then as you add more adsorbate the number of "islands" will stay constant but the size of your islands will increase proportional to θ and so the intensity should increase as θ^2. On the other hand if there is some mechanism which causes the layer to break up into islands anyway, then the number of islands will increase with θ and the intensity should increase proportional to θ.

M. Lagally:

We have considered the peak intensity of superlattice reflections as a function of coverage, as have others (see Ref. 3). In both cases a roughly θ^2 dependence is found below $\theta = 0.3$. Above $\theta = 0.3$, the increase in intensity flattens out and the intensity becomes nearly constant between $\theta = 0.4$ and $\theta = \theta_s = 0.5$. The interpretation of peak intensity vs. coverage is not as simple as indicated in the question. Because the substrate is coherent even if it has steps, there will be interference between atoms on different terraces even if they are thermodynamically isolated. At low coverages, where adsorbate islands are widely separated, it can be shown (Ref. 20) that the θ^2 dependence of the peak intensity is valid, as the questioner points out. Near saturation coverage, however, antiphase domain effects may become important, and the intensity is no longer simply proportional to θ^2. Also, above the coverage where the overlayer exists as a single phase, the intensity will rise only as θ as the random vacancies in the layer become filled.

Sinha, ed. Ordering in Two Dimensions

THEORIES OF COMMENSURATE-INCOMMENSURATE TRANSITIONS ON SURFACES

J. VILLAIN

Centre d'Etudes Nucléaires de Grenoble
Département de Recherche Fondamentale, Laboratoire de Diffraction Neutronique
85 X - 38041 Grenoble Cédex, France

ABSTRACT

The simplest model of commensurate-incommensurate transition [1] may be related to the statistical mechanics of a system of rigid parallel walls at distance ℓ submitted to an indirect interaction $V(\ell) \sim \exp -\kappa\ell$ and to a variable chemical potential δ. This model yields a continuous transition and a fairly good fit to experiment except very near the transition. In two-dimensional cases, wall fluctuations should have a dramatic effect near the transition. Additional complications are : i) The orientational instability of Novaco and Mc Tague. ii) The effect of wall intersections. The effect of substrate imperfections is briefly studied. Dislocations are neglected in most of theoretical models and may have dramatic effects - possibly destroy the C-I transition.

1. INTRODUCTION

In 1949 Frank and Van der Merwe [1] proposed a one-dimensional model of epitaxy of a "solid" adsorbed film on a crystalline substrate (Fig. 1). This model is dominated by the conflict between two forces acting on the adatoms : the periodic potential of the substrate and the interaction between adatoms.

Twenty years later [2,3,4] quantitative experiments became possible in physisorption. Krypton monolayers on the cleavage face of graphite proved especially adequate for a test of the Frank-Van der Merwe (FVdM) theory because they exhibit both commensurate (C) and incommensurate (I) phases (2,3,4). The interatomic distance varies steeply near the C-I transition and its variation is a crucial test for the FVdM theory. This test is favourable [3]. However certain discrepancies were found [5] in the vicinity of the C-I transition. An improvement of the FVdM theory is therefore necessary.

FVdM calculated exactly the ground state of a one-dimensional adsorbate on an ideal substrate. One wishes to treat a two-dimensional system at non-vanishing temperature, eventually in the presence of substrate defects. This can generally not be done exactly.

A very useful concept near a C-I transition is the concept of domain walls. It may be easily introduced in the model of Fig. 1. The size of adatoms is such that the only simple possible C phase corresponds to the occupation of one half of the easy adsorption sites. Indeed the atoms are too big to occupy two neighboring sites, and too small to allow the occupation of one third of the sites. There are therefore two kinds of commensurate phases, which can be called A and B and correspond to the occupation of two different sublattices.

Near the C-I transition, the I phase can be shown [6,7,8] to correspond to a succession of broad domains in which the system is almost commensurate. Any two consecutive domains are separated by walls of fixed width. A and B domains alternate regularly.

The two-dimensional system Kr on graphite may be similarly described in terms of domains and walls (Fig. 2) near the C-I transition [9, 8,10].

In diffraction patterns satellites should testify the domain structure. In three-dimensional systems such satellites are often observed. Their observation is more difficult, though

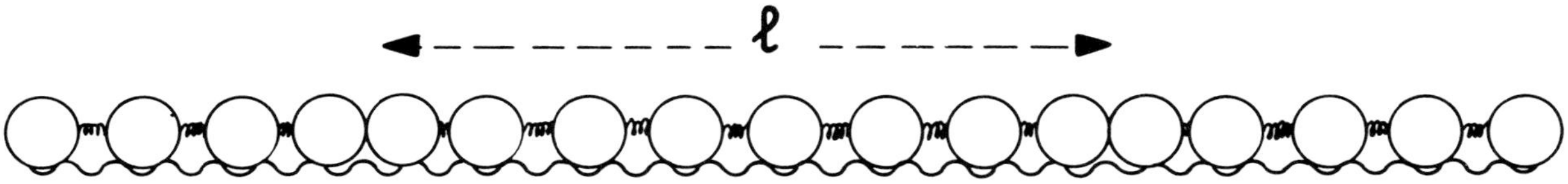

Fig. 1 : The 1-D, FVdM model. The circles represent adatoms, springs are the interactions between them and the wavy line simulates the substrate potential.

possible [5] in two-dimensional (2-D) systems.

C-I transitions appear to be qualitatively different from order-disorder transitions described by the Landau-Wilson theory [11]. C-I transitions are indeed produced by localized fluctuations (walls). The critical behaviour is not described by usual power laws, at least at $T = 0$ (see eq. II-3 below). However, there are attempts to describe C-I transitions by a Landau-Wilson theory [12].

2. THE FVdM THEORY

The 1-D model introduced in the introduction can be simplified by the following approximations : i) The substrate potential, which is a periodic function of the displacement u of the adatoms from their position in the C phase, is replaced by a single Fourier component. ii) The interaction between adatoms is a quadratic function of the displacements u and interactions beyond nearest neighbours are neglected. iii) The system is treated as a continuum. The energy to be minimized is :

$$W = \frac{1}{2} A \int_0^N dx[(\xi \frac{\partial u}{\partial x} + \frac{1}{2}\delta)^2 - \cos^2 u] \qquad \text{(II-1)}$$

where A and the wall width $\xi = 1/\kappa$ are constants, N is the size of the system and the chemical potential δ depends on pressure. The model (II-1) is not appropriate to describe chemisorption on metals, where various long-range forces are to be considered.

Energy (II-1) can be minimized exactly [1]. The ground state satisfies the time-independent Sine Gordon equation :

$$\xi^2 \partial^2 u/\partial x^2 = \sin u \cos u \qquad \text{(II-2)}$$

The commensurate phase $u = 0$ is found to be stable for $\delta > \delta_c$ with :

$$\delta_c = -4/\pi \qquad \text{(II-3)}$$

Just below δ_c the period ℓ of $\partial u/\partial x$ is given approximately by :

$$2\kappa\ell \exp -\kappa\ell = (\delta_c - \delta)/\delta_c \qquad \text{(II-4)}$$

and the system is found to have the domain structure described in the Introduction : $\partial u/\partial x$ is almost zero except in relatively narrow "walls". The distance ℓ between walls is determined by their interaction which is mediated by the atomic displacement u : a wall displaces the neighboring atom which in turn displaces the next atom, etc., and finally the effect reaches the next wall. Mathematically it is possible to eliminate the continuous variables u and this procedure introduces an effective interaction U(r) between walls at distance r [6,13,14,10,8,7]. The mechanism is reminiscent of the phonon-mediated interaction between electrons in superconductors. At large distance x from a wall, sin u is small and eq. (II-2) may be linearized, yielding an exponential decay $\sin u(x) \sim \exp - \kappa x$. Therefore the interaction U(r) is also exponential :

$$U(r) = U_o \exp - \kappa x \qquad \text{(II-5)}$$

The approximate law (II-4) may be derived from (II-5) as well as from the exact solution [6,13,7,8,10].

In 3-D systems walls are no more points, but planes. The FVdM theory may still be applied provided the atomic displacements from the C position have only one non-vanishing component u and depend on only one coordinate x. This implies a uniaxial modulation which may occur, for instance, in an anisotropic medium. This condition is however not sufficient to apply the FVdM theory. Walls must also remain flat. They do remain flat in 3-D systems if the temperature T is below their roughening transition temperature T_R [15].

In 2-D systems, walls are lines and $T_R = 0$. Walls are flat at $T = 0$ only. The FVdM theory may be applied at $T = 0$ only, and only for an anisotropic substrate with an easy wall direction. In other cases an appropriate generalization is necessary. This is the object of the forthcoming Sections.

3. ISOTROPIC SYSTEMS AT T = 0

a) The BMVW theory [16]

In the case of hexagonal or square symmetry, walls have respectively 3 and 2 easy directions. Therefore they can cross. A wall intersection should have some energy Λ which, in principle, can be calculated from the interatomic potential. However Bak et al. [16] by-passed this tedious numerical calculation and introduced Λ as a phenomenological parameter. There are 2 cases :

i) $\Lambda > 0$. The number of wall intersections should be as small as possible. Therefore, near the C-I transition all walls are parallel because the exponentially weak repulsion (II-5) is more easily overcome than the larger repulsion Λ between crossing walls. Since walls are parallel the FVdM theory may be applied and the C-I transition is continuous according to (II-4). Away from the transition the repulsion (II-5) becomes stronger and wall intersections are expected to appear, giving rise to a new phase transition.

ii) $\Lambda < 0$. The number of wall intersections should be as large as possible. This probably implies an equal number of walls in all easy directions. The energy W as a function of the distance ℓ between walls is :

$$W = (N/\ell)(\delta^* + \Lambda/\ell + U_o \exp - \kappa\ell) \qquad \text{(III-1)}$$

where the "natural misfit" δ^* is proportional

to $(\delta - \delta_c)$. If $1/\ell$ is considered as the square of an order parameter η, the term proportional to η^4 is negative. In this case the transition is first order as well known [17].

Conclusion : either the C-I transition is first order or the (cubic or hexagonal) symmetry is destroyed in the I phase near the transition.

There is no well established experimental confirmation of this theoretical prediction. Krypton monolayers on graphite have been most carefully studied by various experimental groups [2,3,4,5] and seem to provide a counter-example : apparently the C-I transition is continuous and hexagonal symmetry is preserved. The reason may be [20] that the distorted phase has a narrow range of stability. On the other hand experiments are done at $T \neq 0$ on imperfect substrates and these features should be taken into account. This will be done in Section 5, and substrate imperfections will be found to provide a possible explanation of the discrepancy between theory and experiment. On the other hand available experimental data correspond to pretty high temperatures (not far from melting) and it would be interesting to have data at low temperatures which are more accessible to theory.

b) The Novaco-Mc Tague (NmT) distortion [18,10,19,20,21]

Another difference between 1-D media and 2-D media is that 2-D media have transverse strains in addition to longitudinal strains. Since transverse strains have lower energy, they can favor a tilt of the adsorbate with respect to the substrate [18]. Near the C-I transition the NmT tilt may be interpreted as a tilt of the walls [10]. Thus at T = 0, the tilt is expected to appear as soon as walls appear, i.e. just at the C-I transition. This is (so far as we understand) in agreement with predictions of Pokrovskii and Talapov [21] but in disagreement with Shiba [20]. Experiments by Fain et al. [19] show a non-tilted phase near the C-I transition in agreement with Shiba, but this may also be due to thermal effects. Experiments at lower temperature would be of interest. The NmT tilt takes place for a sufficiently large value of the Poisson ratio σ of the adsorbate. For a reasonable choice of the substrate potential (Novaco, private communication) the critical value is just 0, so that the NmT effect is expected to take place rather generally. If Cauchy relations are satisfied, $\sigma = 1/4$.

For the sake of simplicity the NmT effect will be neglected in the next Sections.

4. THE POKROVSKII-TALAPOV (PT) THEORY [21,22]

Pokrovskii and Talapov studied the effect of temperature in the case of a single easy direction for walls. Such a situation occurs for instance for an anisotropic substrate such as the (110) face of a metal, though the calculations of ref. 21 are in fact carried out for an isotropic substrate. However the PT model may essentially be applied to a substrate of uniaxial or lower symmetry, and also for square symmetry. The more complicated case of hexagonal symmetry will be discussed in Section 5.

For $T \neq 0$ walls may slightly deviate from their easy direction. Therefore they have many allowed states and a finite entropy S. The quantity to be minimized is no more the energy W but the free energy F = W - TS, which is again a function of the average distance between walls. The entropy per unit length contains a term independent of ℓ and a negative term which expresses the fact that the number of available states for a given wall is limited by the neighboring walls. The corresponding term in the free energy is therefore positive. It is a repulsion between walls. As in § 3-b this repulsion produces a continuous transition. The calculation [21,8] shows that the correction per unit wall length is proportional to $1/\ell^2$ and therefore it overwhelms the exponential repulsion (II-3) for large ℓ. The free energy may be written as :

$$F = (N/\ell)(\mu^{*} + C/\ell^2) \qquad \text{(IV-1)}$$

and its minimisation with respect to leads to the following law [22,21,8,12] :

$$1/\ell \sim [\delta_c(T) - \delta]^{1/2} \qquad \text{(IV-2)}$$

at constant temperature. A similar square root behavior is found at constant pressure as a function of temperature.

For smaller values of ℓ the FVdM formula (II-3) may be applied but its derivation has to be modified in order to take fluctuations into account.

The range of validity of the Pokrovskii-Talapov formula (IV-2) is narrow at low temperature when walls are stiff. On the other hand, when the wall stiffness decreases it may become necessary to take dislocations into account, while they are neglected in the PT theory. This point will be discussed in Section 6.

Let a word be said about the temperature variation of the transition value $\delta_c(T)$. According to ref. [22] it is given for $A^c << T << A\xi^2$ by :

$$\delta_c(T)/\delta_c(o) = (V_o a^2/T)^{T/2(T_o - T)}$$
$$\simeq 1 - \frac{T}{2T_o} \mathrm{Log}(V_o a^2/T) + \ldots$$

where $V_o a^2 \simeq A$ and $T_o \simeq A\xi^2$. The variation of μ_c with temperature is essentially produced by the following mechanism : the energy of adatoms in the commensurate phase or within domains in the incommensurate phase is increased at $T \neq 0$, by an amount essentially proportional to T. Therefore the wall energy is reduced.

Correlation functions in the I phase exhibit [8,21,32] power law singularities characteristic of a "floating solid" [23,24].

5. HEXAGONAL SUBSTRATES AT LOW TEMPERATURE

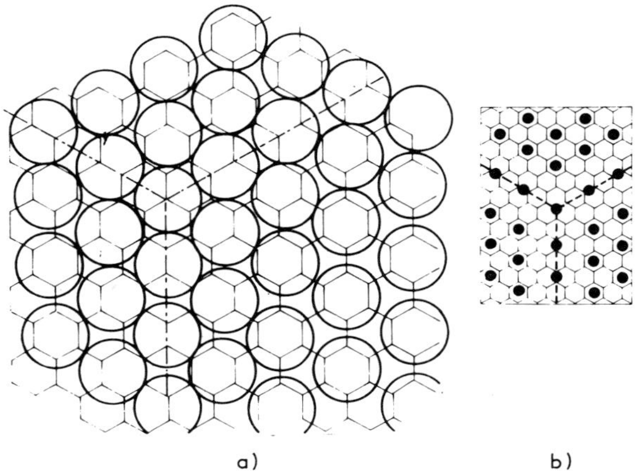

Fig. 2 : a) Semi-realistic picture of a krypton monolayer on graphite in its I phase near the C-I transition. Dot-dashed lines are the cores of "heavy" walls. NnT distortion is not represented. b) Schematic representation of "light" walls.

Rare gas monolayers adsorbed on hexagonal substrates are especially appropriate for experiments on phase transitions. Their hexagonally symmetric I phase may be described [9,8] as a network of walls with the topology of the honeycomb lattice. As in § 3-a the energy contains a term proportional to the total wall length L and a term proportional to the number 2ν of vertices. It is easily checked [25,26] that L and ν are not modified by a "dilatation" of a hexagon as displayed by Fig. 3. The only force which can stabilize a regular honeycomb network (of walls) is the exponential repulsion (II-5) which may be neglected when the average hexagon side ℓ is large. In this case the walls form an "irregular honeycomb network" (Fig. 3). Since any of the ν hexagons may be deformed without energy change, there are ν "soft" degrees of freedom which produce a very large entropy. This property is typical of the honeycomb network. In the case of square symmetry the PT theory would apply.

The maximum amplitude of the deformation of each hexagon is about ℓ. The entropy of an irregular honeycomb network of ν hexagons is therefore about :

$$K_B \nu \, \mathrm{Log}\ell \simeq K_B N \, \ell^{-2} \, \mathrm{Log}\ell \qquad \text{(V-1)}$$

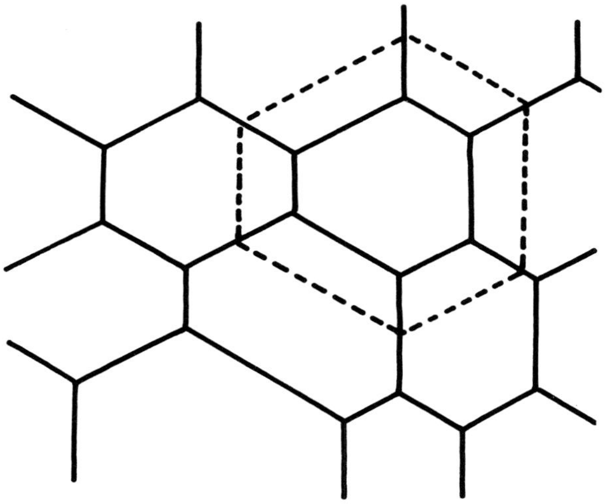

Fig. 3 : Wall arrangement near C-I transition. Dashed lines show a dilatation of a hexagon, which does not modify the energy except for exponentially weak terms.

where N is the number of sites. This entropy dominates both the BMVW interaction $\Lambda\nu$ of Section 3 and the PT term $NC\ell^{-3}$ of expression (IV-1). The resulting effect in the free energy is a negative term $-K_B TN\ell^{-2}\mathrm{Log}\ell$, which has essentially the same effect as a negative interaction $\Lambda\ell^{-2}$ in the BMVW theory of Section 3, namely : a first order C-I transition which does not break hexagonal symmetry [25]. According to experiment the C-I transition is continuous at least in Kr on graphite [2,3,4,5].

This discrepancy may be attributed to substrate defects [26] although their effect has not been completely clarified yet. Substrate defects, like thermal effects, prevent the walls to form a regular honeycomb network. The energy gain due to the deformation into an "irregular honeycomb network" (Fig. 3) is proportional [27,28] to the number of hexagons multiplied by the square root $\sqrt{N/\nu}$ of the average area per hexagon. This effect favors hexagonal symmetry and discards the striped structure predicted by BMVW for an ideal substrate with a continuous transition at T = 0. A crucial question of course is whether the C-I transition on an imperfect substrate is of first or second order. A more precise treatment would be necessary to get the answer.

Stephens et al. [5] made very accurate experiments on the C-I transition of Kr monolayers on graphite. They can be tentatively interpreted as follows [26] : i) Far from the C-I transition the FVdM law is obeyed [3]. ii) The transition region is broadened by finite size effects. iii) The intermediate region is dominated by a positive BMVW term which alters the FVdM behavior, and by substrate imperfections which prevent the BMVW distortion. The apparent power law observed by Stephens et al. would be accidental. An alternative interpretation in terms of "restricted area" effects has been given by Pokrovskii and Talapov [21].

6. HIGH TEMPERATURES

a) The Halperin-Nelson (HN) point

In the previous Sections walls were assumed to introduce either an excess of matter ("heavy" walls) or a lack of matter ("light" walls). Either heavy walls or light walls are favored according to the atomic size of the substrate and the adsorbate. However there may be a point on the C-I transition line in the P-T plane, where both heavy and light walls are present with the same free energy per unit length. This is the HN point, called B in ref. 24. The behavior as a function of T near this point is well understood, but the HN point is very special. The slightest pressure variation is in principle sufficient to exclude either heavy or light walls except of sufficiently short length.

The HN point can only exist [24] if the number p of sublattices is large enough. For Kr on graphite p = 3 and the HN point does not exist. Instead, the solid melts before reaching the HN point.

The HN point may exist in the case of long period structures which are rather common in chemisorption. See also Ref. 8, Fig. 3 and Ref. 21, Fig. 3.

It is of interest to consider the case of a large wall width ξ (see Section 2). This situation is very popular among theoreticians [12,22]. In this case the C phase may be stable up to temperatures of order $A\xi^2/p^2$ which are much higher than the wall energy per unit length which is about $A\xi$. This somewhat surprising result may be derived from renormalization group methods [27,24]. The reason is essentially that the wall entropy per unit length is not so large as might be expected. More generally the concept of rather stiff walls with preferred directions, which has been used in the previous Sections, is questionable at such high temperatures and it is of interest to have better mathematical treatments.

b) Yamamoto [28] and Okwamoto [29] have treated a pressure induced C-I transition at a special temperature near the HN point. Their model excludes dislocations and has the same anisotropy as the PT model. Their mathematical treatment, however, is more accurate at high temperature. Okwamoto finds a continuous transition and a square root law in agreement with (IV-2).

c) Effect of dislocations

As seen above the HN point does not exist in the popular instance of Kr monolayers on graphite, and also in many other cases when p is small. The reason is that free dislocations appear and destroy the solid phase. Dislocations are also present at lower temperature [30] and even if they are paired [31,27] they alter the picture given in the previous Sections. For p = 1 they transform the C-I transition (which exists at T = 0 and follows the FVdM theory) into an ordinary order disorder transition, of the universality class of the XY model. For p = 2, they allow the formation of closed wall encircling inclusions even in the C phase. Inclusions may also exist (with a triangular shape in the simplest case) for hexagonal symmetry with p = 3. Their effect on C-I transitions has not been seriously investigated. It is not excluded that a narrow liquid zone separates the C and I regions in the phase diagram, thus suppressing the C-I transition. If this transition is preserved it may be difficult to find an appropriate order parameter since the presence of inclusions at the C-I transition seems to imply a finite value of the average distance ℓ between walls.

7. CONCLUSION

Two-dimensional solids exhibit fascinating features, a part of which has been considered in this brief review. For instance, nothing or little has been said about dynamics and about absence of conventional order [23,8,32]. Furthermore, we have only investigated here the neighborhood of the C-I transition. The various mechanisms which have been discussed may give rise to crossover effects and complicated phase diagrams with several transitions [16,25,26].

So far, theorists tried to identify basic mechanisms and used qualitative models for this purpose. It is remarkable that mean field calculations [1,20], wall theories [16,8], renormalization group techniques [12,24,27], Fermion theories [33,29,32] are in fact concerned with similar systems, similar problems, but treated with very different techniques and approximations. Among the various mechanisms which have been identified, which of them are really important in practice and in what range of the phase diagram ? This is not yet clear.

It is a pleasure to thank Bob Birgeneau and Sam Fain for many informations on the experimental situation, and Mirta B. Gordon for helpful discussions.

REFERENCES

1. F. C. Frank, J. H. Van der Merwe, Proc. Roy. Soc. London, 198, 205 (1949).
2. A. Thomy, X. Duval, J. Chim. Phys. 66, 1966 (1969) and 67, 286-1101 (1970).
3. M. D. Chinn, S. C. Fain, Phys. Rev. Lett. 39, 146 (1977).
4. Y. Larher, J. Chem. Phys. 68, 2257 (1978).
5. P. W. Stephens, P. Heiney, R. J. Birgeneau, P. M. Horn, Phys. Rev. Lett. 43, 47 (1979).
6. M. L. McMillan, Phys. Rev. B 14, 1496 (1976).
7. A. D. Bruce, R. A. Cowley, J. Phys. C 11, 3577-3591-3609 (1978).
8. J. Villain, in "Ordering in Strongly Fluctuating Condensed Matter Systems" ed. T. Riste, p. 221 (Plenum, 1980).
9. J. A. Venables, P. S. Shabes-Retchkiman, Surface Sci. 71, 27 (1978).
10. J. Villain, Phys. Rev. Lett. 41, 36 (1978).
11. P. G. de Gennes, in "Fluctuations, Instabilities and Phase Transitions" ed. T. Riste, p. 1 (Plenum, New York 1979).
12. T. Natterman, J. Phys. C 13, L265 (1980).
13. P. Bak, V. J. Emery, Phys. Rev. Lett. 36, 978 (1976).
14. J. Villain, M. Gordon, J. Phys. C 13, to be published (1980).
15. J. D. Weeks, in "Ordering in Strongly Fluctuating Condensed Matter Systems" ed. T. Riste, p. 293 (1980).
16. P. Bak, D. Mukamel, J. Villain, K. Wentowska, Phys. Rev. B 19, 1610 (1979).
17. L. D. Landau, E. M. Lifshitz, Statistical Physics, § 135, p. 435 (Pergamon Press, 1959).
18. A. D. Novaco, J. P. McTague, Phys. Rev. Lett. 38, 1286 (1977).
19. S. C. Fain, M. D. Chinn, R. D. Diehl, Phys. Rev. B 12, May 1 (1980, to appear).
20. H. Shiba, J. Phys. Soc. Japan 48, 211 (1980).
21. V. L. Pokrovskii, A. L. Talapov, Zh. E.T.F. 78, 269 (1980).
22. V. L. Pokrovskii, A. L. Talapov, Phys. Rev. Lett. 42, 66 (1979).
23. B. Jancovici, Phys. Rev. Lett. 19, 20 (1967).
24. D. R. Nelson, B. I. Halperin, Phys. Rev. B 19, 2457 (1979).
25. J. Villain, Surface Science, 97, 219 (1980).
26. J. Villain, J. Physique Lettres, 41, L-267 (1980).
27. J. V. Jose, L. P. Kadanoff, S. Kirkpatrick, D. R. Nelson, Phys. Rev. B 16, 1217 (1977).
28. H. Yamamoto, Prog. Theor. Phys. 61, 1295 (1979).
29. A. Luther, J. Timonen, V. Pokrovskii in "Phase Transitions in Surface Films" edited by J. G. Dash and J. Ruvalds (Plenum Press New York 1980) p. 115. See also Y. Okwamoto, Preprint (1980).
30. J. A. Venables, P. S. Schabes-Retchkiman, J. Phys. C 11, L913 (1978).
31. J. M. Kosterlitz, D. J. Thouless, J. Phys. C 6, 1181 (1973).
32. H. J. Schulz, preprint (1980).
33. A. Luther, Phys. Rev. B 15, 403 (1977).

Refs. 3, 9, and 18 are also available, together with other works on adsorbed films, in the Supplément au Journal de Physique 38, fascicule 10-C4 (1977).

DISCUSSION

W. F. Brinkman:

As the wall-wall interaction increases, do you expect a transition in which there will be long range order of some type in the honeycomb lattice?

J. Villain:

I do not expect any transition. Near the C-I transition the walls form a rather soft honeycomb lattice which progressively becomes more harmonic when ℓ decreases and the exponential repulsion becomes more effective. An analogous behaviour is expected in the anisotropic model of Pokrovskii and Talapov. In the hexagonal case, however, we predict a first order transition and the exponential repulsion is never completely negligible. The soft degrees of freedom are never completely soft in practice.

E. K. Riedel:

Doesn't your treatment contain a mean-field argument in that the order of the commensurate-incommensurate transition is determined by minimizing the free energy $F = F(1/\ell)$? For a two-dimensional system, fluctuation effects may render this argument invalid.

J. Villain:

I do not think the expression "mean field" is appropriate. In the treatment outlined in my talk, I did neglect certain fluctuations since the walls were assumed to be rigid. But the other degrees of freedom were treated properly. On the other hand, wall fluctuations can be approximately accounted for [25] and the results are not very different. Minimization of the free energy $F(\ell)$ as a function of a macroscopic parameter ℓ is of course not a mean field approximation, but a general method in statistical mechanics. Generally the effective calculation of $F(\ell)$ is not possible. In the present case we could derive exact upper and lower bounds of $F(\ell)$ which are sufficiently accurate to derive definite conclusions.

DISCUSSION

R. R. Sharma:

The first-order term arising from the defects has been shown by you to be proportional to $(-\ell) \times 1/\ell^2$ which has the behavior similar to the first term of the free-energy. Therefore it can not give information about the effect of the defects. Evidently one needs to consider the second order term. In what way would the type of the phase-transition change if you consider the effect of this term--the second order term?

J. Villain:

Impurities or similar substrate defects destabilize the striped structure for large ℓ. I cannot say anything else. I agree that it would be essential, but probably difficult, to calculate the next term.

R. J. Birgeneau:

I should like to comment that in my view too much emphasis is put on whether or not the commensurate-incommensurate transition is first or second order. From our most recent S.S.R.L. synchotron experiments for krypton on graphite we would conclude that the longest possible first order jump in the misfit is 0.3% at 92°K. Without a quantitative theory, it is difficult to judge whether or not this represents a serious discrepancy. It seems to me that the unusual lineshapes we observe as well as the universal behavior of the misfit versus chemical potential difference represent the most important features of the C.-I.T. I think that your approach contains the correct physical ideas and it would be interesting to see a more quantitative application to the krypton on graphite system.

Published 1980 by Elsevier North Holland, Inc.
Sinha, ed. Ordering in Two Dimensions

OBSERVATION OF A ONE DIMENSIONALLY INCOMMENSURATE CHARGE-DENSITY WAVE STRUCTURE IN 2H-$TaSe_2$

R. M. FLEMING, D. E. MONCTON, D. B. McWHAN and F. J. DiSALVO

Bell Laboratories, Murray Hill, NJ 07974, USA

ABSTRACT

High resolution X-ray scattering studies have revealed a phase transition at 112K between a one-dimensionally incommensurate or "striped" charge-density wave (CDW) structure and a fully incommensurate hexagonal CDW structure. In the striped phase, which is seen on warming between 93K and 112K, the structure is characterized by a triple-$\vec{q}$ CDW with one commensurate and two incommensurate wave vectors. This results in a lattice modulation which is commensurate in one basal plane direction but incommensurate in the orthogonal direction.

Charge-density waves (CDWs) [1], as well as rare gases physisorbed onto surfaces [2,3] are examples of systems which possess a periodicity which may be incommensurate with the underlying host lattice. There has been widespread interest in these systems and recently questions have arisen concerning the possibility of lowering the energy be modulating the phase or the amplitude of the incommensurate superlattice, a process which results in the formation of domains [4,6]. Generally it is felt that phase modulation will dominate far from the normal-incommensurate (NI) transition. Therefore, near the commensurate-incommensurate (CI) transition the domains are thought to consist of regions of commensurate CDWs separated by domain walls where the superlattice phase varies rapidly. The domain wall separation is a function of temperature and diverges as the superlattice approaches lock-in. The domain walls may be ordered in a hexagonal array [5,6], a one-dimensionally incommensurate striped array [5], or they may be disordered [7]. Examples of ordered domain walls are shown in Fig. 1. Phase transitions may occur between the ordered states or from the disordered to the ordered states. Bak, et al [5] have considered within the Landau theory the case of CDWs in 2H-$TaSe_2$. They find that in the case of a continuous CI transition, hexagonal symmetry is broken in the incommensurate phase. This results in the formation of striped CDW domains. Near the NI transition, symmetry favors hexagonal domains. Consequently, in addition to a NI and a CI transition, Bak et al predict a third phase transition where the CDW changes from the striped to the hexagonal phase. Thus far the only experimental evidence for the existance of domain structures in incommensurate systems comes from the observation of higher diffraction harmonics in 2H-$TaSe_2$ [8] and from satellite intensity variations attributed to ordered domains in the krypton/pyrolytic graphite system [2].

CDW DOMAIN STRUCTURE (DIRECT SPACE)

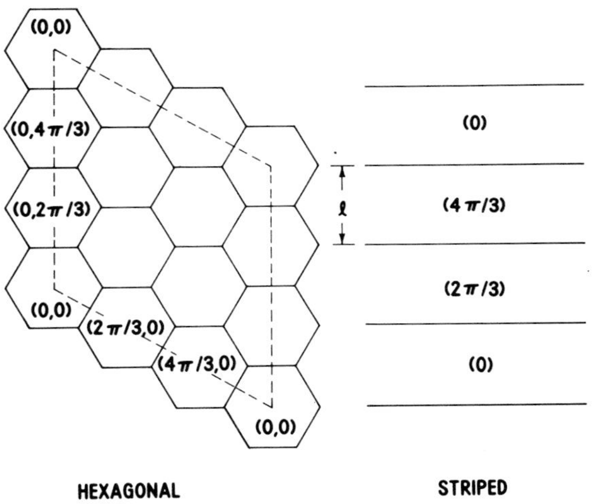

Fig. 1 Examples of ordered domain structures.

We have investigated the incommensurate CDW state in 2H-$TaSe_2$ using high resolution X-ray scattering. We have found new evidence supporting the existence of CDW domains by observing a CDW phase with the same structure as the striped domain phase of Bak et al [5] and also by observing a phase transition at 112 K between the striped and the hexagonal phases. The 112 K transition was originally observed in a dilatometry experiment by Steinitz et al [9].

Fig. 2 shows a measurement of the incommensurability of the CDW, $\delta(T) = \vec{q}_{obs} - \vec{q}_c$ where in this case $\vec{q}_c = \frac{8}{3}a^*$ and $a^* = 4\pi/\sqrt{3}a_o$. On cooling the data are in agreement with earlier neutron scattering data of Moncton et al [8]. We find a normal-incommensurate transition at T_{NI} = 123 K and a commensurate-incommensurate transition at T_{CI} = 85 K. There is no evidence of a third CDW phase transition between onset and lock-in. Contrary to previous reports of a first-order transition at T_{CI} [9-12], we find that the lock-in is continuous.

The behavoir of $\delta(T)$ on warming is dramatically different from the cooling data. First, T_{CI} = 93 K instead of 85 K. Second, scans along [h00] as shown in Fig. 3 show two superlattice peaks, one commensurate and one incommensurate, instead of the single incommensurate peak seen on cooling. Also at 112 K there is clear evidence of a first-order phase transition where the commensurate peak vanishes abruptly. Although $\delta(T)$ approaches zero continuously for both heating and cooling, the existence of a metastable phase between 93 and 112 K as well as hysteresis suggests that T_{CI} may be first order.

Analysis shows that the two peak structure seen near $(\frac{8}{3}\ 0\ 0)$ on warming results from a new CDW phase with a one dimensionally incommensurate or striped geometry. A schematic comparision of the hexagonal and the striped CDW phases is shown in Fig. 4. Fig. 4(a) shows the structure of the symmetric, hexagonal CDW. Each Bragg peak of the host lattice is surrounded by six superlattice peaks with hexagonal symmetry. In the striped phase shown in Fig. 4(b), hexagonal symmetry is broken and each of the six primary superlattice peaks breaks into three peaks. The triple-q CDW is now characterized by one commensurate and two incommensurate wave vectors. This results in a lattice modulation which is commensurate parallel to one wave vector, but incommensurate in the orthogonal direction. Because the resolution function is broad perpendicular to the diffraction plane, scans along [h00] show two peaks as in Fig. 3.

Of course, the observation of multiple peaks near the primary superlattice position does not by itself identify the striped phase uniquely. For a unique determination of the structure one must also observe scattering at second order positions which are the sum of two non-colinear wave vectors such as $\vec{q} = \vec{q}_1 - \vec{q}_2$. In Fig. 4 second order peaks of this type are identified by open symbols. Scans along [hh0] with our resolution are expected to show two peaks near a second order position in the hexagonal phase and three peaks in the striped phase. In no case is a second order peak expected at the <u>commensurate</u> position marked by the "X". A commensurate second order peak would be the signature of spatially separate regions containing commensurate and incommensurate CDWs with hexagonal symmetry.

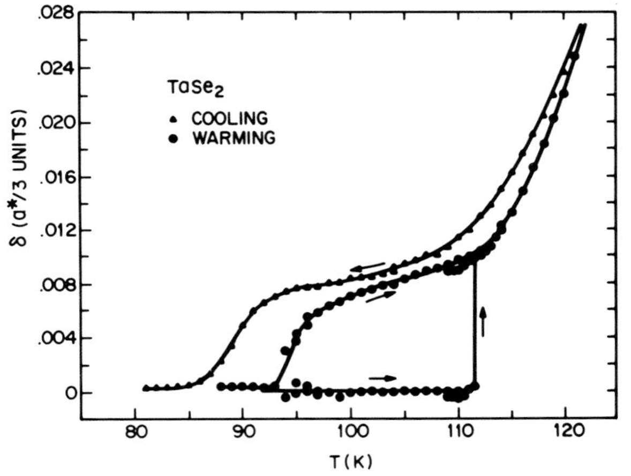

Fig. 2 Incommensurability as a function of temperature. On warming the striped CDW phase is seen the range 93 - 112 K.

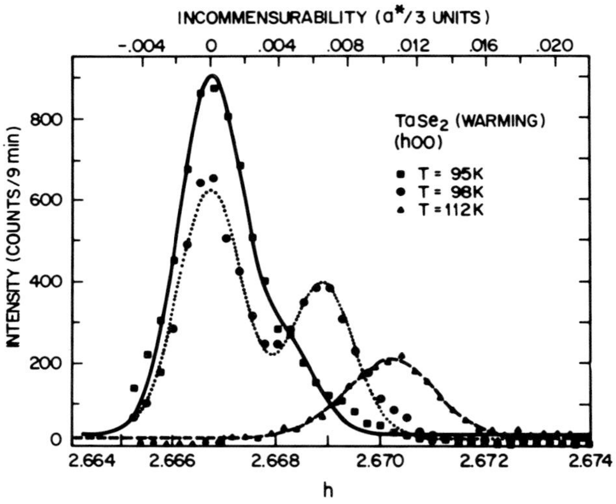

Fig. 3 Scans taken along [h00] through the $(\frac{8}{3}00)$ superlattice position. The two peaks are due to the striped CDW phase where one wave vector is commensurate and two wave vectors are incommensurate.

Fig. 5 shows scans along [h00] through the $[\frac{4}{3}\ \frac{4}{3}\ 0]$ superlattice position. For 100 K (warming) the three peak structure characteristic of the striped CDW phase is seen. For 114 K (warming or cooling) the two-peak structure of the hexagonal CDW phase is seen. No intensity is seen the commensurate second order position for any temperature above T_{CI}. Furthermore, the magnitudes of the in-plane components of momentum transfer along [hh0] are in agreement with the values predicted by the structure shown in Fig. 4. We emphasize that although the peaks perpendicular to the diffraction plane are not individually resolved, the hexagonal and striped structures are uniquely determined by the scans along [h00] (Fig. 3) and [hh0] (Fig. 5).

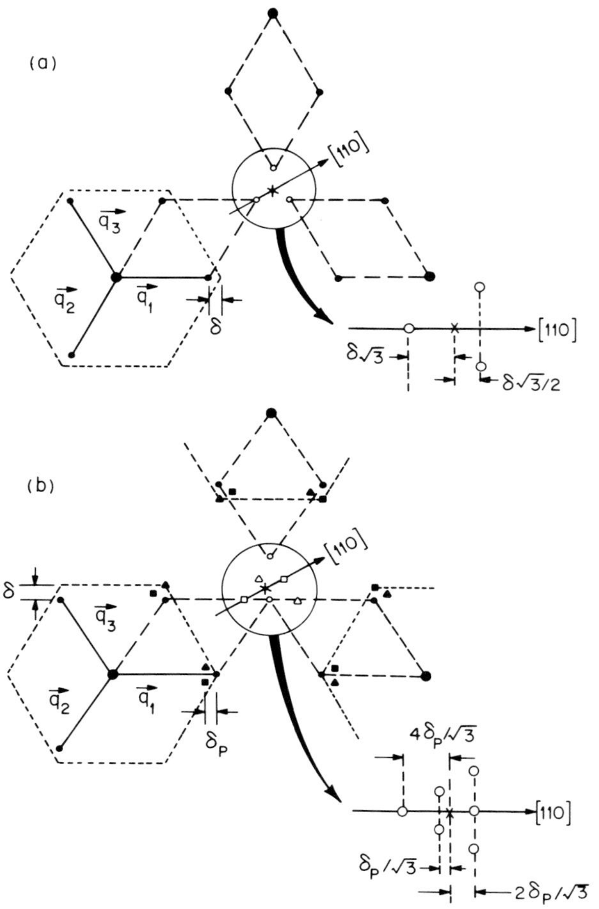

Fig. 4 Schematic plot of reciprocal space showing (a) the hexagonal CDW phase and (b) the striped CDW phase. The large circles Bragg peaks of the host lattice. The small filled symbols are the primary superlattice and the open symbols are the second-order CDW peaks.

A remarkable feature of the striped and the hexagonal CDW phases is the extraordinarily long length scale needed to establish a physical difference between the two phases. The extremely small separation of the two peaks in Fig. 3 suggests that the difference between the striped and the hexagonal structures are only apparent over lengths in excess of 300 A. The fact that we observe superlattice peaks with instrumental widths indicates that the domain wall pattern is coherent on a scale 3000 A. Since the differences between the two phases is so subtle, one might be temped to argue that the hysteresis near lock-in as well as the fact that the striped phase is only seen on warming may be a result of extrinsic effects. However previous disagreements concerning the temperature of the lock-in transition now appear to be resolved by the observation of the 112 K transition. This together with the long coherence of the CDW structures suggest that the hysteresis and the metastability are intrinsic effects.

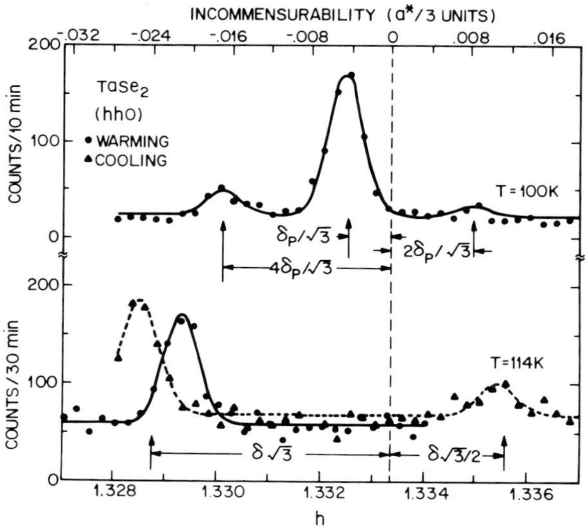

Fig. 5 Scans taken along [hh0] through the second order superlattice position, $(\frac{4}{3}\ \frac{4}{3}\ 0)$. For T = 100 K (warming) three incommensurate peaks characteristic of the striped CDW structure are seen. For T = 114 K (warming or cooling) two peaks characteristic of the hexagonal CDW phase are seen.

REFERENCES

1. For a review see F. J. DiSalvo in Electron-Phonon Interactions and Phase Transitions, edited by Tormod Riste, (Plenum, NY, 1977) p 107.
2. P. W. Stephens, R. J. Birgeneau and P. M. Horn, Phys. Rev. Lett. 43, 47 (1979).
3. J. B. Hastings and D. E. Moncton, Bull. Am. Phys. Soc. 24, 406 (1979).
4. W. L. McMillan, Phys. Rev. B 14, 1496 (1976).
5. Per Bak and David Mukamel, Phys. Rev. B 19, 1604 (1979) and P. Bak, D. Mukamel, J. Villain and K. Wentowska, Phys. Rev. B 19, 1610 (1979).
6. Kazuo Nakanishi, and Hiroyuki Shiba, J. Phys. Soc. Jpn. 44, 1465 (1978).
7. Jacques Villain in Ordering in Strongly Fluctuating Condensed Matter Systems, edited by Tormod Riste, (Plenum, NY, 1980) p. 221.
8. D. E. Moncton, J. D. Axe and F. J. DiSalvo, Phys. Rev. B 16, 801 (1977).
9. M. O. Steinitz and J. Grunzweig-Genossar, Solid State Commun. 29, 519 (1979).
10. M. Barmatz, L. R. Testardi and F. J. DiSalvo, Phys. Rev. B 12, 4367 (1975).
11. M. Barmatz, L. R. Testardi, F. J. DiSalvo and J. M. E. Harper, Phys. Rev. B 13, 4637 (1976).
12. R. A. Craven and S. F. Meyer, Phys. Rev. B 16, 4583 (1977).

DISCUSSION

S. C. Moss:

To what extent does the solution of your diffraction pattern require (striped) domains as opposed simply to permitting a domain interpretation? In other words is there a more-or-less rigorous energy argument which demands domains given your full reciprocal lattice satellite distribution?

R. M. Fleming:

Your question raises an interesting issue. To my knowledge,a rigorous energy argument which shows that the striped structure requires domains has not yet been made. One could in principle have a sinusoidal modulation of the charge density which was incommensurate in ony one direction. The result would be a diffraction pattern similar to the one we report, however the appearance of second harmonic intensity in our data is suggestive of domain wall formation.

D. E. Moncton:

In the neutron experiments of 1975, we solved the CDW crystal structure for the first harmonics. This structure predicts significantly less second harmonic intensity than is observed. We concluded that there must be a second harmonic in the actual CDW displacement wave. The natural implication of this is the tendency to form discommensuration or domain walls.

Roy Clarke:

What is the origin of the upward concave form of the incommensurability, δ, near the onset of the transition?

R. M. Fleming:

The upward curvature in δ (T) was discussed at length in the neutron scattering experiment of Moncton, Axe and DiSalvo. At that time the cause of the upward curvature was not understood. We reproduce the upward curvature in our high revolution x-ray scattering experiment but we do not provide any additional information as to its cause.

S. Fain

1) Is there interlayer coupling in 2H-$TaSe_2$?

2) Hysteresis was predicted by Bak for such a transition; is his mechanism applicable to your system?

R. M. Fleming:

1) Yes, there is weak interlayer coupling in 2H-$TaSe_2$. As a result the charge-density wave is in phase from layer to layer ($q_z = 0$). However, even though the CDW is three dimensional, many of the same domain arguments used in the case of gases physisorbed on surfaces may also be used for incommensurate CDWs in layered compounds.

2) Possibly. I believe the hysteresis and metastability are intrinsic to 2H-$TaSe_2$.

Sinha, ed, Ordering in Two Dimensions

POWDER DIFFRACTION PATTERNS FROM SUPERSTRUCTURES IN ADSORBED MONOLAYERS NEAR COMMENSURATE–INCOMMENSURATE PHASE TRANSITIONS

M. NIELSEN and J. ALS-NIELSEN
Risø National Laboratory,
DK-4000 Roskilde,
Denmark

J.P. McTAGUE
University of California,
Los Angeles, CA, U.S.A.

We discuss possible interpretation of the structured diffraction profiles observed near the commensurate to incommensurate phase transition of the $\sqrt{3}$-structure on (002) surfaces of graphite. These profiles are then related to the various models for the C-I transition. Our calculations show the difficulties in distinguishing between the different proposed models on the basis of the diffraction line profiles.

I. INTRODUCTION

At sufficiently low temperature and coverage simple gases such as He, H_2, Ar, Kr and Xe physisorb on the (002) surfaces of graphite, forming ordered 2-D structures [1-3]. The lattices are either incommensurate with the substrate surface or they take up the registered $\sqrt{3}$-structure [1]. Figure 1 shows schematically a density-temperature phase diagram typical for an adsorbed film for which the $\sqrt{3}$-structure exists. At low temperature this structure (C) coexists with a dilute gas (F) for densities up to $\rho \simeq 1$, where the total substrate surface is covered. Further increase of the coverage will bring us through a narrow region of the homogenous $\sqrt{3}$-phase, and then the monolayer is squeezed out of registry in a commensurate incommensurate (C-I) phase transition. At still higher coverages the density of the monolayer increases and ultimately second layer condensation sets in.

The commensurate $\sqrt{3}$-structure has an equilateral triangular lattice with a nearest neighbour distance of a = 4.26 Å. The substrate has three times as many identical adsorption sites as the number of ad-particles (see Fig. 1b), and we label them, A, B and C. The long range order parameter of the $\sqrt{3}$-phase is the excess density of occupied A sites $\rho_A - \frac{1}{2}(\rho_B + \rho_C)$, meaning that in the ideal $\sqrt{3}$-structure all of the say A-sites are occupied (and none of the B or C-sites).

Several models have been proposed for the C-I transition (for a recent review see Ref. 6). It is the purpose of the present article to give the basis for interpreting the neutron and X-ray diffraction results by comparing the scattering profiles which we can calculate from the different proposed models. This is done in section II. In section III we discuss some experimental results.

II. SCATTERING PROFILES FROM ABC-SUPERSTRUCTURES

HEXAGONAL DOMAINS

The scattering profiles of the $\sqrt{3}$-phase are given by the modified Warren line shape (1) and the first reflection is located at $\tau_{10} = 4\pi/(\sqrt{3}\,a) = 1.703\ Å^{-1}$. It is shown as a broken line in figure 3. Such curves are very commonly observed and show that we are within the C+F or the C region of the phase diagram of Fig. 1a.

When the density of the adsorbed monolayer is increased beyond $\rho = 1$ there are more ad-particles than can be accommodated in the pure $\sqrt{3}$-structure. The resulting competition between the forces within the adsorbed film and the in-plane forces between the film and the substrate may give rise to modulated structures as illustrated in Fig. 2, where three different configurations are shown. We shall only discuss here the diffraction line profiles of the domain structures and refer to Ref. [6] for the discussion of the physics behind the occurrence of domains.

In Fig. 2a and 2b the domain walls form a honeycomb structure; inside each hexagonal domain the ad-particles are adsorbed on one of the kind, A, B or C sites. Near the domain boundaries the local density increases, displacing the particles from the centre of the substrate hexagons as shown by the X-Y diagrams, where Y is the displacement of the ad-particles relative to the ideal A-sites along the line X. The repeat distance between particles along this line is $a\sqrt{3}$ for the pure $\sqrt{3}$ structure and in Fig. 2a the B domain is obtained by displacing the particles $\frac{1}{3}a\sqrt{3}$ relative to the A sites in the -X direction. The X Y plot shows three different ranges of relaxation of the particles around the domain walls [7]. The step function corresponds to no relaxation at all, and this situation is depicted in the shown ABC domains. The S-shaped

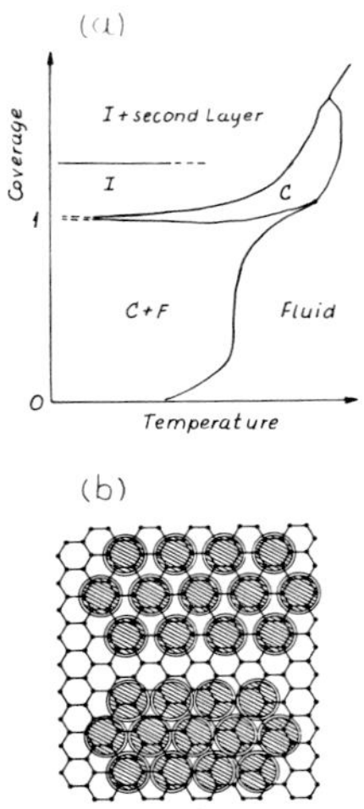

Fig. 1. (a) Schematic phase diagram for a monolayer of a simple gas adsorbed on the (002) surface of graphite.
(b) Schematic representation of the $\sqrt{3}\times\sqrt{3}$, 30° registered phase (top) and of an incommensurate dense monolayer structure (bottom).

curve indicates relaxation within a certain distance ℓ from the wall. We have chosen $\ell = 2\sqrt{3}a$ and assumed the relaxation to vary quadratically. We have chosen this wall to accentuate the effect of narrow walls. A more realistic value may be $\ell = 5\sqrt{3}a$. (Ref. 4). Finally, the straight line in the X Y plot of Fig. 2a gives a homogenous density and thus no superstructures at all. In the last case the (10) Bragg peak is just shifted in accordance with the density. The configuration of Fig. 2a will arise if we, starting with a pure $\sqrt{3}$-phase, first homogenously compress the adsorbed layer and then let every ad-particle relax towards the nearest A, B or C adsorption site. Of course only discrete values of the density will give the exact configuration.

The domains shown in Fig. 2b are obtained by displacing the B-domain-particles $\frac{1}{6}a\sqrt{3}$ along -X-direction and $\frac{1}{2}a$ perpendicular to the X-direction.

The domain structures have triangular lattices with lattice vectors given by the centre of the domains. The unit cell is a hexagonal domain and in order to find the structure factor we must sum the phase factor $\exp(-i\vec{Q}\cdot\vec{r}_j)$ over the particles within a domain. Let the distance from the center to the corner of a domain be $Na+\Delta a$, so the new reciprocal lattice vectors have the length

$$T_{(10)} = \frac{4\pi}{a\sqrt{3}} \, \frac{1}{\sqrt{3}(N+\Delta a/a)} = \tau_o \, \frac{1}{\sqrt{3}(N+\Delta a/a)}$$

where τ_o is the (10) reciprocal lattice vector of the pure $\sqrt{3}$-structure. Further the new reciprocal lattice is rotated relative to the $\vec{\tau}_o$ lattice by 30°. Fig. 4 shows how the lattice points are situated around $\vec{\tau}_o$ for the configurations of Fig. 2a ($\Delta a/a = 2/3$) and Fig. 2b ($\Delta a/a = 1/3$). $\vec{T}_1$ and $\vec{T}_2$ are the (10) and (01) reciprocal lattice vectors of the domain structure and we calculate $F(\vec{Q}=m_1\vec{T}_1+m_2\vec{T}_2) = F(m_1, m_2)$ for all the lattice points near $\vec{\tau}_o$. The summation over the hexagonal domain may be done by first summing over hexagonal rings like that shown in the top part of Fig. 4, and then adding the contribution from each ring. The n'th ring from the center contributes

$$F_n = 2I_{1,n}\cos w_{1,n} + 2I_{2,n}\cos w_{2,n} + 2I_{3,n}\cos w_{3,n}$$

where
$$I_{1,n} = \sin(\tfrac{1}{2}n\phi_j)/\sin(\tfrac{1}{2}\phi_j)$$
$$I_{2,n} = \sin(\tfrac{1}{2}n\phi_i)/\sin(\tfrac{1}{2}\phi_i)$$
$$I_{3,n} = \sin(\tfrac{1}{2}n\phi_k)/\sin(\tfrac{1}{2}\phi_k)$$
$$w_{1,n} = \tfrac{1}{2}(n-1)\phi_j + n\phi_i$$
$$w_{2,n} = -\tfrac{1}{2}(n-1)\phi_i - n\phi_k$$
$$w_{3,n} = \tfrac{1}{2}(n-1)\phi_k + n\phi_j$$

and
$$\phi_i = |\vec{T}_1| \; B_n(m_1 - \tfrac{1}{2}m_2) \; 1/n$$
$$\phi_j = |\vec{T}_1| \; B_n(-m_1 + \tfrac{1}{2}m_2) \; 1/n$$
$$\phi_k = |\vec{T}_1| \; B_n(-\tfrac{1}{2}m_1 - \tfrac{1}{2}m_2) \; 1/n$$

B_n is the distance from the center of the domain to the corner of the n'th ring. Near the domain wall this depends on the wall relaxation Y. Only the reciprocal lattice points (m_1, m_2) which are close to the reciprocal lattice points of the undistorted $\sqrt{3}$ phase will get appreciable intensity. We shall only discuss the reflections close to $\vec{\tau}_o$.

In order to find the powder diffraction intensity we must add the intensity from reflections

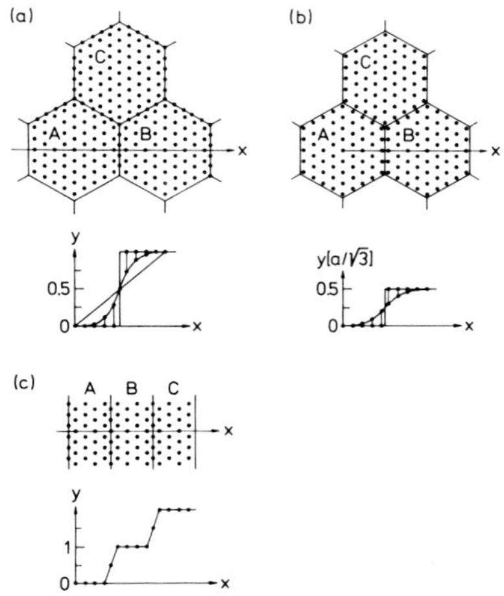

Fig. 2. (a), (b), and (c) show three different domain configurations. Y indicates the displacement along the -X-direction of the particles relative to the ideal A-positions. The (a) configuration is consistent with superstructures observed in Kr.

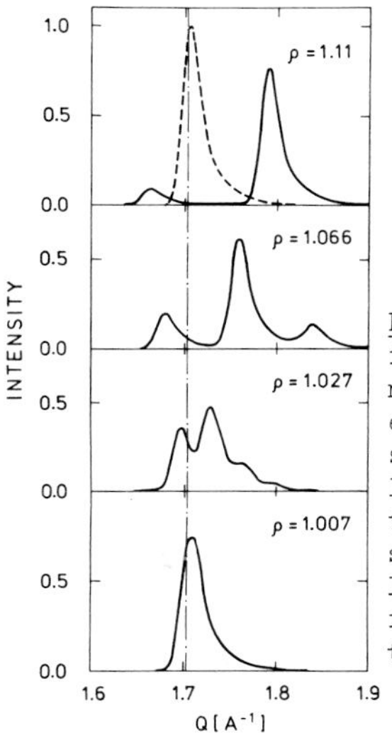

Fig. 3. Calculated diffraction profiles from monolayers adsorbed on an exfoliated graphite substrate. ρ is the monolayer density relative to the density of the $\sqrt{3}$-structure. The broken line curve is calculated for the ideal $\sqrt{3}$-structure.

with identical $|\vec{Q}|$. For large N we can add the intensities along the vertical lines in the Fig. 4. As seen from the figure we will get a series of powder diffraction lines and the two nearest to τ_o are shifted to higher and lower values of Q in the ratio 2:1 for $\Delta a/a = 2/3$ and 1:2 for $\Delta a/a = 1/3$.

In diffraction experiments using exfoliated graphite substrates finite size effects and the non-parallelism of the scattering planes can be taken into account by replacing each of the above calculated powder lines with the Warren line shape with an intensity proportional to that of the powder line and with τ_{hk} given by its position. In practice the line shape is known from measurements at $\rho \leq 1$, where we have a pure $\sqrt{3}$ phase. In using this procedure we assume that the finite size effects are not dependent on the structure we are measuring. Further, close to $\rho = 1$ where only a few domains can be placed within the coherent areas of typical extension L, we must assume that the finite size effects do not favor particular domain wall configurations. We do not know to what extent these conditions are fulfilled.

Fig. 3 shows the calculated scattering profiles for the domain configurations of the type shown in Fig. 2a where $\Delta a/a = 2/3$. The density ρ can be calculated for each discrete set of domain sizes given by N: $\rho = [3(N+1)^2+1]/(3(N+\Delta a/a)^2)$. The full curves are the calculated profiles. At $\rho = 1.11$ we have two well-resolved peaks. For this density the coherent adsorbing area (L = 350 Å) can accommodate about 50 domains. As ρ decreases the number of domains goes down, at $\rho = 1.066$ to about 20, at $\rho = 1.027$ to 4 and at $\rho = 1.007$ to less than one. In Fig. 5a and b we compare the scattering profiles from the domain structures in Fig. 2a and 2b. The shifts of the peaks relative to the 1.703 $Å^{-1}$ position are different for the two cases and are in accordance with the lattice shown in Fig. 4.

STRIPE DOMAIN STRUCTURES

For the stripe domain structure shown in Fig. 2c we can calculate F(Q) by summing the phase factor $\exp-(\vec{Q}\cdot\vec{r}_j)$ over the unit cell shown in Fig. 6. $\vec{D}_1$ and $\vec{D}_2$ are the direct lattice vectors of the domain structure and $\vec{T}_1$ and $\vec{T}_2$ the reciprocal lattice vectors. For the type of domain wall chosen, which is the same as in Fig. 2a, but without wall-relaxation, we have

$$D_1 = a\sqrt{3}/2 \quad (N+2\Delta a/a); \; D_2 = a$$

$$T_1 = \tau_o/(N+2\Delta a/a); \qquad T_2 = \sqrt{3}/2 \; \tau_o$$

Each of the τ_o points of the $\sqrt{3}$ structure is replaced by a series of points, shown in Fig. 6 as filled circles, and lying in the direction perpendicular to the domain walls. In a powder diffraction measurement the position of the reflection lines is given by the distance from these points to the origin. As above we place a Warren line shape profile at each of the reflection lines and finally add the intensities. The result is shown in Fig. 5c for the mean density $\rho = 1.07$, corresponding to N = 8. When N is odd the calculation is a little different but the result quite similar.

The scattering profile shown in Fig. 5d is calculated for the stripe domain structure but with the maximum wall relaxation which means the layer is homogenously compressed in the directions perpendicular to the domain walls. So the figures 5c and 5d give the result in the two extreme situations with none and with maximum wall relaxation; in practice we must expect the intensity of the three lines to have values in between what is shown in the figures. The position of the lines does not change with the wall

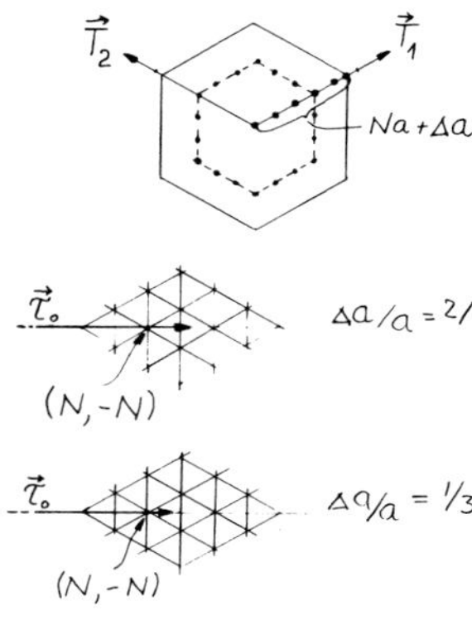

Fig. 4. Unit cell and reciprocal lattice for the hexagonal domain structure shown in Fig. 2(a) and 2(b).

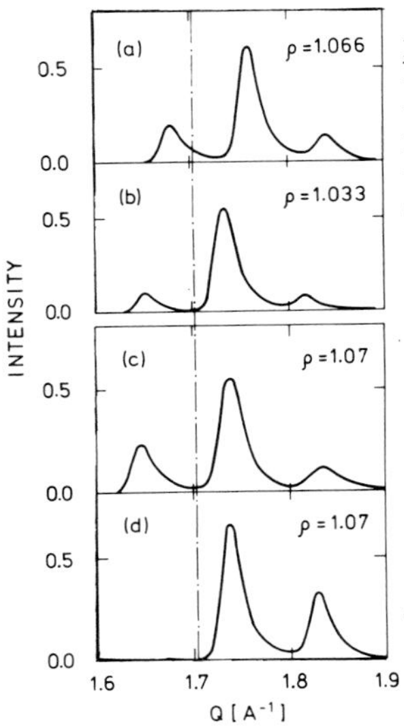

Fig. 5. (a) and (b) show the calculated diffraction profiles from the hexagonal domain structures of Fig. 2(a) and 2(b). For both cases N=10 and for (a) ρ = 1.066; for (b) ρ = 1.033. (c) and (d) show the calculated diffraction profiles for the stripe domain structure (Fig. 2 (c)). In case (c) all particles are located on ideal A, B or C sites except on the walls whereas in case (d) the adsorbate is homogenously compressed axially, along the direction X of Fig. 2.

relaxation. Of course when the density changes we will find a behaviour analogous to what is shown in Fig. 2.

ROTATED STRUCTURES

So far we have only considered domain walls having the (11) direction of the $\sqrt{3}$-structure. To discuss the result in spectroscopic measurements from rotated structures which are both theoretically expected [8] and directly seen in LEED measurements [9] we will use the model of Novaco and McTague [8]. They assume that the interactions between the ad-particles are dominant so that the in-plane substrate-ad-particle forces give rise to harmonic density waves. Thus we should in a powder diffraction measurement expect to get the main Bragg peaks determined by the mean density of the adsorbed monolayer and in addition a set of satellites around each of these from the density waves. The position of the satellites depends however on the relative rotation of substrate and adsorbate. This is illustrated in Fig. 7. The monolayer is compressed relative to the $\sqrt{3}$-structure to the density $\rho = (1+\delta)^2$ so that the reciprocal lattice is expanded linearly by $1+\delta$. This is then assumed to be rotated relative to the substrate by the angle ω. The in-plane potential acting on the ad-particles from the substrate has the periodicity of the graphite honeycomb structure and this has reciprocal lattice vectors coincident with the (11) vectors of reciprocal lattice of the $\sqrt{3}$-structure. For small values of δ and ω the substrate potential will then give density waves in the monolayer with wave vectors which join the (11) reciprocal lattice point of the $\sqrt{3}$-structure with the (11) point of the compressed and rotated film. These wave vectors determine the position of the satellites to be observed around each reciprocal lattice point of the film. In the figure two of them are shown around the $\tau(10)$ position and in the lower graph the same is shown in enlarged scale. As filled circles we show the three reflections which are nearest to τ_o, the (10) vector of the $\sqrt{3}$-structure. When we go beyond the harmonic approximation and let the particles move to the nearest A, B or C site, then the three reflections closest to τ_o become the strongest.

From figure 7 it can be seen that as the adsorbed monolayer rotates relative to the substrate (ω) the pattern of reflections around $\vec{\tau}_o$ rotates through the angle θ, given by $tg\theta = \omega/\delta$. The hexagonal domain pattern of Fig. 2a and 2b can be identified as special cases of this model where for Fig. 2a $\theta = \omega = 0$ and for Fig. 2b $\theta = \pm 60^o$.

The powder diffraction response from rotated structures which do not correspond to domain structures with walls in high symmetry directions are quite complicated, and we have not attempted any model calculation of the intensities.

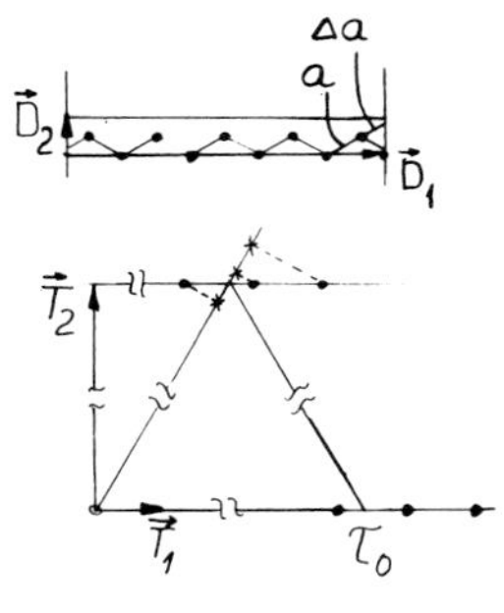

Fig. 6. Unit cell and reciprocal lattice vectors of a stripe domain structure.

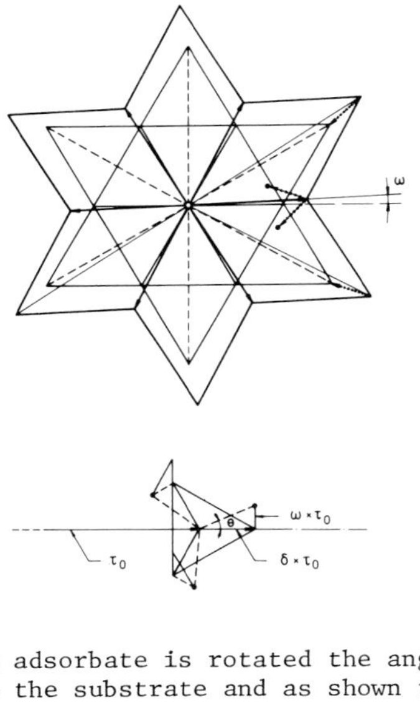

Fig. 7. The adsorbate is rotated the angle ω relative to the substrate and as shown the wave vectors of the harmonic density waves rotate the angle θ where $tg\theta = \omega/\delta$.

III. EXPERIMENTAL RESULTS AND DISCUSSION

Theoretically the following behaviour is expected for the C-I transition [6]: The transition from the pure $\sqrt{3}$-structure to the hexagonal domain structure (Fig. 2a and 2b) is a first order phase transition. This model differs from the case of parallel domain walls (Fig. 2c) by the presence of the crossing points of the domain walls. If the free energy of the crossing points is positive the stripe domains are favoured, and this may imply a continuous C-I transition, followed however at higher densities by a first order transition from the stripe to the hexagonal domain structure. The model of Novaco and McTague which gives harmonic density waves in the adsorbate is only valid close to the ideal hexagonal incommensurate structure of the adsorbate at $\rho > 1$. Thus there is expected always to be a first order phase transition involved somewhere in going from the $\sqrt{3}$-phase to the pure hexagonal incommensurate phase. The conclusions are based on low temperature approximations and may change at higher temperatures [6]. Experimentally the most detailed results are obtained on Kr-layers adsorbed on graphite. The main conclusion from the LEED studies [9] is that for $T \sim 55$ K the Kr-monolayer undergoes a continuous phase transition into a denser hexagonal phase and at about $\delta = 0.02$ it starts to rotate. The X-ray measurements by Stephens et al. [5] used UCAR-ZYX exfoliated graphite as substrate and Fig. 8 shows some of their results. The pressure (P) of the bulk vapor in

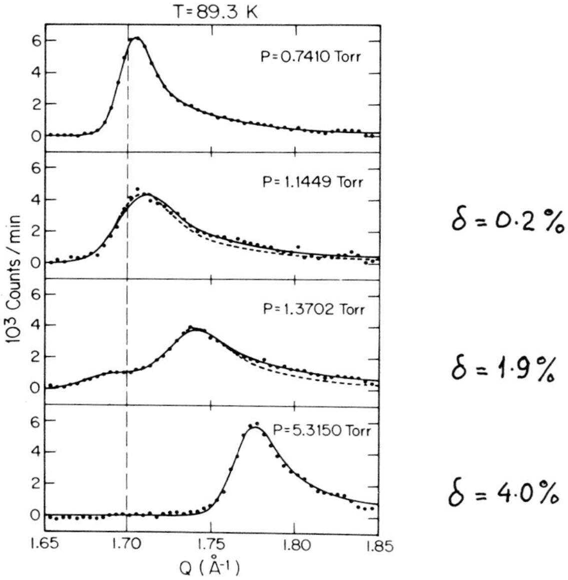

Fig. 8. X-ray data from Stephens et al. reference 5. The figures show the Kr(10) line corrected for the graphite background. The solid lines are their fitted curves.

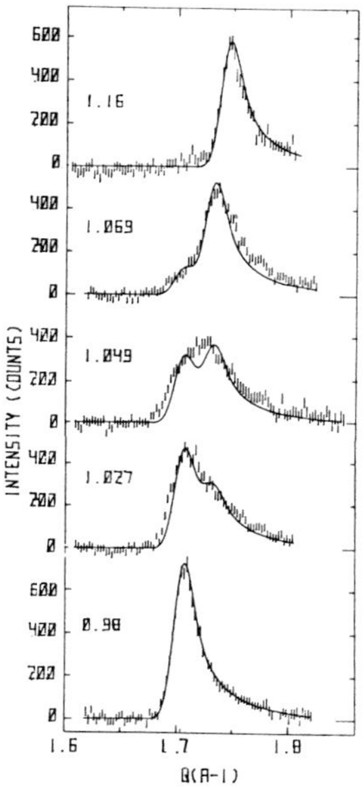

Fig. 9. Neutron diffraction data from CD_4 adsorbed on UCAR-ZYX, from reference 10. The solid lines are calculated profiles. The numbers give the density ρ and for $\rho = 1.027$, 1.049 and 1.069 it is assumed that two phases with $\tau_{10} = 1.703$ Å^{-1} and $\tau_{10} = 1.730$ Å^{-1} coexist.

equilibrium with the adsorbed monolayer is used as independent variable and the temperature is kept fixed at 89.3 K and 90.0 K. Stephens et al. have used a hexagonal domain model with $\Delta a/a = 2/3$ to interpret the scattering line profiles of Fig. 8 and the full curves are their fitted line profiles, which may be compared to those of Fig. 3. Extra broadening of the profiles is allowed in the fit but the general behaviour, a splitting of the (10) peak into a major peak at $\tau_{10}+\varepsilon$ and a smaller peak at $\tau_{10}-\varepsilon/2$, seems well established.

The rotation of the adsorbate relative to the substrate which was seen in the LEED measurement will change the powder diffraction pattern. However from the data of Fain et al. we find that at $\delta = 0.04$ the rotation is $\omega \simeq 0.4^{\circ}$ which gives a rotation of the reciprocal lattice of the domain structure of about $\theta = 10^{\circ}$. Although this contributes to the broadening of the experimental groups it is too small to influence the analysis significantly.

Deuterium monolayers adsorbed on Grafoil have been studied by neutron diffraction near $\rho = 1$ [2]. At low temperature, $T = 1.5$ K, a result rather similar to that of Fig. 8 for Kr-layers was found but no superstructures were observed.

The C-I transition has also been studied for CD_4-monolayers by neutron diffraction, using a UCAR-ZYX substrate [10]. A result very different from that of Kr- and D_2 layers was observed. Fig. 9 shows some scattering profiles measured near $Q = 1.703$ Å^{-1}, at $T = 5$ K. At $\rho \leq 1$ an ideal $\tau(10) = 1.703$ Å^{-1} line profile is observed and it is well fitted by a Warren line shape.

At $\rho > 1$ there is a region of ρ where two contributions to the scattering profile are identified, one at $Q = 1.703\ Å^{-1}$ and the other at $Q = 1.730\ Å^{-1}$ and there is a rapid spill-over of intensity from the first to the second peak. The four lower curves of Fig. 9 are calculated for the two fixed Q-positions and for the same parameter values as used above, only the intensities are adjusted. Exactly this behavior is expected for a transition involving coexisting phases and it is concluded that the transition is of first order.

IV. CONCLUSIONS

Characterization of the C-I phase tranition, in the system described here, means in the context of Ref. [6] and [11] an identification of the domain structures as function of temperature and density or 3-D vapor pressure. We are faced with the contradiction that the continuous C-I transition observed in Kr, and at low temperatures in D_2 layers, theoretically are expected to be of first order, if the transition is from the $\sqrt{3}$-structure to another hexagonal structure. The structured scattering profiles observed for Kr monolayers by Stephens et al. [5] could in fact be identified with the stripe domain structure choosing a suitable type of domain wall but for higher densities this model would predict distinct double peaks (Fig. 5d) which are not seen. For both the Kr and the D_2 data we must conclude that the C-I transition is continuous or at least only weakly of first order, and the transition is not described by the stripe domain model. In order to see if the C-I transition of Kr-layers remains continuous down to low temperatures new X-ray measurements have been initiated [12] and preliminary results indicate that the transition does change to be of first order.

The diffraction result on CD_4 monolayers at low temperatures illustrates the coexisting-phases-response of what is believed to be a first order C-I transition. It is not possible to model these scattering profiles (Fig. 9) using the domain structures without making unrealistic assumptions of rapid changing over from one model to the other.

In this article we have attempted to illustrate the problems involved in interpreting the structured diffraction profiles which are measured near $\rho = 1$ from monolayers adsorbed on exfoliated graphite substrates by neutron and X-ray scattering. No other technique gives the necessary Q-resolution to resolve the domain superstructures proposed. The mixing of the information which is inevitable in the powder diffraction technique is a severe complication which can only be overcome when data from other sources like LEED measurements are taken into account in the interpretation. As the techniques are further developed more information can be obtained by studying the higher order reflections (20) and (21) in the powder diffraction, in greater details than has been possible so far due to high background scattering and low intensity.

REFERENCES

1. M. Nielsen, J.P. McTague and L. Passell, Proceedings of Nato Advanced Study Institute on Phase Transitions in Surface Films, held at Ettore Majorana Centre for Scientific Culture, Erice, Sicily, Italy, June 11-25, 1979, p. 127, Plenum Press.
2. J.K. Kjems, L. Passell, H. Taub, J.G. Dash, and A.D. Novaco, Phys. Rev. 13, 1446 (1976).
3. M. Nielsen, W.D. Ellenson, and J.P. McTague, in "Neutron Inelastic Scattering" 1977 (IAEA, Vienna), Vol. II, p. 433 (1978).
4. M.D. Chinn and S.C. Fain, Jr., Phys. Rev. Lett. 39, 146 (1977). J.A. Venables, and P.S. Schabes-Retchkiman, J. Physique Colloq. 38 (1977) C4-105
5. P.W. Stephens, P. Heiney, R.J. Birgeneau, and P.M. Horn, Phys. Rev. Lett. 43, 47 (1979).
6. J. Villain, Proceedings of Nato Advanced Study Institute on Strongly Fluctuating Condensed Matter Systems, held in Geilo, Norway, April 16-27 (1979) p. 211, Plenum Press.
7. We use the notion "relaxation" to characterize the displacement of the ad-particles away from the ideal A, B or C-sites, forced by the interaction between the ad-particles.
8. A.D. Novaco and J.P. McTague, Phys. Rev. Lett. 38, 1286 (1977). J.P. McTague and A.D. Novaco, Phys. Rev. B19, 5299 (1979).
9. S.C. Fain, Jr., M.D. Chinn, and R.D. Diehl, Phys. Rev. B, to be published.
10. M. Nielsen, S. Sinha, P. Dutta, and M. Bretz, to be published.
11. P. Bak, D. Mukamal, J. Villain, and K. Wentowska, Phys. Rev. B19, 1610 (1979).
12. J. Bohr, M. Nielsen, and J. Als-Nielsen, unpublished.

DISCUSSION

H. Zabel:

What can one say from the diffraction data about the relative orientation of the domains in monolayers of D_2, Kr and CD_4 on graphite?

M. Nielsen:

In X-ray and neutron diffraction technique we can not see orientational epitaxy directly as in LEED. Of the adsorbates you mention only Kr has been studied by LEED near the C-I transition. Sam Fain found that Kr layers do rotate relative to the graphite above $\rho = 1$ but the angle it rotates away from the symmetry direction is small, something like 0.4° for an 8 percent decrease of lattice parameter relative to the value for the $\sqrt{3} \times \sqrt{3}$ phase. This is not enough to change the satellite group seen by x-ray diffraction on Kr layers.

S. Fain:

In answer to the queston from H. Zabel, our 52K LEED data for Kr on graphite indicate that the incommensurate overlayer does not start to rotate until the misfit is greater that 2% [Fain, Chinn, and Diehl, Phys. Rev. B21 (May 1980)]. The maximum amount of rotation is less than 0.5°.

A. D. Novaco:

Very preliminary theoretical results for D_2 on Grafoil indicate that the D_2 lattice is oriented at a very small angle relative to the $\sqrt{3} \times \sqrt{3}$ structure when the system is close to the transition but is still incommensurate.

M. Nielsen:

Experimentally we have no information about the rotation of D_2 or CD_4 layers in the incommensurate regions.

Sinha, ed. Ordering in Two Dimensions

MELTING AND DYNAMICS OF TWO-DIMENSIONAL SOLIDS AND LIQUID-CRYSTALS

B. I. HALPERIN

Harvard University, Cambridge, Massachusetts 02138

ABSTRACT

The theory of dislocation-mediated melting in two dimensions has yielded predictions about the variety of intermediate liquid-crystal phases that may occur, the nature of transitions between the phases, dynamic properties, and the possible effects of an incommensurate crystalline substrate.

During the last few years, the ideas of Kosterlitz and Thouless [1] have been applied to the analysis of melting and the classification of liquid-crystal phases in two dimensions. Although the basic theory is concerned with free suspended layers, or with a layer on a smooth substrate, many features are unchanged when one considers melting of a layer adsorbed on a crystalline substrate, provided that the adsorbate period is "incommensurate" with the period of the substrate. Since this work has been reviewed in several places [2,3], and since the details are now available in a series of published papers and preprints, it does not seem worthwhile to include a full review in these conference proceedings. However, an outline of the subject and a guide to some recent publications, may be useful.

MELTING OF A REGULAR TRIANGULAR LATTICE ON A SMOOTH SUBSTRATE

If melting takes place by the Kosterlitz-Thouless mechanism, i.e., by the appearance above T_m of a small number of free dislocations, then the melted phase retains a quasi-longrange order for the orientation of the "bonds" between neighboring atoms [4]. This phase is a new kind of liquid-crystal, with a six-fold orientational symmetry, that we have called the "hexatic" phase. A transition from the hexatic phase to an ordinary liquid, having only short range orientational and translational order, is driven by the appearance of free disclinations, at a second transtiion temperature T_i. No observable singularity is expected in the specific heat or other thermodynamic functions at either T_m or T_i. Predictions have been made for the divergence of correlation lengths at the transition temperatures, and for the discontinuities in the elastic constants [4,5].

EFFECTS OF CRYSTALLINE SUBSTRATE

The most dramatic effect of a crystalline substrate is the tendency to lock the period of the adsorbate lattice to a simple rational multiple of the substrate periodicity. Nevertheless, in many systems, it is possible to have an adsorbate lattice whose period can vary freely as a function of temperature and coverage [6]. This situation is loosely described as an incommensurate solid phase of the absorbate, but there is actually no difference in the physical properties of this phase if the lattice constants are accidentally commensurate, at sufficiently high order [4].

Potentially the most important effect of an incommensurate crystalline substrate is the establishment of certain favored orientations for the adsorbate lattice. This orientational coupling will have little effect on the nature of the phase transition at T_m, but it will have a significant effect on the nature of the phase above T_m, and a drastic effect on the transition which occurs at the upper temperature T_i for the case of a smooth substrate [4].

Consider a triangular lattice adsorbed on a hexagonal substrate. If the favored orientation for the adsorbate bonds is along a symmetry direction of the substrate, then there will be a six-fold orientational symmetry at all temperatures, there will be no distinction between the "isotropic" fluid and hexatic phases, and there is no necessity for a second transition at T_i. On the other hand, Novaco and McTague [7] have shown that in typical cases, the favored orientations of the incommensurate solid are at an angle $\pm\phi$ from a symmetry direction so that there is a broken orientational symmetry in the solid phase. If melting occurs by the dislocation mechanism, we expect the broken orientational symmetry to persist above T_m, until a second transition temperature T_i' where the hexagonal symmetry is restored. Although T_i' may be close to the disclination-unbinding temperature T_i for the adsorbate on a smooth substrate, the transition at T_i' should be quite different in character. In particular, if the T_i' transition is not first-order, it should have exponents characteristic of the two-dimensional Ising model, such as an order parameter vanishing as $(T_i'-T)^{1/8}$, and a specific heat singularity (probably very weak) proportional to $\ell n\ |T-T_i'|$.

DYNAMIC PROPERTIES

A hydrodynamic description has been developed for the hexatic phase, and the long-wavelength low-frequency modes have been identified [8]. The behavior of transport coefficients in the various phases, and the breakdown of hydrodynamics near the phase transitions T_m and T_i, have been discussed. It has been shown, also [9], that there are (weak) logarithmic divergences in two-dimensional fluids.

TILTED MOLECULES AND MELTING OF ANISOTROPIC LATTICES

In applications, one frequently deals with molecules oriented at an angle away from the normal to the plane. Cooperative effects tend to line up the molecules. The possible phases that can occur in such a system have been discussed with consideration of the coupling between the tilt orientation and the orientation of "bonds" between molecules in the plane [10]. The melting of a uniaxial solid has been considered in some detail [11,12], and the possibility of a state with the properties of a two-dimensional smectic has been discussed [11-13].

CAVEAT

Although the theories discussed above explore the consequences of a dislocation-mediated melting of the Kosterlitz-Thouless type, the theories cannot rule out the possibility of a phase transition by another mechanism, such as a large first order transition directly from the solid to the isotropic fluid. Computer simulations, carried out on several systems by various research groups attempting to decide between these possibilities, have resulted in considerable controversy. Relaxation in the vicinity of the melting temperature is probably a very slow and complicated process, however, proceeding on a hierarchy of time scales involving propagation of phonons, diffusion of vacancies and dislocations, and finally processes such as dislocation climb and grain-boundary diffusion. These long relaxation times greatly complicate the interpretation of computer experiments.

ACKNOWLEDGMENT

The work reviewed here involved collaborations and discussions with D. R. Nelson, A. Zippelius, S. Ostlund, J. Toner, and R. H. Morf. It was supported in part by the NSF through the Harvard Materials Research Laboratory, and Grant No. DMR-77-10210.

REFERENCES

1. J. M. Kosterlitz and D. J. Thouless, J. Phys. C6, 1181 (1973); J. M. Kosterlitz, J. Phys. C7, 1046 (1974); J. M. Kosterlitz and D. J. Thouless, Prog. Low. Temp. Phys., Vol. VII-B, edited by D. F. Brewer (North-Holland, 1978).
2. B. I. Halperin, in Proceedings of the Kyoto Summer Institute 1979 -- Physics of Low Dimensional Systems edited by Y. Nagaoka and S. Hikami, (Publications Office, Progress of Theoretical Physics, Kyoto, 1979) p. 53.
3. B. I. Halperin and D. R. Nelson, in Light Scattering in Solids, edited by J. L. Birman, H. Z. Cummins, and K. K. Rebane, (Plenum Press, N.Y., 1979) p. 47.
4. B. I. Halperin and D. R. Nelson, Phys. Rev. Lett. 41, 121 (1978); E41, 519 (1978); D. R. Nelson and B. I. Halperin, Phys. Rev. B19, 2457 (1979).
5. A, P. Young, Phys. Rev. B19, 1855 (1979); D. R. Nelson, Phys. Rev. B18, 2318 (1978).
6. For reviews of the theory of the commensurate-incommensurate transition in two-dimensions, see V. L. Pokrovsky and A. L. Talapov, preprint; and J. Villain, in Proceedings of the NATO Advanced Study Institute, Geilo, April 1979.
7. A. D. Novaco and J. P. McTague, Phys. Rev. Lett. 38, 1286 (1977).
8. A. Zippelius, B. I. Halperin, and D. R. Nelson, preprint, submitted to Phys. Rev.
9. A. Zippelius, preprint, submitted to Phys. Rev.
10. D. R. Nelson and B. I. Halperin, Phys. Rev. B (in press).
11. S. Ostlund and B. I. Halperin, manuscript in preparation.
12. S. Ostlund and B. I. Halperin, contributed paper at this conference.
13. J. Toner and D. R. Nelson, preprint.
14. R. Morf, Phys. Rev. Lett. 43, 931 (1979); D. Frenkel and J. P. McTague, Phys. Rev. Lett. 42, 1632 (1979); F. F. Abraham, Phys. Rev. Lett. 44, 463 (1980); S. Toxvaerd, J. Chem. Phys. 69, 4750 (1978); F. VanSwol, L. V. Woodcock, and J. W. Cape, (preprint); S. Toxvaerd, Phys. Rev. Lett. 44, 1002 (1980); J. Tobochnik and G. V. Chester, preprint.

DISCUSSION

R. B. Griffiths:

Is it possible for the solid-to-hexatic and hexatic-to-liquid phase transitions to be first order? If so, can you suggest which potentials would favor a first-order transition?

B. I. Halperin:

I do not know of anything that would exclude the possibility of a first order transition from solid to hexatic, or hexatic to liquid. I have no idea which potentials would favor such transitions, however.

M. Schick:

Concerning the Ising transition you described between the phase in which the orientational order of the overlayer is aligned with the symmetry axes of the substrate to that in which it is not, why must one strain a crystal in order to observe it by LEED?

B. I. Halperin:

The order parameter for the transition, Im $<\psi>$, is proportional to the difference of the LEED intensity at angles θ and $-\theta$, relative to the symmetry direction. In order to measure this properly, of course, one must obtain a single domain sample, or at least a multidomain sample in which the domains are predominantly of one sign, due to the effects of a weak symmetry breaking field. (By comparison, to measure the magnetization of a ferromagnet one applies a magnetic field which is very weak compared to the exchange energy, but strong enough to align the domains.) The observation of a symmetric double-peaked LEED intensity, such as would arise from a multi-domain sample without a symmetry breaking perturbation, is strongly suggestive of a non-zero value of Im $<\psi>$, but it is not conclusive evidence. A double peaked structure could also result if there were regions the size of the translational correlation length, each aligned at an angle $\pm\theta_0$ but with no long range correlations in the sign of the orientation.

A. N. Berker:

A follow-up to Griffith's question: Wouldn't a microscopic interaction which lowers the dislocation core energy (or the energy of some other short-wavelength fluctuation) tend to make the transition first-order?

B. I. Halperin:

Quite likely.

A. Holz:

What type of motion do you admit for your dislocation structures. In particular it can be assumed that the Kosterlitz-Thouless theory applies essentially (if it applies at all) only to conservative motion. Non conservative motion requires mass transport which may be rate controlling and in addition introduces a new conservation law into the theory which is not included in the Kosterlitz-Thouless type of approach.

B. I. Halperin:

In the equilibrium theory dislocation climb and glide are both included. Since there will be a finite density of interstitials and vacancies, near the melting temperature, climb can take place at a finite rate. When calculating the dynamic behavior of the system, of course, one must take into account the very different rates of climb and glide, as well as the requirements of particle conservation. This has been done in the preprint by Zippelius, Nelson and myself, cited above.

D. J. Thouless:

(1). How do the results of submonolayer helium experiments fit in with the theory of vortex-mediated melting?

(2). There is a significant contrast between the rather high proportion of defects present at the melting temperature of the solid and the low proportion of defects at the transition temperature in Monte Carlo studies of the planar spin model.

B. I. Halperin:

I do not understand the submonolayer helium experiments.

SIMULATION STUDIES OF THE 2-D MELTING MECHANISMS

J. P. McTAGUE, D. FRENKEL, and M. P. ALLEN

University of California, Los Angeles, California 90024

ABSTRACT

Computer simulations can reveal details of the thermal properties of 2-D systems not observable by other techniques. We report here studies of r^{-6} repulsive systems (256 and 2500 particles) which show the role of dislocations and grain boundaries in the 2-D melting process. Three regions are observed; solid, oriented fluid, and isotropic fluid. The solid is characterized by algebraic decay of g(r), but long range orientational correlations. The oriented fluid has exponential decay for g(r) and algebraic dependence for orientation, while the isotropic fluid phase shows the normal exponential behavior for both properties. The qualitative features of 2-D melting in the systems studied thus are consistent with the Kosterlitz-Thouless-Halperin-Nelson mechanism, although the defect structure in both the oriented and isotropic fluids is rather complex.

Although the melting-freezing transition pair in 3-D is arguably the most important of all phase transitions, we have remarkably little knowledge of the mechanism(s) involved. This ignorance is related to the hard first-order nature of the transition; without a continuous change in order parameter, or at least a significant degree of pretransitional fluctuations, there is no observable signature to test a proposed mechanism. The 2-D melting phenomenon, however, offers more hope: fluctuations are more important in lower dimensions, and there is evidence that, in some cases at least, 2-D melting can be continuous [1].

We report here a combined Monte Carlo (MC) and molecular dynamics (MD) study of melting in a particularly simple system, namely softly repulsive discs interacting pairwise by the potential $V(r_{12}) = \varepsilon(\sigma/r_{12})^6$. The overall density $\rho^* = \rho\sigma^2 = 0.8$ ($\rho \equiv N/A$) was maintained by periodic boundary conditions. Two simplifications of an r^{-n} potential are the impossibility of a gas-liquid transition, and the universal ρ-T phase behavior characterized by a single dimensionless scaling parameter $\phi = \rho^*(\varepsilon/kT)^{2/n}$.

An obvious purpose of these simulations is to test the Kosterlitz-Thouless (K-T) [2], Halperin-Nelson (H-N) [3] theory of dislocation-disclination mediated melting for this potential. Crucial to testing the mechanism rather than just some of its predictions for correlation functions, etc., is determining an unambiguous method for directly observing the defect structure as a function of temperature.

The usual Burgers vector method of dislocation characterization is quite useful for a system with only a few defects, since almost everywhere the underlying lattice structure is apparent. Difficulties appear, however, when the system becomes highly defected.

COMPUTATIONAL DETAILS

All runs reported here were on a 2500 particle system system in a rectangular box of aspect ratio $\sqrt{3}/2$, chosen to accommodate a perfect triangular lattice at low temperature. The standard Metropolis MC method was used. Very long runs were made; in the transition region more than 10^8 moves (4×10^4 per particle) were required, while at least 4×10^7 moves were made away from the transition region. Local equilibrium was tested by examining block averages over 5×10^6 moves. The usual thermodynamic properties were monitored, while the orientational order parameter

$$\psi_6 = N^{-1} \sum_{i=1}^{N} \delta(\vec{r} - \vec{r}_i) \{\tfrac{1}{6} \sum_{j=1}^{6} \exp(i6\theta_{ij})\}$$

was used to determine the four point orientational correlation function $g_6(r) \equiv \langle\psi_6(r)\, \psi_6^*(0)\rangle$. Here j labels the 6 nearest neighbors of a given particle, and θ_{ij} is the angle between some fixed axis and the line joining particles i and j.

The defect structure of a 2-D system is conveniently characterized by the coordination numbers (CN) of each atom. An unbiased definition of nearest neighbors is given by the Wigner-Seitz (or Dirichlet Domain) construction, as shown in figure 1. The lines drawn are the perpendicular

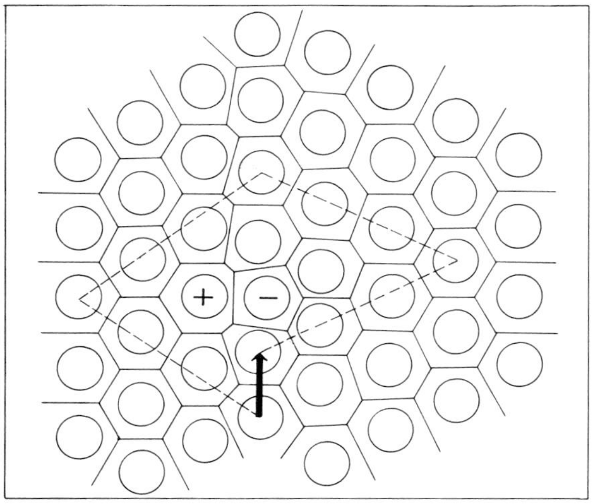

Fig. 1: The Dirichlet domain construction. A 7-5 pair is illustrated, with the path around the dislocation defining the Burgers vector.

bisectors of vectors to neighboring atoms. The smallest polygon around a central particle then uniquely specifies the number of nearest neighbors. For a perfect, T = 0 triangular lattice, all such polygons are regular hexagons, giving CN = 6 for each particle. It is a remarkable property of 2-D, however, that the average coordination number is a conserved variable regardless of the degree of disorder (this is not true in 3-D) [4]. (This statement is true except for a vanishingly small fraction of configurations, of which, unfortunately, the square lattice is one: see the Appendix.)

This conservation principle suggests that the "wrongly" coordinated particles may play a central role as elementary excitations in 2-D disorder. Such is indeed the case. As illustrated in fig. 1, a dislocation of Burgers vector 1 consists of a pair of 7- and 5-coordinated particles, while each non-6 coordinate particle can be viewed as a disclination. Thus a dislocation is a pair of (usually) 7 and 5 coordinate atoms, while a dislocation pair is a 7-5, 7-5 quadrupole. A string of 7-5, 7-5, 7-5... entities is a grain boundary. Vacancies and interstitials can also be identified in terms of clusters of non-6's [5].

RESULTS

A typical configuration of a high temperature ($T^* \equiv k_B T/\varepsilon = 0.15$) solid is shown in fig. 2a. This figure illustrates the great difficulty in identifying the defect structure by eye. However, the Dirichlet domain construction readily reveals the defect structure, as shown in fig. 2b. For clarity, only non-6 coordinate particles are shown, with 7's

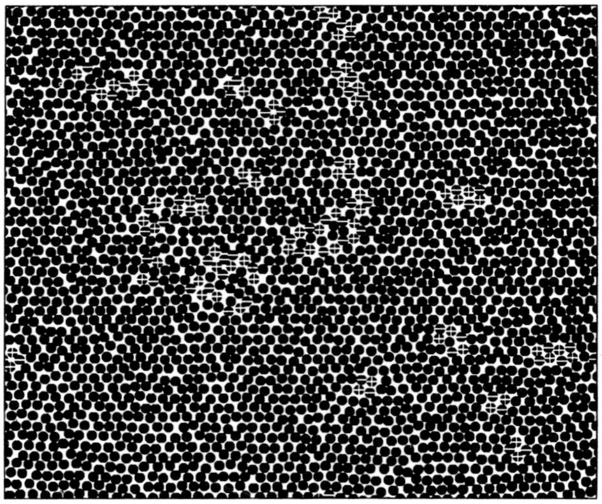

Fig. 2a: Typical configuration for the hot solid at T* = 0.150. Note the misoriented regions within the dislocation loops.

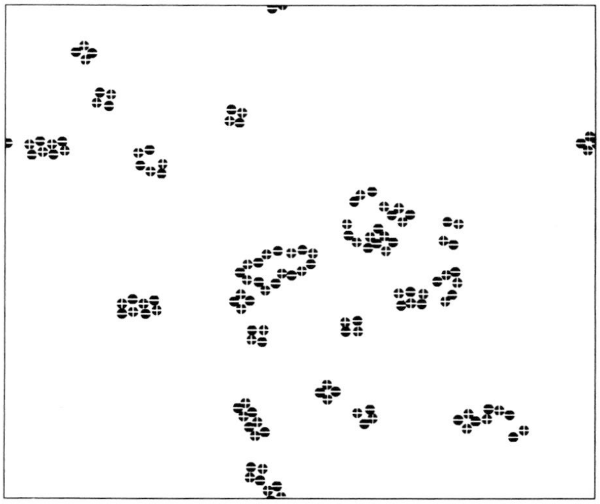

Fig. 2b: Defect structure for the configuration in fig. 2a.

labelled by +, 5's by –, 8's by #, etc. At low temperature ($T^* = 0.1$, Fig. 3) only 7-5, 7-5 quadrupoles (i.e., dislocation pairs) are observed, but at $T^* = 0.15$, just on the edge of the melting transition, more complex structure is apparent, including grain boundary loops and dislocation triplets.

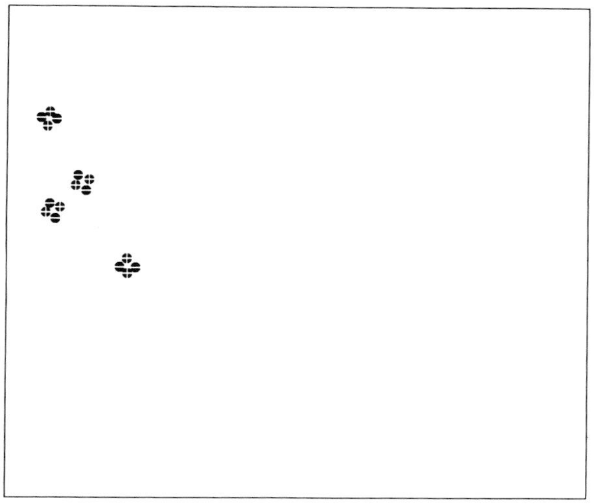

Fig. 3: Defect structure for the moderately hot solid at $T^* = 0.10$. Four dislocation pairs are apparent.

In fig. 4 we display the fraction of defects as a function of T^{-1}. The low temperature behavior yields a dislocation core energy $E_c = 1.5\ \varepsilon \approx 10\ k_B T_m$, where T_m is the melting temperature. The defect density changes rapidly in the region $0.15 < T^* < 0.15625$. At $T^* = 0.15$ no free dislocations are apparent, while at $T^* = 0.1525$ a few can be seen in each configuration (fig. 5). MD studies on a smaller (256 particle) but otherwise identical system show that this appearance of free dislocations correlates with a solid-fluid transition. Although some free dislocations are apparent at $T^* = 0.1525$, the majority are in collective loop-like and cluster-like configurations. There is, however, no sign of macroscopic phase separation, although the structural scale is of order a few hundred particles, so smaller systems might lead one to infer a phase separation [6].

The behavior of the correlation functions g(r) and $g_6(r)$ is shown in figs. 6 and 7. For $T^* \leq 0.15$ examination of ln g vs. ln r plots confirms that the envelope of g(r) decays algebraically, as do

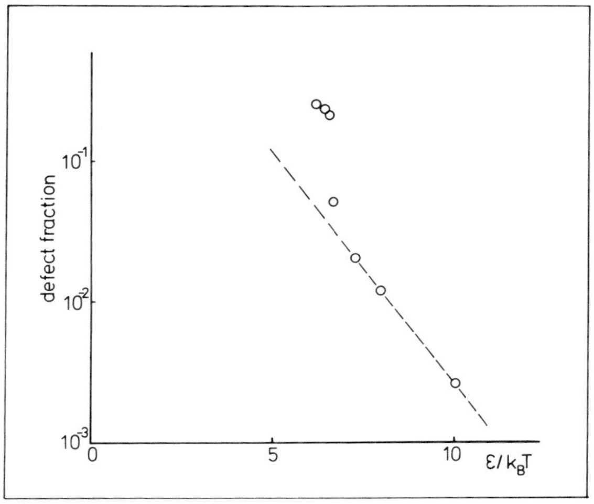

Fig. 4: Ln (defect fraction) vs. T^{*-1}.

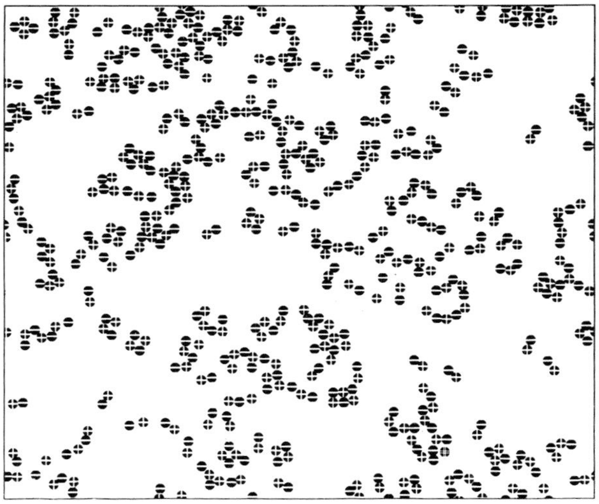

Fig. 5: Defect structure in the oriented fluid at $T^* = 0.1525$. Both free dislocations and many grain boundaries are apparent.

the Bragg peaks $S(\tau_{kh} - q)$. However, for $T^* \geqslant 0.1525$ g(r) decays exponentially. The orientational correlation function $g_6(r)$ shows sensibly infinite range for $T^* \leqslant 0.15$, as expected, while at $T^* = 0.1525$ it decays at $r^{-\eta_6}$, with $\eta_6\ (T^* = 0.1525) \approx 1.13$. At $T^* = 0.15625$ and above $g_6(r)$

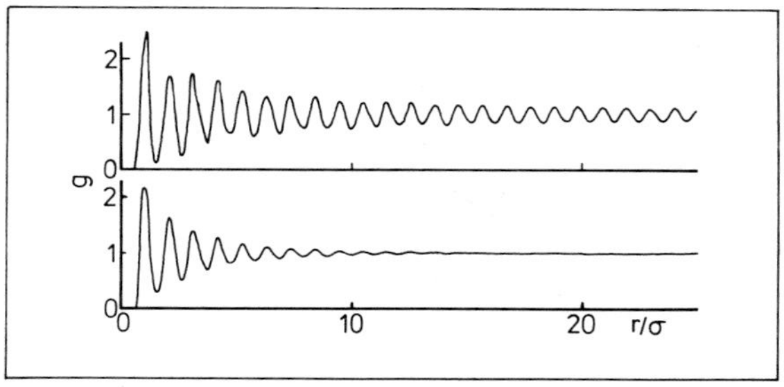

Fig. 6: g(r) vs. r at $T^* = 0.150$ (upper) and 0.1525 (lower). The envelope of the decay of [g(r)-1](0.150) is algebraic, with decay constant $\eta \approx 0.27$. The envelope for $T^* = 0.1525$ is exponential.

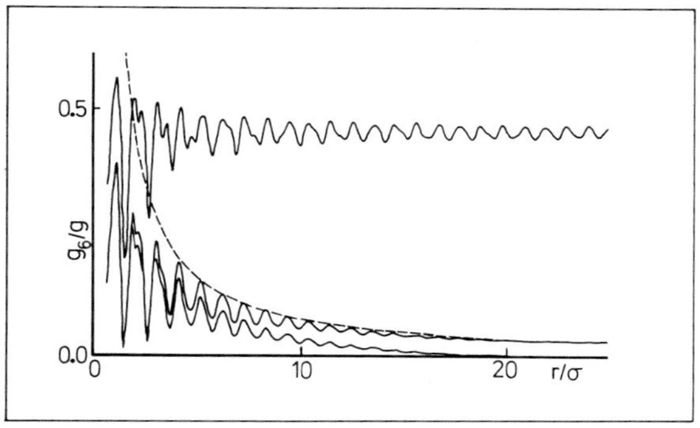

Fig. 7: $g_6(r)/g(r)$ at $T^* = 0.150$ (upper), 0.1525 (middle) and 0.15625 (lower). The dashed line is of the form $r^{-1.13}$. $g_6(r)/g(r)$ represents the orientational order correlation per particle vs. r.

decays exponentially as in a normal isotropic fluid.

According to the K-T theory of melting, T_m is given by the relation $K(T_m^-) = 16\pi$ (≈ 50). Here $K = (k_BT)^{-1} a_o^2 \mu(\mu + \lambda)/(2\mu + \lambda)$ and μ, λ are the Lame coefficients, and a_o is the lattice plane spacing. We find the following values for $K(T^*)$: 101 (0.10); 74 (0.125); 75 (0.137); 56 (0.15); 0 (0.1525). These results are clearly consistent with K-T theory, but the numerical results are preliminary and are probably reliable to only ± 25%. H-N predict that the fluid above this transition should have $g_6(r) \sim r^{-\eta_6(T)}$, with $\eta_6(T_m^+) = 0$, and increasing to $\eta_6(T_i^-) = 1/4$, where T_i is the temperature of decomposition of dislocations into disclinations. We presume that the numerical discrepancy between H-N and the present results is due to the predominance of dislocation loops (grain boundary loops), which provide an additional disorienting mechanism.

Despite the extreme length of our runs we are unable to determine unequivocally whether the melting transition at T_m is continuous or first order. Even after 4×10^4 moves per particle there is still a discrepancy in the energy in coming from higher vs. lower T, and there is still a slow drift in E. As pointed out by Knak Jensen and Mouritsen [7] it is necessary to show that the energy distribution becomes bimodal (first order) or unimodal (continuous) regardless of the direction of approach. What we have shown is that the transition is extremely sluggish, and that even longer runs are required to determine the order. In the region of T_m the energy changes rather rapidly (fig. 8).

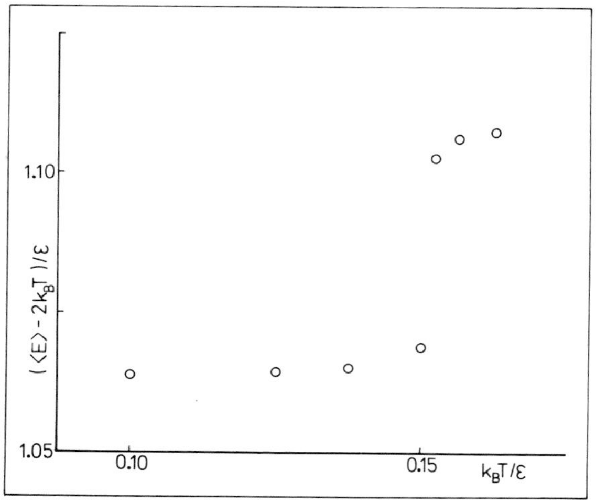

Fig. 8: Temperature dependence of the internal energy.

CONCLUSIONS

By use of the Dirichlet domain construction we have been able to identify the defect structure connected with the melting of a system of soft repulsive discs. The transition from crystal to oriented fluid is both qualitatively and quantitatively given by the K-T theory of dislocation mediated melting, although we are unable to confirm whether the transition is continuous. However, other defect structures become apparent around this transition; in particular, grain boundary loops are quite significant in both the oriented and isotropic fluids. Above the first transition, the oriented fluid has a large number of clustered defects, including grain boundary loops. The transition to isotropic fluid in the system studied appears to involve microscopic grain boundary structure. Isolated disclinations have only rarely been identified. Where comparable, our results are quite similar to those on the one-component plasma ($V(r) \sim r^{-1}$) [8] and 2-D Lennard-Jones ($V(r) \sim (\sigma/r)^{12} - (\sigma/r)^{6}$) [9], and suggest that dislocations play a major role in 2-D melting for a wide variety of potentials, as originally suggested by Cotterill [10].

APPENDIX

We present here a simple proof that the average coordination number is a conserved variable. As illustrated in Fig. 1 the Dirichlet domain construction defines a connected graph each face of which is a Wigner-Seitz cell containing one particle.

Each vertex of the graph is equidistant from three nearby particles and hence is of degree 3: only in a vanishing fraction of degenerate cases (such as the square lattice) are vertices of higher degree found. Consequently the total number of edges E is related to the number of vertices V by the equation

$$E = 3V/2$$

The conservation law is obtained by applying the Euler formula for a connected graph

$$V - E + N = 2 - 2\gamma$$

which relates the numbers of vertices, edges and faces (particles) to the genus γ of the surface in which the graph may be embedded. In our case, the employment of periodic boundary conditions is topologically equivalent to embedding in a toroidal surface, which is of genus $\gamma = 1$. The above equations then yield

$$E = 3N$$

Since each edge separates two particles the average coordination number is exactly 6.

ACKNOWLEDGEMENTS

This research was supported in part by NSF grant CHE 79-15180. We have profited from discussions with David Nelson and Michael Schick.

REFERENCES

1. J. P. McTague, M. Nielsen, and L. Passell in Ordering In Strongly Fluctuating Systems ed. T. Riste Page 195 (Plenum Press 1980).
2. J. M. Kosterlitz and D. J. Thouless, J. Phys. C6, 118 (1973).
3. D. R. Nelson and B. Halperin, Phys. Rev. B 19, 2457 (1979).
4. A. L. Loeb, Space Structure-Their Harmony and Counterpoint, Addison-Wesley 1976.
5. D. S. Fisher, B. I. Halperin, and R. Morf, Phys. Rev. B 20, 4692 (1979).
6. See, e.g., S. Toxvaerd, Phys. Rev. Letters 44, 1002 (1980).
7. S. J. Knak Jensen and O. G. Mouritsen, Phys. Rev. Letters 43, 1736 (1979).
8. R. C. Gann, S. Chakravarty, and G. V. Chester, Phys. Rev. B 20, 326 (1979).
9. D. Frenkel and J. P. McTague, Phys. Rev. Letters 42, 1632 (1979); J. Tobochnik and G. V. Chester, (preprint).
10. R. M. J. Cotterill in Ordering in Strongly Fluctuating Systems ed. T. Riste page 261 (Plenum Press 1980).

DISCUSSION

B. I. Halperin:

The observation of an exponent of greater than 1/4, for the hexatic phase, is very difficult for me to accept. If one has truly reached thermal equilibrium, and if the distance scale is large enough, then the exponent η is directly related to the Frank constant K_A; if η is greater than 1/4, then the system should be unstable to the formation of disclinations. (Of course, if the core energy is high enough, there would not be many disclinations present in a finite size system.) In any case, I feel that the measured η cannot be simply interpreted as being characteristic of a macroscopic hexatic phase.

J. P. McTague:

I have no estimate of the disclination core energy so I cannot estimate whether we are in the asymptotic region, other than to note that our results are similar for 256 and 2500 particle systems, and that similarly large values of η are found for Lennard-Jones systems, where the intermediate phase is stable over a considerably wider temperature region.

The question of thermal equilibrium is always a thorny one near a transition. We have made exceedingly long runs by simulation standards, but the equivalent times are still very short on a laboratory scale. The results in Lennard-Jones systems near the middle of the intermediate phase should not be subject to as much uncertainty.

D. R. Nelson:

I would like to suggest that you clean up your 5-7 pictures further by blocking the system out, and computing the net "charge" and "dipole moment" within each cell. The net dipole moment rotated by 90° could be interpreted as a cell Burger's vector, and the net "charge" as a cell disclination charge.

J. P. McTague:

This is in progress at present.

E. K. Riedel:

Do you find a Kosterlitz-Thouless like melting transition for all densities?

J. P. McTague:

Yes, there is only one parameter in the theory. However, for $u > 6$ the transition may be first order. Already for $u = 6$ the hexatic phase region is very narrow.

D. S. Fisher:
(a)
What is the energy of vacancy and interstitials?

(b)

Grain boundaries can form as metastable structures if the time scale is not long enough for dislocations to climb. This seems to be a worry as far as equilibrium is concerned.

J. P. McTague:

The energy of a vacancy should be about 3 E, while T_m is about 0.15 E, so it is a very high energy excitation for this system. Interstitials should also have a high energy.

A. Holz:

From your slides I had the impression that below the melting transition there exists essentially closed bound pairs of dislocations and above melting there still exist lots of closed bound pairs, grain boundaries and only a few free dislocations. Is it possible that the fluidity of the sample is caused by other mechanisms, say grain boundary sliding? In addition what is the influence of the boundary by means of mirror faces on the generation of free dislocations?

J. P. McTague:

The grain boundary loops can soften the shear modulus but not to zero, so finite sized loops can not influence the $k = 0$ flow, but will influence the short time, short distance motion.

The periodic boundary conditions do not appear to be affecting the generation of free dislocations in any essential way in the large systems (2500 particles) but probably do play a role in the small systems at low temperatures.

J. G. Dash:

The limited number of particles in your computer simulation might be too small for reliable comparisons with theory, but it could be quite adequate for comparison with experiment. In most of the current studies the typical sizes of uniform domains in the monolayer films on graphite are on the order of $\sim$100 Å, and these sizes can contain no more than about 10^3 molecules. One qualification could be important, however. In most of the determinations of experimental domain sizes the films have been in registered phases, and it may well be that the characteristic sizes of incommensurate phases are considerably greater. The differences may in fact be relevant to the tendency that registered films seem to undergo less sharp "melting" transitions.

J. P. McTague:

With regard to Griffith's question about whether the correlation functions observed in the intermediate phase could be modeled by a sum of small crystallite and liquid correlation functions, the answer for g(r) is no, since a good exponential is seen ($g(r) \sim \exp(-r/\xi)$, with ξ of order a few atomic diameters.

D. J. Thouless:

Vacancies and interstitials are very important in equilibrating dislocations. Have you studied the vacancies in your systems?

J. P. McTague:

Vacancies and interstitials are not as easily identified in our defect constructions as are dislocation pairs. However, some interstitials are observable as tight loops of 3 7-5's surrounding a single 6 - coordinate atom.

A. N. Berker:

The Kosterlitz-Thouless-Halperin-Nelson-Young mechanism for melting involves a low-temperature phase of bound dislocation pairs, the average separation between two dislocations forming a pair increasing with temperature and becoming infinite at the transition. The number of dislocations remains small. However, slides from your simulations show small pair separations up to melting and a substantial increase in the number of dislocations at melting. Can't this be interpreted as a preemption by a standard first-order transition?

J. P. McTague:

Although the average distance of separation within a dislocation pair does increase slightly with temperature, it does not go continuously to infinity. At the transition there is a discontinuity in the screening factor which causes a spontaneous creation of unbound pairs.

With regard to the order of the transition, as stated in the text, we can make no statement from the Monte Carlo data concerning the events at the transition. We can state that, away from this region, dislocations and dislocation pairs play a significant role.

Published 1980 by Elsevier North Holland, Inc.
Sinha, ed. Ordering in Two Dimensions

MELTING IN TWO DIMENSIONS IS FIRST ORDER: AN ISOTHERMAL-ISOBARIC MONTE CARLO STUDY

FARID F. ABRAHAM

IBM Research Laboratory
San Jose, California 95193

ABSTRACT: Isothermal-isobaric Monte Carlo computer experiments on melting in a two-dimensional Lennard-Jones system indicate that the transition is first order, in contrast to the two-stage, second-order melting behavior suggested as a possibility by Halperin and Nelson.

"*So let us melt, and make no noise... Let others freeze with angling...*" Donne, 1571-1631

Expanding on the proposals by Kosterlitz and Thouless [1] and by Feynman [2], Halperin and Nelson [3,4] have developed a detailed theory of dislocation-mediated melting for a two-dimensional "crystal." One important feature of the Halperin-Nelson theory is the possibility that the transition from two-dimensional solid to two-dimensional liquid takes place by two second-order transitions with increasing temperature. At some temperature T_m, dissociation of dislocation pairs gives rise to a second-order transition from a solid phase, with algebraic decay of translational order and long-range orientational order, to a "liquid-crystal" ("hexatic") phase, with exponential decay of translational order but algebraic decay of sixfold orientational order. At a higher temperature, $T_i > T_m$, dissociation of dislocations into disclinations gives rise to another second-order phase transition from the hexatic phase to the isotropic fluid phase. Halperin and Nelson do emphasize that this particular melting mechanism is only one possibility. They cannot rule out the possibility of a first-order melting transition.

Attracted by the unusual predictions of the theory and the apparent confirmation by computer experiment [5], I decided to look more closely at the thermodynamics of a two-dimensional melting phenomenon by doing a different type of computer experiment: By employing the isothermal-isobaric Monte Carlo simulation method of classical statistical mechanics [6,7], we may calculate the equilibrium enthalpy and density of the existing phase (whether solid, liquid, or hexatic) for a chosen temperature and pressure. Hence, by performing a series of experiments at fixed pressure and for a range of temperatures, we should observe (a) discontinuities in enthalpy and density at T_m if the melting transition is first-order, or (b) no discontinuities in the enthalpy and density at T_m and T_i if the melting transitions are second order. In particular, Halperin [4] states that the phase transitions in the Halperin-Nelson theory should not be apparent in the free energy or its derivatives at T_m and T_i. Also, there should (should not) be hysteresis if the system passes back through the apparent melting temperature when the phase transition is first order (second order). I have performed such experiments and conclude that two-dimensional melting is a first-order phase transition [8].

Before describing my investigation, I will comment on studying a <u>first-order</u> melting transition by using molecular dynamics [5] or Monte Carlo [9], where the number of particles, the area and, hence, the density is constrained to remain constant. In Fig. 1, the dependence of the pressure, heat capacity and energy on temperature is shown, assuming equilibrium as the system passes through the two-phase region. We note that these features were observed in the fixed density M.D. [5] and M.C. [9] experiments, but the investigators interpreted the experiments as evidence for two consecutive second order melting transitions. Employing the constant density (N,A,T) Monte Carlo method, we have calculated the equation of state for Lennard-Jonesium at a reduced density of 0.84 [10]. Briefly, we see features of the (P,T) equation of state consistent with Fig. 1 and with Refs. [5] and [9]. In Fig. 2, a trajectory plot of the 256 atoms over 1.4×10^6 configurations is presented for the reduced temperature of 0.7, this density/temperature position being bounded by the two "kinks" in the (P,T) equation of state. From this trajectory plot, we feel that a reasonable interpretation is that the system is in the two-phase, solid-liquid region and that envoking the existence of a new phase (e.g., the hexatic phase) is not needed [18]. However, the definitive demonstration lies with the single-phase thermodynamics attainable by doing isobaric-isothermal experiments, and I will now describe such computer experiments.

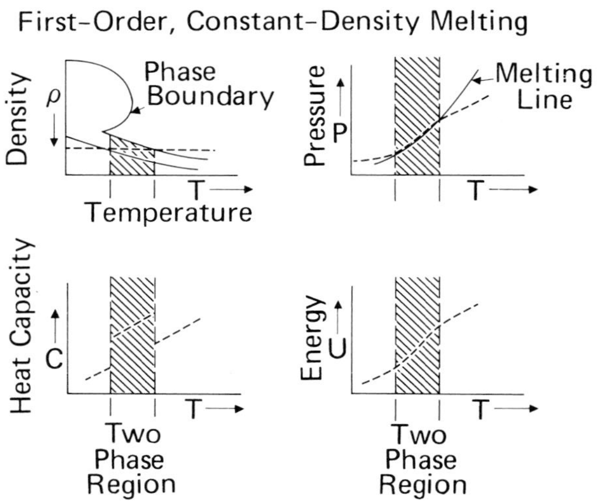

Figure 1. The dependence of pressure, heat capacity and energy on temperature as a system in equilibrium passes through the two-phase (solid-liquid) region.

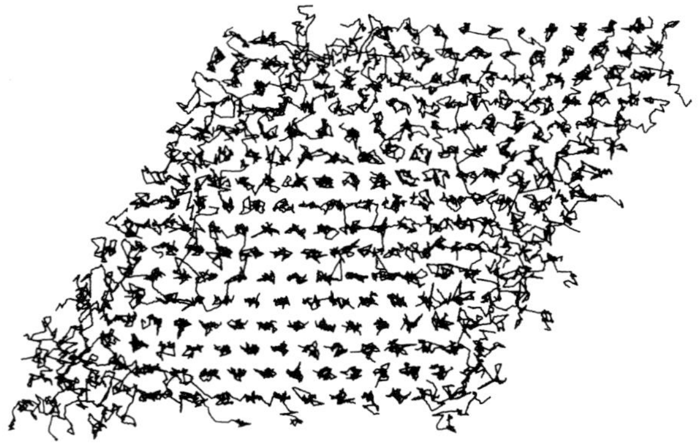

Figure 2. An atomic trajectory plot over 1.4 million constant-density Monte Carlo configurations of a 2-D L-J system in two-phase equilibrium for a reduced density and temperature of (0.84, 0.7), respectively. The mean pressure is P*=2.55.

The (N,P,T) Monte Carlo Calculations were performed on N=256 and 529 atom systems with periodic boundary conditions to simulate the bulk. The interatomic force law was taken to be L-J 12:6 with (ϵ,σ) denoting the well-depth and size parameters, respectively. The phase diagram for 2-D Lennard-Jonesium is given in Fig. 3. In a typical simulation "experiment," the system was initialized using an equilibrium configuration at a neighboring temperature and "equilibrated" through Ne Monte Carlo moves with a 50% acceptance ratio. Each move is comprised of an atomic displacement and an area scaling. After equilibration, Na further moves were performed to obtain the enthalpy H, the average density ρ, the pair distribution function g(r), and other quantities, the total number of moves being greater near the melting transition and in the liquid region. For the 256 atom system, Ne and Na equaled 10^6 and 2 to 6×10^6 configurations, respectively. For the 529 atom system, Ne and Na equaled 2×10^6 and 9×10^6 configurations, respectively. Great care was practice to determine that the system was in "local equilibrium" (stable or metastable) when taking the statistics for the quantities of interest; i.e., the distribution and running mean of density, enthalpy and virial pressure were monitored.

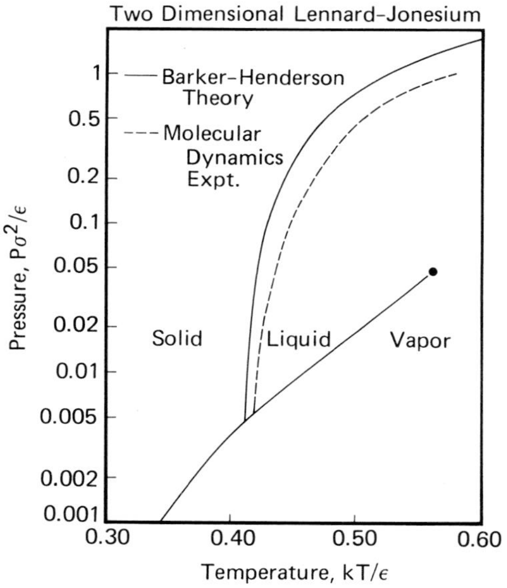

Figure 3. The phase diagram for two-dimensional Lennard-Jonesium as predicted by theory [10]. The melting line denoted by a dashed line is an <u>extrapolation</u> from Toxvaerd's [13] high pressure, molecular dynamics experiments.

In Figs. 4-6, I present the equilibrium density $\rho^*=\rho\sigma^2$ and the enthalpy per atom $h^*=H/N\epsilon$ for the L-J system of 256 atoms as a function of temperature $T^*=kT/\epsilon$ and for a fixed pressure $P^*=P\sigma^2/\epsilon=0.01$, 0.05 and 1.0, respectively. The first two pressures were intentionally chosen to be low since Nelson and Halperin have speculated that the melting transition could be first order at high pressures but second order at low pressures [3], and since it was suggested [3,5] that earlier studies [11-13] were examining a high-(temperature, density) regime where melting becomes first order. I now discuss Fig. 5. Considering first the density behavior, note that the solid density decreases smoothly as I sequentially increase the temperature and equilibrate the system up to the temperature T*=0.46. From T*=0.40 to 0.45, 4×10^6 configurations are generated at each temperature. At T*=0.46, the solid melts into a liquid after 5×10^6 moves with a dramatic decrease in equilibrium density $\rho_s^*-\rho_\ell^*=0.11$ (see Fig. 7).

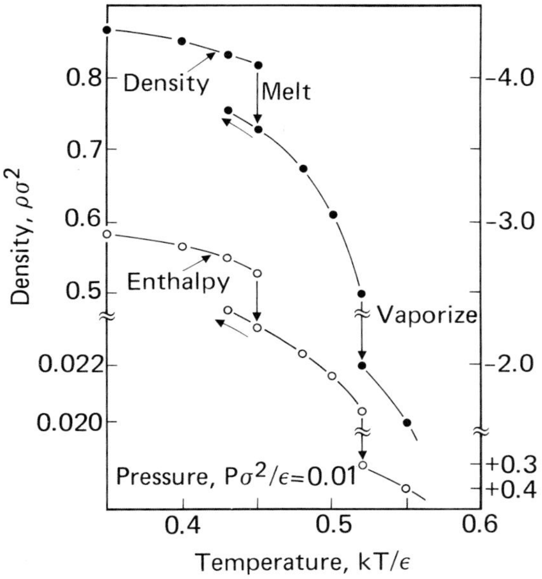

Figure 4. The equilibrium density and enthalpy per atom as a function of temperature for the 256 L-J atom system at fixed pressure P*=0.01. Note that the system both melts and vaporizes.

The pair distribution functions and "snap-shot" pictures of the atomic configurations (see Figs. 8 and 9) show the change from crystalline order to liquid disorder. At higher temperatures, the liquid density decreases smoothly. By sequential decrease of temperature and equilibration, the system passes through T*=0.46 and remains a liquid with smoothly increasing density down to a temperature T*=0.43 (again, the pair distribution functions reflect the fact that the undercooled states are liquid). This establishes hysteresis when passing back through the apparent melting temperature. At T*=0.4, the liquid solidifies with a sharp increase in density $\rho_s^*=0.8$; however, this density is lower than the original solid density at that temperature, $\rho_s^*=0.85$. Examination of the atomic structure of this solid, which was nucleated from the liquid phase, revealed significant defect structure which accounts for the density difference for the two solid systems. By slowly cooling a three-dimensional (3-D) L-J fluid using the (N,P,T) M.C. method, I have also simulated nucleation and crystallization of a super-cooled liquid and have again observed a final defect solid-state structure with a density ~midway between the liquid and solid densities. Of course, it would be unlikely that the supercooled liquid would solidify directly into a defect-free crystal structure.

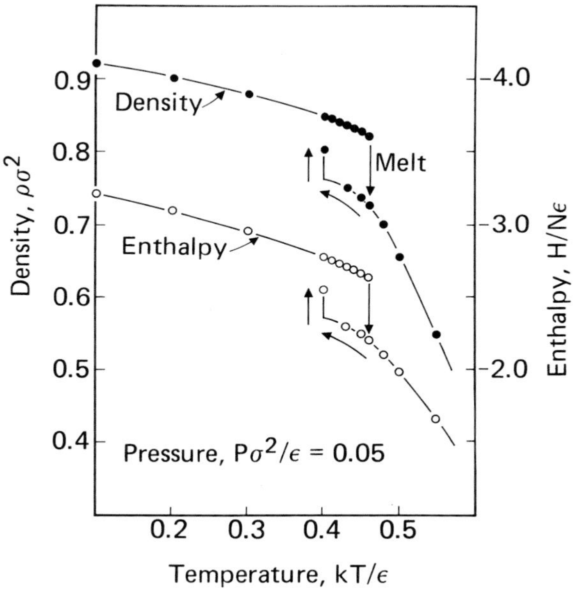

Figure 5. The equilibrium density and enthalpy per atom as a function of temperature for the 256 L-J atom system at fixed pressure P*=0.05. This is an extension of the study reported in Ref. [8].

We also note in Fig. 5 that the enthalpy of the system has the same behavior as the density and yields a latent heat of ~0.44ε/atom.

Similar features are seen in Fig. 4 for the low pressure of P*=0.01 and in Fig. 6 for the high pressure of P*=1.0. It is particularly interesting to note in Fig. 4 that the system vaporizes at T*=0.52, with a very large change in density and enthalpy. Hence, by traversing the phase diagram at constant pressure P*=0.01 for a wide temperature range, I have shown that the 2-D Lennard-Jones system undergoes two first-order phase transitions; at T*=0.46 the 2-D system melts, and at T*=0.52 the 2-D system vaporizes.

Using the 529 L-J atom system, I have studied 2-D melting at P*=0.05 and for T*=0.40, 0.42, 0.44, 0.45 and 0.46 (Fig. 10). In going

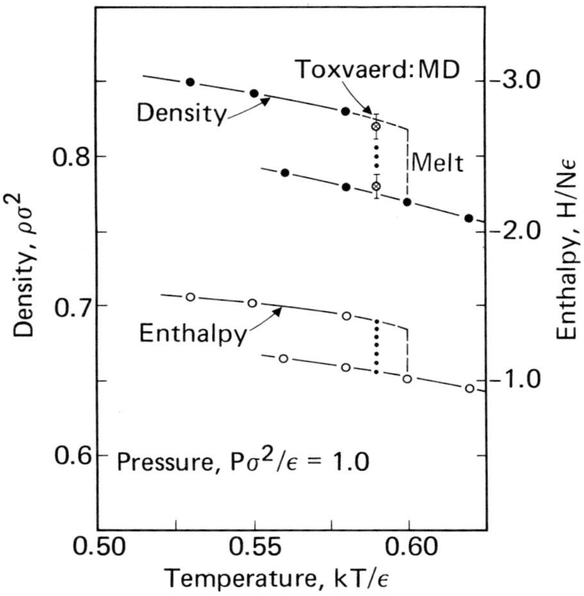

Figure 6. The equilibrium density and enthalpy per atom as a function of temperature for the 256 L-J atom system at fixed pressure P*=1.0. The data of Toxvaerd is taken from Ref. [13].

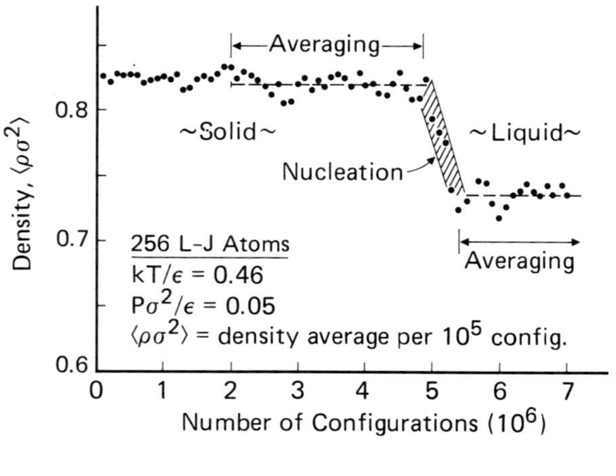

Figure 7. The density, averaged over blocks of 10^5 configurations, as a function of the total number of configurations for the L-J system at T*=0.46 and P=0.05. Note the sudden drop in density when the system nucleates from the solid to the liquid phase. Each configuration (or move) is comprised of an atomic displacement and an area scaling.

directly from T*=0.44 to 0.46, the system began to melt after 7×10^5 configurations. Therefore, I also performed an experiment by going from T*=0.44 to 0.45 and observed the onset of melting at this lower temperature after 5×10^6 configurations (Fig. 11). The density and enthalpy dependence on temperature are essentially identical to the values found using the 256 atom system. The sequential trajectory plots in Fig. 11 graphically depict the melting process.

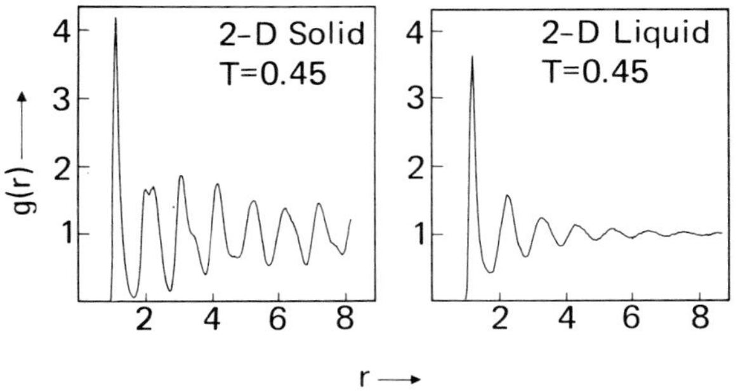

Figure 8. The pair distribution function of the L-J system at T*=0.45 when it is a solid and a liquid, respectively.

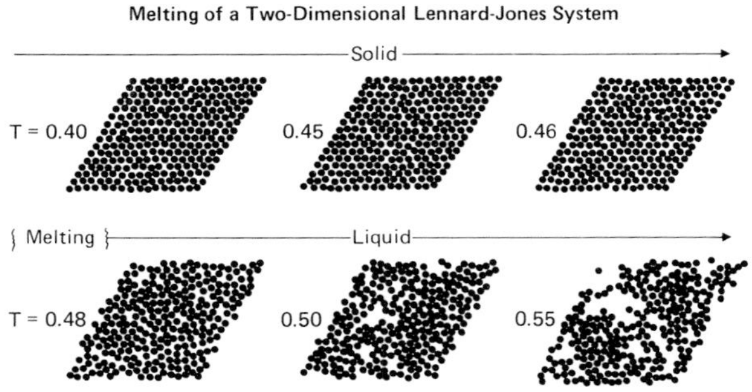

Figure 9. Snapshot pictures of the 256-atom L-J system simulated by the isothermal-isobaric Monte Carlo method for the denoted temperatures and a fixed pressure of P*=0.05.

I did one further experiment. Starting with a configuration consistent with 2-D solid-liquid coexistence obtained using the (N,A,T) M.C. method (Fig. 2), I did a constant pressure-temperature experiment using a pressure equal to the mean pressure of the two-phase system and at the same temperature. The system immediately relaxed to the crystalline state with a density ρ*=0.867.

This shows that (1) the "Before" state in Fig. 12 is not the lowest free energy state and (2) that it is not difficult to crystallize to a perfect solid from the two-phase region.

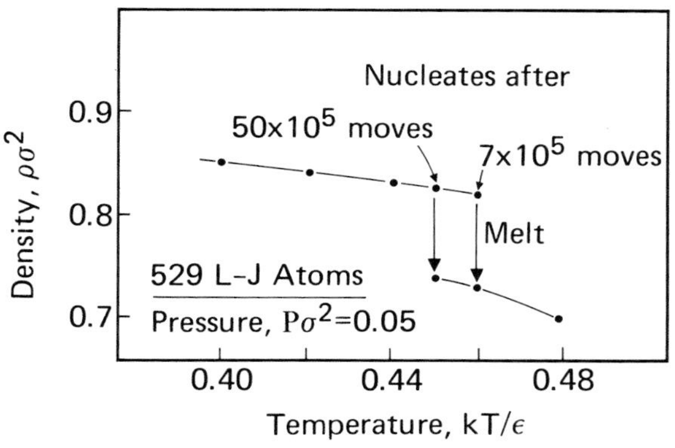

Figure 10. The equilibrium density as a function of temperature for the 529 L-J atom system at fixed pressure P*=0.05.

The density discontinuity between solid and liquid, the evident latent heat, and the existence of metastability certainly demonstrate that melting of a two-dimensional solid is a first-order phase transition.

I conclude with the following comment. One should not be too concerned with the statement that the existence of the crystalline solid state in two dimensions is impossible [14,15]. I explain my viewpoint. In contrast to a 3-D (three-dimensional) crystal, the atomic equilibrium positions themselves in a 2-D crystal become uncorrelated at large separations [14-16]. A quantitative measure of the loss of long-range crystalline order is the difference δ between the average separation between two atoms and the distance corresponding to the proper number of lattice spacings. It is found that δ^2 diverges slowly with N, the number of atoms in the 2-D system, the dependence on N being only logarithmic. Using a relation found by Hoover, Ashurst, and Olness [17], I estimate that to lose crystalline correlation equal to one lattice spacing (δ~3Å) near the melting point, the area of the 2-D crystal should be ~10^{27} cm^2, very large indeed. At the uppermost extreme, a crystal of dimension equal to the size of the universe (~10^{10} light-years) would have a correlation loss of ~6Å [17]. I can only conclude that for real situations, the crystalline solid state in two dimensions can exist. The fact that "for practical purposes" the 2-D crystalline state exists may suggest that Landau's argument that "transitions between bodies of different symmetry (in

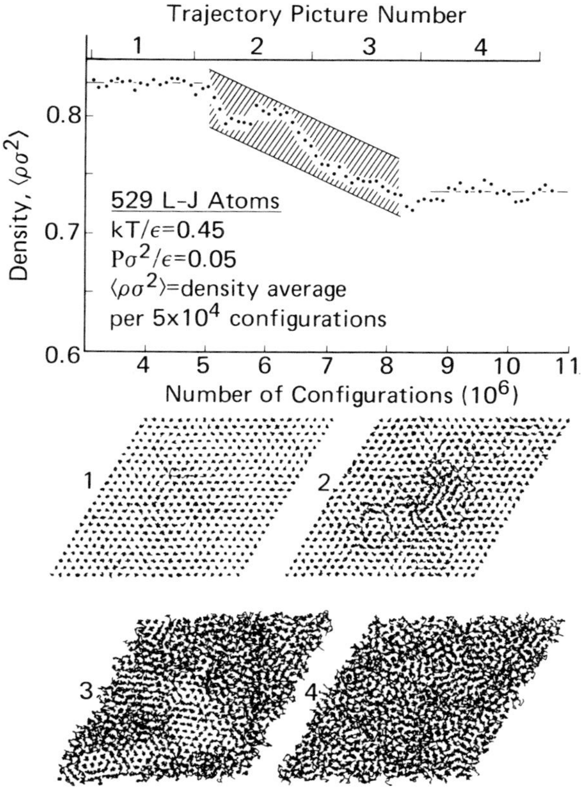

Figure 11. The density, averaged over blocks of 5×10^4 configurations, as a function of the number of configurations for the 529 L-J atom system at T*=0.45 and P*=0.05. The sequential trajectory plots show the melting process.

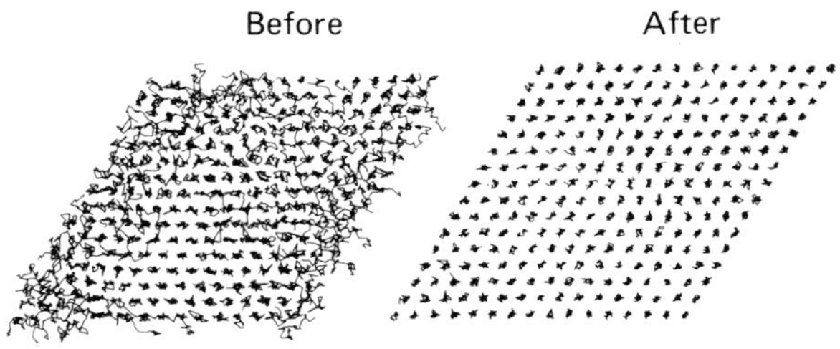

Figure 12. "Before" is the constant density similation of Fig. 2. "After" is obtained by starting from a configuration of "Before" and performing a constant pressure experiment at the same temperature T*=0.7 and pressure P*=2.55. The system becomes entirely crystalline with a density equal to 0.867.

particular between a liquid and a crystal) cannot happen continuously" [15] may well be applicable in two dimensions.

ACKNOWLEDGMENT

I am indebted to John Barker and Doug Henderson for many valuable discussions and for critically reading this manuscript.

REFERENCES

1. J. M. Kosterlitz and D. J. Thouless, J. Phys. C6, 1181 (1973).
2. R. P. Feynman, unpublished. An outline of this theory is given by R. L. Elgin and D. L. Goodstein, Phys. Rev. A9, 2657 (1974), and by J. G. Dash, Phys. Rep. 38C, 177 (1978).
3. B. I. Halperin and D. R. Nelson, Phys. Rev. Lett. 41, 121, 519(E) (1978). Also, D. R. Nelson and B. I. Halperin, Phys. Rev. B19, 2457 (1979).
4. B. I. Halperin, in Physics of Low-Dimensional Systems, edited by Y. Nagaoka and S. Hikami (Proc. Theor. Phys., Kyoto University, Kyoto, Japan, 1979).
5. D. Frenkel and J. P. McTague, Phys. Rev. Lett. 42, 1632 (1979).
6. W. W. Wood, in Physics of Simple Fluids, edited by H. N. V. Temperley, G. S. Rushbrooke and J. S. Rowlinson (North-Holland, Amsterdam, 1968), Chap. 5.
7. I. R. McDonald, Mol. Phys. 23, 41 (1972).
8. F. F. Abraham, Phys. Rev. Lett. 44, 463 (1980).
9. J. Tobochnik and G. V. Chester, preprint.
10. J. A. Barker, D. Henderson and F. F. Abraham, to be published.
11. B. J. Alder and T. E. Wainwright, Phys. Rev. 127, 359 (1962).
12. F. Tsien and J. P. Valleau, Mol. Phys. 27, 177 (1974).
13. S. Toxvaerd, J. Chem. Phys. 69, 4750 (1978).
14. R. E. Peierls, Ann. Inst. Henri Poincaré 5, 1771 (1935).
15. L. D. Landau, in Collected Papers of L. D. Landau, edited by D. ter Haar (Pergamon, Oxford, 1965), Chap. 29.
16. J. G. Dash, Phy. Rep. 38C, 177 (1978), pp. 179-183.
17. W. G. Hoover, W. T. Ashurst and R. J. Olness, J. Chem. Phys. 60, 4043 (1974).
18. At the time of completion of this manuscript, the molecular dynamics study of 2-D melting by S. Toxvaerd appeared in Phys. Rev. Lett. 44, 1002 (1980). By adopting the fixed density used by Frenkel and McTague [5], Toxvaerd demonstrates solid-liquid phase coexistence using a trajectory plot like our Fig. 2 and concludes that the dynamical behavior of the (finite) molecular dynamics system is compatible with a first-order phase transition. Furthermore, in agreement with my recent Monte Carlo study [F. F. Abraham, J. Chem. Phys. 72, 1412 (1980)], Toxvaerd demonstrates liquid-vapor co-existence for the 2-D system, in contradiction to the Frenkel-McTague suggestion [5].

DISCUSSION

J. Sokoloff:

Couldn't you get a fake metastable state and hence a first-order transition because there are long relaxation times near a second-order phase transition?

F. Abraham:

I was very careful to establish that the L-J system was not continually relaxing to some unachieved state near the melting transition by monitoring the various thermodynamic and structure measurements in the configurational averaging. Unlike McTague (previous paper), I did not notice any "slow drift" in the enthalpy (or energy) near the melting transition! In fact, I took on excessive number of configurations near the melting transition (in relation to obtaining good averages) in order to establish that there was no peculiar behavior in the convergence to equilibrium (such as a "critical slowing down" of the system). Both for the melting and for the vaporization, I could see no evidence for being near or at the phase transition point until the system nucleated, as I have demonstrated in my presentation.

J. Sokoloff:

How do you know that the apparent coexisting regions are not simply fluctuations?

F. Abraham:

Following the two-phase region over several million configurations in the constant density Monte Carlo simulation, we have established that the ratio of solid area to liquid area is essentially constant and consistent with being at the appropriate point in the -T phase diagram, while the line interface between the liquid and solid may move throughout the total simulated area (as shown during my talk). Of course, it is not difficult to establish the solid and liquid portions of the system, the diffusivity of the individual atoms being a good signature (as shown in Figure 2).

R.B. Griffiths:

This question is addressed to all of the previous speakers (Abraham, McTague, and Halperin). There are two proposals present for the Lennard-Jones system in two-dimensions: a hexatic phase or a region of two-phase coexistence. Could the latter produce correlation functions which mimic the former-either position or orientation correlation functions?

F. Abraham:

I believe so. Doing a configurational average to obtain a "correlation function" of the two-phase region means that we are averaging liquid-state atoms with short-range order and absence of long-range order (provided we don't cross an interface to the crystal region which, of course, we are likely to do for any respectable ratio of solid area to liquid area) with crystal-state atoms with short and long range order (again provided we don't cross an interface to the liquid region, which, of course, we are, etc...). Also, liquid in contact with a solid gives rise to a layering of the liquid normal to the solid face for several atomic diameters into the liquid (e.g., see F. F. Abraham, J. Chem. Phys. 68, 3713 (1978), an added complication. With some imagination, we can see that a solid-liquid mixture of atoms would retain the short-range order but dampen out any long-range order. However, the existence of solid regions in coexistence with liquid regions would give the appearance that "this fluid" would have a quasi-long-range angular correlation (remember, the solid regions do not rotate in the simulation). A more meaningful measure of correlation is to do it for the individual regions of liquid and of solid.

Sinha, ed. Ordering in Two Dimensions

2-D MELTING IN PHYSISORBED MONOLAYERS OF Ar, Kr, Xe AND CH_4

C. TESSIER and Y. LARHER

Département de Physico-Chimie, Centre de'Etudes Nucléaires de Saclay
Boite Postale No. 2, 91190 Gif sur Yvette, France

ABSTRACT

Sets of adsorption isotherms show that under certain conditions of lattice incompatibility between the adsorbate and the substrate, monolayers of Xe or CH_4 adsorbed on the basal face of lamellar halides can exist under the form of three distinct 2-D phases: one gaseous and the two others dense. Since the reduced temperature of their triple point is constant and equal to a computer simulation estimate for the gas-liquid-solid coexistence of Lennard Jones disks, it is conjectured that the two dense phases are a liquid and an incommensurate solid whose thermodynamic properties are little affected by the surface structure. Our data support the idea of first order melting on a smooth surface.

INTRODUCTION

The melting transition has given rise to numerous studies in recent years. Although interesting ideas concerning the molecular explanation of this phenomenon have already emerged, it does not seem that a unified philosophy meeting a large consensus would as yet exist (1).

Many theoretical and computer simulation studies have been carried out in the 2-D (two-dimensional) space. Working in spaces of various dimensions is a very usual procedure in statistical mechanics which has proved most fruitful. General ideas, important for the understanding of our 3-D world, have been revealed in this way. Now, we know that 2-D statistical mechanics can also be of a more direct interest, since it applies to real physical systems and particularly to monolayers adsorbed on uniform surfaces.

Indeed many experimental findings have already demonstrated the 2-D character of these adlayers (2). We quote here a most simple justification of this statement (3) : the experimental values of the reduced critical temperatures of 2-D condensation for the first monolayers of Ar, Kr and Xe adsorbed on the basal face of graphite are well accounted for by a theoretical estimate based on a virial expansion for a 2-D assembly of 6-12 Lennard Jones particles. Since this calculation implies a smooth surface, it appears that, at least under certain conditions, the properties of adlayers are little affected by the surface structure.

But this is far from general. It has been very clearly shown that on surfaces such as the basal face of lamellar halides, on which the potential wells are much deeper than on graphite (4,5), the surface structure manifests itself even at the critical point of 2-D condensation (6).

Of course one expects that under conditions of higher density or lower temperature, and consequently at the melting transition, this effect would be more pronounced. At first this appears as a troublesome complication, but in the end it is largely compensated by the richer variety of transitions which can be observed. Moreover the lack of experimental results concerning melting on smooth surfaces, which might be considered as very frustrating at the outset, is now compensated by the quasi-experimental data obtained from computer simulations.

Numerous experimental studies of two-dimensional melting of adlayers have already been published. A single adsorbent, graphite, has been used in these studies. In this paper we will present results concerning melting on other substrates, the lamellar halides. Their main interest is that they enable a better assessment of the way in which the melting transition is altered by the surface structure. Only adlayers of classical atoms (Ar, Kr, Xe) or quasi-spherical molecules (CH_4) will be considered, so that neither the orientation of the admolecules, nor quantum effects will have to be taken into account.

THE MELTING TRANSITION OF Ar, Kr, Xe and CH_4 ON THE BASAL FACE OF GRAPHITE.

In this section we will summarize a number of already known data concerning the melting of monolayers of Ar, Kr, Xe and CH_4 adsorbed onto the basal face of graphite.

Monolayers in simple registry have

been observed on this surface, although its potential wells are shallow. They can appear because the substrate possesses a sublattice of sites close to the 2-D lattice that the adsorbate would take if adsorbed on a smooth surface. For the adsorbates considered here (Ar, Kr, Xe and CH_4), it proves convenient to take as a reference for such a solid monolayer, the (111) plane of their bulk crystal. It may differ a little from the former but has the advantage of having a precisely known crystalline parameter d . The three $\sqrt{3} \times \sqrt{3}$, R30° sublattices of sites of the basal face of graphite have a crystalline parameter a not very different from d . Of course only at low values of the relative difference between a and d , $i = (a - d) / d$, which we call the dimensional incompatibility, are registered solids expected to occur. This point is indeed supported by the information concerning the melting transition of monolayers of Xe, CH_4, Kr and Ar on graphite contained in Table I.

Let us remark first that the optimum lateral stability of an adlayer is expected at an incompatibility somewhat higher than i = 0. Indeed, in the expression $i = (a - d) / d$, d corresponds to the optimum stability of a 3-D solid. Now if the potential energy of 2-D or 3-D crystals of 6-12 Lennard-Jones particles is minimized, one obtains in 2-D an interparticle distance 2 % higher than in 3-D. Moreover three-body forces can increase this difference. Consequently registry is favorable for both Kr and CH_4. For Kr this has been observed in numerous structural studies (11-13). For CH_4 there might be a small deviation (0.5 to 1 %) from this situation (14). But for Ar and Xe the solid monolayer is incommensurate (15, 16). For both these gases, either the compression or the expansion necessary for registry involves a loss of lateral stability too important to be compensated by the energy gained in putting the adatoms above the potential wells.

For Xe, CH_4 and Kr the adsorption isotherms clearly indicate that melting is first order (7,17,18). For Ar, neutron diffraction studies (19), heat capacity (20) and adsorption isotherms measurements (17,10) would rather suggest a continuous transition.

THE MELTING TRANSITION OF Xe AND CH_4 ON THE BASAL FACE OF LAMELLAR HALIDES.

The cleavage face of lamellar halides is constituted by a compact arrangment of halide ions, so that the sites form a honeycomb lattice or equivalently two triangular sublattices having a crystalline parameter a of the same order of magnitude as d the crystalline parameter of the (111) plane of bulk adsorbate. Again we define a dimensional incompatibility by $i = (a - d)/d$. The existence of a family of substrates with a varying from 3.48 to 4.54 Å presents the interest that i can be given a reasonably dense although discrete set of values between a minimum of - 21 % for Xe-$NiCl_2$ and a maximum of + 19 % for Ar-PbI_2. From the striking variations of many thermodynamic properties of the adlayers (entropy, energy, critical temperature of 2-D condensation) with the dimensional incompatibility a number of interesting conjectures have already been made. In particular, the results strongly suggested for $0 \lesssim i \lesssim 12$ % the existence of a 1 x 1 registered monolayer (4,5). This was effectively observed later for CD_4 on PbI_2 (i = 8 %) and for ^{36}Ar on MnI_2 (i = 8 %) in neutron diffraction studies (21). Since such a registered monolayer is obtained directly from the gaseous 2-D phase through a condensation, we have a behaviour resembling that of a lattice gas model. This model indeed accounts qualitatively for the variations of the critical temperatures of 2-D condensation with the dimensional incompatibility (6, 22). However the experimental values of these temperatures are lower than the predictions of the lattice gas model. They are intermediate between the latter and those calculated for a continuum model. This reflects well the physical situation : in the dense phase the lateral interactions contribute, under certain conditions, to stabilize the registered phase thus favoring the lattice character of the transition ; but the potential wells, although deeper than on graphite, are insufficient to give a lattice character to the low-density phase.

The situation is even more complicated since we must take into account the existence of two sublattices of sites. If adsorption is sitewise at any density, theoretical calculations predict an ordering transition (23, 24) showing a weak singularity in the isothermal compressibility above a tricritical point. This we never observed. We could be tempted to consider the absence of such a singularity as an indication of the inadequacy of the model, the adsorption being mobile at low density. However a weak singularity may be difficult to demonstrate on an adsorption isotherm and we feel that more precise measurements are needed before its existence could be definitely ruled out.

We have described the simplest situation observed for Ar, Kr, Xe or CH_4 on lamellar halides : the direct first order transition from a 2-D gas to a 1 x 1 registered solid. Our thermodynamic data led us to conjecture that it occurred only for $0 \lesssim i \lesssim 12$ %, which means that outside this range of incompatibilities the loss in lateral stability is too important to be compensated by the energy gained in putting the atoms or molecules above the potential wells. An important point is that, until 1978, the adsorption isotherms which we measured at all incompatibilities showed a single first order

TABLE I. FLUID-SOLID TRANSITION OF MONOLAYERS OF Xe, CH_4, Kr AND Ar ADSORBED ON THE BASAL FACE OF GRAPHITE.

Gas	Xe	CH_4	Kr	Ar
Incompatibility[x]	- 3.5 %	1.7 %	4.6 %	12.3 %
Nature of the solid[xx]	I	I ∿ C	C	I
Order of the transition	1rst	1rst	1rst	higher order
T_t(2-D)	99 K (7)	56 K (8)	84.7 K (9)	50 K (10)
T_t(2-D)/T_t(3-D)	0.61	0.62	0.73	0.60

x Estimated at the triple point temperature.
xx C means : commensurate √3 x √3 R 30° ; I means : incommensurate

transition, a 2-D condensation from a dilute monolayer to a dense one. We considered that outside the range 0 ≲ i ≲ 12 % this monolayer was a solid either incommensurate or commensurate with a large coincidence unit cell (4,5). Since recent calculations (25) suggest that both kinds of structures have neighbouring thermodynamic properties, we will not distinguish between them.

The situation changed in 1978, when Robert and Larher (26) found a first order transition within a dense monolayer of Xe on $NiCl_2$. Since this system corresponds to the lowest dimensional incompatibility (- 21%) within the class studied here, we decided to look for other such transitions particularly at low dimensional incompatibilities.

Table II contains the list of the systems studied and a number of characteristics of their behaviour, in particular triple point and critical point temperatures. Figure 1 shows a set of adsorption isotherms for Xe on $FeCl_2$ which is typical of systems showing two successive first order transitions. Our discussion in the next section will lead us to conjecture that the 2nd one, occuring between two dense monolayers, corresponds to coexistence between a liquid and a solid. It proves convenient to term it melting from now on.

That melting is first order is not obvious on the scale of the figure. But looking at it on a larger scale shows that the pressure variation from the bottom to the top of the step is approximately the same as for the 2-D condensation which is a well characterized first order transition. In one case however, for CH_4 on $NiCl_2$, the order of the transition is doubtful. It seems to become first order very close to the triple point, say from T_t to T_t + 1.5 K. But this is only a tentative statement since, in that temperature range, the pressures being very low (about 10^{-4} torr), are measured with a poor accuracy.

Table II seems to suggest that melting in a monolayer adsorbed on a lamellar dihalide occurs only at dimensional incompatibilities below about - 17 % .This upper boundary is still unprecise, since the absence of melting in the case of Xe on $CoBr_2$ is not yet an entirely warranted result. The work on this system is still in hand. Table II contains only the systems which have been studied recently. We must add that for the numerous other systems at dimensional incompatibilities above - 13 %, which have been studied earlier by Larher, Nardon, and Millot (4-6) , no first order melting was ever detected. For about one half of these systems, the necessary conditions are fulfilled (sufficient uniformity of the substrate, correct temperature and pressure range, sufficient accuracy in the measurements) for the observation of a first order melting if it existed. On the other hand the data contained in these studies are not sufficiently precise to guarantee that no second order transition had been missed. In conclusion it appears probable that a first order transition in a monolayer adsorbed on lamellar halides occurs only for dimensional incompatibilities below about - 17%.

TABLE II. TRIPLE POINT AND CRITICAL POINT TEMPERATURES FOR MONOLAYERS OF Xe AND CH_4 ADSORBED ON THE BASAL FACE OF LAMELLAR HALIDES.

Systems	Dimensional incompatibility	T_c(2-D)/T_c(3-D)	T_t(2-D)/T_t(3-D)
Xe-$NiCl_2$	- 21.0 %	0.39	0.61
Xe-$CoCl_2$	- 19.6 %	0.39	0.61
Xe-$FeCl_2$	- 18.5 %	0.39	0.61
Xe-$CoBr_2$	- 16.6 %	0.39	melting unlikely
Xe-$CdCl_2$	- 12.8 %	0.40	no melting transition
CH_4-$NiCl_2$	- 17.8 %	0.39	0.62

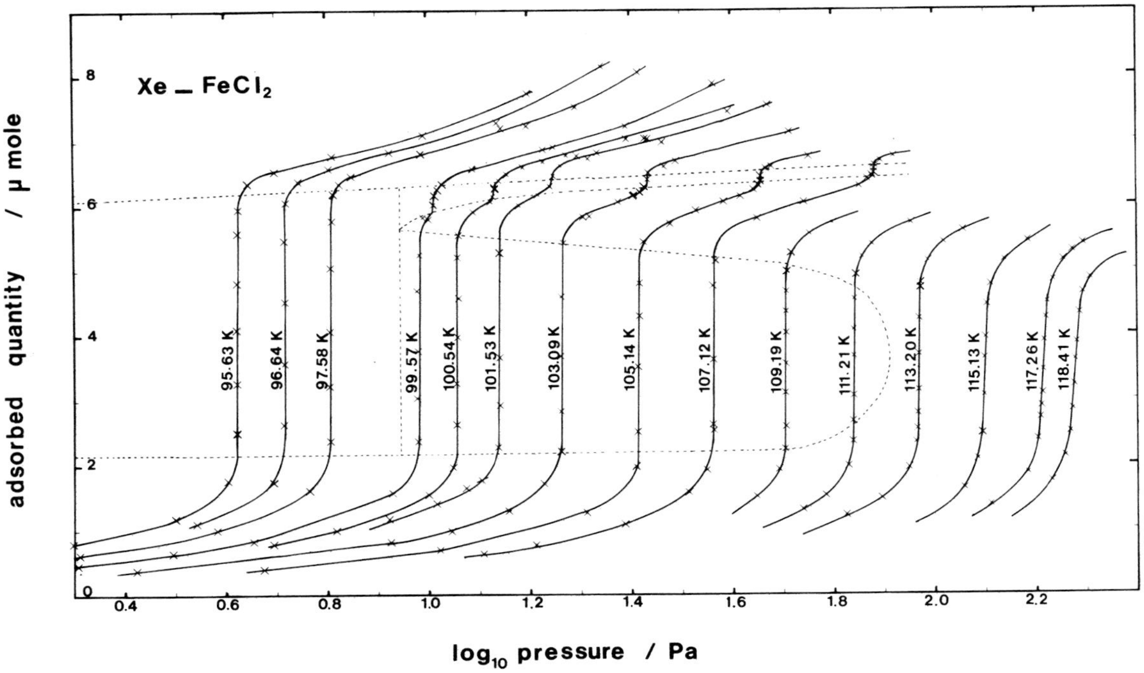

FIG. 1. SET OF ADSORPTION ISOTHERMS OF Xe ON $FeCl_2$.

DISCUSSION

The structure of simple 3-D liquids and solids near melting is known to be mainly determined by the repulsive forces between atoms (1). Thus an assembly of hard spheres already displays the important features of melting, so that statistical theories taking as a starting point the properties of such a system, derived from computer simulations, and using perturbation methods, have proved very efficient (27). The same point of view does not seem to have met a large consensus in 2-D. Indeed, although Alder and Wainright (28) showed as early as 1962 that an assembly of hard disks exhibited a first order melting transition, the idea that melting in 2-D is a continuous transition has been long favored by many physicists (2,29). The absence of long range positional order in 2-D solids, the guess that the melting of a monolayer of He at the surface of graphite is continuous (29) and the scarcity of experimental results constitute probably the main reasons for the reticence to recognize melting in 2-D as a first order transition.

It is now generally accepted that the absence of true long range order in 2-D solid is of no practical importance (29). On the other hand 2-D solids possess long range directional order (30) and it seems natural to associate melting with the loss of this order. Halperin and Nelson (31) have shown that this could happen continuously. Their theory is based on a mechanism of dissociation of dislocations pairs (32). But the authors quote that they cannot rule out another mechanism leading to a first order transition.

Since no first order 2-D melting theory exists at present, the best we can do is to compare the experimental results with computer simulation data. We know that even simple atoms interact according to multiparameter pair potentials and that the contribution from three body forces is far from negligible (33). However comparison in 3-D of computer and experimental results has shown that the 6-12 Lennard-Jones potential can be considered as a good effective potential (34). Many simulations of 2-D assemblies of 6-12 Lennard-Jones particles have already been carried out (35-40). Curiously enough, one amongst these does not support the idea of a first order melting (39). Among the others, the most informative is the molecular dynamics study of Toxvaerd (38) who has calculated the melting line and estimated the triple point temperature for an assembly of 6-12 Lennard-Jones disks. He obtains $kT_t(2D)/\varepsilon = 0.41$. A similar earlier study in 3-D gave $kT_t(3D)/\varepsilon = 0.67$ (41) so that $T_t(2D) / T_t(3D) = 0.61$.

Toxvaerd had already noticed that the experimental values of this ratio for Xe on $NiCl_2$ or graphite agreed well with this simulation estimate. Looking at Tables I and

II we see that the agreement extends to 7 out of the 8 systems for which a 2-D triple point has been estimated from experimental results. That Kr on graphite is an exception can be easily understood (9) : the registered monolayer of Kr, being expanded as compared to a monolayer adsorbed on a smooth surface will have thermodynamic properties different from those of the latter. The agreement for the other seven systems suggests, on the contrary, that the monolayers, gaseous, liquid or solid, mostly behave, as far as their thermodynamic properties are concerned, as if adsorbed on a smooth surface .

For adsorption on graphite this is not surprising. We already know that the gaseous and liquid monolayers are little affected by the surface structure (3). Moreover calculations of Novaco and McTague (42) suggest that the surface structure of the basal face of graphite does not change very much the energy of an incommensurate adlayer. Lastly, the properties of a registered monolayer of CH_4 should not be much different from those of an incommensurate one, since the dimensional incompatibility is very close to 2 %.

On the contrary for adsorption on lamellar halides our results are surprizing. We had been used to observing large variations of the thermodynamic properties of solid monolayers with the dimensional incompatibility (4-6), and we never had any indication of a liquid monolayer. Consequently, having found for Xe on $NiCl_2$ the same triple point temperature as for Xe on graphite, Robert and Larher (26) thought that this might be fortuitous. Given our new results for Xe on $NiCl_2$, $CoCl_2$ and $FeCl_2$, and for CH_4 on $NiCl_2$, and the computer simulation result of Toxvaerd (38), it seems difficult now to escape the conclusion that below a dimensional incompatibility of about - 17 %, a gaseous, a liquid and a solid monolayer of Xe and CH_4 can coexist at the surface of lamellar halides. Moreover their thermodynamic properties should not differ much from those of monolayers adsorbed on a smooth surface. For the gaseous and liquid monolayers this point is supported by the experimental value of $T_c(2D) / T_c(3D) = 0.39$ (see Table II), in good agreement with a theoretical estimate of 0.38 for a continuum model (3).

The melting transition observed being first order, except perhaps in the case of CH_4 on $NiCl_2$, our data strongly support the idea that melting on a smooth surface is first order.

Of course structural studies of these phases are desirable. Meanwhile we present a very qualitative analysis with the aim of trying to explain the apparent smoothness of the basal face of lamellar halides at low dimensional incompatibilities. We use Steele's notation (43) for the three distinct positions above the surface, i.e. S = site, SP = saddle point (midway between 2 sites on the basal face of lamellar halides) and A = above an atom (here an anion). Semi empirical calculations of potential energies, ϕ , of one adatom above the surface, carried out for Ar on Xe (111) (43), which simulates well our systems (5), yield $\phi_A - \phi_S = 288$ K and $\phi_{SP} - \phi_S = 74$ K, to be compared to $\phi_A - \phi_S \simeq \phi_{SP} - \phi_S \simeq 35$ K for Ar on graphite (43). For years we considered the relatively high value of $\phi_A - \phi_S$ as responsible for the absence of a liquid monolayer on the surface of lamellar halides. Now Ar on Xe corresponds to i = 15 %. We expect a decrease of potential barriers to translation as i decreases. But rough calculations indicate that this is not important for $\phi_A - \phi_S$, which explains the reluctance of Robert and Larher (26) to accept a liquid monolayer for Xe on $NiCl_2$.

On the other hand $\phi_{SP} - \phi_S$ can become of the same order of magnitude as on graphite. Consequently adatoms or admolecules could sit not only above the sites but also in any position close to the path formed by the honeycomb lattice of sites. Suppose we take a snap-shot of a crystalline or a liquid monolayer adsorbed on a smooth surface and we superpose it above the lattice of the substrate in trying to avoid the vicinity of the unfavorable A positions for the adatoms by straining the adlayer. It seems intuitive that as soon as we are definitely below i = 0, say $i < -10\%$, the necessary strain will decrease as i decreases. Also the diffusive motion of atoms or molecules in the liquid layer, avoiding the A position will become easier. This is only because the 2-D space above which the atoms can sit becomes closer if compared to the diameter of the adatoms. Thus we can imagine that below $i \simeq -17$ % the liquid and solid monolayer could be comparable to what they are on a smooth surface. This tentative explanation of our data is of course very qualitative. Our feeling is that a realistic theory could be as yet very difficult, while a computer simulation of our systems could be most instructive.

The case of Ar on graphite is exceptional since the melting transition seems to be second order. This suggests that an incommensurate solid might behave differently according to the sign of the dimensional incompatibility. LEED measurements (15) have shown that the solid monolayer has an unusually large expansion as compared to the (111) crystallographic plane ($\Delta d/d = 4.1$ % at 50 K), which brings its density closer to that of a liquid monolayer and might facilitate a continuous melting transition.

REFERENCES

1. S.M. Stishov, Sov. Phys. Usp. 17, 625 (1975).
2. See for example J.G. Sash, Films on Solid Surfaces (Academic Press 1975).
3. Y. Larher and B. Gilquin, Phys. Rev. A20, 1599 (1979).
4. Y. Larher, J. Colloïd. Interface Sci. 37, 836 (1971).
5. Y. Larher and F. Millot, J. Phys. (Paris) Colloq. 38, C4-189 (1977).
6. Y. Nardon and Y. Larher Surf. Sci. 42, 299 (1974).
7. A. Thomy and X. Duval, J. Chim. Phys. 67, 1101 (1970).
8. A. Glachant, J.P. Coulomb, M. Bienfait, and J.G. Dash, J. Phys. (Paris) Lett. 40, 543 (1979).
9. Y. Larher, J. Chem. Phys. Faraday Trans. I70, 320 (1974).
10. F. Millot, J. Phys. (Paris) Lett. 40, 9 (1979).
11. M.D. Chinn and S.C. Fain, Phys. Rev. Lett. 39, 146 (1977).
12. T. Ceva and C. Marti, J. Phys. (Paris) Lett. 39, 221 (1978).
13. P.M. Horn, R.J. Birgeneau, P. Heiney, and E.M. Hammonds, Phys. Rev. Lett. 41, 961 (1978).
14. A. Glachant, J.P. Coulomb, M. Bienfait, P. Thorel, C. Marti, and J.G. Dash, Proceedings of this Conference.
15. C.G. Shaw, S.C. Fain, and M.D. Chinn, Phys. Rev. Lett. 41, 955 (1978).
16. J.A. Venables, H.M. Cramer, and G.L. Price, Surf. Sci. 55, 373 (1976).
17. Y. Larher, J. Chem. Phys. 68, 2257 (1978).
18. Y. Larher and A. Terlain, J. Chem. Phys. 72, 1052 (1980).
19. H. Taub, K. Carneiro, J.K. Kjems, L. Passell, and J.P. Mc Tague, Phys. Rev. B 16, 4551 (1977).
20. T.T. Chung, Surf. Sci. 87, 348 (1979).
21. P. Thorel, Y. Larher et al. unpublished data.
22. F. Millot, thesis, Nancy, 1975, Commissariat à l'Energie Atomique Note n° 1865, 1976 (unpublished).
23. L.K. Runnels in Phase Transitions and Critical Phenomena ed. C. Danb and M.S. Green Vol. II page 305 (Academic Press, 1965).
24. M.W. Springgate and D. Poland, Phys. Rev. A 20, 1267 (1979).
25. C.R. Fuselier, J.C. Raich, and N.S. Gillis, Surf. Sci. 92, 667 (1980).
26. P. Robert and Y. Larher, Phys. Rev. Lett. 40,1275 (1978).
27. H.C. Longuet-Higgins and B. Widom, Mol. Phys. 8, 549 (1964).
28. B.J. Alder and T.E. Wainwright, Phys. Rev. 127, 359 (1962).
29. J.G. Dash, Phys. Rep. 38, 177 (1978).
30. N.D. Mermin, Phys. Rev. 176, 250 (1968).
31. B.I. Halperin and D.R. Nelson, Phys. Rev. Lett. 41, 121 (1978).
32. J.M. Kosterlitz and D.J. Thouless, J. Phys. C 5, L 124 (1972).
33. J.A. Barker, R.A. Fisher, and R.O. Watts, Mol. Phys. 21, 657 (1971).
34. I.R. Mc Donald and K. Singer, Mol. Phys. 23, 29 (1972).
35. R.M.J. Cotterill and L.B. Pedersen, Sol. Stat. Comm. 10, 439 (1972).
36. F. Tsien and J.P. Valleau, Mol. Phys. 27, 177 (1974).
37. S. Toxvaerd, Mol. Phys. 29, 373 (1975).
38. S. Toxvaerd, J. Chem. Phys. 69, 4750 (1978).
39. D. Frenkel and J.P. Mc Tague, Phys. Rev. Lett. 42, 1632 (1979).
40. F.F. Abraham, Phys. Rev. Lett. 44, 463 (1980).
41. J.P. Hansen and I.R. Mc Donald, Theory of simple liquids (Academic Press, London, 1976).
42. A.D. Novaco and J.P. Mc Tague, J. Phys. (Paris) Colloq. 38, C4-116 (1977).
43. W.A. Steele, Adsorption of Gases on Solids (Pergamon, Oxford, 1974).

DISCUSSION

J. P. McTague:

You quoted triple points and critical point values for Ar, Kr, and Xe on graphite. The best capacity studies for these systems show only a single peak. Would you or Alex Stewart like to comment on these different sets of data?

Y. Larher:

Critical points were discovered in 1874 but it was not until 1963 that Voronel first gave evidence of an infinite singularity in the heat capacity at constant density for bulk Ar and O_2 at their critical point. These are known to be weak singularities. Obviously they are even more difficult to demonstrate for ad films since the film signal is a small part of the total signal (film + calorimeter). The second part decreases as the temperature is lowered and this could explain that Prof. Lerner has been able recently (unpublished data) to detect a peak in the heat capacity of a monolayer of Ne adsorbed on graphite on the gas-liquid coexistence boundary up to the critical point.

J. A. Zollweg:

The isotherms which you showed for noble gases adsorbed on Lamellar halides seem to show very large 2-dimensional compressibility before the apparent transition is actually reached. This is unusual for a first order transition in that the compressibility normally remains approximately constant within a phase up to the transition. Would you care to comment on this?

Y. Larher:

Imperfections of the substrate can account for the apparent large compressibilities outside the coexistence domain. Since they appear mainly above the liquid-solid coexistence boundary, they are probably due to small size effects resulting from the presence in the adsorbing powder of a number of very small crystals.

NEUTRON DIFFRACTION STUDIES OF MELTING ON PHYSISORBED MONOLAYERS OF CD_4 ON GRAPHITE

P. Dutta, S. K. Sinha and P. Vora, Argonne National Laboratory, Argonne, IL 60439; M. Nielsen, Riso National Laboratory, DK-4000 Roskilde, Denmark; L. Passell, Brookhaven National Laboratory, Upton, NY 11973; and M. Bretz, Univ. Michigan, Ann Arbor, MI 48109.

ABSTRACT

The system of methane on grafoil has the unusual property that melting always takes place from an incommensurate phase - an expanded solid phase at coverages less than the full commensurate monolayer coverage, and the familiar compressed solid phase at higher coverages. Detailed line-shape analyses in the fluid and incommensurate solid phases and across the melting transitions show that (a) Lorentzian structure factors account well for the observed fluid lineshapes, with correlation lengths $\sim$ 30A well within the fluid phase; (b) "tails" in the incommensurate solid lineshapes indicate the presence of substantial 'thermal diffuse scattering' and can be fitted with power-law structure factors; (c) The expanded solid has only been observed to melt in a first-order way, whereas melting of the compressed solid is indistinguishable from continuous. However, outside the coexistence region, the approach to the transition region from either side is in all cases indicative of Kosterlitz-Thouless melting.

In an incommensurate phase of a monolayer physisorbed on graphite, the substrate potential in the plane may reasonably be approximated by its constant mean value, and such a system may therefore closely simulate a true two-dimensional system. Melting in two dimensions has been the focus of much theoretical activity. In particular, Kosterlitz and Thouless(1) have proposed that the transition is continuous and proceeds via the unbinding of dislocation and antidislocation pairs. Their work has been extended more recently by Halperin and Nelson(2) using renormalization group theory. We have experimentally studied the melting of monolayers of deuterated methane (CD_4) on exfoliated graphite; CD_4 has a large coherent scattering cross-section for neutrons, and moreover it turns out to have the unusual property of always melting from an incommensurate phase (see below and Fig. (1).

The lineshape calculated by Warren(3) for diffraction from a randomly oriented array of finite two-dimensional crystallites can be modified to account for the known partial orientation of exfoliated graphite surface planes, and has been widely used to characterise the neutron and x-ray diffraction spectra from monolayers physisorbed on these surfaces. The line-shapes are fitted in terms of three parameters: L, the linear dimension of the domains; G, the magnitude of the appropriate reciprocal lattice vector; and an overall multiplicative constant; long-range crystalline order is assumed. The system of methane on grafoil was studied in this fashion by Vora, Sinha and Crawford(4) who reported the phase diagram shown in Fig. 1.

From the value of G it was determined that a $\sqrt{3} \times \sqrt{3}$ commensurate structure is formed at low temperatures and coverages. At low temperatures, above a coverage defined as $\rho = 1.0$, the methane monolayer compresses into a triangular incommensurate phase. (The commensurate-incommensurate transition in methane has been discussed by M. Nielsen at this conference). Both these phases are familiar from studies of many other physisorbed systems, as is the liquid phase obtained for all coverages at high enough temperatures. However, with increasing temperature, the commensurate solid unexpectedly expands before melting and it appears that the expanded solid is always an intermediate stage in the melting of monolayers with $0.6 < \rho < 1.0$.

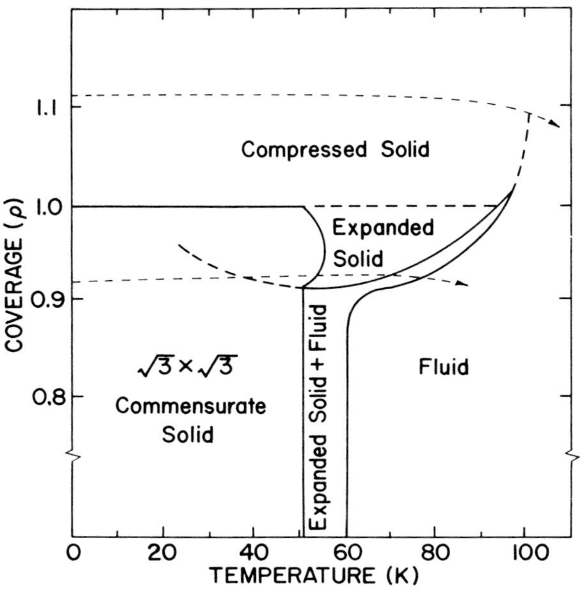

Fig. 1. Phase diagram for CD_4 on graphite (from Ref. 4). The units of coverage are defined such that ρ = 1.0 is the highest commensurate coverage at low T. The horizontal dotted lines indicate the trajectories along which the melting transitions were studied (Figs. 4, 6 and 7); the slight downward curves of these trajectories are due to increasing evaporation into the "dead space" in the graphite cell.

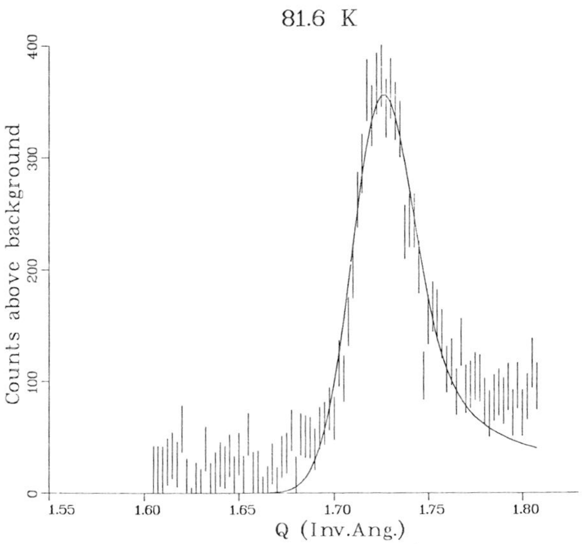

Fig. 2. Lineshape for incommensurate solid CD_4 on ZYX exfoliated graphite at 81.6K. The line is a Warren lineshape fit to the data (i.e. assuming 3D-like correlations and finite size). Note that it is unable to fit the "tail" in the data (in attempting to do so, it has become slightly too wide in the main body of the peak).

The Warren expression for the lineshape is of course inapplicable to fluid phases and solid-liquid coexistence regions, and identifications of such regions in the phase diagram are of necessity made by appropriately interpreting the behavior of the Warren parameters upon entering these regions. Indeed, even an infinite two-dimensional solid is not in principle characterizable in terms of the Warren parameters since long-range order is predicted not to exist. Fig. 2 shows a typical diffuse scan for a compressed solid monolayer of methane on ZYX graphite at 81.6°K and a coverage of $\rho \simeq 1.03$; the Warren lineshape does an adequate job near the peak but does not account for the "tail" at low Q values. These tails are observed in all solid phases (including the registered phase), are substantially larger (in relation to the peak) than is familiar for thermal diffuse scattering in three-dimensional crystals, and indicate the presence of large phonon fluctuations in these quasi-two-dimensional monolayers. The fact that tails exist also in the commensurate solid suggests that the graphite potential wells are particularly weak for methane, and is probably related to the fact that the commensurate monolayer always breaks out of registry before it melts.

In order to analyze lineshapes in cases where the Warren assumption of long-range order is (or may be) incorrect, we have reformulated the Warren expression in terms of a general structure factor S(q), taking into account the fact that the normals to the grafoil surfaces have a distribution $P(\Phi)$ that is peaked around $\Phi = 0$ (partial misorientation) but are completely randomly distributed with regard to rotation about the normals. The structure factor must thus be powder-averaged in the plane:

$$\bar{S}(K') = (2\pi)^{-1}\, f^2(K') \int_0^{2\pi} d\psi\, S(q) \qquad (1)$$

where

$$\vec{q} = \vec{K}' - \vec{G}$$

$\vec{G}$ is the reciprocal lattice vector; the averaging over orientations of G is represented by the integral over ψ(the angle between $\vec{K}'$ and $\vec{G}$); f(K) is the molecular form factor for methane (which rotates freely at all temperatures relevant here).

In terms of $\bar{S}(K')$, the observed intensity for a given scattering wave vector $\vec{K}$ may be shown to be

$$I(K) = \int dK'\, K'\, \bar{S}(K') F(K,K') \qquad (2)$$

where

$$F(K,K') = \int dK'' \frac{R(K - K'')}{K''^2} \frac{\theta(K'' - K')}{\sqrt{1 - (K'/K'')^2}}$$

$$\int_{\beta}^{\pi - \beta} d\Phi \frac{P(\Phi)}{\sqrt{(K'/K'')^2 - \cos^2\Phi}} \qquad (3)$$

Here, $R(\Delta K)$ is the instrumental resolution, $\theta(x)$ is a step-function, and

$$\beta \equiv \sin^{-1}\sqrt{1 - (K'/K'')^2}.$$

The misorientation function $P(\Phi)$ is well approximated by a gaussian with a FWHM of about 30° for grafoil and 15° for ZYX exfoliated graphite.

The Warren lineshape emerges from these expressions if

$$S(q) \propto \exp(-L^2q^2/4\pi)$$

which represents the effects of finite size on a delta-function structure factor. The effect of finite size on other structure factors is far less clear. (A detailed analysis of this effect for two-dimensional systems is in progress and will be reported separately). Nevertheless, we have used theoretical predictions for the infinite-size structure factors in the solid and liquid phases to fit lineshapes obtained in two temperature scans across the melting transitions performed at coverages $\rho = 0.92$ and $\rho = 1.09$. The values of the parameters thus determined, while not quantitatively accurate, may be expected to show illuminating trends.

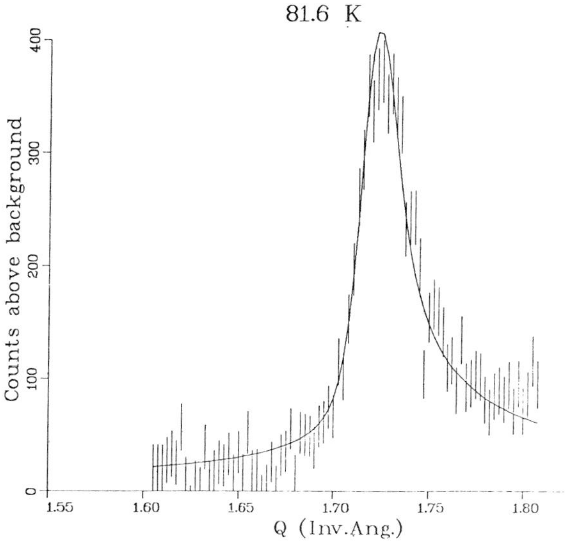

Fig. 3. Same data as in Fig. 2; the line is a fit assuming a power-law structure factor - $S(q) \sim 1/q^{2-\eta}$.

In the case of the solid phases using the power-law structure factors predicted for two-dimensional systems:

$$S(q) \propto \eta\left(\frac{2}{qa}\right)^{2-\eta} \qquad (4)$$

(where a is the lattice constant), we are able to obtain excellent fits to both the "body" and the "tail" of each line (e.g. Fig. 3). As already pointed out, the values of η obtained from the fits (Fig. 4) are not exact. However we can say qualitatively that at $\rho = 1.09$, η stays fairly small and suddenly increases for T>95K just near the melting temperature. At $\rho = 0.92$, in the pure expanded solid phase, is considerably larger (which is reasonable since the density is lower) and increases with temperature. (Above 60K, in the coexistence region, the fits are not meaningful).

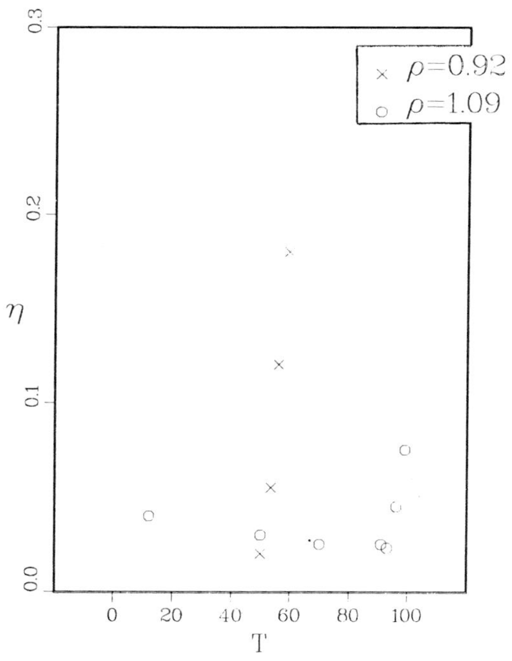

Fig. 4. η vs. T for expanded ($\rho = 0.92$) and compressed ($\rho = 1.09$) incommensurate solid CD_4 on grafoil.

Turning now to the liquid phase, the structure factor is predicted by Halperin and Nelson(2) to be approximately

$$S(q) \propto \frac{\xi^{2 - \eta^*}}{(q^2\xi^2 + 1)^{1 - \eta^*/2}} \qquad (5)$$

Here η^* is the value of η for the solid at melting and ξ is a correlation length (at $T = T_c$, $\xi \to \infty$ and S(q) reduces to Eq. 4 with $\eta = \eta^*$). The description in terms of the distance q from a reciprocal lattice point is applicable because of the postulated existence of a hexatic phase. The temperature dependence of ξ is predicted by these authors to be

$$\xi = \xi_o \exp [A(T - T_o)^{-\nu}] \qquad (6)$$

where ν is calculated by Halperin and Nelson to be 0.369...and T_o is the transition temperature.

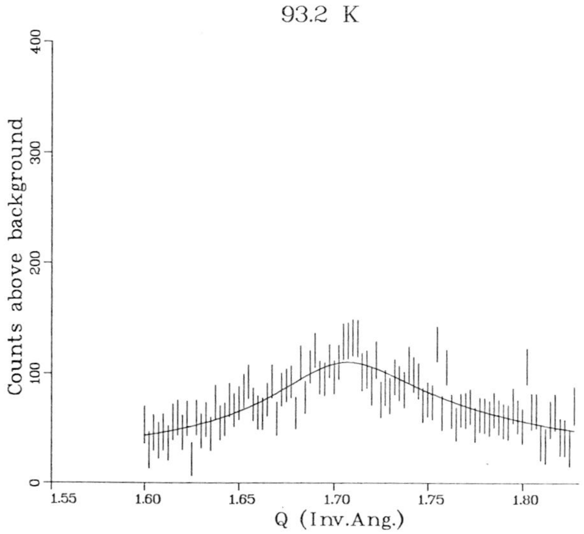

Fig. 5. Lineshape at 93.2K for the same system as in Figs. 2 and 3 (CD_4 on ZYX exfoliated graphite); the line is a fit assuming a Lorentzian structure factor $S(q) \propto A/(1 + q^2\xi^2)^{1-\eta^*/2}$.

We have fitted liquid lineshapes to Lorentzian structure factors (corresponding to the approximation $1-\eta^*/2 \simeq 1$) and find good fits (eg., Fig. 5). The values of ξ well inside the liquid phase are quite large ($\sim$ 30 Å). Further, we find that

(a) for ρ = .92 the lineshapes may be plausibly fitted near the melting transition with a sum of solid and liquid structure factors, with relative weights changing across the transition in such a way as to keep the total amount of material constant. Thus our identification in Ref. 4 of a solid-liquid coexistence region (indicative of a first-order transition) is consistent with our fits.

(b) The melting of the ρ = 1.09 monolayer is considerably sharper and we were not able to fit the rapidly changing lineshape to a sum of solid and liquid structure factors with constant total amount.

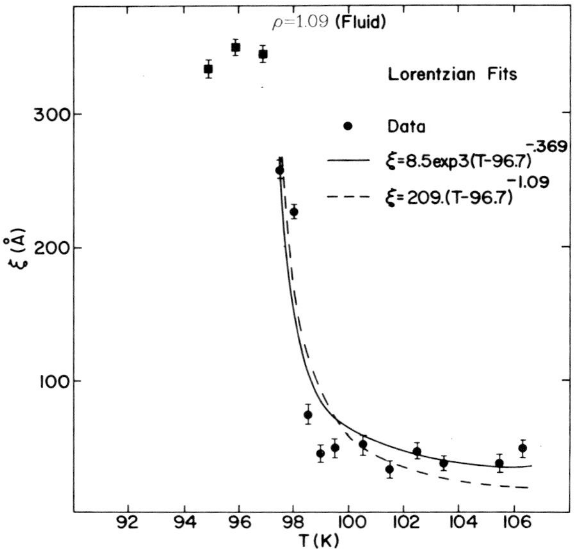

Fig. 6(a). Temperature dependence of ξ for ρ = 0.92.

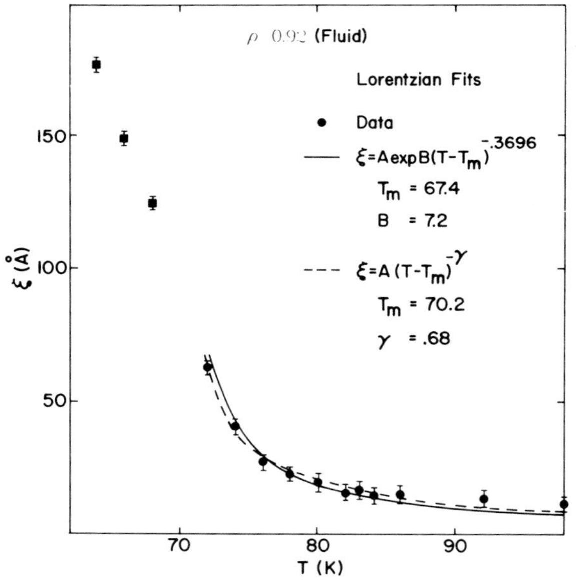

Fig. 6(b). Temperature dependence of ξ for ρ = 1.09.

Fig. 6 (a) and (b) show the behavior of ξ in the liquid as a function of temperature. In both cases ξ is fairly constant at $\sim$ 30Å and then rapidly increases with decreasing temperature. When ξ reaches a few hundred angstroms it tends to level off; we are limited here both by the not-fully-understood effects of finite size and by the spectrometer resolution ($S(q)$ is independent of ξ for $q \gg \xi^{-1}$). Fig. 6 (a) and (b) also show fits of the Halperin-Nelson prediction (Eq. 6) for $\xi(T)$ using ξ_o and T_c as parameters; also shown is a fit to

$$\xi \sim A\,(T - T_c)^{-\nu}$$

The largest values of ξ have been excluded from these fits for the reasons mentioned above. It can be seen that owing to the relatively small number of points, it is not really possible to distinguish between the models and to demonstrate the validity of Eq.(1); however, ξ does show strong 'precursor' behavior above the transition that is unknown in three dimensional fluids and is consistent with Eq. (6). Fig. 7 shows a plot of ln I vs ln ξ, where I is the amplitude of the Lorentzian. In both cases a reasonable straight line is observed with a slope slightly less than 2 as predicted by the theory [see Eq. (5)]. Also shown are points corresponding to temperatures clearly in the solid (for $\rho = 1.09$) or coexisting solid/fluid (for $\rho = 0.92$) phases, for which the Lorentzian structure factor is presumably not valid. In each case, a departure of the plot from linearity is observed which also helps to identify the transition. It may be recalled (Fig. 4) that in the incommensurate solid phase, the $\rho = 0.92$ coverage had a higher η; from the slopes in Fig. 7, this coverage can be also seen to have the higher η^*. Note that this is very qualitative since there is an intervening coexistence region for this coverage.

In conclusion, our detailed lineshape analyses show that:

(1) The substantial excess scattering in the solid phases can be explained in terms of power-law structure factors.

(2) While methane always melts from an incommensurate phase, there is an identifiable solid-liquid coexistence region at low coverages but not at high coverages. Thus the compressed solid may be melting in a continuous way.

(3) In both cases, however, the precursor behavior in the liquid correlation length, and the fact that the coverage with the higher η below the transition has the higher η^* above it, are qualitatively consistent with theoretical predictions. Thus it may be that dislocation unbinding has identifiable effects whether or not it is the primary process very close to the transition temperature.

This work was partially supported by the U.S. Department of Energy.

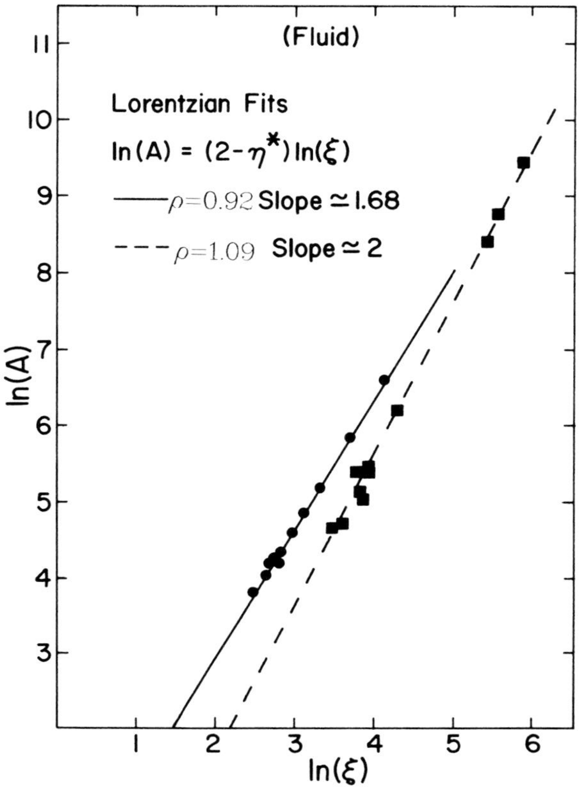

Fig. 7. Dependence of the scale factor A on for the two coverages, showing the different slopes (i.e. different values of η^*) in the two cases.

References:

1. J. M. Kosterlitz and D. J. Thouless, J. Phys. C6, 1181 (1973); J. M. Kosterlitz, J. Phys. C7, 1046 (1974).

2. B. J. Halperin and D. R. Nelson, Phys. Rev. B 19, 2457 (1979).

3. B. E. Warren, Phys. Rev. 59, 693 (1941).

4. P. Vora, S. K. Sinha and R. K. Crawford, Phys. Rev. Lett. 43, 704 (1979).

5. See, e.g. Y. Imry, CRC Crit. Rev. in Sol. State and Mat. Sciences 8, 157 (1979).

DISCUSSION

O. Vilches:

Could you comment on the discrepancies between

the French neutron scattering data and the isotherm work and the phase diagram you showed?

The isotherms show that there is a liquid-vapor coexistence region. Why not put it in your phase diagram?

P. Dutta:

Early data reported by Bienfait and coworkers did not show an expanded phase. They have now studied the system further (see Glachant et al., these proceedings) and have identified the expanded phase. Since Vora et al. did not attempt to distinguish the liquid from the vapor phase, there are now no discrepancies regarding the qualitive features of the phase diagram.

Published 1980 by Elsevier North Holland, Inc.
Sinha, ed. Ordering in Two Dimensions

MECHANICAL PROPERTIES OF FREELY-SUSPENDED LIQUID CRYSTAL FILMS

K. MIYANO

Argonne National Laboratory, Argonne, Illinois 60439

J. C. TARCZON

Argonne National Laboratory, Argonne, Illinois 60439

and

Department of Physics and Astronomy, Northwestern University

Evanston, Illinois 60201

ABSTRACT

The in-plane shear response of thin, freely-suspended films of a liquid crystal, butyloxybenzylidene octylaniline (40.8), driven at sub-audio frequencies (0.5 ∽ 4 Hz), has been studied at temperatures around the bulk smectic A-B phase transition. The amplitude and phase responses were measured as a function of temperature and film thickness. From these, the shear modulus μ and shear viscosity η were calculated. The mechanical properties of thin films are drastically different from the mechanical properties of the bulk material. More than 6 K above the bulk A-B transition temperature, μ was found to be non-zero and η extremely large. The anomalous behavior extends to higher temperature for thinner films, which suggests the effect of the 3rd dimension.

*Work supported by the U. S. Department of Energy

INTRODUCTION

The idea of phase transitions in two-dimensional (2D) systems due to topological ordering proposed by Kosterlitz and Thouless [1] has inspired many theoretical and experimental studies. The theory, later expanded by many workers, contains a number of specific predictions, some of which seem to be verified by recent experiments on 2D systems (e.g., superfluid ^{4}He films [2], superconducting films [3] and electrons on liquid ^{4}He [4]).

The study of the mechanical properties of freely-suspended liquid crystal films [5] provides an additional experimental test of 2D theory. The desirable features of the liquid crystal films are the following: (1) One can study the shear modulus, the quantity of interest in the theory, by literally shearing the film externally. (2) The film thickness can be varied, thereby providing information on the importance of the third dimension. (3) Experimental results can be compared directly with theory without the complicating effect of a substrate interaction. (4) Relative ease of the experiment.

In this paper, we will describe the details of our experiment, present some preliminary data, and discuss the significance of the results.

EXPERIMENTAL

Material

We used butyloxybenzylidene octylaniline (40.8 or BBOA) for our experiment, since this material has been extensively studied and the material constants are well known. It exhibits both the smectic A and B phases, and the bulk A-B transition occurs at ∽49°C, which is a convenient temperature to work around experimentally.

Torsion Head

The experimental setup is shown schematically in Fig. 1. A torsion wire W was stretched between two points under a tension of ∽ 4 x 10^4 dynes. The wire was 50 μm in diameter and made of Pt 8%W alloy. This material was chosen for its low internal friction at low frequencies [6]. Attached to the wire were a small mirror M_2 and a BeCu disc. The disc fits into a concentric hole made in a stainless steel sheet. The edges of the disc and the hole were ground to a sharp knife edge.
The edge of the hole was further serrated to provide a proper boundary condition, which will be discussed later. A liquid crystal film was spread in the annular gap between the disc and

the hole. The top end of the wire was attached to a pivot mechanism made of BeCu leaf springs which allows a turning motion only. This mechanism was twisted sinusoidally by a piezoelectric bender (Bimorph) B, which was driven by an audio frequency generator. The rotations of the top of the wire and the disc were monitored by measuring the positions of laser beams reflected from M_1 and M_2 with two position sensing photodiodes PD_1 and PD_2. The amplitude and phase of the disc rotation were recorded as the temperature of the sample was swept. The moment of inertia I of the disc assembly was $\sim 6 \times 10^{-3}$ g-cm^2 and the torsion constant of the wire was $\sim$ 5.5 dyne-cm. Its natural frequency without the film was thus $\sim$ 5 Hz. The quality factor was $\sim$300.

The temperature of the oven was monitored to 0.01 K using a temperature-sensing transistor. Temperatures quoted in this paper are those of the oven. Although it is impossible to measure the temperature of the film, it is estimated to be close to the oven temperature. A check of this approximation was made by measuring temperature of the disc with no film present. This was done by placing a glass bead thermistor on the disc. The thermistor leads did not touch any part of the oven, thus maximizing the heat loss through the leads. Under this most unfavorable condition, the disc temperature was $\lesssim$0.1 K lower than the oven temperature. From this observation and the consideration that the disc is heated by convection and radiation only, we concluded that the deviation of the film temperature from the oven temperature was substantially less than 0.1 K.

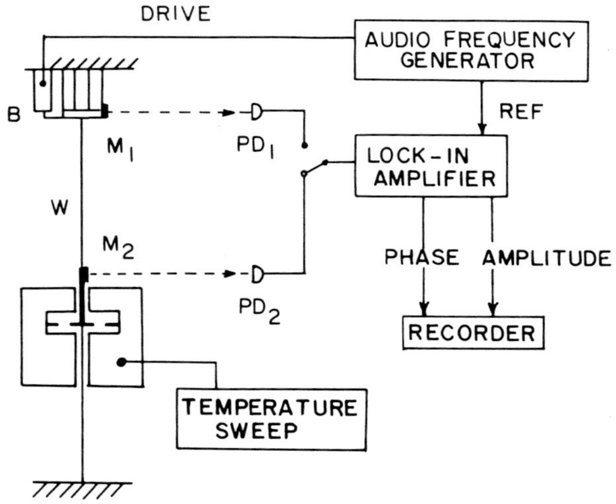

Fig. 1 Block diagram of the apparatus.

CALCULATION

We assume that the liquid crystal film is rigidly attached to the disc and hole. As the disc rotates, the film is deformed. From the symmetry of the system, the deformation vector is written as

$$\vec{U} = (0, U_\phi\ (r), 0) \tag{1}$$

in cylindrical coordinates where the z-axis is along the film normal. We assume that the mechanical behavior of the film can be described by a linearized hydrodynamics as

$$\rho \frac{\partial^2}{\partial t^2} U_\phi = (\mu + \eta \frac{\partial}{\partial t}) (\frac{\partial^2}{\partial r^2} + \frac{1}{r}\frac{\partial}{\partial r} - \frac{1}{r^2}) U_\phi, \tag{2}$$

where μ is the shear modulus and η shear viscosity. Assuming a sinusoidally varying displacement

$$U_\phi\ (r) = R(r)e^{i\omega t}, \tag{3}$$

Eq. (2) reduces to

$$\frac{d^2R}{dr^2} + \frac{1}{r}\frac{dR}{dr} - \frac{R}{r^2} + \frac{\rho\omega^2}{i\omega\eta+\mu} R = 0 \tag{4}$$

The solution is thus

$$R(r) = AJ_1(br) + BN_1(br),\ b^2 = \frac{\rho\omega^2}{i\omega\eta+\mu}, \tag{5}$$

where J_1 and N_1 are Bessel and Neumann function of order one, respectively. The constants A and B are determined from the boundary conditions

$$R(r) = 0 \quad \text{at } r = r_o$$

and

$$R(r) = \alpha r_i \quad \text{at } r = r_i \tag{6}$$

where r_i and r_o are inner and outer radii of the film, respectively. The equation of motion of the torsion pendulum is then

$$e^{i\omega t}[I(-\omega^2\alpha) + D\ (i\omega\alpha) + K_1\ (\alpha - \beta e^{i\varphi}) + K_2\alpha] = \tau e^{i\omega t} \tag{7}$$

where I: moment of inertia of the disc,
D: damping due to all frictional forces except the film viscosity
K_1: torsion constant of upper half of the wire
K_2: torsion constant of lower half of the wire
$\beta e^{i(\omega t+\varphi)}$: angular motion of the top of the wire
$\tau e^{i\omega t}$: torque due to the film
α: angular amplitude of the disc.

The torque due to the film is given by

$$\tau\, e^{i\omega t} = (\sigma_{r\phi})2\pi h r^2 , \tag{8}$$

$$= (\mu+i\omega\eta)\left(\frac{dR}{dr} - \frac{R}{r}\right)2\pi r^2 h\, e^{i\omega t}$$

at $r = r_i$,

where h is the thickness of the film. Although Eq. (8) contains Bessel and Neumann functions of order zero and unity, for $[\mu^2+\omega^2\eta^2]^{1/2} \gg \rho\omega^2$, it can be shown that

$$\frac{\tau}{\alpha} = \frac{4\pi h r_i^2 r_o^2}{r_i^2 - r_o^2}(\mu + i\omega\eta) - \frac{2\pi h r_i^4 r_o^2}{r_i^2 - r_o^2}\rho\omega^2 \ln\left(\frac{r_o}{r_i}\right) + \ldots \tag{9}$$

In our experiment, all terms except the first are negligible. In this case, Eq. (7) can be separated into real and imaginary parts and μ and η are written in a simple closed form as

$$\mu = \frac{1}{A}\left[K - I\omega^2 - \frac{\alpha_o}{\alpha}\sqrt{(K - I\omega^2)^2 + D^2\omega^2}\cos\varphi\right] \tag{10}$$

$$\eta = \frac{1}{A\omega}\left[D\omega - \frac{\alpha_o}{\alpha}\sqrt{(K - I\omega^2)^2 + D^2\omega^2}\sin\varphi\right],$$

where

$$A = \frac{4\pi h r_i^2 r_o^2}{r_i^2 - r_o^2}$$

and α_o/α is the ratio of the angular amplitude of the disc when the film is absent to that when the film is present.

DATA

We spread the film near the nematic - smectic phase transition temperature. In most cases the temperature was then lowered at rates of 3/4 ∽3K/hr to 45°C and the experiment was terminated. α and φ were recorded continuously as the sample temperature was swept. In some cases, the temperature was then increased at the same rate and α and φ recorded. Some raw data are shown in Fig. 2. The calculated μ and η through Eq. (10) are shown in Fig. 3.

DISCUSSION

A striking feature is immediately clear from Fig. 3. At 56°C the viscosity is enormously large. For the sake of comparsion the shear viscosity of a bulk sample of BBOA measured by a capillary viscometer is show in Fig. 4 (From Ref. [7]). Note the viscosity is on a linear

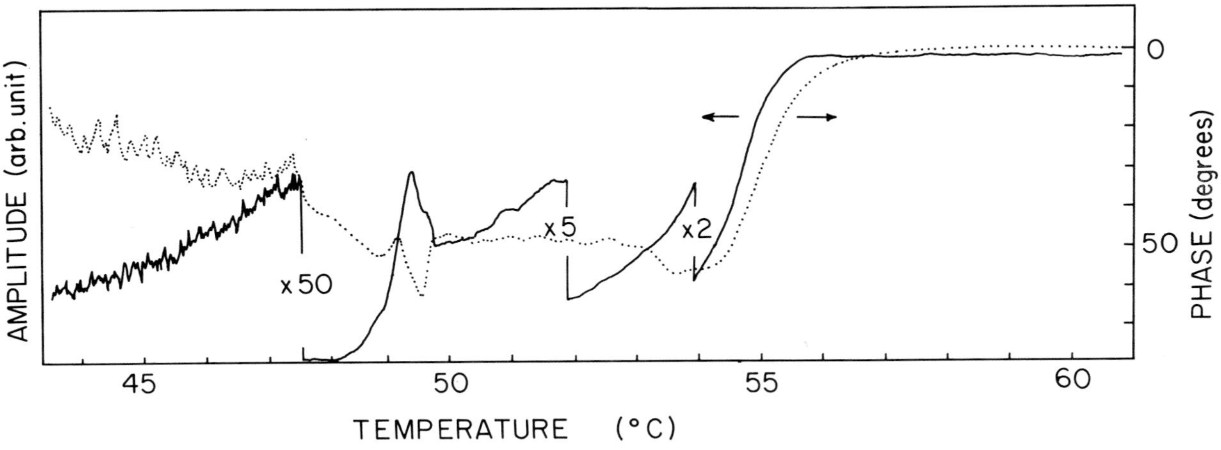

Fig. 2 Amplitude α and phase φ of the disc oscillation for a 10 molecule thick film at 2 Hz.

scale here. Well above the bulk A-B transition temperature the shear modulus is non-zero. A similar behavior was also seen by Pindak et al. [8] at ∽500 Hz. The anomalous behavior of η and μ extends toward higher temperatures for thinner films. This suggests that the anomalies are closely connected to the low dimensionality of the films. There is also a small feature at the bulk A-B transition.

Because the observed behavior is so unusual a few remarks are in order. In the initial phase of this program we experienced irreproducibility problems. Low sample purity due to degradation was first suspected. This idea was, however, quickly dismissed; in some cases repeated heat - cool cycles gave nearly identical results with a small shift of the bulk A-B transition temperature (taken as a small feature around 49°C) as the sample degraded while in others the traces of two successive runs were completely different. During these experiments the film thickness was repeatedly measured. The only other possible cause of irreproducibility was the boundary condition. Becuase the viscosity and shear modulus of the film were many orders of magnitude larger than the bulk values, a boundary region which can have the bulk properties will dominate the apparent shear mechanical property even if the region is quite narrow. We attached the disc to the film after the film was spread in the hole by lowering the film. This disc can be clean, free of excessive liquid crystal, and thus the boundary problem is minimal at the disc edge. However, an excess of material was inevitably deposited at the hole edge when the film was spread. To provide a

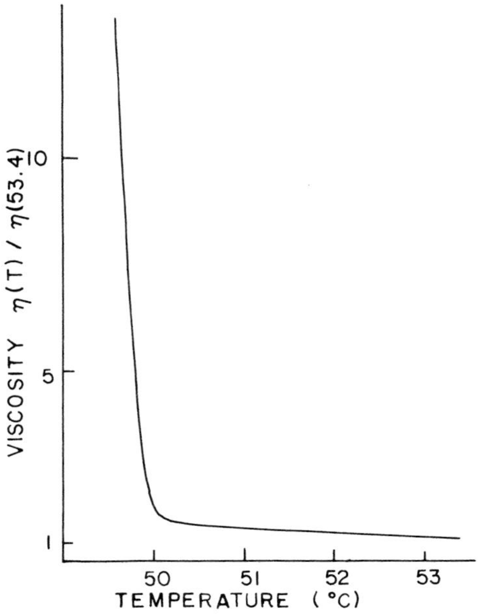

Fig. 4 "Bulk" shear viscosity (taken from Ref. [7]).

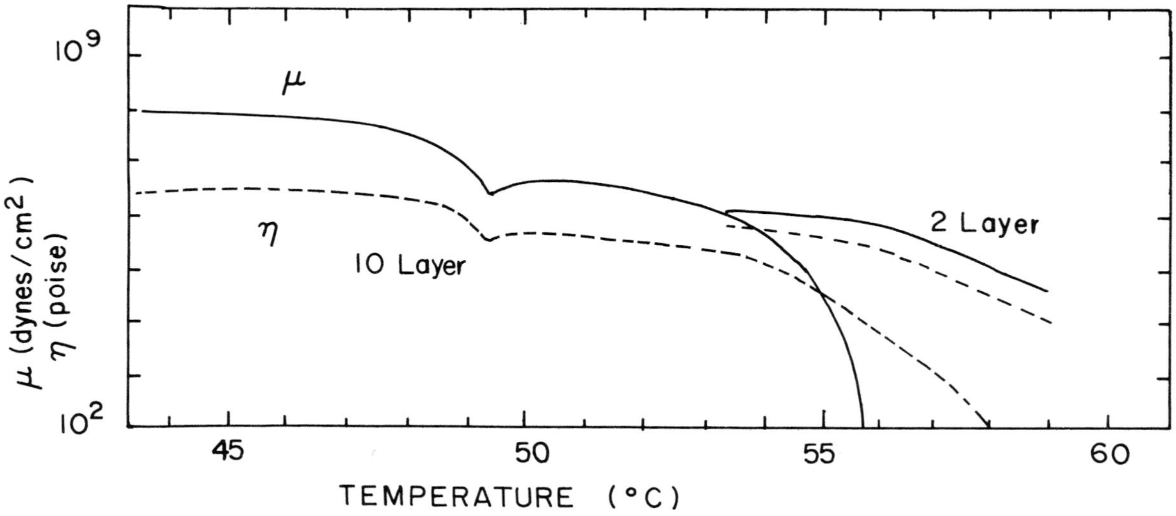

Fig. 3 Calculated μ and η for a 10 molecule thick film (from Fig. 2) and a 2 molecule thick film at 2Hz.

cleaner outer boundary we cut small teeth around the perimeter of the hole. Excess material tends to collect at the bottom of the teeth. The teeth also serve as a reservoir of additional material to maintain the film pressure when the film shrinks as it is cooled.

Our initial intention of this experiment was to search for the discontinous change in μ predicted by theory:

$$\mu_D \geq \frac{4\pi K T_m}{a_o^2} \tag{11}$$

where the equality would hold in the limit that the other Lamé coefficient λ is >> μ. Here, μ_D is the jump in μ at the 2D melting temperature T_m, and a_o is the lattice constant. Using the X-ray result [9] of $a_o = 5 \times 10^{-8}$cm and assuming $T_m \sim 322$ K, we get $\mu_D = 224$ ergs/cm^2. This is the shear modulus per two-dimensional entity. If each layer is considered to be such a unit entity, the shear modulus per thickness jumps from 0 to 7.8×10^8 dynes/cm^2 at T_m. Note that our value for μ at low temperature (∽45°C) is only ∽10^8 dynes/cm^2. Moreover, μ starts to increase continuously from very small values. Thus we did not find the type of phase transition predicted by theory. The phase transition seems rather different from what one would expect in a simple solid-liquid transition. Further study of frequency dependence and the effect of the boundary material and shape is underway.

ACKNOWLEDGEMENT

We thank J. B. Ketterson and B. M. Abraham for valuable suggestions.

REFERENCES

1. J. M. Kosterlitz and D. J. Thouless, J. Phys. C6, 1181 (1973).

2. I. Rudnick, Phys. Rev. Lett. 40, 1454 (1978) and D. J. Bishop and J. D. Reppy, Phys. Rev. Lett. 40, 1727 (1978).

3. A. F. Hebard and A. T. Fiory, Phys. Rev. Lett. 44, 291 (1980).

4. C. C. Grimes and G. A. Adams, Phys. Rev. Lett. 42, 795 (1979).

5. C. Rosenblatt, R. Pindak, N A. Clark, and R. B. Meyer, Phys. Rev. Lett. 42, 122 (1979).

6. K. Tørklep and H. A. Øye, J. Phys. E12,875 (1979).

7. S. Bhattacharya and S. V. Letcher, Phys. Rev. Lett. 42, 458 (1979). The viscosity around 54°C is about 1 poise (S. Bhattacharya, private communication).

8. R. Pindak, D. J. Bishop, and W. O Sprenger, Preprint.

9. D. E. Moncton and R. Pindak, Phys. Rev. Lett. 43, 701 (1979).

DISCUSSION

M. Pomerantz:

Have you been able to check the method by going to the limit of thick films?

K. Miyano:

The responses of the films of thickness ranging from 3 to 15 layers were all quite similar. Much thicker films turned out to be non-uniform and we could not try the limit of thick films.

M. Pomerantz:

How do you measure the number of monolayers in your films?

K. Miyano:

By optical reflectivity measurements.

S. Doniach

Can you say something about stress in the films, and in particular do you observe hysteresis on warm up-cool down recycling?

K. Miyano:

The maximum torque that the system can generate is about 5×10^{-4} dynes-cm. Therefore, the maximum stress, say for 7 layer film, is 5×10^2 dynes/cm^2 when it is in the B phase. The linearity has been confirmed down to at least two orders of magnitude lower stress. Detailed heating experiments have not been done.

D. E. Moncton:

As Pindak et al. point out in their recent paper (see also these proceedings) on identical experiments, the persistence of a solid surface layer above the bulk A→B transitions is also observed in our x-ray experiments. There is no doubt about the existence of a finite shear modulus above 49°C. I would like to emphasize the potential interest in this type of surface transition as discussed by M. Wortis at this conference (I-2).

K. Miyano:

Anomalous behavior of the 2-layer films seems to further support the idea.

B. I. Halperin:

I was worried about the boundary condition in the thin samples. How do you know that you do not have problems similar to the ones you mention in the thicker case?

K. Miyano:

We provided a "drain" for the excessive material on the boundary by cutting teeth around the perimeter. This technique produced a very clean boundary. We could not spread thick films in this manner, however, because the teeth spontaneously drain the film thin.

Published 1980 by Elsevier North Holland, Inc.
Sinha, ed. Ordering in Two Dimensions

Recent Experimental Results on Vortex Processes in Thin-Film Superconductors

A. F. HEBARD and A. T. FIORY

Bell Laboratories, Murray Hill, New Jersey 07974

By measuring the complex a.c. impedance in the frequency range 700 Hz to 80 MHz of a thin granular aluminum film, we have been able to obtain information about the dynamics of vortex motion which is unique to two-dimensional superconductors. Data taken in zero magnetic field reveal the existence of thermally-excited vortex pairs which dissociate at a frequency-dependent temperature where there is a jump in the superfluid density. This observed transition is broadened logarithmically with frequency and the zero-frequency transition temperature is determined by extrapolation. At lower temperatures, vortices created by an external magnetic field exhibit thermally-activated mobilities. We find that vortex mobilities change rapidly at temperatures which depend on the measuring frequency and the vortex areal density. These observations may be associated with a crossover from solid-like to fluid-like behavior in the vortex medium.

1. INTRODUCTION

The Kosterlitz-Thouless theory of phase transitions in two dimensions [1-3] has recently been experimentally verified for superfluid films of ^{4}He on mylar substrates [4] and for granular aluminum thin film superconductors [5]. The theory for these superfluid systems is based on the idea that it is the fluctuations in the phase rather than in the magnitude of the order parameter which determine the details of the transition. These phase fluctuations are driven by thermally excited vortex pairs, of opposite circulation, which unbind at a transition temperature T_c. The applicability of the theory to superconductors has been presented by several authors [6-9].

The transition temperature is given as a universal function of the superfluid areal density n_s and mass m^* ($m^* = 2m$ for super-electron pairs) as

$$T_c = \pi \hbar^2 n_s / 2m^* k. \quad (1.1)$$

The correlation length of the superconductor above T_c is

$$\xi_+ = \xi_c \exp\{[b(T/T_c - 1)]^{-1/2}\}, \quad (1.2)$$

where $\xi_c \simeq \xi_{GL}(T_c)$, the Ginzburg-Landau coherence length, approximately equal to the vortex core radius, and b is a constant of the order of unity. The maximum separation of bound vortex pairs is ξ_+ and the density of free vortices is given by ξ_+^{-2}.

The connection with experiments was made by Ambegaokar *et al* [10-12], who treated the diffusive motion of the vortices under the influence of an oscillating driving force, which in the superconductor is supplied by a transport current. Taking into account the diffusivity of the vortices, these authors show that experiments carried out at finite frequency effectively probe the response of the bound pairs separated by a length $r_\omega = (14D/\omega)^{1/2}$, where D is the vortex diffusivity and the factor of 14 a calculated constant [11]. The result is that the film will exhibit some superfluid density above T_c at finite frequencies satisfying $r_\omega < \xi_+$.

The theory is presumably applicable to superconductors even though the $\ln(r/\Lambda)$ vortex-pair interaction potential is valid only for $r < \Lambda$, where $\Lambda = mc^2/2\pi n_s e^2$ is the magnetic screening length for thin films [13]. Thus one may expect deviations from ideal behavior over lengths greater than Λ, as the theory is formulated only for logarithmic interactions. The influence of sample edges, the finite size of Λ and sample inhomogeneities are subjects where additional work remains to be done.

In the following sections we first develop a formalism (Sec.2) which treats the vortex contribution to the electical impedance of a superconductor which is then followed by a review of the experimental technique used to measure this impedance (Sec.3) [14]. We then review and elaborate upon the experimental evidence [5] for vortex unbinding (Sec.4) and the effect of pinning on this transition (Sec.5). We conclude in Section 6 with a discussion of preliminary data showing a pinning transition which may be related to recent theories of vortex lattice melting [15,16].

2. VORTEX DYNAMICS - EQUIVALENT CIRCUIT

In a consideration of the response of a superconductor to external currents and fields, we include contributions from the superfluid background, normal currents and the dissipative flow of vortices which may be present in the sample. It is our purpose here to present in some detail the electrical response of a two-dimensional superconductor and identify those contributions which can be related to the experimentally measured impedance.

We start with the fundamental assertion that the pair wave function ψ describing the superconducting state satisfies the equation $i\hbar \partial\psi/\partial t = 2\mu\psi$, where the electrochemical potential for a single electron,

$$\mu = \mu_c + eV, \tag{2.1}$$

is written as the sum of a chemical potential μ_c and an electric potential V. This wave equation has solutions $\psi = |\psi| e^{i\phi}$ which satisfy the Josephson relation [17]

$$-\hbar \partial \phi / \partial t = 2\mu. \tag{2.2}$$

The phase ϕ of the order parameter is related to the vector potential $\mathbf{A}$ and the superfluid current density $\mathbf{J}_s$ by the well-known Ginzburg-Landau [17] equation

$$\frac{4\pi\lambda^2}{c}\mathbf{J}_s + \mathbf{A} = \frac{\Phi_o}{2\pi}\nabla\phi, \tag{2.3}$$

where λ is the London penetration depth and Φ_o the flux quantum.

From straightforward algebraic manipulation of the above three equations one calculates the electric field

$$\mathbf{E} = -\frac{1}{c}\frac{\partial \mathbf{A}}{\partial t} - \nabla V = \frac{4\pi\lambda^2}{c^2}\frac{\partial \mathbf{J}_s}{\partial t} + \nabla(\mu_c/e). \tag{2.4}$$

For thin films of thickness $d_F << \lambda$, it is it is convenient to define a surface current $\mathbf{K}_s = d_F \mathbf{J}_s$ and a sheet inductance

$$L_K = 2\pi\Lambda / c^2, \tag{2.5}$$

where $\Lambda = 2\lambda^2/d_F$ is the effective magnetic screening length. Equation (2.4) now becomes

$$\mathbf{E} = L_K \partial \mathbf{K}_s / \partial t + \nabla(\mu_c/e), \tag{2.6}$$

which explicitly characterizes the inductive superfluid response in terms of L_K, the sheet kinetic inductance of the superfluid. In terms of microscopic quantities

$$L_K = m^*/n_s(e^*)^2. \tag{2.7}$$

It is well known that electric fields are associated with the dissipative motion of vortices in thin film superconductors [18]. These vortex contributions arise explicitly from a consideration of the $\nabla\mu_c$ term of Eq.(2.6). We proceed to show this connection by introducing the concept of a vortex current $\mathbf{K}_v$, restricted to the plane of the film, satisfying the conservation equation

$$\nabla \cdot \mathbf{K}_v + \partial \rho_v / \partial t = 0, \tag{2.8}$$

where the vortex areal charge density $\rho_v = q_v n_v$ is equal to the product of the vortex areal number density n_v and the vortex charge q_v [19]. It is a well-established fact that the average rate of change of phase between two points in a superconducting film is equal to 2π times the rate at which vortices cross a boundary between them [20]. This is expressed by the gauge-invariant relation

$$-\frac{\Phi_o}{2\pi c}\frac{\partial}{\partial t}(\nabla\phi) = \frac{\Phi_o}{q_v c}\hat{z}\times\mathbf{K}_v + \nabla V, \tag{2.9}$$

where $\hat{z}$ is a unit vector normal to the film.

Combining equations (2.1), (2.2) and (2.9) we can write Eq.(2.6) in the final form

$$\mathbf{E} = L_K \,\partial \mathbf{K}_s/\partial t + (\Phi_o/q_v c)\,\hat{z}\times\mathbf{K}_v. \tag{2.10}$$

The right-hand side reveals the decomposition of the electric field into respective contributions from the superfluid background and flowing vortices. To include the contribution of normal currents, we follow the suggestion of A. Schmid [21] and assume that the total electric current $\mathbf{K}$ can be written as the sum of a supercurrent $\mathbf{K}_s$ and a normal current obeying Ohm's law $\mathbf{K}_N = R_N^{-1} L_K \,\partial \mathbf{K}_s/\partial t$, where R_N is the normal state sheet resistance. All of the data discussed in this paper were taken at frequencies $\omega << R_N/L_K$ so that $\mathbf{K} \simeq \mathbf{K}_s$ and the normal current contributions could be ignored.

In order to make the final connection with experiment, we assume a local constitutive relation between $\mathbf{A}(\mathbf{r})$ and $\mathbf{K}(\mathbf{r})$ of the form

$$\mathbf{K}(\mathbf{r}) = -i\omega\,\mathbf{A}(\mathbf{r})/cZ, \tag{2.11}$$

where Z is the complex sheet impedance of the film. As electric charging effects are negligible, it is convenient to set $\nabla V = 0$, so that the electric field can be written as

$$\mathbf{E} = -c^{-1}\,\partial\mathbf{A}/\partial t = -i\omega\mathbf{A}/c = Z\mathbf{K}. \tag{2.12}$$

Since the vortex current is proportional to the electric current we immediately see from equations (2.10) and (2.12) that the complex sheet impedance can be written in the form

$$Z = i\omega\, L_K + Z_v, \tag{2.13}$$

where Z_v, the vortex impedance, is associated with contributions due to flowing vortices and gives rise to an electric field $(\Phi_o/q_v c)\,\hat{z}\times\mathbf{K}_v$. We will show in the next section how Z is determined from experiment.

The dynamical theory [10-12] describing the Kosterlitz-Thouless transition is discussed in terms of a frequency-dependent dielectric constant $\epsilon(\omega)$ which is the sum of two parts: a contribution ϵ_f, due to the motion of free vortices, and a frequency-dependent contribution ϵ_b, due to the motion of bound pairs. In zero applied magnetic field and for temperatures $T<T_c$, where there are no thermally excited free vortices ($\epsilon_f = 0$), contributions are made to Z_v at finite frequencies by the vortex currents associated with the reorientation of bound pairs in response to applied electric currents. At zero temperature the attractive vortex-antivortex interaction annihilates all bound pairs, the total impedance Z of Eq.(2.13) becomes $i\omega L_K$, and by Eq.(2.11), $\mathbf{K} = -\mathbf{A}/cL_K$, the well-known London response for a superfluid with "bare" superfluid density. For temperatures $T_c<T<T_{co}$ there is a flood of free vortices which makes dominant contributions to the real (dissipation) and imaginary (inductive) components of Z_v as will be discussed in the following sections. In our notation, the complex dielectric constant appears in a modified form of Eq.(2.13) [9]

$$Z = i\omega L_K\,\epsilon(\omega). \tag{2.14}$$

3. EXPERIMENTAL PROCEDURE

The films for this study were prepared by evaporating aluminum in an oxygen atmosphere onto fire-polished Corning 7059 glass slides. Square samples, 3mm on an edge, were scribed and contacts were applied to the corners. For the rf impedance measurements each film was placed in a perpendicular plane between two coaxial coils. An rf current was applied to the drive coil and a signal voltage from the receive coil was phase-sensitively detected. The circuit diagram which we used was given previously [14]. All measurements were taken at rf amplitudes where the voltage in the receive coil scaled linearly with the current in the drive coil. We present results here obtained for two films, 43 and 4130 ohms per square ($\Omega/\Box$) sheet resistances in the normal state, measured at 4.2K and denoted by R_N.

In our analysis, we use the constitutive relation between the sheet current density in the film and the local vector potential as given in Eq.(2.11) to define a sheet impedance Z. We solve the following equation appropriate to our case where the currents in the film are driven by an external vector potential $\mathbf{A}_e$ from the drive coil;

$$\mathbf{K}(\mathbf{r}) = -\frac{i\omega}{cZ}\left[\mathbf{A}_e(\mathbf{r}) + \frac{1}{c}\int_{film}\frac{\mathbf{K}(\mathbf{r}')d\mathbf{r}'}{|\mathbf{r}-\mathbf{r}'|}\right]. \quad (3.1)$$

The second term is the vector potential arising from currents in the film. The signal voltage in the receive coil, V_r, is then given by an integration of the time derivative of the vector potential around the turns of this coil. Equation (3.1) was solved numerically for our specific geometry. Since we require a solution giving the impedance as a function of V_r, it is convenient to use the following calibration formula:

$$Z = i\omega\,[L_1/\beta - L_2(\beta)], \quad (3.2)$$

where L_1 and $L_2(\beta)$ are numerically- determined parameters. We calculate β from the signal-voltage data as follows:

$$\beta = 1 - V_r/V_{ro}, \quad (3.3)$$

where V_{ro} is the mutual inductance voltage in the absence of the film. We obtain V_{ro} from the receive-coil signal when the film is in the normal state. The calibration constant L_1 was calculated numerically to be 0.36 nH. The $L_2(\beta)$ term is comparable to L_1 and accounts for the effective geometrical inductance of the film. We find that $L_2(\beta)$ is only weakly dependent on β, varying by about 10%.

In presenting the data in the following sections, we partition the measured impedance into its series-equivalent sheet-resistance and sheet-inductance components:

$$Z = R + i\omega L. \quad (3.4)$$

4. SUPERCONDUCTING TRANSITION

In this section we describe the superconducting transition of the aluminum film with the higher normal-state sheet resistance, $R_N = 4130\,\Omega/\Box$. At low temperatures the film's impedance is dominated by the kinetic inductance of the superfluid. Since L_K^{-1} is proportional to the superfluid density at low temperatures, we present the data as plots of L^{-1} vs temperature as shown in Fig. 1 for external magnetic field values of 0, 0.12 and 0.58 G. The measuring frequency is 300 kHz. The corresponding rf resistance data are shown in Fig. 2.

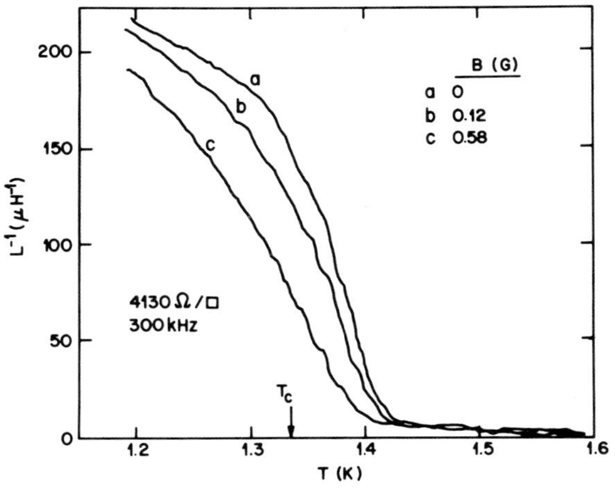

Fig. 1 Temperature dependence of inverse sheet inductance of a granular aluminum film. The extrapolated transition temperature is $T_c = 1.34$K.

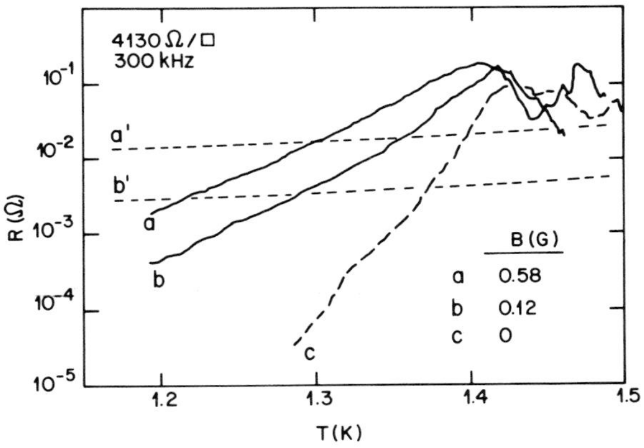

Fig. 2 Resistance transitions for same film as in Fig. 1. Curves a' and b' are theoretical flux-flow resistances in magnetic fields of 0.12 and 0.58G, respectively.

Considering first the zero-field data, we observe that the resistance becomes appreciable for temperatures above 1.4K, where L^{-1} rapidly decreases. This is the region where the motion of thermally-excited vortices makes a measurable increase in the film impedance. At lower temperatures in the range 1.2 to 1.3K this contribution is negligible, so we take L equal to the kinetic inductance, which we find to be independent of frequency from 5 kHz to 40 MHz, as expected theoretically. When fitted to the form $L^{-1} = L_K^{-1} \propto (T_{co} - T)$, we obtain $T_{co} = 1.85 \pm 0.02$K, which is also a temperature near the midpoint of the d.c. resistance transition of the film.

A study of the frequency dependence has been reported previously [5] and will be summarized here. The transition in L^{-1} *vs* T broadens and the bend in the curve (at about 1.42K at 300 kHz, Fig. 1) at small L^{-1} shifts to higher temperatures. This frequency dependence is similar to the prediction for the dynamical response of thermally-excited vortex pairs. The frequency-dependent resistance is attributed to the polarization of bound vortex-pair dipoles, the dominant pairs being those separated by the vortex diffusion length r_ω discussed in Sec. 1.

At higher temperatures the resistance should become frequency independent, owing to the dissociation of the vortex pairs, given by resistance of a free plasma of thermally excited vortices. Our experimental method gives the sensitivity needed to observe this effect in the resistance for frequencies above about 7 MHz. Theoretically, the polarization of the bound vortices adds to the inductance of the film. In practice, we expect some additional contribution from pinning forces. We associate the frequency-dependent temperature T_ω where L^{-1} approaches zero as the temperature at which r_ω becomes comparable to the maximum separation of bound pairs ξ_+. This is the temperature where pairs of separation $r \to r_\omega$ dissociate, so we equate r_ω to the vortex correlation length ξ_+. In ref. 5 we presented a procedure for extrapolating T_ω to zero frequency, to find the effective d.c. dissociation temperature T_c. We assume that ξ_+ has the theoretical temperature dependence given in Eq.(1.2). The diffusivity that enters into the expression for r_ω is expressed in terms of the vortex mobility μ and viscosity coefficient η in the usual manner:

$$D = \mu kT = \eta^{-1} kT. \tag{4.1}$$

The viscosity coefficient is given in terms of the flux-flow resistance R_f of vortices produced by an external magnetic field as follows [22]:

$$R_f = B\Phi_o/\eta c^2. \tag{4.2}$$

Equation (4.2) is presumably valid in the low-temperature regime where R_f is linear in B and vanishes at $B = 0$. The mobility of free vortices was calculated by Bardeen and Stephen [18] and is used to calculate the curves a' and b' in Fig. 2, to be compared with experimental curves a and b. The data show that $R = R_f$ is given to order-of-magnitude accuracy by the theory, but the temperature dependence is more pronounced. In particular, R_f decreases at low temperatures, probably because of the onset of pinning effects. This is discussed in the next section. The behavior above T_c is complicated because of the thermally excited vortices. Since the measured resistances are close to the theoretical curves at high temperatures, we use Eq.(4.1) for computing r_ω. We analyze the data by plotting $\ln^{-2}(r_\omega/\xi_{GL})$ vs T_ω and making a linear fit. The result, referring to Eq.(1.2), is $T_c = 1.34 \pm 0.01$K and $b = 0.24 \pm 0.02$. The value of $L(T_c)$ averaged over frequency is 6.7 ± 0.5 nH, so that $\Lambda(T_c)T_c = c^2L(T_c)T_c/2\pi = 1.4 \pm 0.1$ cm-K [23]. The theoretical relation, Eq.(1.1), can be written $\Lambda(T_c)T_c = 1.96$ cm-K.

The presence of an external magnetic field has the effect of shifting the frequency-dependent transition in L^{-1} *vs* T to lower temperature by stimulating the unbinding of bound pairs. We see from Fig. 1 that at $B = 0.58$ G the dissociation temperature is about 1.40K, as compared to about 1.42K in zero field at this frequency. One may expect that with an externally applied field L^{-1} would approach zero at a lower temperature, where the separation between the dominant pairs, r_ω, becomes comparable to the separation $a = \sqrt{\Phi_o/B}$ between the vortices created by the external field, and where $\xi_+(T) > a$. At temperatures above this point, where $r_\omega > a$, the response is dominated by the free vortices created by the external magnetic field. Since the transition is associated with $r_\omega = a$, the threshold sensitivity to magnetic field is frequency dependent, occuring as a function of the ratio B/ω, which is in accordance with our observations at other frequencies. For the $B = 0.58$ G data at 300 kHz, these lengths are $a = 6.0\mu$m and $r_\omega = 5.1\mu$m at $T = 1.4$K. This result, that $r_\omega \sim a$ at $L^{-1} \to 0$, serves as a good consistency check on our calculation of r_ω.

5. PINNING CONTRIBUTIONS

If the assumption is made that the motion of a given vortex in a harmonic pinning potential is uncorrelated with the motion of similar vortices in neighboring potentials, then it is straightforward to show [14] that the complex vortex impedance Z_v is proportional to B and can be modeled as a parallel combination of a pinning inductance, inversely proportional to the strength of the pinning, and a flux flow resistance R_f of Eq.(4.2). The linear dependence of this pinning impedance with B has been experimentally verified for our $R_N = 4130\,\Omega/\square$ film at 1.2K and 10 kHz [14]. However, we might expect this simple description to break down either at temperatures where there is a contribution from free vortices or for lower R_N films where correlations between vortices are expected to become important [15,16]. It is therefore advantageous to be more general and define a complex mobility

$$\mu = \mathbf{v}/\mathbf{F}_L = c^2 Z_v / B\Phi_o \tag{5.3}$$

relating the velocity $\mathbf{v}$ to the Lorentz force $\mathbf{F}_L$. As a vortex moves in a viscous medium over a sea of pins one would expect $\mathbf{v}(\mathbf{r})$ to lag $\mathbf{F}_L(\mathbf{r})$ just as $\mathbf{E}$ (proportional to $\mathbf{v}$) lags $\mathbf{K}$ (proportional to $\mathbf{F}_L$) for an inductive response. The phase angle $\theta = \arg\{Z_v\}$ therefore becomes an important measure of the relative importance of the pinning forces compared to the velocity-dependent viscous forces which give rise to dissipation.

Figure 3 is a plot at four different frequencies of the phase angle θ as a function of field for the $4130\ \Omega/\square$ -film at 1.20 K. The ωL_K contribution at $B = 0$ to Z has been subtracted out prior to the evaluation of θ. It is apparent from this plot that there is a slowly increasing contribution to the pinning with increasing field at each frequency as well as with increasing frequency at a given field. We infer that at $\omega = 0$ the pinning is negligible ($k \simeq 0$) and any zero-voltage critical current due to pinning should also be negligible. This checks with our dc current-voltage measurements on this same film and measurements of others [24] and is also in accord with

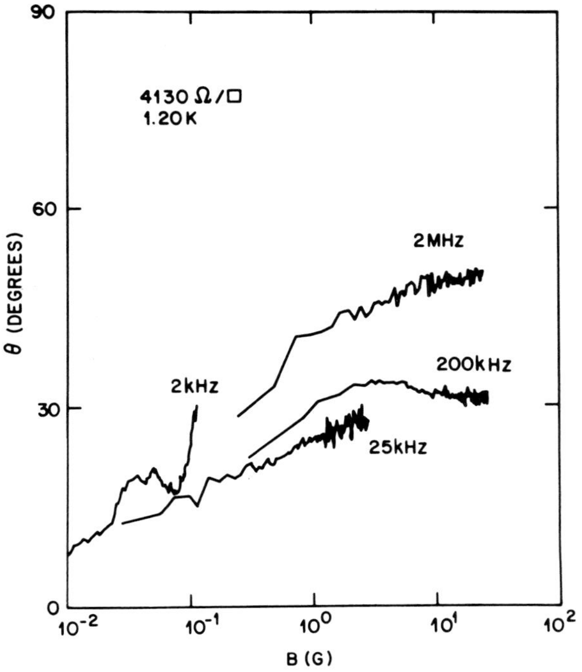

Fig. 3 Magnetic field dependence of the vortex impedance phase angle for several frequencies.

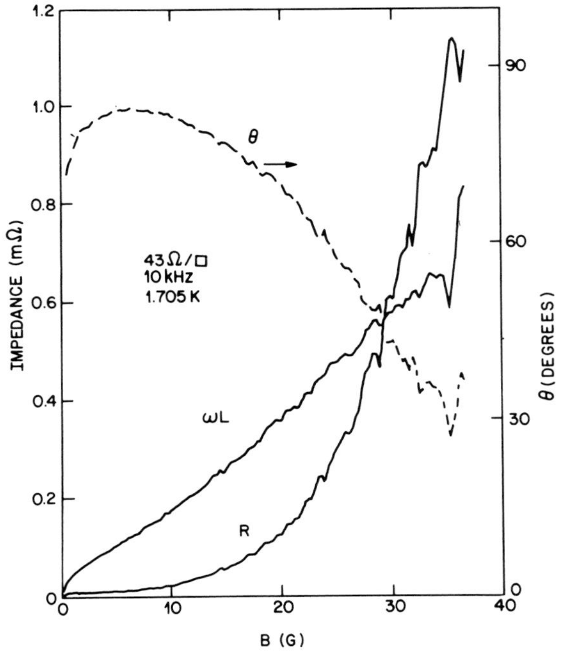

Fig. 4 Impedance components and vortex impedance phase angle versus magnetic field.

theoretical expectations [15,16] that the vortices are in a fluid-like uncorrelated state in this relatively high R_N film.

In Fig. 4 we have plotted $\theta = \tan^{-1}[\omega(L-L_K)/R]$ as a function of B at 10 kHz and 1.705K for the film with substantially lower sheet resistance, $R_N = 43\,\Omega/\square$. Here we expect stronger vortex correlations to order the vortices into a more solid-like phase which would interact strongly with the randomly distributed pinning sites [25]. This behavior is suggested in the plot of Fig. 4 where the phase angle approaches 90° at low B but then rapidly decreases to zero at higher fields. This field-induced pinning transition will be discussed in the following section and is associated with a pronounced increase in dissipation (R of Fig. 4) which has an onset at $B \simeq 10$G.

6. VORTEX-LATTICE TRANSITION

We have shown in the previous sections that the vortices exhibit high, nearly-free mobilities and the effects of pinning are weak in the 4130 - Ω/□ film. By comparison, strong pinning and low mobilities were found in the 43 - Ω/□ film. In this section we present additional results for the 43- Ω/□ film in an external magnetic field at temperatures well below the two-dimensional transition temperature T_c and our preliminary conclusions about the effects of vortex-lattice melting.

Doniach and Huberman [15] and Fisher [16] have used the Kosterlitz criterion to calculate the vortex-lattice melting temperature T_M. Fisher has included the renormalization of the shear modulus by lattice vibrations and defects, finding

$$T_M = \Phi_o^2 A_1 / 64\sqrt{3}\pi^3 k \Lambda(T_M), \qquad (6.1)$$

where he estimates that $0.40 < A_1 < 0.75$. Equation (6.1) is valid in the intermediate field range $\xi_{GL} < a_o < \Lambda$ appropriate to this work, where $a_o = (2\Phi_o/\sqrt{3}B)^{1/2}$ is the nearest neighbor spacing in a triangular vortex lattice. Comparison of Eqs. (1.1) and (6.1) shows that the melting temperature is below T_c. Noting that in the limit of high sheet resistance Λ scales with R_N [6], we find that all of the data for the 4130 - Ω/□ film were taken above T_M, for which theory gives $0.18\text{K} < T_M < 0.30\text{K}$, while for the 43−Ω/□ film, theory gives $1.73\text{K} < T_M < 1.81\text{K}$. A low melting temperature for the 4130 - Ω/□ film is consistent with the data.

Data for the temperature dependence of the phase angle of the vortex impedance is shown for several external fields for the 43 - Ω/□ film in Fig. 5, where the measuring frequency is 10 kHz. The zero-field transition occurs over a narrow temperature region near 1.90K. For all of the data displayed in Fig. 5, the temperature is well below T_c, so that the contribution from thermally-excited vortices should be negligible. Below about 1.6 K the vortices exhibit strong pinning and a phase angle approaching 90°. This is a temperature region where the conventionally-measured 4-probe current-voltage curves show a zero-voltage critical current. The critical current is attributed to pinning of the vortex lattice on film inhomogeneities.

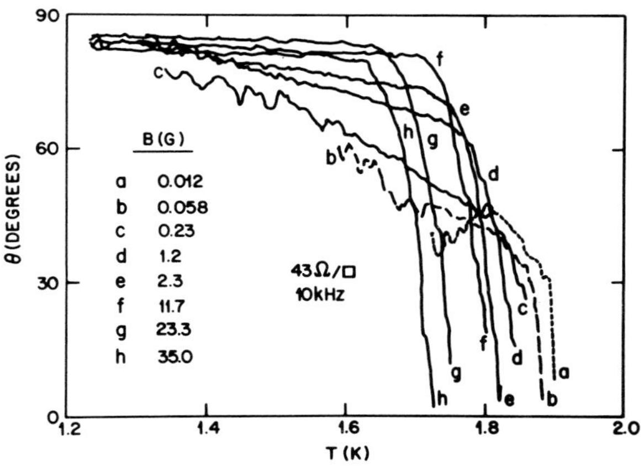

Fig. 5 Temperature dependence of vortex impedance phase angle for several applied magnetic fields.

In these aluminum films, the inhomogeneities are comparatively weak, so that pinning is dominated by the excitation of long-wavelength shear fluctuations in the lattice [25,26]. The resistance component of the vortex impedance is thermally activated. From measurements carried out at frequencies from 700 Hz to 50 MHz we find that the rf resistance extrapolates to zero at zero frequency, which is consistent with the dc result at low current levels.

At higher temperatures we observe that the phase angle approaches zero at a field-dependent temperature which our data taken at other frequencies shows to be also frequency-dependent. These temperatures correspond to the onset of free-vortex response and the apparent disappearance of vortex pinning. Since the fields applied are well below the critical field and the resistances are also less than R_N, we assume that in the high-temperature region the vortices are in an uncorrelated fluid-like state.

At an intermediate temperature, again depending on field and frequency, there is an abrupt change in the phase angle, a knee-like feature which is most pronounced at the higher magnetic fields. Pinning strengths rapidly decrease with temperature in this region and vortex mobilities increase. The resistance in the region between the knee and the $\theta \to 0$ point increases rapidly with temperature and also approaches frequency independence. Qualitatively, these results are as expected from the dislocation theory of melting [1-3,27]. The appearance of mobile thermally-excited dislocations would be expected to reduce the pinning and increase the average vortex mobility and resistance.

The authors acknowledge useful discussions with D. S. Fisher, B. A. Huberman, D. R. Nelson and L. A. Turkevich and appreciate the technical assistance of R. P. Minnich.

REFERENCES

[1] J. M. Kosterlitz and D. J. Thouless, *J. Phys. C* **6,** 1181 (1973).

[2] D. R. Nelson and J. M. Kosterlitz, *Phys. Rev. Lett.* **39,** 1201 (1977).

[3] J. M. Kosterlitz, *J. Phys. C* **7,** 1046 (1974).

[4] D. J. Bishop and J. D. Reppy, *Phys. Rev. Lett.* **40,** 1727 (1978); D. J. Bishop, *Thesis,* Cornell University (1978), unpublished.

[5] A. F. Hebard and A. T. Fiory, *Phys. Rev. Lett.* **44,** 291 (1980).

[6] M. R. Beasley, J. E. Mooij, and T. P. Orlando, *Phys. Rev. Lett.* **42,** 1165 (1979).

[7] S. Doniach and B. A. Huberman, *Phys. Rev. Lett.* **42,** 1169 (1979).

[8] L. A. Turkevich, *J. Phys. C* **12,** L385 (1979).

[9] B. I. Halperin and D. R. Nelson, *J. Low Temp. Phys.* **36,** 599 (1979).

[10] V. Ambegaokar, B. I. Halperin, D. R. Nelson and E. D. Siggia, *Phys. Rev. Lett.* **40,** 783 (1978).

[11] V. Ambegaokar and S. Teitel, *Phys. Rev. B* **19,** 1667 (1979).

[12] V. Ambegaokar, B. I. Halperin, D. R. Nelson and E. D. Siggia, *Phys. Rev. B,* to be published.

[13] J. Pearl, *Low Temperature Physics - LT 9,* edited by J. G. Daunt, D. O. Edwards, F. J. Milford and Yagüb (Plenum, New York, 1965), pg. 506; *Thesis,* Polytechnic Institute of Brooklyn (1965), unpublished.

[14] A. T. Fiory and A. F. Hebard, *AIP Conf. Proc.* **58,** 293 (1980).

[15] B. A. Huberman and S. Doniach, *Phys. Rev. Lett.* **43,** 950 (1979); *AIP Conf. Proc.* **58,** 87 (1980).

[16] D. S. Fisher, *AIP Conf. Proc.* **58,** 95 (1980); preprint: "Flux Lattice Melting in Thin Film Superconductors."

[17] M. Tinkham, *Introduction to Superconductivity* (McGraw-Hill, New York, 1975).

[18] J. Bardeen and M. J. Stephen, *Phys. Rev.* **140,** A1197 (1965).

[19] The vortex charge is defined as $q_v = \Phi_o^2/8\pi^2\Lambda$, in the analogy to the 2-D coulomb gas. See also [7].

[20] P. W. Anderson, *Rev. Mod. Phys.* **38,** 298 (1966).

[21] A. Schmid, *Phys. Kondens. Materie* **5,** 302 (1966).

[22] Y. B. Kim and M. J. Stephen, in *Superconductivity* edited by R. D. Parks (Marcel-Dekker, New York, 1969), Vol. 2, Chap. 19.

[23] The value for $L(T_c)$ quoted in Ref. 5 differs slightly because we had omitted the correction term $L_2(\beta) \approx 0.38$nH given in Eq. (3.2).

[24] P. M. Horn and R. D. Parks, *Phys. Rev.* **4,** 2178 (1971).

[25] A. Schmid and W. Hauger, *J. Low Temp. Phys.* **11,** 667 (1973).

[26] A. T. Fiory, *Phys. Rev. B* **7,** 1881 (1973); **8,** 5039 (1973).

[27] D. R. Nelson and B. I. Halperin, *Phys. Rev. B* **19,** 2457 (1979).

DISCUSSION

K. E. Gray:

Granular aluminum is known to be inhomogeneous and this can explain the broad resistive transition and kinetic inductance. What evidence is there in your results which cannot be explained by an inhomogeneous sample?

A. F. Hebard:

The frequency dependencies of our data are in agreement with the predictions of the dynamical theory and provide strong evidence for a Kosterlitz-Thouless interpretation. It is clear however that inhomogeneities do play a role and are most likely responsible for the saturation effects observed at low frequencies.

S. Doniach:

Could you say where the predicted melting temperature would appear in your last plot (phase angle versus temperature) and in particular is it consistent with the temperature at which the large change of pinning is observed?

A. F. Hebard:

The melting temperature for the R_N = 43 Ω/□ film is predicted to appear in the range 1.7K to 1.8K and is consistent with those temperatures where we observe a rapid change in the pinning behavior.

Sinha, ed. Ordering in Two Dimensions

TWO-DIMENSIONAL WIGNER CRYSTALLIZATION AND ELECTRONS ON HELIUM

DANIEL S. FISHER

Bell Laboratories
Murray Hill, New Jersey 07974

Recent experiments on a layer of electrons floating on liquid helium indicate a phase transition at a temperature, T_M, which scales as the square root of the areal electron density. These experiments are interpreted in terms of 2-dimensional electron crystallization below T_M. Current theoretical and experimental understanding of the transition is summarized with emphasis on comparison with the Kosterlitz-Thouless theory of dislocation mediated melting.

I. INTRODUCTION

In systems which are candidates for Kosterlitz-Thouless (KT) transitions [1], it has proven difficult to observe sharp phase transitions. There are expected to be no observable singularities in thermodynamic measurements [1,2] and even experiments which measure quantities which are predicted to have universal jumps, (such as ρ_s in ^{4}He films [3,4]) show considerable rounding due to finite frequency effects.

In this paper we briefly review the current theoretical understanding of electron crystallization on the surface of liquid helium. In particular, we will discuss the analysis of a recent experiment by Grimes and Adams (GA) [5]. This experiment, rather surprisingly, observed a relatively sharp appearance of several coupled longitudinal modes of the system beneath a temperature T_M. The appearance of these modes was interpreted as being associated with the formation of a hexagonal electron crystal for $T < T_M$ [6].

For a classical system of electrons (the appropriate limit for electrons on helium) there is a single dimensionless parameter $\Gamma = \frac{\sqrt{\pi n_s}}{T} e^2$, (where n_s is the areal electron density which lies in the range $10^8 - 10^9\ \mathrm{cm}^{-2}$). The melting curve is found, as expected, to be well parameterized by [5]

$$\Gamma(T_M) = \Gamma_M = 131 \pm 7\,. \tag{1}$$

In addition to having a single dimensionless parameter, electrons on helium have several other advantages for studying two dimensional melting.

1) The hamiltonian is known exactly, hence numerical and analytical results can be directly compared with experiment. In particular non-universal quantities such as T_M can be calculated. (Section II).

2) The helium substrate (unlike crystalline or amorphous substrates) has a negligible effect on the *static* properties of the electrons.

3) The coupling of the electrons to the helium substrate provides a means of effectively "time averaging" over many microscopic times of the electron system [6]. This effect (see below) gives rise to the relatively sharp transition observed by GA [5]. (Section III)

4) The long wavelength shear modulus can in principle be measured via transverse sound (see Section IV).

II. ESTIMATE OF Γ_M

The KT theory predicts that the melting temperature of a two dimensional isotropic solid will be given by

$$\frac{\mu(T_M)\,(\mu(T_M) + \lambda(T_M))}{2\mu(T_M) + \lambda(T_M)} \frac{a_o^2}{T_M} = 4\pi \tag{2}$$

where $\mu(T)$ (shear modulus) and $\lambda(T)$ are the temperature dependent Lame coefficients of elasticity and a_o is the lattice spacing [1,2]. Because of the long range $1/r^2$ forces, the longitudinal coefficient λ is infinite at T=0 for the case of electrons [7]. The shear modulus, however, is finite and easily calculable. Thouless [7] used the zero temperature values for μ and λ to estimate Γ_M. He found $\Gamma_M = 79$, corresponding to a melting temperature considerably higher than observed experimentally ($\Gamma_M \sim 131$). Two computer simulations, on the other hand, have observed transitions which are apparently second order at temperatures in good agreement with experiment. (Morf [8] $\Gamma_M \sim 130$, Gann et. al. [9] $\Gamma_M \sim 125$).

At any non-zero temperature, the shear modulus will be renormalized by phonon anharmonicities and the presence of dislocation pairs [1,2,10]. (Note: λ is also renormalized to a finite value by dislocation pairs [8]. This effect is small and hardly affects the melting temperature. We ignore it here.)

Morf [8] assumed an empirical form for the anharmonic phonon contribution (later verified by this author [10]) and used the calculated dislocation core energy [11] and the renormalization group equations of Halperin and Nelson [2] to find $\mu(T)$ and hence T_M. He found a temperature dependent shear modulus in excellent agreement with his molecular dynamics simulation and $\Gamma_M \sim 130$.

This remarkable agreement between the experimentally and numerically observed melting temperature and a calculation with no adjustable parameters based on dislocation-mediated melting, strongly suggests that the melting of a 2-d electron crystal may in fact be a KT transition.

III. DIMPLES AND COUPLED MODES

In the absence of a substrate, a two dimensional hexagonal electron crystal has a simple phonon spectrum: a transverse sound-like mode with

$$\omega_{ot}^2(q) = \frac{\mu}{m}q^2 \tag{3}$$

and a 2-d plasma-like longitudinal mode

$$\omega_{o\ell}^2(q) = 2\pi\frac{n_s e^2}{m}q\ . \tag{4}$$

(Note: in the experimental geometry of GA, the longitudinal mode is actually sound-like at very long wavelengths due to screening from image charges in the metal bases of the container [5].) The bare dispersion relations are drastically modified in the presence of helium due to coupling to capillary waves (or "ripplons") which is caused by a vertical electric field $E_\perp$ which presses the electrons down onto the helium surface [6]. This coupling mixes the long wavelength phonon modes of the electron crystal with ripplons of wavevector Q and frequency $\Omega^2(Q) = \frac{\sigma}{\rho}Q^{3/2}$ (where σ and ρ are the helium surface tension and density respectively) with Q near electron lattice reciprocal lattice vectors, G_i. At low temperatures, an array of depressions or "dimples" will be formed in the helium surface (typically .01 to .1 Å deep) under the electron lattice sites [6,12]. These dimples are mostly sinusoidal (i.e. consist of ripplons of only the lowest few reciprocal lattice vectors) for $T \lesssim T_M$. The strongly-coupled modes in the presence of these dimples were analyzed by Fisher, Halperin, and Platzman [6].

The characteristic frequency scale for the coupled modes is set by $\Omega_R \equiv \Omega(G_1) = 1\cdot10^8\ \mathrm{sec}^{-1}$ for $n_s = 5\cdot10^8\ \mathrm{cm}^{-2}$, where G_1 is the first reciprocal lattice vector. At frequencies low compared to Ω_R, the dimples follow the motion of the electrons and the phonon modes will be modified only by an effective mass $m^* \sim 60m$ of the electron plus dimple [6]:

$$\omega_{\ell,t}^2(q) = \frac{m}{m^*}\,\omega_{o\ell,t}^2(q) \tag{5}$$

For $\omega \gg \Omega_R$, on the other hand, the dimples are effectively static and the dispersion relations become

$$\omega_{\ell,t}^2(q) = \omega_d^2 + \omega_{o\ell,t}^2(q) \tag{6}$$

where ω_d is the oscillation frequency for a single electron in a dimple.

At intermediate frequencies, there are several branches of the spectrum: one mode at $\Omega(G)$, for each $|G|$ involving uniform vertical electronic motion; three others at each $\Omega(G)$ involving no electronic motion; and two modes shifted slightly below each $\Omega(G)$ involving transverse and longitudinal horizontal electron motion respectively and partially out of phase ripplon motion [6].

Grimes and Adams [5] observed four branches of the intermediate frequency coupled longitudinal modes at frequencies just below $\Omega(G)$ for the lowest four reciprocal lattice vectors. These modes are resonant modes of the cylindrical experimental container with wavelengths determined by the geometry. Three different resonances have been observed for the branch with $\omega \lesssim \Omega_R$ corresponding to the three longest wavelengths of the experimental chamber.

The dependences on n_s, $E_\perp$ and T of all of the resonant frequencies observed, are in excellent agreement with theory which uses a self-consistently determined effective Hamiltonian to account for the high frequency thermal motion of the electrons [6]. The effective interaction between a ripplon of wavevector near a low order reciprocal lattice vector (which for definitiveness we take to be G_1) and the long-wavelength modes of the electron lattice, is determined by the pressure due to the electrons averaged over a time on the order of $\tau_R = 1/\Omega_R$ [6]. This time is much longer than the characteristic time for electron motion $\tau_e = 1/\omega_e$ where $\omega_e = 2\cdot10^{11}\ \mathrm{sec}^{-1}$ for $n_s = 5\cdot10^8\ \mathrm{cm}^{-2}$.

For $T < T_M$, the electrons are localized near lattice sites even on long time scales. There will thus be a non-uniform pressure on the helium surface on the time scale τ_R causing the dimples to form. (Note: effects of the logarithmically divergent mean-square displacement due to long wavelength phonons are not important unless the "averaging time" is extremely long).

For $T \gg T_M$, or the other hand, the electrons will wander much further and the electron pressure will be uniform on any time scale much longer than τ_e. There will thus be no array of dimples, and the electron and ripplon modes will decouple. There will be a "transition" between these two regimes at a temperature $T_M^{app} \gtrsim T_M$ with a width determined by the small ratio of the characteristic time scales: $\tau_e/\tau_R \sim 5\cdot10^{-4}$. Thus, even though the longitudinal coupled modes observed by GA do not directly measure the solid ordering of the electron system, the observed modes do disappear relatively sharply at the apparent melting temperature T_M^{app}.

The simplest picture of the crossover region (which is in terms particularly applicable to KT melting) assumes that for $T > T_M$ the solid like ordering is destroyed by the motion of free dislocations which are present only above T_M. The solid-like ordering (and hence the dimples) will be destroyed on a time scale τ_R if free dislocations can diffuse a distance on the order of their average separation, $\xi(T)$, in time τ_R. The apparent melting temperature will then be determined by

$$\Omega_R \sim D_{dis}\,\xi^{-2}(T_M^{app}) \tag{7}$$

where $D_{dis} \sim \omega_e a_o^2$ is the diffusion coefficient for dislocation glide. (Climb will be much slower.) It then follows that

$$\frac{\xi(T_M^{app})}{a_o} = C\left[\frac{\omega_e}{\Omega_R}\right]^{\frac{1}{2}} \tag{8}$$

where C is a dimensionless number probably considerably less than unity. Since the correlation length $\xi(T)$ diverges exponentially as $T \to T_M$, T_M^{app} will be logarithmically dependent on Ω_R/ω_e and will approach T_M as $\Omega_R/\omega_e \to 0$. The width of the observed transition $(\sim T_n^{app} - T_M)$ will also decrease as $\Omega_R/\omega_e \to 0$.

The experimentally observed width [5] ($\sim 5-10\%$) of the transition at a frequency $\sim \Omega_R$, is roughly consistent with the criteria (Eq. 8) from this rather oversimplified picture. The actual melting temperature is probably slightly below where the amplitude of the resonances starts to decrease dramatically; within error bars this is still consistent with the theory and computer simulations [8,9] discussed in Section II. It is interesting to note that the criteria for the apparent transition temperature (Eq. 8) is analogous to the criteria for the disappearance of superfluidity at finite frequency in thin helium films observed by Bishop and Reppy. [4,5]

At this point it is worth making a comment on the order of the melting transition for electrons on helium. While a small first order jump cannot be ruled out, the absence of an observed jump in either the frequency or the amplitude of the GA modes [5], probably rules out a large first order melting transition such as is found in three dimensions. This conclusion is in agreement with the computer simulations by Morf [8] and Gann et. al. [9]

IV. FUTURE EXERIMENTS AND CONCLUSIONS

Currently, the agreement between the KT theory applied to electrons, computer simulations, and experiments with electrons on helium, is rather good. However the most important predictions of the KT theory have not yet been tested experimentally and are just beginning to be tested in computer simulations [8].

As mentioned previously, the KT theory predicts an (almost) universal jump in the ratio of the shear modulus to the temperature at T_M (See Eq. 2 and discussion in Section II) [1,2,7]. In principle, the temperature dependent shear modulus of the electron lattice on helium can be measured by exciting and measuring the velocity of transverse phonons. There are, however, several theoretical difficulties. Firstly, one would like to be able to measure the transverse sound velocity, c_t, at as long wavelengths as possible since the jump in μ at T_M will occur only in the limit of infinite wavelength. At any finite wavelength, L, the liquid at $T > T_M$ will support shear if $L < \xi(T)$ [2]. While it should be possible to excite long wavelength transverse modes, the interpretation will be clouded by the effective mass corrections to c_t as in (Eq. 5). Fortunately, the shear modulus and effective mass can be extracted independently (at least approximately) by performing the experiment in the presence of a vertical magnetic field [13].

The second, and currently more serious difficulty, is that it appears quite likely that the long wavelength transverse mode will be overdamped (or at best very broad) in the range of densities and temperatures studied by GA. This problem can be circumvented if the experimental temperatures can be reduced to the dilution refrigerator range. Since, $T_M \propto n_s^{\frac{1}{2}}$, it should then be possible to freeze the electron system at lower densities. The coupling to the ripplons, and hence the damping, can then be reduced by reducing the pressing electric field $E_\perp$. Thus it should be possible, at least in principle, to measure the shear modulus of the electron crystal and test the KT prediction directly.

Other consequences of the KT theory are probably much more difficult to test. The structure factor, S(q) might be measurable below T_M by scattering light from the array of dimples. However this is not possible above T_M, even in the region where there is predicted to be an exponentially diverging correlation length [1,2], except perhaps for $T < T_M^{app}$.

As mentioned earlier, there is expected to be no observable specific heat singularity at a KT transition [1,2]. Measurements of the specific heat, which could test this prediction as well as settle questions on the order of the transition, are probably not possible for electrons on helium.

In spite of the difficulties discussed above, it appears that the advantages of being able to measure the shear modulus and make direct comparisons with theory, coupled with the absence of substrate effects on static properties, will make electrons on helium one of the best systems for studying two dimensional melting.

ACKNOWLEDGEMENTS

The author's understanding of the theory and experiments summarized here has benefited greatly from interaction with colleagues too numerous to mention. I would like to thank especially Mike Grimes, Bert Halperin and Rudolf Morf for countless fruitful discussions.

REFERENCES

1. J. M. Kosterlitz and D. J. Thouless, J. Phys. *C6*, 1181 (1973); J. M. Kosterlitz, J. Phys. *C7*, 1046 (1974); J. M. Kosterlitz and D. J. Thouless, Prog. Low. Temp. Phys. Vol. VII-B edited by D. F. Brewer (North-Holland, 1978).
2. B. I. Halpèrin and D. R. Nelson, Phys. Rev. Lett. *41*, 121 (1978); E*41*, 519 (1978); D. R. Nelson and B. I. Halperin, Phys. Rev. *B19*, 2457 (1979); See also A. P. Young, Phys. Rev. *B19*, 1855 (1979).
3. D. J. Bishop and J. D. Reppy, Phys. Rev. Lett. *40*, 1727 (1978) and to be published.
4. V. Ambegaokar, B. I. Halperin, D. R. Nelson and E. D. Siggia, Phys. Rev. Lett. *40*, 783 (1978) and Phys. Rev. *B21*, 1806 (1980).
5. C. C. Grimes and G. Adams, Phys. Rev. Lett. *42*, 795 (1979), and private communication.
6. D. S. Fisher, B. I. Halperin and P. M. Platzman, Phys. Rev. Lett. *42*, 798 (1979), and D. S. Fisher to be published.
7. D. J. Thouless, J. Phys. *C11*, L189 (1978).
8. R. Morf, Phys. Rev. Lett. *43*, 931 (1979), and private communication.
9. R. C. Gann, S. Chakravarty and G. V. Chester, Phys. Rev. *B20*, 326 (1979).
10. D. S. Fisher, unpublished.
11. D. S. Fisher, B. I. Halperin, and R. Morf Phys. Rev. *B20*, 4692 (1979).
12. Yu. P. Monarkha and V. B. Shikin, Zh. Eksp. Teor. Fiz. *68*, 1423 (1975) [Sov. Phys. JET P*41*, 710 (1976)].
13. D. S. Fisher and V. B. Shikin, Pis'ma Zh. Eksp. Teor. Fiz. *31*, 238 (1980).

DISCUSSION

A. Holz:

What is actually the significance of your approach when you represent the system by one where interactions are short range and the Cauchy relations are not satisfied because the Poisson modulus $\gamma = 1/2$ imples that. If, however, the incompressibility is taken seriously then only conservative motion of dislocations is possible and a reduction of phase space implies that Γ is reduced by a factor of 3/4. In the latter case your $\Gamma \simeq 130$ is reduced to $\Gamma' \simeq 100$.

D. S. Fisher:

The suggestion (due to D. J. Thouless) that the limit $\lambda \to \infty$ of elasticity theory is appropriate for the electron system has been verified numerically by calculating the energy of dislocation pairs and other defects by Fisher, Halperin and Morf.

R. R. Sharma:

1) How do the experimental results change as a function of the electron density?

2) Are there phase transitions observed experimentally for high electron densities?

D. S. Fisher:

1) The transition temperature scales as the square root of the electron density. In addition, the ripplon-electron lattice coupling becomes weaker at lower densities.

2) The electron density is limited by a surface instability of the helium to be less than $\sim 2 \cdot 10^9$ cm^{-2}. This density is too small for quantum mechanical effects to be important.

A. Czachor:

1) Is there any more direct evidence for 2D Wigner crystallization (one presented here involves intermediate concepts - dimplers, ripplons, and is far from being direct)?

2) What is the nature of the screening which changes the $\omega \sim \sqrt{q}$ behavior of plasmons in 2D electron gas to the type $\omega \sim q$?

D. S. Fisher

1. The most direct evidence for a 2D crystal is the observation of coupled modes very near to the ripplon frequencies at several reciprocal lattice vectors of a triangular lattice. The evidence is certainly not very direct. However, the agreement with theory is excellent. The most that can directly be claimed is evidence for relatively well defined crystalline order on rather long time scales.

2) The interaction between the electrons is screened at distance on order of 1 mm by image charges in metal plates on the top and bottom of the container. This softens the 2D plasmon dispersion relation to $\omega_p(q) \propto q$.

B. Halperin:

It may be worth pointing out that there is a closely related electron system, namely, electrons at the interface between He^3 and He^4 which has been studied by Leiderer. In this case the electrons are in bubbles, so that they move slowly even above the melting transition. It may be possible to measure S(q) using light scattering in this system.

SECTION II

A. PHYSISORBED AND CHEMISORBED SYSTEMS

Sinha, ed. Ordering in Two Dimensions

ADSORBED OVERLAYER CRITICAL PHENOMENA BY LOW-ENERGY ELECTRON DIFFRACTION*

T. M. LU

Metallurgical and Mineral Engineering Department and Materials Science Center

University of Wisconsin, Madison, Wisconsin 53706

A general expression is obtained for the low-energy electron diffraction (LEED) superlattice beam intensity from an adsorbed overlayer that undergoes a continuous order-disorder phase transition. The low-temperature phase may possess more than one component in the order parameter, and the transition may belong to one of the many 2-D universality classes. Based on an Ising-lattice gas analog, we show that the intensity can be written as the sum of two terms, a Bragg diffraction term, giving the temperature dependence of the square of the order parameter (long range order), and a diffuse scattering, describing the critical fluctuations (short range order) near the transition temperature. The effect of multiple scattering is also discussed.

Recently, it has been shown [1,2] that continuous order-disorder phase transitions for localized adsorption on surfaces (or equivalently, that of two-dimensional (2-D) Ising models) may exhibit many different classes of critical behavior such as Ising-like, X-Y like (with cubic anisotropy), Heisenberg-like, q-state Potts model, etc. The symmetry properties of the order parameter play an important role in characterizing the nature of the transitions. In this paper we derive a general expression for the temperature dependence of the intensity of a LEED superlattice beam when an adsorbed overlayer undergoes a continuous order-disorder phase transition. This result will allow a study in detail of the critical properties near the transition temperature. The special case of a square lattice with nearest-neighbor repulsive interactions, which has a single-component (Ising-like) order parameter, has been treated previously [3,4].

Let G_o be the space group of a 2-D lattice. The lattice may be any of the common Bravais lattices. The system of localized interacting adsorbed atoms is assumed to have a Hamiltonian of the form

$$H = H_o - \vec{\eta}\cdot\vec{\psi}. \qquad (1)$$

H_o is the "bare" Hamiltonian, which may contain an arbitrary set of neighboring interactions, and is invariant under the group G_o. The symmetry-breaking term, $-\vec{\eta}\cdot\vec{\psi}$, is the product of the n-component order parameter $(\vec{\psi} \equiv (\psi_1, \psi_2, \ldots, \psi_n))$ and its conjugate field $\vec{\eta}$. If a continuous order-disorder phase transition occurs, then according to the Landau Theory [5], $(\psi_1, \psi_2, \ldots, \psi_n)$ should transform as the basis of an n-dimensional irreducible representation of G_0. The n-component order parameter is related to the "star of $\vec{k}$", $\vec{k} \equiv (\vec{k}_1, \vec{k}_2, \ldots, \vec{k}_n)$. Let $\vec{k}_1$ be a vector in the first Brillouin zone. The "star of $\vec{k}$" is a s[illegible] of n independent vectors which are obtained by applying all the symmetry elements of G_o on $\vec{k}_1$. (One includes only those vectors that may not be obtained by connecting a reciprocal-lattice vector to any of the $\vec{k}_p$'s.) The order parameter may then be defined as

$$\psi_p \equiv \sum_i b_i\, e^{i\vec{k}_p\cdot\vec{r}_i}, \quad p=1,\ldots,n, \qquad (2)$$

where $\vec{r}_i$ is the position of the ith site, and b_i is the occupation number, which may be 0 or 1.

The kinematic LEED intensity from the overlayer atoms is given by [6]

$$I(\vec{K}_{\|}, T) = \sum_{i,j} \langle b_i b_j \rangle\, e^{i\vec{K}_{\|}\cdot(\vec{r}_i - \vec{r}_j)}, \qquad (3)$$

where $\vec{K}_{\|}$ is the momentum transfer parallel to the surface, and <> indicates the thermal average. The system of a lattice gas can be mapped into the corresponding 2-D Ising system according to the transformation [7]

$$b_i = \frac{\sigma_i + 1}{2}, \quad (4)$$

where σ_i, the spin variable in the Ising system, may take the value of ± 1. Therefore, a spin up (down) implies an occupied (empty) site. By substituting Eq. (4) into Eqs. (2) and (3) and recognizing that for a superlattice beam,

$$\sum_i e^{i\vec{K}_{\parallel}\cdot\vec{r}_i} \simeq \sum_i e^{i\vec{k}_p\cdot\vec{r}_i} \equiv 0, \quad (5)$$

one can reduce Eqs. (2) and (3) to the following forms:

$$\psi_p = \frac{1}{2}\sum_i \sigma_i\, e^{i\vec{k}_p\cdot\vec{r}_i} \equiv \sum_i S_i(p) \quad (6)$$

$$I(\vec{K}_{\parallel}, T) = \frac{1}{4}\sum_{i,j} \langle\sigma_i\, \sigma_j\rangle\, e^{i\vec{K}_{\parallel}\cdot(\vec{r}_i - \vec{r}_j)} . \quad (7)$$

Here, $\langle\sigma_i\, \sigma_j\rangle$ is the spin-spin correlation function, and we have defined $S_i(p)$ as $\frac{1}{2}\sigma_i\, e^{i\vec{k}_p\cdot\vec{r}_i}$. Equation (7) can be written in a more convenient form by looking at a particular superlattice beam with wavevector $\vec{k}_p$:

$$I_p(\vec{K}_{\parallel},T) = \langle\psi_p\rangle^2\, \delta(\vec{K}_{\parallel} - \vec{k}_p)$$
$$+ \sum_{i,j} [\, \langle S_i(p)S_j^*(p)\rangle - \langle S_i(p)\rangle\, \langle S_j^*(p)\rangle]$$
$$\times\, e^{i(\vec{K}_{\parallel} - \vec{k}_p)\cdot(\vec{r}_i - \vec{r}_j)} . \quad (8)$$

This is achieved by letting the conjugate field $\vec{\eta} = (\vec{\eta}_1 = 0, \vec{\eta}_2 = 0, \ldots, \vec{\eta}_p \to 0^+, \ldots, \vec{\eta}_n = 0)$. The average values of $S_i(p)$ and ψ_p are given respectively by

$$\langle S_i(p)\rangle = \frac{1}{2}\langle\sigma_i\rangle\, e^{i\vec{k}_p\cdot\vec{r}_i}$$

and

$$\langle\psi_p\rangle = \sum_i \langle S_i(p)\rangle .$$

The first term in Eq. (8) is the sharp Bragg diffraction, resulting from the long-range order in the system. The summation describes the short-range correlation near the critical temperature T_c. The critical properties, which may belong to one of the universality classes mentioned above, can be deduced from the temperature dependence of the peak value and angular distribution of the superlattice beam intensity in LEED from the overlayer by using Eq. (8). For instance, for $T<T_c$, the intensity is dominated by the Bragg diffraction and the critical exponent β associated with the long-range order can be measured. Different universality classes should possess different values for the exponent. (The X-Y model with cubic anisotropy is nonuniversal and therefore does not possess a definite value for the exponent.) For $T>T_c$, $\langle S_i(p)\rangle = 0$, and the long-range order vanishes. However, Short-range order still exists and in general gives a diffuse and broader profile. Again one expects that different universality classes should fluctuate quite differently. The temperature dependence of the peak intensity ($T>T_c$, $\vec{K}_{\parallel} = \vec{k}_p$) allows us to measure the critical exponent γ. The correlation length can be determined from the entire superlattice beam profile ($T>T_c$, $\vec{K}_{\parallel} \neq \vec{k}_p$).

For the special case of a square lattice with nearest-neighbor repulsive interactions (at a coverage of 1/2), Eq. (8) can be reduced to a form analogous to the scattering of neutrons from a 3-D binary alloy undergoing a substitutional order-disorder phase transition [8].

In LEED multiple scattering may make important contributions to the diffracted intensity. It has been shown recently [9,10] that the superlattice beam intensity can also include, in addition to the back-scattering from the overlayer, a multiple-scattering contribution arising from a "sequential scattering" that consists of a deflection from an overlayer atom [11] with $\vec{K}_{\parallel}$ equal to a superlattice wavevector in the nearly forward direction, followed by a nearly back-scattered reflection from the substrate. Multiple scattering events may also occur in the opposite order. In such a "sequential scattering," there is only one deflection from an overlayer atom, and therefore, it possesses the same temperature dependence as the kinematic (single-scattering) intensity--namely, that of Eq. (8). Events involving two or more scatterings in the overlayer have been shown experimentally to be negligible [9]. We therefore believe that the simple result of Eq. (8) is likely to be sufficient for analyzing overlayer order-disorder phase transitions.

In conclusion, LEED is not only a powerful tool for characterizing adsorbed overlayer phase transitions, but is also potentially useful for studying quantitatively many very interesting 2-D critical phenomena. Some experimental progress has already been made in this direction. Examples include

(1) Oxygen chemisorbed on a W(110) surface [12] at 1/2 coverage (X-Y model with cubic anisotropy).

(2) He physisorbed on the basal plane of graphite [13] at 1/3 coverage (3-state Potts model), and

(3) Oxygen chemisorbed on a Ni(111) surface [14] at 1/4 coverage (4-state Potts Model).

ACKNOWLEDGEMENTS

The author would like to thank M. B. Webb, P. A. Bennett, D. L. Huber, L. D. Roelofs, A. R. Kortan, G. C. Wang, and M. G. Lagally for valuable discussions.

REFERENCES

*Research supported by ONR.

1. S. Krinsky and D. Mukamel, Phys. Rev. B16, 2313 (1977).
2. E. Domany, M. Schick, and J. S. Walker, Phys. Rev. Lett. 38, 1148 (1977); E. Domany, M. Schick, J. S. Walker, and R. B. Griffiths, Phys. Rev. B18, 2209 (1978).
3. G. Doyen, G. Ertl, and M. Plancher, J. Chem. Phys. 62, 2957 (1975).
4. T. M. Lu, Surface Sci. 93, L111 (1980).
5. L. D. Landau and E. M. Lifshitz, Statistical Physics, Pergamon, New York (1968), 2nd ed. Chapter XIV; D. Mukamel and S. Krinsky, Phys. Rev. B13, 5065 (1976).
6. Note that for a complete description of the temperature dependence of the intensity, one should include the Debye-Waller factor which describes the thermal vibrations.
7. T. D. Lee and C. N. Yang, Phys. Rev. 87, 410 (1952).
8. J. Als-Nielsen, Phase Transitions and Critical Phenomena, C. Domb and M. S. Green, eds., Academic Press, (1976), vol. 5A, P. 88.
9. M. B. Webb and P. A. Bennett, private communication; P. A. Bennett, Ph. D. dissertation, University of Wisconsin, Madison, (1980, unpublished).
10. L. D. Roelofs and A. R. Kortan, private communication.
11. In Ref. (9), the "overlayer" refers to a reconstructed clean surface layer.
12. T. Engel, H. Niehus and Bauer, Surface Sci. 52, 237 (1975); J. C. Buchholz and M. G. Lagally, Phys. Rev. Lett. 35, 442 (1975); T. M. Lu, G. C. Wang, and M. G. Lagally, Phys. Rev. Lett. 39, 411 (1977); G. C. Wang, Ph.D. dissertation, University of Wisconsin, Madison, (1978, unpublished); E. Domany and E. K. Riedel, Phys. Rev. Lett. 40, 561 (1978).
13. M. Bretz and J. G. Dash, Phys. Rev. Lett. 27, 647 (1971); J. G. Dash, Films on Solid Surfaces, Academic Press, New York (1975), Chapter 8; S. Alexander, Phys. Lett. A54, 353 (1975); M. Bretz, Phys. Rev. Lett. 38, 501 (1977).
14. A. R. Kortan, P. I. Cohen, and R. L. Park, J. Vac. Sci. Technol. 16, 541 (1979); A. R. Kortan, Ph. D. dissertation, University of Maryland, College Park, (1980, unpublished).

Published 1980 by Elsevier North Holland, Inc.
Sinha, ed. Ordering in Two Dimensions

LOW-ENERGY ELECTRON DIFFRACTION RESULTS FOR PHYSISORBED NITROGEN ON GRAPHITE

RENEE D. DIEHL, CHRISTOPHER G. SHAW[@], SAMUEL C. FAIN, JR., and MICHAEL F. TONEY

Department of Physics, University of Washington, Seattle, Washington 98195

ABSTRACT

Low-energy diffraction (LEED) observations of N_2 physically adsorbed on the basal plane of graphite single crystals are reported for 34K < T < 54K and P < 10^{-4} Torr. Photometric measurement of the attenuation of graphite first-order LEED spots is used to determine vapor-pressure isotherms for the submonolayer fluid-to-solid transition. For T < 47K, these isotherms indicate fluid-solid coexistence. Above 47K, the isotherms become progressively less vertical and show no evidence for phase coexistence. LEED patterns near 34K are analyzed to investigate the commensurate-incommensurate phase transition. These results are combined with published data to obtain a vapor-pressure versus inverse temperature phase diagram for 33K < T < 100K.

I. INTRODUCTION

The density-temperature phase diagram of N_2 adsorbed on graphite has been partially determined by neutron scattering [1-3], heat capacity [4], and volumetric isotherm [5] measurements. Both neutron scattering and heat capacity measurements indicate a submonolayer gas-solid coexistence region which terminates near 47K [2-4]. Volumetric isotherms have not been reported near this phase boundary due to the very low equilibrium vapor pressure there. This paper presents isotherms near 47K which were obtained from low-energy electron diffraction (LEED) intensity measurements, preliminary LEED measurements of the commensurate-incommensurate transition near 34K, and a tentative pressure-temperature phase diagram for 33K < T < 100K.

II. EXPERIMENTAL PROCEDURE

The two different graphite crystals used were about 3 mm in diameter and 0.2 mm in thickness and were kindly provided by Dr. T.S. Noggle of Oak Ridge National Laboratories. Each crystal was mounted on a copper sample holder and the surface layers pulled off in air with Scotch Magic Transparent Tape. After the sample was mounted in the vacuum system and the system baked at 160°C for about 24 hours, a base pressure of less than 5 x 10^{-11} Torr was attained. Initially both crystals showed a 2 x 2 LEED pattern; no ordered N_2 monolayer adsorption was observed when the 2 x 2 pattern was present. The origin of this 2 x 2 pattern is unknown; no such superstructure has been observed in recent molecular beam experiments on crystals that were obtained from the same mine in Ticonderoga, N.Y. and were prepared using methods similar to ours [6,7]. In our experiments the 2 x 2 pattern was eliminated by bombarding the crystal with a 100 nA, 144 eV electron beam for several hours. After this treatment, N_2 monolayers would order on the surface. Typical electron beam parameters for the adsorption experiments reported here were 2 nA and 64 to 144 eV.

The sample holder used was identical to that described previously [8] with the exception of the thermometer. A silicon diode temperature sensor [9] mounted to the copper sample holder was used instead of a thermocouple. The sensor was calibrated against a platinum resistance thermometer; the absolute calibration is believed accurate to ± 0.5K. The flow of cold helium gas to the holder was manually regulated to provide temperature control which was better than ± 0.1 K for the measurements reported here. The room temperature N_2 pressure was determined to an absolute accuracy of ± 20% from a glass Bayard-Alpert ionization gauge that was calibrated for nitrogen at Boeing Technology Services in Seattle. The effective pressure at the sample was corrected by the thermomolecular factor $(T/295K)^{1/2}$ to account for the difference between the temperature of the N_2 gas incident on the sample and the temperature T of the sample [10,11].

The composition of gas in the vacuum system was monitored with a quadrupole residual gas analyzer. The pressures of Ar and He increased noticeably when the ionization pumps were turned off during measurements. The maximum rise in Ar pressure was to about 10^{-7} Torr approximately ½ hour after turning off the ionization pumps. This rise in Ar pressure may have affected the data presented here for the commensurate-incommensurate transition near 34K due to the ½ hour duration of that experiment, and to the low temperature, which readily condenses Ar onto the sample. The effect on the higher temperature photometric isotherms is expected to be small as a result of the shorter duration of each measurement (<10 minutes) and the higher temperatures involved. The pressure of CH_4 was minimized by use of a 60 K cold surface [12] and was below detectability for these experiments.

The spot photometer used for LEED intensity measurements was the same as in previous experiments [11] except for the use of a smaller entrance aperture to minimize the contribution from the diffuse background near the graphite (10) LEED spots. A small (about 10%) contribution to the photometer intensity measurements can result from the second order ($\sqrt{3} \times \sqrt{3}$) 30° N_2 LEED spots from the ordered overlayer. However the predominant effect on the graphite (10) intensity of an increase in N_2 pressure near 47K is an attenuation due to elastic and inelastic scattering by adsorbed N_2 molecules.

The mean lattice constant was measured for incommensurate N_2 overlayers observed near 34K by methods similar to those used for Kr [13,14]. Photographs of the LEED pattern were taken at an electron energy of either 64 eV or 144 eV to maximize the multiple-scattering contribution to the extra LEED spots observed for the incommensurate monolayer [8,13]. The misfit $(d_o-d)/d_o$, where d is the mean nearest neighbor distance inferred from the photographs by assuming the commensurate ($\sqrt{3} \times \sqrt{3}$)30° distance d_o=4.26 Å, was measured from the photographs. For misfits greater than 2.5%, the distances between the centers of the split spots were measured to determine misfit. For smaller misfits, where the split spots could not be visually resolved, the patterns were analyzed by comparing them to drawings that represented the overlapping spots at small misfits.

III. SUBMONOLAYER PHASE TRANSITION RESULTS

The LEED pattern produced upon adsorption of N_2 on the basal plane of graphite single crystals at $P < 10^{-4}$ Torr and 44K < T < 54K was that of the commensurate ($\sqrt{3} \times \sqrt{3}$)30° structure inferred from N_2 neutron diffraction experiments [1-3]. No additional superstructure from orientational ordering of the N_2 molecular axes was observed, in agreement with recent neutron diffraction measurements which located the molecular ordering transition at 30K for monolayers [15].

Five LEED intensity isotherms are shown in Fig. 1, and the normalized inverse slope at the inflection point of twenty such isotherms is shown in Fig. 2. Isotherms at 47K and below show an almost-vertical linear segment which extends over at least 25% of the total intensity decrease shown. Coexistence of a 2-D fluid and a 2-D commensurate solid inferred from neutron scattering [2,3] and heat capacity [4] studies would produce such a region. The finite width of the linear segment in the isotherms is similar to that seen in LEED isotherms from Ar [11], Kr [16], and Xe [17] at temperatures expected to produce submonolayer gas-solid coexistence. Isotherms taken for decreasing pressure (not shown in Fig. 1) were usually displaced slightly

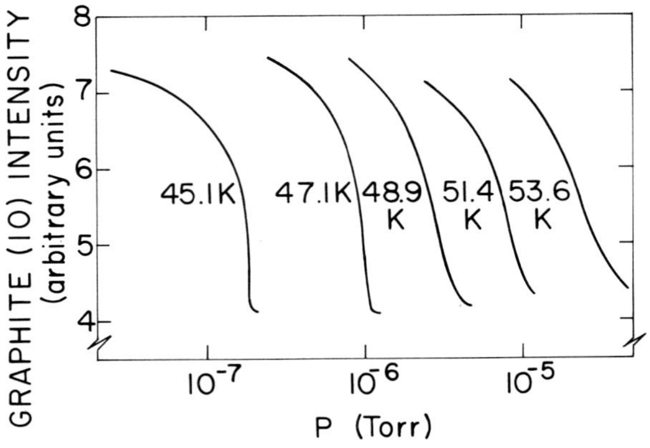

FIG. 1. Graphite (10) LEED intensity at 141eV versus increasing N_2 vapor pressure.

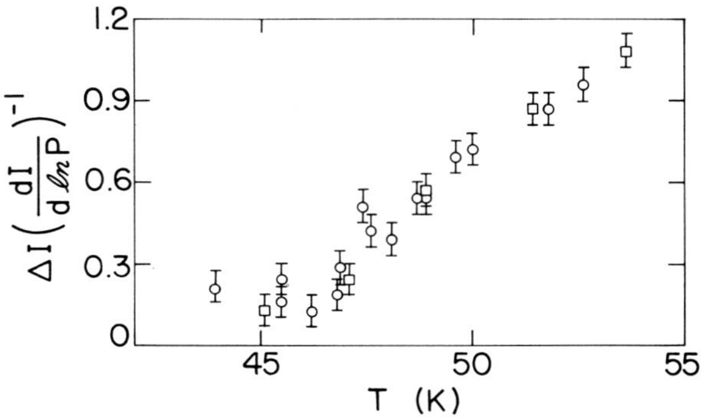

FIG. 2. The normalized inverse slope $(\Delta I)(dI/d\ln P)^{-1}$ at the inflection point of 20 LEED isotherms including the 5 isotherms from Fig. 1 (square points). The normalization ΔI was chosen to be 3.0 in the intensity units of Fig. 1. Similar plots have been used by Larher to determine critical points from isotherms [23].

to the left of those for increasing pressure as also for Ar [11] and Xe [17] on graphite. In this experiment the temperature sensitivity is sufficient to rule out temperature drift as the cause of the shift in pressure.

A maximum in the heat capacity of submonolayer N_2 on Grafoil occurs near 48K and has been identified as indicating a triple point at 48K [4]. A similar maximum has been observed for submonolayer Kr on Grafoil at 85 to 86K and interpreted as an incipient triple point [18]. Neither our N_2 LEED isotherms nor Kr volumetric isotherms [19] show any direct evidence for two distinct fluid phases. Our N_2 LEED isotherms also show no evidence for fluid-solid phase coexistence above 47K such as that observed for Kr above 85K [19]. It is possible that the small contribution (mentioned in the previous section) of the overlayer ordering to the intensity measurements in Fig. 1 makes observation of a fluid-solid phase transition with a small density change difficult to see in the type of measurements presented here. However, Ostlund and Berker have proposed that the fluid-solid phase transition above 48K for N_2 (85K for Kr) is intrinsically second order, but is influenced by size effects to appear first order [20]. This raises the intriguing possibility that substrates such as the single crystals used here might show quite different behavior from a polycrystalline aggregate such as used in previous studies of N_2 on graphite [1-5]. Measurements are underway to test this hypothesis.

IV. COMMENSURATE-INCOMMENSURATE TRANSITION RESULTS

At lower temperatures we were able to observe the commensurate-incommensurate transition for $P < 10^{-4}$ torr. Results at 34.3K are shown in Fig. 3. The accuracy of these measurements is not sufficient to rule out a first-order transition with a misfit change of less than 2%. Neutron scattering results near 78K have been interpreted as indicating a first-order commensurate-incommensurate transition [2,3], but the data are not detailed enough to prove this. There can be a change in the nature of the transition due to the large difference in temperature of the two observations. The results in Fig. 3 do provide an approximate location of the onset of this transition for the phase diagram in the next section.

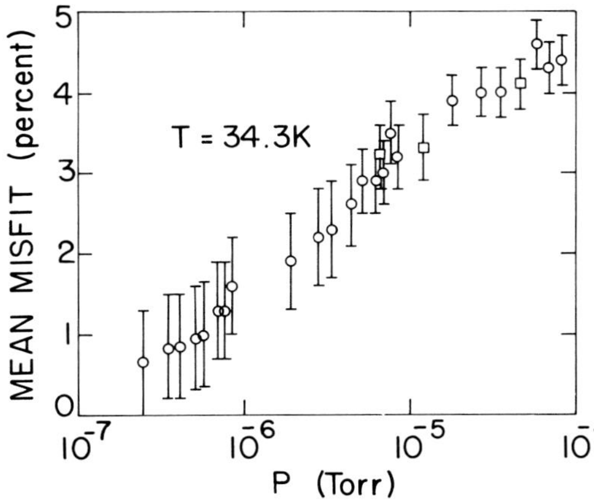

FIG. 3. Mean misfit versus N_2 vapor pressure at 34.3K. Circles are increasing pressure; squares, decreasing.

V. VAPOR PRESSURE-INVERSE TEMPERATURE PHASE DIAGRAM

The pressure-temperature locations of our isotherm inflection points above 47K are shown in Fig. 4 as open triangles. The long-dashed line, labeled "a" is the best fit line for the open

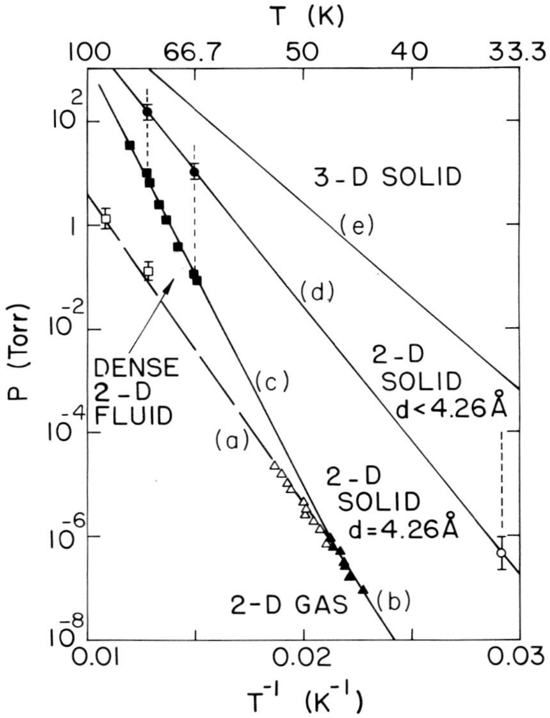

FIG. 4. Vapor pressure versus inverse temperature phase diagram for N_2 on graphite discussed in text. The equations for lines a-e are given in Table I.

TABLE I

Constants in equation $\log_{10}P(\text{Torr}) = B - A/T(\text{K})$ and isosteric heats q for N_2 on graphite.

	T RANGE	A	B	q(kJ/mole)
Fig.4a	47-54K	571 ± 20	6.06	10.9 ± 0.4
Fig.4b	43-47K	673 ± 49	8.26	12.9 ± 0.9
Fig.4c	47-83K	814 ± 30	11.23	15.6 ± 0.6
Fig.4d	34-78K	523 ± 30	8.87	10.0 ± 0.6
Fig.4e	35-63K	356.3	7.61	6.82

triangles (see Table I). This line is extrapolated through the open squares at higher temperatures, which mark approximately half-monolayer coverage from Ross and Winkler's volumetric isotherms on graphitized carbon black [21]. Although no phase transition is observed along this line, it provides a reference line for half-monolayer coverage. The locations of our almost vertical isotherms below 47K are shown as closed triangles. The best fit line for these points is labeled "b" and has a greater slope than line "a" (see Table I).

The solid line "c" has been drawn to intersect our data at 47K and to go through Larher's points at higher temperatures (solid squares) [5]. (Larher's measurements indicate that the slope of the line through his points changes in the region of his measurements, so it is difficult to extrapolate his measurements to low temperatures.) Larher's identification of his events with a disordered fluid to commensurate solid transition is in agreement with neutron diffraction results at 78K [1]. As mentioned in section III, no transition in the LEED isotherms for 47K < T < 54K was observed.

The onset of the commensurate-incommensurate (C-I) transition we observed is shown in Fig.4 as the open circle. The upper solid circle marks the C-I transition location observed in neutron-diffraction measurements at 78K [2]. The lower solid circle marks the higher pressure inflection point in Larher's 66.5K isotherm [5] which looks similar to the C-I transition point in the isotherm taken in the neutron experiments [2]. The vertical dashed lines show the extent of the isotherms at 78K [1,2], 66.5K [5], and 34.3K [Fig. 3]. The uppermost solid line "e" is the saturation vapor pressure P_0 of bulk N_2 [22]. The proximity of the upper solid circles to line "e" is consistent with previously suggested second-layer effects on the C-I transition at high temperatures [2].

VI. SUMMARY AND ACKNOWLEDGMENTS

It has been proposed [18,20] that the submonolayer phase diagrams of N_2 and Kr on graphite are qualitatively similar. The major difference between the LEED isotherms presented here and Kr volumetric isotherms near 85K is the absence of any evidence for a dense fluid-commensurate solid phase coexistence for N_2 above 47K. This may be due to the differences in the measurement techniques, the difference in substrate quality, and/or the difference between N_2 and Kr. Although LEED investigations of this region of the phase diagram for Kr are not possible due to its high vapor pressure, more extensive LEED studies for N_2 are planned.

We wish to especially thank T.S. Noggle for providing graphite crystals, A.N. Berker and M. Schick for explaining Ref. 20, D.M. Butler, J.G. Dash, G.D. Halsey, Y. Larher, and A.G. Stewart for other helpful discussions, D.M. Bylander and C.E. Platt for experimental assistance, and NSF Grant DMR77-26931 for financial support.

REFERENCES

@ Current address: Boeing Aerospace Co., Seattle, WA 98124.

1. J.K. Kjems, L. Passell, H. Taub, and J.G. Dash, Phys. Rev. Lett. 32, 724 (1974).
2. J.K. Kjems, L. Passell, H. Taub, J.G. Dash, and A.D. Novaco, Phys. Rev. B13, 1446 (1976).
3. W.F. Brooks, Brookhaven National Laboratory Informal Report 22617 (1977).
4. T.T. Chung and J.G. Dash, Surface Sci. 66, 559 (1977).
5. Y. Larher, J. Chem. Phys. 68, 2257 (1978).
6. G. Boato, P. Cantini, C. Guidi, R. Tatarek, G.P. Felcher, Phys. Rev. B20, 3957 (1979).
7. D. Wesner, G. Derry, G. Vidali, T. Thwaites, D.R. Frankl, to be published.
8. M.D. Chinn and S.C. Fain, Jr., J. Vac. Sci. Technol. 14, 314 (1977).
9. Lake Shore Cryotronics, Inc., Westerville, Ohio 43081, silicon sensor type DT-500.
10. T. Edmonds and J.P. Hobson, J. Vac. Sci. Technol. 2, 182 (1965).
11. C.G. Shaw and S.C. Fain, Jr., Surface Sci. 83, 1 (1979).
12. C.G. Shaw, S.C. Fain, Jr., and M.D. Chinn, Phys. Rev. Lett. 41, 955 (1978).
13. M.D. Chinn and S.C. Fain, Jr., Phys. Rev. Lett. 39, 146 (1977).
14. S.C. Fain, Jr., M.D. Chinn, and R.D. Diehl, Phys. Rev. B21 (in press for 1 May 1980).
15. J. Eckert, W.D. Ellenson, J.B. Hastings, and L. Passell, Phys. Rev. Lett. 43, 1329 (1979).
16. S.C. Fain, Jr. and M.D. Chinn, J. Phys. (Paris) Colloq. 38, C4-99 (1977).
17. J. Suzanne, Thesis, L'Universite d'Aix-Marseille II (1974, unpublished), Fig.4-19.
18. D.M. Butler, J.A. Litzinger, G.A. Stewart, and R.B. Griffiths, Phys. Rev. Lett. 42, 1289 (1979).
19. Y. Larher, J. Chem. Soc. Faraday Trans. I 70, 320 (1974).
20. S. Ostlund and A.N. Berker, Phys. Rev. Lett. 42, 843 (1979).
21. S. Ross and W. Winkler, J. Colloid Sci. 10, 319 (1955).
22. W. Frels, D.R. Smith, and T. Ashworth, Cryogenics 14, 3 (1974).
23. Y. Larher, Molecular Physics 38, 789 (1979).

Sinha, ed. Ordering in Two Dimensions

NEW NEUTRON SCATTERING RESULTS ON METHANE ADSORBED ON GRAPHITE

A. GLACHANT, J.P. COULOMB AND M. BIENFAIT

Département de Physique, Faculté des Sciences de Luminy
13288 Marseille, France

P. THOREL and C. MARTI

I.L.L., B.P. 156 - 38042 Grenoble Cédex

J.G. DASH

Department of Physics, University of Washington
Seattle, Washington 98195, USA

Recent neutron scattering observations obtained on the CH_4/graphite system are controversial [1,2]. The disagreement bears on the triple point temperature and the commensurability of 2D solids and liquids. New neutron diffraction and quasi-elastic scattering experiments were carried out to resolve the discordance. Our temperature scale was checked against our previous one by a measurement of the triple point of a monolayer of N_2 adsorbed on graphite. Both diffraction and mobility measurements confirm our triple point at 56 K, but the new elastic scattering experiment shows that the former resolution was too poor to detect a 0.5 to 1 % expansion of the $\sqrt{3}$ solid. In fact we now observe an expansion from a $\sqrt{3}$ structure to an out-of-registry solid around 48 K, in qualitative agreement with the Argonne study [1]. Finally, we propose the following stability domains of a CH_4 submonolayer/graphite (coverage 0.6 and 0.8). Below $\sim$ 48 K, a $\sqrt{3}$ solid occurs and transforms itself into an expanded solid above this temperature. The triple point is 56 K. The 0.6 layer melts into an expanded out-of-registry liquid. The liquid and the incommensurate solid coexist in the 0.8 layer between 57 and 69 K.

1. INTRODUCTION

Two recent letters dealt with the study by neutron diffraction experiments of the phase transitions occurring in a monolayer of methane adsorbed on the (0001) plane of graphite. The most complete [1] bore on a detailed analysis of the various solid phases stable in this surface film. According to temperature and coverage, one could observe one in-registry $\sqrt{3}\times\sqrt{3}$ solid, one out-of-registry compressed solid. The melting of those solids was also studied.

In the second letter [2], the same system was analyzed but on more narrow coverage and temperature ranges. The authors were mainly interested in the melting of the methane film. They found that this solid-liquid transition was first order and were able to determine the monolayer triple point.

The two works overlapped partly and it appeared that a few results were at variance. The disagreement bore on the triple point temperature (60 and 56 K in [1] and [2] respectively) and on the commensurability of 2D solid and liquid. Glachant et al. claimed that just before melting, the 2D solid at coverage 0.6 and 0.8 was commensurate with the graphite substrate. At melting the layer expanded by about 1.5 %. Whereas Vora et al. showed that before the 2D triple point, the solid layer is incommensurate with a lattice parameter $\sim$ 1 % larger than a $\sqrt{3}$ in-registry solid. At melting the layer contracted and became a "lattice liquid".

In order to remove these discrepancies, we performed new neutron elastic and quasi-elastic scattering experiments on submonolayers of methane adsorbed on graphite. We chose papyex as a substrate. This recompressed exfoliated graphite [2] is very close to grafoil [1] which is of common use in surface studies.

DIFFRACTION EXPERIMENTS

We first calibrated the sample cell temperature by measuring the triple point of a monolayer of N_2 adsorbed on graphite. We found that a 0.6 layer of nitrogen melted at ≃ 49 K, a temperature in good agreement with previous calorimetry (T ≃ 48 K) [3] and neutron diffraction (T ≃ 49 K) [4] measurements.

The diffraction results for 0.6 and 0.8 layers of CD_4 are summarized in Table I. Figs. 1 and 2 present the diffraction scans of the 0.8 layer coverage film at 15 K and 49.9 K, showing the (10) peak of the $\sqrt{3}\times\sqrt{3}$ registered phase.

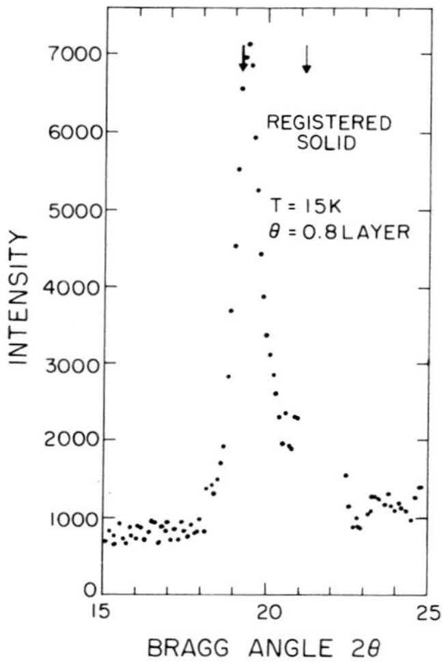

Fig. 1

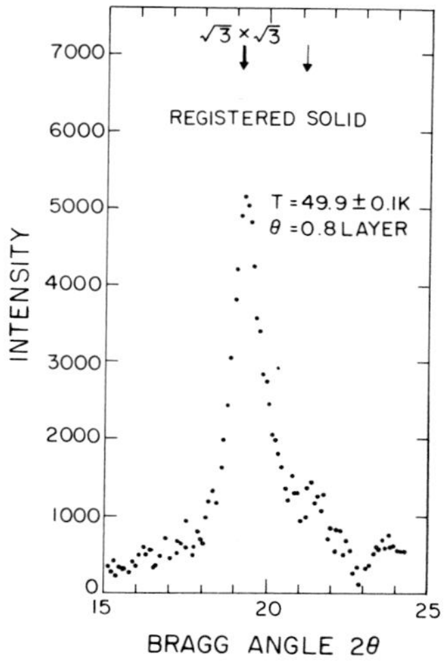

Fig. 2

The broad peak observed at coverage 0.6 and T > 57.2 K indicates that the layer is fully melted under these experimental conditions. At 46.5 K, the sharp line characterizes a 2D registered solid. At coverage 0.8, the solid and the liquid coexist at and above 57.7 K whereas the layer is solidified between 2 and 53.6 K. All these results are consistent with a triple point temperature around 56 K [2]. However just before melting, the lattice parameter of the solid is slightly larger (by 0.5 %) than that given in our previous paper. At the time, we did not pay enought attention to the wavelength uncertainty (∿ 1 %) and to the instrumental resolution (two counters are 0.1° apart). Here, the wavelength uncertainty is only 0.3 % and we observe before melting an expanded out-of-registry solid in agreement with the results of Vora et al. [1]. Fig. 3 shows the diffraction of the 0.8 layer expanded solid at 53.6 K.

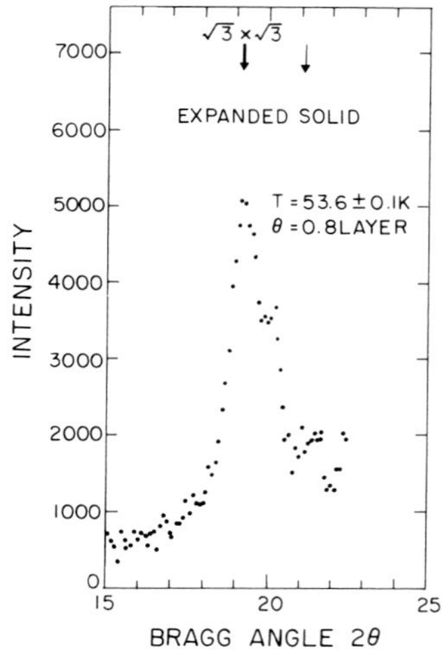

Fig. 3

Furthermore, when cooling down the system from 49.9 K to 35 K, we detect a small contraction of the layer, associated with a rise of the intensity of the line. The new lattice parameter is compatible with a $\sqrt{3}$ in-registry solid. This new transition is in qualitative agreement with the Argonne results. Taking account of Table I, we believe that the commensurate-incommensurate transitions occurs between 47 and 49 K.

Finally, at melting, the incommensurate solid transforms into an out-of-registry liquid whose mean nearest neighbor distance is probably larger than that of a registered structure. Fig. 4 displays results for the 0.8 layer film in the solid-fluid coexistence region.

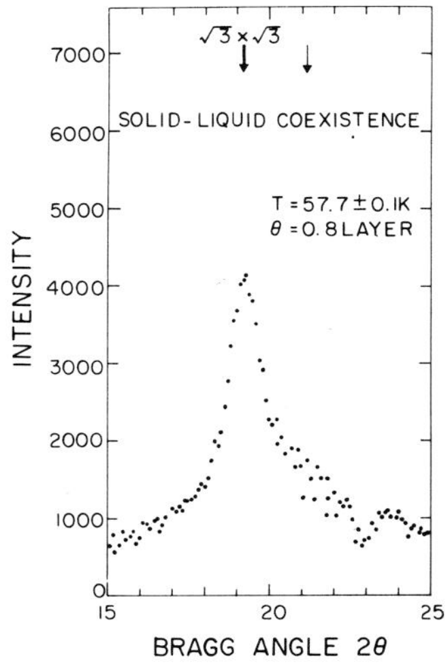

Figure 4

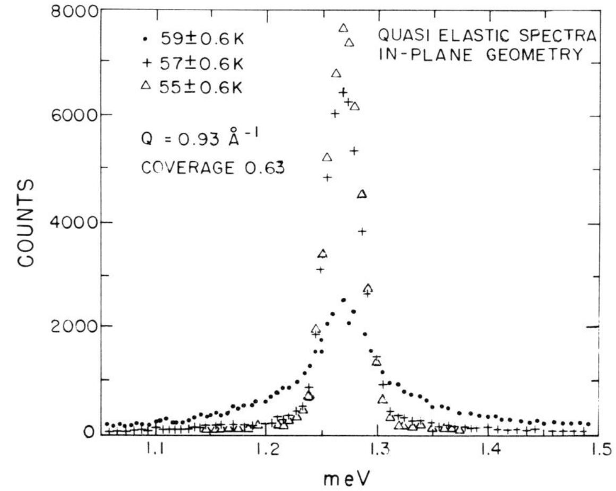

Fig. 5 Quasi-elastic spectra, with in-plane geometry, of 0.63 monolayer film at three temperatures, indicating fluid, fluid + solid, and solid phases.

QUASI-ELASTIC SCATTERING

The broadening in energy of a scattered incoherent neutron beam can be used to determine the mobility of a surface fluid [5,6]. Solid phases exhibit narrow scattering spectra showing the instrumental resolution, whereas fluids display broad peaks whose width yields the diffusion coefficient of the mobile phases. We extended our previous measurements [5], performed on the CH_4/graphite system towards the temperature of the triple point. A 0.63 layer is mobile at 59, 62, 67, 69 and 70.5 K. It is solidified at 55 K. At 57 ± 0.6 K, we observe a superposition of a narrow peak due to a solid and of a broad spectrum due to a liquid-like phase. During the recording (∿ 12 h), the drift of our temperature regulation probably caused the system to go across the triple line. This observation supports again our melting point at ∿ 56 K. Fig. 5 summarizes the results for the 0.8 layer film at 3 temperatures spanning the solid-fluid transition.

All these new results allow us to propose the following transitions of the methane submonolayer adsorbed on graphite (coverage 0.6 and 0.8): T = 47-49 K, transition between an in-registry solid and an out-of-registry solid.

The solid-fluid transition at coverage 0.63 is quite sharp, being completed within about 0.7 K. The experimental relative width $(\delta T/T) \gtrsim 0.013$ strongly suggests that the transition corresponds to a 2D triple point. This fact alone indicates the existence of a bound liquid phase, and is therefore consistent with the line width results of both the elastic and quasi-elastic measurements. The liquid phase, since it has a density very close to the registered value, may well have a structure influenced by the substrate, and will be an extremely interesting phase for further study.

TABLE I

Diffraction results of CD_4 submonolayers adsorbed on graphite. The intensity profiles are analyzed after subtracting the intensity of scattering from the empty substrate. I_{max} is the maximum intensity of the line at $2\theta_{max}$; d is the distance between rows of CD_4 molecules ($d = a\sqrt{3}/2$ with a the nearest neighbor distance) and L is the mean size of the diffracting arrays. The uncertainty in d is not negligible because of the lack of precision in the wavelength and the limited instrumental resolution, but the variation of $2\theta_{max}$ is significant.

$\lambda = 1.230 \pm .004$ Å ; $\theta = 0.6$

T_k(± 0.1 K)	46.5	57.2	65.3	75.2
I_{max}(a.u)	21000	9000	7200	6600
$2\theta_{max}$(°)	19.4	∿ 19	∿ 18.8	∿ 19
shape	sharp line	broad line	broad line	broad line
$d_{Å}$	3.69 ± 0.02	∿ 3.73	∿ 3.76	∿ 3.73
$L_{Å}$	∿ 180	∿ 20	< 20	< 20

$\lambda = 1.230 \pm .004$ Å ; $\theta = 0.8$

T_k(± 0.1 K)	∿ 2	∿ 15	∿ 25	∿ 35	49.9	53.6	57.7	61.7	69.2
I_{max}(a.u)	34000	33700	33000	32000	24000	24000	19000	16000	13000
$2\theta_{max}$(°)	19.4	19.4	19.4	19.4	19.3	19.3	19.3	19.2	19.2
shape	sharp line	sharp line	sharp line	sharp line	sharp line	sharp line	sharp line + wings	sharp line+ wings	bump +small peak
$d_{Å}$(± 0.02)	3.69	3.69	3.69	3.69	3.71	3.71			
$L_{Å}$	∿ 200	∿ 200	∿ 200	∿ 200	∿ 150	∿ 150			

REFERENCES

(1) P. Vora, S.K. Sinha and R.K. Crawford, Phys. Rev. Letters 43, 704 (1979).
(2) A. Glachant, J.P. Coulomb, M. Bienfait and J.G. Dash, J. Physique Lettres 40, L-543 (1979).
(3) T.T. Chung and J.G. Dash, Surf. Science 66, 559 (1977).
(4) J.K. Kjems, L. Passell, H. Taub, J.G. Dash and A.D. Novaco, Phys. Rev. B13, 1446 (1976).
(5) J.P. Coulomb, M. Bienfait and P. Thorel, Phys. Rev. Lett. 49, 733 (1979).
(6) J.P. Coulomb, M. Bienfait and P. Thorel, to be published.

Sinha, ed, Ordering in Two Dimensions

SUBSTRATE EFFECTS ON LONG-RANGE ORDER AND SCATTERING FROM LOW-DIMENSIONAL SYSTEMS*

Charles L. CLEVELAND, Charles S. BROWN and Uzi LANDMAN

School of Physics, Georgia Institute of Technology, Atlanta, Ga. 30332

ABSTRACT

Using solvable models it is shown that coupling 1D and 2D systems to substrates produces significant alterations in their long-range order and scattering characteristics, even if the coupling is very weak. Expressions for Peierls's long-range-order parameter, $<\delta_n^2>$, are obtained, with their asymptotic forms, and static structure factors, $S(\vec{Q})$, are evaluated.

Most if not all systems exhibiting one-dimensional (1D) behavior (such as organic [1] and inorganic complexes [2]) and those reported to show two-dimensional (2D) behavior (such as adsorbed layers [3], electrons trapped on a liquid Helium surface [4] and thin "soap bubble" films [5]) are coupled to skeletal or substrate environments. Consequently they may be better characterized as quasi-low-dimensional systems. Particularly intriguing are questions of ordering (degree and type) and stability in such systems [6-10]. Since dimensionality influences radically the properties of physical systems, and in view of physical arguments [6] and rigorous proofs [7] that true long-range order does not exist in strict 1D and 2D systems, it is of interest to investigate the behavior and degree of long-range order in approximately lower dimensional systems. Such coupling effects have been observed in recent neutron scattering studies of $Hg_{3-\delta}AsF_6$ [2,11], and of phases of CD_4 monolayer films on graphite [12] for which the scattered neutron line shapes could not be interpreted, even for the registered (commensurate) phase, on the basis of strict 2-D theories.

Our purpose is to show that coupling to a substrate significantly affects the degree of long-range order and scattering characteristics in certain 1D and 2D model systems. To elucidate our discussion we limit our considerations to certain solvable models employing simple coupling schemes.

A measure of the long-range order in an N-particle system is provided by the function $<\delta^2_{n\hat{\kappa}}> \equiv <((\vec{u}_n-\vec{u}_o)\cdot\hat{\kappa})^2>$ given by

$$<\delta^2_{n\hat{\kappa}}> = 4(Nm)^{-1}\sum_{\vec{q}} <|U_{\vec{q}}\cdot\hat{\kappa}|^2> \sin^2(\tfrac{1}{2}\vec{q}\cdot\vec{R}_n) \qquad (1)$$

where $\vec{u}_n$ is the deviation of particle n of mass m from its equilibrium position $\vec{R}_n$, $\hat{\kappa}$ is an arbitrary direction in the lattice and $U_{\vec{q}}$ is the normal mode amplitude. The angular brackets denote temperature ensemble averaging. At sufficiently high temperatures, T, (typically larger than the Debye temperature) equipartition can be used [6,8] to write $<|U_{\vec{q}}\cdot\hat{\kappa}|^2> = k_BT\omega_q^{-2}$, where ω_q is the normal mode frequency.

Consider first a 1D chain of atoms of lattice spacing a and interparticle n.n. force constants K. Let it be coupled via n.n. and next n.n. force constants $K_\perp$ and K_D, respectively, to a 1D parallel substrate chain of heavy masses each a distance a, for simplicity, below a lattice site of the first chain. In the harmonic approximation, and for a stationary substrate, the longitudinal normal mode frequency (describing motions along the chain axis) is given by $\omega_q^2 = (4K/m)[R^2+\sin^2(\frac{qa}{2})]$, where $2R^2 = K_D^2/K^2$ is a measure of the interchain relative coupling strength. Notice that this mode possesses a q=0 gap, equal to $4KR^2/m$. The existence of this gap, due to the inclusion of interactions between the subsystems beyond n.n.'s, is essential for the modified behavior discussed below. Using the high-T approximation and the above ω_q, transformation of the sum in Eq. (1) over $\vec{q}$ to an integral and converting to a contour integral in the complex plane yields the following closed form result

$$<\delta_n^2>/a^2 = \sigma\left[\frac{1-(2R^2+1-2R\sqrt{R^2+1})^n}{R\sqrt{R^2+1}}\right]$$

$$\equiv C_1[1-e^{-\alpha n}] \qquad (2)$$

where $\sigma = k_BT/2Ka^2$, (typically [11] of the order $10^{-3} - 10^{-4}$). In the limit of vanishing coupling, R=0, and for large n the previously known result [11], $<\delta_n^2>/a^2 \sim (2\sigma)n$ is recovered. With a criterion that long-range order exists when

$<\delta_n^2>/a^2 < 1$ as $n \to \infty$, it follows that there is long-range order if $R \gtrsim \sigma$, so that even weak coupling to a substrate restores long-range order. The modified behavior upon coupling is shown in Fig. (1a).

Using the expression given in Eq. (2), the frequency integrated dynamical structure factor $S(\vec{Q})$ (for $\vec{Q}$ parallel to the chain) can be evaluated yielding

$$S'(Q) \equiv S(Q)-Ne^{-f^2}\delta(Q,G) = \sum_{\ell=1}^{\infty} \frac{(-f)^{2\ell}}{\ell!} Z(\ell), \quad (3a)$$

$$Z(\ell) = \sum_{j=0}^{\ell} (-1)^j \binom{\ell}{j} \left[\frac{e^{-\alpha j}-\cos(Qa)}{\cosh(\alpha j)-\cos(Qa)}\right], \quad (3b)$$

where $2f^2 = Q^2C_1$ and G is a reciprocal lattice vector. In the limit of vanishing interchain coupling [11] S'(Q) (and S(Q)) consists of a series of narrow peaks centered upon the reciprocal lattice vectors. For non-vanishing coupling strengths a broadening of the peaks accompanied by a pronounced asymmetry occurs as shown in Fig. (1b) (note change in scales). Sufficient accuracy is obtained by truncating the sum over ℓ in Eq. (3a) typically at ℓ=3-5. The above could provide practical functional form for fitting purposes.

We turn next to the evaluation of $<\delta_{||}{}^2(\rho)>$ (where $\rho = r/a$ and $r_{||}$ is an interparticle distance in an arbitrary direction) for a 2D square lattice, of lattice constant a which is coupled to a stationary square substrate layer via n.n. and next n.n. force constants $K_\perp$ and K_D. Following arguments similar to the above, we obtain

$$<\delta_{||}{}^2(\rho)>/a^2 = (\sigma/2\pi)\,\rho^{-2} \times \int_0^{2\sqrt{\pi}\rho} dy\; y\,\frac{1-J_o(y)}{R^2+\sin^2(y/2\rho)} \quad (4)$$

where the Debye cutoff has been employed and J_o is the Bessel function of the first kind. For vanishing interplane coupling, R=0, the previously [3,8] derived asymptotic logarithmic divergence of $<\delta_{||}{}^2(\rho)>$ is observed. For finite coupling asymptotic analysis yields non-logarithmic asymptotic behavior,

$$<\delta_{||}{}^2(\rho)>/a^2 \sim (\sigma/\pi)\left[F(R) - \frac{\sqrt{\pi}J_1(2\pi\rho)/\rho}{R^2+\sin^2\sqrt{\pi}}\right], \quad (5a)$$

where

$$F(R) = \int_0^{2\sqrt{\pi}} dy\; y[2R^2+1-\cos y]^{-1}. \quad (5b)$$

Numerical evaluation of Eq. (4) for various values of R indicated that good fits to $<\delta_{||}^2(\rho)>$ are given by the form $A-B\exp(-\gamma\rho^{1/2})$ where A,B and γ are constants dependent upon R. Sample results are shown in Fig. (2a). Using the above form, an expression for $S'(\vec{Q})$ (for $\vec{Q}$ parallel to the plane) can be derived, which for various values of R yields results shown in Fig. (2b). It is of interest to comment that for a strict 2D lattice i.e., R=0 the peaks in $S(\vec{Q})$ near reciprocal lattice vectors G are given by power law singularities [10], $S(\vec{Q}) \sim |\vec{Q}-\vec{G}|^{-2+\eta_{\vec{G}}(T)}$, where the bounded exponents, $\eta_{\vec{G}}(T)$ are related to the elastic moduli of the lattice. We note that for both the 1D and 2D cases, the $\delta(\vec{Q},\vec{G})$ term has been subtracted in $S'(\vec{Q})$, (e.g., Eq. (3a)). This term which is absent in the R=0 limit [11] increases with R, i.e., increasing coherent scattering intensity at $\vec{Q}=\vec{G}$. Correspondingly, the residual $S'(\vec{Q})$ decreases in amplitude and broadens upon increased coupling to the substrate (note scales in Figs. 1b and 2b).

It is important to note that for both the 1D and 2D coupled systems the long-range-order parameters exhibit an altered asymptotic behavior, deviating significantly from the uncoupled results (R=0) even for small substrate coupling strengths (Fig. 1a, 2a). In fact for both cases $<\delta_n^2>$ converges to a limit at microscopic distances even for small R values. Consequently, even for small couplings to the substrate strict 1D or 2D behavior is lost. This is due to the fact that by turning on the couplings to the substrate (finite R) the number of possible paths for linkage between any two atoms increases. Thus the tendency to maintain long-range-order increases upon coupling. These characteristics are exhibited in the integrated scattering functions (Figs. 1b, 2b).

While we recognize that the above model calculations employed simplifying assumptions, such as a particular geometry, range of interaction, classical description and a stationary substrate, the essential results pertaining to the salient effects of the dimensionality of the system on the degree of long-range-order and scattering characteristics should remain valid in more general circumstances. Moreover, the first three assumptions can be easily relaxed (for commensurate arrangements) and do not modify the main conclusions. Non-commensurate configurations and couplings to extend non-stationary substrates remain the subjects of further investigations. Finally, in analyzing scattering data in addition to the coupling effects discussed above, finite size effects which become significant for small surface layer crystallites, and corrections due to random orientation effects should be considered.

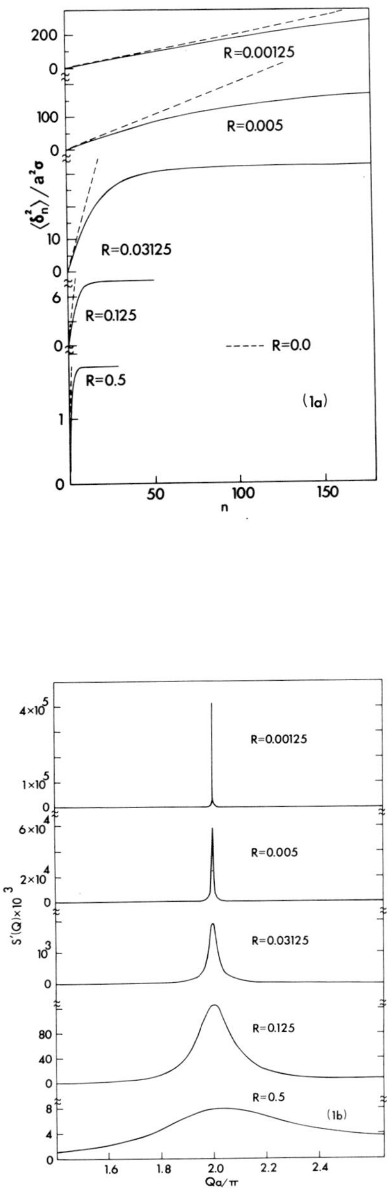

Figure 1: 1D chain coupled to a stationary substrate chain. (a) $<\delta_n^2$ vs. n, for various values of relative coupling strength, R. Solid lines after Eq. 2, dashed lines correspond to the R=0 case. (b) Subtracted static structure factors, S'(Q), around the first Bragg peak for various coupling strengths, $\sigma = 10^{-4}$. Note changes in scale.

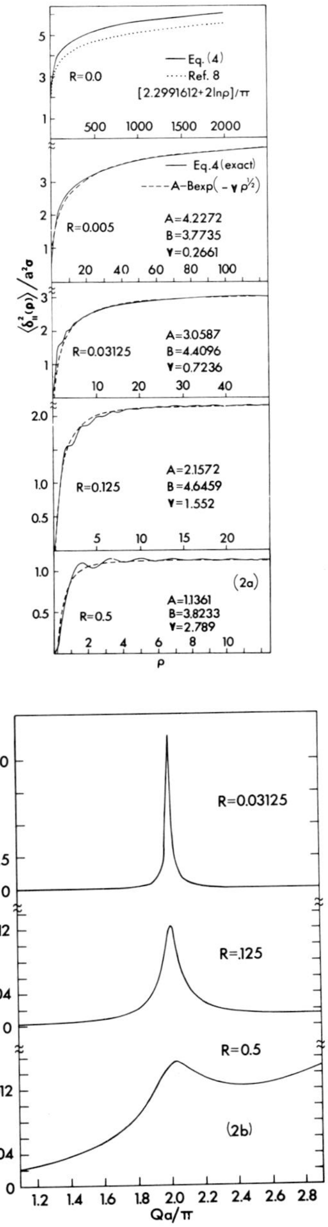

Figure 2: 2D layer coupled to a stationary substrate layer. (a) $<\delta^2(\rho)>$ vs. ρ=r /a for various values of R. For R=0, results obtained with (dotted) and without (solid) the long-wavelength approximation are shown. For finite R values, results are shown using Eq. (4), (solid) and via the indicated fits (dashed). (b) Subtracted static structure factors, S'($\vec{Q}$) around the (10) Bragg peak for several coupling strengths, R, $\sigma=10^{-4}$. Note changes in scale.

REFERENCES

*Work supported by U.S. DOE Contract No. EG-S-05-5489.

1. Chemistry and Physics of One-Dimensional Metals, edited by H. J. Keller (Plenum, New York, 1977), Vol. 25B.
2. I. U. Heilmann, J. D. Axe, J. M. Hastings, G. Shirane, A. J. Heeger and A. G. MacDiarmid, Phys. Rev. B 20, 751 (1979).
3. J. G. Dash, Films on Solid Surfaces (Academic, New York, 1975).
4. P. M. Platzman and H. Fukuyama, Phys. Rev. B 10, 3150 (1974); C. C. Grimes and G. Adams, Phys. Rev. Letts. 42, 795 (1979).
5. R. J. Birgeneau and J. D. Lister, J. Phys. Letts. (Paris) 39, L399 (1978).
6. R. E. Peierls, Ann. Inst. Henri Poincare 5, 177 (1935).
7. a) N. D. Mermin, Phys. Rev. 176, 250 (1968) b) B. Jancovici, Phys. Rev. Letts. 19, 20 (1967); c) Y. Imry and L. Gunther, Phys. Rev. B3, 3939 (1971).
8. J. G. Dash and M. Bretz, J. Low Temp. Phys. 9, 291 (1972).
9. J. M. Kosterlitz and D. J. Thouless, J. Phys. C 6, 1181 (1973).
10. D. R. Nelson and B. I. Halperin, Phys. Rev. B 19, 2457 (1979); and references cited therein.
11. V. J. Emery and J. D. Axe, Phys. Rev. Letts. 40, 1507 (1978).
12. P. Vora, S. K. Sinha and R. K. Crawford, Phys. Rev. Letts. 43, 704 (1979), and private communications.

Sinha, ed. Ordering in Two Dimensions

STRUCTURES OF MONOLAYERS OF CF_4 ON GRAPHITE

H.J. LAUTER, B. CROSET, C. MARTI

Institut Laue-Langevin, 156X, 38042 Grenoble Cédex, France

P. THOREL

Département de Recherche Fondamentale, CEN Grenoble,

85X, 38041 Grenoble Cédex, France

ABSTRACT

Adsorption isotherms [11] made at T > 87K suggested a phase diagram with extrapolated values of the triple point of a coverage x > 0.73 and a temperature T << 77K. By neutron diffraction we found the triple point at <u>x < 0.5</u> and T ∿ 74K. This leads to a quite original phase diagram. The 2d-solid has a triangular cell, expanding fast near the triple point, where it is commensurate with the graphite (2 x 2 superstructure). The 2d-liquid is unique near 74K: high coherence length and localization with respect to the graphite; we describe it as a "sublattice liquid", partly localized but disordered among four graphite sublattices. It is verified that the second layer is formed only above 65K.

The experiments were performed on the D1B and D2 spectrometers at the HFR in Grenoble. We used a wavelength of 2.52 Å and 1.69 Å respectively.

The sample consisted of a 6 cm high stack (diameter 2 cm) of Papyex foils, which were oriented parallel to the scattering plane. The monolayer completion (x=1) has been determined

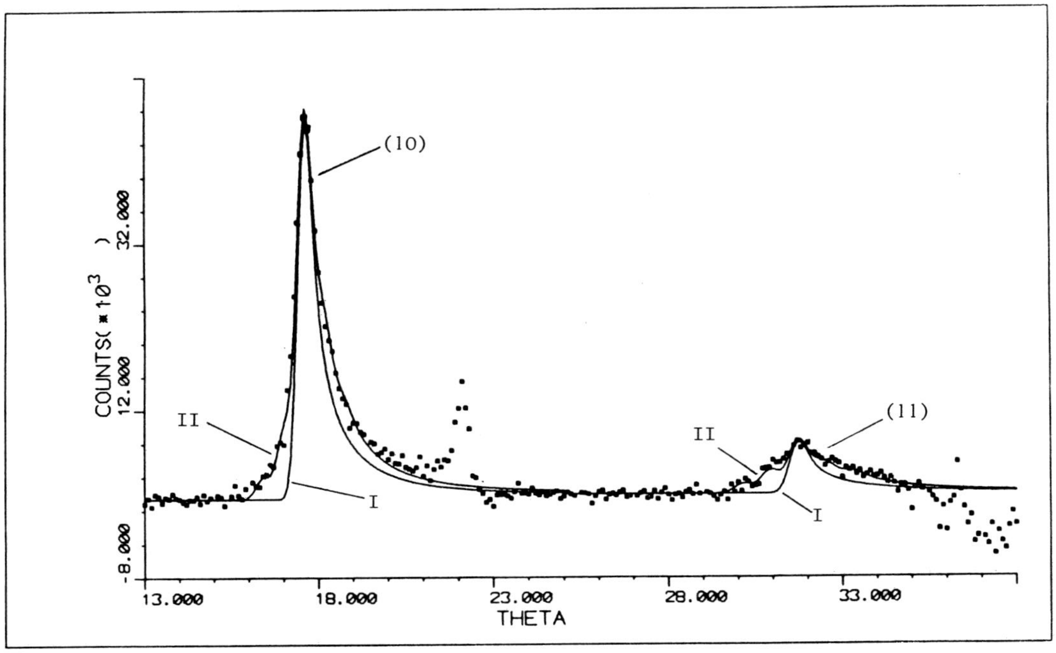

Fig. 1 Diffraction pattern of adsorbed CF_4 at X=1 and T=40K. Lines are fits without (I) and with (II) satellite peaks.

by an adsorption isotherm with CF_4 at T=107K.

The spectrum of the first adsorbed layer (x=1) is shown in fig. 1. The background spectrum, which consists of the scattering of the cryostat with sample cell and bare graphite, has already been subtracted. The intensity at $\theta \sim 22°$ shows that the (002)-peak of the graphite has been influenced by the adsorbed layer [1]. The existence of the (10) and (11) reflections proves a triangular structure for the adsorbed layer, which is nearly the 2x2 superstructure.

The curve (I), which fits the experimental points only aside the peaks and at the points with the highest intensity, has been derived using Warren's formula [2] with a coherence length of ∿200 Å as for other adsorbates on Papyex [3,4,5]. The formula has been modified to take into account the mosaic distribution of the graphite and has been convoluted with the instrumental resolution. This formula does not fit our data although it fits the data resulting from measurements of adsorbed C_2D_6[3], C_2D_2[4] and Ar [5].

Long range order should not exist in two dimensions [6];this effect could cause a broad foot of the peaks. But because this broad foot has only been seen with CF_4 and not with the other mentioned adsorbates, we made a fit including satellites . The satellites arise because of only a small difference (q) between the (20)-CF_4 and the (10)-graphite reciprocal lattice point (fig. 2). The instrumental resolution was not sufficient to resolve the different satellites. So we could not deduce anything about the orientation of the CF_4 lattice with respect to the graphite lattice and assumed the configuration shown in fig. 2. We took into account two satellites of first order and one of second order for each side of the (10) peaks and one first order and one second order satellite for each side of the (11) peak as foreseen by the theoretical model (fig. 2), due to the different relative orientations of the satellite star. The agreement with the data is quite quantitative. But high resolution experiments will be made to separate the satellites from each other and from the main peak.

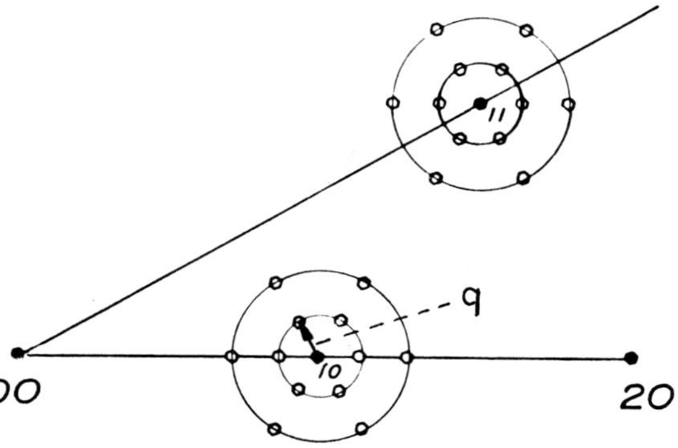

Fig. 2 Satellite stars around the (10) and (11) reflex in reciprocal space; (●) CF_4 reciprocal points; (0) satellite spots.

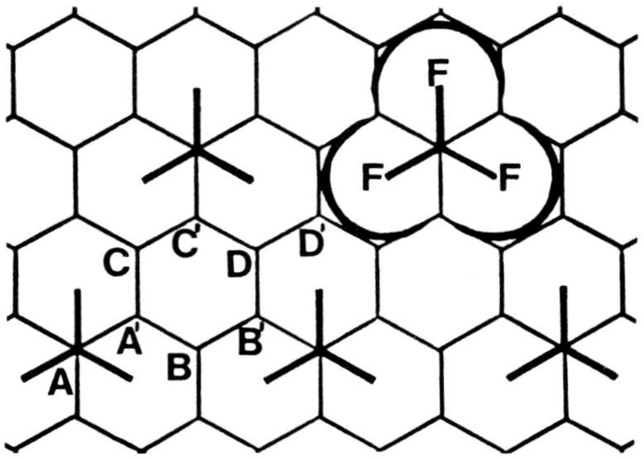

Fig. 3 Sketch of the (2x2) superstructure of CF_4 tripodes on graphite. ABCD are equivalent adsorption sites and A'B'C'D' is a second set but with a 180° turned CF_4 molecule.

In fig. 3 the graphite lattice is seen with adsorbed CF_4 in the 2x2 structure. The CF_4 molecules are described by the projection of the tetrahedron, which is assumed to sit with one F-F-F plane on the graphite and each F nearly on an adsorption site of the graphite (F-F distance 2.22 Å). Therefore already small deviations from the 2x2 structure will cause a rotation of the molecules which could deform the CF_4-lattice and create the satellites with more intensity than seen with Kr [7]. It should be noted that the relative height of the satellites of the (11) peak is greater than those of the (10) peak [8]. This problem will be further discussed [9].

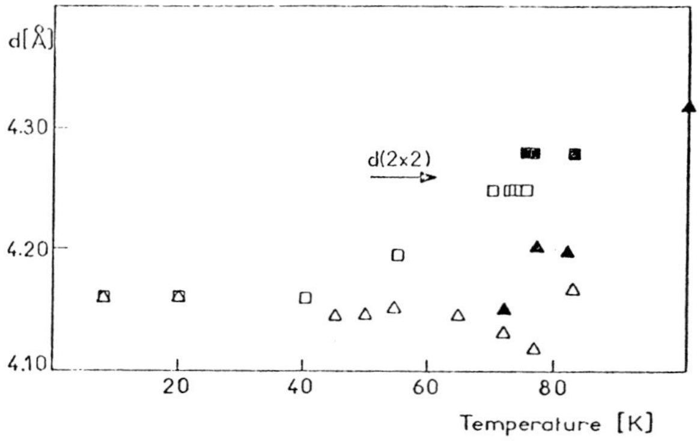

Fig. 4 d-spacing of the triangular CF_4-lattice ((10)-reflex) in dependence of temperature, (□) X=0.6, (Δ) X=0.9, open symbols indicate a solid and full symbols a fluid.

The figure 4 shows the evolution of nearest neighbour distance a_{nn}. If two points exist at the same temperature and coverage, it indicates that the line shape of the (10) peak was fitted by the sum of two contributions :

- a solid part without satellites with a coherence length of 200 Å ; - a liquid part fitted by the same formula as for the 2D-solid but with the long coherence length of ≈30 Å.

It should be pointed out that for x = 0.6 the (10) peak is in the (2x2) position just before and during melting and does not show satellites as expected. Furthermore the layer at x=0.9 melts at a lower temperature than at 0.6. This leads to a tentative phase diagram as shown in fig. 5. It shows a critical point for the transition 2x2 solid-liquid. The two triple points are at higher temperatures than the triple point extrapolated by P. Dolle et al. [11] (T_t<<77). This disagreement can be explained by the pinning of the CF_4 molecules in a 2x2 structure around 70K, which was not seen with the isotherm measurements. More experiments will precise the phase diagram.

In fig. 4 is seen that at low temperatures (T<40K) the nearest neighbour distance a_{nn} stays constant. This is valid until monolayer completion. For x>1 a_{nn} becomes smaller. The densities are ρ = 0.0500 m/Å^2 for x<1, ρ = 0.0515 m/Å^2 for x>1 and ρ = 0.0544 m/Å^2 in the dense a-b plane of bulk CF_4 [10]. Thus for x<1 adsorbed CF_4 forms 2-D clusters in equilibrium with 2-D gas. The density of this solid remains constant on adding more CF_4 until a complete solid monolayer is formed. Then a further compression takes place.

For x=1.9 and T=40K the fig. 6 shows the

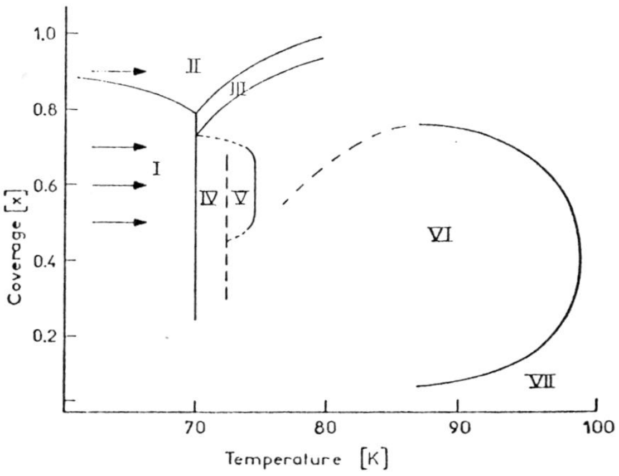

Fig. 5 Tentative phase diagram: (I) solid + gas; (II) solid; (III) solid + liquid; (IV) 2x2 phase + gas; (V) 2x2 phase + liquid; (VI) liquid + gas; (VII) gas; transition line (VI)-(VII) from Ref. [11]. The arrows indicate the investigated coverages.

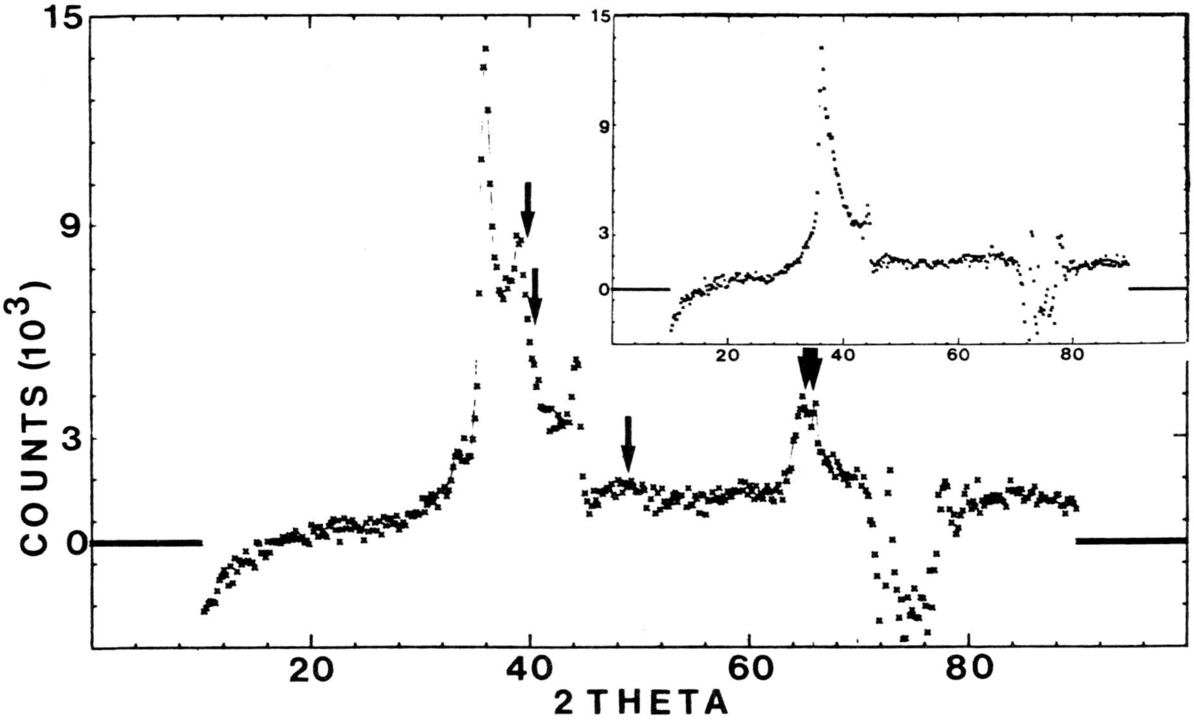

Fig. 6 Diffraction pattern of CF_4 at X=1.9 and 40K. The arrows indicate the angular position where the reflections in bulk-CF_4 [10] have been seen. The inset shows the diffraction pattern at X=1.9 and T=78K.

(10) and (11) peak and in addition some small peaks which are very close to the arrows which indicate the position of the diffraction peaks of bulk CF_4 [11]. The (10) peak has the same intensity as at x=1. The (11) is greater than for x=1 but only because the diffraction of the bulk takes place at about the same angle. With the nucleation of 3-D crystallites the monolayer is further compressed to a constant value of $a_{nn} \sim 4.73$ Å ; this has been verified until x=4. The small peak at $2\theta=33.4$ could not be attributed.

The inset in fig. 6 shows the diffraction pattern for X=1.9 at T=78K. All 3-D crystals have melted and the (10) peak consists now of a solid part and a liquid like part, which enlarges the foot of the peak. Careful measurements at X=1.5 show that the bulk has already disappeared at T=67K in agreement with T=65K given by Doll et al. [11] by isotherm measurements.

When the molecules from the 3D crystallites move to the second layer the peak intensity of the (10) peak remains unchanged and no diffraction of a liquid or other peak appears. So at T= 67K the second layer is a 2D solid and the molecules of the second layer are supposed to be in the center of the triangles formed by the first layer as suggested by the unchanged structure factor. This is quite different from bulk CF_4 where the molecules of each dense a b plane are supposed to be stacked nearly one over each other.

The inset in fig. 6 also shows that the (11) peak has disappeared. It disappears already at about 40K in the region in fig. 4, where a_{nn} becomes dependent on the temperature.

In conclusion our results show that the existence of the periodic graphite potential is very important for the structure of CF_4 adsorbed on graphite. It perturbs the incommensurate solid, it changes the phase diagram by creation of a 2x2 commensurate solid and liquid . We think that the great influence of the graphite potential is due to the good agreement between the triangle of one CF_4-teträheder plane and the distances between the adsorption sites of the graphite.

The nucleation of 3D-crystallites at low temperatures instead of a second layer could be due to the different stacking of the second layer,seen at T>65K,with respect to the bulk along the c-axis.

REFERENCES

1. C. Marti, B. Croset, P. Thorel, J.P. Coulomb, Surf. Sci. 65,532 (1977)
2. B.E. Warren, Phys. Rev. 59, 693 (1941)
3. J.P. Coulomb, Priv. Com.
4. P. Thorel, C. Marti, G. Bomchil, J.M. Alloneau, Proc. ECOSS 3 (1980) to be published
5. B. Croset, C. Marti, T. Ceva, P. Thorel, Proc. ECOSS 1 , 316, Amsterdam (1978)
6. P.C. Hohenberg, Phys. Rev. 158, 383 (1967)
7. P.W. Stephens, P. Heiney, R.J. Birgenau, P.M. Horn, Phys. Rev. Lett. 43, 47 (1979)
8. for 3 D : A. Guinier, X-Ray Diffraction, e.d. H.M. Foley and M.A. Ruderman, W.H. Freeman and Comp., 1963, p. 216
9. B. Croset, H.J. Lauter, C. Marti, P. Thorel Proc. ECOSS 3, to be published
10. D.N. Bol'shutkin, V.M. Gasan, A.I. Prokhvatilov, A.I. Erenburg, Acta Cryst. B28, 3542 (1972)
11. P. Dolle, M. Matecki, A. Thomy, Surf. Sci. 91, 271 (1980)

Sinha, ed. Ordering in Two Dimensions

NMR STUDY OF ETHYLENE ON GRAFOIL*

J.Z. LARESE and R.J. ROLLEFSON

Wesleyan University, Middletown, Connecticut 06457

ABSTRACT

Using NMR spectroscopy at 24 MHz we have measured spin-lattice relaxation times and spin-echoes for ethylene monolayers adsorbed on Grafoil. Temperatures ranged from 63K to 120K. The spin-lattice relaxation time has a minimum at about 87K and a discontinuous jump to larger values at about 89K. We interpret this as signaling the melting of the monolayer solid. The spin-echo data show a periodic modulation of the echo envelope for varying pulse delays, suggesting a double peaked structure in the absorption spectrum. The width of the absorption spectrum varies from 2.3kHz at 63K to 0.6kHz at 86K. The narrowness of the spectrum indicates that significant molecular reorientation is occurring, narrowing the dipolar interaction.
*Supported by Research Corporation

INTRODUCTION

In recent years a large variety of monatomic and diatomic gases adsorbed on Graphite have been the focus of various thermodynamic and spectroscopic studies. A small but increasing number of larger molecules have also received attention. These latter, most of which contain several protons, form particularly attractive systems for nuclear magnetic resonance spectroscopy. NMR is sensitive to the state of motion of the adsorbed molecules, in particular it easily detects rotation in which the center of mass of the molecule is stationary. The large gyromagnetic ratio of the proton allows studies at higher temperatures and with adsorbents of lower specific area than would be possible with other nuclei. The possibility of replacing a fraction of the protons with deuterons further expands the experimental horizons.

We report here some initial measurements in a study we have undertaken of ethylene molecules (C_2H_4) adsorbed on Grafoil, a commercial form of exfoliated graphite. Ethylene is a planar molecule consisting of two double-bonded carbon atoms surrounded by a rectangle of hydrogen atoms, forming a fairly rigid structure. Neutron scattering studies [1] indicate that the ethylene monolayer forms a registered solid, and it is the melting of this solid that has been the subject of our initial work.

EXPERIMENTAL PROCEDURE

The piece of Grafoil (4.7 gm.) used as the substrate material was folded in a zig-zag manner such that the finished sample was in the form of a cylinder with the grafoil sheets parallel to the cylinder axis. A large number of holes were punched with a pin to facilitate the flow of the ethylene molecules into the sample. The grafoil was then baked in a quartz tube under vacuum at 930°C for 17 hours. Following the bake-out it was transferred to a nitrogen filled glove bag. Strips of 12.7 micron teflon sheet were interleaved between the grafoil layers to provide electrical insulation. The sample chamber, constructed of Macor machinable ceramic, had previously been baked in vacuum at 900°C. A Macor cover was epoxied on using Stycast 2850 epoxy. Care was taken to ensure that no epoxy entered the sample volume. The Stycast seal was placed a distance of one diameter away from the end of the RF coil (see fig. 1) so that the protons of the epoxy made a negligible contribution to the NMR signal. A carbon thermometer on the inner surface of the cover pressed against the edges of the Grafoil. To ensure thermal equilibrium the sample was surrounded by two vacuum cans with the inner can maintained at the sample temperature.

The primary thermometer, a Rosemount model 146 MA platinum thermometer, was mounted at the top of the inner vacuum can at a position out of the magnetic field. Calibration of the sample thermometer was done with helium exchange gas in the inner vacuum can and the sample chamber to improve the thermal contact. Under these conditions the thermal relaxation time between the sample thermometer and the platinum thermometer was about 15 minutes.

Following assembly the sample chamber was evacuated at room temperature and a 77 K nitrogen isotherm measured to determine monolayer capacity and check surface uniformity. A Baratron capacitive transducer was used. The data are shown in fig. 1. The small feature at an adsorbed volume of 20.0 cm^3 STP has been identified as a deregistry transition [2]. Several isotherms were taken during the course of the experiments and this feature reproduced very well. Using this volume as the measure of the registered monolayer capacity a surface area of 84.5 m^2 is calculated.

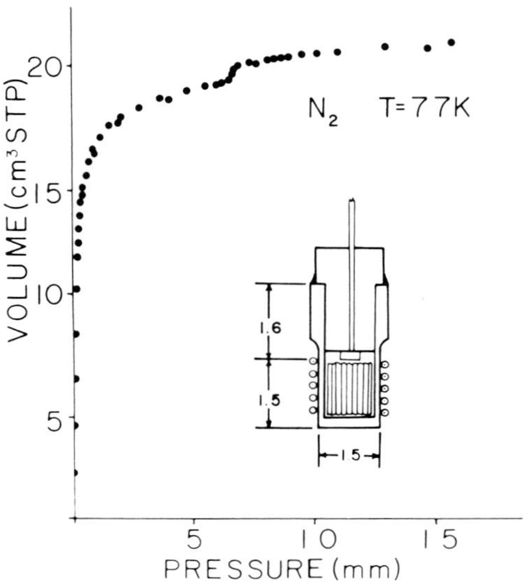

Fig. 1. Adsorption isotherm of nitrogen on grafoil. Inset shows cross-section of sample cell, dimensions in centimeters.

The ethylene sample was prepared from Matheson CP grade gas. To remove remaining impurities (oxygen is a particular concern) about 250 cm^3 STP of the gas was removed from the main storage container and put into a stainless steel storage can. The end of this can was immersed in LN_2 to freeze out the ethylene and the can was then pumped for ~60 sec., following which it was returned to room temperature. This procedure was repeated until the pressure above the frozen ethylene was less than one micron.

The purified ethylene was admitted to the sample cell with the cell at 160K. The sample was then annealed at this temperature for at least 6 hours and cooled to 77K over a period of 12 to 18 hours.

The NMR data were taken using a coherent pulsed spectrometer described elsewhere [3]. A 180° pulse length of 40 μsec. was obtained. Measurements of the spin-lattice relaxation were made with a 180° - τ - 90° spin inversion method, while the spin-spin interaction was measured using a 90° - τ - 180° spin echo. The highly anisotropic susceptibility of the graphite has the effect of making the magnetic field rather non-uniform [4]. In an effort to minimize this the sample was oriented with the Grafoil sheets parallel to the applied field.

DATA

Relaxation data at 24 MHz were taken as a function of temperature between 63K and 120 K at a coverage corresponding to registry. A small amount of data was also obtained at a coverage of 0.75 monolayers.

The T_1 data for the monolayer coverage is shown in fig. 2. As the temperature increases the relaxation time is seen to pass through a minimum at about 87K and immediately above this minimum to jump discontinuously to longer relaxation times at about 89K. This behavior was observed in two passages through the transition, as shown in the figure. Between these two data runs the sample was re-annealed.

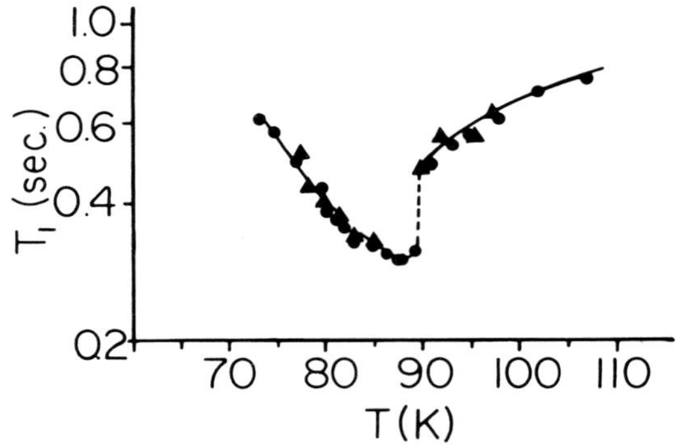

Fig. 2. Longitudinal relaxation time vs. temperature for monolayer coverage. Circles and triangles are for two different passes through the transition.

The recovery of the longitudinal magnetization was observed to be exponential at all temperatures except in the immediate vicinity of the discontinuity. There it seemed to follow a sum of two exponentials with relaxation times approximately equal to those just above and just below the discontinuity. The data for which the recovery was non-exponential are not shown in this figure.

The spin echo data showed a monotonic decrease with increasing pulse spacing at the highest temperatures, while at lower temperatures a modulation in echo amplitude was superimposed on the overall decrease. The period of the modulation decreased as the temperature decreased. Figure 3 shows several representative traces. The lower coverage data also showed this modulated echo envelope. Because of this behavior it was impossible to assign a value of T_2.

DISCUSSION

The most striking feature of the present data is the discontinuous jump in T_1 at about 89K. Ocuring as it does on the high temperature side of the T_1 minimum it indicates an increase in the mobility of the film. We interpret this as the melting of the monolayer solid. The fact that the change is discontinuous indicates an abrupt change in mobility and therefore an abrupt rather than gradual melting transition. A qualitatively similar discontinuity is observed for the 3-D solid when it melts at 104K [5]. In the vicinity of the discontinuity the apparent decay of the magnetization as the sum of 2 exponentials suggests the coexistence of both a solid and fluid phase, each with its distinct mobility, as would be expected for melting at constant area.

The modulation of the spin-echo amplitude is characteristic of a spectrum that is dominated by coupling between a small number of resonant nuclei, in this case presumably dipolar coupling of the four protons in the ethylene molecule. The modulation arises from the fact that the 180° pulse flips all of the interacting nuclei. Thus the component of the local field at a given nucleus contributed by its resonant neighbors is reversed, so that fast and slow nuclei interchange. The dephasing caused by this local field component is therefore not reversed by the 180° pulse and the echo amplitude is modulated as the Fourier transform of the absorption spectrum. A similar effect has been observed in liquid crystals [6].

In the present experiments the signal to noise ratio was not sufficient to allow a determination of the absorption spectrum by a Fourier transform of the echo amplitudes. However, some qualitative information can be obtained. In most of the traces below 86K it was possible to follow the echoes out far enough that 2 or in some cases 3 maxima were observed. In all cases the location of these maxima were found to be quite accurately given by a multiple of the location of the first maximum. This suggests a roughly double humped spectrum resulting in beats between the two dominant frequencies. The separation in time between the beats is equal to the inverse of the frequency splitting. To get a measure of the temperature dependence of this splitting the pulse spacing, τ, at which the first maximum in the echo amplitude occurs is plotted in fig. 4. A general increase with temperature is evident which we interpret as motional narrowing. At the temperature where T_1 shows a discontinuity there is also a sharp change in the time at which this first maximum occurs. The observed increase in time indicates a decrease in the width, as would be expected for an increase in mobility. At slightly higher temperatures it was no longer possible to find evidence of increased echo amplitude at longer τ and the decrease in echo amplitude at short τ was more gradual. This suggests that the double humped spectrum has merged into a single, somewhat broadened line.

The width of the splitting can be calculated from the spacing of the maxima in the echo amplitudes, noting that the echoes occur at a time $t = 2\tau$. The results range from 1.2 kHz (0.29 gauss) to 0.6 kHz (0.15 gauss) for the data in fig. 4 below the transition. Even at 63 K a width of only 2.3 kHz (0.54 gauss) was found. These values are considerably smaller than would be expected if

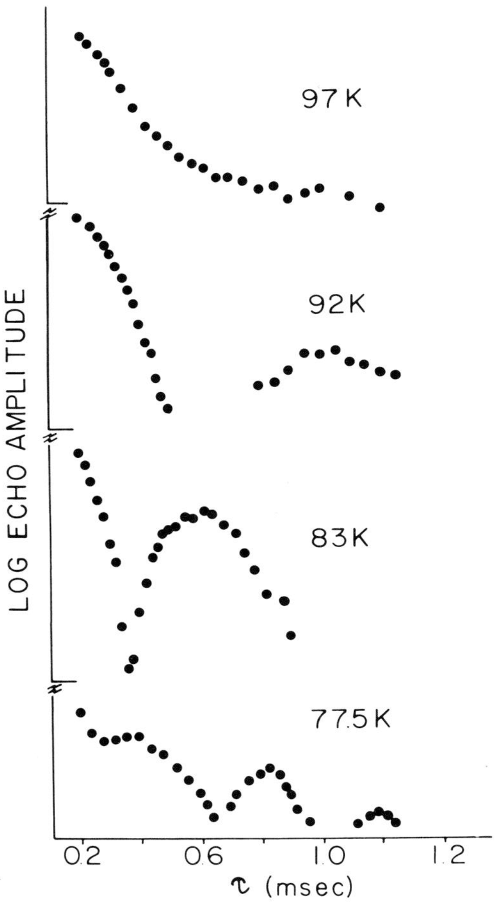

Fig. 3. Echo amplitude vs. pulse spacing, τ, for monolayer coverage.

the molecules were fixed rigidly. For two protons separated by a distance of 1.92 Å, corresponding to the protons at one end of the ethylene molecule, a rigid lattice splitting of $\Delta H \simeq 3\mu/r^3 \simeq 5.9$ gauss is calculated. The other protons in the same and surrounding molecules would increase this value (for bulk solid ethylene at 90K, 14K below melting, a value of about 8 gauss is observed [7]). Rotation of the molecule will notionally narrow the line. However, rotation about a single axis can no more than halve the width [8]. Thus the narrow splitting deduced for the adsorbed ethylene indicates rotation about more than one axis.

For the data of figures 2-4 the samples were carefully annealed, cooled to 77K, and the data taken as the sample temperature was stepped upward. The results of two such runs showed good reproducibility. One set of data, not shown in the figures, was taken after the sample had been

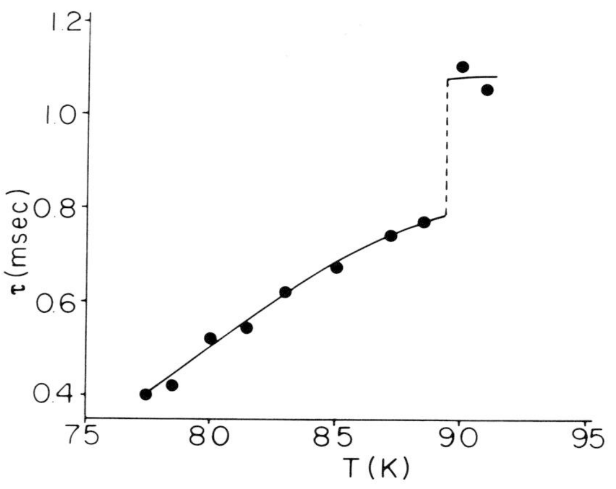

Fig. 4. Pulse spacing giving first maximum in echo amplitude vs. temperature. Note that the echo actually occurs at a time equal to 2 τ.

cold for a considerable length of time and a number of passes had been made up and down through the melting transition. While the same qualitative behavior was observed, including the T_1 discontinuity, the measured values of T_1 and the echo maxima separations were somewhat different. We conjecture that in passing through the melting transition strains may have developed which in time led to irreproducibility. Further study will be required to confirm this point.

CONCLUSION

Both longitudinal and transverse relaxation of adsorbed ethylene monolayers show that a discontinuous increase in mobility occurs at a temperature of 89K. We interpret this as the melting of the monolayer solid. Spin-echo measurements show evidence of a double-humped absorption spectrum with a splitting which decreases as the temperature is increased. Shortly above the melting transition the spectrum appears to merge into a single line. The width of the spectrum is an order of magnitude narrower than would be observed if the molecules in the solid monolayer were rigidly fixed, indicating substantial rotational motion. We are continuing measurements to lower temperatures in an effort to observe the orientational ordering of the ethylene monolayer.

REFERENCES

1. L. Passell, private communication.
2. D.M. Butler, G.B. Huff, R.W. Toth, and G.A. Stewart, Phys. Rev. Lett. 35, 1718 (1975); T.T. Chung and J.G. Dash, Surf. Sci. 66, 559 (1977).
3. R.F. Buzerak and R.J. Rollefson, J. Low Temp. Phys. 38, 105 (1980).
4. D.L. Husa, D.C. Hickernell, and J.E. Piott, in Monolayer and Submonolayer Helium Films, J.G. Daunt and E. Lerner, eds. (Plenum Press, New York, 1973), p. 133; D.C. Hickernell, D.L. Husa, J.G. Daunt, and J.E. Piott, J. Low Temp. Phys. 15, 29 (1974).
5. N.J. Trappeniers and F.A.S. Lighthart, Chem. Phys. Lett. 19, 465 (1973).
6. R.L. Vold and S.O. Chan, J. Chem. Phys. 53, 449 (1970).
7. H.S. Gutowsky, G.B. Kistiakowski, G.E. Pake, and E.M. Purcell, J. Chem. Phys. 17, 972 (1949).
8. A. Abragam, The Principles of Nuclear Magnetism (Oxford University Press, England, 1961).

Sinha, ed. Ordering in Two Dimensions

STUDY OF THE DYNAMICS OF HYDROCARBON FILMS ADSORBED ON GRAPHITE

R. WANG, H.R. DANNER and H. TAUB

University of Missouri-Columbia, Columbia, Missouri 65211, U.S.A.

ABSTRACT

Surface vibratory modes have been identified in the inelastic incoherent neutron spectra of monolayer films of butane (C_4H_{10}), pentane (C_5H_{12}) and hexane (C_6H_{14}) adsorbed on graphite. A new dynamic model with more reasonable molecule-surface force constants has been deduced for butane assuming the equilibrium orientation of the molecule is tilted 30° with respect to the graphite basal plane, rather than parallel as previously assumed. When this model is used to calculate the inelastic spectra of pentane and hexane monolayers, better agreement with observation is found for hexane than for pentane. Finally, a dramatic broadening of the surface vibratory bands of monolayer butane is observed above the 2D melting point at 116 K.

INTRODUCTION

In general, when a molecule is adsorbed on a solid surface, additional vibrational modes will appear resulting from hindered translational and rotational motion of the molecule against the surface. Inelastic neutron scattering was first used to observe surface vibratory modes of this type in experiments with ^{36}Ar monolayers adsorbed on Grafoil [1]. The results suggested the existence of a dispersionless Einstein mode in which the argon atoms vibrate normal to the surface. Similar modes were later identified in the inelastic neutron spectra of adsorbed monolayers of H_2 and D_2 [2]. A review of other studies of the vibrational spectra of adsorbed molecules by inelastic neutron scattering can be found elsewhere [3].

Probably the clearest evidence for a surface vibratory mode involving hindered rotational motion was obtained with butane (C_4H_{10}) adsorbed on a graphitized carbon powder [4]. An intense peak in the inelastic neutron spectrum was assigned to a libration of the molecule about its long axis aligned parallel to the surface. Another broader band at lower frequency was attributed to two unresolved surface vibratory modes: one a vibrational mode normal to the surface and the other a libration of the molecule about its short axis parallel to the surface. These assignments were made on the basis of calculated spectra for the adsorbed molecule using a simple force constant model. The calculations also suggested that the butane molecule adsorbed with the plane of its carbon skeleton parallel to the substrate surface.

In this paper we report further studies of the dynamics of n-alkane molecules adsorbed on graphite. We have observed the temperature dependence of the inelastic neutron spectrum from a butane monolayer and recalculated the vibrational spectrum for the adsorbed molecule. This work was motivated by recent elastic neutron diffraction experiments with deuterated butane (C_4D_{10}) adsorbed on Grafoil [5] which showed an abrupt melting of the submonolayer at 116 K. It was therefore of interest to determine the effect of melting on the surface vibratory modes. Also the structure deduced from an analysis of the low temperature diffraction pattern indicated a small tilting of the butane carbon skeleton with respect to the graphite basal plane. This result suggested new model calculations to determine whether the fit to the observed spectrum could be improved by assuming a tilted orientation for the molecule.

We have also measured the inelastic neutron spectra of monolayers of pentane (C_5H_{12}) and hexane (C_6H_{14}) adsorbed on the same substrate. These longer molecules directly follow butane in the n-alkane series. Because of their structural similarity, it was thought that the dynamical model developed for adsorbed butane might be applicable to adsorbed monolayers of these molecules as well. The paper concludes with the results of such an analysis.

EXPERIMENT

The substrate used in these experiments is a graphitized carbon powder known as Carbopack B [6]. It was selected over Grofoil [1] because of its large surface area ($\sim$80 m^2/g versus $\sim$30 m^2/g) and yet high homogeneity. The particles have no preferred orientation and are $\sim$50 Å thick along the graphite c-direction. The random particle orientation simplifies the calculation of the vibrational spectra, since no additional parameters are needed to describe the particle orientation distribution function.

The aluminum sample cell was in the shape of a circular plate 15 cm in diameter and 1 cm deep. A complete description of the sample and handling techniques is contained in Ref.4.

In order to determine the coverage corresponding to monolayer completion, vapor pressure isotherm measurements were performed at room temperature with both pentane and hexane, and compared with the butane isotherm previously obtained at 273 K. The systematic decrease observed in the adsorbate volume required for monolayer completion suggests that all three of the molecules adsorb with their long axes parallel to the surface.

All of the inelastic neutron spectra reported here were obtained on the beryllium-filter, time-of-flight spectrometer at the University of Missouri Research Reactor. The spectrometer, previously described [3], operates in a neutron energy-gain mode with a scattering angle of 27.4° selected to minimize Doppler broadening of the vibrational bands and the multiphonon contribution to the spectra. The neutron energy transfer ΔE varied from 6 mev (48 cm^{-1}) to 40 mev (320 cm^{-1}) with a resolution of 10 cm^{-1}. The momentum transfer $\vec{Q}$ varied from 1 $\mathring{A}^{-1}$ to 3.5 $\mathring{A}^{-1}$ in this energy range.

RESULTS

Figure 1 shows the inelastic spectrum of a butane monolayer adsorbed on Carbopack B measured at several temperatures. Except for an increase in the thermal population of the higher frequency modes, the spectrum at 98 K is essentially identical to that measured at 77 K. These two low-temperature spectra reproduce very well the butane film spectra previously measured with the same cell [4].

Our initial analysis [4] identified the peaks, labeled A and B, as the intramolecular torsions of the end-methyl groups and peak C as the internal CH_2-CH_2 torsion. Band D at 112 cm^{-1} is not observed in the spectrum of the bulk solid and was assigned to a surface vibratory mode in which the adsorbed butane molecule librates about its long axis aligned parallel to the graphite basal plane. Two additional surface vibratory modes - one (E), a libration of the butane molecule about its short axis parallel to the surface, and the other (F), a uniform vibration of the entire molecule normal to the surface - are unresolved in the broad band near 50 cm^{-1}.

The above assignments were based on a spectrum calculated assuming the butane molecule was oriented with the plane of its carbon skeleton parallel to graphite basal plane. We have now recalculated the vibrational spectrum of adsorbed butane assuming the carbon skeletal plane is tilted 30° with respect to the surface as suggested by elastic diffraction studies of deuterated butane monolayers [5]. As before, the model incorporates a set of seven fixed intramolecular force constants [7], and simulates the presence of the surface by introducing additional force constants between atoms in the molecule and infintely massive atoms located in the surface plane. Butane-butane interactions are neglected. The molecule-substrate force constants are the parameters adjusted to give the best agreement with the frequencies and relative intensities of the observed modes.

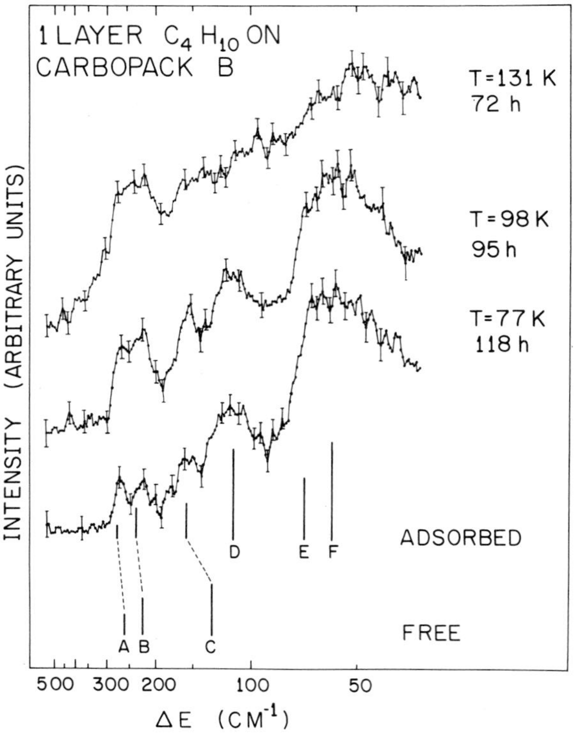

Fig. 1. The neutron inelastic spectrum of a butane ($CH_3(CH_2)_2CH_3$) monolayer adsorbed on Carbopack B for three temperatures. In each case the background inelastic scattering from the substrate has been subtracted. The vertical bars labeled ADSORBED represent the monolayer spectrum calculated from the simple force constant model described in the text. The spectrum calculated for the intramolecular modes of the free molecule is shown for comparison.

The vertical bars, labeled ADSORBED in Fig. 1, show the recalculated frequencies and relative intensities of the intramolecular torsions and surface vibratory modes at 98 K. Due to the difficulty in estimating the multiphonon contribution to the inelastic spectrum, it was not believed worthwhile to perform a quantitative fit to the mode intensities by folding the instrumental resolution with this spectrum. Nevertheless, we find that the agreement with the mode frequencies and, at least qualitatively, the intensities is as good as obtained previously [4] assuming the parallel equilibrium orientation. The only difference is that the present calculation reverses the order of the two lowest surface vibratory modes, i.e., the "bouncing" mode now has the lowest frequency.

Although the two calculations describe the observed spectrum equally well, we believe that the present calculation represents an improvement in that more resonable values of the force constants are obtained. In our initial model only the four coplanar hydrogen atoms (one from each CH_3 and CH_2 group) nearest the surface were bound; to reproduce the principal features of the spectrum it was necessary to assume a larger force constant between the CH_2 hydrogens and the surface (k_2 = 0.065 mdyne/Å) than for the CH_3 hydrogens (k_3 = 0.02 mdyne/Å) even though both were thought to be at nearly equivalent adsorption sites. Moreover, the fact that any binding, however weak, of the carbon atoms to the surface worsened the agreement seemed unreasonable. The present model incorporates three film-to-surface force constants binding the carbon atoms (k'), the "nearest" hydrogen atoms (k'') and the "next nearest" hydrogen atoms (k'''), respectively. The best fit, represented by the vertical bars in Fig. 2, was obtained with k' = 0.003 mdyne/Å, k'' = 0.04 mdyne/Å and k''' = 0.002 mdyne/Å. A more complete description of this model will be published later [8]. The fact that all the hydrogen atoms at the same height above the surface have the same force constant and the value of this constant diminishes with increasing distance from the surface would appear to be an improvement over the original model.

The surface vibratory modes observed at the two lower temperatures are dramatically broadened in the spectrum at 131 K. The methyl torsions are broadened to a lesser extent although the two modes are no longer resolved. We attribute these effects to the 2D melting transition which is observed to occur abruptly at 116 K in the elastic diffraction experiments [5]. Since the bulk melting point of butane is at 135 K, the broadening of the surface vibratory modes presumably results from a 2D liquid state.

As a further test of our dynamical model, we measured the inelastic spectrum (shown in Fig. 2) of a monolayer of hexane (C_6H_{14}) adsorbed on Carbopack B at 98 K. Although hexane has two additional CH_2 groups it has the same molecular symmetry as butane. Therefore, it seemed resonable to expect hexane to adsorb on the graphite basal plane in the same orientation as butane and that the same force constant model could then be applied.

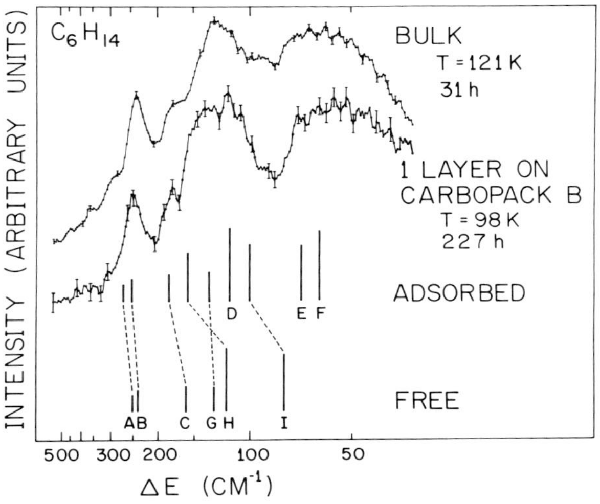

Fig. 2. The neutron inelastic spectrum of a hexane ($CH_3(CH_2)_4CH_3$) monolayer adsorbed on Carbopack B at 98 K. The background scattering from the substrate has been subtracted. The calculated spectrum labeled ADSORBED is based on the same molecular orientation and force constant model used for the butane monolayer. D, E, and F are the surface vibratory modes. The spectrum calculated for the free molecule and the spectrum of solid bulk hexane at 121 K are shown for comparison.

The low frequency spectrum of the hexane monolayer is more complicated than that of butane due to contributions from three (rather than one) intramolecular CH_2-CH_2 torsion modes and a low frequency bending mode. The vertical bars, labeled ADSORBED in Fig. 2, show the calculated frequencies and relative intensities of the six lowest intramolecular modes and the three surface vibratory modes using the same molecular orientation and force constants as for butane. The agreement with the observed spectrum is encouraging. The calculated frequencies of the end-methyl torsions (A,B) are separated by only 11 cm^{-1} and are unresolved in both the film and bulk spectra, while the highest frequency internal torsion (C) is better resolved in the film spectrum. The remaining two internal torsions (H,I) and the bending mode (G) all contribute to a broad band near 135 cm^{-1}. Superimposed on this band is a well-defined peak (D) at the frequency calculated for the most intense surface vibratory mode--the libration of the molecule about its long axis parallel to the surface. As in the case of butane, the two surface vibratory modes appear unresolved in a broad band below 70 cm^{-1}.

We have also measured the inelastic spectrum of a pentane monolayer under the same experimental conditions. However, the agreement with the vibrational spectrum calculated from our model is not as good as for butane and hexane. It is possible that the lower molecular symmetry of pentane favors a different orientation of the adsorbed molecule. Consequently, we have postponed any further analysis of pentane until after elastic diffraction experiments have been performed to determine the monolayer structure.

SUMMARY AND CONCLUSION

The inelastic neutron spectra of monolayers of butane, pentane, and hexane adsorbed on a graphitized carbon substrate all show evidence of surface vibratory modes associated with hindered translation and rotation. The vibrational spectrum of adsorbed butane has been recalculated assuming the plane of its carbon skeleton tilted at an angle of 30° to the surface as supported by recent elastic diffraction experiments. The fit between calculated and observed spectra is as good as that obtained previously with the carbon skeletal plane assumed parallel to the surface. Finally, the surface vibratory modes of the butane monolayer have been observed to broaden dramatically above the 2D melting point.

ACKNOWLEDGEMENT

We acknowledge partial support of this research by the National Science Foundation under Grant No. DMR-7905958, the Petroleum Research Fund of the American Chemical Society, a Dow Chemical Company Grant of the Research Corporation, the University of Missouri Research Council, and the University of Missouri Research Reactor.

REFERENCES

1. H. Taub, K. Carneiro, J.K. Kjems, L. Passell and J.P. McTague, Phys. Rev. B 16, 4551 (1977).
2. M. Nielsen, J.P. McTague and W.D. Ellenson, J. Phys. (Paris) C-4, 10 (1977).
3. H. Taub, Am. Chem. Soc. Symposium Series (in press).
4. H. Taub, H.R. Danner, Y.P. Sharma, H.L. McMurry and R.M. Brugger, Phys. Rev. Letters 39, 215 (1977); Surf. Sci. 76 50 (1978)
5. G.J. Trott, H. Taub, F.Y. Hansen and H.R. Danner, to be published.
6. Carbopack B is the trade name of a graphitized carbon powder manufactured by Supelco, Inc., Bellefonte, PA 16823.
7. K.W. Logan, H.R. Danner, J.D. Gault and H. Kim, J. Chem. Phys. 59, 2305 (1973).
8. R. Wang, H.R. Danner and H. Taub, to be published.

Sinha, ed. Ordering in Two Dimensions

NEUTRON DIFFRACTION STUDY OF BENZENE AND TOLUENE ON GRAFOIL

M. MONKENBUSCH and R. STOCKMEYER

Institut für Festkörperforschung, Kernforschungsanlage Jülich,
5170 Jülich, West Germany

ABSTRACT

Benzene and toluene layers adsorbed on Grafoil have been investigated in the temperature region between 40 K and 270 K by thermal neutron diffraction (λ_o = 4.78 Å). The coverage was varied between 0.5 and 1.5 BET monolayers. Below a coverage dependent transition temperature T_{tr} in the region of 150 K benzene forms a registered $\sqrt{7} \times \sqrt{7}$ triangular lattice with the molecules "lying" parallel to the surface. Above T_{tr} a highly compressible 2D fluid has been observed. Toluene does not transform directly from a registered 2D solid to a fluid, but forms an incommensurable lattice with high thermal expansion in the intermediate region. Below 70 K (Θ = 0.9) it forms a 3 x 3 registered phase, and above 150 K it is fluid.

INTRODUCTION

In the recent years neutron scattering techniques have been used to investigate adsorbed layers |1,2|. Even though neutrons are not a surface sensitive probe because of their weak interaction with matter, adsorbed layers can be observed if a reasonable large part of the sample volume consists of adsorbed layers, and if the scattering length of the layer is not too small.

The combination graphite (= substrate) and hydrocarbons (= adlayer) is a choice which fits these requirements because there are commercially available graphites with high homogeneous (mainly 0001 planes exposed for adsorption) surfaces with the surfaces partially oriented (Grafoil) |3|. Fortunately the graphite basal planes exhibit only a weak lateral variation of the surface-potential for physisorbed species, thus being a suitable substrate to model nearly 2D systems, and to study their physical properties. Hydrocarbons as adlayers offer the possibility to choose between deuterated and hydrogeneous samples showing mainly coherent or incoherent scattering respectively. The substitution of D by H in diffraction work gives valuable information because of the different scattering lengths (H = -3.74 fm, D = 6.674 fm). Theoretical works profit from the availability of reasonable empirical interaction potentials between hydrocarbons and graphite |4| and between two hydrocarbon molecules |5,6|. In this paper we present the first diffraction measurement of benzene (C_6D_6, C_6H_6) and toluene (C_6D_5-CD_3, C_6H_5-CH_3) layers in the monolayer regime on Grafoil.

EXPERIMENTAL

All our measurements were done at the multiple (64)-detector TOF instrument SV5c at the FRJ2. The energy resolution has been relaxed to $\Delta\hbar\omega$ = 1.5 meV at ω = 0 by using a very rough chopper, installed to get rid of problems with higher order contamination. The incoming neutron wavelength was λ_o = 4.78 Å, the 64 detectors cover the angular region of 21° < Θ < 159°; (0.5 Å^{-1} < Q < 2.6 Å^{-1}).

The sample consists of 44 g Grafoil sheets (Ø = 25 m) stacked in a thinwalled stainless steel can. Every 5 mm a gadolinium plate 0.1 mm thick has been inserted into the stack to reduce multiple scattering, which is mainly caused by double (002) graphite Bragg diffraction. The preferential orientation of the basal planes is parallel to the scattering plane. The temperature is controlled with an accuracy of ± 1 K by two heaters on the ends of an Al tube surrounding the steel can, and the whole system is cooled by a closed cycle cryostat with rough preregulation of the temperature. We measured the diffraction patterns from the graphite samples with and without adlayer at temperatures T_v = 40 K + v·Δt between 40 K and 250 K with temperature intervals ΔT between 5 K and 10 K. The BET monolayer capacity of our sample has been determined at 21° C to be 0.48 cm^3 liquid benzene and 0.43 cm^3 liquid toluene respectively. In the following these values are referred to as coverage Θ = 1. Measurements were done with Θ = 0.5, 1, 1.5 for benzene and Θ = 0.9 for toluene.

OBSERVATIONS

All data shown in the following are gained by subtracting intensities from the degassed sample from those of the loaded sample, after normalising to the incident neutron flux and correcting each data-file for self-absorption.
At the lowest temperatures all diffraction patterns exhibit sharp sawtooth shaped peaks as should be expected from an ordered 2D solid phase. At the highest temperatures all diffraction patterns show a very broad liquidlike structure peak.
For benzene the transition between these two cases is sharp within our 10 K step. Fig. 1 shows a comparison of the high temperature to the low temperature pattern for $\theta = 0.8$. For $\theta = 0.5$ and 1.5 the character of the pattern is the same, the transition temperature rises slightly with θ and the liquidlike structure peak shifts to higher scattering angles with increasing θ.

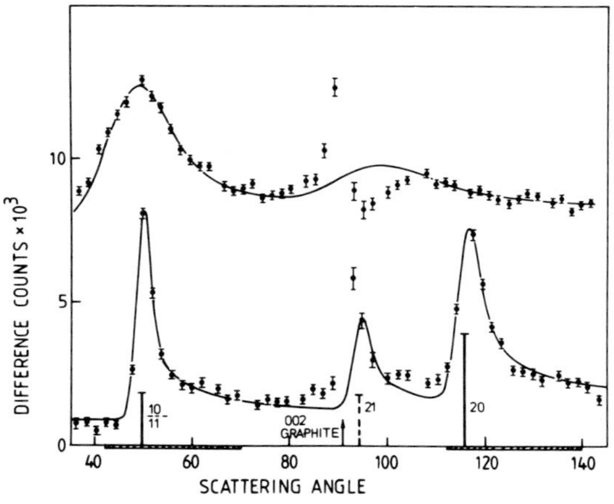

Fig. 1:
Diffraction pattern from a C_6D_6 layer on Grafoil with coverage $\theta = 0.8$. Upper data from the fluid phase is the sum of all data taken in the temperature interval 150 K $\leqslant$ T $\leqslant$ 240 K. Lower data from the solid phase summed over the temperature interval 40 K $\leqslant$ T $\leqslant$ 140 K. There is no significant change of the diffraction pattern within these temperature intervals. The region around the graphite 002 reflection suffers from interference effects and large systematical errors.

Toluene exhibits an additional phase between the two extremes where the peak (only one intense peak observed) shifts to lower scattering angles and becomes slightly broader, as illustrated by Fig. 2. Coherent elastic ($\Delta\hbar\omega = 0.1$ meV) scattering from the hydrogeneous samples yields only the first peak for benzene and besides the first peak for toluene the beginning of a peak at about 160° unobservable in the former measurement because of instrumental reasons.

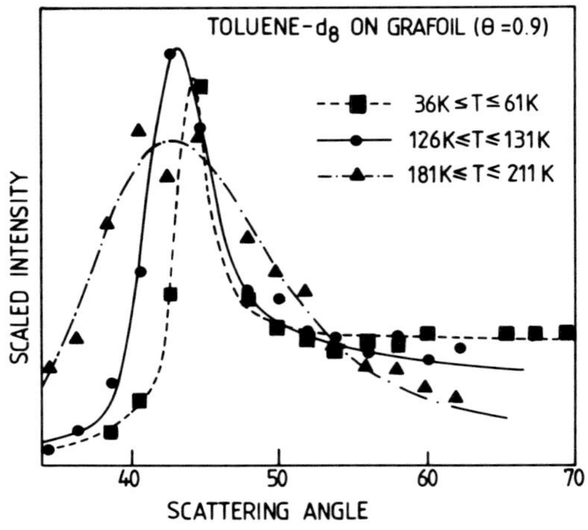

Fig. 2:
Diffraction patterns from a C_6D_5-CD_3 layer ($\theta = 0.9$) on Grafoil showing the peak forms and positions for the registered solid, incommensurate solid and fluid phase respectively.

DATA REDUCTION

To get more precise information about the different phases and the transitions between them we fitted the observed peaks to a theoretical lineshape containing the mean size of coherently ordered regions L in the adlayer and the true position Q_{hk} of the Bragg reflection as parameters. The scattered intensity from a plane lattice consists in the 3D reciprocal space of a series of rods perpendicular to the plane of the lattice, intersecting it at the points of the 2D reciprocal lattice. A finite size L of the coherent regions results in a broadening of these rods approximately describable by |7|:

$$I_{hk}(\vec{Q}) = N^2 \exp\left(-\left(L(\vec{Q}_{hk}-\vec{Q}_{||})/2\sqrt{\pi}\right)^2\right) \quad (1)$$

$\vec{Q}_{||}$ is the momentum transfer parallel to the lattice plane. The observed intensity from (hk) is then

$$I_{hk}^{Det} = \frac{1}{Q^2} \int_S I_{hk}(\vec{Q}) F(\vec{Q}) \hat{w}(\theta)\, do \quad (2)$$

S = sphere with radius Q around the origin

$F(\vec{Q})$ = structure factor of the unit cell
θ = angle between $\vec{Q}$ and the direction normal to the planes
$\hat{w}(\theta)$ = angular distribution function

The function $\hat{w}(\theta)$ depends on the actual angular distribution $w(\theta)$ of the normals of the basal planes in the substrate $w(\theta)$ as

$$\hat{w}(\theta - \pi/2) = \int_{o}^{2\pi} w(\arccos(\cos\theta \cos\psi))\, d\psi \quad (3)$$

We approximated $\hat{w}(\theta)$ by

$$\hat{w}(\theta) = p_1 + p_2 \exp(-(\theta - \pi/2)^2/\sigma^2) \quad (4)$$

with $p_2/p_1 = 1.3$ and $\sigma = 0.23$.

These values were determined by fitting the solid curve in Fig. 1 for the low T data and then kept fixed for the rest of the work.

In Fig. 3 the results of fitting simultaneously the lineshapes of the 1st and 3rd peak in the regions indicated by the hatched bars in Fig. 1 for benzene are shown. The size of coherent regions L is in the order of 100 Å a value typical for the substrate Grafoil. For high coverage L is slightly larger than for low Θ. At a Θ-dependent temperature between 120 K and 160 K L decreases abruptly (within our stepwidth of 10 K) to values below 20 Å.

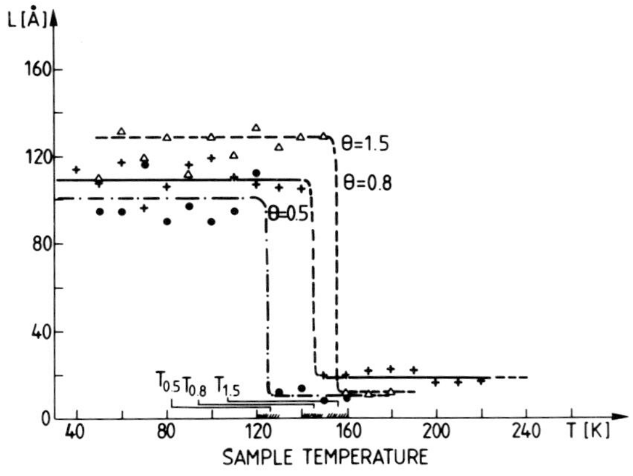

Fig. 3:
Temperature variation of the size of coherent structure regions for C_6D_6 on Grafoil. Phase transitions occur at temperature $T_{0.5}$, $T_{0.8}$ and $T_{1.5}$ for the corresponding coverages.

Up to this transition temperature T_{tr} the position of the peak is constant and Θ-independent, while above T_{tr} the position of the peak shifts from $\theta = 36^o$ for $\Theta = 0.5$ to $\theta = 54^o$ for $\Theta = 1.5$.

The corresponding data for toluene are presented in Fig. 4. One observes two abrupt changes in L at 70 K and 140 K. Below 70 K the position of the peak is constant at 70 K there is a jump, and at 100 K it starts to decrease continuously, the jump of L at 140 K has no corresponding feature in the peak position which can be distinguished from statistical scatter of the points.

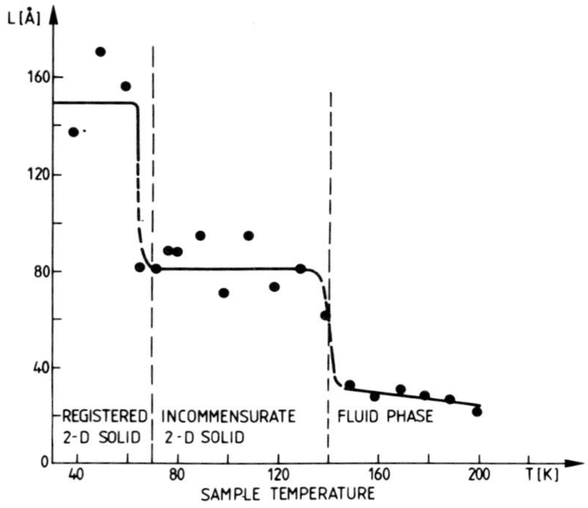

Fig. 4:
Temperature variation of the size of coherent structure regions for C_6D_5-CD_3 on Grafoil ($\Theta = 0.9$). Phase transitions from 3x3 registered solid to a incommensurate solid at 70 K and from incommensurate solid to fluid at 140 K reveal themselves by jumps of L.

CONCLUSIONS

From the temperature and coverage independence of the peak positions for the low temperature phase of benzene on Grafoil we conclude that this phase consists of a registered lattice. The absolute values of the peak positions are in accordance with the assumption of a $\sqrt{7} \times \sqrt{7}$ triangular lattice within the experimental error. In Fig. 5 we show the structure we propose for this phase. The solid line through the data points is calculated with a structure factor of the proposed structure. The elastic coherent scattering from a C_6H_6 layer is also in accordance with this model. While the potential energy of the layer as calculated with potentials from |4,5| has a minimum for this model, the relative position on the graphite surface cannot be deduced from the data and has been taken from the energy minimalization. The distance between the surface and the adlayer is 3.2 Å.

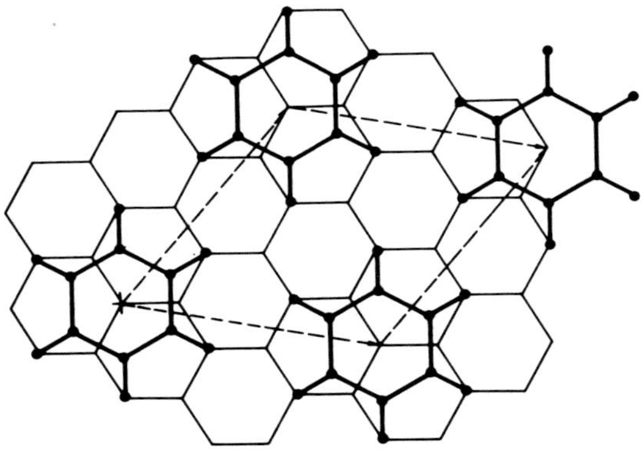

Fig. 5:
Structure model for benzene on the basal planes of graphite. The layer forms a $\sqrt{7} \times \sqrt{7}$ superlattice, the distance from the surface is 3.2 Å (A_{Molec} = 36.7 Å^2).

A model assuming the benzene ring standing on the surface |8| can be discarded because it cannot account for the intense 3rd peak in the C_6D_6 data.

The phase transition between 140 K and 150 K for $\Theta = 0.8$ seems to lead directly from a registered solid to a fluid phase.

The fluid phase is highly compressible as can be deduced from the Θ-dependence of the position of the structure peak.

Toluene on Grafoil ($\Theta = 0.9$) shows three different phases. The first phase below 70 K exhibits the same properties as the benzene low temperature phase and can thus be attributed to a 3x3 registered lattice. The second phase between 70 K and 140 K is an incommensurate solid, the apparent size of the coherent region has dropped to about 80 Å. This may be due to modulations in the layer as expected for incommensurate structures. At 140 K the size L drops to below 40 Å. This phase seems to be a fluid phase similar to that observed for benzene.

Inelastic incoherent scattering experiments which will be reported in detail elsewhere give additional information on the different phases:

a) The fluids are highly mobile (comparable or higher than the corresponding bulk liquids at ambient temperature).
b) The registered solids exhibit phonon-like scattering as expected from 3D systems.
c) The incommensurate phase of toluene shows a scattering near $\omega = 0$ which can be attributed to a 2D Debye solid. Because of its incommensurability, this phase is quasi-free and is a very good model for a real two dimensional system.

Acknowledgement

The authors would like to thank Prof. H. Stiller for helpful discussions and R. Wagner for his assistance with the experiments.

REFERENCES

1. J.P. Mc Tague, M. Nielsen and L. Passell, CRC Critical Reviews in Solid State and Material Sciences 8(2), 135 (1978).
2. J.W. White in Topics in Current Physics 3, Dynamics of Solids and Liquids by Neutron Scattering, Ed. SW. Lovesey, T. Springer, Springer Verlag, Berlin (1977).
3. Grafoil is a product of Union Carbide.
4. L. Battezzati, C. Pisani and F. Ricca, J. Chem. Soc. Faraday Trans. II 71, 1629 (1975).
5. G. Taddei, H. Bonadeo, M.P. Marzocki and S. Califano, J. Chem. Phys. 58, 966 (1973).
6. D.E. Williams, J. Chem. Phys. 45, 3770 (1966).
7. B.E. Warren, Phys. Rev. 59, 693 (1941).
8. B. Boddenberg, J.A. Moreno, Journal de Physique 38, C4-52 (1977).

Published 1980 by Elsevier North Holland, Inc.
Sinha, ed. Ordering in Two Dimensions

MÖSSBAUER CROSS SECTION ISOTHERMS OF ADSORBED FILMS

H. Shechter and R. Brener

Physics and Solid State Institute, Technion I.I.T., Haifa, Israel

J. Suzanne

Départment de Physique, Université d'Aix-Marseille II, Luminy, Marseille
France

ABSTRACT

Mössbauer spectroscopy can be used to study adsorbed phase boundaries through their dynamical properties. It is particularly useful for cases where there is no measurable pressure in equilibrium with the adsorbed phases, in the interesting T-range. One can obtain information on the 2D phases by plotting experimental resonance cross section for $\vec{k}_\gamma$ parallel (σ'') and perpendicular ($\sigma^{\perp}$) to the film surface, for various temperatures and coverages. (σ''; x; T) surface is represented to demonstrate the method and possible deviations from this model are discussed.

INTRODUCTION

It has been known for a long time that adsorption vapor pressure isotherms resemble strikingly the vapor pressure isotherms of bulk matter [1,]]. Even though many other methods are employed today in the attempts to understand the 2D world, still much can be learned from these isotherms. There are longstanding debates concerning the order of transitions in the two dimensional phases. Depending on the theoretical model, the melting can be a sharp (first order) or continuous (higher order) transition [3-5]. It has been extensively studied experimentally with numerous adsorbates using graphite as a substrate [6,7,8]. The results indicate that solid-fluid transitions are more or less abrupt depending on the system. When the transition is continuous, one does not know, from most of the above techniques if it is due to intrinsic properties of the system (film) or a broadening produced by surface heterogeneities or size effects.

Until recently it was believed that first order melting would occur in registered layers only. Our results on butadiene iron tricarbonyl [9] and tetramethyltin [10] and other studies [7] have shown that this rule does not hold because coexistence was observed between a 2D liquid and a 2D incommensurate solid.

The phase boundaries of adsorbed films are still interesting, and conventional adsorption isotherms are employed. In most of the adsorbed gases studied by isotherms (mostly rare, and few light molecules like N_2, CH_4, NO, etc.), the auxiliary experiment was done within the existence temperature of the vapor pressure isotherms. For some heavier molecules, however, the situation is different: For the BIT [9] and TMT [10] the bulk T_r-pressure ≃ 1 torr and 92 torr, respectively. The TMT bulk melts at 218 K and below this T the vapor pressure is barely measurable. Its 2D phases, however, are expected to be well below ∿ 150 K. Thus, the isotherm technique fails. The surface behavior of heavy molecules (probably unregistered) is interesting. Besides its technological importance, one would not expect any positional order if the substrate mesh is fine relative to the molecular size.

We summarize in this report the introduction of the Mössbauer spectroscopy to this problem. We describe how some of the dynamical properties of these phases may affect the Mössbauer resonance cross section. A relation between a "cross section-coverage-temperature surface" and a possible phase diagram is suggested and some recent experimental results are mentioned.

MÖSSBAUER ISOTHERMS

The Mössbauer effect was used in the past to study diffusion in bulk glasses and supercooled liquids [11]. In typical liquids the technique fails due to the high diffusion rate. It was demonstrated, however, that for adsorbed films there is still an observable effect [9]: In same cases the Mössbauer f-factor in a direction perpendicular to the surface is larger than that in bulk. When aligned substrates are used, such as Grafoil or Papyex, the molecule can be followed in the diffusion region (2D-fluid) when observed in direction other than parallel to the film plane.

For low temperatures the mobility of the adsorbate is negligible. It forms either a registered or an unregistered solid which is characterized by many factors and the surface density as well. Each atom (molecule) experiences a mean square displacement (MSD) and one can characterize it by an Einstein frequency in the perpendicular direction and by a 2D-Debye model for the lateral direction. In both cases, the recoil free fractions ($f_\perp$ and $f_{\parallel}$) are not too small, $f_\perp > f_{\parallel}$ and there is a measurable effect [12,13], which yields through this dynamical state the desired information on the 2D phases. For elevated temperatures, surface diffusion may start.

The effect of diffusion on the Mössbauer cross section was first treated theoretically by Singwi and Sjölander two decades ago [14] where they adopted the concepts used in the case of quasi-elastic neutrons in bulks. Later the theory was extended to the 2D case [9]. A solution for the 2D diffusion equation was represented and substituted in the general expression of the resonance cross section of the Mössbauer atom in a classical liquid [14]: For perfectly aligned absorbers one find that the on-resonance cross section for the 2D case depends on the direction of $\vec{k}_\gamma$ relative to the film:

$$\sigma_{res}(\theta,T) = \frac{\sigma_o\Gamma}{4h}\, e^{-\hbar^2k^2/8mk_BT}\, F(D) \qquad (1)$$

$$F(D) \equiv (1 + 2\hbar Dk^2\sin^2\theta/\Gamma)^{-1/2}$$

where θ is the angle between $\vec{k}_\gamma$ and the normal to the plane, D is the coefficient of diffusion, Γ is the natural width of the Mössbauer level. For Grafoil with its partial alignment ($\sim$ 50% random, 50% narrow peaked oriented crystals [15], one obtains

$$\overline{\Delta\sigma}^{\perp}/\sigma_o = 1 - (1/\sqrt{8\hbar Dk^2/\Gamma})(\pi/2 + 1/\delta\theta)$$
$$\overline{\Delta\sigma}''/\sigma_o = 1 - 1/\sqrt{8\hbar Dk^2/\Gamma} - \pi/\sqrt{32\hbar Dk^2/\Gamma} \qquad (2)$$

where $\overline{\Delta\sigma} \equiv \sigma\,(2\hbar Dk^2/\Gamma<<1) - \sigma\,(2\hbar Dk^2/\Gamma>>1)$ is the average reduction in the Mössbauer intensity. $\delta\theta$ is the angle of the sample-irradiation set by the experimental conditions and is, in our case, 0.05. $\overline{\Delta\sigma}^{\perp}$ and $\overline{\Delta\sigma}''$ are, respectively, for θ = 0° and 90°.

Figure 1 illustrates the dependence of $\Delta\sigma''/\sigma_o$ on D, for Fe-57, Sn-119 and Kr. Notice that for relatively slow diffusion rates, D $\sim 10^{-6}$ or slower, one can distinguish between parallel and perpendicular directions.

One can construct $\sigma''_T(x)$ (Mössbauer isotherms) guided by the following considerations: Suppose we reduce the surface density. The remaining molecules can either resettle in the whole available area, or form clusters. If the T is low enough, the 2D solid occupies the entire "space" and by reducing the coverage x, the MSD is expected to increase. σ" will follow a concaved line (Fig. 2). If by further reduction of x the isotherm enters into a coexistence region of two phases, σ" will follow a straight line ending at σ" = 0 for x = 0. The isotherms in Fig. 2 demonstrate a simple case. Practically $\sigma''_x(T)$ is more conveniently measured and one can construct then the isotherms. Figure 2 then illustrates how (σ",x,T) reflects thephase diagram.

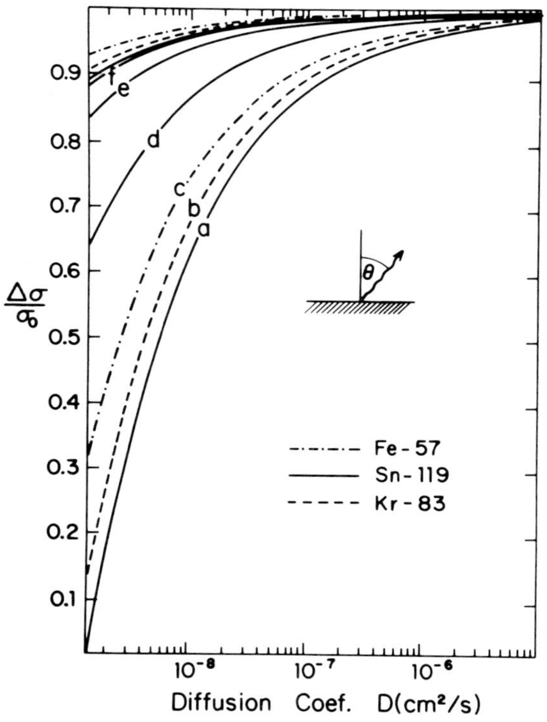

Fig. 1 The relative average change in the spectral line intensity $\Delta\overline{\sigma}''/\sigma_o$ as a function of the diffusion coefficient D, for Fe-57, Sn-119 and Kr-83; a) θ = 0; b) 10°; c) 30°; d) 60°; e) 90°.

Experimentally, a stack of $\sim$ 10 g of cleaned Grafoil sheets were located in a cryo-oven cell with through-mylar windows to allow $\vec{k}_\gamma$ to pass perpendicular or parallel to the sheets. The Grafoil is then loaded following adsorption vapor pressure isotherm, to the appropriate coverage. Then the spectra are recorded vs. temperature. The coverage is then changed. $\sigma''_T(x)$ is plotted after $\sigma''_x(T)$ Unfortunately, except from other difficulties these experiments are limited to materials which contain a Mössbauer isotope as a constituent [9,12,13].

Such is the case of the tetramethyltin

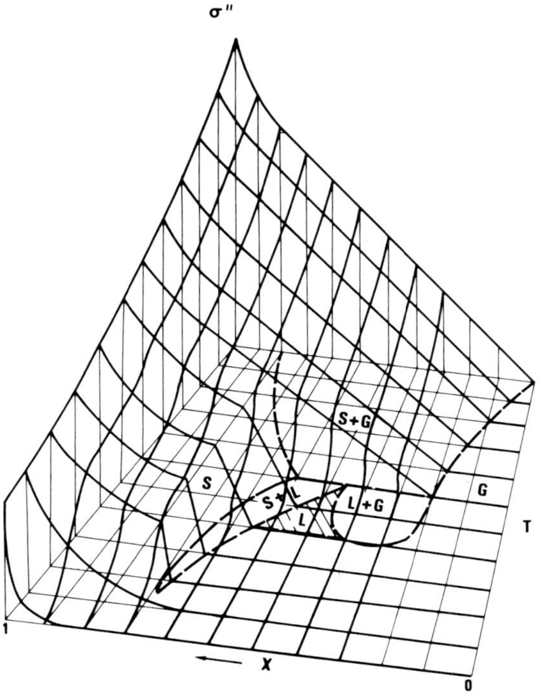

Fig. 2 A schematic view of a (σ''; x; T) surface. x is the coverage and σ, the cross section for θ = 90° direction.

$Sn(CH_3)_4$ adsorbed on Grafoil [16]. The spherical molecule has a diameter of ∿ 7 Å; it cannot form a commensurate solid on graphite (0001) plane. Preliminary X-ray measurements suggest that below 85 K the patches are close to the substrate (∿ 5 Å).

A first order "melting" is observed at 85 K (there is no gradual increase in the line width). One would, however, expect a melting at ∿ $T_m(3D)/2$ ∿ 109 K and indeed a second melting is observed above 100 K.

A break in $\sigma''_T(x)$ is noticed around x ≅ 0.67 for temperatures below 85 K; $\sigma''_T(x)$ increases monotonically with x and at this coverage there is a regular increase from the new value. In this T-range one expects a solid for higher coverages and S + G at intermediate surface densities. The transition from the first region to the second is continuous; such an irregular change in $\sigma''_T(x)$ at $T \leq 85$ K, $0.65 < x < 0.7$ cannot describe a regular phase transition (cf. Fig. 2). It is argued [16] that at lower coverages clusters start to form. The conventional isotherms show that only one layer is formed up to the saturation vapor pressure; thus the thickness of the clusters is monomolecular. The patches are not strongly bound to the surface ($f_{film} < f_{bulk}$). These islands may swivel. This happens with a slow rate. The effective σ'' is reduced (Eq. 1). At the critical coverage of ∿ 0.67 percolation results in a formation of a "continent" which due to its size practically stops. This fraction of a material will contribute to σ''. Above 85 K this "continent" is broken, probably at dislocations, and the irregularity in the Mössbauer isotherms disappears.

These results demonstrate some of the possibilities one finds in this method. It should, however, be stated that the structure of the solid cannot be revealed by this method and it must be accompanied by elastic neutron scattering or LEED experiments.

REFERENCES

1. J. G. Dash, "Films on Solid Surfaces," Academic Press, N.Y. 1975.
2. A. Thomy and X. Duval, J. Chim. Phys. 67, 1101 (1979); Y. Larher, J. Chem. Soc. Faraday Trans. 170, 320 (1974); F. A. Putnam and T. Fort, Jr., J. Phys. Chem. 79, 459 (1975); 81, 2164 (1977); F. A. Putnam, T. Fort, Jr., and R. B. Griffiths, J. Phys. Chem. 81, 2171 (1977).
3. A. Holz and J. T. N. Madeiros, Phys. Rev. B17, 1161 (1978).
4. D. R. Nelson and B. I. Halperin, Phys. Rev. B5, 2457 (1979).
5. E. Domany, M. Schick, J. S. Walker, and R. B. Griffiths, Phys. Rev. B18, 2209 (1978).
6. J. P. Coulomb, J. P. Biberian, J. Suzanne, A. Thomy, G. J. Trott, H. Taub, H. R. Danner, and F. Y. Hansen, Phys. Rev. Lett. 43, 1878 (1979).
7. J. Suzanne, J. P. Coulomb, M. Matecki, A. Thomy, B. Croset, and C. Marti, Phys. Rev. Lett. 41, 760 (1978).
8. M. Bretz, J. G. Dash, D. C. Hickernell, E. O. McLean, and O. E. Vilches, Phys. Rev. A8, 1589 (1973); G. B. Huff and J. G. Dash, J. Low Temp. Phys. 24, 155 (1976); see also reports by M. Nielsen et al, I. Marlow et al, Y. Grillet et al, J. Suzanne et al, in J. Physique (Paris) C4 (1977); T. T. Chung and J. G. Dash, Surf. Sc. 66, 559 (1977); Y. Larher, J. Chim. Phys. 68, 2257 (1978); A. Glachant, J. P. Coulomb, M. Bienfait, and J. G. Dash, J. Phys. (Paris) Lett.; A. Widom, J. R. Owers-Bradly, and M. G. Richards, Phys. Rev. Lett. 43, 1340 (1979).
9. H. Shechter, J. Suzanne, and J. G. Dash, Phys. Rev. Lett. 37, 706 (1976); J. Phys. (Paris) 40, 467, (1979).
10. H. Shechter, R. Brener, J. Suzanne, and S. Bukshpan (to be published).
11. D. St. Bunbary, J. A. Elliott, H. E. Hall,

and J. M. Williams, Phys. Lett. 6, 34, (1963); D. P. Craig and M. Sutin, Phys. Rev. Lett. 11, 460, (1963); S. L. Ruby, J. C. Love, P. A. Flynn, and B. J. Zubrowsky, Appl. Phys. Lett. 27, 322, (1975).
12. S. Bukshpan, T. Sonnino, and J. G. Dash, Surf. Sc. 52, 466 (1975).
13. H. Shechter, J. G. Dash, M. Mor, R. Ingalls, and S. Bukshpan, Phys. Rev. B14, 1876, 1976).
14. K. S. Singwi and A. Sjölander, Phys. Rev. 120, 1039, (1960).
15. J. K. Kjems, L. Passell, H. Taub, and J. G. Dash, Phys. Rev. Lett. 32, 724, (1974).
16. H. Shechter, R. Brener, and J. Suzanne (to be published).

Published 1980 by Elsevier North Holland, Inc.
Sinha, ed. Ordering in Two Dimensions

PAIR POTENTIAL OF KRYPTON ADSORBED ON GRAPHITE

F. A. PUTNAM

Massachusetts Institute of Technology
Cambridge, Mass. 02139

ABSTRACT

An improved pair potential for Kr adsorbed on graphite is obtained by adding a substrate-mediated repulsion to the known pair potential for bulk Kr. The magnitude of the substrate-mediated repulsion is adjusted to fit two-dimensional second virial coefficients obtained from low coverage adsorption isotherm data. The resulting potential has well depth $\varepsilon/k = 171$ K and is steeper than both the Lennard-Jones 6-12 potential and the bulk Kr potential.

In view of the intense current interest in the adsorption of krypton on graphite [1], it is advantageous to determine as accurately as possible the pair potential for physisorbed Kr atoms. It is well known that the adsorbent modifies the pair potential of adsorbed molecules [2-5]. Detailed calculations for this and other nonadditive interactions have been carried out recently for Xe and Kr adsorption on Ag (111) and graphite [6,7]. The dominant correction to the vapor phase pair potential is the <u>substrate-mediated interaction</u>. This is given by

$$U_s = C_{s1}\left\{\frac{4}{3} - [4L^2/(r^2 + 4L^2)]\right\}/[(r^2 + 4L^2)^{3/2}\, r^3] - C_{s2}(r^2 + 4L^2)^{-3} \qquad (1)$$

in the theory of McLachlan [3], where r is the atom separation and L is the distance of both atoms from the imaging plane. The Sinanogolu and Pitzer theory [2] gives a simpler expression, $U_s = C/r^3$. For the rare gas-graphite systems, several workers have attempeted to find C by comparison with experiment [8,9]. A persistent difficulty in these works was the uncertainty as to the bulk (unadsorbed) pair potential. However, more recently Barker, Watts, Lee, Shaefer, and Lee (BWLSL) [10] have arrived at a very accurate bulk Kr potential. This allows us to determine the potential for adsorbed Kr much more accurately.

The primary experimental information regarding the pair interaction of Kr on graphite is from low coverage adsorption data. There are three sets of appropriate data in the literature, the data of Putnam and Fort (PF) which were taken near 100 K [11], and the data of Sams, Constabaris, and Halsey (SCH) and Edmonds (E) which were taken near 270 K [12]. The information contained in the adsorption data is a weighted volume of the potential energy well, the two-dimensional second virial coefficient,

$$B_{2d} = -\pi\int_0^{\infty}\{\exp[U(r)/kT] - 1\}\, r\,dr \qquad (2)$$

where U(r) is the adsorbate-adsorbate potential. This is determined from the data by fitting to the virial coefficient adsorption equation, the two-dimensional form of which is [11]

$$\ln\left(\frac{n^{\sigma}}{pK_H}\right) = 2B_{2d}\left(\frac{n^{\sigma}}{\mathscr{A}}\right) + \frac{3}{2}\,C_{2d}\left(\frac{n^{\sigma}}{\mathscr{A}}\right)^2 + \dots \qquad (3)$$

where n^{σ} is the amount adsorbed, p is the vapor pressure, K_H is the Henry's Law constant, $\mathscr{A}$ is the adsorbent surface area, and B_{2d}, C_{2d} are the two-dimensional virial coefficients. As can be seen from (2), the weighted well volume B_{2d} is determined by the initial curvature of the isotherm. Also, in fitting (3) to data the parameter B_{2d} is highly colinear with $\mathscr{A}$ so that an accurate value of the surface area is needed in order to determine B_{2d}.

The SCH-E adsorption data is most reliably used by fitting to the two-dimensional Boyle temperature, $T_{B_{2d}}$, because this temperature is given by the conditions that $B_{2d} = 0$ which eliminates the $\mathscr{A}$ collinearity [4]. Everett [5] analyzed the SCH-E data and determined that $T_{B_{2d}} = 230 \pm 10$ K for the Kr-graphite system. The constant C_{s1} or C is evaluated by iteration to make the potential $U(r) = U_{BWLSL}(r) + U_s(r)$ yield $B_{2d} = 0$ at 230 K, where U_{BWLSL} is the K2 potential of Barker et al [10]. For the fitting, the second term in eq.(1) is neglected, so that the correction due to this term is folded into this term. The C_{s1}, C_{s2} values of Bruch and Watanabe [7] imply that the second term is about one fifth the magnitude of the first. Calculations using both terms in eq.(1) yield essentially the same potential as with one term. Following Bruch and Phillips [6], the position of the imaging plane is taken from Zaremba and Kohn [13], who predict it to be one half the interlayer spacing above the centers of the surface atom layer. Z is thus approximately one-half the atom-atom spacing in crystalline Kr, 2.05 Å.

The PF adsorption data was collected far below $T_{B_{2d}}$. For this data the collinearity cannot be eliminated. Fortunately, the surface area for the PF sample is accurately known from the identification of the coverage at which the Kr registers with the graphite lattice [11]. The data has been analyzed [11] by the virial coefficient theory. In this analysis, a Lennard-Jones (LJ) 6-12 adsorbate-adsorbate potential was directly fit to the data via eq.(2).

The LJ distance parameter σ_{gg} was fixed and the LJ well depth ε_{gg} was adjusted to fit the data. In this case, the LJ potential is an effective potential in which a single parameter is adjusted to give the correct B_{2d}. For σ_{gg} fixed at 0.360 nm, ε_{gg} was 145 $\pm$ 3 K and for σ_{gg} fixed at 0.373 nm, ε_{gg} was 143 $\pm$ 3 K. To fit the PF data, C_{s1} or C is adjusted to yield the same B_{2d} as the above effective potentials at the center of the temperature range of the PF data (T = 99.337 K, B_{2d} = - 4.02 x 10^{19} m^2). The result is given in Table 1.

Table 1. Values of the Substrate-Mediated Interaction Constants.

r^{-3} repulsion	C/k, m^3K
Experiment:	
Fitting to SCH data	$(1.0 \pm .2) \times 10^{-27}$
Fitting to PF data	$(1.5 \pm .1) \times 10^{-27}$
Theory:	
Yaris [16]	1.57×10^{-27}
Sinanogulu and Pitzer [2]	2.11×10^{-27}
McLachlan [3] repulsion	C_{s1}/k, m^3K
Experiment:	
Fitting to SCH data	$(1.9 \pm .4) \times 10^{-27}$
Fitting to PF data	$(2.4 \pm .2) \times 10^{-27}$
Theory:	
Bruch and Watanabe [7]	1.1×10^{-27}

DISCUSSION

The results in Table 1 indicate that the McLachlan repulsion is to be preferred over that of Sinanoglu and Pitzer in that the two sets of experimental data give C_{s1} values which are in agreement to within experimental error. The C values for the two sets of data disagree significantly, however. The difference is due to the inverse cube law having a more significant long range repulsive effect that eq.(1). Our best estimate of the krypton pair potential on graphite is thus given by eq.(1), with C_{s1}/k = 2.2 x 10^{27} m^3 K, C_{s2} = 0 (the weighted average of the value obtained from the two sets of data).

The pair potential arrived at above is shown in Figure 1 along with two other potentials which have been used for bulk and adsorbed krypton. The potential has well depth ε/k = 171.48 K, slow collision diameter 0.361 nm, and minimum at 0.40368 nm. At r > 0.5 nm, it decays toward zero faster than the other Kr potentials. Like the BWLSL potential, the new potential is steeper than the LJ 6-12 potentials. The new potential is actually even steeper than the BWLSL potential, its width at half depth being 0.106 nm, compared to 0.116 nm for the BWLSL potential and 0.125 nm for the LJ potentials.

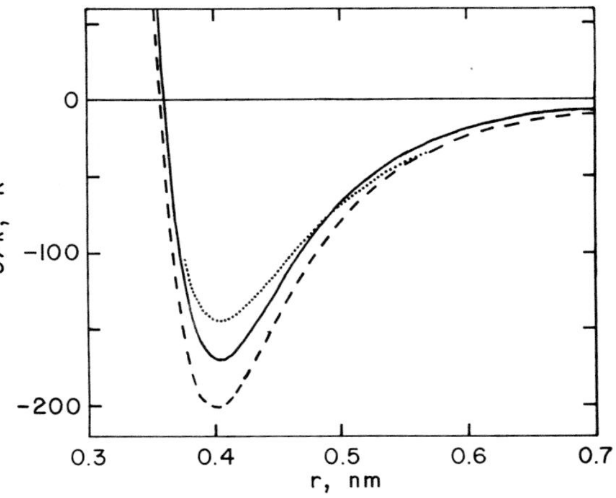

Figure 1. Kr pair potentials. The dashed line is the BWLSL potential for bulk Kr [10]. The solid and dotted lines are both potentials for Kr on graphite obtained by fitting to experimental 2d second virial coefficients. Dotted line: Lennard-Jones 6-12 potential [11]. Solid line: this work.

It is of interest that the new potential has a well depth which is very close to the depth of some of the bulk LJ 6-12 potentials which have been used to model the adsorption of Kr on graphite. Spurling and Lane's potential has $\varepsilon/k = 168.8$ K, $\sigma = 0.367$ nm.

The new potential is compared to other potentials which have been used for bulk and adsorbed krypton 2d "Lennard-Jonesium" in Table 2.

Table 2. Lennard-Jones Interatomic Potentials for Krypton on Graphite.

Reference	Parameters	
	σ, nm	ε/k, K
Toxvaerd [16]	.3827	164
Lane and Spurling [1]	.367	168.8
Hanson and McTague [1]	.36	170
Putnam and Fort [11]	.36	145
Wolfe and Sams [8]	.373	142.8
Glandt et al [1]	.3573	152

The constant C_{s1} is approximately twice as large as the C_{s1} calculated theoretically by Bruch and Watanabe. The disagreement could well be due to the failure of the McLachlan dipole image theory for small distances from the solid. It is well known that the Z^{-3} atom-surface long-range interaction potential fails for distances operative in submonolayer adsorption, so that the attraction is larger and varies more like Z^{-4} [14]. The McLachlan interaction is likely similarly perturbed for small Z. For this reason, eq. (1) may be taken only as a first approximation to the substrate-mediated interaction for submonolayer adsorption.

REFERENCES

1. P.M. Horn, R.J. Brigeneau, P. Heiney, and E.M. Hammonds, Phys. Rev. Lett.m 41, 961 (1978); D.M. Butler, J.A. Litzinger, and G.A. Stewart, Phys. Rev. Lett., 44, 466 (1980); J. Regnier, J. Roguerol, and A. Thomy, J. Chem, 31, 465 (1978); F.E. Hanson and J.P. McTague, J. Chem. Phys., 32, 1279 (1960).
2. O. Sinanglu, and K.S. Pitzer, J. Chem. Phys., 32, 1279 (1960).
3. A.D. McLachlan, Mol. Phys., 7, 381 (1964).
4. T. Takaishi, Prog. in Surf. Sci, 6, 43 (1975).
5. D.H. Everett, Disc. Faraday Soc., 40, 177 (1965.
6. L.W. Bruch and J.M. Philips, Surf. Sci., 91, 1 (1980).
7. L.W. Bruch and H. Watanabe, Surf. Sci., 65, 619 (1977).
8. R. Wolfe and J.R. Sams, J. Chem. Phys., 44, 2181 (1966).
9. J.D. Johnson and M.L. Klien, Trans. Faraday Soc., 60, 1964 (1964).
10. J.A. Barker, R.O. Watts, J.K. Lee, J.P. Schafer, and Y.T. Lee. J. Chem. Phys., 61, 3081 (1974).
11. F.A. Putnam and T. Fort, Jr., J. Chem. Phys., 81, 2164 (1977).
12. T. Edmons, Ph.D. Thesis, Bristol University, 1962.
13. E. Zaremba and W. Kohn, Phys. Rev. B., 13, 2270 (1976).
14. W.A. Steele, "The Interaction of Gases with Solid Surfaces", Pergamon Press, 1974.
15. R. Yaris, Ph.D. Thesis, Univeristy of Washington, Seattle, 1962.
16. S. Toxvaerd, Mol. Phys., 29, 373 (1975).
17. E.D. Glandt, A.L. Myers, and D. Fitts, Chem. Eng. Sci., 33, 1659 (1978); J. Chem. Phys., 70, 4243 (1979).

Published 1980 by Elsevier North Holland, Inc.
Sinha, ed. Ordering in Two Dimensions

EVIDENCE FOR A PHASE TRANSITION IN THE HYDROGEN/W(110) SYSTEM

R. DIFOGGIO and R. GOMER

The James Franck Institute, The University of Chicago, Chicago, Illinois 60637

ABSTRACT

A study of the surface diffusion of ^1H and ^2H on the (110) plane of tungsten by the field emission of fluctuation method has revealed the existence of a first order phase transition over the coverage range $\theta = 0.3$ to $\theta = 0.9$. The transition is revealed by a sharp oscillation in the mean square density fluctuation and by dips in the diffusion coefficient at the transition temperature. The dynamics of the transition can be followed by changes in the mean field emission current on supercooling or superheating and indicate that it is first order.

The field emission current fluctuation method of studying surface diffusion of adsorbates [1] consists of measuring the time autocorrelation functions of the emitted current from a small region (50-100 Å in radius r_o) of a single crystal plane of a field emitter uniformly covered with adsorbate, when the latter is mobile. The current fluctuations are related (in fact nearly proportional) to adsorbate density fluctuations, which build up and decay with a relaxation time $\tau_o \sim r_o^2/4D$ where D is the adsorbate surface diffusion coefficient. Thus, a comparison of the experimental and theoretical correlation functions f(t) permits determination of D.

It was recognized at the outset that the autocorrelation function at zero time, f(0), (the mean square density fluctuation) also contained thermodynamic information since it is given by

$$\langle \Delta n^2 \rangle / \bar{n} = (\bar{n}/A)\ kT\ K \qquad (1)$$

where K is the 2-dimensional compressibility of the adsorbed phase. In particular, phase transitions in the adsorbed layer should manifest themselves by divergences [2] in f(0). Until the present work we had been unsuccessful in seeing such transitions, possibly because of insufficient mobility at the transition temperature, or because of the occurrence of prediffusive flip-flop, which could mask the short time behavior of the diffusional part of the correlation function.

In the case of ^1H and ^2H on W(110) diffusion becomes temperature independent below 140 and 120 K respectively because of adsorbate tunneling [3], and clear evidence for a phase transition is seen. Figure 1 indicates the behavior of f(0) for ^1H at $\theta = 0.6\ \theta_{max}$. A sharp fluctuation occurs at 89K. Similar behavior is seen for ^2H. Concomitantly there is a dip in the diffusion coefficient, shown in Figure 1 of Reference 3. This dip is presumably the first order analogue of critical slowing. The magnitudes of the fluctuation in f(0) and the dip in D may be limited by the finite size of the plane (~1000 Å) on which observations are being made.

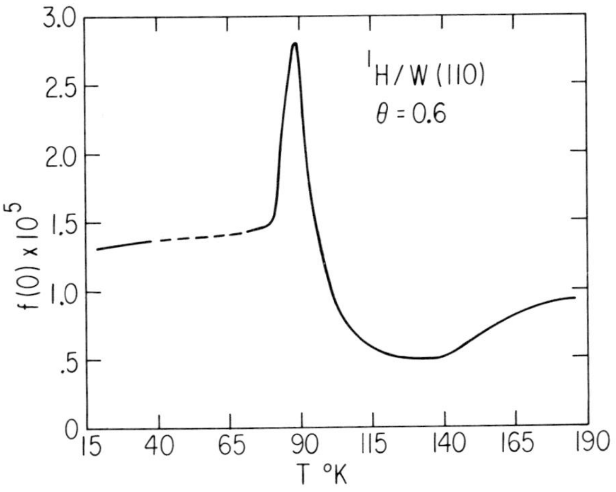

[1] Zero-time current fluctuation for ^1H at $\theta = 0.6\ \theta_{max}$ vs. temperature.

The mean field emission current above and below the transition differs in magnitude by a few percent. This effect was exploited to study the dynamics of the transition, by rapidly cooling from above to just below the transition temperature and following the change in current with time. This was most easily done with ^2H since it diffuses more slowly and therefore takes longer to reach equilibrium at the new temperature. Figure 2 shows the results for $\theta = 0.6\theta_{max}$. There is clear indication of a plateau, i.e., the dynamic halt typical of first order processes. Similar behavior was found on going from below to above the transition temperature.

To date the transition has been observed at three coverages. The resulting phase diagram is shown in Figure 3. Since He diffraction data for this system are lacking the nature of the ordered phase is not known. Very probably it corresponds to a (1x1) overlayer. It could also, by analogy to oxygen on this plane, correspond to a p(2x1) structure. In any case it seems clear that we are observing a lattice gas-ordered lattice

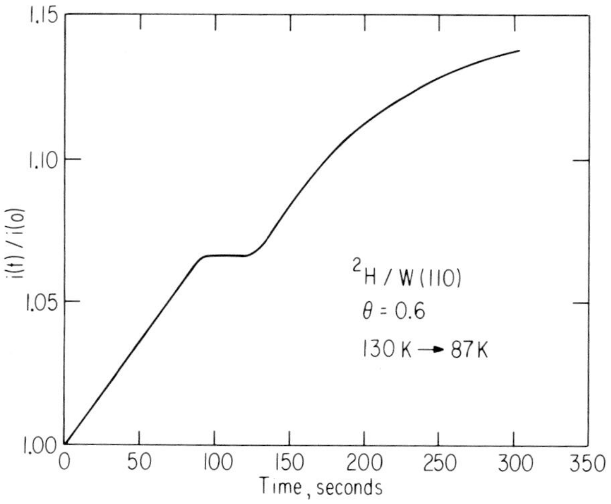

[2] Time dependence of field emission current for ^{2}H at $\theta = 0.6\theta_{max}$ when the emitter is rapidly cooled from 130K to 87 K, i.e., just below the transition temperature. Time is measured from the moment the emitter reached 87 K. Note the plateau.

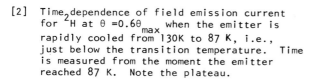

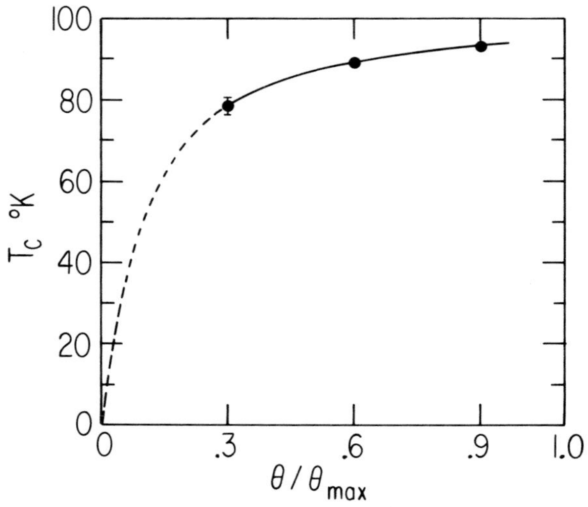

[3] Phase Diagram for H/W (110). The region to the left of the phase boundary represents the single phase disordered (lattice gas) region, that to the right the 2 phase (lattice gas plus ordered solid) region.

solid transition. The method holds promise of providing information not easily obtained by other means, and should be applicable to any system showing adequate mobility at its transition temperature(s).

ACKNOWLEDGMENT

This work was supported in part by NSF Grant CHE 77-08328 and also by the Materials Research Laboratory Program of the National Science Foundation at the University of Chicago.

REFERENCES

[1] J.-R. Chen and R. Gomer, Surf. Sci. 79,413 (1979)

[2] B. Bell, R. Gomer, and H. Reiss, Surf. Sci. 55, 494 (1976)

[3] R. DiFoggio and R. Gomer, Phys. Rev. Lett. 44, 1258 (1980)

Sinha, ed. Ordering in Two Dimensions

CHEMISORPTION ON STEPPED SURFACES AND ADATOM BINDING ENERGIES*

P. KLEBAN

Department of Physics and Astronomy
University of Maine
Orono, Maine 04469

*supported in part by the Office of Naval Research and a Stauffer Chemical Company grant of Research Corporation

We report the results of a theory of chemisorption on stepped surfaces for systems which may be modeled by a lattice gas in thermodynamic equilibrium. We mention several new results, including a general expression (in single-scattering approximation) for the LEED (Low Energy Electron Diffraction) intensity for an overlayer on a stepped substrate including the effects of disorder. In addition we argue that a comparison of LEED results for flat and stepped substrates can determine the sign of the change in adatom adsorption energy at terrace edge sites. This is illustrated by a Monte Carlo calculation of O on a certain stepped W(110) surface, where we find the binding energy to be less strong on either terrace edge site than on the flat surface. We also argue that in general the same information can be gleaned by a very simple comparison of appropriate LEED features.

I. Introduction

This paper is a review of the main results[1,2] of a new theory of chemisorption on stepped surfaces and its application to O/stepped W(110). Full details of this work are too long for inclusion here. They may be found in references 1 and 2.

The influence of defects on surface properties is a question of fundamental importance in surface science. In this work we consider the effects of one kind of surface defect - a step-on chemisorbed overlayers. In particular we present a detailed theory of LEED results for certain chemisorbed systems on stepped surfaces. We also show how information about changes in adsorption energies at terrace edge sites may be extracted from those results.

Section II reviews the general theory of LEED scattering from overlayers on stepped surfaces, including the effects of disorder in the overlayer. We restrict ourselves to systems for which a lattice gas model is appropriate. We consider the overlayer spot splitting and integrated intensity in the kinematic (single-scattering) approximation and also point out that qualitatively new LEED features may appear at low or high coverages.

In Section III we describe an application of the theory to existing experimental results for O adsorbed on a certain stepped W(110) surface. Integrated LEED intensity as a function of coverage θ and spot splitting results for the two adlayer domains are considered. A Monte Carlo program was employed to calculate the statistical averages of interest. As input we used adatom-adatom (AA) interaction energies for O/W(110) derived by comparison of experimental studies on the flat surface and theoretical lattice gas models. Using the signs of the changes in adsorption energy at terrace edge sites as unknown parameters, we find that the observed LEED features are reproduced most accurately when the O adsorption is less strong at either terrace edge site than on the flat surface.

Section IV contains some suggestions for future work. In particular we point out that, in general, one should be able to determine the sign of the change in adsorption energy at terrace edge sites by a simple comparison of appropriate overlayer LEED results for the flat and stepped surfaces without detailed modeling.

The lattice gas theory employed here makes use of the properties of thermodynamic equilibrium. This has some important advantages in surface systems that have not been fully exploited, especially in chemisorption studies. First, properly chosen systems are characterized by only a few parameters. Since there exist several well-developed methods for calculating thermodynamic quantities, one can determine these parameters with considerable confidence. For instance, in an adsorption system many observable effects, such as LEED results, depend only on the adsorption energy and adatom adatom (AA) interactions. This is a consequence of the general fact that all kinetic effects are irrelevant to equilibrium properties. This situation is especially advantageous when one is dealing with a lattice gas model with finite range AA forces. There are only a few energy parameters in the problem, since the adatoms sit at discrete sites and are only influenced by a few near neighbors. Second, although equilibrium studies certainly cannot solve all the important questions in surface physics, they can provide a secure starting place from which to attack other more difficult problems in complicated systems. One such application of equilibrium studies on a flat surface is demonstrated here for 0 on stepped W(110).

II. Theory of LEED Scattering from Overlayers on Stepped Surfaces

In this section we review the main points of the theory developed in reference 1. There we assumed the stepped surface was composed of identical terraces, with equivalent points on neighboring terraces separated by a fixed vector $\underline{g}$, and that N_s terraces were within the coherence length ξ of the LEED apparatus. We assumed the adatoms on a given terrace sit at N fixed sites $\underline{R}_\ell$, with occupation variable $n_\ell = 0(1)$ for an unoccupied (occupied) site. If the adatoms are in thermal equilibrium, and the AA forces of not too long range, it is reasonable to assume that the overlayer on each terrace is statistically independent of that on the others. Another important point is that for strong chemisorption the change $\delta\varepsilon$ in adsorption energy at terrace edge sites will generally be such that $|\delta\varepsilon| >> kT$. Hence edge sites are either empty ($\delta\varepsilon > 0$) or occupied ($\delta\varepsilon < 0$) given a sufficient value of θ.

The result of all this is that in the kinematic (single-scattering) approximation the LEED intensity for scattering vector $\underline{k}$ is

$$I(\underline{k}) = N_s(\langle|\rho(\underline{k})|^2\rangle - |\langle\rho(\underline{k})\rangle|^2) + h(N_s,\underline{k}\cdot\underline{g})\,|\langle\rho(\underline{k})\rangle|^2 \qquad (1)$$

Here

$$\rho(\underline{k}) \equiv \sum_{\ell=1}^{N} n_\ell \exp(i\,\underline{k}\cdot\underline{R}_\ell), \qquad (2)$$

$$h(N_s,\chi) = \sin^2(N_s\chi/2) \,/\, \sin^2(\chi/2) \qquad (3)$$

N is the number of adsorption sites on a terrace and the angular brackets denote a statistical average on a single terrace. The first term in Eq. (1) includes effects of statistical disorder on a single terrace, while the second term includes effects of interterrace interference via the function h. Thus for a given stepped chemisorption system and set of LEED scattering parameters, the intensity I is determined by the temperature T, coverage θ, AA interactions, and signs of the changes $\delta\varepsilon$ in adsorption energy at terrace edge sites. All of these quantities except $\delta\varepsilon$ may be measured or determined by calculations for the corresponding flat surface adsorption system. Thus the sign of $\delta\varepsilon$ can be found by studying experimental results for the stepped surface. This is shown by various model calculations in references 1 and 2.

In reference 1 we also consider several features of the LEED scattering in more detail. In particular, we show in a model calculation that the overlayer beam intensity $I(\underline{k}_o')$, where $\underline{k}_o'$ is an appropriately chosen scattering vector, has a maximum at a coverage θ_m that is characteristic of the change in adsorption energies $\delta\varepsilon$. This result is only due to the change in effective number of adsorption sites available on the stepped vs. flat surface. Thus one would expect a similar trend in the maximum of the integrated intensity $\mathcal{I}$ vs. θ, where

$$\mathcal{I} \equiv \iint_a I(\underline{k}) \frac{d(\underline{k}\cdot\underline{a})}{2\pi} \frac{d(\underline{k}\cdot\underline{b})}{2\pi} \qquad (4)$$

In Eq. (4), $\underline{a}$ and $\underline{b}$ are unit mesh vectors. $\mathcal{I}$ is an experimentally measurable quantity. The reciprocal space integration area a is determined by instrumental parameters and beam energy. This expectation is confirmed by the results for O/stepped W(110) in reference 2.

In reference 1 we also consider the splitting of LEED spots, which is due to interterrace interference via the function h in Eq. (1). We show by means of general arguments and specific model calculations at T=0 that the splitting strength depends on coverage, the signs of the $\delta\varepsilon$, and whether the terrace width is odd or even. The more elaborate finite temperature calculation in reference 2 verifies these results in part.

We also point out the interesting possibility of new LEED features appearing in the stepped chemisorption systems we are considering. Suppose $\delta\varepsilon < 0$ for a certain edge. Then the sites along that edge will fill up first, while all others on the terrace remain empty. Since the substrate configuration is quite different at the edge than elsewhere in the terrace, one would expect the adatom-adatom (AA) interaction to be different as well. If the AA edge energies favor a particular one-dimensional order and the temperature is not too large, new LEED spots will appear. If one edge has $\delta\varepsilon < 0$ and the terrace is D sites wide, these spots will reach a maximum intensity for some coverage $0 < \theta < 1/D$. The spots will be streaked in the direction perpendicular to the step edge, since they are due to one-dimensional order. There should also be a spot splitting in this direction due to the fixed phase relation between the edge sites. For a repulsive edge ($\delta\varepsilon>0$) similar effect can occur for coverages $1 - 1/D < \theta < 1$. Note that the two terrace edges are different, hence one can have two sets of AA interactions. The corresponding

LEED features will occur simultaneously at low θ if both $\delta\varepsilon<0$, at high θ if both $\delta\varepsilon>0$, or separately at low and high θ, if $\delta\varepsilon<0$ for one edge and $\delta\varepsilon>0$ for the other.

III. Application to O/stepped W(110)

The system O/W(110) has been extensively studied[3]. For $\theta <$ 0.5, and $T<T_c(\theta)$ the adlayer shows (2x1) ordering. For $T> T_c$ it is in a disordered state. AA (adatom-adatom) interaction energies have been derived by several authors[4] by comparing experimental and theoretical T_c results at a few θ values. We use the values determined by Ching et al.

Note that in the flat surface system there are two degenerate (2x1) structures which are completely equivalent in the absence of defects. Hence the (1/2,1/2) and $(\overline{1/2},1/2)$ LEED beams have the same behavior as functions of θ and T.

An experiment for O on stepped W(110) has been performed by Engel et al.[5] The W crystals were cut so that the average terrace width D on one was D=10 adsorption sites and the other D=24. In the nomenclature of Lang et al.[6] these can be indexed as W-(S) [10(110)x(011)] and W-(S) [24(110)x(011)]. Thus the terrace edges are in the (close packed) $[\bar{1}11]$ direction, as illustrated in Fig. 1. This means that the two (2x1) overlayer domains become crystallographically inequivalent. Domain I refers to that with the [0,1] direction parallel to the terrace edge, domain II to the other, as shown in Fig. 1. The (1/2,1/2) $((\overline{1/2},1/2))$ LEED beam corresponding to domain I (II) was monitored in this experiment.

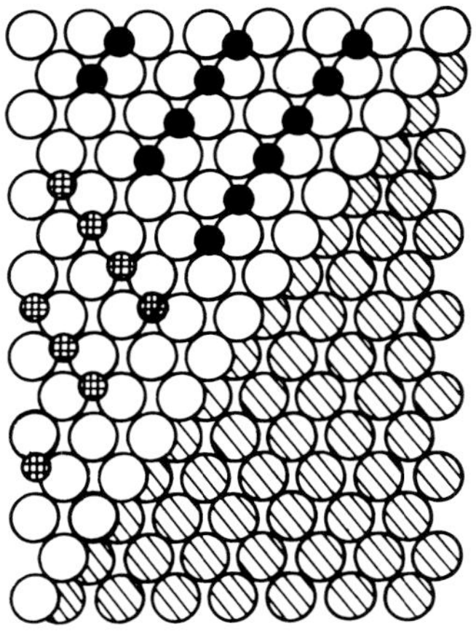

Fig. 1. The stepped W(110) surface used in Ref. 5. Large Circles: W substrate atoms, Small Circles: O adatoms. Domain I shown dark, Domain II cross-hatched.

The experimental results[5] included (a) the integrated LEED intensity $\mathcal{I}$ and (b) the LEED overlayer spot shapes. Under (a) it was observed that while the integrated intensities $\mathcal{I}_I$ (for domain I) and $\mathcal{I}_{II}$ (for domain II) are roughly equal at low coverage ($\theta<0.1$), $\mathcal{I}_I$ is considerably larger than $\mathcal{I}_{II}$ for larger values of θ, as one would expect from the terrace orientation. Further, the maximum value of $\mathcal{I}_I$ is reached not at $\theta = 0.5$, which would correspond to a complete (2x1) overlayer on a flat surface, but at $\theta \cong 0.32$. (for D=24 the corresponding maximum is at $\theta \cong 0.44$). $\mathcal{I}_{II}$ has a minimum at a coverage slightly below the maximum in $\mathcal{I}_I$ in both cases.

Under (b), the $(1/2,\bar{1}/2)$ (domain I) beam shows a splitting in the $[\bar{1}\ 1\ 2]$ direction at characteristic energies, from which terrace width and step heights were deduced. The (1/2,1/2) beam (domain II) showed no splitting. All overlayer features were streaked in the $[\bar{1}\ 12]$ direction, which is consistent with a distribution of terrace widths and the clean surface LEED pattern.

For reasons outlined in reference 2, we considered the case D=10 only. We used the Monte Carlo technique on a finite terrace of size Dx30, with periodic boundary conditions in the $[\bar{1}\ 11]$ direction. LEED beam intensities I(k) and integrated intensities $\mathcal{I}$ were calculated using the theory described in section II and reference 1. We had to assume reasonable values for certain experimental parameters - see reference 2 for details.

In applying our theory we made use of the AA interaction energies derived by Ching et al in a lattice gas treatment[4] of the flat surface chemisorption system. These parameters were employed as input in a statistical mechanical calculation of LEED scattering on the stepped surface chemisorption system with the signs of the $\delta\varepsilon$ as variables. The results of this calculation were compared with experiment. We found agreement only for $\delta\varepsilon>$ 0 on both terrace edges. Reference 2 contains a full discussion of the behavior of $\mathcal{I}_I$, $\mathcal{I}_{II}$ and the overlayer spot splitting in both theory and experiment. All observed LEED features were consistent with our model calculation when both $\delta\varepsilon$ were >0 (both terrace edge sites less attractive). Note there are generally only three possibilities for the two $\delta\varepsilon$: both $>$ 0 (++), both$<$ 0 (−−) or one of each (+ −). The only LEED feature that distinguished between them was the location θ_m of the peak in $\mathcal{I}_I$ vs. coverage θ. The observed $\theta_m \cong 0.32$ (for D=10) on the stepped surface, while $\theta_m = 0.5$ for a complete (2x1) overlayer on a flat surface. The model calculation gave a peak for $\theta_m< 0.5$ only for the (++) case. The numerical value was $\theta_m \cong 0.4$, somewhat larger than observed. Several possible reasons for this difference are mentioned in reference 2. However, the reason for the shift to lower θ values is quite simple - excluding 2/D of the adsorption sites means that a complete (2x1) adlayer is formed at a lower value of θ . The evidence for the (++) case is strongly buttressed by the experimental results[5] for D=24. Here the peak is at a higher θ value ($\theta_m \cong 0.44$), but still less than 0.5, which is consistent with our interpretation.

IV. Prospects for Further Work

We have considered the general theory of LEED scattering from certain types of overlayers on stepped surfaces including the effects of disorder in the overlayer for the first time. By comparison with experimental results of O on a stepped W(110) surface, we have demonstrated the adsorption at terrace edge sites on this particular surface is less strong than elsewhere on the terrace. This conclusion rests most strongly on a single experimental fact: that the peak intensity for LEED scattering from the I domain occurs at a coverage θ less than that on the flat surface. Referring to the discussion in Section IV, it seems clear that the three possibilities for terrace edge site adsorption energy changes, (+ +), (+ -) and (--) should be easily distinguishable by a simple comparison of experimental results on the stepped and flat surfaces. This indeed seems to be the case, and we hope to spell out the full details elsewhere. Basically, the idea is that one should look at the change in θ_m, the coverage of a particular overlayer LEED feature (e.g. the maximum in $\mathcal{I}_I$ in section II), from the flat to stepped surface. By simple stearic arguments one expects that θ_m will decrease for (+ +) (as seen herein), increase for (- -) and stay about the same for (+ -).

REFERENCES

1. P. Kleban, submitted to Surface Sci.
2. P. Kleban and R. Flagg, submitted to Surface Sci.
3. T. Engel, H. Niehus and E. Bauer, Surface Sci. 52 (1975) 237; G.-C. Wang, T.-L. Lu and M.G. Lagally, Phys. Rev. Letters (1977) 411; J. Chem. Phys. 69 (1978) 479.
4. W.Y. Ching et al., Surface Sci. 77 (1978) 550; E. Williams, S. Cunningham, and W.H. Weinberg, J. Chem. Phys. 68 (1978) 4688.
5. T. Engel, T. von dem Hagen, and E. Bauer, Surface Sci. 62 (1977) 361.
6. B. Lang, R.W. Joyner and G.A. Somorjai, Surface Sci. 30 (1972) 454.

Sinha, ed. Ordering in Two Dimensions

PREPARATION OF SURFACE FILMS FAR FROM THERMODYNAMIC EQUILIBRIUM BY SURFACE CHEMICAL REACTIONS

D. G. HOWARD and R. H. NUSSBAUM

Portland State University, Portland, Oregon 97207

Strong cohesive forces promote clustering in surface films of many materials, especially of refractory materials and metals. We report a method of preparation of dispersed layers by surface reaction which proceeds at sufficiently low temperatures that diffusion and thus clustering is inhibited. Such deposits may therefore be in a state far from thermodynamic equilibrium. In particular, we prepared highly dispersed iron oxide deposits on graphite by oxidizing monolayer films of $Fe(CO)_5$ at room temperature. Mössbauer measurements indicate the layer to be amorphous Fe_2O_3, in which the cohesive forces are indeed larger than the forces of adsorption at room temperature. Reduction of the oxide in hydrogen at reasonably low temperatures produced highly dispersed metallic iron on the graphite.

INTRODUCTION

The recent commercial development of porous, exfoliated graphite (Papyex, Grafoil) has made available a unique type of high surface area substrate for adsorption studies. The nature of the surfaces has been investigated by x-ray and neutron diffraction techniques [1,2]. They are found to be composed primarily of homogeneous basal planes of the order of 100 Å in linear dimensions, of which about half are oriented to within a few degrees of lying parallel to the external plane of the sheets. The basal plane of carbon has been found to exert relatively weak forces of adsorption on most molecules, and can therefore be easily cleaned by heating in a rather modest vacuum. In addition to these advantages, graphite is particularly suitable for Mössbauer studies because of its high transparency to low-energy gamma rays.

To date, several techniques have been used to study a variety of molecules adsorbed on graphite surfaces. In all cases, the samples are prepared by condensation of molecules from the vapor phase. While this procedure often produces monolayers of high uniformity, it deposits molecules on the surfaces at temperatures where surface diffusion is rapid: thus, if the molecules possess relatively strong cohesive forces (as compared to the forces of adhesion), clustering will result. Moreover, many interesting systems cannot be prepared in this manner because the molecules cannot be easily prepared in the vapor phase. For this reason we have been investigating alternative techniques for the preparation of surface films. We report in this paper a method of preparing films of Fe_2O_3 and Fe by surface chemical reactions.

CHEMICAL CONVERSION OF MONOLAYER FILMS

The starting material in these studies is iron pentacarbonyl adsorbed onto a graphite substrate that has previously been cleaned by heating to 1000° C in a 10^{-5} Torr vacuum. Monolayer completion, as determined by vapor pressure isotherms at room temperature, corresponds to a surface area per molecule of about 45 $Å^2$. Since about 50% of the substrate surface is aligned parallel to the plane of the Papyex sheet, the intensity of Mössbauer adsorption spectra, measured perpendicular and parallel to the sheets, can be interpreted to give mean-squared displacements of the molecules in the two directions. At low temperatures, the mean-squared displacements of the $Fe(CO)_5$ molecules parallel to the surfaces are found to be considerably larger than those perpendicular, indicating that the cohesive forces are weaker than those of adsorption. This behavior is qualitatively similar to that of other molecules adsorbed as monolayers onto graphite surfaces [3,4]. Above 180 K, the increasing mobility of the adsorbed molecules reduces the Mössbauer intensity to background levels. However, we found no indication of a sharp transition in the mean-squared displacement parallel to the substrate surface as was reported for another organo-metallic surface layer [3]. The observed quadrupole splitting in the spectra of the adsorbed molecules is identical to that observed in bulk and in frozen solutions of $Fe(CO)_5$, indicating that the molecule is not appreciably distorted by the surface.

If the substrate is heated above room temperature, thermal decomposition of the $Fe(CO)_5$ appears to be preceeded by desorption, so that the reaction takes place in the gas phase. The resulting deposition of metallic iron under these conditions tends to favor the formation of relatively large clusters of Fe atoms rather than dispersed layers. The resulting Mössbauer spectrum is that of a Zeeman-split pattern characterized by the full 330 kOe. field of bulk metallic iron. The spectra taken perpendicular and parallel have approximately the same total intensity; the only major difference is in the relative intensities of the component lines of the Zeeman pattern, indicating that the direction of magnetization in individual clusters is always parallel to the substrate. This parallel orientation of the magnetization is consistent with that found in Fe films of more than 30 Å thickness [5].

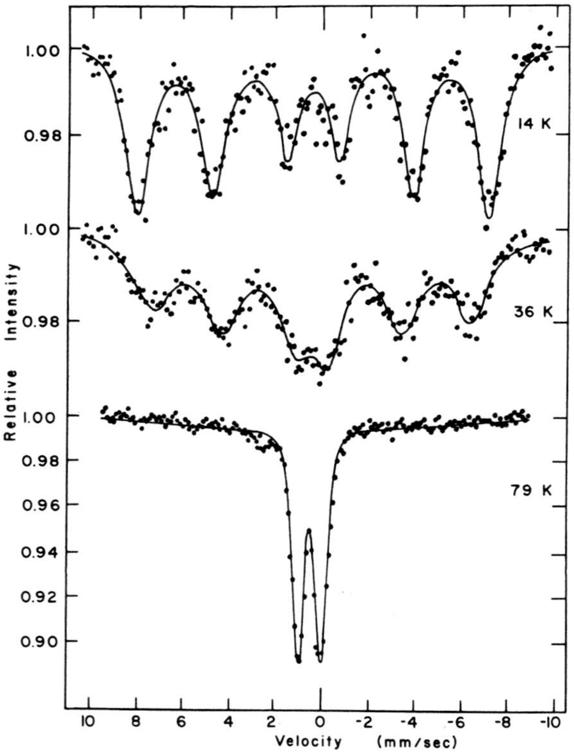

Fig. 1. Mössbauer spectra of ultra-thin films of amorphous Fe_2O_3 at three temperatures. The source is Co^{57}:Rh, but the velocity scale is referred to natural Fe. In all three spectra, the radiation is normal to the plane of the partially oriented surfaces. The solid line in the 14 K and 36 K spectra represent an unrestricted fit using six Lorentzians; at 79 K, the solid line represents a superposition of Lorentzians of natural width (corrected for source width) representing a Gaussian distribution of values of the electric field gradient.

In contrast, when the adsorbed layer of $Fe(CO)_5$ is exposed to oxygen at room temperature, the chemical decomposition now appears to occur on the surface rather than after desorption. Oxidation proceeds rapidly and completely. Mössbauer spectra of the oxide film formed in this way, taken at room temperature or 79 K, reveal a symmetric, quadrupole-split doublet of energy spearation 0.95 ± 0.05 mm/sec. The equality of the components indicates no preferred orientation of the crystal field axis. The room temperature isomer shift is 0.35 ± 0.02 mm/sec with respect to metallic iron. Magnetic hyperfine splitting is, however, observed below 79 K. From preliminary measurements, we estimate the critical temperature to lie in the range 60 to 75 K. Based on these data, we identify the oxide layer as an amorphous form of Fe_2O_3, similar to a layer of about 10μm thick prepared from the decomposition of an aerosol on a heated substrate [6].

STRUCTURAL PROPERTIES OF A SURFACE Fe_2O_3 LAYER

A striking feature of the oxide film is that at room temperature the mean-squared displacement of Fe parallel to the substrate surface is decidedly less than that perpendicular, in contrast to the behavior of the $Fe(CO)_5$ layer. If the Fe_2O_3 layer were clustered, it is expected to show a near-isotopic mean-squared displacement. The definite anisotropy which we found suggests, therefore, that the material resembles more closley a dispersed film than clusters or particles. The equal intensity of the two components of the quadrupole doublet indicates a random orientation of the crystal field axis at the Fe sites which is consistent with our interpretation of an amorphous structure of the dispersed layer, although it could also result from randomly oriented clusters. However, our conclusion that the material is highly dispersed rather than clustered is further supported by the observation that such an oxide layer on the graphite substrate will adsorb only about half as much $Fe(CO)_5$ as the clean graphite surface, indicating that a substantial fraction of the basal plane surfaces are covered by the oxide layer, which apparently impedes adsorption.

The reduced mean-squared displacement parallel to the surface relative to that perpendicular is consistent with the assumption that for this material, the force of cohesion within the layer is stronger than that of adhesion to the substrate. Such a condition should lead to clustering. We submit that clustering does not occur because the layer is formed at a temperature low enough (300 K) that diffusion mobility is strongly inhibited. We find that the structure of the oxide is not altered by heating up to about 200° C. Above this temperature, however, the amorphous film will crystalize to α-Fe_2O_3 in the presence of excess oxygen, or to a mixture of Fe_3O_4 and metallic iron if the heating is carried out in vacuum (presumably due to the reducing properties of the graphite). Similarly the bulk amorphous oxide, prepared from an aerosol, was found to transform to α-Fe_2O_3 at 250° C or above [6].

ELECTRIC AND MAGNETIC HYPERFINE INTERACTIONS

The line widths of all Mössbauer spectra of the oxide films are considerably larger than natural (after correction for source line width). In keeping with our identification of the system as amorphous, we would expect variations in local environments of the Fe atoms to lead to a distribution of values for the local electric field gradient (EFG). Assuming natural width doublets and a Gaussian distribution of splitting values centering around 0.95 mm/sec significantly improves the computer fit to our experimental spectra compared to a fit with a single Lorentzian doublet with broadened components. The required full width at half maximum for the Gaussian distribution is approximately the same as the average value for the EFG. A similar analysis can be applied to the published spectra of the bulk material formed from an aerosol [6], yielding similar parameters except that the Gaussian width in that case is only about three quarters of that for our films. This seems to indicate that our material has lower homogeniety as might be expected for an ultra-thin film as compared to larger particles or bulk amorphous material.

Spectra of the amorphous Fe_2O_3 films at 14 K and 36 K exhibit Zeeman splittings characteristic of a saturation magnetization of about 470 kOe, slightly less than for α-Fe_2O_3. The line widths of these spectra are broadened even more than the quadrupole doublets, and in particular the broadenings increase with increasing temperature. The bulk material formed from the aerosol also exhibits broadened lines in the magnetically split spectra below 80 K, but those line widths do not appear to vary with temperature [6]. The effect we found in our films, therefore, cannot be due entirely to inhomogeniety in the magnetic fields. The nature of the variation is just what could be expected of small assemblies of spins exhibiting superparamagnetism.

SUMMARY OF THE STRUCTURE OF THE OXIDE FILMS

For a film sample prepared from a monolayer coverage of $Fe(CO)_5$, we estimate a total amount of Fe_2O_3 of about 5.8 mg per gram of graphite. Based on estimates from bulk Fe_2O_3 densities, we can deduce that this is insufficient to completely cover the exposed substrate surfaces. We therefore propose that the film is made up of small film patches which exhibit superparamagnetism, and that their thickness is less than 10 Å on the average, consistent with our finding that the vibration amplitudes of Fe parallel and perpendicular to the substrate surface are different. This picture is substantiated by the observation that an oxide film prepared from a sample originally given only half monolayer coverage of $Fe(CO)_5$ exhibits the same shift and quadrupole splitting as the films prepared from full monolayers. Ultimate confirmation of the correctness of our proposed structure of the Fe_2O_3 films must derive from direct surface probes, e.g. neutron scattering.

REDUCTION TO THIN IRON LAYERS

In a preliminary investigation we have succeeded in completely reducing the oxide to metallic iron at 327°C. Under these conditions, about 75% of the resulting iron is in small enough particles to exhibit superparamagnetic narrowing in a room temperature spectrum. Hydrogen reduction at even lower temperatures, currently under investigation, could provide still more finely dispersed systems.

REFERENCES

1. J. K. Kjems, L. Passell, H. Taub, J.G. Dash and A.D. Novaco, Phys. Rev. B13, 1446 (1976).
2. M. Nielson and J. P. McTague, Phys. Rev. B19 (1979).
3. H. Shechter, J. Suzanne and J. G. Dash, J. Physique 40, 467 (1975).
4. S. Bukshpan and G. J. Kemerink, to be published.
5. S. W. Duncan, R. J. Semper, A. H. Owens and J. C. Walker, J. Physique 41, Suppl. Colloque C1, 213 (1979).
6. A. M. VanDiepen and Th. J. A. Popma, Solid State Commun. 27, 121 (1978).

MODEL DISPERSION RELATIONS OF SURFACE PHONONS FROM ATOMIC SCATTERING

RICARDO AVILA*

EEAP Dept., Case Western Reserve University, Cleveland, Ohio 44106

MIGUEL LAGOS

Inst. de Fisica, Univ. Catolica de Chile, Cas 114-D, Santiago, Chile

ABSTRACT

We consider the dynamics of inelastic atomic scattering from crystal surfaces in the single-phonon approximation. A geometrical picture of the process displays the conditions such that i) creation and annihilation events are completely separated on either side of the elastic peak's polar angle, ii) the azimuthal spread of the inelastic processes is bounded, and iii) a dynamical focussing effect produces sharp observable inelastic streaks along the azimuthal boundaries. Use of measurements of these inelastic streaks allows for a fitting of a model surface phonon dispersion relation needing no energy analysis of the inelastic intensity. Application of this method to He/LiF and He/NaF data gives encouraging results. The search for inelastic streaks produced by scattering from a selective adsorption channel is proposed.

1. INTRODUCTION

One interesting possible application of atomic scattering from crystal surfaces is towards the study of surface dynamics, much the same way as neutron scattering discloses the bulk vibrational spectrum. The straightforward use of the similarities of both techniques calls for the energy analysis of the scattered beam; this approach is being developed both by time of flight [1,2] and by triple axis spectrometry [3]. Another technique peculiar to atom-surface scattering is the use of selective adsorption processes [4] that carry a built in energy analysis of the inelastic events. Two major difficulties thus far encountered are the superposition of surface and bulk contributions and the smallness of the resulting signal. These obstacles are not insurmountable, and great progress has been made on the way to overcome them.

In a previous work [5], we have proposed an alternative, no-energy analysis method which should be specially useful to obtain the roughly linear part of the dispersion relation for acoustic surface phonons. This method relies on the existence of a Frontier of the Allowed Zone (FAZ) in detector space for scattering by both surface and bulk phonons, the bulk allowed zone being interior to the surface allowed zone. On the FAZ for surface phonons the density of the projections from the surface first Brillouin zone onto the space of the detector (DDP) is divergent.

In the present article we present in Section 2 a geometrical picture of the scattering process that should help to develop an intuitive grasp of the FAZ and the DDP; i.e., their meaning in terms of the conservation equations and the necessary conditions for their existence. Section 3 contains an exact expression for the aforementioned density of allowed projections $(Q_x, Q_y) \rightarrow (\Theta, \phi)$. This expression provides a method by which one can fit the parameters of a model dispersion surface by using experimental data on the divergence of the density of projections. Finally, in Section 4 we consider the applicability of the previous sections to the inelastic scattering from a closed channel.

2. GEOMETRICAL PICTURE

A useful geometrical picture of the dynamics of inelastic atom-surface scattering can be obtained by means of a simple generalization of an expression written by G. Benedek [6] for the planar scattering case.

Consider an atom of mass m and wavevector $\vec{k}_i$ incident on a crystalline surface. Let Θ_i be the polar angle of the specular diffraction beam and take a coordinate system with its $\hat{x}$-axis parallel to $\vec{K}_i$ (uppercase letters represent the surface parallel component of vectors) and $\hat{z}$ pointing outward from the surface.

The energy and pseudomomentum shifts $\hbar\omega^*$ and Q^* of the gas atom in an inelastic process are

$$\hbar\omega^* = \frac{\hbar^2}{2m}(k^2 - k_i^2) \qquad (1)$$

$$Q^* = \vec{K} - (\vec{K}_i + \vec{G}) \qquad (2)$$

where $\vec{k}$ is the scattered atom wavevector

and $\vec{G}$ is a surface reciprocal lattice vector representing a concurrent elastic process.

From (1) and (2) we easily obtain a two dimensional extension of Benedek's eq. (2) in ref. [6]

$$\frac{\omega^*}{\omega_i} = \frac{1}{K_i^2} \frac{\sin^2\Theta_i}{\sin^2\Theta} [(K_i+G_x+Q_x^*)^2 + (G_y+Q_y^*)^2] - 1 \quad (3)$$

where $\hbar\omega_i$ is the incident atom kinetic energy and Θ is the polar angle of $\vec{k}$.

Equation (3) represents a paraboloid in (Q_x^*, Q_y^*, ω) space opening upward from $(-\vec{K}_i-\vec{G}, -\omega_i)$.

Equations (1) and (2) represent an allowed inelastic process if the energy and pseudomomentum shifts of the gas atom are matched by the inverse shifts in the crystal phonon field. In the single phonon approximation, this means $\vec{Q}^* - \xi\vec{Q} = 0$, $\omega^* - \xi\omega = 0$ (where $\xi = \pm 1$ for phonon annihilation/creation), that is

$$\omega^* = \xi\omega(\xi\vec{Q}^*) \quad (4)$$

ω and $\vec{Q}$ being the surface phonon frequency and pseudomomentum.

Consider now the case of planar scattering ($Q_y^* = 0$) around the diffraction peak $\vec{G}$. From eq. (3) we obtain

$$\left.\frac{\partial\omega^*}{\partial Q_x^*}\right|_{\omega^*=0} = \frac{v}{\sin\Theta} \quad (5)$$

where v is the scattered atom velocity. This means as illustrated in Fig. 1, that the parabola $\omega(Q^*)$ at $\Theta = \Theta_{\vec{G}}$ becomes tangent to the dispersion relation $\omega(Q)$ of Acoustic Surface Phonons (ASP) at $Q = 0$ if

$$\left.\frac{\partial\omega}{\partial Q}\right|_{Q=0} = v_s = \frac{v_i}{\sin\Theta_{\vec{G}}} . \quad (6)$$

That is when $\beta_s \sin\Theta_{\vec{G}} = 1$, where $\beta = \omega/(Qv_i)$ and $\beta_s = v_s/v_i$.

In this circumstance, a large amount of both phonon creation and annihilation processes appear superimposed on the elastic peak. We also see from Fig. 1 that, when $\beta_s \sin\Theta_{\vec{G}} < 1$, there is one solution of eqs. (3) and (4) for each $\xi = \pm 1$ and for each $\Theta \gtrless \Theta_{\vec{G}}$. For

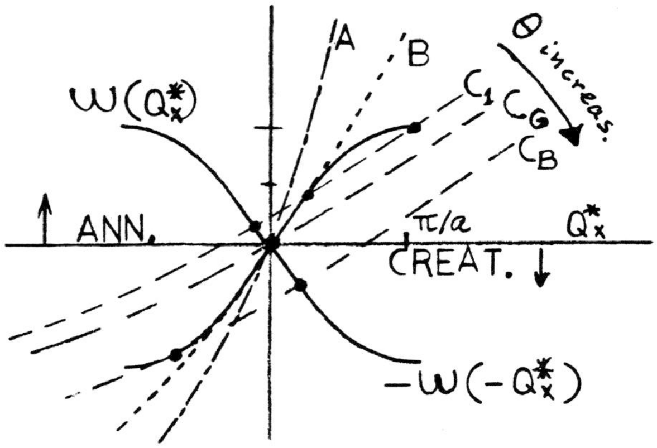

Fig. 1. The conservation equations on the incidence plane. The parabolas representing eq. (3) are drawn for $\Theta = \Theta_{\vec{G}}$ in A, B and C_G; $\Theta < \Theta_{\vec{G}}$ in C_1 and $\Theta > \Theta_{\vec{G}}$ in C_B. Also $\beta_s \sin\Theta_{\vec{G}}$ is < 1 in A, $= 1$ in B and > 1 in C. The surface phonon dispersion curves are drawn according to eq. (4). Dots indicate solutions to eqs. (3) and (4).

$\beta_s \sin\Theta_{\vec{G}} > 1$, there are two solutions for annihilation only for $\Theta < \Theta_{\vec{G}}$ and there are two solutions for creation only for $\Theta > \Theta_{\vec{G}}$. The relation $\beta_s \sin\Theta_{\vec{G}} \lessgtr 1$ is thus the criterion for the creation and annihilation events to appear overlapped or completely separated on either side of $\Theta_{\vec{G}}$. In what follows we shall always assume $\beta_s \sin\Theta_{\vec{G}} > 1$.

We turn now to the out of incidence plane behavior of the sets of solutions to eqs. (3) and (4). For this purpose we define the critical polar angles Θ_1 at which the paraboloid (eq. (3)) first touches the dispersion surface (see curve C_1 in Fig. 1) and Θ_B at which $\omega^*(Q^*)$ becomes tangent to $\xi\omega(\xi Q^*)$, which is the case already studied by G. Benedek (see curve C_B in Fig. 1). Stretching our imagination (or by numerical methods) we obtain now the full intersection of $\omega^*(Q^*)$ and $-\omega(-Q^*)$ for the phonon creation case. The result is depicted in Fig. 2. The phonon annihilation case is very much alike.

We note in Fig. 2 that, for $\Theta_{\vec{G}} < \Theta < \Theta_B$, there are two solutions to the conservation equations for each angle ϕ up to the limiting angle ϕ_F. No single ASP scattering is allowed farther away from the incidence plane. The locus (Θ, ϕ_F) will consequently be called the Frontier of the Allowed Zone (FAZ). On this locus (Θ, ϕ_F), the surface component $\vec{K}$ lies tangent to the set of solutions, and the density of allowed surface contributions $D_{\vec{k}s}$ becomes singular. A similar behavior produces a singularity on the incidence plane at $\Theta = \Theta_B$. Moving to higher values of Θ, this (Benedek's) singularity approaches the one running along the FAZ until they meet at a certain angle $\Theta_c > \Theta_B$. For $\Theta > \Theta_c$, $D_{\vec{k}s}$ is regular all over the ϕ range. Let us note finally that our assumption $\beta_s \sin\Theta_G > 1$ is necessary for the existence of these singularities of $D_{\vec{k}s}$. Otherwise, the solution curves are open and ϕ has no turning points on them.

As we expect [7], the probability of single phonon scattering to be a rather smooth function of (Θ, ϕ), we expect the divergencies of $D_{\vec{k}s}$ to appear as sharp streaks on the total inelastic intensity. We show in ref. [5] that these structures are not interfered with by the contribution from bulk phonons. In the next section we will see how to calculate a model $\omega(\vec{Q})$ from experimental data.

3. DENSITY OF PROJECTIONS

We develop now a quantitative measure of the density of allowed inelastic processes in a given direction (Θ, ϕ). From this result, we will obtain a method to calculate an acoustic surface phonon dispersion relation from inelastic intensity structures.

Let us write the conservation equations in the standard form

$$\vec{K} = \vec{K}_{\vec{G}} + \xi Q \tag{7a}$$

$$\frac{\hbar^2 k^2}{2m} = \frac{\hbar^2 k_i^2}{2m} + \xi\hbar\omega_{\vec{Q}} \tag{7b}$$

where $\vec{K}_{\vec{G}} = \vec{K}_i + \vec{G}$. The direction of the scattered wavevector $\vec{k}$ is then given by

$$\tan\phi = \frac{K_{\vec{G}y} + \xi Q_y}{K_{\vec{G}x} + \xi Q_x} \tag{8a}$$

$$\sin\Theta = \frac{K_{\vec{G}x} + \xi Q_x}{k\cos\phi} . \tag{8b}$$

The equations (8a, b) can be considered as a projection from the surface First Brillouin Zone onto the space of the detector. The density of such projections is given by

$D_{\vec{k}s} \sim |\sin\Theta J(\Theta, \phi)|^{-1}$ where $J(\Theta, \phi) = \partial(\Theta, \phi)/\partial(Q_x, Q_y)$. A look at Fig. 2 shows that, at each scattering direction, there are either none or up to three solutions to eqs. (7a, b). The total density of projections is then

$$D_{\vec{k}s} = d_s \Sigma_i |\sin\Theta\, J_i(\Theta, \phi)|^{-1} \tag{9}$$

where $d_s = (L_x L_y)/(2\pi)^2$ is the density of surface modes.

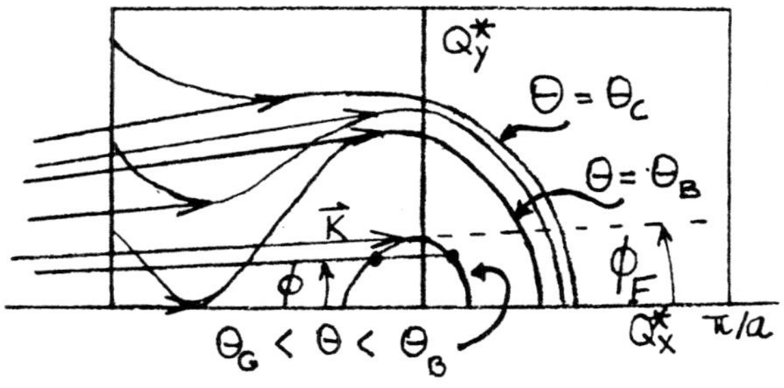

Fig. 2. Surface parallel projections of solution sets to eq. (4) for phonon creation events. We assume a square surface Brillouin Zone and $\beta_s \sin\Theta_{\vec{G}} > 1$ incidence conditions. Dots indicate solutions to eq. (4) in fixed direction (Θ, ϕ). Arrow heads indicate turning points for ϕ on these iso-Θ curves that produce divergencies of the density of solutions in (Θ, ϕ) space.

The Jacobian of the projection (8) can be put into the form

$$J = \frac{1}{k^2 \sin\Theta\cos\Theta}\left[1 - \frac{\sin\Theta_{\vec{G}}}{v^2}\vec{\nabla}\omega\cdot\vec{v}\right] \tag{10}$$

Where $\vec{v}$ is the scattered atom velocity. This expression reminds us of the case of neutron scattering by bulk exitations [8].

The use of the expression (10) in eq. (9) is certainly quite involved, but in dealing with the divergencies of $D_{\vec{k}s}$ at the tangency points in Fig. 3, eq. (10) is most useful as these divergencies occur precisely at the zeros of J. Given a dispersion relation, the straightforward numerical determination of the roots (Q_x, Q_y) of J yields through eqs. (8a, b) the locus (Θ, ϕ) of the divergencies of $D_{\vec{k}s}$.

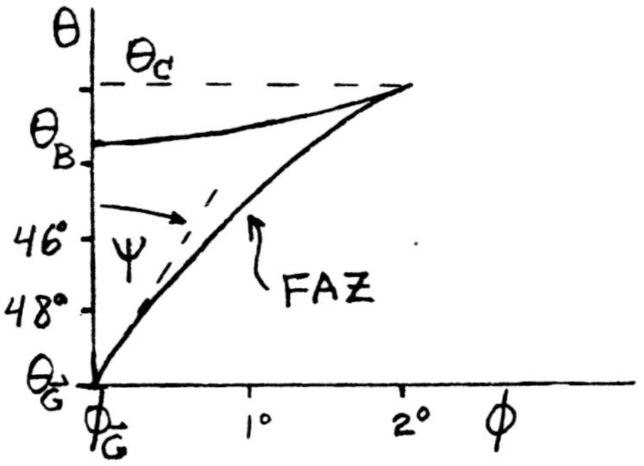

Fig. 3. Calculated locus of the Divergence of the Density of Projections $(\Theta_x, Q_y) \to (\Theta, \phi)$. The angle $\psi \to \pi/2$ as $\beta_s \sin \Theta_{\vec{G}} \to 1^+$. Also $\Theta_G - \Theta_B$ increases as $\omega(Q)$ becomes more linear and $\beta_s \sin \Theta_{\vec{G}}$ increases.

The DDP graph, shown in Fig. 3, has been calculated for the model dispersion relation of the form

$$\omega = (v_{so} + v_{sa} \cos 4\eta)\,(1 - r^p/(p+1))\,Q \qquad (11)$$

where we have used a square Brillouin zone, where $r = Q/Q_{max}(\eta)$, $p = p_o + p_a \cos 4\eta$, and the angle η is measured from the $\vec{G} = (1,0)$ direction. This model dispersion relation has the necessary symmetry and has four adjustable parameters, v_{so}, v_{sa}, p_o, and p_a, which can be used to fit experimental DDP curves.

We have developed an iterative solution procedure to minimize $\sum_i (\Delta\phi_i)^2 = \sum_i (\phi_i^{theor} - \phi_i^{exp.})^2$ by varying these adjustable parameters. Convergence is found to be quite rapid with some 10 iterations required to fit the model parameters within 1%.

We have applied the DDP method to analyze the experimental data on He/LiF [9] and He/NaF [10]. The data is insufficient to analyze the anisotropy of the dispersion curves so we have set $v_{sa} = p_a = 0$ in this analysis. The results of fitting the experimental data by varying v_{so} and p_o are shown in Table I. In addition, the Table gives the theoretical values of v_s for Rayleigh modes [11] calculated from the bulk elastic constants [12], the theoretical v_s from the slab dynamics calculations of Chen et al [13], Williams and Mason's estimates of v_s from their experimental data [9, 10], and our earlier estimates using the FAZ method [5].

Table 1. Surface speed of sound for LiF and NaF. Units are 10^5 cm/s. The angle η indicates the direction for the FAZ and DDP values and is measured from the <100> direction.

		Rayleigh	Chen	Williams
LiF	$v_{s<100>}$	3.96	3.62	----
	$v_{s<110>}$	4.17	4.10	3.85
NaF	$v_{s<100>}$	2.99	3.11	----
	$v_{s<110>}$	2.96	2.91	2.55

	FAZ	DDP		η
	v_s	v_{so}	p_o	
LiF	4.2	4.3	7.	30°
NaF	3.4	3.2	9.	10°

A good general accord is apparent between the different approaches to v_s in Table I. As pointed out by Williams and Mason, their estimates seem to be too low; this appears to result from their $Q_x = 0$ assumption for the inelastic streak phonons which is not quite right according to our analysis [5]. We believe our values for v_s to be a little too large. This results from the shift of the DDP maxima towards the Θ axis produced by a finite detector size and non-monochromatic incidence. We have made no attempt to incorporate this effect in our calculations.

We have compared the shape of our dispersion curves to those calculated by Chen et al and find that our values of p_o are within experimental error for NaF but slightly too large for LiF. The uncertainty of our fitted parameters is less than 20% for v_{so} and 50% for p_o.

4. INELASTIC SCATTERING FROM A RESONANT STATE

Up to this point we have kept in mind the picture of inelastic scattering around an allowed elastic peak. All of sections 2 and 3 can be extended though, with minor restrictions, to the case where the diffraction channel $\vec{G}$ is closed.

We can identify a closed channel by the corresponding imaginary $k_{\vec{G}z}$ component of the diffracted wavevector $k_{\vec{G}z} = i(-2m\varepsilon/\hbar^2)^{1/2}$ where

$\varepsilon < 0$. The closed channel is a resonant or selective adsorption channel if ε matches the energy ε_n of a bound state of the atom-surface potential. In this state, the atom is expected to have an enhanced probability of undergoing inelastic scattering, but the actual relevance of inelastic scattering from a resonant state is a matter of controversy [14].

The arguments in the previous sections are applicable to closed channels provided i) we define a complex polar angle $\Theta_{\vec{G}} = \pi/2 + i\zeta_{\vec{G}}$ for the closed channel such that $\sin\Theta_{\vec{G}} = \cosh\zeta_{\vec{G}} = (\vec{K}_i + \vec{G})/k_i = [1 + |\varepsilon|/\varepsilon_i]^{1/2}$ where $\varepsilon_i = \hbar^2 k_i^2/2m$ and ii), we restrict the inelastic events to phonon annihilation processes for which $k_z^2 > 0$. This means $k_z^2 = 2\beta k_i Q - 2\vec{Q}\cdot\vec{K}_{\vec{G}} - Q^2 - k_i^2|\varepsilon|/\varepsilon_i > 0$. The main consequence of this requirement is that low frequency phonons do not contribute to the inelastic scattering and that the FAZ and DDP structures are truncated at an angle $\Theta < \pi/2$.

The detection of streaks from a resonant state would not only support the use of the single phonon approximation but it would also lend further evidence of the selective adsorption phenomenon by allowing us to peek into the closed channel just below the crystal surface.

5. CONCLUSIONS

We conclude that observed intensity streaks in atomic scattering are the result of single surface phonon scattering, and that the angular positions of these streaks can be used to determine model parameters for the surface phonon dispersion relation. Observation of streaks may be especially prominent for closed channels under bound state resonance conditions. Such observations would have profound consequences for 2-D structures as they indicate energy and momentum exchange between the adsorbed species and the substrate, particularly under low coverage conditions.

REFERENCES

1. J. M. Horne and D. R. Miller, Phys. Rev. Lett. 41, 511 (1978).
2. G. Brusdeylins, R. B. Doak and J. P. Toennies, private communication.
3. B. F. Mason and B. R. Williams, Surf. Sci. 77, pp. 385-399 (1978).
4. P. Cantini, R. Tatarek and G. P. Felcher, Surf. Sci. 63, 104 (1977).
5. R. Avila and M. Lagos, submitted to Phys. Rev. Letters.
6. G. Benedek, Phys. Rev. Lett. 35, 234 (1975).
7. M. Lagos and L. Birstein, Surf. Sci. 51, 469 (1975).
8. I. Waller and P. O. Froman, Arkiv Pysik 4, 183 (1952).
9. B. R. Williams, J. Chem. Phys. 55, 3220 (1971).
10. B. F. Mason and B. R. Williams, J. Chem. Phys. 61, 2765 (1974).
11. D. N. Gazis, R. Herman and R. F. Wallis, Phys. Rev. 119, 533 (1960).
12. American Institute of Phys. Handbook, McGraw Hill, 1963, p. 2-52.
13. T. S. Chen, F. W. deWette and G. P. Alldredge, Phys. Rev. B 15, 1167 (1977).
14. H. Chow and E. D. Thompson, Surface Science 82, 1 (1979).

† Research supported in part by the NSF under Grant No. DMR 79-01658

It is a pleasure to thank Prof. Eric Thompson for his creative reading of the manuscript and Dr. B. R. Williams for providing us with his inelastic streaks data.

Sinha, ed. Ordering in Two Dimensions

EVIDENCE FOR A DISORDERED $\sqrt{19} \times \sqrt{19}$ STRUCTURE FOR THE QUENCHED CLEAN Si(111) SURFACE

Y. J. CHABAL and J. E. ROWE

Bell Laboratories, Murray Hill, NJ 07974

The ability to produce clean Si(111) surfaces with different well-characterized structures is important for further understanding of various chemisorption mechanisms. We report the LEED observation and UPS study of a structure obtained by quenching a Si(111) surface. The streaky background of the LEED pictures is attributed to the presence of small domains with $\sqrt{19} \times \sqrt{19}$ symmetry and is to be distinguished from either the sharp 1×1 structure obtained by laser annealing or the sharp $\sqrt{19} \times \sqrt{19}$ structure obtained when 10% of a Ni monolayer is present. The transition to a 7×7 structure is reversible. UV Photoemission spectroscopy indicates large differences in the electronic structure between the two surfaces.

INTRODUCTION

The Si(111) surface has several metastable and stable surface reconstruction structures: upon cleaving, the surface produces a 2×1 LEED pattern; heating the surface to about 400°C induces a disordered 1×1 structure while further heating to 700°C gives the well-known 7×7 structure(1). Above 870°C, a disordered 1×1 pattern appears(2). Recently, it has also been shown that a sharp 1×1 LEED pattern can be obtained by laser annealing(3). Besides the reconstruction patterns of the clean Si(111) surface, several structures are induced by trace amounts (less than 10% monolayer) of various absorbates. The so-called Si(111) 1* pattern is caused by trace amounts of Cl(4) while Te will give rise to a sharp 1×1 structure(5). A sharp $\sqrt{19} \times \sqrt{19}$ pattern can be obtained by quenching a Si(111) sample with trace amount of Ni present(6).

We report here the observation of a structure, obtained by quenching a nominally clean Si(111) sample from 1000°C, which we attribute to disordered domains with $\sqrt{19} \times \sqrt{19}$ symmetry. The $k_{//}$ dependence of the streaks with respect to the integral order spots is studied in order to arrive at a structural model. The presence of metal impurities such as Ni is monitored both by AES and Rutherford Backscattering Spectroscopy (RBS). The photoemission spectra of the quenched and 7×7 surfaces indicate large differences in the electronic structure close to the Fermi energy.

EXPERIMENTS AND RESULTS

The samples used were polished Si(111) crystals very lightly doped [p-type, $\sim 10^{15}$ - 10^{16} cm^{-3}, $\rho \sim 1$ to 20 Ω cm]. Each crystal [2 × 0.5 × 0.02 cm^3] was held by two Ta clips and could be heated resistively to 1100°C by passing 10 Amp through. The Ta clips were in thermal contact with a liquid N_2 reservoir allowing the sample to be cooled down to about 170K. The cleaning procedure consisted in bombarding the sample for a few hours with 10 μA/cm^2 of 1 KV Ar^+ ions to remove the native oxide and the carbon present and in thermal annealing at 900°C for several minutes to restore the crystallinity of the surface layer. The cleanliness was checked with a Phi single Pass Auger Analyzer capable of detecting about 0.3% of an impurity monolayer. When the sample was cleaned and annealed, the LEED pattern showed sharp 7×7 features characteristic of the annealed Si(111) surface as shown in Fig. 1a. When the sample was heated briefly to 1000°C and suddenly cooled by throwing the heater switch while keeping the liquid N_2 reservoir full, the resulting LEED pattern changed drastically as shown in Fig. 1b. The integral order spots of the quenched pattern are strong without evidence of broadening. The 7th order spots are completely absent. Instead a moderately strong and structured background is present. As shown in Fig. 1b for E = 50 eV and present for most electron energies used (30 → 100 eV), slightly asymmetric but straight streaks occur at the 3/5 and 2/5 of the integral order reciprocal vector. The 7×7 pattern could readily be recovered by a slow anneal: as the sample is heated, the 7th order spots start to appear at about 450°C. A short anneal to 800°C followed by a relatively slow cooling restored the sharp 7×7 pattern. The quenched structure could always be obtained by heating briefly to T > 900°C and cooling at a rate estimated to be 100°C/sec around 900°C. For quantitative analysis, the LEED pictures were recorded with a PAR Optical Multichannel Analyzer (OMA) Vidicon camera which digitizes the intensity/position information and transfers it to a Nova 3 minicomputer where it is stored(7). An example is given in Fig. 1c.

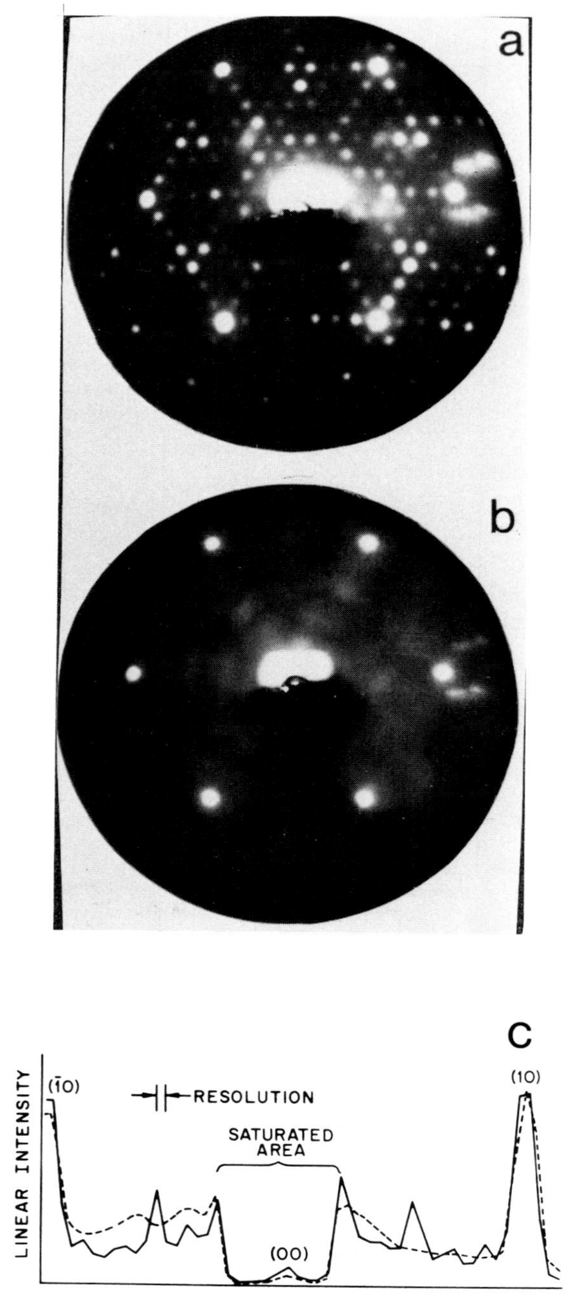

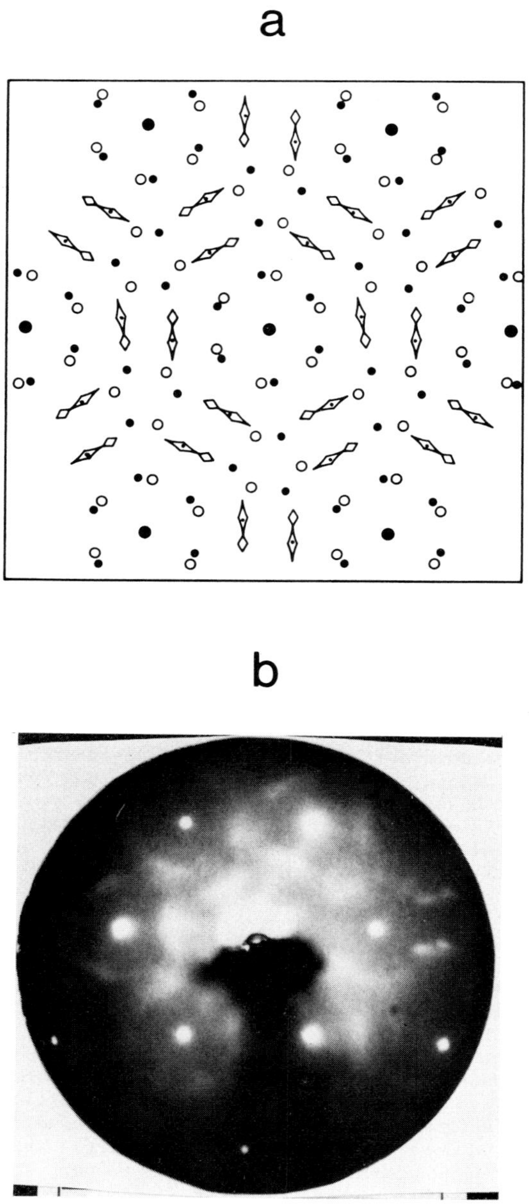

FIGURE 2

Figure 1: (a) LEED picture of the annealed Si(111) 7×7 surface for 50 eV electrons.
(b) LEED picture of the quenched Si(111) surface for 50 eV electrons.
(c) LEED intensity versus $k_{//}$ along the (10) direction for 50 eV electrons. The solid line is for the 7×7 surface and the dashed line for the quenched surface. The (00) spot is reduced due to saturation of the vidicon camera by the visible reflected light.

Figure 2: (a) Diffraction pattern from the $\sqrt{19}\times\sqrt{19}$ structure. The dots and open circles correspond to the 2 equivalent domains. The most intense streaks are sketched with different weight for the 2 different domains to bring out the asymmetry with respect to the reciprocal vectors associated with the 1×1 structure.
(b) LEED picture of the quenched Si(111) surface for 80 eV electrons.

Several structures were investigated as possible candidates for this disordered background: 2×1, 7×7, 3×3, $\sqrt{3}\times\sqrt{3}$, 5×5 and $\sqrt{19}\times\sqrt{19}$. The first four could readily be eliminated since both the streak position ($k_{//}$ dependence) and direction could not be matched. The 5×5 structure however was an interesting candidate since it had been obtained by Bauer and Poppa on Si(111) with a few percent of a Ni monolayer[8]. However, the inherent asymmetry of the observed streaks most noticeable for 40<E<60 eV cannot be explained by a symmetric 5×5 reconstruction. Instead, we find that disordered domains of $\sqrt{19}\times\sqrt{19}$ surface reconstruction fit the data well. Fig. 2 shows the $\sqrt{19}\times\sqrt{19}$ structure with the most intense streaks sketched. The pattern is made up of 2 equivalent $\sqrt{19}\times\sqrt{19}$ patterns with unit meshes rotated by $\theta = \pm \tan^{-1}(\sqrt{3}/4) = \pm 23.4°$ relative to the substrate lattice, i.e. rotated by 46.8° with respect to each other. The most intense streaks are associated with the spots located at a distance of $\sqrt{3}/\sqrt{19}$ from the integral order spots but with a small rotation off the integral reciprocal vectors of $\pm[\frac{\pi}{6} - \tan^{-1}(3/4)]$. In the $\sqrt{19}\times\sqrt{19}$ superlattice space they correspond to the (1 1) and (1 2) reciprocal vectors. If one of the two equivalent domains dominates, the streaks will appear slightly asymmetric with respect to the integral order spots. In Fig. 3, a possible surface structure is shown where long and narrow domains of $\sqrt{19}\times\sqrt{19}$ symmetry are present. The overall direction of these domains is given by the reciprocal vectors associated with the (1 1) and (1 2) 2d-"lines" of the superlattice structure. The domains with a given symmetry are rotated 60° from each other and separated by antiphase $\sqrt{19}\times\sqrt{19}$ domains with the other possible symmetry. The width of those domains is about 4 lattice vectors of the $\sqrt{19}$ unit mesh, i.e. $4\sqrt{19} \times 3.84Å \simeq 70Å$ with a length about twice as long (∿ 140Å). To check the validity of the assignment, we prepared a Si(111) surface with 10% of a Nickel monolayer. Upon quenching, a sharp $\sqrt{19}\times\sqrt{19}$ pattern was obtained and the $k_{//}$ dependence of the quenched structure for the clean Si(111) could be reconciled nicely with the sharp pattern. The presence of Ni on the nominally clean quenched surface was investigated by AES in situ and RBS after the runs. Ni was occasionally found but never in excess of about 1×10^{-1} of a monolayer ($<8\times10^{13}$ Ni/cm^2).

The electronic structure of the surface was studied by means of UPS. A typical spectrum obtained with 11.7 eV radiation is shown in Fig. 4. The peak at -4.70 eV below E_F for the quenched surface corresponds to a peak in the bulk silicon energy distribution of the joint density of states and occurs at -4.80 eV below E_F for the 7×7 surface. This indicates a change in the relative position of E_F and the valence band near the surface. The quenching of the (111) surface therefore modifies the surface band bending due to dangling bonds by about 0.10 eV. The difference in the dangling

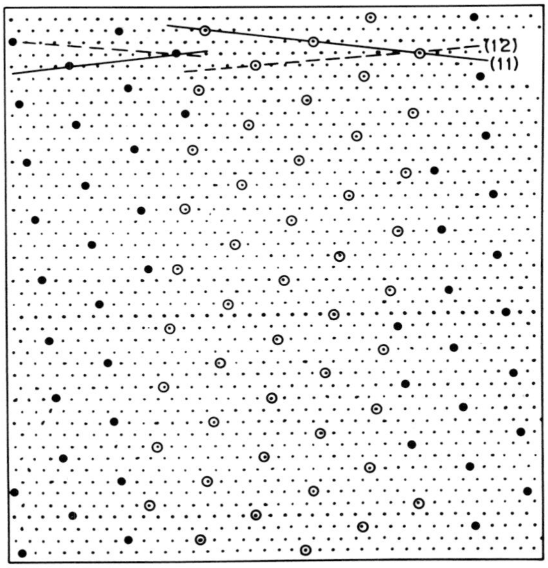

Figure 3: Possible real space configuration. The open circle represent one of the small domains with $\sqrt{19}\times\sqrt{19}$ symmetry. The large dots represent the domains with the other possible $\sqrt{19}\times\sqrt{19}$ symmetry. The small dots represent the underlying Si(111) surface.

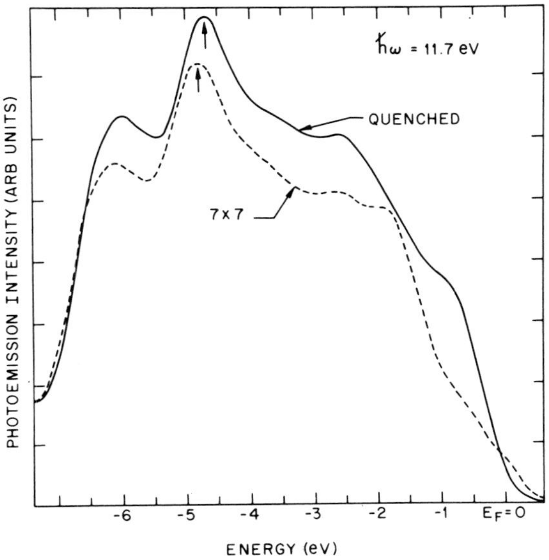

Figure 4: Photoemission spectra of both the 7×7 and quenched surface for $\hbar\omega$ = 11.7 eV.

band surface state is further seen in the region within 1 eV below E_F. The metallic character of the surface peak on the 7×7 surface disappears as the surface peak moves further away from E_F for the quenched surface as was observed in the case of the 2×1 surface(1).

CONCLUSION

A disordered $\sqrt{19}\times\sqrt{19}$ structure has been reproducibly obtained on a nominally clean Si(111) surfaces by quenching. The same surface give rise to a sharp 7×7 structure if the annealing is done slowly. The presence of Ni is always less than 10 % of a monolayer when detected but, although no direct evidence exists at present, we cannot eliminate the possibility that Ni stabilizes this quenched structure. The position of the dangling bond surface peak is moved away from the Fermi energy leaving a gap as in the case of the 2×1 surface. However, chemisorption of Cl and H indicates that there are no longer well defined chemisorption sites for the quenched structure in contrast to the 2×1 surface(9).

REFERENCES

(1) J. E. Rowe and J. C. Phillips, Phys. Rev. Lett. 32, 1315 (1974).
(2) P. Bennett and M. B. Webb, 39th Annual Conference on Physical Electronics; University of Maryland, June 1979.
(3) D. M. Zehner, C. W. White and G. W. Ownby Surf. Sci. 92, L67 (1980).
(4) J. V. Florio and W. D. Robertson, Surf. Sci. 24, 173 (1971).
(5) H. D. Shih, F. Jona, D. W. Jepsen and R. M. Marcus, Phys. Rev. Lett. 37, 1622 (1976).
(6) A. J. Van Bommel and F. Meyer, Surf. Sci. 8, 467 (1967).
(7) S. P. Weeks, J. E. Rowe, S. B. Christman and E. E. Chaban, Rev. Sci. Instr. 50, 1249 (1979).
(8) E. Bauer, Vacuum 22, 539 (1972).
(9) M. Schluter, J. E. Rowe, G. Margaritondo, K. M. Ho and M. L. Cohen, Phys. Rev. Lett. 37, 1632 (1976).

Sinha, ed. Ordering in Two Dimensions

POLARIZATION OF RAMAN MODES OF CO ADSORBED ON Ni(111)†

E. B. BRADLEY J. M. STENCEL*

Department of Electrical Engineering, University of Kentucky
Lexington, Kentucky 40506, U.S.A.

We have measured the activity of the xy, yy and xx components of the polarizability tensor for Raman bands at 2181 cm^{-1} and 81 cm^{-1} for CO + H_2 adsorbed on Ni(111). Also, the effects upon the intensities of these bands from Ar^+ milling of the surface shows that the bands cannot be eliminated by milling, but only by repolishing the crystal. It is thought that the Raman active adsorbed species is graphitic.

1. INTRODUCTION

Experiments in Raman scattering from CO adsorbed on Ni(111), Ni(110) and Ni(100) have shown the utility of Raman spectroscopy as a probe for the bonding and structure of adsorbed molecular species [1]. This technique can also be used for studies of adsorbed species at high pressures where LEED can no longer provide useful information. The Ni crystal was purchased from Materials Research Corporation, oriented, then cut to needed dimension with a diamond-edged saw. The crystal was then polished in a water-alumina slurry which had successive particle sizes of 1.0, 0.3 and 0.05μ. Chemical etching was performed in a solution of HNO_3:H_3PO_4:HCH_3COO; the X-ray diffraction patterns showed well-defined Laue spots indicative of the (111) plane to within ± 1.5° of the sample surface. This sample was placed in a UHV chamber for study by laser Raman spectroscopy of the adsorption of CO + H_2 on the surface. With the chamber pressure at 5 x 10^{-9} torr the Ni crystal was heated to 500°C and H_2 was flowed at 5 x 10^{-1} torr for one hour while a spectral scan was run. No Raman bands were observed. Afterward the sample was cooled to 30°C, CO was introduced at 1 torr, then H_2 was admitted to obtain a total pressure of CO + H_2 of 4 torr. The following experiments were performed.

2. EXPERIMENTAL

After this exposure a strong Raman band was found at 69 cm^{-1} using 1.5 W of the 5145 Å line with the detector sensitivity at 5 x 10^2 cps. The sample chamber was evacuated for a two-week period with a Ni crystal temperature of 25°C. The center frequency of the band shifted approximately 12 cm^{-1} to 81 cm^{-1} and the band intensity increased by 300%.

The effects of Ar^+ milling were investigated next to determine whether it would creat observable spectral changes on the 81 cm^{-1} band. This milling and heating of the sample to 400°C decreased the 81 cm^{-1} band intensity by 25%. It also showed that the broad 81 cm^{-1} band is a composite of other bands; the shoulder at 69 cm^{-1} is at the same frequency as the band observed originally. The sputtering and heating may have reordered the adsorbed species. The sample was then heated to 440°C for a two-hour period and a survey scan acquired. Besides the 81 cm^{-1} band, a new peak was observed which was centered at 2181 cm^{-1} and which had approximately 10% the intensity of that at 81 cm^{-1}. In Figure 1 is shown the 2160 -

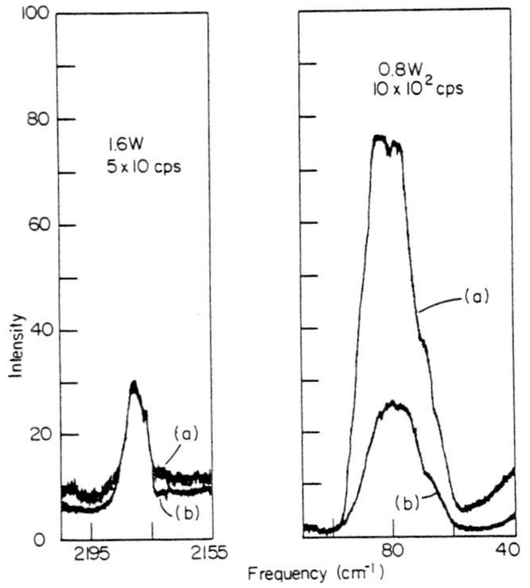

Figure 1. The 2180 cm^{-1} and 81 cm^{-1} spectral regions using 4880 Å excitation with 1.5 cm^{-1} spectral slitwidths. (a) H_2 exposure, and (b) CO exposure.

†Work supported by Department of Energy, Office of Basic Research, Contract #DEAS05-79ER10447.

*Present address: Pittsburgh Energy Technology Center, Pittsburgh, Pennsylvania 15213, U.S.A.

2210 cm^{-1} and 50 - 100 cm^{-1} spectral regions after the Ni crystal was exposed to CO + H_2 and the chamber evacuated to 6×10^{-8} torr with the sample at 350°C. The 81 cm^{-1} band is shown to have complex structure which consists of at least three modes at 69, 78 and 85 cm^{-1}. Polarization studies were also performed to characterize the 2181 and 81 cm^{-1} bands. The z-direction is perpendicular to the crystal surface. In Figure 2 is shown the results of this analy-

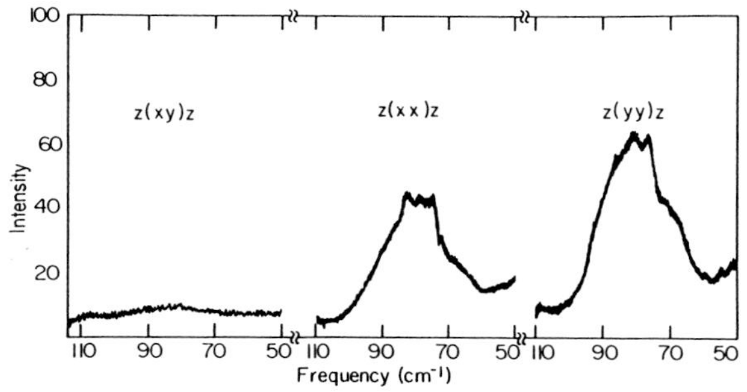

Figure 2. The activity of the xy, yy, and xx polarizability components for the 81 cm^{-1} band.

sis for the 81 cm^{-1} region. The xx and yy polarizability components are active for the 81 cm^{-1} band and give approximately the same band shape; the intensity differences are due to the polarization efficiencies of the half-wave plate and analyzer. The xy component does not contribute to the 81 cm^{-1} region. The 2181 cm^{-1} band is present for all polarizations which were studied.

Next, the effects of severe milling were investigated using Ar^+ sputter gun settings of 2 KV accelerating voltage and 30 mA filament current. The 81 cm^{-1} band in Fig. 2 shows side-band structure as a result of this sputtering. However the intense sputtering would decrease the 81 cm^{-1} band intensity but would not completely remove the surface species which is responsible for this band. Only repolishing of the crystal surface would eliminate the band. As shown in Figure 3 the exposure of the sputtered Ni surface to O_2 at 5×10^{-1} torr at a 350°C temperature decreased the intensity to 13% of its original value. At the same time, the intensity of the 2181 cm^{-1} band increased substantially. Furthermore, a survey scan over the 40 - 4000 cm^{-1} region showed that two more Raman bands existed at 3160 and 3223 cm^{-1}.

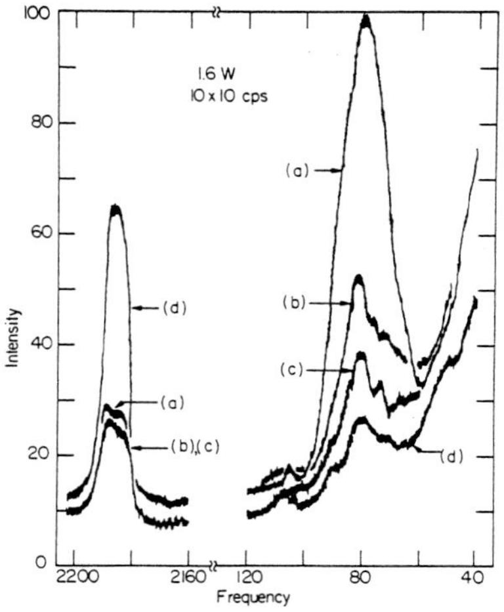

Figure 3. The intensity of the 2180 cm^{-1} and 81 cm^{-1} regions (a) before Ar^+ sputtering, (b) after 1.5 hours Ar^+ sputtering, (c) after 3.0 hours Ar^+ sputtering and (d) after O_2 exposure at 350°.

3. DISCUSSION

The Raman data are not easily explained by characteristic Raman frequencies and intensities of known compounds. CO + H_2 exposure causes the 81 cm^{-1} band to appear but its frequency is well below infrared bands for CO (2090 - 1050 cm^{-1}) and H_2 (4300 cm^{-1}). The lack of Raman data on low-frequency modes (< 100 cm^{-1}) and the fact that few Raman surface studies have been performed complicate further the spectral interpretation. The low frequency of the 81 cm^{-1} band is characteristic of lattice modes in crystalline solids but Raman intensities due to intra-molecular vibrations are normally more intense. Other Raman bands which were observed remain approximately constant in intensity during O_2, CO and H_2 exposure. For this reason these bands (2181, 3160 and 3223 cm^{-1}) are not expected to be due to an active surface species as is the 81 cm^{-1} band; however, the 2181 cm^{-1} band may be due to a surface species in equilibrium with that which creates the 81 cm^{-1} band.

Some studies have shown that the low-frequency modes of hydrocarbons have significant intensities. For example, the Raman spectrum of p-diphenyl benzene (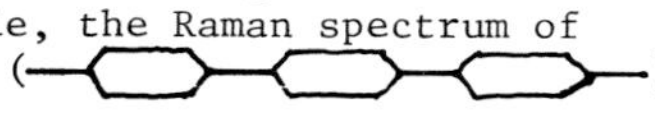)

shows it to have intense bands in the 50 - 100 cm^{-1} region [2]. Spectra we have obtained of this compound show that these low-frequency modes are more intense than the higher-frequency intra-molecular vibrations. This intensity may be due to crystal orientation with respect to the impringing radiation, and if so, similar arguments would be applicable to the spectra of the adsorbed species on Ni.

Spectral studies which have been important in providing a possible explanation for these results are the polarization measurements. Besides those which are shown in Fig. 2, preliminary measurements have been obtained to determine the activity of the zy and zx polarizability components. These measurements show that the zy component is 10 times more intense than zx for the 81 cm^{-1} band. Thus the xx, yy and zx components produce appoximate-equal intensities whereas xy and zy are very close to zero. This activity is similar to what is observed for lattice vibrations in naphthalene ($C_{10}H_8$) and anthracene ($C_{14}H_{10}$) [3]. Six librational modes for the monoclinic unit cells of these compounds are located between 125 - 46 cm^{-1} ($C_{10}H_8$) and 125 - 39 cm^{-1} ($C_{14}H_{10}$). Unpolarized measurements obtained in our laboratory show that three of these modes are 1 1/2 - 8 times more intense than higher frequency modes. The C/H ratio for the cyclic hydrocarbons compounds which have been investigated in our laboratory is between 1.25 and 2.0. Tetracene (C/H = 1.5) and perylene (C/H = 2.0) have an orange-yellow color which introduced intense fluorescence obscuring the Raman modes. As the C/H ratio becomes larger, the CH band intensities should decrease relative to those due to CC modes, and the number of Raman modes will change because of the shift in the symmetry of the compound. For example, graphite (C/H = ∞) has two Raman active modes; the intraplanar mode at 1582 cm^{-1} is approximately 10^2 more intense than the interplanar mode at 140 cm^{-1} [4]. If the adsorbed species which produces the 81 cm^{-1} band has cyclic-carbon structure, the initial layers adjacent to the Ni surface would be distorted as a result of strong adsorbent-adsorbate bonding. Nevertheless, the C-C bond distances could be nearly equal to those in cyclic hydrocarbons, e.g., the 1.42 Å Ni-Ni distances (Ni(111)); the C-C distance in graphite is 1.42 Å while it varies between 1.39 - 1.45 Å for pyrene ($C_{16}H_{10}$). Recent semiempirical model calculations show that the energy profile of Ni(111) is relatively constant with respect to that calculated for Ni(100) or Ni(110) [5]. The ± 3 Kcal/mole profile for CO adsorption on Ni(111), as compared to ± 15 Kcal/mole on Ni(110), could ease lateral dimensional requirements for the carbon atom positioning. At the same time, intense π-bonding forces would allow the adsorbate to be very stable. Such stability is evident in our adsorption-evacuation studies which show that the time to completely remove the adsorbed species is greater than six days at 550°C. Strong π-bonding for C_6H_6 and other unsaturated hydrocarbons on Ni orient the admolecules parallel to the surface. Above 200°C, at which temperature oxidation effects are observed in our experiments, C_6H_6 dissociates and forms a 'carbonaceous' surface species [6,7]. A 'carbonaceous' species is also formed when CO dissociates over Ni at temperatures above 100°C. Similarly, it has been shown that clean Pt surfaces are not catalytically active whereas a 'carbonaceous' adlayer introduces activity [8.9]. Thus, our adsorption results do not preclude a cyclic hydrocarbon with a high C/H ratio from being able to explain the present experimental data. Therefore, we assume that the Raman active adsorbed species is graphitic in nature.

REFERENCES

1. J. M. Stencel and E. B. Bradley, J. Raman Spectrosc. 8, 203 (1979).
2. S. K. Mukerji and Laksman Singh, Phil. Mag. 37, 874 (1946).
3. M. Suzuki, T. Yokoyama and M. Ito, Spectrochim. Acta 24A, 1091 (1968).
4. J. J. Song, D. D. L. Chung, P. C. Eklund and M. S. Dresselhaus, Solid State Comm. 20, 1111 (1976).
5. G. Doyen and G. Ertl, Surf. Sci. 69, 157 (1977).
6. J. E. Demuth and D. E. Eastman, Phys. Rev. Letters 32, 1123 (1974).
7. T. E. Fischer, S. R. Keleman and H. P. Bonzel, Surf. Sci. 64, 157 (1977).
8. G. A. Somorjai, Catalysis Rev. 7, 87 (1972).
9. K. Baron, D. W. Blakely and G. A. Somorjai, Surf. Sci. 41, 45 (1975).

EFFECT OF H ON THE COMMENSURATE-INCOMMENSURATE TRANSITION OF W(001) SURFACE+

K. H. LAU AND S. C. YING

Department of Physics, Brown University, Providence, Rhode Island 02912

ABSTRACT

The lattice gas model with the possibility of lattice displacement is used to describe the H/W(001) system. We extend our earlier discussions for low coverages to higher coverages and focus in particular on the commensurate-incommensurate transition. It is found that H-W and H-H interactions have significant influences on this transition. We illustrate with a particular choice of parameters how increasing coverage of H adatoms drive the system from a commensurate to an incommensurate phase.

+Supported in part by a N.S.F. Grant No. DMR76-10379 and by the Materials Research Program at Brown University funded through the National Science Foundation.

Hydrogen adsorbed on W(001) is one of the most studied chemisorption systems[1]. In the LEED study of H/W(001) [2], it has been well established that a c(2×2) pattern is formed at low coverages, followed by a transition to an incommensurate phase at higher coverages as indicated by the splitting of the half-order spots. In an earlier paper [3], we introduced a theoretical model to study the effect of H adsorption on the displacive transition of W(001) surface. We showed that at low coverages, H enhances the W surface distortion and causes the distortion to switch direction at some finite coverage. In this paper, we shall focus attention on higher coverages where the commensurate-incommensurate transition takes place. In our model, the Hamiltonian describing the H adatom in the presence of a substrate distortion is written as

$$H_{ad}=\frac{1}{2}\sum_{i',j'} J_{i'j'}n_{i'}n_{j'}+\sum_{i,i'} n_{i'}v(\vec{R}_i^0+\vec{u}(\vec{R}_i^0)-\vec{R}_{i'}) \quad (1)$$

where $\vec{R}_{i'}$ is the coordinate of the bridge-binding site for H and $n_{i'}$ is the occupation number. $J_{i'j'}$ is the H-H interaction with strengths J_1 and J_2 between nearest and next-nearest neighbors. The second term in (1) describes the H-W interaction as a sum of pair potentials v, with the possibility of a substrate atom displacement from the periodic position $\vec{R}^0$ to $\vec{R}^0+\vec{u}$.

As in [3], the W atom displacement is taken to be in the surface plane and is described by two order parameters ϕ_1, ϕ_2:

$$\vec{u}(\vec{R})=\hat{K}_1\mathrm{Re}(\psi_1(\vec{R}))+\hat{K}_2\mathrm{Re}(\psi_2(\vec{R}))$$
$$=\hat{K}_1\mathrm{Re}(e^{i\vec{K}_1\cdot\vec{R}}\phi_1(\vec{R}))+\hat{K}_2\mathrm{Re}(e^{i\vec{K}_2\cdot\vec{R}}\phi_2(\vec{R})) \quad (2)$$

where $\vec{K}_{1,2}=(\pi,\pm\pi)$ and $\hat{K}_{1,2}$ are unit vectors along $\vec{K}_{1,2}$. The unit of length has been chosen such that the lattice constant is unity. In the commensurate phase ϕ_1 and ϕ_2 are constants while in the incommensurate phase, they are functions of position.

In the presence of a periodic lattice distortion described by (2), the binding energy of H over the bridge sites is also modulated periodically with the same wavevector. Hence, the average occupation number $n_{i'}$ takes the form

$$\bar{n}_{i'}=\frac{\theta}{2}+\mathrm{Re}(n'(\vec{R}_{i'})e^{i\vec{G}\cdot\vec{R}_{i'}}$$
$$-im_1(\vec{R}_{i'})e^{i\vec{K}_1\cdot\vec{R}_{i'}}-im_2(\vec{R}_{i'})e^{i\vec{K}_2\cdot\vec{R}_{i'}}), \quad (3)$$

where $\vec{G}=\vec{K}_1+\vec{K}_2$, and θ is the coverage with $\theta=1$ defined as one H atom per surface W atom.

In the mean field approximation, the free energy for the H adatoms in the presence of lattice displacement can be expressed as

$$F_{ad}=-2\lambda\int\{\mathrm{Re}(m_1(\vec{R}))\mathrm{Re}(\phi_1(\vec{R}))+\mathrm{Re}(m_2(\vec{R}))\mathrm{Re}(\phi_2(\vec{R}))\}d^2\vec{R}$$
$$+\frac{1}{2}\sum_{i',j'} J_{i'j'}\bar{n}_{i'}\bar{n}_{j'}$$
$$+T\sum_{i'}\{\bar{n}_{i'}\ln\bar{n}_{i'}+(1-\bar{n}_{i'})\ln(1-\bar{n}_{i'})\}, \quad (4)$$

where T is the temperature, $\bar{n}_{i'}$ is given by (3) and λ is the coupling constant related to the derivative of the pair potential v.

Recent electronic structure calculations

suggest that the driving mechanism for the displacive transition for the W surface is either a Fermi-surface instability leading to a charge density wave [4,5], or an intrinsic atomic instability with wavevector determined by the Fermi-surface structure [6]. In either case the free energy for the substrate may be described phenomenologically by a Landau-Ginzburg expansion similar to that used by McMillan [7]:

$$F_s=\int d^2\vec{R}[r\{(Re\psi_1)^2+(Re\psi_2)^2\}+u\{(Re\psi_1)^2+(Re\psi_2)^2\}^2$$
$$+v(Re\psi_1)^2(Re\psi_2)^2+\sum_{i=1,2}\{c|\vec{k}_i\cdot(\vec{\nabla}-i\vec{k}_i)\psi_i|^2$$
$$+d|\vec{k}_i\times\vec{\nabla}\psi_i|^2\}]. \qquad (5)$$

where $\vec{k}_1$, $\vec{k}_2$ are nesting vectors for the W Fermi surface. For $\vec{k}_{1,2}\neq\vec{K}_{1,2}$ the gradient terms in (5) favor the incommensurate phase, which is minimized when the order parameter ψ_i has wavevector $\vec{k}_i$. The coefficients r, u, etc., exhibit the periodicity of the substrate. For example

$$u=u_0+u_1\sum_\alpha e^{i\vec{G}_\alpha\cdot\vec{R}}+u_2\sum_\beta e^{i\vec{G}_\beta\cdot\vec{R}}+\ldots \qquad (6)$$

where $\vec{G}_\alpha$ is a reciprocal lattice vector of length $2|\vec{K}_1|$ and $|\vec{G}_\beta|=4|\vec{K}_1|$.

The total free energy F is the sum of (4) and (5). As discussed in [3], to minimize F, we must have either (i) $\phi_1=0$ or $\phi_2=0$ or (ii) $\phi_1=\pm\phi_2$. For simplicity, we shall discuss the 1-component case here, so that $\phi_2=m_2=n'=0$ and

$$\vec{u}(\vec{R})=\hat{K}_1 Re(e^{i\vec{K}_1\cdot\vec{R}}\phi_1(\vec{R})) \qquad (7)$$
$$=\hat{K}_1\phi_0 Re(e^{i\vec{K}_1\cdot\vec{R}}e^{i\alpha(\vec{R})}),$$

$$m_1(\vec{R})=m_0e^{i\alpha(\vec{R})} \qquad (8)$$

In writing down (8), we already anticipate that for the free energy to be minimized, the phase factors for the W atom displacement $\vec{u}$ and for the H order parameter m_1 have to be the same.

To fourth order in the order parameters, the total free energy takes the form

$$F=\int d^2\vec{R}[\tfrac{1}{2}(r_0-r_1\cos 2\alpha)\phi_0^2+(\tfrac{3}{8}u_0+\tfrac{1}{2}u_1\cos 2\alpha+\tfrac{1}{8}u_2\cos 4\alpha)\phi_0^4$$
$$+\tfrac{1}{2}\{c_0|\hat{K}_1\cdot(\vec{\nabla}\alpha-\vec{Q})|^2+d_0|\hat{K}_1\times(\vec{\nabla}\alpha-\vec{Q})|^2\}\phi_0^2$$
$$+r'm_0^2\cos^2\alpha+u'm_0^4\cos^4\alpha-2\lambda m_0\phi_0\cos^2\alpha]$$
$$+\sum_{\vec{q}}[\tfrac{1}{2}J_1\{-\cos(\vec{q}\cdot\hat{K}_1)+\cos(\vec{q}\cdot\hat{K}_2)\}$$
$$+J_2\{1-\cos(\vec{q}\cdot\hat{K}_1)\cos(\vec{q}\cdot\hat{K}_2)\}]\times m(\vec{q})(m(\vec{q}))^* \qquad (9)$$

where

$$\vec{Q}=\vec{k}_1-\vec{K}_1 \qquad (10)$$

$$r'=4T/\theta(2-\theta)-4J_2 \qquad (11)$$

$$u'=32T(1-\tfrac{3}{2}\theta+\tfrac{3}{4}\theta^2)/3\theta^3(2-\theta)^3 \qquad (12)$$

and

$$m(\vec{q})=m_0\int d^2\vec{R}e^{i\alpha(\vec{R})}e^{-i\vec{q}\cdot\vec{R}} \qquad (13)$$

is the Fourier transform of $m_1(\vec{R})$.

In (9), the umklapp term r_1(taken to be positive without loss of generality) and the coupling term $-2\lambda_0 m_0\cos^2\alpha$ favor the commensurate phase, while the gradient term and the quadratic and quartic terms in m_0 both favor the incommensurate phase. The last term in (9) may favor the commensurate or the incommensurate phase, depending on the magnitudes and signs of the interactions J_1 and J_2. For $|J_1|>2J_2$ or $J_2<0$, the expression inside the parenthesis is not positive definite, and the free energy will be lowered when the order parameter $m_1(\vec{R})$ developes a long wavelength spatial modulation.

For the commensurate phase, we can set $\alpha=0$ in (9), leading to the expression

$$F_c=\tfrac{1}{2}(r_o-r_1+c_0Q^2)\phi_0^2+(\tfrac{3}{8}u_0+\tfrac{1}{2}u_1+\tfrac{1}{8}u_2)\phi_0^4$$
$$+r'm_0^2+u'm_0^4-2\lambda m_0\phi_0. \qquad (14)$$

For the incommensurate phase we shall consider in this paper only the simple sinusoidal form for the phase factor, i.e. $\alpha(\vec{R})=\vec{q}\cdot\vec{R}$, neglecting all higher harmonics. This is known to be incorrect near the commensurate-incommensurate transition where the effect of higher harmonics in $\alpha(\vec{R})$ is important leading to a domain wall structure [7]. However, our main goal in this paper is to discuss the location of this transition as a function of the coverage. The single harmonic approximation is quite adequate for this purpose since the region for which the higher harmonics are important is very narrow. For this choice of phase factor, the free energy function in the incommensurate phase is

$$F_I=\tfrac{1}{2}r_0\phi_0^2+\tfrac{3}{8}u_0\phi_0^4+\tfrac{1}{2}c_0\phi_0^2(\vec{q}-\vec{Q})^2+\tfrac{1}{2}r'm_0^2+\tfrac{3}{8}u'm_0^4$$

$$+\frac{1}{2}\{\frac{1}{2}J_1(-\cos(\vec{q}\cdot\hat{K}_1)+\cos(\vec{q}\cdot\hat{K}_2))$$

$$+J_2(1-\cos(\vec{q}\cdot\hat{K}_1)\cos(\vec{q}\cdot\hat{K}_2))\}m_0^2-\lambda m_0\phi_0. \quad (15)$$

Experimentally, only the normal-commensurate transition is observed for the W(001) surface and the H/W(001) system at low coverages. This implies that $r_1-c_0Q^2\geq 0$. At higher coverages, there are various competing factors in the coupled system as discussed earlier, each tending to enhance or suppress the commensurate or the incommensurate phase. Thus, the final result depends very much on the coupling constants appearing in the free energy expression (9). This is in sharp contrast to the generally parameter independent consequence of H adsorption on the displacive transition at low coverages. To make all these ideas more concrete we have calculated the free energy for the commensurate state F_c and that for the incommensurate state F_I for a particular choice of parameters. For the commensurate phase, we put $\frac{1}{2}(r_0-r_1+c_0Q^2)=A(T-T_s)$ where T_s is the transition temperature for the clean W surface. We take (in units of degree K) A=160, T_s=290, $\frac{3}{8}u_0+\frac{1}{2}u_1+\frac{1}{8}u_2=2.0\times10^6$ and $\lambda=1.1\times10^5$. The values of r' and u' are given in (11) and (12) with J_2= -400[8]. For the incommensurate phase, we choose $r_1-c_0Q^2\simeq 0, u_0\simeq 0$ and $\vec{Q}=0.1\times(\vec{K}_1+\vec{K}_2)$. The free energy for the commensurate state and the incommensurate state are obtained by minimizing with respect to ϕ_0, m_0 and $\vec{q}$, and are plotted as functions of coverage for T=350K in Fig. 1. This illustrates that for this particular choice of parameters, the adsorbed H can drive the system to the incommensurate state at a coverage $\theta\simeq 0.35$ and $\vec{q}=\vec{Q}$.

The effect of different choices of parameters on the commensurate-incommensurate phase transition will be considered in more detail in a future publication.

REFERENCES

1. See, for example, L. D. Schmidt, in Interactions on Metal Surfaces, edited by R. Gomer (Springer, New York, 1975) p. 63, and E. W. Plummer, ibid., p. 143.

2. R. A. Barker and P. J. Estrup, Phys. Rev. Lett. 41, 1307 (1978).

3. K. H. Lau and S. C. Ying, Phys. Rev. Lett. 44 1222 (1980).

4. E. Tosatti, Solid State Commun. 25, 637 (1978); in Karpacz Winter School of Theoretical Physics, edited by A. Rekalski and J. Pryzstava (Springer 1979).

5. H. Krakauer, M. Posternak and A. J. Freeman, Phys. Rev. Lett. 43, 1885 (1979).

6. J. E. Inglesfield, J. Phys. C12, 149 (1979).

7. W. L. McMillan, Phys. Rev. B14, 1496 (1976).

8. These values differ somewhat from those used in ref. [3], but they also give qualitative agreement with LEED data.

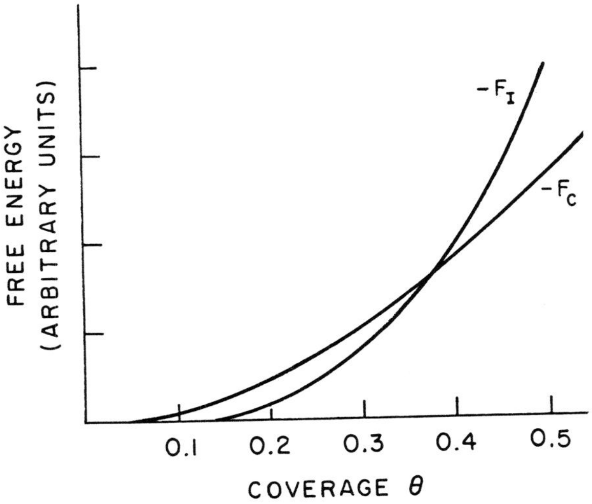

Fig. 1. Plots of free energy for the commensurate state (F_C) and the incommensurate state (F_I) versus coverage, at T=350K.

Published 1980 by Elsevier North Holland, Inc.
Sinha, ed. Ordering in Two Dimensions

THERMODYNAMIC IMPLICATIONS OF BAND STRUCTURE EFFECTS FOR RARE GASES ON GRAPHITE

W.E. CARLOS,[a] M.W. COLE,[b] S. RAUBER and G. VIDALI[c]

The Pennsylvania State University, University Park, Pennsylvania 16802

A.F. SILVA-MOREIRA,[d,e] J.L. CODONA and D.L. GOODSTEIN[f]

California Institute of Technology, Pasadena, California 91125

Recent analyses of atomic beam scattering data have suggested that an anisotropic He-C pair interaction is appropriate to the problem of He on graphite. This results in considerably more corrugated equipotential surfaces than previously assumed, and correspondingly manifest band structure effects. These have been observed for He/graphite in the specific heat for temperature T > 3 K. The implications for other gases and temperatures and for the effective adatom-adatom interaction are discussed.

INTRODUCTION

Calculations of film properties often assume for simplicity the strictly two-dimensional (2D) approximation, thus neglecting the lateral variation of the potential energy $V(\vec{r})$ and adatom motion perpendicular to the substrate [1]. However, surface corrugation should cause even high temperature, low coverage He on graphite to exhibit marked deviation from 2D ideal gas behavior. We describe here the basis for this prediction [2] and its subsequent experimental confirmation [3].

Early work leading to the 2D model [4,5] started by choosing

$$V(\vec{r}) = \sum_i U(\vec{r} - \vec{R}_i) \quad , \tag{1}$$

and assuming the pair potential $U(\vec{x})$ to depend only on the <u>magnitude</u> $x = |\vec{r} - \vec{R}_i|$ of the adatom-substrate separation [6]. As described below, the results of He scattering experiments [7,8] have led us to conclude instead [9,10] that an anisotropic pair potential, varying with the orientation of $\vec{x}$, is appropriate to rare gases on graphite. The resulting potential is less uniform than was calculated previously, resulting in larger band structure effects and more favorable energies of epitaxial phases [11].

The predicted band structure governs the density of states for adsorbed atoms in the low coverage limit. The result is revealed in the heat capacity of ^{3}He and ^{4}He after a thermodynamic analysis which eliminates both the effects of substrate inhomogeneity and He-He interactions [3].

POTENTIAL AND BAND STRUCTURE

Atomic beam scattering is a remarkably direct way of investigating the gas-surface interaction. [12]. A bound state resonance (BSR), which occurs when the incident state is coupled by the periodic potential to a surface bound state shows up dramatically in the reflected beam intensity. The resonance condition is degeneracy of the two states: $E_{inc} = E_{bound}$. In the <u>approximation</u> of free propagation along the surface, this becomes

$$\hbar^2(k_z^2 + \vec{K}_{||}^2)/2m \simeq \varepsilon_{nz} + \hbar^2(\vec{K}_{||} + \vec{G})^2/2m \quad , \tag{2}$$

where the wave vectors are those of the incident beam, $\vec{G}$ is a 2D reciprocal lattice vector, $\varepsilon_{nz} < 0$ is a discrete eigenvalue of the laterally averaged potential $V_o(z)$. The latter is the leading term in the Fourier expansion of the potential

$$V(\vec{r}) = V_o(z) + \sum_{\vec{G}\neq 0} V_{\vec{G}}(z) \exp(i\vec{G}\cdot\vec{r}) \quad . \tag{3}$$

Because graphite provides a relatively smooth potential (as compared with LiF, for example) the periodic part of $V(\vec{r})$ is relatively small in the spatial region sampled by the bound states (see Fig. 1). Consequently deviations from Eq. (2), i.e., band structure corrections, can be treated as a perturbation in analyzing the data [10,13]. The observed deviations then yield matrix elements of the perturbation between the unperturbed eigenfunctions $\phi_n(z)$:

$$V_{\vec{G}}^{mn} = \int dz\, \phi_m^*(z)\, V_{\vec{G}}(z)\, \phi_n(z) \tag{4}$$

Ref. [10] demonstrated how the 19 values of ε_{nz} and $V_{\vec{G}}^{mn}$ obtained by Boato <u>et al</u>. [7] and Derry <u>et al</u>. [8] could be used to assess hypothetical forms of $V(\vec{r})$. In particular, isotropic pair potentials $U(|\vec{r} - \vec{R}_i|)$ were found to yield corrugations (i.e., $V_{\vec{G}}^{mn}$ values) smaller than were observed experimentally. Consistency was obtained, however, with anisotropic potentials, e.g.,

$$U(\vec{x}) = 4\varepsilon\{(\sigma/x)^{12}[1+\gamma_R(1-\tfrac{6}{5}\mu^2)] - (\sigma/x)^6[1+\gamma_A(1-\tfrac{3}{2}\mu^2)]\} , \quad (5)$$

where μ is the cosine of the angle between $\vec{x} = \vec{r} - \vec{R}_i$ and the surface normal. The parameter $\gamma_A = 0.4$ was determined _ab initio_ from the polarizability anisotropy of graphite. Its effect is to decrease the attraction when the He atom is above a C atom - a configuration for which the van der Waals interaction is small [10,14].

Fig. 1 shows the potential derived from Eq. (5) with parameters optimized to yield good agreement with the scattering data. It is less deep and less smooth than previous potentials [4]. For example, well depths above the C atom and hexagon center are 15.5 and 19.1 meV, respectively, compared to earlier values of 20.3 and 22 meV. The previous potentials, which were derived by rough estimation of isotropic potentials, are inconsistent with the present data [10].

A similar conclusion concerning the corrugation follows from an "empirical" band structure calculation of Carlos and Cole [2]. Their technique employed the experimental values of ε_{nz} and $V_{\vec{G}}^{mn}$ in the Hamiltonian matrix; this obviated any assumptions, e.g., Eq. (1), about the form of $V(\vec{r})$. The computed band gaps are approximately twice as large as those derived previously, corresponding to considerably less free translation across the surface. The effective mass is enhanced by a factor 1.06 for ^{4}He and 1.03 for ^{3}He; the latter isotope's lower mass increases its delocalization.

A further test of these calculations comes from a recent neutron scattering study. By measuring the adsorbate enhancement of the (002) Bragg reflection from the graphite, Passell and co-workers [15] have derived a preliminary value of the adatom-substrate separation in favorable agreement with $\langle z \rangle = 2.92$ Å computed for the potential of Fig. 1.

In closing this Section, we remark that the derivation [9,10] of anisotropy in $U(\vec{r})$ applies to other rare gases on graphite, in fact, the value $\gamma_A = 0.4$ is the same for all.

BAND STRUCTURE EFFECTS IN THE HEAT CAPACITY

The isosteric heat capacity of the film, C_N, is a function of the single particle density of states $n(\varepsilon)$. For a smooth substrate, strictly 2D motion leads to $n(\varepsilon)$ = constant and thereby to the prediction $C_N/Nk_B = 1$ in the limit of low density. Since, however, the substrate is periodic, the relevant energies ε are those of the band states. One would thus expect C_N/Nk_B to deviate from its ideal value. Unfortunately, if the coverage N is not small, C_N also depends on the effects of He-He interactions and quantum degeneracy, while at very low N, C_N is difficult to measure, and in any case is dominated by the influence of unavoidable substrate inhomogeneities. These factors have in the past obscured the influence of band structure on C_N.
Elgin and Goodstein [16] have shown, however, how a knowledge of the distribution of inhomogeneities on the surface, together with sufficiently complete thermodynamic data, may be used to correct C_N to find the values it would have on an ideally homogeneous graphite surface. We have performed the necessary analysis for their ^{3}He and ^{4}He data to obtain values of $C_N^o(T)$ where C_N^o is the $N \to 0$ limit of C_N on an ideal graphite substrate [3]. Values of C_N^o/Nk_B are obtained from the corrected data by extrapolating to N=0

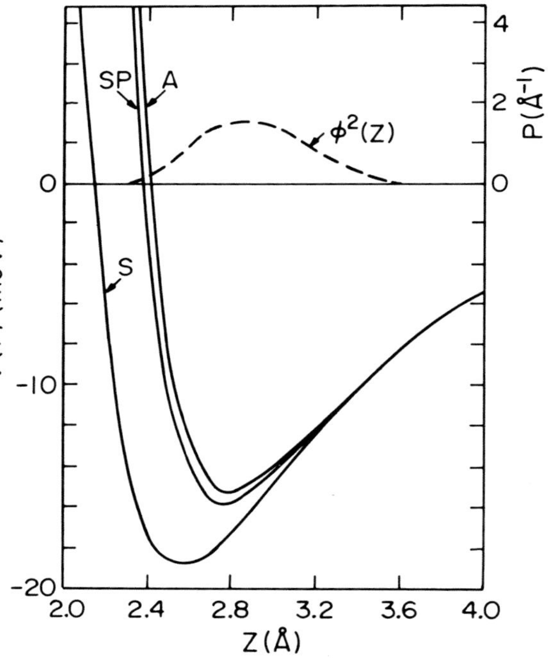

Fig. 1. Full curve represents the potential energy of a He atom as a function of distance z above symmetry points in the basal plane (hexagon center (S), C atom (A) and potential point (SP)), obtained from ref. [10]. Dashed curve is the probability density $P = \phi_o^2(z)$ for a ^{4}He atom in the laterally averaged potential $V_o(z)$.

plots of C_N/Nk_B versus ρ, the 2D density. This procedure is standard for analyzing a virial gas. One would expect the plots to yield straight lines intercepting at 1. Results for C_N^o/Nk_B are shown in Fig. 2 where they are compared to the band structure predictions of Carlos and Cole [2]. The effects of band structure are strikingly evident in the data, particularly in the case of ^3He, where more complete data are available.

The shape of the curves in Fig. 2 is a consequence of the gap of forbidden states introduced by the 2D periodicity of the substrate. As T rises, the heat capacity falls as the missing states in the gap are encountered, then rises again as the next band begins to be populated.

We wish to describe briefly our method of correcting the C_N data. The distribution of binding energies for helium on the Grafoil substrate used for the measurements has been shown [3] to be fitted by the form

$$\varepsilon(N) = -E_b[1 + (1 + N/N_o)^{-3}] \quad , \qquad (6)$$

where N_o, crudely the number of sites of binding energy substantially different from the ideal binding energy, E_b, corresponds to approximately 0.025 monolayers, and E_b/k_B = 142 K for ^4He [3], and 136 K for ^3He [6].

The procedure involves the division of the system into a large but finite number of subsystems with binding energies ε_i and number of sites N^i designed to reproduce closely Eq. (6). The equilibrium amount adsorbed on each subsystem, $N_i = \rho_i N^i$, where ρ_i is the equilibrium density, is determined by the condition that the chemical potential, μ, of the entire system be uniform. An activated site has an increased binding energy compensated by an increased density. The increase in density needed to compensate a given increase in binding energy is taken directly from the experimental data for the chemical potential as a function of density. The empirical chemical potential is modified slightly to prevent double occupancy of highly activated sites.

The picture, then, is of a large unactivated subsystem, whose density $\rho_o = N_o/N^o$ is lower than average, in equilibrium with a known distribution of activated subsystems with higher densities. Each subsystem is assumed to have the thermodynamic properties that the entire film is measured to have at the same temperature when the overall density is equal to the subsystem density.

For the system as a whole, the energy U may be written as $dU = TdS + \mu dN$; for the subsystems, one has $dU_i = TdS_i + \mu_i dN_i$. Expressing dU as

$$dU = \left(\frac{\partial U}{\partial T}\right)_{\{N_i\}} dT + \sum_i \left(\frac{\partial U}{\partial N_i}\right)_{T,\{N_i-1\}} dN_i$$

where in $\left(\frac{\partial U}{\partial N_i}\right)_{T,\{N_i-1\}}$ one N_i is changed while the others are kept fixed, leads to

$$C_N = \sum_i C_{N_i} - T \sum_i \left(\frac{\partial \mu_i}{\partial T}\right)_{N_i} \left(\frac{\partial N_i}{\partial T}\right)_N \qquad (7)$$

where C_{N_i} is the heat capacity of each subsystem. All quantities in Eq. (7) except C_{N_o} are known by the procedure outlined above. It is thus possible to deduce C_{N_o} at the correct coverage N_o using Eq. (7).

We have found that the procedure breaks down for coverages below about 0.2 monolayers, because the corrections then become sensitive to the way in which second layer formation has been prevented and to the fact that we have used a discrete rather than continuous distribution of subsystems. Only data above 0.2 monolayers were therefore used in the analysis leading to Fig. 2.

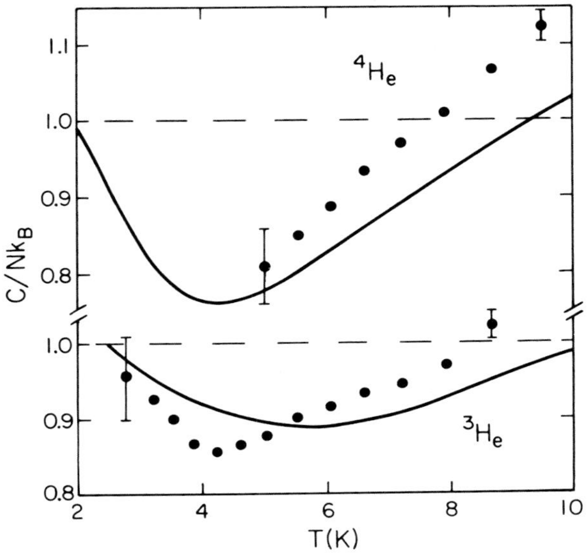

Fig. 2. Specific heat of He isotopes in the limit of zero coverage. Full curve from density of states of ref. [2]. Data from ref. [3]. The dashed curve represents the 2D result $C = Nk_B$. Note the axis labels.

CONCLUSION

The earliest scattering results for ^{4}He on graphite [7] were found to give a ground state energy in remarkable agreement with the thermodynamically determined binding energy of that system [16]. The ensuing interplay of scattering experiments [8], thermodynamic experiment [17] and theory [9,10,13] has led to a highly detailed understanding of the interactions of He with a graphite surface, culminating in the predictions [2] and confirmation [3] of band structure for 2D motion of He in the graphite basal plane.

REFERENCES

(a) Present address: Naval Research Lab., Code 5377, Washington, DC 20375.
(b) Research supported in part by DOE contract DE-AS02-79ER10454.
(c) Research supported by the Consiglio Nazionale delle Ricerche (Italy) and in part by NSF grant DMR77-22961.
(d) On leave from University of Campinas, Brazil.
(e) Research supported in part by the Brazilian agency FAPESP, contract 78/0303.
(f) Research supported in part by NSF grant DMR77-00036.

1. J.G. Dash and M. Schick, in The Physics of Liquid and Solid Helium, edited by K.H. Bennemann and J.B. Ketterson (Wiley, NY, 1978), Part II.
2. W.E. Carlos and M.W. Cole, Phys. Rev. B, 21, 3713 (1980).
3. A.F. Silva-Moreira, J.L. Codona, and D.L. Goodstein, Phys. Lett., to be published.
4. D.E. Hagen, A.D. Novaco, and F.J. Milford, in Adsorption-Desorption Phenomena, ed. by F. Ricca (Academic Press, London, 1972), p. 49.
5. W.A. Steele, Surf. Sci. 36, 317 (1973).
6. The further assumption in Eq. (1) of pairwise additivity is currently under investigation by M.W. Cole, B. Garrison and W.A. Steele.
7. G. Boato, P. Cantini, C. Guidi, R. Tatarek, and G.P. Felcher, Phys. Rev. B 20, 3957 (1979).
8. G. Derry, D. Wesner, W.E. Carlos, and D.R. Frankl, Surf. Sci. 87, 629 (1979).
9. W.E. Carlos and M.W. Cole, Phys. Rev. Lett. 43, 697 (1979).
10. W.E. Carlos and M.W. Cole, Surf. Sci. 91, 339 (1980).
11. See e.g., A.D. Novaco, Phys. Rev. A 7, 1653 (1973).
12. See e.g., F.O. Goodman, Crit. Rev. Solid St. and Mater. Sci. 7, 33 (1977); or M.W. Cole and D.R. Frankl, Surf. Sci. 70, 585 (1978).
13. H. Chow, Surf. Sci. 79, 157 (1979); N. Garcia, W.E. Carlos, M.W. Cole, and V. Celli, Phys. Rev. B 21, 1636 (1980).
14. G. Bonino, C. Pisani, F. Ricca, and C. Roetti, Surf. Sci. 50, 379 (1979) took this polarizability anisotropy to be infinite.
15. L. Passell and H. Taub, private communication.
16. R.L. Elgin and D.L. Goodstein, Phys. Rev. A 9, 2657 (1974). The analysis also incorporates data from M. Bretz et al., Phys. Rev. A 8, 1589 (1973).
17. R. Elgin, J.M. Greif, and D.L. Goodstein, Phys. Rev. Lett. 41, 1723 (1978).

THE HEAT CAPACITY OF MONOLAYER XENON ON GRAPHITE

J. A. LITZINGER and G. A. STEWART

Department of Physics, University of Pittsburgh, Pittsburgh, Pa. 15260

Results of heat capacity measurements for xenon adsorbed on Grafoil are presented. Near delta-function anomalies and an essentially vertical density-temperature phase boundary at 100K are observed for coverages between 0.2 and 0.7 monolayer. The heat capacity shapes in combination with the nearly vertical character of the associated phase boundary suggest a triple point interpretation. The xenon data, and the possible triple point interpretation, are in marked contrast with previous heat capacity studies of krypton which suggested a single fluid phase. Although triple point characteristics are observed in the xenon heat capacity, no explicit anomaly associated with a critical point for T >100K is observed.

The study of phase transitions in heavy noble gas monolayers physisorbed onto graphite has been an active area of research following the vapor pressure work of Thomy and Duval [1]. Their provocative isotherm experiments suggested that submonolayer krypton and xenon are isomorphic to the bulk. In particular, they suggested that there were two disordered fluid phases consisting of monolayer liquids and gases, a quasi-two dimensional solid phase at the highest densities, and accompanying triple and critical points. Indeed, the solid phases have been observed in both these systems using diffraction techniques [2-6].

However, for krypton on the basis of heat capacity experiments [7,8], an alternative to the triple and critical point interpretation has been proposed. This alternative interpretation, involving a first order transition (phase coexistence) between the registered solid and a single fluid phase, is without a normal triple and critical point. Recent theoretical treatments [9-11] have also yielded submonolayer phase diagrams with a single fluid phase.

Heat capacity experiments [12] at constant density--because their thermodynamic path in the density-temperature plane crosses orthogonally those phase boundaries containing triple and critical points--are particularly sensitive probes of these features. In this paper we present the first measurements of the heat capacity of xenon monolayers on graphite with specific reference to the submonolayer region containing the conjectured triple and critical points.

The heat capacity of xenon on Grafoil has been measured for coverages between 0.19 and 1.05 monolayers in the temperature range 65-160K.

The experimental system is the same as that previously used for krypton studies [7,8]. The calorimeter has an approximate xenon monolayer capacity of 98 CC STP as obtained from a 109K isotherm.

The characteristic feature of the xenon heat capacities is a single line of anomalies whose locus is plotted in the density-temperature phase diagram of Fig. 1. At the lowest coverage the anomaly is a low, broad, peak centered at 99K. With increasing coverage the peak rapidly evolves into a near delta-function whose experimental amplitude increases approximately linearly with coverage to 0.64 monolayer as

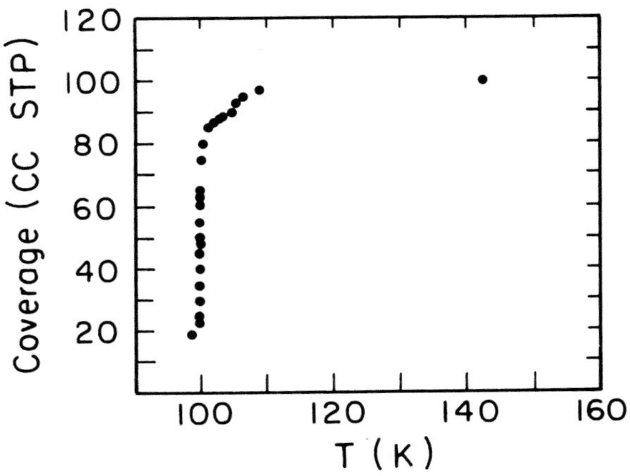

Fig 1: Locus of heat capacity peaks for Xe on Grafoil. Monolayer coverage is approximately 98 CC STP as determined from isotherm measurements.

shown in Fig. 2. Over a range in coverage of 0.42 monolayer, the peak position is nearly constant at a mean temperature of 99.95K, suggesting a triple point phase boundary. The position of this boundary varies by no more than 100 mK as the coverage ranges from 0.25 to 0.70 monolayer. As the coverage increases beyond 0.7 monolayer, the peaks broaden and move to high temperature as shown in Fig. 1. As the monolayer is approached, the peak temperature rises abruptly, increasing from 105K to 142K as the coverage changes from 0.95 to 1.02 monolayer.

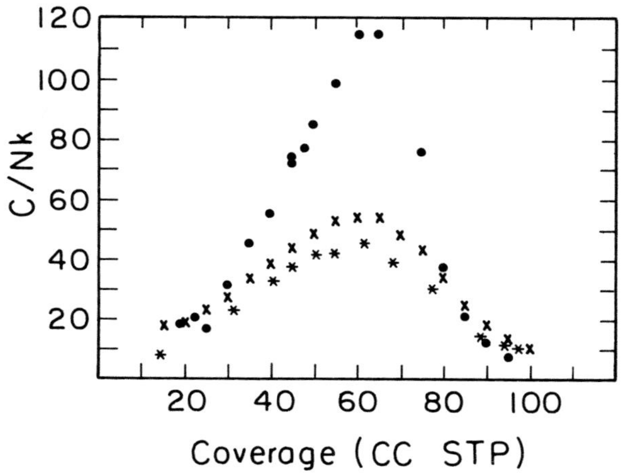

Fig. 2: Heat capacity per atom peak heights versus coverage for xenon ●, krypton X [7], and nitrogen * [13]. The nitrogen coverages are normalized to our calorimeter by N_2 isotherm comparisons.

The peak at 0.61 monolayer is shown in Fig. 3. Using heating pulse widths of 90 mK we are unable to resolve the top of this peak. Experimentally the peak appears as a near delta-function singularity, the expected signature for a triple point. The fractional width at half maximum (FWHM) may be taken as a measure of the sharpness of the anomaly. Because the peak height is resolution limited, this is an upper limit. From the data of Fig. 3, FWHM < 0.003. This is comparable with the neon delta function anomaly observed by Huff and Dash [14] for which FWHM = 0.006 for a (resolved) peak at 13.5K.

Although the peak shapes and near independence of peak position on density are qualitatively consistent with a triple point, there are a number of deviations from ideal behavior that should be noted. First, there is no detectable discontinuity in the high temperature tails which would confirm liquid-vapor or liquid-solid phase boundaries. However, these discontinuities could be quite small--a few boltzmanns--and difficult to observe since the film signals in the tails are typically less than 1% of the total (film plus calorimeter) measured signal. Second, no explicit anomaly that could be identified as a liquid-vapor critical point is observed at any coverage for temperatures above 99.95 K. We note, however, that critical points may be extremely sensitive to substrate uniformity. For example, inhomogeneities may be expected to appear as effective variations in chemical potential Δu. These chemical potential variations would in turn produce variations in the mean density near the critical point through coupling between the density and compressibility via

$$\frac{\Delta\rho}{\rho} = \rho\kappa\ \Delta\mu \tag{1}$$

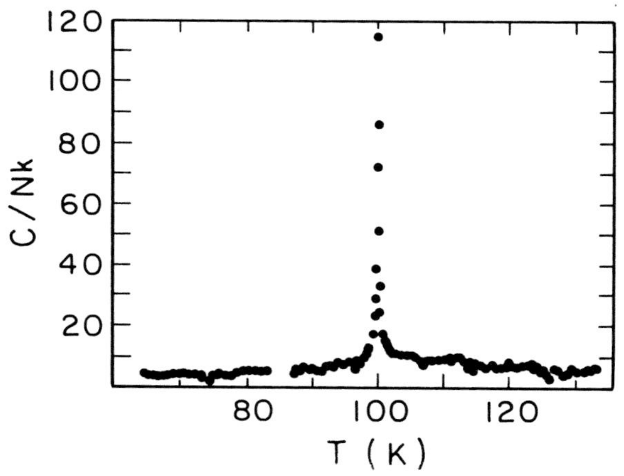

Fig. 3: The heat capacity per atom of 0.61 monolayer xenon on Grafoil.

Such a density variation could smear out the critical point and make it experimentally unobservable. In contrast, the triple point would be relatively insensitive to the same effect since the relevant compressibilities are not divergent. Even in the presence of some density variation the observation of the triple point may be relatively unaffected since, in contrast to the critical point, it exists over a wide range of densities.

We also note that the linear dependence of the per atom peak height in Fig. 2 is a deviation from expected behavior for an ideal triple point. The maximum heat capacity C_L in crossing a triple line in the coverage-temperature plane occurs at that coverage N_L where the system passes from solid-vapor coexistence to pure liquid. For the present system this maximum corresponds to the peak of Fig. 3 at 0.6 monolayer. For an ideal triple point the heat capacity per atom for coverages $N_V < N < N_L$, where N_V is the low coverage terminus of the triple line, is given by

$$\frac{C}{Nk} = \frac{(1 - \frac{N_V}{N})}{(1 - \frac{N_V}{N_L})} \frac{C_L}{N_L k}\ \delta(T - T_t) \tag{2}$$

which is not linear in N as observed experimentally in Fig. 2. However, caution must be exercised in detailed quantitative interpretation of peak heights versus coverage since the peak heights are not fully resolved.

In Fig. 2 we note an intriguing similarity in the peak heights of the three heavy atom systems (N_2, Kr, Xe) which have been studied by heat capacity techniques. Nitrogen [13] and krypton [7], systems with a commensurate submonolayer solid phase, exhibit a single line of peaks for coverages below that corresponding to a maximum peak height. These peaks, however, are relatively broad, are not fixed in temperature, and do not correspond to near delta-function singularities as observed in xenon, a system with an incommensurate solid phase. Indeed, the commensurate systems have been associated with an "incipient" vis-a-vis "normal" triple point [7]. Nevertheless, the maximum peak heights for all three adsorbates occur at the same number density, suggesting a substrate effect. In bulk krypton and xenon, for example, the number densities at the maximum in the triple point heat capacities differ by 24%.

At present, from heat capacity data for the noble gases on graphite, only neon and xenon would suggest a triple point interpretation at submonolayer coverage. Yet neither of these systems has revealed a critical point which would confirm the anticipated liquid-vapor coexistence region. The extreme interpretation of this fact would suggest an extremely steep but non-vertical phase boundary. On the high temperature side of this boundary would be a highly compressible fluid. On the low temperature side would be solid-fluid phase coexistence with a solid-fluid conversion term in the heat capacity given by [12]

$$C_{conv} = T(\overline{S}_f - \overline{S}_s)\left(\frac{dn_f}{dT}\right)_{eq} \qquad (3)$$

In equation 3 the entropies are partial molar and the equilibrium derivative is the rate of appearance of fluid with increasing temperature. For a nearly constant solid density and an extremely steep phase boundary like that of Fig. 1, the derivative in equation 3 (by the lever rule) is proportional to the slope of the phase boundary. Further, as the solid-fluid system approaches the boundary with increasing temperature, the derivative is large only in the vicinity of that boundary. By making the boundary arbitrarily steep, almost all the conversion occurs immediately adjacent to the boundary. This yields an abrupt heat capacity rise terminated by a discontinuity at the boundary. Resolution limited heat capacities may not be able to distinguish between such a rapid conversion to a single fluid phase or a resolution limited delta function characteristic of liquid-vapor coexistence.

An experiment that can distinguish between one or two fluid phases above the transition temperature is required. Such an experiment requires examination of density dependences at fixed temperature above the boundary. For example, heat capacity isotherms should be linear in coverage in a region of two fluid coexistence. This, however, is but a necessary condition and is difficult to verify experimentally because of the small heat capacity magnitudes above the transition. Alternatively, a diffraction experiment at fixed temperature above the transition, if possible, could unambiguously resolve any question regarding the number of fluid phases. For a system with a triple point, the position of the liquid short-range order peak should be independent of mean density. For a system with an extremely sharp phase boundary corresponding to a single fluid phase, the peak should shift to larger angles with increasing density.

The authors are grateful to P. K. Das for experimental assistance, to D. M. Butler and R. B. Griffiths for discussions, and to the National Science Foundation (Grant No. DMR78-20603) for financial support.

REFERENCES

1. A. Thomy and X. Duval, J. Chem. Phys. 67, 1101 (1970).
2. M. D. Chinn and S. C. Fain, Jr., J. Phys. (Paris) Colloq. 38, C4-99 (1977).
3. P. M. Horn, R. J. Birgeneau, P. Heiney, and E. M. Hammonds, Phys. Rev. Lett. 41, 961 (1978).
4. J. A. Venables, H. M. Kramer and G. L. Price, Surf. Sci. 55, 373 (1976).
5. E. M. Hammonds, P. Heiney, P. W. Stephens, R. J. Birgeneau, and P. Horn, TBP.
6. J. J. Lander and J. Morrison, Surf. Sci. 6, 1 (1967).
7. D. M. Butler, J. A. Litzinger, G. A. Stewart, and R. B. Griffiths, Phys. Rev. Lett. 42, 1289 (1979).
8. D. M. Butler, J. A. Litzinger, and G. A. Stewart, Phys. Rev. Lett. 44, 466 (1980).
9. A. N. Berker, S. Ostlund, and F. A. Putnam, Phys. Rev. B 17, 3650 (1978).
10. S. Ostlund and A. N. Berker, Phys. Rev. Lett. 42, 843 (1979).
11. S. Ostlund and A. N. Berker, TBP.
12. J. G. Dash, Films on Solid Surfaces, Academic Press, 1975.
13. T. T. Chung and J. G. Dash, Surf. Sci. 66, 559 (1977).
14. G. B. Huff and J. G. Dash, J. Low Temp. Phys. 24, 155 (1976).

SUPERFLOW ONSET: ^4He ON GRAFOIL

J. MAPS and R.B. HALLOCK

Laboratory for Low Temperature Physics
Department of Physics and Astronomy, University of Massachusetts
Amherst, MA 01003

Thermal conductance is used to signal the onset of superflow on a Grafoil ribbon with a constriction at the center. Onset data is presented in terms of helium coverage vs. onset temperature. Further analysis reveals that the data are reasonably consistent with the predictions of Huberman and Dash.

INTRODUCTION

Phase transitions in systems of restricted geometry have been of interest for many years. These general studies have recently experienced an explosive renewed interest stimulated in part by the work of Kosterlitz and Thouless [1]. In the case of superfluid helium the onset transition to superflow has been studied by a variety of techniques on a number of surfaces [2-8] and the nature of the transition is the subject of current debate [9].

The present work results from a desire to simultaneously study both third sound and thermal transport in a thin ^4He film on a variety of surfaces. Preliminary results of thermal transport measurements in the case of Grafoil [10] are presented here.

APPARATUS AND PROCEDURE

The apparatus consists of a brass chamber which contains both the Grafoil substrate of interest as well as $A\ell_2O_3$ powder. The $A\ell_2O_3$ serves to provide a large reservoir to stabilize the thickness of the helium film. The Grafoil [11] used for the thermal transport measurements reported here is of dimension 1 mm x 1 cm x 8 cm with a constriction cut at the center. In the region of the constriction the width of the Grafoil is reduced to approximately 1 mm. After fabrication the Grafoil was baked to 800°C under high vacuum ($P < 10^{-6}$ Torr) for four days. All subsequent handling of the Grafoil was accomplished in a ^4He environment. Located along the Grafoil are nominal 100Ω Allyn-Bradley resistors which have been sanded flat on one side [12]. The resistors are attached to the Grafoil using a small amount of General Electric 7031 insulating varnish. A heater consisting of 500Ω of manganin wire (3.4 Ω/cm) is wound around one end of the Grafoil and uniform contact is made to the Grafoil through use of the 7031 varnish. The opposite end of the Grafoil is affixed to a copper plate and the entire assembly is placed in the brass chamber (Fig. 1).

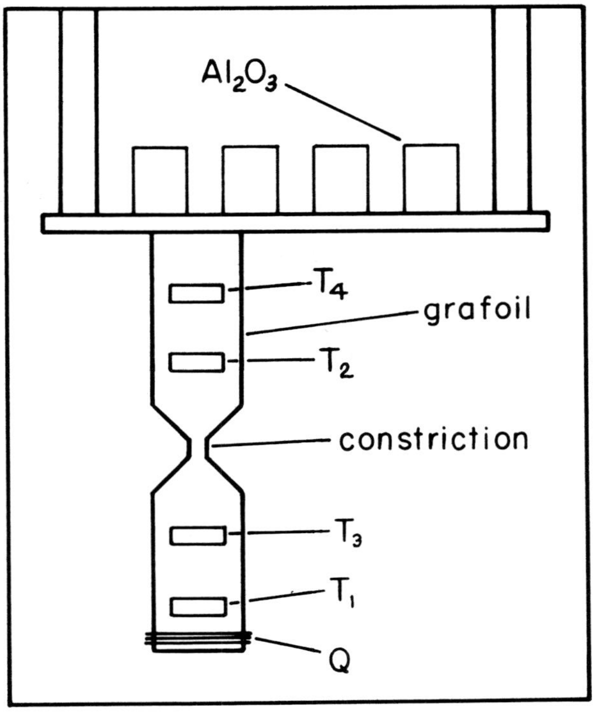

Figure 1. Schematic representation of the Grafoil in the brass chamber. Heat is applied to heater Q and the temperature is measured at various points along the Grafoil.

Under normal operation power is supplied to the heater Q and the temperature gradient along the Grafoil substrate is monitored as a function of the amount of ^4He added to the chamber by use of the four carbon thermometers. The purpose of the constriction is to localize the temperature gradient imposed by the heater Q. The resistors located along the Grafoil confirm that the constriction serves this purpose.

In a typical experiment heat is supplied to heater Q with very little helium present in the chamber. For fixed heater power and fixed bath temperature the temperature gradient across the constriction is monitored as a function of increasing amounts of helium gas in the chamber. In the present apparatus the helium is added to the experimental chamber by an unplanned (but in retrospect rather fortunate) small superleak in the chamber. The pressure, P, in the experimental chamber is simultaneously measured relative to the pressure, P_o, above the main dewar helium reservoir by the use of a MKS Baratron pressure gauge (0-10 Torr range).

The pressure measurements allow a determination of the coverage of ^4He on the Grafoil in terms of equivalent helium layers. For the present report we adopt the thickness scale obtained from the approximate relation

$$- d^3 = \alpha/(T \ln P/P_o) \qquad (1)$$

where we have taken [8] α = 44 layers^3 - K. Given the existence of new isotherm measurements for Grafoil [13] we view this thickness scale as tentative.

RESULTS

Data representative of that we have collected is presented in figure 2. The upper figures show the measured temperatures at thermometers T_1 and T_2 located on either side of the Grafoil constriction as a function of the average ^4He film thickness on the Grafoil. The precipitous drop in the temperature recorded at thermometer T_1 records the onset of superflow in the region of the constriction. The lower thermometer, T_2, shows a less dramatic reduction in its temperature at a slightly smaller film thickness. This too is a signal for onset and it occurs at a smaller film thickness since the thermometer T_2 is at a temperature somewhat lower than the thermometer T_1.

We choose to call onset the value of the film thickness at which the thermometer T_1 begins to record the rapid drop in temperature as a function of film thickness. This location is marked by an arrow on the figure.

The lower figure in figure 2 represents the temperature difference observed across the constriction in the Grafoil as a function of the ^4He film thickness. The peak in the curve is due to the fact that the thermometer T_2 records onset at a slightly smaller value of the film thickness than does the thermometer T_1. Although we have not done so, one might conveniently choose the location of the peak as a reasonable criterion for onset.

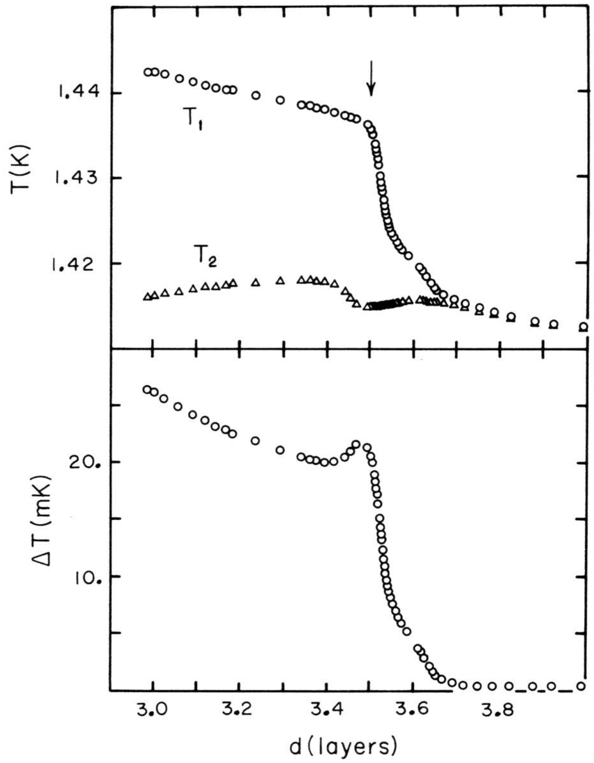

Figure 2. a) Observed temperatures as measured on thermometers T_1 and T_2. The precipitous change in T_1 signals onset in the region of of the constriction. The arrow illustrates our choice for the meaning of onset. b) The observed difference $\Delta T = T_1 - T_2$ in temperature across the constriction.

To date we have made measurements of this sort over the temperature range $1.3 < T < 1.95$K and have used heater powers of typically 20 μw. In general our results are not strongly dependent on the heater power used.

DISCUSSION

In spite of the tentative nature of our film thickness scale we have proceeded to analyze the data in more detail. In figure 3 we display the onset thickness vs. temperature plot which results from the measurements we have conducted to date. Shown as the solid line are measurements made elsewhere for glass [14]. We observe the onset curve for helium on Grafoil to be smooth within our experimental error and remarkably similar to glass in appearance. In particular we see no clear evidence for the anomaly (flat spot) in the onset curve as observed by Polanco and Bretz [7] in the case of ZYX. This difference may be due to the dif-

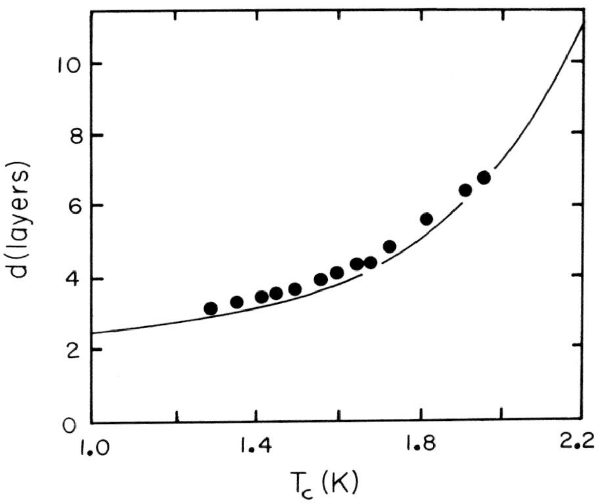

Figure 3. Onset thickness as a function of temperature.

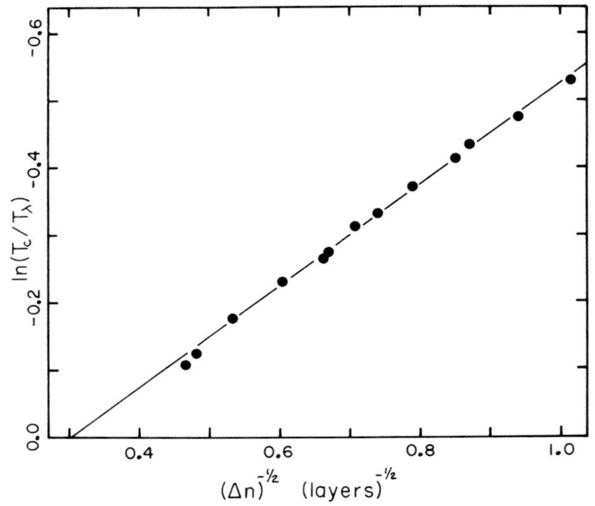

Figure 4. Comparison to the Huberman-Dash prediction [15]. The line shown is a fit to the data.

ference in substrate; it may also arise due to the difference in the measurement technique. In any case, for the Grafoil used in the present measurements the onset curve appears smooth and shows no unusual features.

Use of the tentative film thickness scale also allows us to compare the present results with the predictions of the Huberman and Dash droplet model [15]. In this model the onset of superflow in the helium system is considered to be coincident with a percolation transition in the phase angles of the order parameters of the various helium droplets. In the model the onset temperature T_c is related to the bulk superfluid transition temperature T_λ by the expression

$$\frac{T_c}{T_\lambda} = \frac{\exp[-\gamma(\Delta n)^{-1/2}]}{\exp[-\gamma(\Delta n_\lambda)^{-1/2}]} \qquad (2)$$

where γ is a constant. Here n is a measure of the film thickness and $\Delta n = n - n_o$ and $\Delta n_\lambda = n_\lambda - n_o$ with n_o the minimum thickness for droplet formation and n_λ the film thickness above which the droplets merge to form a "bulk-like" film across the entire sample. For the data we report here the choice $n_o \sim 2.1$ layers results in the graph presented in figure 4. From this we determine $n_\lambda \sim 13.3$ layers a value somewhat smaller than that reported previously by others [7,15]. This difference may well be due to our choice of a thickness scale.

The data shown in figure 4 are in generally good agreement with the qualitative predictions of Huberman and Dash [15]. We view this in terms of the data being consistent with the theory rather than constituting a test of the theory.

ACKNOWLEDGMENTS

This work was supported by the National Science Foundation through DMR78-07762 and DMR79-09248.

REFERENCES

[1] J.M. Kosterlitz and D.J. Thouless, J. Phys. C6, 1181 (1973).
[2] I. Rudnick, Phys. Rev. Lett. 40, 1454 (1978).
[3] D.J. Bishop and J.D. Reppy, Phys. Rev. Lett. 40, 1727 (1978).
[4] J.A. Herb and J.G. Dash, Phys. Rev. Lett. 29, 846 (1972).
[5] M. Chester and L.C. Yang, Phys. Rev. Lett. 31, 1377 (1973).
[6] D.T. Ekholm and R.B. Hallock, Phys. Rev. Lett. 42, 449 (1979).
[7] S.E. Polanco and M. Bretz (to be published).
[8] J.A. Roth, J. Jelatis and J.D. Maynard, Phys. Rev. Lett. 44, 333 (1980).
[9] J.G. Dash, Phys. Rev. Lett. 41, 1178 (1978).
[10] Union Carbide Corporation
[11] We are grateful to M. Bretz for providing us with a sample of Grafoil.
[12] Allyn Bradley nominal 100Ω, 1/8 watt, after the shape change R ≃ 55Ω.
[13] J.G. Dash (private communication).
[14] See, for example, R.S. Kagiwada, J.C. Fraser, I. Rudnick and D. Bergman, Phys. Rev. Lett. 22, 338 (1969).
[15] B.A. Huberman and J.G. Dash, Phys. Rev. B17, 398 (1978).

SUPERFLUID ONSET IN $^3He - ^4He$ MIXTURE FILMS

F.M. ELLIS, J.S. BROOKS* and R.B. HALLOCK

Laboratory for Low Temperature Physics
Department of Physics and Astronomy, University of Massachusetts
Amherst, MA 01003

We report measurements of superfluid onset in unsaturated $^3He - ^4He$ mixture films. The measurements are made using the techniques of third sound and mass transport in an apparatus designed to allow an unambiguous characterization of the mixture film.

INTRODUCTION

Third sound and mass transport have proven to be extremely useful techniques in helping to elucidate the properties of thin films of superfluid helium [1-4]. In the case of pure ^4He in the vicinity of onset both techniques have been employed to study the transition. In particular, through the use of [5,6],

$$C_3^2 = 3 \frac{\langle\rho_s\rangle}{\rho} \frac{\alpha}{d^3} [1 + TS/L]^2 \qquad (1)$$

it was first experimentally established that $\langle\rho_s\rangle/\rho$ remains finite at onset [7,8]. Here C_3 is the third sound velocity, $\langle\rho_s\rangle/\rho$ is the superfluid fraction in the film, α the strength of the Van der Waal's interaction, d the film thickness, T the temperature, S the entropy and L the latent heat.

The case of $^3He - ^4He$ mixtures has received substantially less attention. At temperatures above 1.2K it has been shown that the theory of Downs [9] is not obeyed and that the observed dependence of C_3 on ^3He concentration can be understood by assuming [10] the ^3He simply modifies the value of $\langle\rho_s\rangle/\rho$. The present work has been undertaken in an effort to study thin mixture films as a function of concentration and thickness at temperatures low enough so that the effects of phase separation might be important. In addition it was desired to study the effect of increasing ^3He on the onset temperature.

In this work we report some of our observations of the effect of ^3He on the velocity of third sound for T > 0.3K and document the decrease in the temperature of superfluid onset as a function of increasing ^3He concentration.

APPARATUS

The apparatus (Figure 1) has been designed so as to provide detailed information on the $^3He - ^4He$ mixture in the apparatus. A BeCu chamber (volume 14.1 cm^3) which can be sealed by means of a superfluid valve is connected to a conventional re-circulating ^3He refrigerator by a link positioned to minimize temperature gradients in the chamber. In order to minimize the effects of capillary condensation but still stabilize the film thickness, the chamber con-

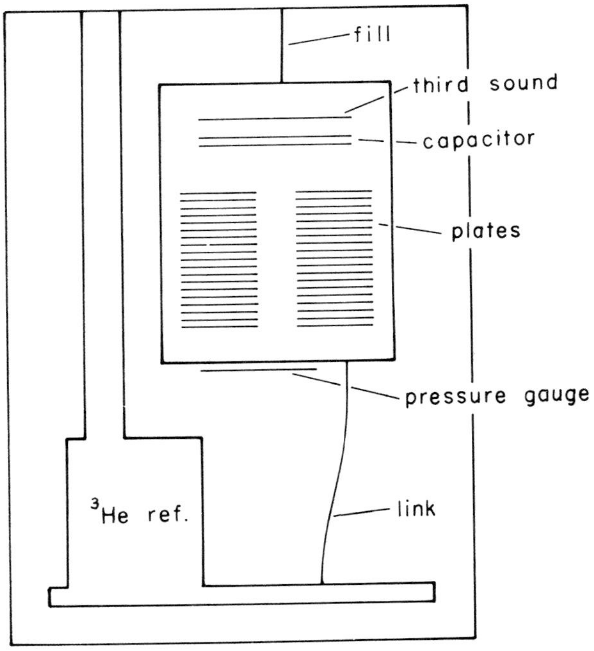

Figure 1. Schematic representation of the apparatus.

tains 798 spaced glass plates 18mm in diameter and 25μm thick rather than the customary Al_2O_3 powder. A flexible BeCu membrane forms an integral part of the experimental chamber and through the use of tunnel diode techniques allows precise measurements of changes in the $^3He - ^4He$ vapor pressure in the chamber. The chamber also contains a glass plate with superconducting strips which serve as interchangable sources and detectors of both third sound in the helium film and ordinary sound in the vapor above the film [10]. A capacitor which provides a direct measure of the amount of helium between its plates is also enclosed in the chamber. The ordinary sound measurements [10] allow a direct measure of the ^{3}He concentration in the vapor above the film. Thus, we are able to deduce the ^{3}He concentration associated with the film directly.

In typical operation a known amount of ^{4}He is admitted to the chamber at 0.4K. For most of the work we report here the ^{4}He thickness was 5.8 layers at low temperatures [11]. The chamber is then sealed and measurements are made as a function of temperature as the chamber is warmed until superfluid onset is reached. Particular care is taken to ensure equilibrium has been reached at each temperature selected for study. Data typical of that available from the apparatus is shown in Figure 2 for pure ^{4}He from earlier work with a thicker film. Various ^{3}He concentrations are then studied in sequence: the apparatus is cooled to 0.4K (to ensure that the ^{4}He remains in the chamber) and a selected increment of ^{3}He is added through the superfluid valve. The valve is then sealed and measurements are taken as the sealed chamber is again warmed in steps.

RESULTS

Figure 3 illustrates the substantial effect the ^{3}He has on the observed third sound velocity for data collected in the manner just described. In all of the situations studied it is observed that the presence of the ^{3}He causes a reduction in the third sound velocity. This is consistent with earlier observations in thicker films at higher temperatures [10]. An interesting phenomena seen in Figure 3 is the nearly linear dependence of the third sound velocity on temperature for all but the highest concentrations. In the low temperature range over which this behavior is observed it is expected that nearly all of the ^{4}He remains in the film. The concentrations which label each data set refer to the total helium initially admitted to the chamber at low temperature. That is, we define $X = N_3/(N_3 + N_4)$ where the N's are the number of respective atoms in the chamber.

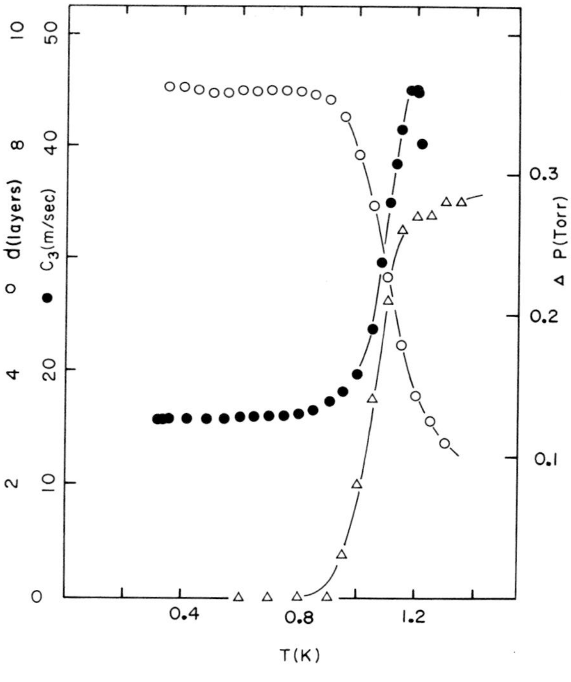

Figure 2. Data for ^{4}He which illustrated the general behavior of the various measurement devices in the experimental chamber. Here d is the film thickness as deduced from the capacitor using a provisional thickness scale and P is the vapor pressure in the sealed chamber.

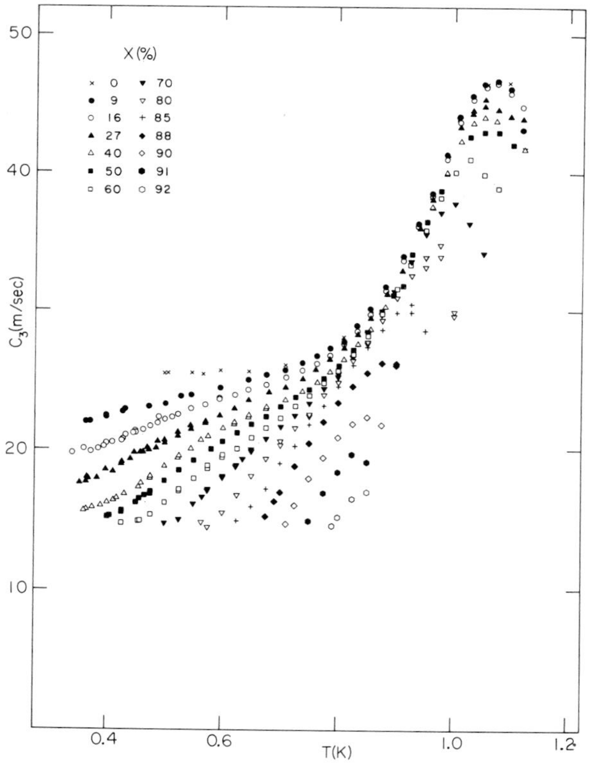

Figure 3. Velocity of third sound as a function of temperature for various values of X.

We observe the superfluid onset temperature to be suppressed by the addition of ^{3}He. The onset temperature T_c is plotted as a function of X in Figure 4. The onset temperatures

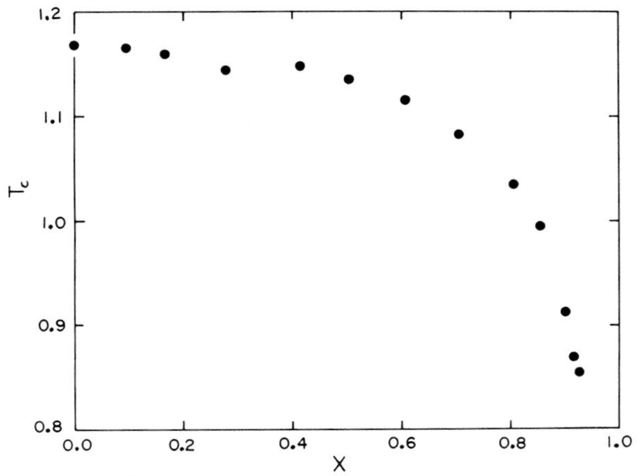

Figure 4. The observed superfluid onset temperature as a function of X.

shown are taken from measurements made with the film thickness capacitor rather than from the data presented in Figure 3. When a small temperature drift is imposed on the chamber a small anomaly in the observed capacitance is seen at a temperature 10 - 30 mK above the temperature at which the third sound signal becomes unmeasurable. We believe the T_c values obtained in this way are more accurate than those determined from the third sound measurements.

DISCUSSION

Although we have taken precautions to minimize the effects of capillary condensation, its effects can be clearly seen. The data presented in Figure 3 have been truncated at low temperatures because of these effects. An example of the effect of capillary condensation in the apparatus is shown in Figure 5 for one of the sets of data shown in Figure 3. The peak in the observed third sound velocity at low temperature is caused by the changing properties of the film in the third sound apparatus due to capillary condensation elsewhere in the apparatus [12]. Changes in the film thickness are simultaneously observed in the capacitor when these effects are observed in the third sound. It is interesting to note that these effects appear to set in at essentially the same third sound velocity for most of the ^{3}He concentrations studied.

One of the initial goals of this set of measurements included the hope that the occurrence of phase separation in these films would be

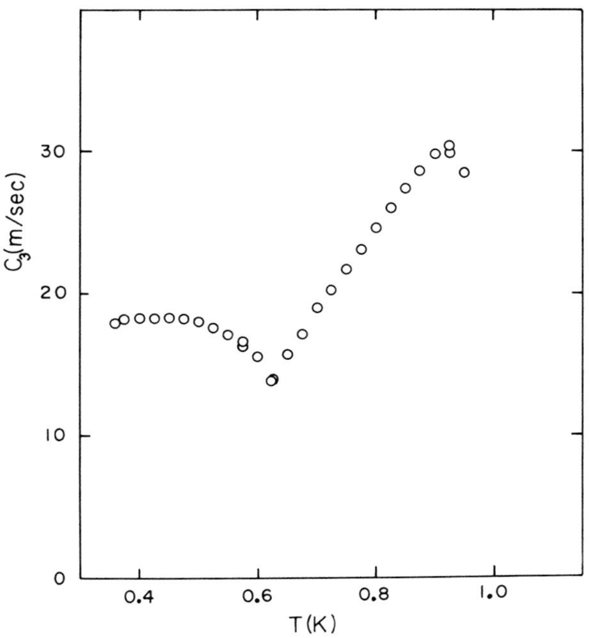

Figure 5. An example of the effect of capillary condensation in the apparatus. In this example X = 0.85.

signaled by a clear signature in the third sound velocity measurements. No sharp signature suggestive of a dramatic change in the properties of the film is observed in the third sound velocity measurements. Rather, in all cases the evolution of the third sound velocity is smooth with no abrupt changes in curvature (except near 0.9K where the ^{4}He begins to leave the film in large amounts per increments of increasing temperature). In the case of bulk helium a low temperature ^{3}He surface state is known to become depopulated above 0.2K [13] but this may not be the case in systems of finite but small film thickness [14] where the area of the film is much larger in comparison to the volume of the fluid than it is in the bulk. These facts suggest that a ^{3}He surface layer may be present on the film over a wider range of temperatures. This layer may grow in a uniform way with a decrease in the temperature. These facts coupled with the linearity mentioned earlier are particularly interesting when it is recognized that much of the data at low temperatures is well within the forbidden region of the bulk phase diagram.

As of this writing the experiments are underway and a careful calibration of the film thickness apparatus is incomplete. We thus are not presently able to reliably compute $\langle\rho_s\rangle d/T_c$ from the data.

ACKNOWLEDGMENTS

We have benefited from numerous conversations with R.A. Guyer and M.D. Miller. This work was supported by the National Science Foundation through DMR78-07762 and DMR79-09248.

REFERENCES

* Boston University, Boston, Massachusetts

[1] See, for example, K.R. Atkins and I. Rudnick, Progress in Low Temperature Physics, ed. C.J. Gorter (North Holland, Amsterdam, 1970), Vol. 6, ch. 2.

[2] J.H. Scholtz, E.O. McLean and I. Rudnick, Phys. Rev. Lett. 32, 147 (1974).

[3] J.A. Herb and J.G. Dash, Phys. Rev. Lett. 29, 846 (1972).

[4] D.T. Ekholm and R.B. Hallock, Phys. Rev. Lett. 42, 449 (1979).

[5] D.J. Bergman, Phys. Rev. 188, 370 (1969); A3, 2058 (1971).

[6] R.K. Galkiewicz, K.L. Telschow and R.B. Hallock, J. Low Temp. Phys. 26, 147 (1977).

[7] I. Rudnick and J.C. Fraser, J. Low Temp. Phys. 3, 225 (1970).

[8] I. Rudnick, Phys. Rev. Lett. 40, 1454 (1978).

[9] J. Downs, Ph.D. Dissertation, USC (1974), (unpublished).

[10] D.T. Ekholm and R.B. Hallock, Phys. Rev. B May (1980).

[11] This thickness was determined from the usual $d^3 = 27(T\ln(P_o/P))^{-1}$ for glass.

[12] Similar effects were seen by J.S. Brooks and R.B. Hallock (J. de Phys. 39, C6, 315 (1978)) but not clearly identifiable at that time as effects of capillary condensation.

[13] D.O. Edwards, P.P. Fatouros, G.G. Ihas and C.P. Tan, Phys. Rev. B (1977).

[14] M.D. Miller (private communication).

He-3 IN He-4 FILMS: A DOUBLY TWO DIMENSIONAL SYSTEM

M. J. DIPIRRO and F. M. GASPARINI

SUNY at Buffalo, Amherst, New York 14260

Measurements of the heat capacity of He-3 in He-4 films of 22 to 53 Å thickness in the temperature range of 40 to 350 mK are reported. Data analysis shows a bound state of He-3 to the free surface of He-4 exists, but with a binding energy smaller than for the bulk He-4 surface. The surface He-3 behaves as a nearly ideal 2D Fermi gas. A small attractive interaction is found, which is in contrast to the bulk surface where the interaction is weakly repulsive. In addition, for the thinnest films studied, the He-3 which remains dissolved within the film at finite temperatures also behaves as a 2D gas.

In 1965 Atkins and Narahara noticed that the surface tension of a dilute solution of He-3 in He-4 decreased anomalously below about 0.7 Kelvin[1]. Andreev postulated that the decrease was caused by He-3 preferentially adsorbing on the He-4 free surface[2]. In other words the He-3 was more strongly bound to the free surface than to the bulk He-4. Since that time more accurate measurements of surface tension[3], as well as surface sound[4], film flow[5], negative ion mobility[6], and evaporation[7] have confirmed that He-3 surface states exists and that the He-3 behaves as a nearly ideal 2D Fermi gas of quasiparticles.

The energy spectrum of the quasiparticles proposed by Andreev is

$$E = -\varepsilon_3 - \varepsilon_o + \frac{p^2}{2M_s} \qquad (1)$$

where ε_3 is the binding energy of He-3 to bulk He-4, ε_o is the excess binding of He-3 to the He-4 surface, p is the two dimensional momentum, and M_s is the surface effective mass. We note that in this picture, the binding energy and effective mass of the He-3 quasiparticle are independent of density. One may extend this model by introducing a quasiparticle interaction in the same way as was done by Bardeen, Baym, and Pines[8]. This interaction in the simplest case can be taken as a constant, V_0^s, and appears as a density dependent term in the chemical potential[4],

$$\mu_3 = -\varepsilon_3 - \varepsilon_o + \frac{1}{2}V_0^s N_s + k_B T \ln(\exp(T_{fs}/T) - 1) \qquad (2)$$

where N_s is the He-3 surface density and the 2D Fermi temperature, T_{fs}, is

$$T_{fs} = \frac{h^2 N_s}{4\pi M_s k_B} \qquad (3)$$

The physical picture of the system is as follows. At the absolute zero of temperature, if the amount of He-3 is small enough that the 2D Fermi energy is less than the surface binding energy, then all the He-3 resides in the surface states. This is the case for densities up to about one layer for the surface of bulk He-4. For any temperature greater than zero a finite amount of He-3 dissolves in the He-4. In thermal equilibrium the amount that dissolves is governed by two constraints. The chemical potential of the surface and dissolved He-3 must be equal and the total number of He-3 atoms must be conserved.

Edwards and Saam[9] report the parameters obtained by fitting surface tension, surface sound, and film flow data to this model. They obtain the values listed in the last column of Table I. Note that if one restricts the temperature to low enough values, the amount of vapor over the liquid is negligible. There would therefore be no conversion of liquid to vapor and vice-versa. In this regime only the energy difference between the surface and bulk states of He-3 would be important. ε_3 would not enter the data analysis.

We utilized adiabatic calorimetery to measure the heat capacity of He-3 in He-4 films of 22 to 53 Å thickness in the temperature range from 40 to 350 mK. He-3 densities reported here are 0.095, 0.19, 0.28, and 0.50 layer, where a one layer density is arbitrarily defined as the two-thirds power of the bulk He-3 density at zero pressure, $6.45\mathrm{X}10^{14}$ cm^{-2}. These films were formed on Nuclepore filters[10] which we had previously characterized through adsorption isotherms and heat capacity measurements of pure He-4 near the lambda transition[11]. These results demonstrated that one may form relatively thick, uniform films on this substrate. The film thicknesses we quote here include a 3 Å first layer of He-4 and subsequent layers of 3.6 Å.

In our data analysis we used the models discussed above and detailed at greater length by Edwards, et al.[4]. The heat capacity consists of three parts: the heat capacities of the surface and dissolved He-3 and the heat absorbed through promotion of the He-3 from the surface states to the film states. Due to this promotion the specific heat rises above its expected value of k_B for an ideal gas (Fig. 1).

For thicker films the He-3 in the film was treated as if it were dissolved in bulk He-4. That is, it was considered to be a 3D ideal gas of Fermi quasiparticles having an effective mass of $2.3M_3$, which is the experimentally

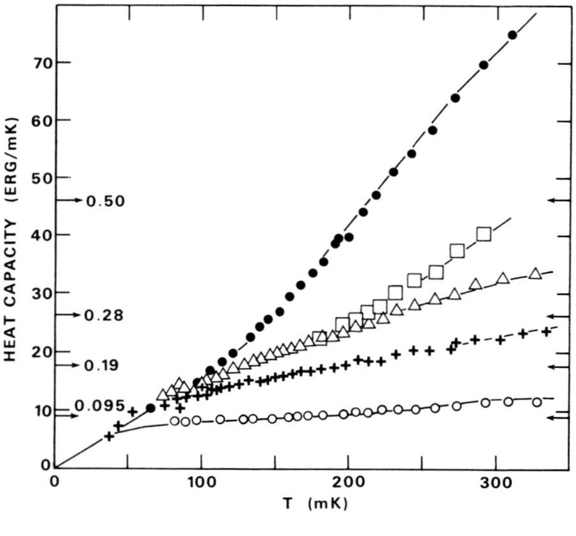

Figure 1. The heat capacities of 0.095 layer He-3 on a 22 Å He-4 film (open circles), 0.19 layer on a 22 Å film (plusses), 0.28 layer on a 22 Å film (triangles), 0.28 layer on a 47 Å film (squares), and 0.50 layer on a 48 Å film (filled circles). The arrows indicate Nk_B for each of the He-3 concentrations. The lines through the data are least squares fits using a 2D model for the He-3 dissolved in the 22 Å film and a 3D model for the He-3 dissolved in the 47-48 Å film.

obtained value for dilute He-3 in bulk He-4 at zero pressure. The concentration gradient within the film induced by the pressure gradient of the van der Waals field was properly taken into account[12]. This has the effect of pushing the He-3 away from the substrate. Our model calculations indicate that the concentration of He-3 dissolved in the film is less than 6%, the phase separation limit at T=0 in bulk helium mixtures (Fig. 2).

High surface concentrations of He-3 yielded data which could not be fit well with either the ideal or interacting quasiparticle models. We will therefore not present the data for 0.69 and 0.99 layer here[12].

The fit of the data for thin He-4 films contained systematic deviations at high temperatures when a 3D model for the dissolved He-3 was used. (See Figure 3). If one considers the size of the thermal (or DeBroglie) wavelength of a dissolved He-3 atom, the reason is obvious; this length becomes of the order of the film thickness. The He-3 within the film must then be more properly treated as a 2D system. Thus this experimental situation involves two 2D systems! As the film thickness is increased the number of 2D states accessible to the He-3 in the film increases, and a gradual crossover to 3D behavior is observed. As a practical matter, once the number of 2D states needed to fit the data reached three (which occured at a film thickness of around 40 Å), the fit was indistinguishable from that obtained with the 3D model.

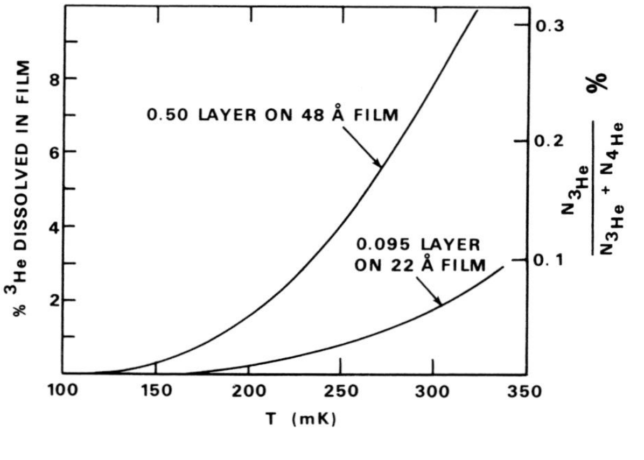

Figure 2. The amount of He-3 dissolved in the He-4 film as a percentage of the total He-3 (left axis) vs. temperature. The right axis corresponds to the average percent of He-3 in He-4 for the 0.50 layer data only. To obtain the proper scale for the 0.095 layer data, multiply by 0.434.

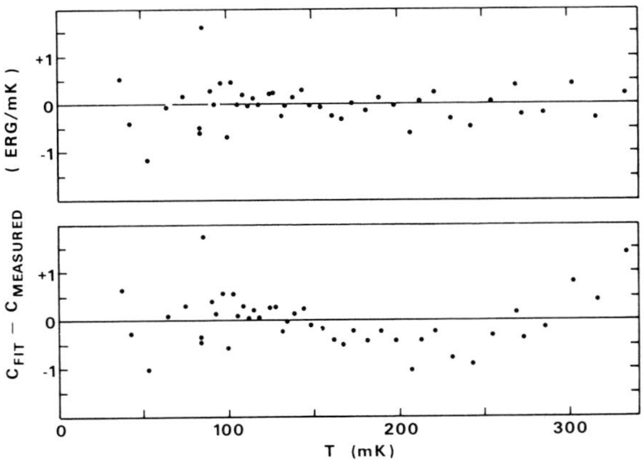

Figure 3. Deviation plots of the 0.19 layer data (22 Å film) for fits with the 2D gas model (top) and the 3D model (bottom) for the dissolved He-3. Systematic deviations for the 3D model only appear at high temperatures where the amount of He-3 within the film becomes appreciable.

A least squares fit of our data yields parameters which are different from those of the bulk surface. Table I contains these results. The surface binding is as much as 1.0 K smaller than for the bulk surface. ε_0 increases with film thickness, tending toward the bulk value. The surface effective mass is somewhat larger than that obtained for the bulk surface, although the difference lies within the combined error bars of the values. Most of the error indicated for the surface effective mass is due to the accuracy of our measurement of the surface area of the substrate. However, the random error in our determination of M_s is only ±.02, and hence the difference between M_s for 22 Å and 48 Å is statistically significant.

Data sets for a single nominal He-3 density (for example, 0.095 layer) are adequately fitted by using the ideal 2D Fermi quasiparticle model. However, if we try to fit data of more than one density with the same parameters, we must extend the model to include the interaction, V_0^s, in order to achieve a satisfactory fit. (See Figure 4). The sign of this required interaction term is negative, indicating that there is an attraction between the He-3 quasiparticles. This is in sharp contrast to the weakly repulsive interaction found for the He-3 quasiparticles on the bulk surface. Table I gives the interaction in two ways, as V_0^s, the energy per particle per unit area, and as v_o^s, which is related to V_0^s by

$$k_B \frac{N_s}{N_o} v_o^s = \frac{1}{2} N_s V_0^s \tag{4}$$

where N_o is the one layer density. In this form one may readily compare the interaction energy with the surface binding energy. For example, we see that for a 22 Å film this interaction is a substantial fraction of the binding energy even for densities of a fraction of a layer.

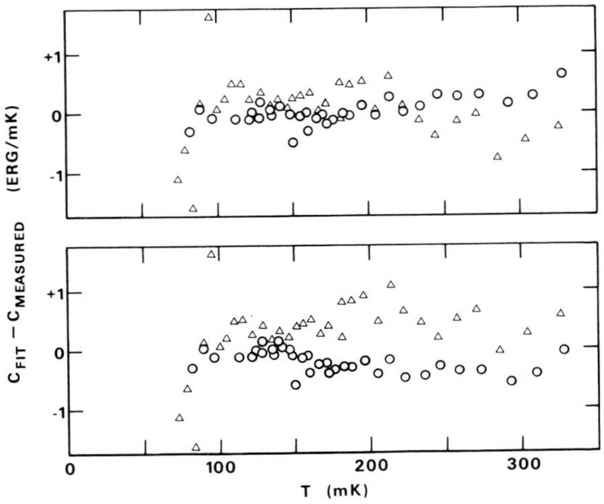

Figure 4. Deviation plots of the 0.095 (circles) and 0.28 (triangles) layer He-3 on a 22 Å He-4 film. The fits are without the interaction, V_0^s (bottom), and with V_0^s (top). The data merge together when fit with the interaction term, although there remains some systematic deviation.

The reason for the differences in the He-3 states on the bulk and on the film is very likely the indirect influence of the substrate. While the direct effect of the van der Waals potential of the substrate on the surface He-3 is small, the surface density profile must somehow be altered. This profile in turn determines the surface binding potential and influences the spectrum of surface excitations, or ripplons, which mediate the surface ^3He-^3He interaction. In addition the cylindrical geometry of the filters changes the ripplon spectrum from that of a planar geometry. For the films we investigated this last effect is probably small. The film thickness is therefore a device which allows us to "tune" the He-3 interaction with the He-4 and with each other.

The physical system we have described differs substantially from other realizations of 2D He-3 such as via direct adsorption on a solid substrate. Clearly, our system will approach this more conventional situation as the He-4 film thickness is decreased. A comparison of our results with the specific heat of He-3 on Grafoil[13] shows that indeed the data are quite different, in overall magnitude, temperature dependence, and coverage dependence. One might also view our system as a two dimensional mixture of He-3 in He-4 on top of a superfluid substrate. Again, comparison with physisorbed mixtures on Grafoil[14] reveals the two systems to be quite different. Viewed as a mixture the surface mole fraction, X_s, of He-3 for the data presented here, spans the range from X_s=0.079 to X_s=0.45. This is a range for which there have been calculations of the 2D phase diagram[15,16] which exhibits both a superfluid transition and a phase separation. The data presented here and in reference 17 show no evidence of such behavior.

Heat capacity measurements on different thickness He-4 films and He-3 concentrations are continuing.

Table I

film thickness (Å)	22	47-48	bulk
M_s (M_3)	1.64±0.08	1.56±0.08	1.45±0.1
ε_o (K)	1.28±0.04	1.68±0.02	2.22±0.03
v_o^s (K/layer)	-1.13±0.13	-0.13±0.05	0.33±0.29
V^s (10^{-31}ergcm^2)	-5.25±0.7	-0.6±0.3	+1.6±1.4

REFERENCES

1. K. R. Atkins and Y. Narahara, Phys. Rev. 138, A437 (1965).
2. A. F. Andreev, Zh. Eksp. Teor. Fiz. 50, 1415 (1966) [Sov. Phys. JETP 23, 939 (1966)].
3. H. M. Guo, D. O. Edwards, R. E. Sarwinski, and J. T. Tough, Phys. Rev. Lett. 27, 1259 (1971).
4. J. R. Eckardt, D. O. Edwards, P. P. Fatouros, F. M. Gasparini, and S. Y. Shen, Phys. Rev. Lett. 32, 706 (1974) and D. O. Edwards, S. Y. Shen, J. R. Eckardt, P. P. Fatouros, and F. M. Gasparini, Phys. Rev. B12, 892 (1975).
5. D. B. Crum, Ph. D. thesis (The Ohio State University, 1973), unpublished.
6. A. J. Dahm, Phys. Rev. 180, 259 (1969).
7. S. Ohta and Y. Sawada, Phys. Rev. B19, 4518 (1979).
8. J. Bardeen, G. Baym, and D. Pines, Phys. Rev. 156, 207 (1967).
9. D. O. Edwards and W. F. Saam, Progress in Low Temperature Physics, Vol. VIIA, ed. by D. F. Brewer (North Holland, New York, 1978).
10. Nuclepore Corporation, 7035 Commerce Circle, Pleasanton, CA 94566.
11. T. P. Chen, M. J. DiPirro, A. A. Gaeta and F. M. Gasparini, J. of Low Temp. Phys. 26, 927 (1977) and T. P. Chen and F. M. Gasparini, Phys. Rev. Lett. 40, 331 (1978).
12. M. J. DiPirro, Ph. D. thesis (The State University of New York at Buffalo, 1979), unpublished. The complete set of data, including 0.28 layer He-3 on intermediate thicknesses of He-4 is contained there.
13. S. V. Hering and O. E. Vilches, Monolayer and Submonolayer Helium Films, edited by J. G. Daunt and E. Lerner (Plenum, New York, 1973), p. 1.
14. D. C. Hickernell, E. O. McLean, and O. E. Vilches, J. Low Temp. Phys. 23, 143 (1976).
15. J. L. Cardy and D. J. Scalapino, Phys. Rev. B19, 1428 (1979).
16. A. N. Berker and David R. Nelson, Phys. Rev. B19, 2488 (1979).
17. M. J. DiPirro and F. M. Gasparini, Phys. Rev. Lett. 44, 269 (1980).

Sinha, ed, Ordering in Two Dimensions

SOLIDON EXCITATIONS IN LIQUID HELIUM FILMS

M. HÉRITIER and G. MONTAMBAUX

Laboratoire de Physique des Solides, Université Paris-Sud, 91405 Orsay, France.

In a nearly solid two-dimensional quantum liquid, a longitudinal phonon can be self-trapped in a solid-like region and form a 2-D "solidon", analogous to a roton with a large solid-like core. If bulk liquid is present on top of a solid-liquid interface, a three-dimensional half-solidon can be bound to the surface. Experimental data on Helium films are briefly discussed.

PHONON SELF-TRAPPING IN QUANTUM LIQUIDS

In a physical system in which an elementary excitation has a lower energy in an excited phase than in the stable phase, a self-trapping of the excitation can occur [1]. Under certain conditions, such an excitation may induce the formation of a finite volume of the excited phase in which it gets self-trapped. The first condition is a favourable balance between, on one hand, the excitation energy lowering, and, on the other hand, the excitation localization energy within the excited volume and the formation thermodynamic potential of this volume (in principle, surface and volume terms are to be included). Another condition for the existence of a well-defined self-trapped excitation is that its lifetime remains large compared to the inverse of its energy and to its time of formation. Superfluid ^{4}He is a good candidate for phonon self-trapping [2] because it is a "nearly solid" quantum liquid [3] (first condition) and because the absence of individual excitation and small cross-sections for scattering ensure a long lifetime (second condition).

In a quantum liquid like ^{4}He, a typical atomic translation kinetic energy $t = \hbar^2/2ma^2$, where m is the atomic effective mass and a the mean interatomic distance, is much lower than a typical longitudinal phonon energy $k_B\theta_D$ (for wave-vector about π/a)

$$t \ll k_B\,\theta_D \qquad (1)\ .$$

Moreover, near enough to the melting curve

$$G_s(p,\ T) - G_\ell(p,\ T) \ll t \qquad (2)$$

where G_s and G_ℓ are the free enthalpies per atom of respectively the solid phase and the liquid phase at a given pressure p and a given temperature T.

Consider a longitudinal phonon with wave vector K belonging to the reciprocal lattice of the metastable solid phase. Its energy, of the order of $k_B\theta_D$ in the liquid, vanishes in an infinite crystal, because of the translational invariance. The phonon excitation reduces to a translation of atoms. In a finite crystal of N atoms, this translation has a momentum $\hbar K$ and a kinetic energy (localization energy)

$$E_{loc}(N) = \frac{\hbar^2K^2}{2Nm} \qquad (3)\ .$$

Therefore, inequalities (1) and (2) ensure a phonon self-trapping in a "solid-like" region. However, flow conservation requires a backflow in the disordered medium, which, at long distance, involves essentially a dipolar velocity field. Moreover this backflow allows periodic boundary properties and, therefore, a true translation invariance. It requires an energy proportional to 1/N, which, in fact, amounts to redefine the localization energy :

$$E'_{loc}(N) = (1+\gamma)\,\frac{\hbar^2K^2}{2Nm} \qquad (4)$$

where γ is a geometrical factor of order unity.

Therefore, the variational estimate of the excitation free enthalpy is :

$$\Gamma(p,T) = (1+\gamma)\frac{\hbar^2K^2}{2Nm} + N\delta G(p,T) + N^{(d-1)/d}\sigma \qquad (5)\ .$$

In (5), the variational parameter N is the number of atoms in the "solid-like" region. $\delta G(p,T) = G^*(p,T) - G_\ell(p,T)$, where $G^*(p,T)$ is the free enthalpy per atom in this region. $G^*(p,T)$ is lowered below $G_s(p,T)$ by relaxation effects. The last term, in which d is the space dimensionality, is a phenomenological expression of the solid-liquid interface energy [2]. Minimizing (5) with respect to N, we obtain the excitation free enthalpy $\Gamma_m(p,\ T)$, and the corresponding number of atoms in the solid-like region $N_m(p,\ T)$.

This self-trapped phonon, which we have called a "solidon", leads to a minimum in the dispersion curve at the shortest reciprocal lattice wave vector of the metastable solid. Our solidon picture is, in a certain way, analogous to the Feynman's description of a roton [4]. However this vortex ring possesses a core: atoms, when penetrating in this core, are strongly correlated so as to preserve the translation symmetry of the metastable solid phase. This semi-phenomenological approach is an attempt to include non-linear effects of the phonon itself on the local order and therefore on the phonon energy. We believe these non-linearities are an essential aspect of collective excitation in liquid helium.

SOLIDON BOUND STATE AT A SOLID LIQUID INTERFACE

Experimental properties pointing out specific aspects of the solidon should be a manifestation of the core: such might be the case of the possible existence of a bound state at a solid-liquid interface. Now, we study the dynamics of a solidon in the vicinity of a surface.

We find it convenient to use the analogy of a vortex ring with a current ring . In a perfect incompressible fluid, the equations of hydrodynamics are quite analogous to Maxwell equations. A rigid surface perturbs the dynamics in a fluid like a metallic surface perturbs a magnetic field. In both cases, the surface imposes the boundary condition :

$$(\vec{n} \cdot \vec{\nabla}\phi(\vec{r}))_{\vec{r} \in \text{surface}} = 0 \qquad (7)$$

where $\vec{n}$ is a unit vector normal to the surface and ϕ is the velocity potential or the scalar magnetic potential. Of course, this is only true for the coherent velocity field of the solidon. In the liquid, atoms can be adsorbed on the surface or desorbed. These random events produce atomic velocities with non-zero normal components and may affect the lifetime of the bound state. To describe the solidon in presence of a surface, it is convenient to use the image method, as in electromagnetism. A ring in the presence of a metallic surface is described by replacing the surface by a ring with a magnetic moment which is the image of the object in a mirror (figure 1). A ring and its image repel themselves. In the same way , a vortex ring is repelled by a surface (with a force $\sim 1/r^4$, at a distance r much larger than the ring radius r_0) Therefore, the solidon energy increases when it gets closer to the surface.

However, half a vortex ring can be bound to the surface (fig.1). When adding the "image flow", the flow is the same as in a vortex ring in free space. Obviously, all the terms in Eq.(5) (kinetic energy and free enthalpy) are divided by two for half a solidon. If we neglect interaction with the surface, we find a bound state free enthalpy half that of a free solidon. The bound state wave vector is parallel to the surface and has the same modulus as in free space.

Due to the finite size of the core, this bound state is stable, even in the presence of small surface irregularities or small displacements from the surface due to incoherent atomic flow scattering. Indeed, when the "ring" is removed from its equilibrium position, first, its kinetic energy increases with the distance r from the surface, as long as r is smaller than the ring radius r_0; then, for $r>r_0$ it decreases. Therefore, a solidon bound state can be stable for smooth enough surface (i.e. on a scale r_0) while a roton bound state is not (the corresponding r_0 is about one interatomic distance).

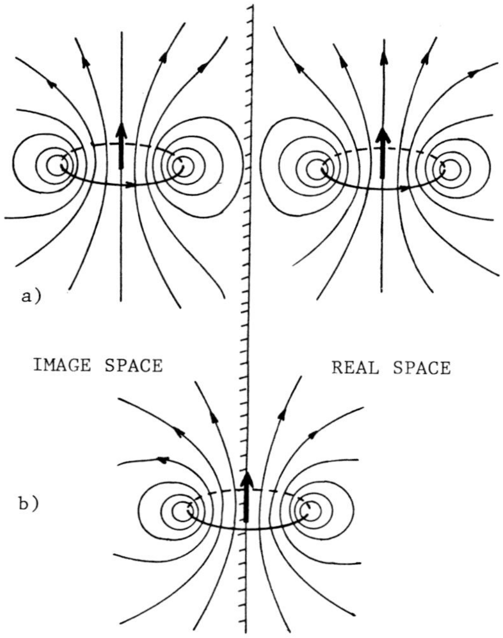

FIG 1: a) repulsion of the solidon by the surface.
b) solidon bound state.

This simple-minded approach disregards a number of difficulties related to the detailed interactions between the liquid and the surface. Friction of the atomic flow on the substrate as well as adsorption or desorption phenomena probably shorten the bound state lifetime. Surface irregularities at the atomic scale should increase the bound state energy. However, the most important effect which we have neglected is probably the strong Van der Waals attraction with the surface, which creates a large pressure gradient in the direction normal to the surface [5]. This should influence both the wave vector and the energy of the bound state. Corresponding to decreasing pressure, the lattice parameters of the metastable solid increase with increasing distance from the surface; this is true, of course, for the lattice spacing in the direction normal to the surface, but also in planes parallel to the surface. In these planes, the areal density decreases with increasing distance to the surface, because of change in the balance between elastic energy and Van der Waals interaction with the substrate [6]. In such a structure, the phonon would not benefit from a translational invariance because of the distorsion due to the large pressure gradient. Therefore, because of inequalities (1) and (2), the phonon is thought to self-trap in a translation invariant structure with a lattice spacing intermediate between the high pressure and low pressure values. Correspondingly, we also expect a bound state wave vector intermediate between the extreme values. Moreover, besides the free enthalpy formation of the metastable solid at intermediate pressure, this structure requires an additional formation energy which we write very roughly as $E=N\frac{\beta}{2}(\frac{\Delta a}{a})^2$. $\frac{\Delta a}{a}$ represents the mean relative lattice deformation necessary to obtain the translation invariant structure from the distorted metastable solid. This expression should be considered as a dimensional argument rather than a rigorous estimation. Without this correction, our simple model predicts a bound state energy of about 4 K. Estimating $\Delta a/a \sim 0.1$, the order of magnitude of the correction is $\sim$ 1 K. Let us mention that the pressure and density gradients induce anisotropy in the backflow. We did not take into account this effect on the solidon energy. Our estimation of the bound state energy is about 5 K.

TWO-DIMENSIONAL SOLIDONS

Truly two-dimensional solidons may also exist. Experimentally, they can be studied in liquid helium films adsorbed on graphite substrates, in which solid-liquid transitions have been detected by various techniques. In the vicinity of these transitions, they certainly provide good examples of "nearly solid" two-dimensional quantum liquids. It is interesting from many respects to study the solidon excitations in these systems. Minimizing Eq.(5) with respect to N in two dimensions, we find the following expressions for the solidon gap $\Gamma_m(p, T)$ and the number of atoms in the solid-like core,

-when the free enthalpy term dominates :

$$\Gamma_m(p, T) = 2\left|\frac{\hbar^2K^2}{2m}(1+\gamma)\,\delta G\right|^{1/2} \qquad (6\ a)$$

$$N_m(p, T) = \left|(1+\gamma)\frac{\hbar^2K^2}{2m}/\,\delta G\right|^{1/2} \qquad (6\ b)$$

-when the interface term dominates :

$$\Gamma_m(p, T) = \frac{3}{2}\left|\frac{\hbar^2K^2}{m}(1+\gamma)\,\sigma^2\right|^{1/3} \qquad (7\ a)$$

$$N_m(p, T) = \left|(1+\gamma)\frac{\hbar^2K^2}{m\,\sigma}\right|^{2/3} \qquad (7\ b)$$

In two-dimensional He layers adsorbed on graphite substrates, two distinct solid phases with different properties exist [7]. At high density, the phase is a solid incommensurate with the substrate, with a close-packed triangular structure [8]. At density corresponding to about 0.6 complete monolayer, a commensurate $\sqrt{3}\times\sqrt{3}$ structure can be formed [9]. The core of a solidon in the liquid at temperature T and pressure p will have the one of these structures with the lowest free enthalpy (Eq.(6 a)) or with the lowest interface tension (Eq.(7 a)). Then, one may speak of a "commensurate" or "incommensurate solidon, depending on its core structure. Different properties are expected for these two kinds of solidons. In both cases, the metastable solid phase has a triangular structure. The solidon wave vector is $4\pi/a\sqrt{3}$, where a is the interatomic distance. For an incommensurate solidon, it varies with external pressure and temperature. For a commensurate solidon, a = 4.20 Å is fixed by the substrate lattice. The solidon wave vector remains constant.

In fact, the above analysis assumes implicitly that the liquid is locally much more disordered than the solid, as in three dimensions. Far from the melting curve, this is certainly true. However, close to the transition, particular aspects of melting in two dimensions should be considered. The problem is different near the liquid-incommensurate solid transition and near the liquid-commensurate solid transition. The experimental situation in liquid He films is still unclear in both cases [7]. Short range order in these liquids is only poorly known. From a theoretical point of view, these two transitions are believed to belong to different university classes. It has been proposed that melting of incommensurate solid He layer can be described by the X-Y theory of melting [10] : melting is due to dissociations of pairs of bound dislocations with opposite Burgers vectors. This transition is accompanied by an essential singularity. Near the transition, the number of free dislocations in the liquid is exponentially small, so the liquid is locally similar to the solid. If this description near the melting line is correct, it is clear that the concept of solidon is meaningless near this line. However, at present time, it is not quite clear that the X-Y model can account for the experimental data in He layers, particularly the specific heat anomaly at the transition [11]. Therefore, we cannot exclude a priori the existence of incommensurate solidons near the melting curve, although we must keep in mind that it might be ruled out by the nature of the liquid phase.

The case of the liquid-commensurate solid is different. Melting of solid He layers in registry with a graphite substrate has been described by a three-state Potts model [12]. In this model, melting is an order-disorder transition with a specific heat singularity in good agreement with experimental data in He layers [13,14]. In such a case, one can really envisage the existence of commensurate solidons near the melting curve.

The liquid-incommensurate solid transition as well as the liquid-commensurate solid transition seems to be continuous in He layers adsorbed on grafoil [7](in accord with the X-Y model and the three-state Potts model). That means that both δG and σ should vanish at the transition. Therefore, in our simple approach, the solidon radius should diverge and the solidon energy should soften at the transition. In fact, difficulties arise in this simple treatment.

First, when the solidon radius diverges, one can no longer consider solidons as independent as in the above treatment. Their interactions will be considered as negligible when the mean distance between solidons is much larger than the solidon radius. This condition can be written :

$$\left(\frac{R}{a}\right)^3 \exp - \frac{\Gamma(K)}{k_B T} << 1 \qquad (8) ,$$

where $\Gamma(K)$ is the softening solidon gap.

Another difficulty arising from Eqs.(6) and (7) is that they imply the occurence of long range order when R diverges, as a result of mean field approximation. Too close to the transition, the solidon radius is no longer determined by minimization of (5). In fact, it is roughly given by the correlation length $\xi \sim t^{-\nu}$, where $t = (T-T_c)/T_c$ and T_c is the critical temperature. Then, the solidon gap behaves like $t^{d\nu}$ (d = 2 is the space dimensionality. Therefore, we expect a cross-over from the classical behaviour described by Eqs.(6 a) and (7 a) in which the critical exponents of the solidon gap are, respectively, $d\nu/2$ and $2y_\sigma/3$ (y_σ is the critical exponent for σ) to a second regime in which the critical exponent is d.

COMPARISON WITH EXPERIMENTS

In discussing the experimental situation, it is difficult at the present time to determine whether the data can be accounted for by surface excitations or by bulk excitations bound to the solid-liquid interface. However, it seems that one can distinguish two cases :

i) "bulk" liquid is present over the substrate (or in the Vycor pores) [15,16] or the He film is thicker than about four layers [6]. In such cases, there is an agreement on a value of about 6 K for the excitation energy in fairly good accord with our estimation of the bound state energy. Neutron data indicate a roton wave vector of 2.0 $\AA^{-1}$ and parallel to the surface. This wave vector value can be accounted for either by the bound solidon model or by the surface solidon model. It does not agree with the value predicted in the surface roton model based on the Feynman-Cohen theory [17,18].

ii) In films thinner than 3 to 4 layers [19,20], the excitation energy seems definitely smaller ($\simeq 5$ K) and the wave vector value is consistent with that of a surface solidon.

It is possible that the case ii) corresponds to surface excitation in the second layer, while the first corresponds to surface excitation in the third layer, the second layer being solidified by the additional pressure. We propose another interpretation : excitations bound to the solid-liquid interface can be detected in case (i), while only surface excitations exist in case (ii).

REFERENCES

1. M. Héritier, P. Lederer and G. Montambaux, to appear.
2. M. Heritier, G. Montambaux and P. Lederer, J. Physique Lett. 40, L493 (1979).
3. B. Castaing and P. Nozières, J. Physique 40, 257 (1979).
4. R. P. Feynman and M. Cohen, Phys. Rev. 102, 1189 (1956).
5. See for example D. Spanjaard, D. L. Mills and M. T. Béal-Monod, J. of Low Temp. Phys. 34, 307 (1979).
6. W. Thomlinson, J. A. Tarvin and L. Passell, Phys. Rev. Lett. 44, 266 (1980).
7. J. G. Dash, Phys. Rep. 38C, 179 (1978).
8. K. Carneiro, W. Ellenson, L. Passell, J. P. McTague and H. Taub, Phys. Rev. Lett. 37, 1695 (1976).
9. M. Nielsen, J. P. McTague and W. Ellenson, J. Phys. C4, 10 (1977).
10. J. M. Kosterlitz and D. J. Thouless, J. Phys. C5, 124 (1975).
11. R. L. Elgin and D. L. Goodstein, Phys. Rev. A9, 2657 (1974).
12. S. Alexander, Phys. Lett. 54A, 353 (1975).
13. M. Bretz, Phys. Rev. Lett. 38, 501 (1977).
14. M. Schick and J. S. Walker, Phys. Rev. B16, 2205 (1977).
15. D. F. Brewer, A. J. Symonds and A. L. Thomson, Phys. Rev. Lett. 15, 182 (1965).
16. C. W. Kicwiet, H. E. Hall and J. D. Reppy, Phys. Rev. Lett. 35, 1286 (1975).
17. T. C. Padmore, Phys. Rev. Lett. 32, 826 (1974).
18. W. Götze and M. Lücke, J. Low Temp. Phys. 25, 671 (1976).
19. J. E. Rutledge, W. L. McMillan, J. M. Mochel and T. E. Washburn, Phys. Rev. B18, 2155 (1978).
20. D. J. Reppy, J. M. Parpia and J. D. Reppy, J. Low Temp. Phys. LT-14, M. Krusius and M. Vuorio eds. (North-Holland, New-York, 1975) Vol.1, p.380 .

Sinha, ed. Ordering in Two Dimensions

SPATIAL ORDERINGS AND ELEMENTARY EXCITATIONS OF HELIUM FILMS*

A. ISIHARA AND L. IORIATTI

Statistical Physics Laboratory, Department of Physics, State University of New York at Buffalo, Buffalo, N. Y., 14260

ABSTRACT

In consideration of collective couplings of the particles, an important formula for the structure factor S(q) of two-dimensional liquid helium is derived for low but finite temperatures. It enables us to determine the population of the elementary excitations. In generalization of the Feynman and Cohen relation, the formula predicts that S(q) approaches a constant for small momentum in proportion to temperature and shows a minimum at high temperatures. Such a behavior has recently been observed for bulk helium. New formulae for the energy and pair distribution function are derived. Superfluidity in actual helium films is analyzed. The roton gap may depend on film coverages.

INTRODUCTION

In an attempt to improve the previous Feynman theory of liquid helium, Feynman and Cohen made a variational approach taking into consideration a backflow of atoms [1]. Their excitation spectrum showed some improvements over the previous one, but the theory was essentially for absolute zero. Since the superfluid properties are sensitive to temperature, Samulski and Isihara [2] have recently developed a new microscopic theory of liquid helium. Treating long distance collective couplings of the real Bose particles at finite temperature, they have shown that the internal energy can be expressed in Landau's form in the first approximation in terms of massless quasiparticles.

In view of the significance of their new approach and high interest in helium films, we shall examine the case of two-dimensional helium in what follows in the present paper, derive a new formula which generalizes the Feynman-Cohen relation and discuss its significant consequences. We shall also treat short range correlations and derive new expressions for the energy and pair distribution functions.

STRUCTURE FACTOR

Spatial order in liquid helium can be represented by the static structure factor S(q) which is related to the pair distribution function $\rho_2(r)$ as follows:

$$S(q) = 1 + n \int e^{i\vec{q}.\vec{r}} g(r)\, d\vec{r}$$
$$\rho_2(r) = n^2 + n^2 g(r) \tag{1}$$

where n is the number density. Samulski and Isihara used the chain diagram approximation for the case of bulk liquid helium. It is possible to express the pair distribution for the two-dimensional case as

$$\rho_2(r) = n^2 - n\delta(r) + \frac{1}{\beta(2\pi)^2} \sum_j \int \frac{\lambda_j \exp(i\vec{q}.\vec{r})\, d\vec{q}}{1 + u(q)\lambda_j(q)} \tag{2}$$

u(q) is the Fourier transform of the potential and the λ_j are the boson eigenvalues of the propagator representing the unit of the chain. For low temperatures, the eigenvalues are given in the first approximation by

$$\lambda_j = \frac{2nq^2}{q^4 + (\frac{2\pi j}{\beta})^2} \tag{3}$$

Use of Eqs. (2) and (3) in (1) gives

$$S(q) = \frac{q^2}{\varepsilon(q)} \{1 + 2f(\varepsilon(q))\} \tag{4}$$

where

$$\varepsilon(q) = q(q^2 + 2nu(q))^{1/2} \tag{5}$$

and where $f(\varepsilon)$ is the Bose distribution function of the quasiparticles of energy ε:

$$f(\varepsilon) = \frac{1}{e^{\beta\varepsilon} - 1} \tag{6}$$

Formula (4) shows that the temperature variation of S(q) is determined by the population of the quasiparticles. It generalizes the Feynman relation for absolute zero:

$$S(q) = \frac{q^2}{\varepsilon(q)} \tag{7}$$

Note that the Feynman relation has been expressed for the energy rather than S(q). For low temperatures, Eq. (5) predicts a phonon spectrum

$$\varepsilon = cq \tag{8}$$

With the sound velocity c given by

$$c = (2nu(0))^{1/2} \tag{9}$$

Hence, according to Feynman's relation we expect

$$S(q) \rightarrow q/c \tag{10}$$

for small q. On the other hand, if q is brought to zero at a fixed temperature, Formula (4) predicts that S(q) approaches a constant given by

$$S(0) = \frac{2}{\beta c^2} \tag{11}$$

The pair distribution function given by Eq. (2) yields the internal energy for low but finite temperatures as follows:

$$u = \frac{1}{2(2\pi)^2} \int d\vec{q}\{\varepsilon(q) - q^2\} - \frac{n^2}{2} \int d\vec{r}\phi(r) + \frac{1}{(2\pi)^2} \int \frac{\varepsilon \, d\vec{q}}{e^{\beta\varepsilon} - 1} \tag{12}$$

From this expression we conclude that ε represents the energy of quasiparticles obeying Bose statistics. Note here that no activity parameter enters the Bose distribution function. Thus, we have confirmed the Landau theory itself not merely his excitation spectrum. We remark that this microscopic justification is based on the eigenvalues of Eq. (3). It is possible to evaluate correction terms to this expression due to finite temperature. Samulski and Isihara have found that the excitation energy for the bulk case becomes a function of temperature.

The excitation spectrum and thermodynamic properties of the system depend on the Fourier transform u(q). If the potential has a hardcore, u(q) can be oscillatory. In such a case, the energy spectrum will be particle-like whenever u(q) vanishes. In order to be more explicit, we have adopted a soft-potential:

$$\phi(r) = \begin{cases} V_0 & r < a \\ \varepsilon^*\{(a/r)^{12} - (a/r)^6\}, & r > a \end{cases}$$

One can show that the excitation energy is of the following form:

$$\varepsilon(q) = cq(1 + \delta_1 q^2 - \delta_2 q^3 + ..) \tag{13}$$

Hence, for small q and low temperatures, the structure factor is

$$S(q)/S(0) = 1 + (\frac{\beta^2 c^2}{12} - 2\delta_1)q^2 + 2\delta_2 q^3 + .. \tag{14}$$

where S(0) is given by Eq. (11). This result indicates that S(q) at elevated temperatures will dip first before reaching the first peak. The temperature at which such a decrease takes place is given by

$$T_I = \frac{c}{(24\delta_1)^{1/2} k} \tag{15}$$

Such a behavior has been found for the bulk case by X-ray and neutron scattering experiments [3].

NEW EXCITATIONS

It has been found experimentally that the excitation spectrum in bulk liquid helium is actually multibranched: There are at least two higher excitation spectra in addition to the Landau spectrum. We remark in this respect that the results in the previous section are based on the eigenvalues given by Eq. (3). Samulski and Isihara have shown that a correction to the eigenvalue expression results in a temperature dependent excitation spectrum. When the momentum transfer representing interaction becomes higher, it becomes necessary to take into consideration short range correlations to the collective couplings such as considered in the previous section. One of the present authors has shown in this respect that what may be called the chain-exchange diagrams are most important [4]. The treatment of such diagrams requires calculating the following new eigenvalues:

$$\lambda_j(q, r) = \frac{1}{(2\pi)^2} \int d\vec{p} \, e^{i\vec{p}.\vec{r}} \times \{ \frac{f(\vec{p}+\vec{q}) - f(p)}{p^2 - (\vec{p}+\vec{q})^2 + \frac{2\pi i j}{\beta}} \} \tag{16}$$

Changing the variable in the first term from $\vec{p}$ to $-(\vec{p}+\vec{q})$ and using the same small p approximation as for Eq. (3) we find

$$\lambda_j(q,r) = n(r)\{\frac{e^{-i\vec{q}.\vec{r}}}{q^2 + \frac{2\pi i j}{\beta}} + \frac{1}{q^2 - \frac{2\pi i j}{\beta}}\} \tag{17}$$

where

$$n(r) = \frac{1}{(2\pi)^2} \int f(p)\, e^{i\vec{p}.\vec{r}}\, d\vec{p} \tag{18}$$

The contribution to the pair distribution function from the chain-exchange diagrams is given by

$$I_{cx} = -\frac{1}{(2\pi)^2} \sum_j \int \frac{u(q)\, \lambda_j(q,r)\, e^{i\vec{q}.\vec{r}}}{1 + \lambda_j u(q)}\, d\vec{q} \tag{19}$$

After some calculations we arrive at

$$I_{cx} = -\frac{I_2(r)}{(2\pi)^2} \int d\vec{q}\, u(q) \Big[\frac{1}{\varepsilon(q)} \coth(\frac{\beta\varepsilon(q)}{2}) + \frac{\cos(\vec{q}.\vec{r})}{\varepsilon(q)^2 - \varepsilon(0)^2} \{2\varepsilon(0) \coth(\frac{\beta\varepsilon(0)}{2}) - \frac{\varepsilon^2(0) + \varepsilon^2(q)}{\varepsilon(q)} \coth(\frac{\beta\varepsilon(q)}{2})\} \Big] \tag{20}$$

where $I_2(r)$ is an ideal gas function defined by

$$I_2(r) = \{G_1(r,\alpha)\}^2 \tag{21}$$

$$G_1(r,\alpha) = K_0(2(\pi\alpha)^{1/2} r/\lambda) \tag{22}$$

α being the ratio of the chemical potential and kT.

Formula (20) gives the chain-exchange diagram contribution to the pair distribution function. It is characterized by the Bose function G_1 of the helium atoms. $I_2(r)$ is short ranged. For three dimensions, a similar Bose function $G_{3/2}$ appears, which is like a screened Coulomb potential above the λ point.

Using the pair distribution function we can find a new energy term due to the chain-exchange diagrams. We obtain

$$u_{cx} = \frac{1}{(2\pi)^2} \int \{g_1 f(\varepsilon(0)) + g_2 f(\varepsilon(q))\} d\vec{q} \tag{23}$$

where

$$g_1 = n(u_1 - u_2) - \frac{2q^2}{u} (\frac{u_1}{2} + \frac{u_2}{3})$$

$$g_2 = -\frac{n\,\varepsilon(q)}{q^2} u_1 + \frac{\varepsilon(q)^3}{3uq^4} u_1 + \frac{\varepsilon(q)}{u} u_2 + \frac{n\,\varepsilon(q)}{q^2} u_2 - \frac{\varepsilon(q)^3}{3uq^4} u_2 \tag{24}$$

and where

$$u_1 = n^{-2} \int \phi(r)\, I_2(r)\, d\vec{r} \tag{25}$$

$$u_2 = n^{-2} \int \phi(r)\, I_2(r) \cos(\vec{q}.\vec{r})\, d\vec{r}$$

Formula (23) indicates that the excitation energy given by Eq. (5) needs corrections. The first term of Eq. (23) may effectively be represented by an effective mass which is momentum dependent. The second term may be represented by a new excitation spectrum plus an effective mass. The details of the new formulae for the pair distribution function and energy will be given in the near future.

SUPERFLUID DEFICIT

In real helium films, a certain portion of the molecules does not participate in superfluidity. It is customary to represent the difference between the actually measured superfluid density and its bulk value as follows:

$$-\delta\rho_s = \rho_{sb} L = \rho_{sb} H + \delta\rho_n \tag{26}$$

Here, the left side, called superfluid deficit, represents the difference, ρ_{sb} is the superfluid mass density of bulk helium and L is the healing length. In the second equation, the first term indicates that within H atomic layers, the helium atoms are immobilized and the second term shows that the deficit results from the excitation to the normal fluid states. It has been observed that H ≅ 1.47 in atomic layers between 0.1 and 0.6 ^{o}K and the healing length L is approximately the same. Therefore, in the low temperature limit, the entire deficit is due to the immobilization of the atoms.

Following Landau one can evaluate the normal fluid density based on Eqs. (5) and (12). The superfluid density of a film of thickness D may be given by $\rho_{sb}(D - L)$ in view of the immobilized atoms. Fig. 1 represents our recent theoretical result (solid curve) [5]. For the data see Scholtz, Mclean and Rudnick [6].

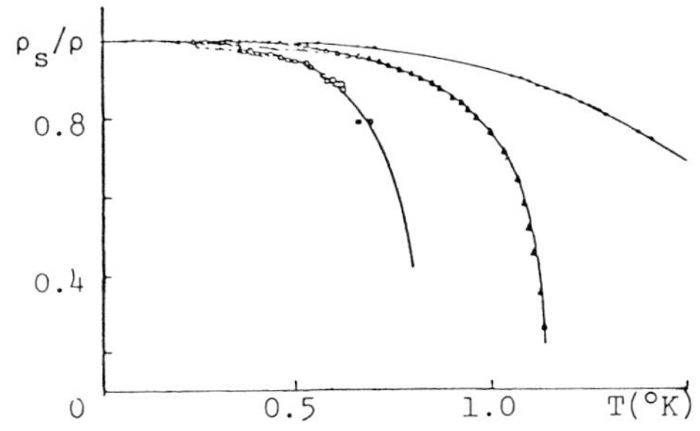

Fig. 1. Superfluid density vs. temperature.

The superfluid density can be evaluated from the normal fluid density as a function of temperature. Fig. 2 compares the theoretical results (solid curves) with the data by Bishop, Parpia and Reppy [7]. Although the theoretical results depend on the choice of molecular parameters, the agreement is good.

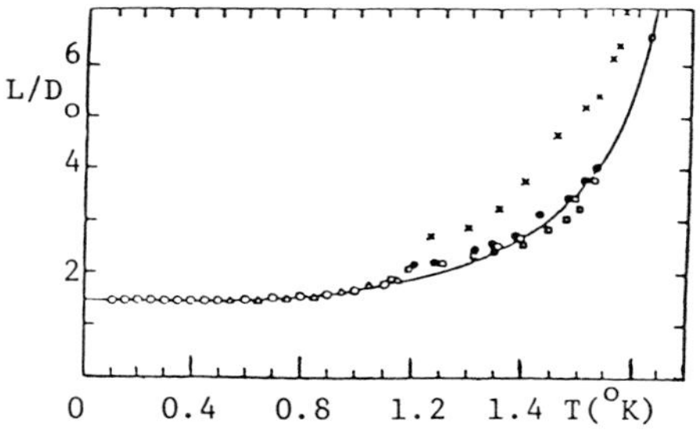

Fig. 2. Healing length divided by one standard layer thickness vs. temperature.

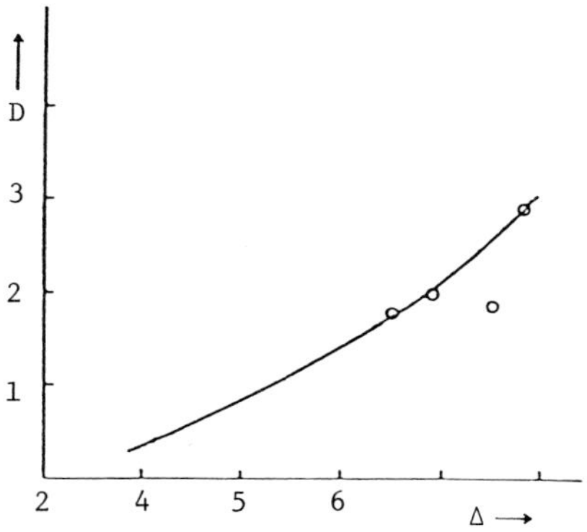

Fig. 3. Roton energy gap vs. film coverage.

In analyzing the films of different coverages, we have found that the roton gap parameter Δ is a function of D. Fig. 3 is our theoretical result (solid curve) compared with the data by Washburn, Rutledge and Mochel [8]. Although the data points may be represented by a horizontal line, a gradual reduction of the gap parameter with the film thickness seems to be also possible.

ACKNOWLEDGEMENT

This work was supported by the ONR under Contract No. N00014-79-0451.

REFERENCES

1. R.P. Feynman, Phys. Rev. 94, 262 (1954), R.P. Feynman and M. Cohen, ibid, 102, 1189 (1956).
2. T. Samulski and A. Isihara, Physica 82A, 294 (1976), A. Isihara and T. Samulski, Phys. Rev. 16B. 1969 (1977).
3. R.B. Hallock, Phys. Rev. 5A, 320 (1972), E.C. Svensson, V.F. Sears, A.B. Woods and P. Martel, Phys. Rev. to be published.
4. A. Isihara, Phys. Rev. 172, 166 (1968), Prog. Theo, Phys. 44, 1 (1969).
5 A.Isihara, Soon-Tahk Choh and Chung-In Um, Phys. Rev. 20B, 4482 (1979), A. Isihara and Chung-In Um, ibid 19B, 5725 (1979).
6. J.H. Scholtz, E.O. Mclean and I. Rudnick, Phys. Rev. L. 32, 147 (1974).
7. D.J. Bishop, J.M. Parpia and J.D. Reppy, in Low Temperature Physics, LT 14, ed. by M. Krusius and M. Vuorio (North-Holland, Amsterdam, 1975), Vol. 1, p. 380.
8. T.E. Washburn, J.E. Rutledge and J.M. Mochel, ibid, Vol. 1, p. 372.

MELTING OF ^{3}He SUBMONOLAYERS AND STRUCTURES OF ADSORBED ^{4}He ON GRAFOIL

H.J. LAUTER

Institut Laue-Langevin, 156X, 38042 Grenoble Cedex, France

H. WIECHERT, R. FEILE

Institut f. Physik, Johannes Gutenberg Universität, 65 Mainz, Germany

ABSTRACT

The melting of dense ^{3}He submonolayers was investigated by neutron scattering. Up to a coverage of x = 0.91 (x = 1 is monolayer completion) a slight compression of the nearest neighbour distance is observed during melting, which can be explained either by the presence of an additional liquid phase or by the creation of dislocations. At x = 0.98 the nearest neighbour distance increases during melting, which originates in the second layer promotion.

With ^{4}He the ($\sqrt{3}$ x $\sqrt{3}$) 30° structure of the first layer at x = 0.55 was confirmed. The diffraction pattern of the second solid layer was observed at T $\sim$0.62 K and x = 2.33.

Neutron scattering has been used to investigate the melting of the incommensurate solid phase of adsorbed ^{3}He on graphite near monolayer completion to get more information about the melting in two dimensions [1,2]. Some structural data of adsorbed ^{4}He on graphite are added.

The experiments were performed on the D16 spectrometer at the HFR in Grenoble. This instrument is placed at the cold source and we used a wavelength of 4.59 Å (with a Be-filter in the beam). A twenty wire multicounter was adjusted to see the (10)-reflection of the triangular lattice of the adsorbed He.

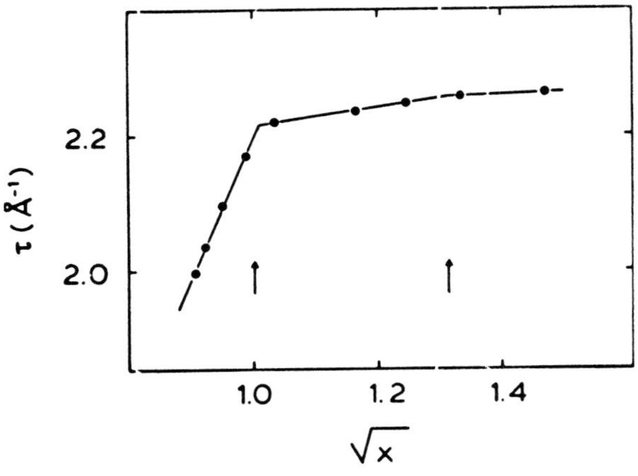

Fig. 1 The length of the reciprocal lattice vector τ in dependence on the root of the coverage at T $\simeq$ 1.06 K.

The sample consisted of a 5.5 cm high stack of Grafoil [3] sheets (6 cm x 0.8 cm), which were oriented parallel to the scattering plane and using reflecting geometry. The monolayer completion (x = 1) has been determined by an adsorption isotherm at 4.2 K [4].

Figure 1 shows the dependence of the length of the reciprocal lattice vector τ of the (10)-reflection in dependence of the square root of the coverage. The straight line (from the origin) to monolayer completion indicates that the atoms are all adsorbed and that they are homogeneously spread over the surface [5].

Above monolayer completion the first layer is further compressed by the influence of the second layer and there is still a change in the compression visible when third layer promotion starts. From the intersections of the straight lines (arrows) the density of the first layer at first and second layer completion was calculated, which gave 0.108 Å^{-2} and 0.112 Å respectively. The total density at second layer completion was determined to be 0.186 Å^{-2}, which is in very good agreement with the data in Ref. 6.

In Fig. 2 it is seen how the (10) peak loses intensity, broadens and shifts to higher angles, when the temperature is raised at constant coverage. The background measurement has already been subtracted, which consists of a measurement at T = 5.57 K (with the same amount of ^{3}He) [5]. The full lines were fitted by a least square fit to the data. It is the result of the convolution of the Warren formula

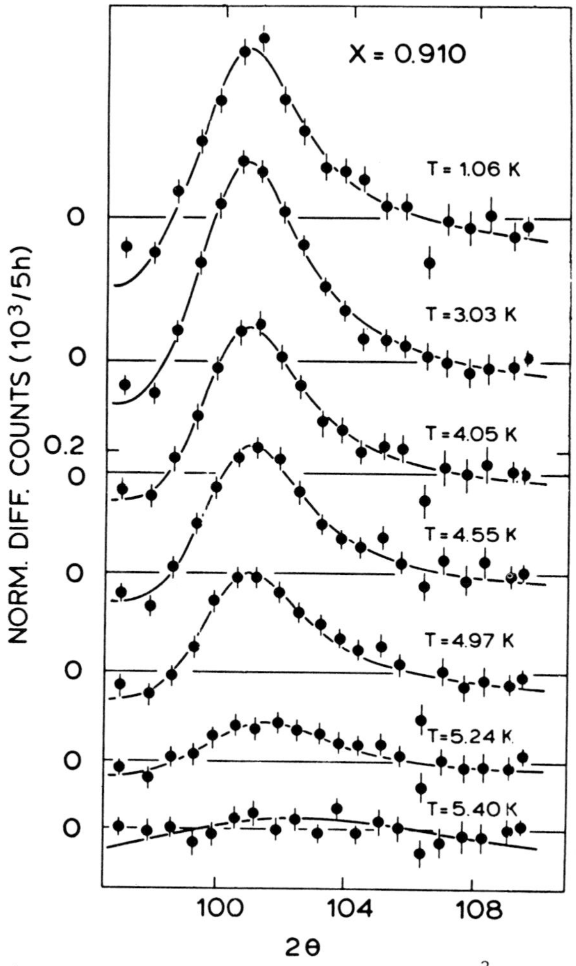

Fig. 2 Diffraction groups of adsorbed ^3He on Grafoil at x = 0.910 for various temperatures.

the surface pressure will increase. But on melting a two phase region could be formed [9]. Thus the liquid phase compresses the solid phase, which is mainly seen by the neutrons. The same result is obtained by introducing a liquid crystal region [2,10] where the solid phase is compressed by the increasing number of dislocations

In addition, the compression of the solid could be favoured by a temperature dependent change of the surface potential, induced by the oscillations perpendicular to the surface [14].

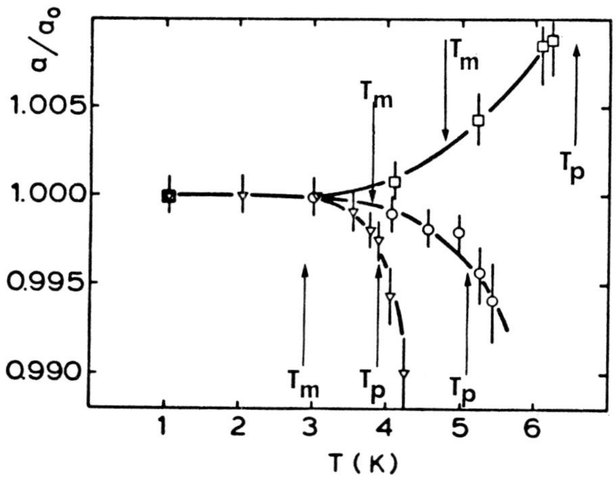

Fig. 3 Temperature dependence of the normalized nearest neighbour distance for x = 0.856 (▽), x = 0.910 (o), x = 0.981 (□); a_o(T = 1.06 K). The lines are drawn as guide for the eye. The arrows (T_p) indicate the melting temperature taken from Ref. 8. T_m marks the transition temperature between the solid and liquid crystal region as taken from the fit in Fig. 4.

including the mosaic distribution of the graphite (e.g. 5) with the resolution function of the instrument. The same behaviour as at x=0.910 is seen at x=0.829 and at x=0.856. But at x=0.981 the peak suddenly shifts to lower angles during melting. This is illustrated in Fig. 3, where the normalized nearest neighbour distance a_{nn} is plotted against temperature. The increase of a_{nn} at x = 0.981 during melting is certainly caused by squeezing atoms out of the first layer into the second and into the 3 D gas as for ^4He [7]. Thus the strong peaks in the specific heat measurements [8], which are far more intense than the ^4He ones, cannot be explained by second layer promotion for x < 0.98.

The decrease of a_{nn} during melting can be explained in two ways. On raising the temperature a_{nn} does not change because the surface area and the coverage are assumed to stay constant and the surface is uniformly filled. Only

The diffraction of the liquid is not seen directly in the spectra. The diffraction group at T = 1.06 K in Fig. 2 shows a negative counting rate on both sides of the peak. This arises because the spectrum taken as background at T=5.57 K includes obviously the diffraction of adsorbed fluid layers. The diffraction of a liquid gives only a flat contribution over the angular region in Fig. 2. At low temperature the intensity concentrates in the diffraction peak and the subtraction gives a negative counting rate at the foot of the peak. On melting the counting rate becomes less and less negative (see Fig. 2) indicating the growth of a fluid phase. But it is impossible to get qualitative data from the limited range in angle to decide whether a liquid is formed by some kind of islands or dislocations are produced uniformly spread over the surface.

The coherence length L, which has been derived from the fit of the Warren formula, consists of two parts, which add as $1/L = 1/L_o + 1/L_t$. At low temperatures it is

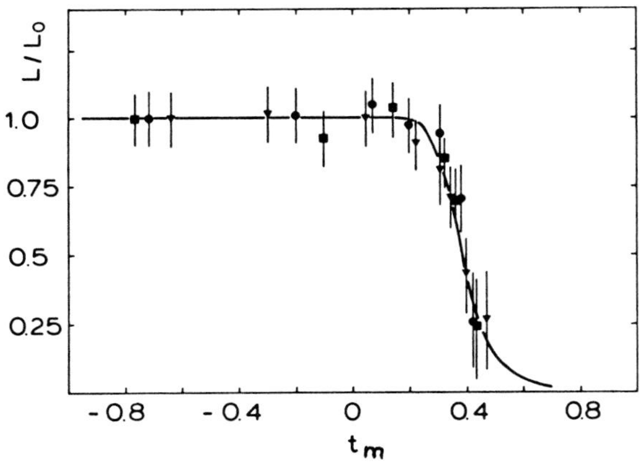

Fig. 4 Temperature dependence of the normalized coherence length, $L_o(T = 1.06\ K) \sim 150$ Å. Symbols as in Fig. 3.

determined by the size of the graphite crystallites ($L_o \sim 150$ Å). Assuming a liquid crystal region, L_t is given by $\exp(C/|t_m|^{0.37})$ [2], $t_m = (T-T_m)/T_m$ with T_m the transition temperature between the solid and the liquid crystal. The fit of the normalized coherence length L/L_o is shown in Fig. 4. T_m is 2.9 K, 3.8 K and 4.6 K for x = 0.856, 0.910 and 0.981, respectively, which is in agreement with Ref. 2, 10. But this good agreement between the fitted curve and the data might be fortuitous because the transition region, $t_m > 0$, could also be determined by a two phase region.

The normalized integrated intensity is shown in Fig. 5. The line serves as guide for the eye. For $t_c \lesssim -0.2$ the intensity is determined by the Debye-Waller factor and for $t_c \gtrsim -0.2$ by the transition region. More detailed calculations will follow.

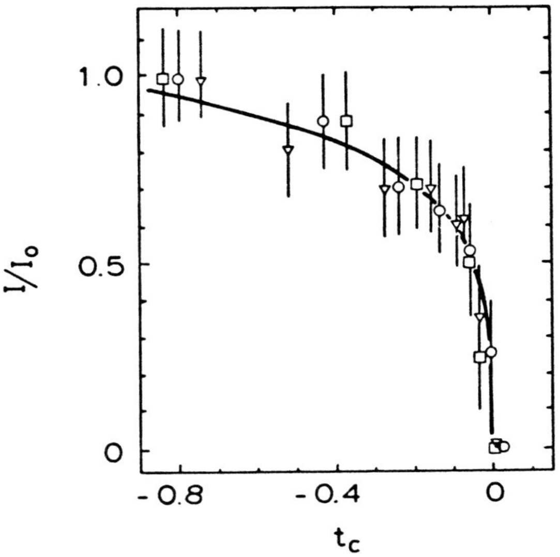

Fig. 5 Temperature dependence of the normalized integrated intensity of the diffraction peak; $I_o(T = 1.06\ K)$. Symbols as in Fig. 3. $t_c = (T-T_c)/T_c$, T_c the assumed melting temperature to the liquid.

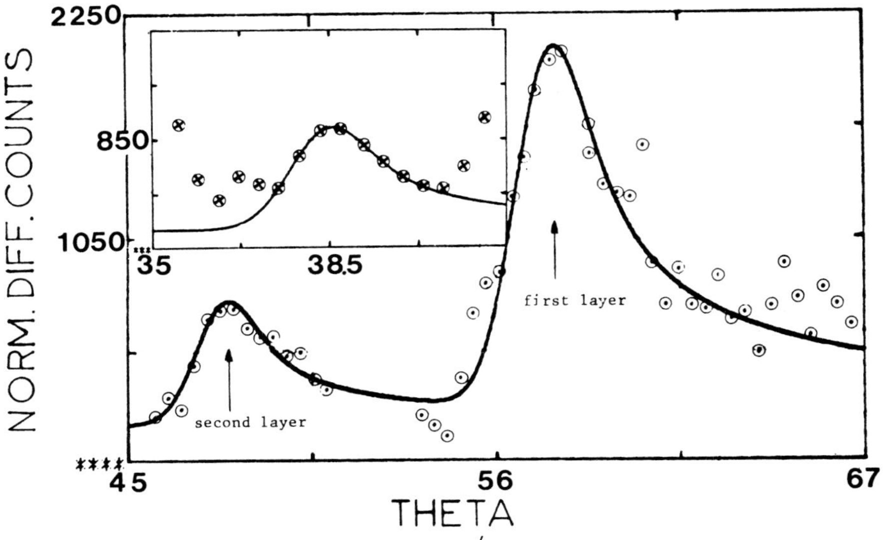

Fig. 6 Diffraction group of adsorbed ^4He on Grafoil at x = 1.977 and at T = 0.62 K. The inset gives the diffraction group at x = 0.508 and at T = 0.67 K.

It should be noted that the values of $\tau = 2.230$ Å^{-1} and $\tau = 1.967$ Å^{-1} (see Fig. 1) correspond within the triangular lattice to a $(\sqrt{7} \times \sqrt{7})R \sim 19°$ and a (3x3) overstructure with four atoms per unit cell (one atom in the middle of a graphite hexagon, the other three between two C atoms). But no deviations are caused in Fig. 1 by these phases, which could affect the nature of the phase transitions.

In an other sample (diameter 2cm) but in the same scattering geometry we measured structures of adsorbed ^{4}He on Grafoil. In Fig. 6 the diffraction pattern of the first and second layer are seen. The nearest neighbour distances have been calculated to be 3.16 Å and 3.59 Å [11], respectively, supposing a triangular lattice structure. Thus the second layer solidifies by the influence of a part of the third layer [6,11] and has a nearest neighbour distance which is only sligthly smaller than 3.66 Å [13], the a_{nn} in the basal plane of hcp helium on the melting curve. The inset in Fig. 6 gives the diffraction pattern at x = 0.508 confirming a $\sqrt{3}$ structure [12].

We are grateful to B. Castaing, G. Dash, H. Godfrin, C. Marti, M. Nielsen, O.E. Vilches and J. Villain for their interest and helpful discussions.

REFERENCES

1. J.M. Kosterlitz, D.J. Thouless, J. Phys. C6, 1181 (1973)
2. B.I. Halperin, D.R. Nelson, Phys. Rev. Lett. 41, 122 (1978).
   D.R. Nelson, B.I. Halperin, Phys. Rev. B19, 2457 (1979).
3. Grafoil is a product of the Union Carbide Corporation.
4. Further details will be published, R. Feile, H. Wiechert and H.J. Lauter.
5. M. Nielsen, J.P. McTague, W.D. Ellenson, J. de Physique C4, supp. 10, C4-10 (1977).
6. S.W. Van Sciver, O.E. Vilches, Phys. Rev. B18, 285 (1978).
7. R.L. Elgin, D.L. Goodstein, Phys. Rev. A9, 2657 (1974).
8. S.W. Hering, S.W. Van Sciver, O.E. Vilches, J. Low Temp. Phys. 25, 793 (1976).
   O.E. Vilches, private communication.
9. S. Toxvaerd, Phys. Rev. Lett. 44, 1002 (1980).
10. A. Widom, J.R. Owers-Bradley, M.G. Richards, Phys. Rev. Lett. 43, 18 (1979).
11. W. Thomlison, J.A. Tarvin, L. Passell, Phys. Rev. Lett. 44, 266 (1980).
12. M. Bretz, J.G. Dash, D.C. Hickernell, E.O. McLean, O.E. Vilches, Phys. Rev. A8, 1589 (1973).
13. H. Wiechert, H.J. Lauter, to be published.
14. L.M. Sander, M. Bretz, M.W. Cole, Phys. Rev. B, 14, 61 (1976).

Published 1980 by Elsevier North Holland, Inc.
Sinha, ed. Ordering in Two Dimensions

HEAT CAPACITY OF ^{4}HE FILMS ON LARGE GRAPHITE CRYSTALS

J.H. CAMPBELL and M. BRETZ

University of Michigan, Ann Arbor, Michigan 48109

M.H.W. CHAN

Pennsylvania State University, University Park, Pennsylvania 16802

ABSTRACT

We have made preliminary measurements of the heat capacity of submonolayer ^{4}He films adsorbed on a single flake of monochromator grade HOPG graphite. AC calorimetry techniques allow us to see a helium difference signal on the order of 1×10^{-9} J/K against a background of 3×10^{-8} J/K at 3K. Given the apparently large crystal domains of our substrates and the excellent temperature resolution of the AC technique, we hope to measure submonolayer phase transition peaks essentially free of rounding. This experiment should then clarify understanding of finite size effects in these 2D phase transitions. Measurements of heat capacity near the 1/3 ordering transition are now in progress.

INTRODUCTION

For some time, various forms of exfoliated graphite have been used in adsorbed film experiments. Exfoliation, however, damages the adsorption surfaces and leads to undesirable geometry effects. These effects complicate data analysis and prevent the observation of long range coherent behavior in 2D. Some experimental techniques promise to not require the large surface area of exfoliated substrates. Among these is the AC calorimetry method of Sullivan and Seidel [1,2]. This method has been applied to thin metal films and other very small samples [2,3,4], but not yet to adsorbed films. We report here our earliest AC heat capacity data on ^{4}He films in the submonolayer region. These films were adsorbed on one side of a small leaf of monochromator grade Highly Ordered Pyrolytic Graphite (HOPG). While full interpretation of the data is not clear, we are satisfied that it shows the potential for future heat capacity studies on our improved substrate.

METHOD

The substrate is a cleaved and cut HOPG square 5μm×7mm×7mm. A heater and a thermometer are placed on one side separated and covered by SiO layers. The lack of any backing material keeps the total heat capacity small, due almost entirely to the graphite (fig. 1). This is necessary to detect the presence of a ^{4}He film for which one would predict a heat capacity of about 4×10^{-9} J/K, based on $c/Nk = 1$ and $A = (.5\ \text{cm})^2$. For this data, a heating power of 1 μW_{rms} @ 600 Hz was used, resulting in a typical temperature modulation of 6 mK rms. Film coverages were determined from a vapor pressure isotherm of an adjacent grafoil reservoir.

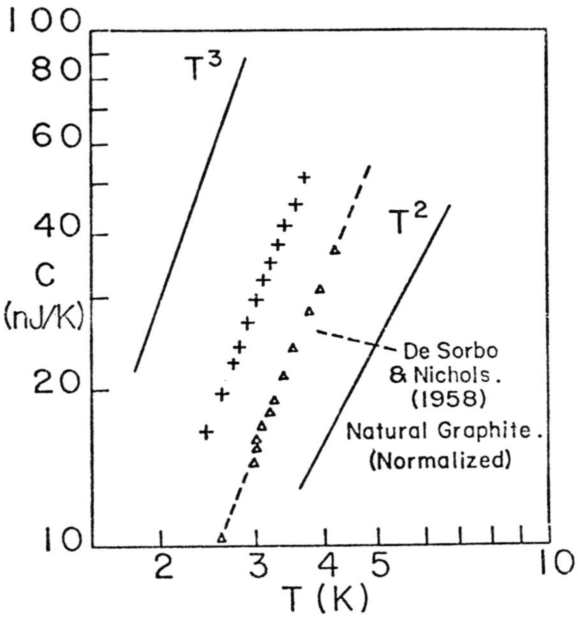

Fig. 1. Total Heat Capacity vs. Temperature

RESULTS

After subtracting the background, the data seems to show a 2D gas signature over our limited experimental range. Δc is nearly independent of temperature (Fig. 2) and linear in coverage after an offset of nearly one half of a monolayer (Fig. 3). This offset might be due to problems of surface contamination, preferential adsorption elsewhere in the cell or temperature differential between substrate and reservoir. We see no definite sign of the 1/3 ordering transition.

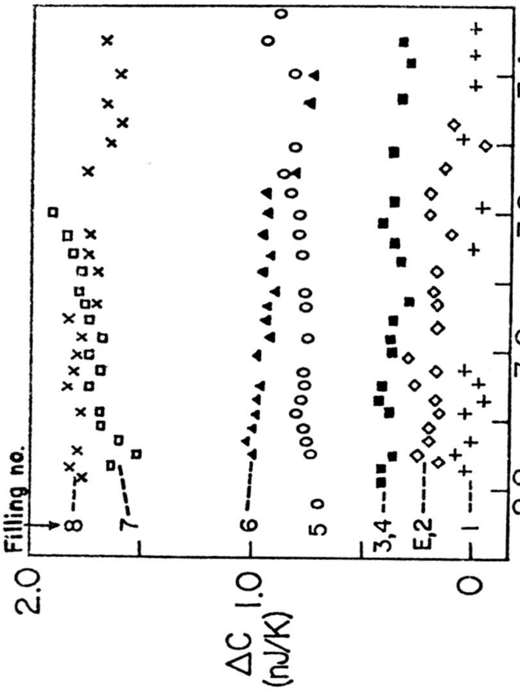

Fig. 2. Difference Heat Capacity vs. Temperature

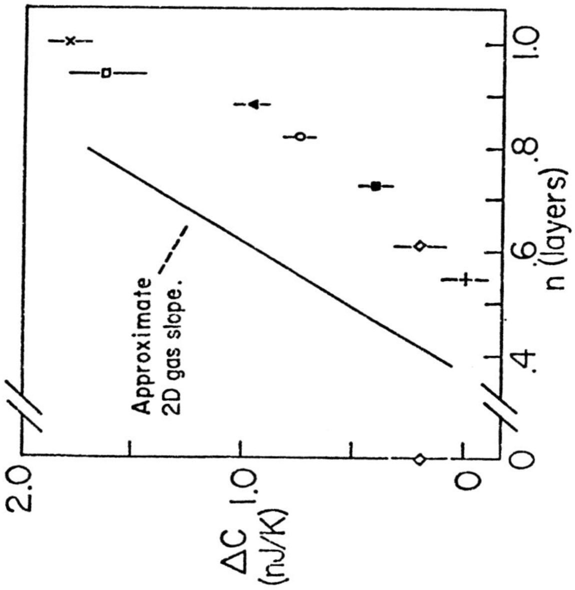

Fig. 3. Difference Heat Capacity vs. Coverage

CONCLUSIONS

After checking several possible systematic errors including changes in frequency response due to gas heat conduction, we feel confident we have succeeded in measuring film heat capacities. Further refinements in substrate preparation, coverage determination and signal resolution are in progress. Once these problems are overcome, we expect to continue our investigation of substrate effects [5] starting with the 1/3 ordering transition.

REFERENCES

1. P.F. Sullivan, PhD Thesis (Brown University Providence, RI, 1968).
2. P.F. Sullivan and G. Seidel, PR 173, 679 (1968).
3. P. Schwarz, PRB 4, 920 (1971).
4. J. E. Smaardyck and J.M. Mochel, RSI 49, 988 (1978).
5. M. Bretz, PRL 38, 501 (1977).

Published 1980 by Elsevier North Holland, Inc.
Sinha, ed. Ordering in Two Dimensions

DO HELIUM MONOLAYERS MELT BY UNBINDING OF DISLOCATIONS?

J.M. GREIF,[a] A.F. SILVA-MOREIRA[b,c] and D.L. GOODSTEIN[a]

California Institute of Technology, Pasadena, California 91125

ABSTRACT

This paper reports progress in a program to compare the predictions of the Nelson and Halperin theory of dislocation-mediated two-dimensional melting to thermodynamic data on ^3He and ^4He monolayers adsorbed on Grafoil. Experiment identifies melting by means of a series of distinct peaks in the heat capacity and other thermodynamic manifestations. The theory, which is supposed to be quite general, predicts that no measureable thermodynamic consequences will be found at the transition. Nevertheless, it is possible that the theory may be relevant. The theory is difficult to evaluate because up to now its predicted consequences have not been thermodynamically testable and because several parameters central to the theory (those describing dislocation cores and measuring the interaction with periodic substrates) are not known experimentally or theoretically. We report the status of new calculations and comparisons with experiment of the predicted dislocation contribution to the monolayer heat capacity above and below the dislocation unbinding transition for physically reasonable values of the dislocation core parameters, and preliminary attempts to use recent data on the adsorbate-substrate potential to include the effects of the substrate periodicity.

(a) Research supported in part by NSF grant DMR77-00036.
(b) On leave from University of Campinas, Brazil.
(c) Research supported in part by the Brazilian agency FAPESP, contract 78/0303.

MELTING DILEMMA

Two-dimensional melting was discovered experimentally in sharp heat capacity peaks observed in ^4He monolayers [1] and later in ^3He films [2]. From the Debye-like heat capacity of the films at low temperature and the fact that long wavelength shear modes propagate in them, as could be determined by comparing the observed 2-D compressibility and Debye temperatures [3], it could soon be shown that the low temperature phase was a solid, and the heat capacity peaks were manifestations of melting. The data, when properly analyzed, showed considerable similarity between the film and bulk solids [4].

Recently, a self-consistent and presumably rigorous analysis of a proposed mechanism for melting of these films has been presented [5]. The essence of the theory is that the resistance to shear of the two-dimensional solid is lost when thermally excited dislocation pairs unbind, removing the translational order of the solid, but not its orientational order. A liquid crystal phase exists until at some higher temperature than the dislocation unbinding, the fluid becomes isotropic. The second transition is not expected if the substrate is periodic. The quantitative predictions of the theory are made by renormalization group (RG) methods. The theory distinguishes films on perfectly smooth substrates from those on periodic substrates and has the power to analyze both registered and unregistered phases.

Does the theory describe the melting of real adsorbed films, in particular, the helium monolayers?

Several important issues are raised in considering this question. First, the theory seems completely inapplicable in bulk three-dimensional matter; yet, the two- and three-dimensional helium solids show similar properties and melting behavior when considered at the same atomic spacing. Second, the theory predicts that most thermodynamic quantities vary smoothly through the dislocation unbinding transition; yet melting was discovered through sharp heat capacity peaks. From these considerations, the theory seems inapplicable to helium monolayers, but that conclusion is not yet proven and the theory may yet turn out to be relevant to these materials.

It is expected that, despite the essential singularity in the heat capacity predicted by the Nelson-Halperin theory at the phase transition, there should be a maximum at higher temperature, owing to a saturation effect when the dislocation density approaches the atomic density of the two-dimensional material. There are no predictions of how much higher in temperature this peak should occur and there is qualitative evidence that it should be a rather broad maximum. Until quantitative calculations of the heat capacity of the dislocations can be made, it will be impossible to tell whether the theory can agree with thermodynamic measurements. Our first results of such calculations are now being studied and checked.

Given the smoothness of the singularities at the phase transition, no thermodynamic measurements can detect the actual melting point. However, the experiments indicate that if the theory is to match them, the melting temperature will occur substantially below the heat capacity peaks. This result can be derived by constructing an "unrenormalized" melting temperature using the elastic constants of the film deduced from the thermodynamic measurements at low temperature. The theory predicts that, on smooth substrates, melting occurs at

$$K_B T_m = \frac{4\mu(\mu + \lambda)a_0^2}{16\pi(2\mu + \lambda)} \quad (1)$$

where μ and λ are Lamé constants and a_0 is the lattice parameter. An estimate for T_m can be gotten by putting the bare elastic constants into the equation. Since the elastic constants are renormalized downward, this temperature must be higher than the actual melting temperature.

Figures 1 and 2 show the heat capacity peak temperatures and these predicted "unrenormalized" melting temperatures for ^3He and ^4He at various densities computed for a substrate that is nearly smooth. These peaks have been corrected for second layer promotion and desorption according to the scheme in Elgin and Goodstein [1]. In all cases, the heat capacity peaks are significantly above the predicted melting temperatures. We do not plot predicted melting temperatures for renormalized elastic constants in the figure even though the renormalization of these parameters can easily be carried out if certain undetermined parameters could be found or estimated well. One of the difficulties of the theory is that it is couched in terms of several parameters which are unknown. It is known, however, that stability criteria imply that the renormalized melting temperature will be pushed downward further when effects of the periodic substrate are accounted for.

UNKNOWN PARAMETERS

The line of starting points of the renormalization group calculation has the equation

$$y_0 = \frac{2}{\sqrt{3}} \left(\frac{a}{a_0}\right)^{2-K/8\pi} e^{-CK/8\pi} \quad (2)$$

which contains two unknown parameters a, the dislocation core radius and C which parametrizes its energy. Actually, only one of the two is arbitrary, as is explained below, but they are not known either theoretically or experimentally for helium solids and are rather difficult to compute. $K = (K_0 a_0^2)/(k_B T)$, $K_0 = 4\mu(\mu+\lambda)/(2\mu+\lambda)$ and y_0 is the initial value of the other RG coupling constant.

The core energy of a dislocation is a partly arbitrary energy designed to remove an infinity from continuum elastic theory and make its predictions match the energy of a true atomic crystal with a dislocation in it. The energy of the strain field of the dislocation is calculated by a continuum integral cut off at some distance from the center of the dislocation. Let this energy be $E_S(a)$ where a is the cutoff distance. If the actual potential energy of the crystal is E and the potential energy of the crystal without a dislocation is E_0 then the core energy $E_C(a)$ is

$$E_C(a) = E - E_0 - E_s(a) \quad (3)$$

so the core energy and radius are interdependent.

Attempts to calculate the dislocation core energy have been made [6] for classical systems in which the zero point energy is ignored. The method consists of setting up a computer model of a dislocation in a perfect crystal of finite size containing atoms interacting by a pairwise potential, allowing the atoms to find the configuration of minimum energy by a relaxation technique and using Eq. (3) for a particular choice of cutoff radius a. For the classical rare gas solids, this procedure may be accurate, but for helium, it will not, since the atoms are effectively smeared out by zero point motion. Averages over many configurations determine the equilibrium average position. As in the case of the substrate-adsorbate orientation ordering, the calculation must be done self-consistently and is thus considerably more complicated than for the other rare gas solids. For the helium solids, the core energy may actually be temperature dependent, since the atomic motions are so large.

In Eq. (1), K_0 is a combination of unrenormalized elastic constants (as would be measured at zero Kelvins) which can be measured, or actually deduced from the thermodynamic measurements of

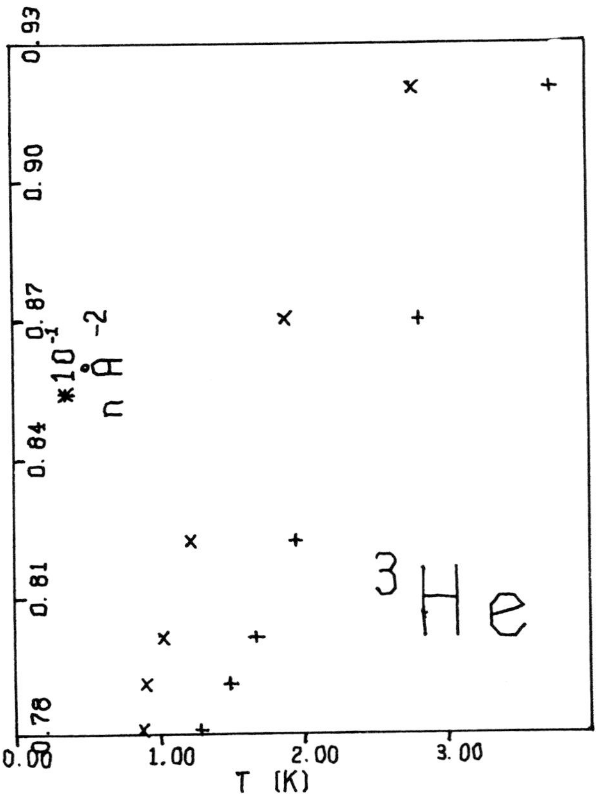

Fig. 1. Predicted unrenormalized melting temperatures (+) calculated for $\gamma << \mu$ and observed heat capacity peak temperatures (x) corrected for second layer promotion and desorption for several coverages of ^3He on Grafoil.

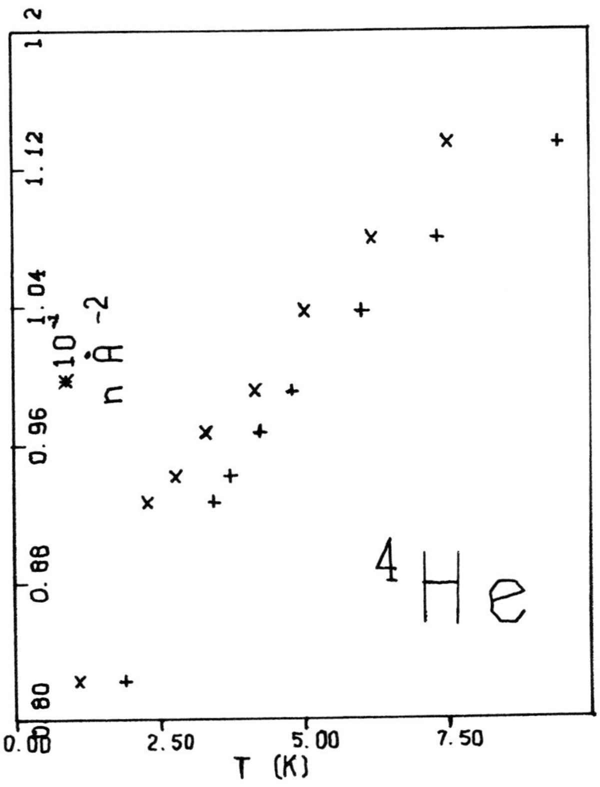

Fig. 2. Predicted unrenormalized melting temperatures (+) calculated for $\gamma << \mu$ and observed heat capacity peak temperatures (x) corrected for second layer promotion and desorption for several coverages of ^4He on Grafoil.

the heat capacity and the vapor pressure of the three dimensional helium gas above the substrate. For this purpose we have used the ^4He data of [1] and previously unpublished ^3He data from our laboratory together with [2]. Thermodynamic analysis of these measurements, as explained in Elgin and Goodstein, can provide two independent elastic constants (usually the bulk modulus and the Debye temperature) which can be combined to give the Lamé constants.

In principle, as the temperature of the film is varied, the renormalization of the elastic constants should start from different points on the line of starting points, assuming the core energy parameter to be independent of temperature. There is no good reason to expect that it will be independent of density, so there may well be a band of starting points over the solid density range, although the band should be narrow if the core energy parameter varies smoothly and slowly with density as would be expected if the cutoff parameter scaled with the lattice spacing.

This result, however, applies only to the solid adsorbed on a smooth substrate. For periodic substrates, there is a new elastic constant which measures the restoring force opposing a twist of the adsorbate with respect to the substrate, even if the lattices are incommensurate. There are no measurements of that elastic constant and no way to deduce it from the thermodynamic data, but calculations have been carried out for the case of argon on graphite [7] and for helium on graphite in this work. We have found that we could not get reasonable values of the third elastic constant γ using the original theory of Novaco and McTague [7], because the linear response calculation overestimates the distorting effects of the substrate on the adsorbate lattice, even though we used accurate data on the anisotropic substrate-adatom potential derived from the helium-graphite scattering experiments [8].

The reason this overestimate occurs is that the zero-point energy of the helium has not been accounted for. Each helium atom in its wide range of motion about its equilibrium position samples regions of the crystal of considerably higher potential energy than the energy at the minimum, and the averaging over many positions partially washes out the energy gain near the local minimum of the substrate-adatom plus adatom-adatom potential. We are attempting to redo the calculation using a self-consistent phonon approach worked out by Novaco [7] which takes this effect into account.

For this case of a periodic substrate, there are three renormalization group couplings, K_1, K_2 and y, and the line of starting points becomes a surface of starting points. Even if the core energy parameter behaves nicely with density, γ varies rather rapidly with density, especially near densities at which the substrate and adsorbate lattices are commensurate [9].

CONCLUSIONS

We have attempted in this paper to emphasize that the Nelson-Halperin theory of melting in solids by the Kosterlitz-Thouless-Feynman [10] mechanism is not yet verified, and that the verification requires the solution of several experimental and theoretical problems. Progress has been made in all of the needed areas, but the final results are not yet complete.

REFERENCES

1. M. Bretz *et al.*, Phys. Rev. A 8, 1589 (1973); R.L. Elgin and D.L. Goodstein, Phys. Rev. A 9, 2657 (1974).
2. S.V. Hering, S.W. Van Sciver and O.E. Vilches, J. Low Temp. Phys. 25, 793 (1976).
3. G.A. Stewart, S. Siegel and D.L. Goodstein, *Proceedings of the Thirteenth International Conference on Low Temperature Physics*, R.H. Kropschot and K.D. Timmerhaus, eds., (University of Colorado Press, Boulder, CO, 1973).
4. G.A. Stewart, Phys. Rev. A 10, 671 (1974).
5. D.R. Nelson and B.I. Halperin, Phys. Rev. B 19, 2457 (1979).
6. Early calculations appear in A. Englert and H. Tompa, J. Phys. Chem. Solids 21, 306 (1961) and 24, 1145 (1963). D.S. Fisher, B.I. Halperin and R. Morf, preprint 1980, contains a calculation for the two-dimensional electron solid.
7. A.D. Novaco, Phys. Rev. B 19, 6493 (1979); J.P. McTague, Phys. Rev. Lett. 38, 1286 (1977).
8. W. Carlos and M. Cole, Phys. Rev. Lett. 43, 697 (1979).
9. It is a peculiar property of this system that the equilibrium orientation angle of the two lattices varies with the density of the adsorbate, so that it is possible to rotate the adsorbate by uniformly compressing it in two-dimensions.
10. J.M. Kosterlitz and D.J. Thouless, J. Phys. C Solid State Phys. 6, 1181 (1973). R.P. Feynman independently predicted the same mechanism, see Elgin and Goodstein, ref. 1.

Sinha, ed. Ordering in Two Dimensions

ROUGHENING TRANSITION IN THE ^4He SOLID-SUPERFLUID INTERFACE

J.E. AVRON, L.S. BALFOUR, C.G. KUPER, J. LANDAU,
S.G. LIPSON AND L.S. SCHULMAN

Department of Physics
Technion-Israel Institute of Technology, Haifa, Israel

Principal planes of the ^4He solid-superfluid interface are expected to undergo roughening transitions at temperatures of about 1K. We describe an experiment in which two such transitions were observed in the hcp-superfluid interface: first for the basal plane at 1.08 K and second for an orthogonal face at 0.85 K.

The roughening transition [RT] is a theoretical concept, known to apply to certain Ising models [1], and to computer simulations of crystal growth [2,3]. In this Letter we discuss the applicability of the RT to a real physical system: the ^4He solid-superfluid interface. We report the observation of morphological transitions in ^4He crystals at 1.08 K and 0.85 K which, we argue, are RT's. Parshin and his coworkers [4,5] have argued that the interface in this system is always rough, but their view is contrary to evidence presented here (and in ref. 16). Balibar [6] has independently proposed the existence of a RT in this system.

The roughening temperature, T_R, is characterized by the vanishing of the step energy [1,2]. Above T_R the interface fluctuates macroscopically, and in the thermodynamic limit, translation invariance is restored [7]. Unfortunately, this aspect of the RT is not expected to be observable in practice. Fluctuations in ordinary macroscopic systems are microscopic; for the two-dimensional interface, the mean-square amplitude fluctuation is of order $\ell n\ N$, where N is the number of lattice points in the interface.

<u>Theory</u>: The interface is modeled by a <u>2-D lattice Hamiltonian</u> with discrete, unbounded Ising spin Z: the vertical coordinate of the interface. A capillary-wave Hamiltonian [8] for the interface is:

$$H = \sum_{\vec{i}} [1 + (\vec{\nabla} Z(\vec{i}))^2]^{1/2} * \{\sigma_0 \Omega + (\rho_o \hbar^2/M^2)(\vec{\nabla}\psi)^2\}. \qquad (1)$$

Here, ψ is the superfluid field, ρ_o its density, $\vec{\nabla}$ the discrete gradient, Ω the area of the 2-D unit cell and $\vec{i}$ a lattice bond. The term proportional to σ_0 (a coupling constant) represents the energy density of the boundary of the solid, and the term proportional to $\hbar^2$ is the energy density of the superfluid layer near the interface [9,10]. In the lowest approximation the ψ field integrates out. Similar conclusions hold for the Villain model [11] and for other models [12] of the interface. The surface tension is the free energy of the Hamiltonian H.

The specific choice of the interface Hamiltonian is of little importance here since the Kosterlitz-Thouless roughening transition [13] is a general feature of 2-D models whose Hamiltonians are invariant under discrete uniform shifts $Z(\vec{i}) \rightarrow Z(\vec{i}) + a$. [13-15].

The quantum aspects of solid He will be taken into account only phenomenologically; σ_0 is taken to be the measured surface tension of ^4He at T = 0 [16]. The roughening temperature is given by $T_R = K\sigma_0\Omega$ where K is a constant of order unity, whose value depends on the details of the interface model (the choice of Hamiltonian), the lattice structure etc. [17]. Taking σ_0 and Ω appropriate to the ^4He interface [16] gives $T_R \simeq 1$ K.

Despite the difficulty of calculating T_R, we can estimate <u>relative</u> values of T_R on different planes and hence establish a <u>hierarchical ordering</u> of plane orientation according to roughening temperature. To accomplish this we

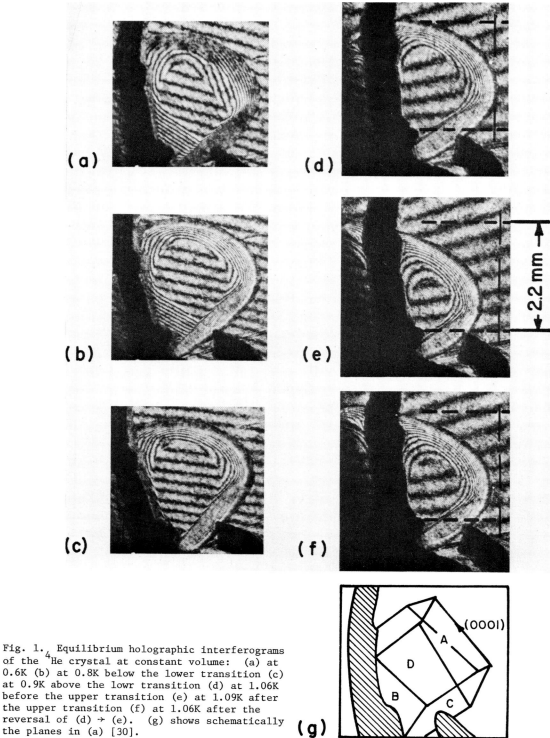

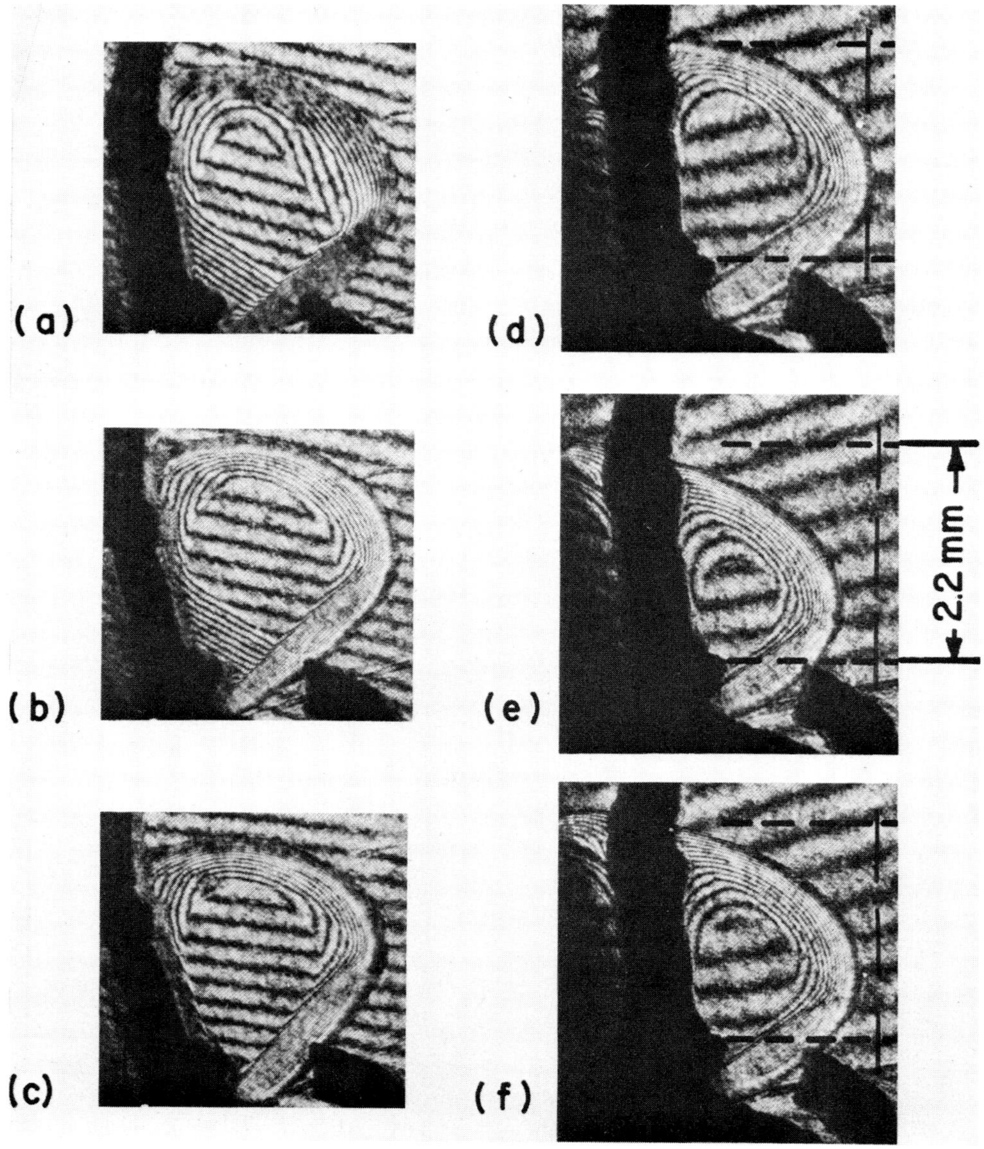

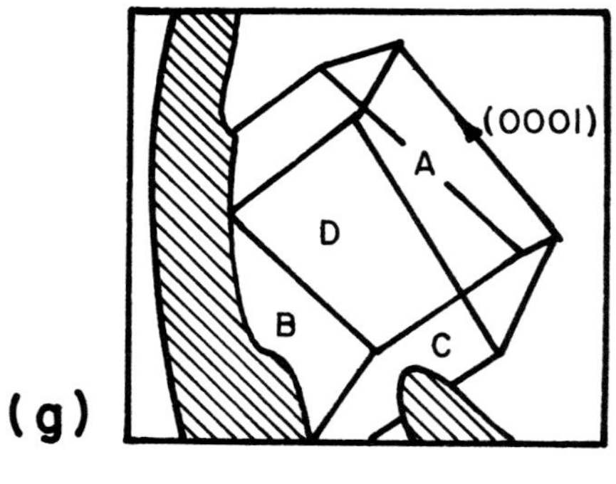

Fig. 1. Equilibrium holographic interferograms of the ^{4}He crystal at constant volume: (a) at 0.6K (b) at 0.8K below the lower transition (c) at 0.9K above the lowr transition (d) at 1.06K before the upper transition (e) at 1.09K after the upper transition (f) at 1.06K after the reversal of (d) → (e). (g) shows schematically the planes in (a) [30].

identify the ^4He interface with the interface of a 3-D Ising Model with Slater-Kirkwood [18] pair interaction and ± boundary conditions. This is reasonable only below 1 K where ^4He, like the Ising model, has essentially no latent heat [19, 20]. To estimate T_R we use a bound of van Beijern as if it were an equality [22]. Van Beijern proved that for a given crystal orientation $T_R \geq T_c$ where T_c is the critical temperature of the 2-D Ising model obtained by considering the restriction of the 3-D Ising model to the plane of the interface. This reduces the problem to that of computing T_c for certain 2-D Ising models, which represent sections of the 3-D structure and therefore have different lattices and interatomic couplings.

We have estimated T_c for these 2-D models with the Slater-Kirkwood interaction both by mean field theory [23] and by reduction to soluble models [24]. The results are summarized in Table I, showing that the basal plane has the largest T_R.

Table I

ESTIMATES OF THE ROUGHENING TEMPERATURE FOR VARIOUS CRYSTAL PLANES. THE TEMPERATURE SCALE WAS CHOSEN SO THAT THE HIGHEST T_R IS UNITY

	T_R Arbitrary Units	
Miller Indices	Mean Field	Nearest Neighbours
(0001)	1	1
$(11\bar{2}0)$	0.5	0.5-1
$(10\bar{1}0)$	0.35	0.2
$(10\bar{1}1)$	0.34	0.16

Experiment: The roughening transition implies [25] that faces which are flat below T_R will become curved (atomically rough) above it [26]. This can happen with an arbitrarily small change in $\sigma(\theta,\phi)$ that smooths a cusp in the Wulff diagram [26].

We have recently been investigating[16] experimentally the surface tension at the solid-superfluid helium interface. In particular, we found that σ is independent of T within a single crystallographic phase, and that large equilibrium crystals have apparently more facets at low temperatures than at high temperatures. Evidence for a RT, and even for a hierarchy of such transitions, comes from observing the shape of a small [28] crystal as it was very slowly warmed up. The size of the crystal was comparable with the capillary length, 1.4 mm. A complete sequence of holographic interferograms [29] was photographed, of which several significant prints are assembled in Fig. 1. At the lowest temperature, 0.6 K, the interferogram (a) shows several plane interfaces A,B,C (g) to be present. The plane A, of type $<11\bar{2}0>$, shrank and disappeared by 0.8 K (b). Shortly afterwards the crystallographically equivalent plane B dramatically disappeared; compare (b) and (c), at 0.9 K. As the temperature continued to rise the facet C started to shrink and the crystal was observed to undergo a sudden change in external dimensions at 1.08 K. This transition reversed as the temperature was reduced again, and occurred a second time at the same temperature as the heating was recommenced. Apart from the changes in extremal dimensions of the crystal (Fig. 1: d,e,f) the transition is less obvious at first glance, because the facet C is almost parallel to the line of sight. To make the effect clearer, we show in Fig. 2 the area of the face C as a function of T, and the transition at 1.08 K is obvious. These observations suggest that the first two transitions in Table I occur at 1.08 ± 0.02 K and 0.85 ± 0.05 K.

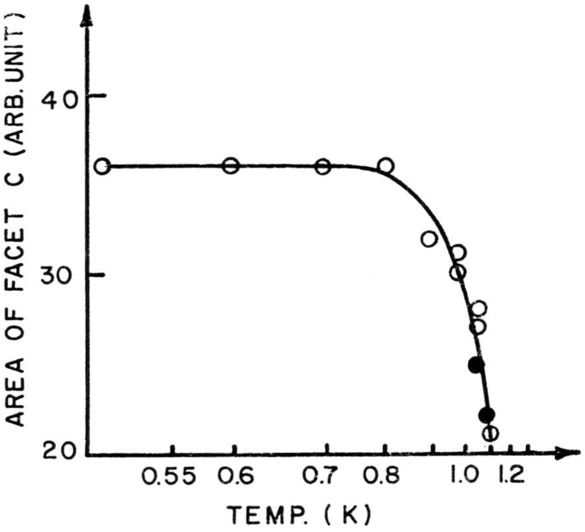

Fig. 2. The area of place C as a function of T: O increasing T; ● decreasing T.

^{3}He Impurities: We found that ^{3}He fractions of the order of 10^{-4} or larger enhanced the flat faces, and eliminated curves in the interface [16]. This presents two problems: (a) How does ^{3}He stabilize facets? (b) How can such small concentrations of impurities affect the interface?

The answer to (a) is that ^{3}He pins the interface. This happens because ^{3}He is preferentially expelled from the solid phase by its zero point energy. A capillary-wave description of the rough surface, following [4], shows that the time-scale for interface fluctuations on an atomic scale is considerably shorter than that for diffusive motion of the ^{3}He impurities in the superfluid. To understand (b), note [12] that the mean square amplitude fluctuation of the rough interface $<Z(0) * Z(r)>$ is of the order $(T/2\pi\sigma_0\Omega)\ln(r/a)$. The interface flattens when the mean distance between the ^{3}He impurities, r, leads to fluctuations small compared to unity. The consequent logarithmic dependence on the concentration explains (b).

In conclusion: The concept of a roughening transition accounts for all the phenomena we have observed in ^{4}He hcp-superfluid interfaces [29].

JEA and JL thank S. Balibar for stimulating correspondence. JL is indebted to D.O. Edwards for suggestions and JEA thanks E. Witten and R. Zia for discussion.

This work was supported in part by the U.S.-Israel Binational Science Foundation, grant numbers 1425 and 1573.

FOOTNOTES AND REFERENCES

1. H. Van Beijern, Phys. Rev. Lett. 38, 992-996 (1977); see also Ref. [21], and R.L. Dobrushin, Theor. Prob. App. 17, 582 (1972).
2. K.A. Jackson, J. Crys. Growth 24/25, 130-136 (1974), H.J. Leamy and G.H. Gilmer, ibid, 499-592.
3. J.D. Weeks, G.H. Gilmer and H.J. Leamy, Phys. Rev. Lett. 31, 549-551 (1973) give evidence for RT based on low temperature expansion for the s.c. Ising model.
4. A.F. Andreev and A. Ya. Parshin, Soviet Phys. JETP 48(4) 763-766 (1978).
5. K.O. Keshishev, A.J. Parshin and A.V. Babkin, JETP Lett. 30, 63 (1979).
6. S. Balibar and B. Castaing, J. de Physique Lett., to appear.
7. M. Aizeman, Phys. Rev. Lett. 43, 407 (1979).
8. D.J. Wallace and R.K.D. Zia, Phys. Rev. Lett. 43, 808-811 (1979) and Ref. therein.
9. D.R. Nelson and J.M. Kosterlitz, Phys. Rev. Lett. 39, 1201 (1977). We neglect superfluid excitations.
10. For lattice superfluid models see D.D. Betts in "Phase Transition and Critical Phenomena, Vol. 3", C. Domb and M.S. Green, Ed. Academic Press (1974).
11. J. Villain, J. de Physique (Paris) 36, 581 (1975).
12. Other choices are $\Sigma(\sigma_0\Omega)|\vec{\nabla}Z|$ etc. See e.g. H.J.F. Knops, Phys. Rev. Lett. 39, 766-769 (1977).
13. There is extensive literature on the Kosterlitz-Thouless transition, see Ref. [15] and Physics Today, August (1978), p. 17 and Ref. therein. The RT is related to the Kosterlitz-Thouless transition by duality.
14. S. Elitzur, R. Pearson and J. Shigemitsu, Phys. Rev. D19, 3698 (1979).
15. D.J. Amit, Y.Y. Goldschmidt and G. Grinstein, J. Phys. A13, 585-620 (1980). J. Jose, L.P. Kadanoff, S. Kirkpatrick and D.R. Nelson, Phys. Rev. B16, 1217 (1977).
16. J. Landau, S.G. Lipson, L.M. Maattanen, L.S. Balfour and D.O. Edwards, Phys. Rev. Lett. - submitted.
17. Rigorous estimates of the Kosterlitz-Thouless transition temperature are given in M. Aizenman and B. Simon, Phys. Lett. - to appear. Accurate, but non-rigorous estimates of T_c are also quoted there.
18. J.C. Slater and J.G. Kirkwood, Phys. Rev. 37, 682 (1931).
19. The latent heat vanishes at T = 0 and T = 0.79 K. E.R. Grilly, J. Low Temp. Phys. 11, 33 (1973).
20. T_c of the Ising model can be pushed to ∞ by imposing a solid-on-solid condition preventing more than a single spin flip in any column. This does not affect our estimates of T_r.
21. H. Van Beijern, Comm. Math. Phys. 40, 1-6 (1975). Extension to interactions with ∞ range is given in J. Bricmont, J.L. Lebowitz, C.E. Pfister and E. Olivieri, Comm. Math. Phys. 66, 1-20 (1979).
22. This means that, e.g. the 3-D, s.c., nearest neighbour Ising model has $T_k \neq 0$ only for the six principal planes. Compare with [1].

23. Mean field is an upper bound on T_c. See R. Griffith, Comm. Math. Phys. 6, 121 (1967). For extensions of these results see, e.g. B. Simon, J. Stat. Phys. (to appear) and references given there.
24. For a given plane, we retain only nearest neighbour interactions. This underestimates T_c. T_c for Ising 2-D models are given by I. Syozi "Phase Transition and Critical Phenomena Vol. 1", C. Domb and M.S. Green, Ed. Academic Press (1972).
25. The large heat conductivity of superfluid ^4He ensures thermodynamic equilibrium.
26. C. Herring in "Structure and Properties of Solid Surfaces", R. Gomer and G.S. Smith, Ed. Univ. of Chicago Press (1955).
27. If σ was T dependent no conclusion could be drawn. For a T-dependent σ, facets may disappear without RT e.g. by adding a constant to σ.
28. Gravity gives the vortex excitation a mass inversely proportional to the capillary length. Finite mass destroys RT in the strict sense. For crystals small compared to the capillary length this is irrelevant.
29. J. Pipman, S.G. Lipson, J. Landau and N. Bochner, J. Low Temp.Phys. 31, 119 (1978).
30. The crystallographic details were determined by observing the crystal during growth, when it has a simple pyramidal shape. The faces overlapping at D in Fig. 1(g) are not exact crystallographic planes, but are in contact with the glass windows.

SECTION II

B. 2-D WIGNER CRYSTALS AND COMPUTER SIMULATION STUDIES

Sinha, ed. Ordering in Two Dimensions

DETECTION OF ORDERING OF ELECTRONS ON LIQUID HELIUM

G. DEVILLE, F. GALLET, D. MARTY, J. POITRENAUD, A. VALDES, F.I.B. WILLIAMS

CEN-Saclay, Orme des Merisiers, 91190 Gif-sur-Yvette, France

ABSTRACT

The ordering is observed for surface densities $1.5\,10^8 < n < 8.10^8\ \mathrm{cm}^{-2}$ at temperatures $0.3 < T < 0.8\,K$ by monitoring the response to a 25 MHz in plane electric field polarized parallel to its wave vector. Ordering influences this susceptibility by the shift in frequency of the plasmon modes as the electrons correlate and so self-trap into the surface deformation they impose on the helium. This experiment is sensitive to the optical mode of the coupled systems (electron and surface deformation move in antiphase), whereas the first observation of ordering by Grimes and Adams[3] was based on the appearance of the acoustical modes (electron + deformation move in phase).

Our results for the ordering temperatures can be expressed by $n_c \sim T_c^{1.8 \pm 0.2}$, or, if one fits the results to $\Gamma_c = (\pi n_c)^{1/2}/T_c$, $\Gamma_c = 118 \pm 10$. It seems that the optical modes establish themselves before the acoustic modes.

We report briefly two experiments on the 2-dimensional system of electrons held onto the vapour side of the liquid-vapour interface of helium. The first is a particularly simple and convenient method for measuring the melting point. The second measures directly the transverse optical mode frequency.

This system is classical (non "Wigner") where the relevant scaling parameter

$$\Gamma = \frac{(\pi n)^{1/2}\, e^2}{T} \qquad (1)$$

measures the ratio of correlation to thermal energy (both $\gg$ degeneracy temperature $n\hbar^2/2m$; n is areal number density, T the temperature in energy units).

The first experiment I detects an anomaly in the longitudinal radio frequency electric susceptibility $\chi_{xx}(\omega, k\hat{x})$ for $\omega/2\pi \approx 25$ MHz, $k \approx 1\ \mathrm{cm}^{-1}$, $10^8 < n < 10^9\ \mathrm{cm}^{-2}$ and $300 < T < 900\,\mathrm{mK}$. From the position (n_c, T_c) of this anomaly one deduces the values of $\Gamma = \Gamma_c$ at which the electrons undergo a phase change.

For electrons constrained to move on a plane non-deformable surface, no change is expected in the long wavelength longitudinal excitation spectrum for constant density. The helium surface, however, can be deformed by the pressure of the electrons. Shikin[1] pointed this out for a single electron and showed that the energy gained $\varepsilon \approx$ a few mK ; such an electron + dimple deformation would not be bound at our experimental temperatures. But for an assembly of N strongly correlated electrons, the total binding energy to an ensemble of dimples is $N\varepsilon$ which must be compared with the thermal energy T of the centre of mass motion ; when $N\varepsilon > T$ there is stability of electrons + dimples. To detect

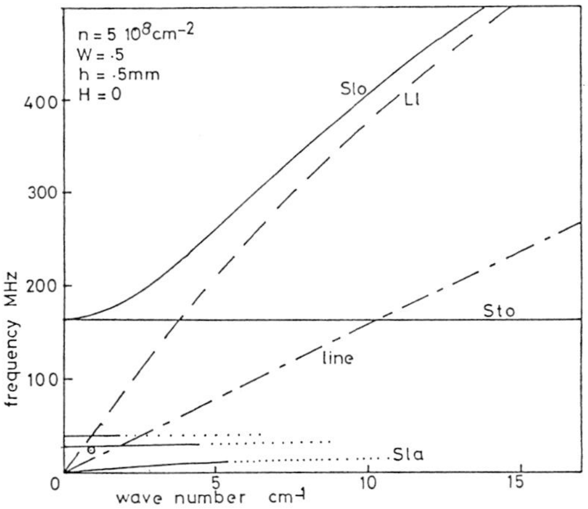

Fig. 1. Dispersion curves of excitations. Electrons in liquid phase : only longitudinal excitations (plasmons) propagate, shown by dashed line labeled Lℓ. Electrons in solid phase : both longitudinal and transverse modes propagate, but binding to surface deformation induced by electron correlation leads to optical and acoustical branches for each polarization shown by solid lines $S\ell_o$, St_o, $S\ell_a$ and St_a, the latter two being dotted in the region of progressive decoupling. S = solid, L = liquid, ℓ = longitudidal, t = transverse, o = optical, a = acoustical. Curves drawn for $n \approx 5\ 10^8\ \mathrm{cm}^{-2}$. The point ⊙ gives the approximate coordinates for the susceptibility measurement of experiment I ; The dot-dashed line the locus of susceptibility measurements of experiment II

the onset of this stability is to detect correlation amongst the electrons.

As Fisher et al[2] pointed out, the dynamics of the bound system must take into account the relative motions of the electrons and the surface deformation. For $n \to 0$, this results in the appearance of optical and acoustical branches for both longitudinal and transverse excitations ; as ω_{ac} increases, there is decoupling and the excitations revert to separate ripplon and electron motions.

The appearance of the acoustical modes was detected by Grimes and Adams[3] and constituted the first experimental evidence of correlation in the electrons.

Here we detect instead the behaviour of the optical branch.

Figure (1) illustrates the low k excitation spectrum for $n = 5\ 10^8\ cm^{-2}$. The (ω,k) point encircled situates the measurement of experiment I. χ at this point is sensitive to the change in the excitation spectrum because of damping (broadening out of the dispersion curves). The absorptive component of $\chi_{xx}(\omega,k\hat{x})$ for $k \approx 1\ cm^{-1}$ takes on the forms of figure (2)

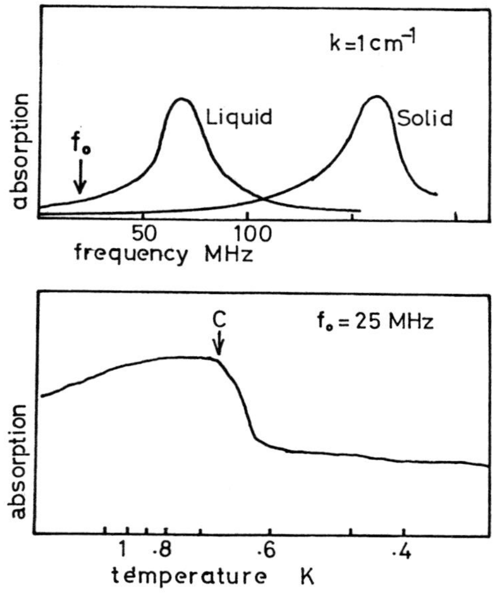

Fig. 2. Absorptive component of $\chi_{xx}(\omega,k\hat{x})$ for $k = 1.1\ cm^{-1}$ deduced from the dispersion curves of figure (1) by including damping ; f_o is the frequency of experiment I. The lower curve shows the observed variation of χ'' at $\omega_o/2\pi = 25$ MHz on going from liquid to solid by lowering the temperature. These curves may be envisaged as sections of the surface $\chi_{xx}(\omega,\Gamma,k\hat{x})$ for fixed Γ and fixed ω.

for the correlated and uncorrelated electron phases. The acoustic mode is usually too weak and narrow to contribute and the experiment detects the appearance of the longitudinal optical mode (a "local" vibration of an electron in its dimple).

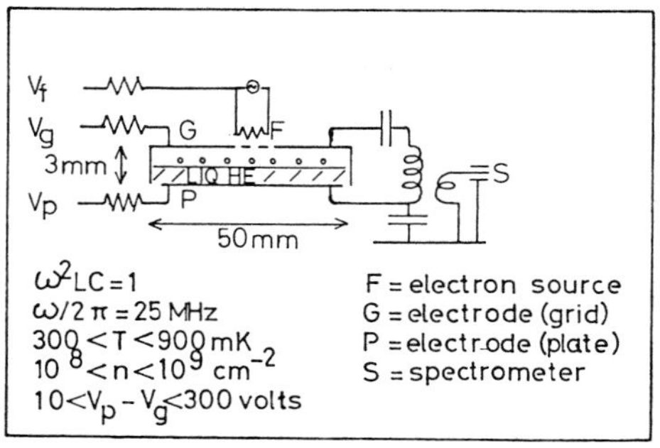

Fig. 3. Schematic diagram of the experimental arrangement of experiment I at fixed $\omega/2\pi = 25$ MHz, $k \approx 1\ cm^{-1}$. For experiment II, P is replaced by a slow wave meander transmission line connected directly to spectrometer line ($\omega/k \approx 10^8\ cm\ s^{-1}$).

The experimental arrangement, shown on figure (3), is essentially an L-C circuit matched at resonance to the transmission line of a reflexion mode 25 MHz R-F spectrometer. The radial fringing fields of the circular parallel plate condenser excite the longitudinal modes. The DC potentials control the electron density ; usually the cell is operated at "saturation" condition where $E = 0$ above and $E = 4\pi n e$ below the electronic pancake. An experimental curve of the absorptive response of the spectrometer vs temperature ($\omega/2\pi = 25$ MHz determined by $\omega^2 LC = 1$ and $k \approx 1\ cm^{-1}$ determined by cell geometry) for fixed $n = 4.3\ 10^8\ cm^{-2}$ is shown on figure 2b. The points "C" of such curves (n_c,T_c) are located on the n,T^2 plot of figure (4). A log.log plot would suggest $n_c \sim T_c^{1.8 \pm 0.2}$, but if one forces a $n_c \sim T_c^2$ fit one finds a mean value of [4]

$$\Gamma_c = 118 \pm 10 \tag{2}$$

Before commenting on this result, we should mention some still provisional results of the other experiment II [5]. Here too we detect $\chi_{xx}(\omega,k\hat{x})$, but now for

$$1 < \omega/2\pi < 350\ \text{MHz}, \quad k = \omega/v \tag{3}$$

situated on figure (1) by the dot-dashed line. The (longitudinal) excitation is produced by the induction field of a slow wave transmission line

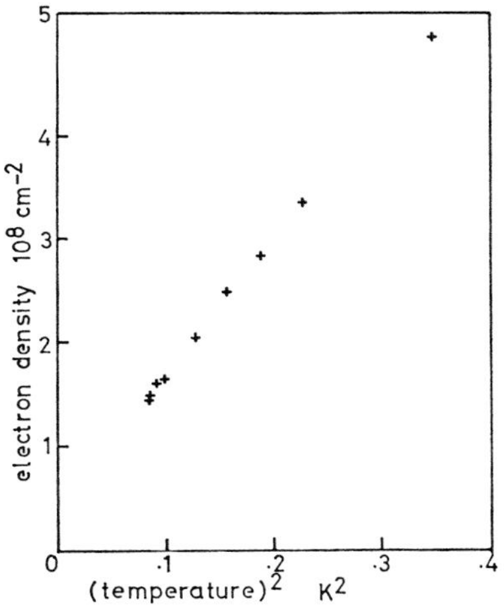

Fig. 4. Coordinates of phase transition as detected by experiment I on a (n,T^2) plot. The best fit to equation (1) gives $\Gamma = 118 \pm 10$.

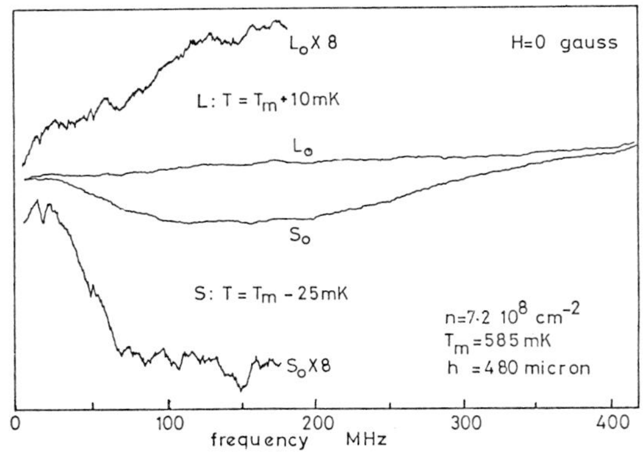

Fig. 5. Experimental curve from experiment II : $\chi''_{xx}(\omega,k\hat{x})$ for $k=\omega/v$, $v=10^8 \mathrm{cm\,s}^{-1}$ phase detected with respect to modulation of holding potential in zero magnetic field. L : high temperature ("Liquid") phase just above T_m. S : low temperature ("Solid") phase just below T_m. The uppermost and lowermost curves are taken at voltage gain X 3 and RF power + 5db with respect to the 2 middle curves. The resonance from the acoustic branch can be distinguished at $f \approx 18$ MHz on the lowest curve. Other low (ripplon) frequency branches appear at lower temperature. h = depth of lower electrode beneath helium surface, n = areal density of electrons, T_m = melting temperature deduced from high frequency $H \neq 0$ behaviour figure (6).

which continues the spectrometer feed line with the same characteristic impedance and replaces the lower plate of figure (3). It is a meander line of half period 100 μ, width 17 mm, length 18 mm and propagation velocity $v \approx 10^8$ cm s^{-1}.

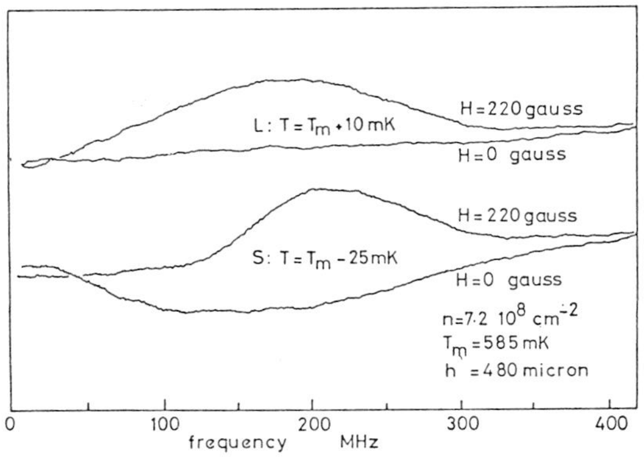

Fig. 6. Experimental curves from experiment II : $\chi''_{xx}(\ ,kx)$ for $k=\omega/v$, $v = 10^8 \mathrm{cm\,s}^{-1}$ phase detected with respect to modulation of the holding potential. Response in absence and presence of normal magnetic field H reveals the "transverse optical" mode of figure (1) for $k \approx 12$ cm^{-1} whose frequency is reduced as H is increased. This frequency extrapolated to H = 0 gives the resonance of an electron vibrating in antiphase with its dimple, the shift due to the electronic shear modulus being too small to detect at such small k, h : depth of lower electrode beneath the helium surface, n = areal density of electrons, T_m = melting temperature deduced from series of such $H \neq 0$ curves.

Both acoustical and optical modes have been detected. The limited data so far indicates that the optical mode appears before but starts to narrow at a value of Γ definable to about 1% at

$$\Gamma = 135 \pm 15 \qquad (4)$$

Typical experimental curves are shown in figures (5) and (6).

These values are to be compared with the

$$\Gamma = 131 \pm 7 \qquad (5)$$

of Grimes and Adams [3].

The differences in the values of Γ are not sufficiently significant to be used as an indication of an intermediate phase [5].

Without going into a detailed comparison with theoretical models, we situate these experimental values of Γ by recalling that a Monté-Carlo calculation of the free energies gives

$\Gamma = 125$ [7], whereas the Kosterlitz-Thouless model of dislocation mediated melting gives $\Gamma = 80$ [8] using the $T = 0$ elastic constants or $\Gamma \approx 128$ using renormalized elastic constant [9].

Although it is not the real point of this paper, it would not be clear how the optical branch is excited if we did not mention that the longitudinal field is coupled to the transverse excitations by applying a normal magnetic field (Lorentz force $f_y \approx \frac{\omega\,\omega_c}{(\omega_\ell^2 - \omega^2)}\, e\, E_x$ where E_x is the longitudinal field amplitude, ω_c = cyclotron frequency of an electron in the normal field and ω_ℓ the longitudinal mode frequency - [9]). It is then the transverse optic branch which is excited, the propagation velocity on the slow wave line being too slow to excite the longitudinal branch.

REFERENCES

[1] V.B. Shikin, Pisma Zh. Eksp. Teor. Fiz. 68, 1423 (1975)[Sov.Phys. JETP 41, 710 (1976)]

[2] D.S. Fisher, B.I. Halperin and P.M. Platzman, Phys. Rev. Lett. 42, 798 (1979)

[3] C.C. Grimes and G. Adams, Phys. Rev. Lett. 42, 795 (1979)+ private communication 1980

[4] D. Marty, J. Poitrenaud, and F.I.B. Williams, J. Phys. Lett. (Paris) à paraitre(july 1980)

[5] G. Deville, F. Gallet, D. Marty, J. Poitrenaud, A. Valdes and F.I.B. Williams (to be published)

[6] D.R. Nelson, B.I. Halperin, Phys. Rev. B19, 2457 (1979)

[7] R.C. Gann, S. Chakravarty and G.V. Chester, Phys. Rev. B70, 326 (1979)

[8] D.J. Thouless, J. Phys. C, 11, L189 (1978)

[9] Morf. Phys. Rev. Lett. 43 (1979) 931.

[10] F.I.B. Williams, J. Phys. Coll. (Paris) 41 C3, 249 (1980) (Congrès de la Société Française de Physique, Toulouse, June 1979).

Sinha, ed. Ordering in Two Dimensions

DEFECT STATES AND PHASE TRANSITION IN THE TWO-DIMENSIONAL WIGNER CRYSTAL

A. HOLZ*

Institut für Theoretische Physik, Universität des Saarlandes, 66 Saarbrücken, BRD

Based on a recently developed formalism to calculate defect states of dislocation type on a microscopic level it is shown that within the "harmonic approximation" Wigner crystals are unstable against formation of dislocations. For the two-dimensional Wigner crystal it is shown that the formation of a pair of dislocations leads to a rotational instability of the crystal in the domain affected by the dislocations. It is then suggested that the long-range Coulomb forces on a triangular lattice are "frustrated" leading to a multitude of degenerate ground states and therefore melting may be considered as a glass transition. Some qualitative features of the melting transition are discussed.

INTRODUCTION

A classical gas of negative charged particles constrained to two-dimensional (2D) motion on a neutralizing background and interacting via the 1/r-law is studied. Electrons moving on the surface of liquid Helium may be a physical realization of such a system. This system is of interest for at least two reasons. First, does it form a Wigner lattice, and what type of lattice is it? Second, what type of phase transition does it show, and what is the nature of the solid-liquid PT? The first point raised above is already clarified to a certain degree by experiment. Grimes and Adams [1] observed for such a system the formation of a Wigner lattice and a solid-liquid transition at a coupling constant $\Gamma=137\pm15$. Here $\Gamma=\pi^{1/2}n^{1/2}e^2/kT$, where n is the density of electrons per area unit, e the charge unit, and T the temperature. The nature of the PT is less clear. Hockney and Brown [2] observed a λ-type PT at $\Gamma=95\pm2$. Surprisingly, however, they found that their ground state is a macroscopic domain structure and disordering takes place over growth and motion of domain boundaries. Recent Monte Carlo calculations by Gann et al. [3] locate a melting transition in the range $110\lesssim\Gamma\lesssim140$ which may be of first order. Furthermore molecular dynamics experiments by Morph [4] show a sharp drop in the shear modulus and onset of electron self diffusion in the range $120\lesssim\Gamma\lesssim140$. Finally Thouless [5] obtained $\Gamma\simeq78.7$ at melting based on the Kosterlitz-Thouless [6] approach to dislocation mediated melting. Thouless [5] simulates the incompressibility of the Wigner crystal taking $\nu = 1/2$ for the Poisson modulus instead of $\nu = 1/4$ which is required by central force interaction. Accordingly the Cauchy relations are not satisfied in his model and non-central force interaction takes place. Such a mapping is therefore hardly justifiable in particular not for the Wigner crystal where interaction is by point charges and where stability of configurations is extremely delicate as will become clear in the following.

The first problem studied in the next section is the calculation of the binding energy of a pair of dislocations formed on a triangular Wigner lattice. Due to the long-range Coulomb interaction this problem cannot be treated by means of the standard dislocation theory. In the final section we present a discussion of the possible sorts of defect states which may drive the melting transition.

GROUNDSTATE AND DISLOCATIONS

In this section we sketch briefly how the formation energy of a pair of dislocations in the Wigner crystal is calculated. Details are given in Reference 7, 8, and 9. The ground state of the Wigner crystal forming a regular 2D Bravais lattice can be written in the form

$$E_o = \Phi_o + \sum_{n\neq m} f(|\vec{R}_o^n-\vec{R}_o^m|). \tag{1}$$

Here the set $\{\vec{R}_o^n\}$ describes the equilibrium configuration of the lattice, n is a particle label, and $f(r)=e^2/r$ represents the Coulomb law. All other terms are contained in ϕ_o. It can now be shown that the second term of Eq. (1) displays the symmetry property

$$\sum_{n\neq m} f(|\vec{R}_o^n-\vec{R}_o^m-\vec{b}^{n,m}|)= \sum_{n\neq m} f(|\vec{R}_o^n-\vec{R}_o^m|) \quad , \tag{2}$$

where each lattice vector $\vec{b}^{n,m}$ is assigned to a pair of sites $\vec{R}_o^n$ and $\vec{R}_o^m$ and is required to satisfy the condition

$$\vec{b}^{n,m} = - \vec{b}^{m,n} \tag{3}$$

and a further so called "transversality condition"

[8]. This latter condition guarantees that number and modulus of a given argument $(\vec{R}_o^n-\vec{R}_o^m-\vec{b}^{n,m})$ on the right and left hand side of Eq. (2) is equal and that this is the case for all possible arguments of this type. A representation of the set of difference vectors $\{\vec{R}_o^n-\vec{R}_o^m-\vec{b}^{n,m}\}$ as difference vectors between lattice points $\{\vec{R}^\kappa\}$ defined in 2D euclidean space is now not possible. Making, however, the replacement

$$\{\vec{R}_o^n-\vec{R}_o^m-\vec{b}^{n,m}\}\to\{\vec{R}_o^n-\vec{R}_o^m+\vec{s}^{n-m}-\vec{b}^{n,m}\} \qquad (4)$$

and subjecting the relative displacement field $\{\vec{s}^{n,m}\}$ to the condition

$$\oint(\vec{s}^{n-m}-\vec{b}^{n,m}) = 0 \ , \qquad (5)$$

which is demanded to hold for all possible closed contours, then the right hand side of Eq. (4) can be embedded in 2D euclidean space. This means and has been shown in Reference 8 that there exists a unique mapping

$$\vec{R}_o^n-\vec{R}_o^m+\vec{s}^{n-m}-\vec{b}^{n,m} = \vec{R}^\kappa-\vec{R}^{\kappa'} \ , \qquad (6)$$

where the set $\{\vec{R}^\kappa\}$ defines the coordinates of N particles in 2D euclidean space. It follows from this that the introduction of the set $\{\vec{b}^{n,m}\}$ amounts to a gauge transformation on the relative displacement field $\{\vec{s}^{n-m}\}$.
The energy of the Wigner crystal can now be written in the form

$$E=\phi_o+\sum_n g((\vec{R}^\kappa-\vec{R}_o^n),\vec{R}_o^n)+\sum_{\kappa\neq\kappa'} f(|\vec{R}^\kappa-\vec{R}^{\kappa'}|) \ , \qquad (7)$$

where the second term represents the interaction with the background and is normalized such that $g(0,\vec{R}_o^n) = 0$ holds. The set $\{\vec{R}^\kappa\}$ is defined by Eq. (6). Eq. (7) can now be expanded in a Taylor series leading to

$$E/kT=E_o/kT-\frac{1}{4}\sum_{n,m}\Phi_{i,k}^{n-m}(s_i^{n-m}-b_i^{n,m})(s_k^{n-m}-b_k^{n,m}) + \text{higher order terms}, \qquad (8)$$

where the coupling constant matrix

$$\Phi_{i,k}^{n-m} = \frac{\Gamma'(1-\delta_{n,m})}{|\vec{R}_o^n-\vec{R}_o^m|^3}(\delta_{i,k}-3(\vec{e}_{n,m})_i(\vec{e}_{n,m})_k) \qquad (9)$$

has been introduced. Here $\delta_{i,k}$ is the Kronecker symbol, $\Gamma'=(\sqrt{3}/2\pi)^{1/2}\Gamma$, and $\vec{e}_{n,m}=(\vec{R}_o^n-\vec{R}_o^m)/|\vec{R}_o^n-\vec{R}_o^m|$. The second term of Eq. (7) contributes only a linear term in $(\vec{R}^\kappa-\vec{R}_o^n)$. This term can be shown to cancel with a corresponding term of the last term of Eq. (7). The first two terms of Eq. (8) constitute the "harmonic approximation" to dislocation theory. The practical use of the set $\{\vec{b}^{n,m}\}$ is therefore that it allows an expansion of the energy in the small quantities $(s_i^{n,m}-b_i^{n,m})$. The energy of a defect structure characterized by the set $\{\vec{b}^{n,m}\}$ is obtained From Eq. (8) by standard methods [7] and yields using Einstein summation convention

$$E^{(2)}(\{\vec{b}^{n,m}\})/kT = -\frac{1}{2N}\sum_q\Big(\sum_{m,n}\Phi_{j,s}^{m-n}b_s^{m,n}e^{i\vec{q}\cdot\vec{R}_o^m}\Big) *\Phi_{j,i'}^{-1}(\vec{q})\Big(\sum_{m',n'}\Phi_{i',s'}^{m'-n'}b_{s'}^{m',n'}e^{-i\vec{q}\cdot\vec{R}_o^{m'}}\Big) - \frac{1}{4}\sum_{m,n}\Phi_{i,k}^{m-n}b_i^{m,n}b_k^{m,n} \ , \qquad (10)$$

where $\Phi_{i,k}(\vec{q}) = -\Sigma^h\,\Phi_{i,k}^h(1-\cos(\vec{q}\cdot\vec{h}))$ represents the inverse phonon propagator. Because the matrix $\Phi^{-1}(\vec{q})$ is positive definite for the triangular lattice the first term of Eq. (10) is therefore positive definite for it is of the form $a^+\Phi^{-1}(\vec{q})a$. Accordingly it gives always a negative contribution to Eq. (10). The last term of Eq. (10) we study for two different sets $\{\vec{b}^{n,m}\}$ defining a pair of dislocations obtained by a transversal slip. In the first case we use the abbreviation

$$I^{(2)}(\kappa,\xi) = -\frac{1}{4}\sum_{n,m}\Phi_{i,k}^{m-n}b_i^{m,n}b_k^{m,n} \qquad (11)$$

for the last term of Eq. (10). Then we take

$$\vec{b}^{m,n} = b\,\vec{e}_{+,-},\ |\vec{R}_o^m-\vec{R}_o^n|\le\kappa \ ,$$
$$= 0 \qquad ,\ |\vec{R}_o^m-\vec{R}_o^n|>\kappa \ , \qquad (12)$$

where $\vec{R}_o^n$ and $\vec{R}_o^m$ are located at opposite sites of the cut line ξ. The latter extends along the direction $\vec{e}_{+,-}$ between the locations $\vec{r}_+$ and $\vec{r}_-$ of the constituents of the pair, which are a distance $R_\xi = |\vec{r}_+-\vec{r}_-|$ apart. Eq. (12) satisfies the "transversality condition" stated earlier and κ is a decay constant measured in lattice units. Simple calculation gives

$$I^{(2)}(1,\xi) = -\Gamma' R_\xi b^2/4 \ , \qquad (13a)$$

$$I^{(2)}(2,\xi) = (1+1/4+1/3\sqrt{3})I^{(2)}(1,\xi) \ , \qquad (13b)$$

$$I_c^{(2)}(\kappa,\xi) = I^{(2)}(1,\xi) \ . \qquad (13c)$$

The last expression applies to continuous integration over the lattice points and is useful for $\kappa>>1$. It follows from this that formation energy of a pair of dislocations described by Eq. (13) is negative.
In the second case we use instead of Eq. (13)

$$\vec{b}^{m,n} = b\vec{e}_{+,-} \text{ for } |(\vec{R}_o^n-\vec{R}_o^m)\cdot\vec{e}^{\,*}_{+,-}|\le\kappa_\perp \ ,$$
$$|(\vec{R}_o^n-\vec{R}_o^m)\cdot\vec{e}_{+,-}|\le\kappa_\parallel \ , \qquad (14)$$
$$= 0, \text{ otherwise},$$

where $\vec{e}^{\,*}_{+,-}\cdot\vec{e}_{+,-} = 0$. If $\kappa_\perp/\kappa_\parallel<<1$ is assumed then one obtains in the continuum approximation for the quantity defined by Eq. (11)

$$I_c^{(2)}(\kappa_\parallel,\kappa_\perp,\xi)\simeq\frac{4}{\sqrt{3}}\Gamma' R_\xi b^2(1+0(\kappa_\perp/\kappa_\parallel)) \ . \qquad (15)$$

Because $I^{(2)}>0$ is a necessary condition for a positive defect energy it is possible that in this case the last term of Eq. (10) is larger in magnitude than the first term of this equation and a positive formation energy arises. This does, however, not prevent the instability of the system in the "harmonic approximation" via the formation of objects defined by Eq. (12).

The displacement field $\{\vec{p}(\vec{r})\}$ can be calculated from [7]

$$p_j(\vec{r}) = -\frac{1}{N}\sum_{\vec{q}}\Phi_{j,k}(\vec{q})\sum_{m,n}\Phi^{m-n}_{k,i}b^{m,n}_i e^{i\vec{q}\cdot(\vec{R}^m_o-\vec{r})} \quad . (16)$$

It should be pointed out that this field displays a discontinuity of order $b\vec{e}_{+,-}$ along the cut line ξ for the case described by Eq. (12). In Reference 9 it is shown that the leading term to the displacement $\vec{p}(\vec{r})$ in the direction $\vec{e}^*_{+,-}$ and for $\vec{r}$ close to the cut line ξ is given by the following expressions for the case that Eq. (12) applies

$$\vec{p}^r(\vec{r}) \simeq -\frac{1}{\pi}\, b\, \ell n \left|\frac{\vec{r}_- - \vec{r}}{\vec{r}_+ - \vec{r}}\right| \vec{e}^*_{+,-} \quad (17a)$$

for $\kappa/R_\xi \ll 1$ and $\kappa=6.55$, and

$$\vec{p}^r(\vec{r}) \simeq -\cdot 65\, b(|\vec{r}_- - \vec{r}|-|\vec{r}_+ - \vec{r}|)\vec{e}^*_{+,-} \quad (17b)$$

for $\kappa/R_\xi \sim 1$. Eq. (17b) corresponds to a uniform rotation of the cut line ξ around its center by an angle of about 52°. The result of the present calculations is therefore that in the "harmonic approximation" of defect theory the triangular lattice undergoes a rotational instability. This effect is certainly amplified if pairs of dislocations are used in an arrangement forming the boundary of a domain, because in this case already for short range interacting systems a rotation occurs.

On account of symmetry and homogeneity arguments it is plausible that perhaps the polydomain structure observed by Hockney and Brown [2] at low temperatures and the irreversibility of the grain boundary and defect structure observed by Morph [4] at high temperatures has its physical cause in the rotational instability. In this case one could say that the long-range Coulomb forces on a regular triangular lattice are "frustrated" and an inhomogeneous structure is energetically prefered. Taking the higher order terms of Eq. (8) into account may, however, modify this result.

THE MELTING TRANSITION

It is generally supposed that the generation of unbound dislocations results in a liquid phase because free motion of dislocations provides easy internal glide of the lattice and screening of transverse shear over short distances. Clearly nonconservative motion like diffusion of particles or groups of particles must also play an important role because it is usually this process one easily identifies in liquid system simulations. As a matter of fact because dislocations can move conservatively and non-conservatively both defect systems must be strongly coupled. In addition it must be taken into account that stacking faults are possible. In the triangular lattice this leads to rows of particles in a relative rectangular lattice position with $\lambda=\sqrt{3}/2$ for the ratio of the lengths of the sides a_1 and a_2 of the elementary cell, i.e., $\lambda=a_1/a_2$. If one uses the values given in Reference 10 and 11 for the ground state energy E_t, E_s, and E_λ for the triangular, square and rectangular lattice respectively one obtains for their energy differences $(E_s-E_t)/kT \simeq \cdot 00585\Gamma$ and $(E_\lambda-E_t)/kT \simeq \cdot 0093\Gamma$. Using $\Gamma=137$ [1] at the PT one obtains $(E_s-E_t)/kT_c \simeq \cdot 80$ and $(E_\lambda-E_t)/kT_c \simeq 1.27$. It follows from this that at the PT the energy differences of the structures considered are already of the order of the thermal energy. Because the stacking fault energy is governed by the energy of the λ-structure plus the energy of two partial dislocations many stacking faults should be present at temperatures of the order of T_c. However, because the λ-structure is unstable against shear the situation is better described by strong anharmonic phonon vibrations where whole rows of atoms slide on top of each other. If the triangular lattice transforms in a certain domain into the square lattice structure this requires diffusion of particles because it is associated with shape transformation which will set up electrical potentials. We conclude from this that the structure of the Wigner lattice around T_c may display a rather complicated pattern of various defect states like domain boundaries etc. and anharmonic lattice motion.

If we assume that the ground state of the Wigner lattice is frustrated as a consequence of the long-range Coulomb interaction as explained earlier then it is also plausible to assume that it is degenerate. This degeneracy may be lifted for finite particle systems through the periodic boundary conditions which may be the physical cause for the qualitatively different results obtained in the computer work of Reference 2, 3, and 4. The interaction law of dislocations (if they are well defined at all) in such a system depends on the mobility properties of the domain boundaries and glass properties may arise. If domain boundary sliding screens the long-range dislocation interaction then even ignoring the structural effects discussed above the "melting" transition should be qualitatively different from the one observed in short-range interacting systems with large domain boundary energies. In the latter case a first order melting transition is predicted in Reference 12 and 13, and if domain boundary formation is a competitive process to dislocation formation, a continuous transition may occur [13].

REFERENCES

*) Work supported by Deutsche Forschungsgemeinschaft under Sonderforschungsbereich 130.

1. C.C. Grimes and G. Adams, Phys. Rev. Lett. 42, 795 (1979).
2. R. W. Hockney and T. R. Brown Jr., J. Phys. C8, 1813 (1975).
3. R. C. Gann, S. Chakravarty, and G.V. Chester, Phys. Rev. B 20, 326 (1979).

4. R. H. Morph, Phys. Rev. Lett 43, 931 (1979).
5. D. J. Thouless, J. Phys. C 11, L 189 (1978).
6. J. M. Kosterlitz and D. J. Thouless, J. Phys. C 6, 1181 (1973).
7. A. Holz, Phys. Rev. A 20, 2521 (1979).
8. A. Holz, Dislocation Formalism in Crystals with long-range Interaction, preprint 1980.
9. A. Holz, Defect States and Phase Transition in the two-dimensional Wigner Crystal, preprint 1980.
10. G. Meissner, H. Namaizawa, and M. Voss, Phys. Rev. B 13, 1370 (1976).
11. L. Bonsall and A. A. Maradudin, Phys. Rev. B 15, 1959 (1977).
12. R. M. J. Cotterill and L. B. Pedersen, Solid State Comm. 10, 439 (1972).
13. A. Holz and J. T. N. Medeiros, Phys. Rev. B 17, 1161 (1978).

Sinha, ed. Ordering in Two Dimensions

ON THE EQUATION OF STATE OF THE 2-D WIGNER MODEL

R. CALINON and Ph. CHOQUARD LPT/EPFL, P.O. Box 1024 CH-Lausanne
E. JAMIN and M. NAVET* CRPE/CNRS-CNET, 45045 Orléans Cedex, France

ABSTRACT

We report the first results of Monte-Carlo simulation of the 2-D Wigner model subject to free boundary conditions, i.e to perfectly reflecting walls. Our computer experiments yield new informations on the equation of state of this one-component Coulomb system, on shape dependent effects, on static equilibrium configurations and on the one-particle distribution function.

I. INTRODUCTION

There is currently considerable interest in two dimensional classical systems of electrons and in particular in the nature of their states |1|, |2|,|3|. This has motivated our search for equations of state for these systems suitable for describing their equilibrium properties and improving our understanding of them. To begin with, we have considered the model of a classical assembly of point charges imbedded in a passive homogeneous background of opposite charge and confined by perfectly reflecting walls |4|,|5|. Unlike periodic boundary conditions (p.b.c.), these conditions, called free boundary conditions (f.b.c.), enable the investigation of the one-particle distribution function, of shape dependent effects, of non strictly neutral systems and of the kinetic pressure.

The kinetic pressure is that due to the transfer of momentum of the particles at the walls of the container. It is a non-negative physical observable, which can be measured in a computer experiment. In |5| we have shown that 1) this pressure is always equal to the virial pressure, 2) the latter is composed of the familiar thermal pressure (known to become negative at low temperature for all one-component Coulomb systems and often erroneously called the virial pressure in the literature) plus an excess pressure, 3) the excess pressure can be analysed in terms of a wall part and a possibly non-vanishing, long range order dependant bulk part, 4) in the ground state the excess pressure compensates exactly the Madelung pressure of the Wigner crystal, the dominant part of this compensation arising from the bulk part of the excess pressure.

The purpose of this paper is to illustrate these features by means of a few modest computer experiments performed with two-dimensional (2-D) systems of particles interacting with $\ln r^{-1}$ and r^{-1} potentials. In the next section we introduce the necessary definitions. In sec.III we review the theory of the virial equation of state of these models for finite systems. In sec.IV we present results of Monte-Carlo simulations with 37 and 32 particles in a disc and in a $\sqrt{3}/1$ rectangle and obtain results for the virial pressure and the internal energy between high and intermediate temperatures. In sec. V we examine the ground states configurations in discs for two sequences of particle numbers: an "ordered" sequence which yields the expected triangular crystal and a "disordered" one which yields dislocated and distorted crystals. In the last section we present a few results of radial densities for 61 and 121 particles in a disc for neutral and sub-neutral systems with $\ln r^{-1}$ and r^{-1} interactions between low and intermediate temperatures.

*presently at L.P.T./E.P.F.L.

II. DEFINITIONS

We consider N classical electrons of charge $-e$, confined in a 2-D domain Λ of volume $V=|\Lambda|$, immersed in a homogeneous background of charge density $e\rho_b$ and in equilibrium at the temperature T $=1/\kappa\beta$. The particles density is $\rho=N/V$ and $\rho_b V=$ $N+s$, s being a non-extensive excess or defect charge (neutrality implies $s=0$).

Let $C(\vec{r})$ be a Coulomb potential. It is instructive to consider in parallel the case where $C=\ln r^{-1}$ and the case where $C=r^{-1}$ with $r=|\vec{r}|$. The first case corresponds to the strictly 2-D Wigner model, the second one to the pseudo 2-D model, most frequently considered |1|,|6|,|7|. With the convention that all integrals occuring below are performed over the domain Λ and that the summations are performed from 1 to N, unless otherwise specified, the potential energy of these systems reads

$$U = U_{pp}+U_{pb}+U_{bb} = \sum_{i<j} e^2 C(\vec{x}_i-\vec{x}_j)$$
$$-\sum_i e^2\rho_b \int d^2y\, C(\vec{x}_i-\vec{y}) + \tfrac{1}{2}e^2\rho_b^2 \int d^2x\, d^2y\, C(\vec{x}-\vec{y}) \quad (2.1)$$

For latter purposes we introduce also $U'=U_{pp}-U_{bb}$ and $U''=U_{pb}+2U_{bb}$. Clearly $U=U'+U''$.

With λ standing for the thermal wavelength of the particles, the canonical partition function reads

$$Q = \frac{1}{N!}\lambda^{-2N}\int d^2x_1..d^2x_N \exp(-\beta U) \tag{2.2}$$

and the Helmholtz free energy is $F_\Lambda(N,\beta,\rho_b,s) = -\kappa T \ln Q$. The expectation value of an observable $A(\vec{x}_1,..\vec{x}_N)$ is defined by

$$\langle A\rangle = \int d^2x_1..d^2x_N A(\vec{x}_1,..\vec{x}_N)\exp\{\beta(F-U)\} \tag{2.3}$$

In particular we have for the one and two particle correlation functions

$$\rho(\vec{x}) = \sum_i \langle\delta(\vec{x}-\vec{x}_i)\rangle \tag{2.4}$$

$$\tau(\vec{x},\vec{y}) = \sum_{i\neq j}\langle\delta(\vec{x}-\vec{x}_i)\delta(\vec{y}-\vec{y}_j)\rangle \tag{2.5}$$

and we define truncated correlation functions through $\bar{\rho}(\vec{x})=\rho(\vec{x})-\rho$ and $\bar{\tau}(\vec{x},\vec{y})=\tau(\vec{x},\vec{y})-\rho(\vec{x})\rho(\vec{y})$. In terms of these quantities the mean potential energy of these systems becomes, for s=0

$$u = \frac{1}{N}\langle U\rangle = \frac{1}{N}\int d^2x d^2y C(\vec{x}-\vec{y})\{\bar{\tau}(\vec{x},\vec{y})+\bar{\rho}(\vec{x})\bar{\rho}(\vec{y})\} \tag{2.6}$$

and it is understood that all mean values and correlation functions occuring in the text depend in general upon β, ρ, s and Λ.

III. EQUATION OF STATE

Let us begin with the virial equation of state. If $\vec{F}_i=-\partial U/\partial x_i$ is the force which acts on the ith particle then

$$2pV = 2N\kappa T + \sum_i\langle\vec{x}_i.\vec{F}_i\rangle \tag{3.1}$$

The sketch of the proof that p equals the kinetic pressure p_k is as follows: we start from the first equation of the B.B.G.K.Y. hierarchy, take the scalar product with x and integrate by parts. Passing the term $-2N\kappa T$ on the right hand side of the equation produces the r.h.s. of eq. (3.1). What is left over is, in dividing by 2V, precisely the kinetic pressure contour integral namely

$$\frac{\kappa T}{2V}\int_{\partial\Lambda} d\vec{\sigma}.\vec{x}\rho(\vec{x}) = p_k \tag{3.2}$$

where $d\vec{\sigma}=|d\vec{\sigma}|\vec{n}$, $\vec{n}$ being the outer normal to Λ. If Λ is a disc shaped domain of radius R, eq. (3.2) becomes simply

$$p_k = \frac{\kappa T}{2\pi}\int_0^{2\pi} d\alpha\rho(R,\alpha)$$

On coming back to eq. (3.1) and using the definitions of sec. I we find, with s=0 (for s≠0 see |4| and |5|)

$$2pV = 2N\kappa T + \int d^2x d^2y \vec{x}.\vec{F}(\vec{x}-\vec{y}).\{\tau(\vec{x},\vec{y})-\rho\rho(\vec{x})\} \tag{3.3}$$

At this point we make the key observation that the quantity in parenthesis of eq. (3.3) is not symmetric in x and y. On adding and substracting the quantity $-\rho\rho(y)+\rho^2$ and using the definition introduced in sec. I we find

$$2pV = 2N\kappa T + \int d^2x d^2y \vec{x}.\vec{F}(\vec{x}-\vec{y})\{\bar{\tau}(\vec{x},\vec{y})+\bar{\rho}(\vec{x})\bar{\rho}(\vec{y})\} + \int d^2x d^2y \vec{x}.F(\vec{x}-\vec{y})\rho\bar{\rho}(\vec{y}) \tag{3.4}$$

The second term of eq. (3.4) can be further symmetrised through $\vec{x}\to\frac{1}{2}(\vec{x}-\vec{y})$ and we identify, in the first two terms of eq. (3.4) precisely 2V times the thermal equation of state defined as usual by

$$p_\theta = -\frac{\partial}{\partial V}F_\Lambda(N,\beta,\rho_b=N/V) \tag{3.5}$$

For the ln r^{-1} potential, and since $\vec{r}.\vec{F}(\vec{r})=1$, we recover Hauge-Hemmer's equation of state

$$p_\theta = \kappa T\rho - \tfrac{1}{4}e^2\rho$$

and for the r^{-1} potential, since $\vec{r}.\vec{F}(\vec{r})=r^{-1}$, we have

$$p_\theta = \kappa T\rho + \tfrac{1}{2}\rho u$$

In both cases we can write

$$p = p_\theta + \Delta p \tag{3.6}$$

with $$\Delta p = \frac{1}{2V}\int d^2x d^2y \vec{x}\vec{F}(\vec{x}-\vec{y})\rho\bar{\rho}(\vec{y}) \tag{3.7}$$

This is the excess pressure announced in sec. I. Note at this point that if the neutralizing background or substrate was inhomogeneous eq. (3.6) would apply on replacing simply $\rho\bar{\rho}(\vec{y})$ by $\rho_b(\vec{x})$. $(\rho(\vec{y})-\rho_b(\vec{y}))$ in the integrand of (3.7) and similarly $\bar{\rho}(\vec{x})$ by $\rho(\vec{x})-\rho_b(\vec{x})$ in p_θ. Some properties of eq. (3.6) in its thermodynamic limit have been examined in |5| and |8| for long and short range potentials. It turns out that the limiting excess pressure can always be analysed in terms of a wall part and a bulk part. This implies the knowledge of the one-particle correlation function in a semi-infinite and in an infinite system. It is clear that the bulk part vanishes for a normal liquid phase. It also follows from our analysis that this limiting p will be shape dependent for the ln r^{-1} interaction whereas its limiting p_θ is known to be shape independent. If Λ is a disc and for the ln r^{-1} interaction equation (3.1) becomes simply

$$\Delta p = -\frac{1}{2}\frac{e^2\rho}{R^2}\int_0^{2\pi} d\alpha\int_0^R dr r^3\bar{\rho}(r,\alpha) \tag{3.8}$$

It has also been shown that in this case $\Delta p=-\frac{\langle U''\rangle}{V}$ and, for the r^{-1} interaction in a disc that Δp can be written in terms of elliptic integrals.

IV. MONTE-CARLO SIMULATION

The principles of the simulation are those described by Metropolis et al. |10|, but compared to other computer experiments, we consider N particles in a finite volume with f.b.c.: if a particle jumps out of the volume, it is returned to its previous position and this configuration is recounted.

We used two types of geometries: rectangular ($L \times L\sqrt{3}$) with 32 particles and circular with 37, 61 or 121 particles. Our codes include the possibility of fixing the center of mass of the system and of removing the rotational degree of freedom of the circular system. Typical orders of magnitude of our simulations are a few 10^4 states for the preliminary relaxation process followed by an average over a few 10^5 states.

For the $\ln r^{-1}$ potential, the results presented in fig. 1 have been obtained essentially from

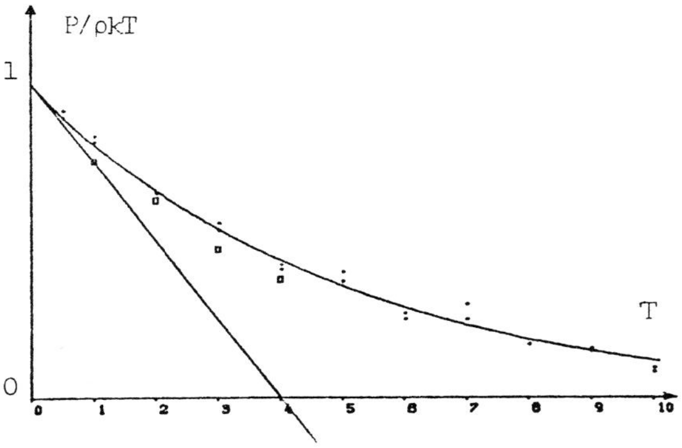

Fig. 1: Points show the Monte-Carlo measured virial kinetic pressure interpolated by the continuous curve for the $\ln r^{-1}$ potential systems, with circular boundary conditions. Squares show the virial pressure obtained in similar experiments with rectangular boundary conditions. The straight line shows the usual thermal pressure.

the circular geometry. Four points correspond to the rectangular geometry: they deviate slightly but significantly from the results of the disc, as expected on theoretical grounds. With Γ designating the plasma parameter βe^2, a good fit for the virial pressure is:

$$\beta p/\rho = 1+\Gamma(-1/4 + 5.82\ 10^{-2}\ \Gamma^{1/2} - 7\ 10^{-4}\ \Gamma^{3/2})$$

For the r^{-1} potential, we worked only with the circular geometry: both pressures decrease faster than in the preceding case (fig. 2) (note that in this case $\Gamma = \beta e^2 a^{-1}$, with $\pi a^2 = \rho_b^{-1}$). The thermal pressure can be interpolated, for $\Gamma < 100$, in the way followed by Gann et al. |6|, by:

$$\beta p_\theta/\rho = 1 + \frac{\Gamma^2}{2}\left\{\frac{a_1}{a_2+\Gamma} + \frac{a_3}{(a_4+\Gamma)^2} + \frac{a_5}{(a_6+\Gamma)^3}\right\}$$

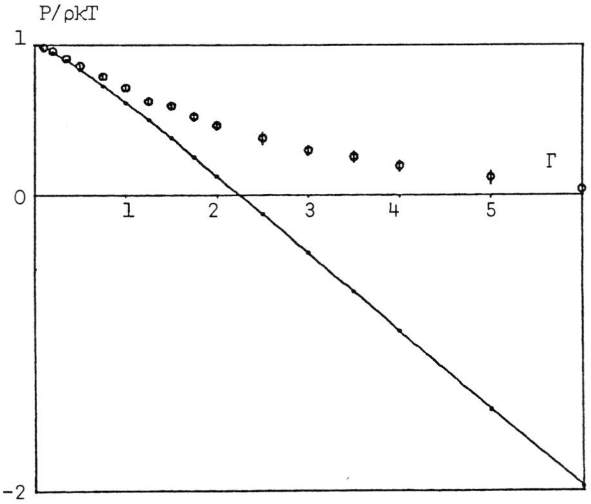

Fig. 2: Circles with error bars show the Monte-Carlo measured virial kinetic pressure for the r^{-1} potential system. Points interpolated by the continuous curve show the thermal pressure obtained in the same simulation.

with the a_i given in table 1.
We found that our values of the thermal pressure are in good agreement with the one given by Gann et al. (in their table 2) in this domain. (Unfortunately there seems to be a misprint in their interpolation coefficients).

We also checked that our results are close to being N independent: the probable maximum error should be smaller than a few percent in spite of the very small number of particles. Note that these results are very similar to those previously found in 1-D (|8|,|9|) and in 3-D (|4|).

Table 1: Fitting parameters for the thermal pressure of the r^{-1} potential system.

a_1 = -1.1131	a_4 = 1.4935
a_2 = .2323	a_5 = -5.941
a_3 = 1.0679	a_6 = 4.1464

It is apparent that the worrysome negative thermal compressibility problem faced if one works in particular with p.b.c., where $\rho=0$ by assumption, no longer occurs. Even in the absence of long range order, a meaningfull equation of state will only be obtained for a semi-infinite

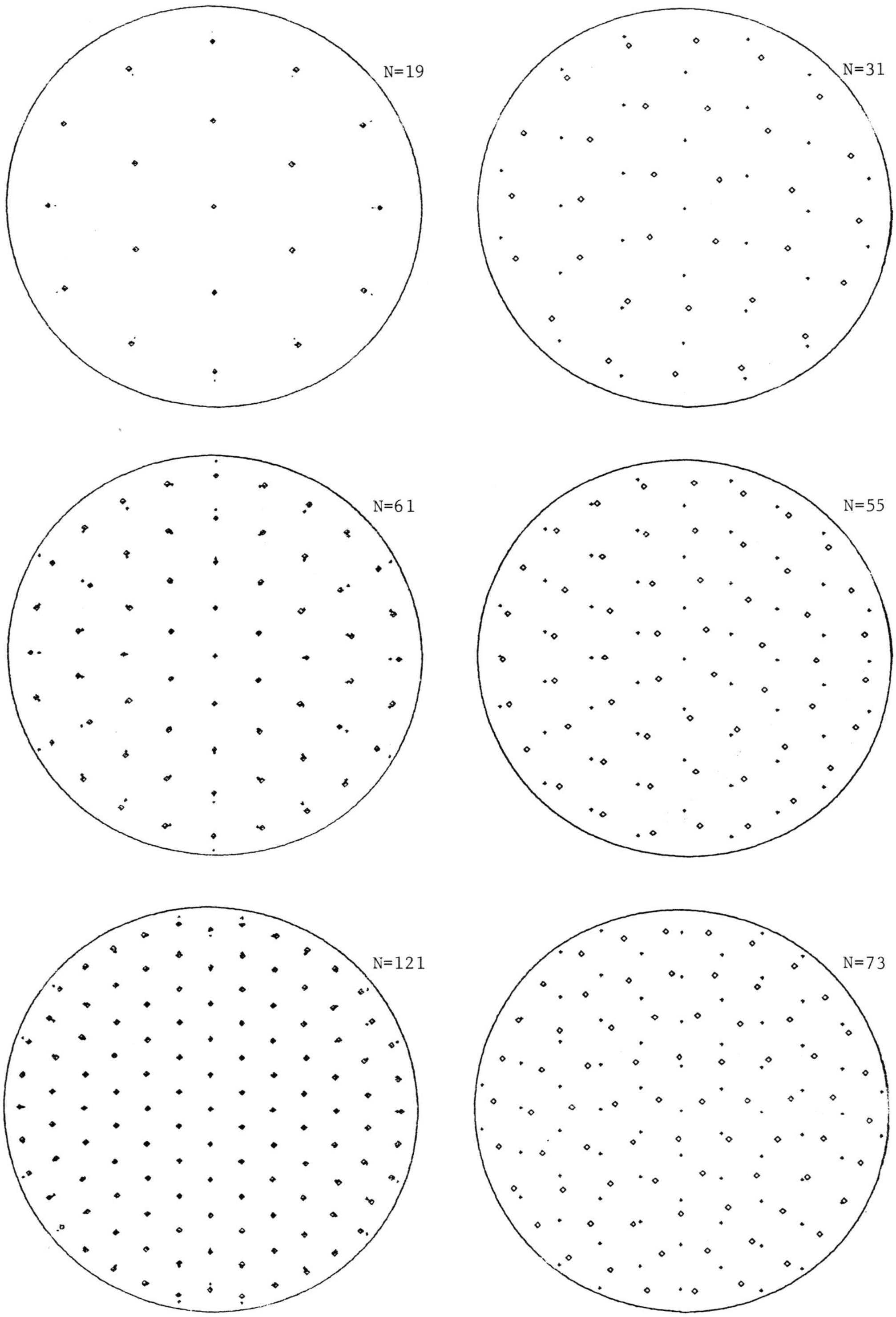

Fig. 3: The states of minimum energy for some ln r^{-1} potential systems (squares). The points show the corresponding triangular lattice.

one-component plasma. This suggests that one should investigate approximate schemes governing ρ and τ in these conditions.

V. MOLECULAR STATICS

We consider the equilibrium (state of minimum energy) of a system of N particles with $\ln r^{-1}$ potential, such that, if one arranges this system in the form of a perfect triangular lattice, it may be included in a circle, centered on a particle, so that the background (the density of which is prescribed by the condition of neutrality of the elementary cell) enclosed in the circle ensures total neutrality.

The method is a combination of standard, gradient-type methods, the initial configuration being either at random or according to the perfect crystal.

For one series of values of the particle number (N=7,19,37,61,85,91,121,151) which we call "ordered", the system keeps at equilibrium the expected triangular lattice structure, the only change with respect to the perfect crystal being a slight adjustment near the boundary. This contraction contributes to the wall part of the excess pressure, the bulk part being given by the moment of inertia of the hexagonal Wigner-Seitz cell |8|. For another series of values (N=13,31,55,73,109,139) which we call "disordered", the system experiences a "catastrophic" change of structure: defects appear (i.e. one or several particles have a number of neighbours not equal to 6 but equal to 5 or 7), but this change corresponds indeed to a decrease of the total energy: the increase of energy due to the defect being over-compensated by a good re-arrangement near the boundary. In both cases the final potential energy is about one percent (for the values of N of the order 10^2) above the one of a perfect crystal ($-.374e^2$). Similar results (in particular the same two ordered and disordered series) have been obtained in the case of the r^{-1} potential. The only difference is that this second system is a little more contracted near the center of the disc than the first one, and also the type of defect in the disordered series may be different. The enhanced contraction is explained by a less effective electrostatic screening (no Newton theorem in this case!). In order to relate the energy per particle of the ordered and disordered sequences of particle numbers, we have determined the energies per particle for a series of numbers N running from 19 to 31. The results are given in table 2.

Table 2: Madelung energy per particle for circular geometry and $\ln r^{-1}$ potential. The value for the infinite crystal is -.374.

N	U_o/N
19	-.36947
20	-.36888
21	-.36983
22	-.36679
23	-.36755
24	-.36811
25	-.36778
26	-.36826
27	-.36925
28	-.36864
29	-.36900
30	-.36976
31	-.36913

VI. ONE-PARTICLE DENSITY

We have shown that the excess pressure is a boundary condition dependent, linear functional of $\bar{\rho}(x)$ and that it may be shape dependent whereas the thermal pressure is shape independent and, within numerical accuracy, independent of boundary conditions. These features suggest investigating the properties of the one-particle density for different geometries and boundary conditions. We have considered the following cases: a rectangular geometry with f.b.c., a similar geometry with semi-p.b.c. (i.e. periodic in one direction only) and a circular geometry.
To begin with, we have chosen the circular geometry since it possesses the highest symmetry compatible with the potential energy given by eq. (2.1). We have investigated the azimuthaly averaged radial density $\rho(r)$ for a subneutral ($s>0$) system of 61 (fig. 4) and 121 (fig. 6) particles interacting via the $\ln r^{-1}$ potential and for a neutral system of 61 particles interacting via the r^{-1} potential (fig. 5). The M.C. experiment of fig. 4 has been performed under the conditions of frozen center of mass and vanishing global rotation. A characteristic feature is the formation of inner shells (three in this case, the two outer shells being due to the boundary conditions). We observe that the density peaks are equally spaced and that the spacing is 10% larger than the height of an equilateral triangle of particles at the same density. The M.C. experiment of fig. 5 has been performed without constraint. A pronounced structure is also observed. Fig. 6 shows the influence of the constraints on $\rho(r)$ for a system of 121 particles with $\ln r^{-1}$ potential. The inner shells are considerably sharpened when the constraints are applied.

In forthcoming experiments we plan to bridge the gap between low and zero temperatures, to investigate the inter-shell angular correlation function and to study the behaviour of $\rho(x)$ and

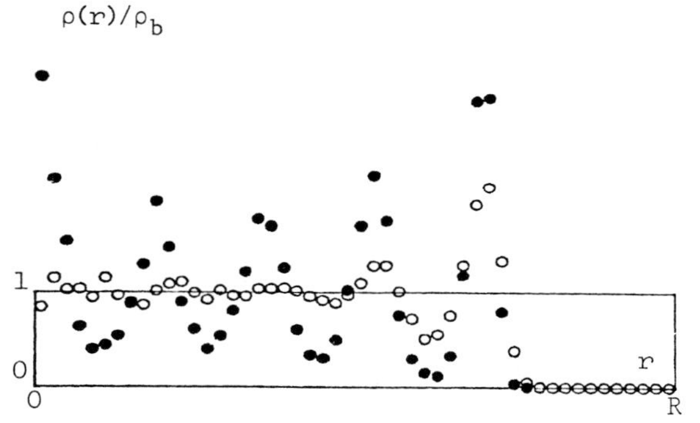

Fig. 4: The density of particles in a disc with ln r^{-1} potential and an excess charge in the background, for Γ=40 (open circles) and Γ=60 (black circles).

$\tau(x,y)$ for the semi-periodic case. The purpose of these experiments is to clarify the nature of the observed order. We expect significant differences in the behaviour of the systems with ln r^{-1} and r^{-1} potentials, owing to the presence of depolarizing effects in the ln r^{-1} case.

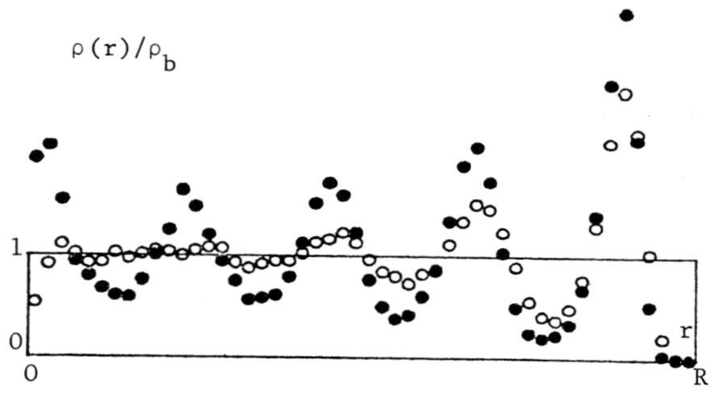

Fig. 5: The density of particles in a disc with r^{-1} potential (N=61), for Γ=30 (open circles) and Γ=55 (black circles).

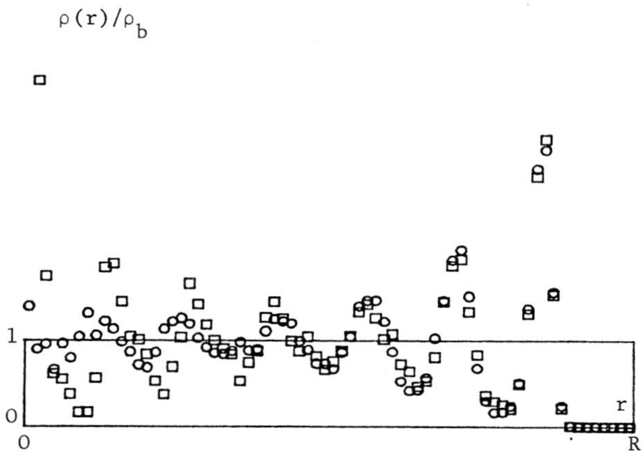

Fig. 6: The density of particles in a disc with ln r^{-1} potential for 121 particles and Γ=60. The circles correspond to a simulation without constraints, the squares to a simulation with frozen center of mass and vanishing global rotation.

REFERENCES

1. R.W. Hockney and T.R. Brown, J. Phys. C 8, 1813 (1975).
2. C.C. Grimes and G. Adams, Phys. Rev. Lett. 42, 795 (1979).
3. D.R. Nelson and B.I. Halperin, Phys. Rev. B 19, 2457 (1979).
4. M. Navet, E. Jamin and M.R. Feix, J. Physique-Lettres 41, L-69 (1980).
5. Ph. Choquard, P. Favre and Ch. Gruber, to appear in J. Stat. Phys. 23 (1980).
6. R.C. Gann, S. Chakravarty and G.V. Chester, Phys. Rev. B 20, 326 (1979).
7. D.S. Fisher, B.I. Halperin and R. Morf, Phys. Rev. B 20, 4692 (1979).
8. Ph. Choquard, to appear in the Proceedings of the 1979 Brasov Summer School on "Recent Advances in Statistical Mechanics" Bucarest 1980.
9. M. Navet, doctoral thesis Orléans 1980.
10. N. Metropolis, A.W. Rosenbluth, M.N. Rosenbluth, A.H. Teller and E. Teller, J. Chem. Phys. 21, 1087 (1953).

Published 1980 by Elsevier North Holland, Inc.
Sinha, ed. Ordering in Two Dimensions

COMPUTER SIMULATION OF THE DIFFUSION BEHAVIOR OF AN ORDERED ADSORBATE

G. E. MURCH

Chemistry Division, Argonne National Laboratory, Argonne, Illinois 60439

ABSTRACT

This paper reports on new Monte Carlo results for the "chemical" diffusivity of an ordered adsorbate. The model used is the square planar lattice gas with repulsive nearest neighbor interactions and attractive next nearest neighbor interactions. The simulation is based on the recently formulated rigorous Darken equation. Within the ordered c (2 x 2) region a strong maximum in the diffusivity as a function of coverage is observed. It is demonstrated that this maximum is a direct consequence of the large gradient in the adsorption isotherm within the ordered region.

INTRODUCTION

It has been usual for many properties of chemisorbed species on metal surfaces to be discussed in the light of the lattice gas model. This model is valid to the extent that adsorption occurs at particular localized sites relative to the substrate atoms and that magnetic and dipole-dipole, etc., interactions are negligible. Recently, the diffusivity of an adsorbed species (O on W {110}) has been studied as a function of coverage at temperatures less than desorption temperatures [1,2]. This work inspired Bowker and King [3] to make a Monte Carlo study of a lattice gas in order to investigate specifically the effect of lateral interactions on the diffusion behavior. Using experimentally derived nearest and next nearest neighbor interaction energies [4], Bowker and King were able to demonstrate semiquantitatively the experimentally observed strong maximum in the diffusivity as a function of coverage [1,2]. We should note, however, that good agreement should not necessarily be expected since the simulation study was in a square planar lattice while the lattice gas appropriate to O on W {110} should be triangular with anisotropic interactions. More importantly, however, the simulation was not sufficiently transparent for any insight to be made into the atomistic reasons for the maximum except that it was reported to be associated with ordering.

In view of the importance of the result for other adsorption systems and the development of a theory of diffusion in ordered structures, two dimensional or otherwise, we have undertaken a new Monte Carlo study on the same model that Bowker and King have employed. The nature of our method, in which the diffusivity is built up from its component parts, permits valuable insight to be made into the reason for the maximum in the diffusivity.

THEORY

The Lattice Gas

We consider a lattice gas of N adsorbed species on B lattice sites where the coverage, θ, is defined by N/B. Introducing the index i for the labeling of lattice sites and a local occupation number c_i such that $c_i = 1$ if the site is occupied and $c_i = 0$ if vacant, we may also designate the coverage by $\theta = \langle c_i \rangle$. We may write down the Hamiltonian of the lattice gas in the following general way [5]:

$$\mathcal{H} = \sum_{i \neq j} c_i c_j \phi(r_i - r_j) + \sum_i \varepsilon_i c_i + \mathcal{H}_o \qquad (1)$$

where ϕ denotes lateral interactions between adatoms, ε_i represents the binding force between the adatom and the substrate, and $\mathcal{H}_o$ represents separable contributions from lattice vibrations of the adsorbed layer, etc. For the case of the square planar lattice gas with nearest neighbor (nn) repulsive interactions and next nearest neighbor (nnn) attractive interactions, one generates a phase diagram with an ordered structure c (2 x 2) centered around $\theta = 0.5$ and two two-phase regions at lower temperatures, see [5].

With these interaction conditions and writing, $c_i = (1 + s_i)/2$; where s_i denotes a spin, we may transform to an Ising model [5]:

$$\mathcal{H} = -\sum_{i \neq j} J_{nn} s_i s_j - \sum_{i \neq j} J_{nnn} s_i s_j - H\sum_i s_i + \mathcal{H}_o' \qquad (2)$$

where $J(r_i - r_j) = -\frac{1}{4} \phi(r_i - r_j)$

and $H(r_i) = -\frac{1}{2}\left[\sum \varepsilon_i + \sum_{j \neq i} \phi(r_i - r_j)\right]$

where H is the imposed magnetic field and J is the exchange coupling between spins. The zero field critical temperature, i.e. at $\theta = 0.5$, for this Ising model has been obtained in an accurate [6] approximate solution by Fan and Wu [7]. They write for the critical condition in the square planar lattice:

$$\exp(|J_{nn}|/kT_c) = \sqrt{2}\exp(-2J_{nnn}/kT_c) + \exp(-4J_{nnn}/kT_c) \quad (3)$$

where T_c is the critical temperature (at θ = 0.5). Now, in their simulation Bowker and King used values of 14.4 kJ mol^{-1} for the nn repulsion interaction and 6.7 kJ mol^{-1} for the nnn attraction interaction with a temperature of 1153 K. These parameters lead to T_c = 1615 K. Thus, the ordered region is indeed entered during the simulation.

Diffusion

We are interested in the calculation of the intrinsic diffusivity, D^i, which is synonymous with the chemical diffusivity, $\tilde{D}$, for the one component lattice gas.

$$D^i = \tilde{D} = -J\left(\frac{d\theta}{dx}\right)^{-1} \quad (4)$$

where J is the diffusive flux and x is the distance. There are three simulation methods available to calculate $\tilde{D}$. The first, an non-steady state method [8], provides a direct means for calculating $\tilde{D}$ from concentration profiles. But this method is prone to statistical uncertainty. The second method makes use of a steady-state technique [9] and is considerably more precise. But both these methods give no real insight into the detailed makeup of $\tilde{D}$. In the third method, however, one can make use of the recently formulated rigorous 'Darken' equation for the one component lattice gas [10].

$$\tilde{D} = D\left(\frac{d\ln a}{d\ln\theta}\right) \quad (5)$$

where D is the self diffusivity of unmarked atoms and a is the thermodynamic activity defined by

$$\mu = kT\ln a + \mu^o \quad (6)$$

and the chemical potential, μ by

$$\mu = \left(\frac{\partial A}{\partial N}\right)_{B,T} \quad (7)$$

where A is the configurational free energy.

D is not measurable experimentally itself, but is related to the familiar tracer diffusivity by

$$D = D^* f_I/f \quad (8)$$

where f_I is the physical correlation factor [10-11] and f is the tracer correlation factor. D may be expanded as [12].

$$D = \frac{1}{2}\lambda^2 f_I[\nu zVW\exp(-E°/kT)] \quad (9)$$

where the adatom jump frequency, [ ] in eqn 9 depends on the surroundings *before* the jump. V is the availability of vacant nn sites to a diffusing atom, $E°_m$ is the diffusion activation energy of an *isolated* atom, λ, ν, and z are the jump distance and vibrational frequency and co-ordination, respectively, and W, the effective frequency factor is of the form

$$W_i = \exp[(4Z^i_{nn}J_{nn} + 4Z^i_{nnn}J_{nnn})/kT)] \quad V_i \neq 0 \quad (10)$$

and Z^i is the number of occupied neighboring sites around an atom i which is in a position to jump.

In a lattice gas treatment where hopping-like motion is assumed, it is sufficient to calculate V, W, f_I and $d\ln a/d\ln\theta$ in order to access $\tilde{D}(\theta,T)$. We used an array of 10^4 sites and divided the calculation into two parts. Specifying J_{nn}, J_{nnn}, and T, we attained equilibrium with the grand canonical ensemble and calculated $V(\theta,T)$, $W(\theta,T)$, and $\mu(\theta,T)$ (the latter is trivially known in this ensemble) by averaging over 10^5 configurations. Then we switched to the petit canonical ensemble in order to calculate f_I for 100 steps per atom [13]. Further details are described elsewhere [14].

RESULTS

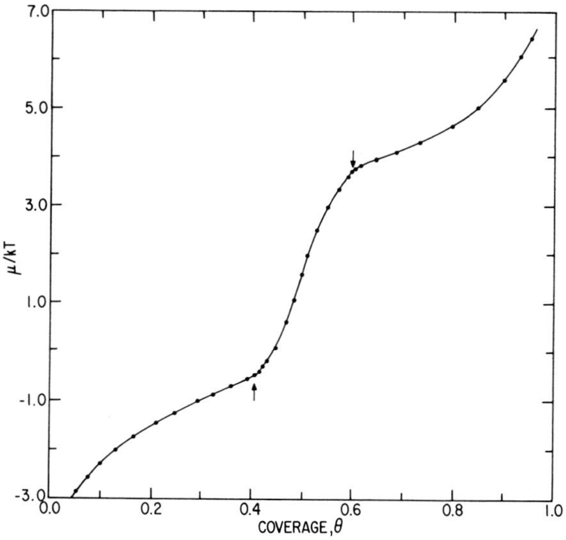

Fig. 1. The adsorption isotherm for the square planar lattice gas with J_{nnn}/J_{nn} = 0.465 and $J_{nn} > 0$.

In Fig. 1 we display the μ/θ isotherm at 1153 K. We have denoted the entry and exit (second order transitions [14]) to the ordered, c (2 x 2) phase by arrows. We note in particular the very steep rise of the isotherm within the ordered region. We will return to this observation below.

For reasons of space, we do not present detailed results for V, W, and f_I. They will be discussed elsewhere [14]. Our results for $\tilde{D}$ (normalized to unity) after numerical differentiation of the μ/θ isotherm are shown in Fig. 2. There is only a rather superficial resemblance with the results of Bowker and King [3]. But the rather large statistical uncertainty in their work, coupled with the unlikelihood of generating well-behaved concentration profiles within the ordered region, probably accounts for any discrepancy. We note in particular that when $\theta \to 1$, then $f_I \to 1$, $V \to (1-\theta)$, $d\ln a/d\ln\theta \to (1-\theta)^{-1}$ and $\tilde{D} \to W$ (θ=1). This is verified in our work, but while Bowker and King do not examine coverages > 0.9, the terminal point just mentioned seems unlikely to be obtained on the basis of any reasonable extrapolation.

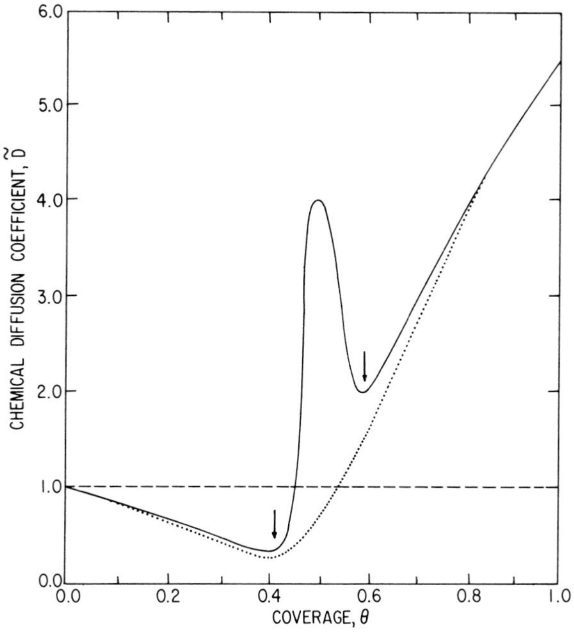

Fig. 2. The chemical diffusivity, $\tilde{D}$, as a function of coverage, θ, for the square planar lattice gas with $J_{nnn}/J_{nn} = 0.465$, $J_{nn} > 0$, and $T/T_c = 0.7136$. Monte Carlo results: ————. Monte Carlo results with $d\ln a/d\ln\theta$ replaced by $(1-\theta)^{-1}$: ....
Ideal mixing case: — — — —.

It is clear that the maximum in $\tilde{D}$ is associated with the ordered region. The dotted line in Fig. 2 was obtained by substituting $(1-\theta)^{-1}$, i.e. the *ideal* mixing case instead of the derivative of Fig. 1. We see now that the maximum completely disappears. This implies that the maximum is dictated only by the very sharp rise in the μ/θ isotherm within the ordered region. That is to say, the ordered region provides a very large thermodynamic driving force over and above that for random mixing alone.

For reference purposes, $\tilde{D}$, for the assumption of ideal mixing, is given by the dashed line in Fig. 2. In this case, $V = 1-\theta$, $W = 1$, $f_I = 1$, and $d\ln a/d\ln\theta = (1-\theta)^{-1}$ and $\tilde{D}$ is accordingly independent of coverage.

As we pointed out above, the present study was intended as an examination of a model system which permitted direct comparison with another simulation. Both arrays were square planar, and, despite the comments of Bowker and King [3], diffusion in this array is unlikely to bear more than a passing resemblance to the undoubtedly *anisotropic* diffusion behavior for O on W [110]. This should be 'modeled' as a triangular lattice gas with anisotropic interactions [4]. Results using our procedure for that lattice gas will be reported elsewhere [14].

REFERENCES

1. R. Butz and G. Wagner, Surface Sci. 63, 448 (1977).
2. J. R. Chen and R. Gomer, Surface Sci. 79, 413 (1979).
3. M. Bowker and D. A. King, Surface Sci. 72, 208 (1978).
4. T. M. Lu, G. C. Wang, and M. G. Lagally, Phys. Rev. Lett. 39, 411 (1977).
5. K. Binder and D. P. Landau, Surface Sci. 61, 577 (1976).
6. D. P. Landau, Phys. Rev. B, 21, 1285 (1980).
7. C. Fan and F. Y. Wu, Phys. Rev. 179, 560 (1969).
8. M. Bowker and D. A. King, Surface Sci. 71, 583 (1978).
9. G. E. Murch, Phil. Mag. to be published.
10. G. E. Murch and R. J. Thorn, Phil. Mag. A, 40, 477 (1979).
11. G. E. Murch, "Atomic Diffusion Theory in Highly Defective Solids," Trans Tech Monograph Series, to be published.
12. H. Sato and R. Kikuchi, J. Chem. Phys. 55, 677, 702 (1971).
13. G. E. Murch, Phil. Mag. to be published.
14. G. E. Murch, to be published.

Sinha, ed. Ordering in Two Dimensions

STUDY OF TWO DIMENSIONAL AND ADSORBED MICROCLUSTERS BY MOLECULAR DYNAMICS

M. WEISSMANN

Departamento de Física, Comisión Nacional de Energía Atómica, Buenos Aires,Argentina, and Departamento de Física,Universidad Simón Bolívar, Caracas, Venezuela.

N. V. COHAN

Departamento de Física,Comisión Nacional de Energía Atómica, Buenos Aires, Argentina.

Clusters of very few rare gas atoms, in two dimensions and adsorbed on graphite and on another rare gas solid, are studied by the molecular dynamics method. Our main interest is to find the structure of the small two-dimensional clusters and to study the effect of the solid surfaces on them, both as a function of temperature and of the strength of the interaction. The results show clearly the existence of two different condensed phases: a solid-like one at the lower temperatures, where each atom vibrates around its equilibrium position, and a liquid-like one at higher temperatures. In it the system remains condensed, but the atoms move considerably more and in very different ways,depending on their number.

INTRODUCTION

The method of molecular dynamics has been widely used to study the bulk properties of liquids and phase transitions, and it has also been applied recently to study adsorbed films [1]. All these calculations use periodic boundary conditions in order to simulate an infinite system. The study of small open systems of microclusters by molecular dynamics started only a few years ago. The first calculations,for three dimensional groups of rare gas atoms [2], show the existence of two condensed phases: a solid-like one and a liquid-like one, but the melting transition is not quite clear.

Our aim in this paper is to study very small groups of rare gas atoms in two dimensions and adsorbed on different surfaces. We wish to compare the structure and thermodynamics of these small systems with the three dimensional ones and also with the infinite adsorbed films when a very low coverage is assumed. Our results show clearly the existence of a solid-like state, where each atom vibrates around its 0°K equilibrium position and a liquid-like state,where the atoms move in very different ways depending essentially on the size and shape of the original 0°K cluster.

CALCULATIONS

We performed calculations for very small groups of atoms (N=4,6,7,8,19) in two dimensions (2D), and also for the three dimensional (3D) case that considers similar groups adsorbed on a smooth surface. We assumed the Lennard-Jones potential among the N rare gas atoms, which in adimensional units is;

$$V(r) = r^{-12} - 2\, r^{-6}$$

and we also set Boltzmann's constant and the mass of the cluster atoms equal to one. In 3D the interaction with the surface was simulated by a function only of the distance of each atom to the surface plane. This interaction was postulated by Steele [3] and does not take into account specific surface sites for adsorption. Therefore, it is an approximation that should be adequate for example in incommensurate adsorbed films and we hope may give a qualitative picture in other cases.

Two different syrface types were considered: one of them models clusters of Ar atoms on the basal plane of graphite, the other models Xe atoms on the 100 face of solid Ne. They represent two extreme cases of strong and weak physisorption. The parameters are taken from

Steele [3].

We used Verlet's algorithm to solve the 2N or 3N differential equations with a time step τ=0.005, that corresponds approximately to 10^{-14} sec. The calculations were started by distorting slightly the most stable 0°K configuration for the N atoms. This configuration is quite easy to guess in 2D or adsorbed (Fig.1) opposite to what happens in three dimensional open systems [4]. From there on, the total energy of the system was slowly increased by multiplying all the velocities by a factor α, slightly larger than one.
Every time the system was left to evolve until it reached a new equilibrium state (the transition region may be an exception, see below), and the time average of the kinetic and potential energies was calculated, $\langle E_k \rangle$ and $\langle E_p \rangle$. The temperature was defined as:

$$\frac{2N}{2N-3}\langle E_k \rangle \quad or \quad \frac{2N}{3N-6}\langle E_k \rangle$$

for 2D and 3D systems, respectively.

Other properties that we averaged for each total energy were: the mean deviation from the equilibrium 0°K configuration

$$\Delta = \sum_{i=1}^{n}\left[(r_i - r_i^o)^2/n\right]^{1/2}$$

where n is N times the number of time steps; the number of pairs of atoms at a certain distance from each other and the mean distance $\langle z \rangle$ to the surface plane. Also, snapshots of the atomic positions were drawn every 1000τ or 5000τ .

In the solid region the averages needed a small number of steps, usually a few thousand, but in the transition region we observed that very slow processes took place and therefore we averaged over 50,000 τ for N=4, over 30,000 τ for N=7, and so on. The dispersion σ_T between averages for different groups of 10,000 τ was also found to be an interesting variable.

RESULTS

The systems behave at low temperatures as a collection of harmonic oscillators, 2N-3 oscillators in 2D and 3N-6 in 3D cases. Fig.2 plots the difference in energy at the same temperature between the 2D and harmonic systems as a function of temperature. The 3D corresponding plots are not shown because they are very similar to these. The curve for N=4 is smooth, with no clear indication of a phase transition, but for N=7 there is a jump at T=.2. The N=6, N=8 groups behave similarly to N=4; and N=19 similarly to N=7. The mean values of T for succesive groups of 10,000 time steps at the same total energy E have a dispersion σ_T of about 0.02 for N=4 and T>.18; but for N=7 the value of σ_T increased significantly in the transition region and decreased again for higher tempertures (5). For this reason we do not know if a real equilibrium state is achieved in the transition region. We believe this difference in behaviour between different groups of atoms is another indication that the "melting" process is also different.

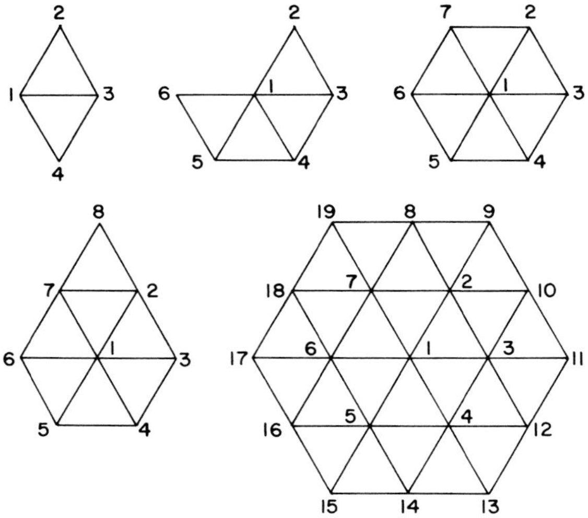

Fig.1. Equilibrium 0°K atomic positions from which the calculations were started. Lines joining pairs of atoms indicate only first neighbors.

Other averaged properties, such as the mean deviation from the 0°K configuration, Δ, show a more similar behavior for all systems studied. At low temperatures (or low total energy E) Δ increases linearly with E but at a certain energy E^* there is an abrupt increase in Δ. By visual inspection of the snapshots one can see that an interchange of particles or a complete change in shape occurs. Therefore, the statistical meaning of Δ is lost for $E>E^*$, as a hexagon with different particle numbering can have as large a value of Δ as any other shape. The system continues to interchange particles and moves in a complicated way, that we call "liquid", for all energies larger then E^*. Calculations were carried out until one particle evaporated from the system. This happened very soon after "melting" for N=19 but smaller clusters remained liquid for a large range of energies.

Fig.3 shows snapshots of liquid-like movements

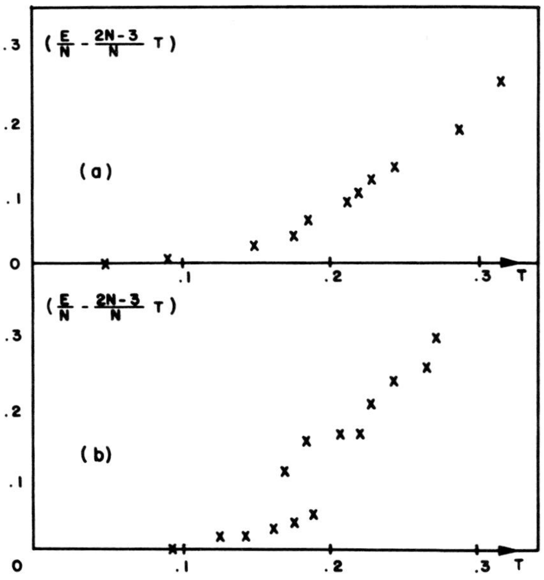

Fig.2. Difference in energy between the two dimensional system and a harmonic one at the same temperature, plotted against temperature. (a) for N=4, temperatures averaged over 50,000 time steps or more; (b) for N=7, temperatures averaged over 30,000 time steps or more.

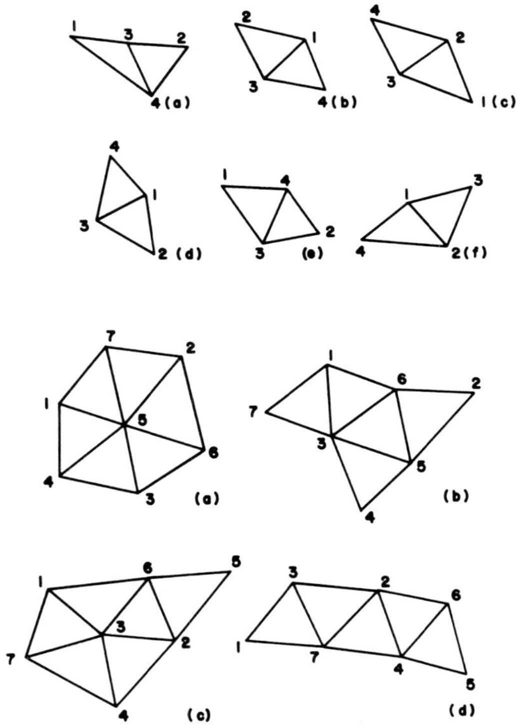

Fig.3. Snapshots of liquid-like movements in 2D. For N=4, E/N = -0.833, T=.243, every 10,000 time steps; for N=7, E/N = -1.174, T=.245, every 5000 time steps.
Letters indicate the time sequence. Lines joining pairs of atoms indicate first neighbors.

for 2D calculations of N=4 and N=7. Particles are numbered so as to make interchanges evident.

Because the transition region is extense,and for a small range of energies the system fluctuates very slowly between solid and liquid-like structures it is difficult to define a precise melting temperature by this procedure. The changes surely occur between T=.15 and T=.25 in all cases and this temperature increases with N, in our finite open systems. Infinite (periodic) adsorbed films in 2D have been studied by Hanson,Mandell and McTague for different coverages [1], the lowest being 15%. However, their curves corresponding to ours of fig.2 are practically superimposable for 15% and 64% and again show a broad feature, but centered at T=.38. It is a well known fact (in3D) that the melting temperature decreases with decreasing particle size, and this may also be the case here. We found no clear evidence of van der Waals loops, such as those reported for 3D open systems [2].

Calculations for the case of strong adsorption such as Ar on graphite, reproduce qualitatively all the previous results. Particles remain at approximately the same distance from the surface for all the energies studied. In weak adsorption, such as Xe on Ne,after the cluster "melts" there are large changes in its configuration,proving that it has lost its 2D character. For example, the average value of z, distance from each atom to the surface, increases significantly because some of the atoms move away from the surface, to a distance about double that at 0°K. We call this a droplet configuration. Thus, there is another transition region, where the system slowly fluctuates between the 2D and the droplet adsorbed configuration for the same range of energies.

CONCLUSIONS

Both 2D and 3D adsorbed microclusters present solid-like and liquid-like states. Systems that have closed shells of neighbors of a central atom at $0^{o}K$ (such as N=7 and N=19) show a clear indication of a phase transition and a transition region with large fluctuations in the average properties. Systems that do not have closed shells of neighbors of a central atom (N=4,6,8) oscillate between different equivalent positions before particle interchanges and real changes of shape occur. For example, for N=4 two equivalent positions are: the one shown in fig.1 and another having the longer distance horizontal (1,3) instead of vertical (2,4). We believe this is the reason for the different types of curves in fig.2. The effect of a surface is negligible for strong adsorption (Ar on graphite) and becomes important for weaker interactions (such as Xe on Ne) and high temperatures. In that case some atoms of the initial 2D cluster prefer to move away from the surface,thus forming an adsorbed cluster.

REFERENCES

1. F.E.Hanson,M.J.Mandell & J.P.McTague, J.de Physique, Colloque C4, 38, C4-76 (1977).
2. reviewed in:
   J.J.Burton & C.L.Briant, Adv. in Coll. and Interface Sci. 7,131,(1977).
3. W.A.Steele, Surf.Sci. 36, 317, (1973).
4. M.R.Hoare & P.Pal, J.Crystal Growth 17, 77, (1972).
5. M.Weissmann & N.V.Cohan, J.Chem.Phys, (in press).

ANALYSIS OF ORIENTATIONAL ORDER IN CONFIGURATIONS OF TWO-DIMENSIONAL SYSTEMS GENERATED BY COMPUTER SIMULATION

J. A. ZOLLWEG

Department of Chemistry, University of Maine, Orono, Maine 04469

A method is described for calculating the orientational order parameter at each molecule in a system for a configuration produced by computer simulation. A Fourier transform method is described for calculating the autocorrelation function of the order parameter and its rate of decay with distance. Fast Fourier transform methods are also described for obtaining the spatial gradient of the order parameter and its autocorrelation function. From this information can be obtained the Frank elastic constant of the system.

Preliminary results are given for the application of these methods to the hard disc system near its melting transition and comparisons are made with the dislocation-mediated melting theory of Halperin and Nelson.

INTRODUCTION

With the development of the dislocation-mediated melting theory of Halperin and Nelson (HN) [1], attention has been drawn to the analysis of orientational order in two-dimensional systems. This paper describes methods which have been developed for analysis of orientational order in the hard disc system, but since the methods depend only on the locations of the centers of the discs, they could be applied to systems of particles interacting with any interparticle potential. The principal question which needs to be answered is whether the melting transition is first order, involving a region of coexistence between solid and fluid phases, or whether there are two second order transitions with a hexatic phase existing between them.

Some evidence concerning the nature of the melting transition has been obtained from observation of the energy, pressure, or shear viscosity for Lennard-Jones systems under various conditions [2-4]. This study was undertaken to look more closely at the orientational order parameter and its correlations because orientational ordering is central to the HN theory. If it is found that the theory does not apply to melting of hard discs, it would be reasonable to expect that the analysis of behavior of the orientational order in the transition region should shed some light on why the theory does not apply to this system.

This paper briefly describes the methods which were used to obtain the orientational order parameter and functions of it and discusses techniques which can be used for determining the nature of the decay of correlations of that order parameter to determine the phase behavior of the system.

METHOD OF ANALYSIS

The orientational order parameter is defined at the location of a particle i by

$$\psi_i = \frac{1}{n} \sum_{j=1}^{n} e^{6i\theta_{ij}} \quad (1)$$

where θ_{ij} is the angle of the line joining the centers of particles i and j relative to an arbitrary reference direction. The value of ψ_i is affected by how many neighbors are included in the sum. I have chosen to include all neighbors within $r\sqrt{7/y}$ of disc i, where y is the ratio of the area of the discs to the total area and r is the radius of a disc. This causes n to be 6 on the average and corresponds to including all of the neighbors within the first peak in the pair correlation function. It differs from the definition used by Frenkel and McTague (FM) who set n = 6 for each particle [2].

Because FM found that the straightforward calculation of the correlation of ψ resulted in a function with deep minima, the correlation of ψ was calculated by first finding the Fourier transform of ψ.

$$\hat{\psi}(\vec{q}_k) = \frac{A}{N} \sum_{i=1}^{N} \psi_i e^{i\vec{q}_k \cdot \vec{r}_i} \quad (2)$$

The desired correlation function is then found by squaring the magnitude of the Fourier coefficients, dividing by A, and taking the inverse transform. So that fast Fourier transform techniques could be used for the inverse transform, the vectors $\vec{q}_k$ were chosen as the lattice vectors of a triangular lattice with nearest neighbor distance $2\pi/L_y$ where L_y is the smaller dimension of the rectangular area of the system. This choice of the $\vec{q}_k$ preserves in the Fourier transform the underlying sixfold symmetry of the two-dimensional solid. It is therefore believed that the Fourier coefficients for a finite system will converge more rapidly and uniformly to the values they have in an infinite system. The correlation functions obtained in this way do not have deep minima at the points where they are calculated.

Since one of the means by which hexatic order is distinguished is by the algebraic decay with distance of the correlation of ψ, techniques were explored for distinguishing between exponential and algebraic decay in a finite system. For an infinite system with exponential decay of correlation, the square of the Fourier coefficients should be proportional to $(\alpha^2 + q^2)^{-3/2}$ where α is the exponential decay constant. For an infinite system with algebraic decay of correlations, $r^{-\eta}$, the square of the Fourier coefficients should fall off as $q^{-3+2\eta}$. Thus one would expect that a plot of $\log|\hat{\psi}(q)|^2$ vs. log q should have a slope of -3 for $q \gtrsim \alpha$ if the decay is exponential. The plot would tend to level off at small q because of the α term. As the density of the liquid increases toward the freezing transition, though, one expects the range of correlation of ψ to increase, and hence α should decrease, making the region with slope -3 longer. If algebraic decay occurs, the maximum slope should be smaller in magnitude. Here also one expects to see a levelling off at small q because the finite size of the system limits the magnitude of the Fourier coefficients.

In the cases where the decay is exponential, the decay constant α can be evaluated, in principle, by plotting $|\hat{\psi}|^{-4/3}$ vs. q^2. This will have slope $(2\pi B\alpha)^{-2/3}$ and intercept $(\alpha^2/2\pi B)^{2/3}$ where B is the amplitude of the exponential correlation function. In practice this is unsatisfactory for small α because the finite size of the system limits the magnitude of $\hat{\psi}$ for small q and the plot curves up, causing α to appear to be too large.

Another indication of the appearance of hexatic order is the discontinuous jump in the Frank elastic constant K_A from 0 in a fluid with exponentially decaying correlations of ψ to $72k_BT/\pi$ in the hexatic phase at the transition. K_A should then increase slowly as the density increases until it diverges strongly at the freezing transition. Of course, in a finite system one cannot expect to see K_A precisely 0 or infinity, but a jump of approximately the correct magnitude has been observed in other systems [5,6]. In this study, K_A was determined by finding the limit of the Fourier transform of the logarithmic gradient of ψ. First ψ was calculated at the vertices r_1 of a triangular lattice by inverse Fourier transformation of $\hat{\psi}$. The gradient of ψ at each point on the lattice was also calculated

$$\nabla\psi(r_1) = \frac{1}{A}\sum_k -i\vec{q}_k\hat{\psi}(\vec{q}_k)e^{-i\vec{q}_k\cdot\vec{r}_1} \qquad (3)$$

Then the gradient of the bond angle field is found:

$$\nabla\theta(r_1) = \frac{\mathrm{Im}(\psi_1^*\nabla\psi_1)}{6|\psi_1|^2} \qquad (4)$$

and finally K_A

$$K_A = \frac{k_BT}{A}\lim_{q\to 0}|\hat{\nabla\theta}(\vec{q})\cdot\hat{\nabla\theta}(-\vec{q})| \qquad (5)$$

RESULTS AND CONCLUSIONS

The configurations which were analyzed were produced by the Monte Carlo method using an algorithm which has been previously described [7]. Systems of two sizes were used, N = 64 and N = 1024. All of the results quoted in this paper are for the 1024 disc system. The density was determined by the size of the rectangular system with height equal $\sqrt{3}/2$ times

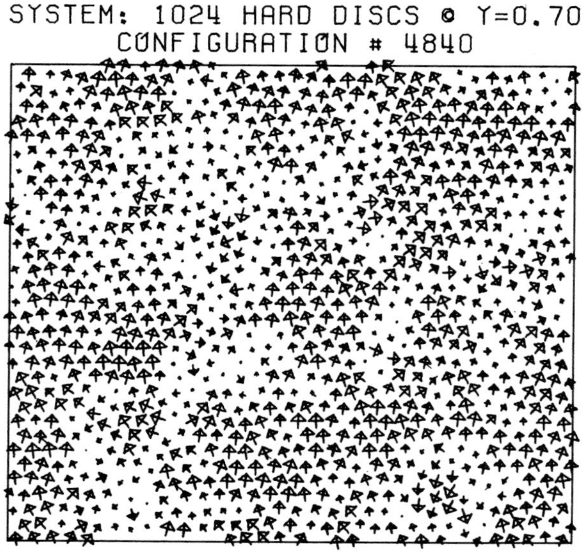

Fig. 1, Order parameter for a hard disc configuration

the width. Periodic boundary conditions were employed in both directions. At least 100,000 moves of each particle in the system were made before any configurations were analyzed. After analysis was begun, a configuration was analyzed after 20 attempts were made to move each particle.

Three densities were used: y = 0.65 which is definitely in the fluid phase; y = 0.70 which is in the transition region between fluid and solid; and y = 0.75 which is definitely in the solid phase. One of the configurations is visualized in Figure 1. The arrows are centered on the discs and are scaled so that ψ of unit magnitude gives an arrow with length equal to the diameter of the disc. The direction is determined by the argument of ψ when it is in complex exponential form. Comparison of this figure with one drawn for configuration #4080 reveals considerable change in the magnitude of ψ at most discs, but the direction is mostly upward for both configurations. A figure drawn for a configuration at y = 0.65 showed only small regions with ψ of substantial magnitude and approximately equal numbers of all orientations of the arrows, while a figure for y = 0.75 shows only a few areas where ψ is not large.

The correlations decay rapidly to zero at y = 0.65, as one would expect for a liquid. At y = 0.70, the correlations decay more slowly, but level off for interparticle distances greater than about 15 disc radii up to half the size of the system, which is about 32 radii. This is caused in part by the periodic boundary conditions which will cause increased correlation at greater distances, but the effect is probably increased by the fact that the system dimensions correspond to a perfect triangular lattice. At y = 0.75, the correlation of ψ decays rapidly to its asymptotic value which is almost 90% of the mean square ψ.

There are two reasons why the correlation functions calculated by this method do not have oscillations of the kind observed by FM. First is that the correlation function is calculated only for interparticle distances which correspond to a regular triangular lattice, and second the Fourier transform interpolates a smoothly varying ψ between particles in an ordered part of the system while FM find contributions only from particles which are in poor orientational relationship to their neighbors when they are at interparticle distances which are far from the spacings of the regular lattice.

The plots of log $|\hat{\psi}|^2$ vs. log q at y = 0.65 and 0.70 are both fit accurately by a line of slope near -3 over the range $0.4 < q^2 < 2.0$. The plot for y = 0.70 is shown in Figure 2. The region of linear slope is shorter at y = 0.65 because of the shorter range of cor-

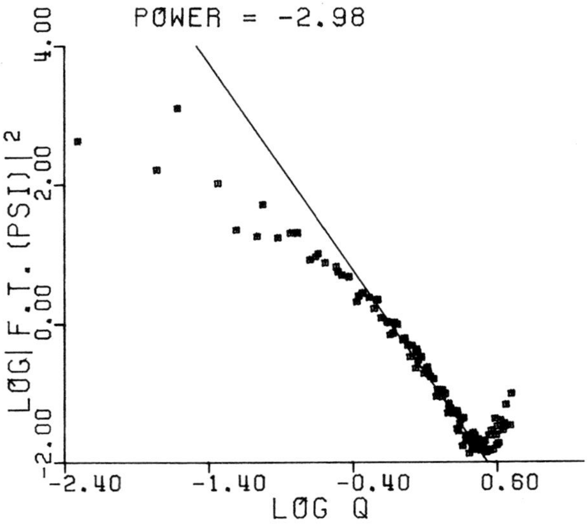

Fig. 2, Logarithm of $|\hat{\psi}|^2$ vs. q^2

relation at that density which corresponds to large α. The fit is not so good at y = 0.75 because α is quite large. The decrease in $|\hat{\psi}|^2$ beyond $q^2 = 2.5$ was seen at all densities.

Figure 3 shows a plot of $|\hat{\psi}|^{-4/3}$ vs. q^2 which should be straight if the decay of cor-

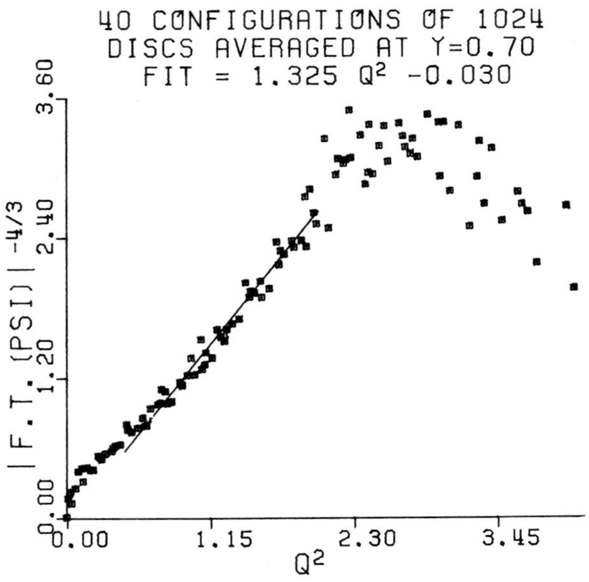

Fig. 3, $|\hat{\psi}|^{-4/3}$ vs. q^2 showing fit of a straight line to the data for $0.4 < q^2 < 2$

relations is purely exponential, and finite size effects are negligible. Apparently those conditions are not met because the plots for the other densities appear similar, with a fairly straight segment for $0 < q^2 < 2$ with extrapolated intercept near zero, and a downward turn for $q^2 > 2.5$. The parameters α and B found from a straight line fit over $0 < q^2 < 1$ do not fit the calculated correlation of ψ well, but a more satisfactory fit is made by combining the slope of the $1 < q^2 < 2$ segment with an estimate of B from the difference between the correlation of ψ at 0 and large separations. The values obtained in this way are: y = 0.65, α = 0.39; y = 0.70, α = 0.32; y = 0.75, α = 1.0.

The plots of $|\hat{\nabla}\theta|^2$ vs. q^2 showed a great deal of scatter but very little trend with q. For both y = 0.65 and 0.70, the value at q = 0 was about twice the average for all other q's. At y = 0.75, there was a slight sinusoidal trend vs. q^2 and the q = 0 value was in line with the values at small q. The Frank constants K_A/k_BT determined from the q = 0 values are 0.57 @ y = 0.65; 8.8 @ y = 0.70, and 87 @ y = 0.75. These are to be compared with $72/\pi$ = 22.9 which suggests that the system at y = 0.70 does not have hexatic ordering.

In summary, the configurations at y = 0.65 and 0.75 show the expected fluid and solid behavior. The configurations studied at y = 0.70 show non-zero correlations of ψ at the largest distances accessible to the simulation, but the approach to that asymptotic value appears to be exponential. The Frank elastic constant at that density appears to be too small for there to be hexatic orientational ordering.

Since the non-zero asymptotic correlations can be explained by the periodic boundary conditions which were used, it seems likely that two phases are in coexistence at y = 0.70. The data presented here were obtained from relatively few configurations, so a definitive answer to the question of the nature of the melting transition in the hard disc system must wait until more work is done, especially in the region between y = 0.691 and y = 0.716 which are thought to be the densities of coexisting fluid and solid for hard discs [8].

The author wishes to thank Prof. David Nelson for several encouraging and helpful conversations and the Computing and Data Processing Center at the University of Maine which made computing time available.

REFERENCES

1. B. I. Halperin and D. R. Nelson, Phys. Rev. Lett. 41, 121 519(E) (1978).
   D. R. Nelson and B. I. Halperin, Phys. Rev. B19, 2457 (1979).
2. D. Frenkel and J. P. McTague, Phys. Rev. Lett. 42, 1632 (1979).
3. F. F. Abraham, Phys. Rev. Lett. 44, 463 (1980).
4. S. Toxvaerd, Phys. Rev. Lett. 44, 1002 (1980).
5. R. Morf, Phys. Rev. Lett. 43, 931 (1979).
6. J. Tobochnik and G. V. Chester, private communication.
7. J. O. Milliken, thesis, University of Maine (1976).
   J. A. Zollweg, J. Chem. Phys. in press.
8. W. W. Wood in Physics of Simple Liquids ed. H. N. V. Temperley, J. S. Rowlinson, and G. S. Rushbrooke (North Holland 1968).

EPITAXIAL CRYSTALLIZATION FROM A MELT: A SURFACE MOLECULAR DYNAMICS STUDY*

U. LANDMAN, C. L. CLEVELAND and C. S. BROWN

School of Physics, Georgia Institute of Technology, Atlanta, GA. 30332

ABSTRACT

A newly developed Surface Molecular Dynamics method is described and is employed to investigate the epitaxial crystallization of a melt supercooled to the same temperature as the substrate. The procedure allows for thermal dissipation via a dynamic bulk reservoir, thus allowing the study of the approach to equilibrium. The early stages of the crystallization involve layering in the fluid, followed by intralayer ordering. The kinetics and dynamic coupling between fluid and solid regions is exhibited in particle number, temperature, potential energy, and orientational order-parameter profiles.

Epitaxial solidification of materials from a melt is an important scientific and technological process. Several theoretical models have been proposed for the description of crystallization at interfaces. The earliest ideas, due to Wilson [1] and Frenkel[2] assume a continuous growth mode. Later models employ a two-dimensional nucleation and growth process [3] and an attempt at a phenomenological unification of the above approaches has been advanced by Cahn [4]. In addition kinetic models [5] based on phenomenological mass transport and atomic incorporation into the crystal and studies of kinetic Ising models and the roughening transition have been reported [6].

While much attention has been devoted to melting little is known about the "reverse" process. In particular the atomic mechanisms occuring in epitaxial crystallization are not satisfactorily known. This may be partly due to the complicated nature of the system in which both disordering and ordering may occur (see below). The molecular dynamics (MD) technique is a powerful tool for the study of atomic properties of condensed matter and of phase transformations [7]. In MD the classical equations of motions of an interacting collection of particles are integrated and the recorded phase-space trajectories $(\vec{r}(t), \vec{v}(t))$ are then analyzed and employed in the calculation of various quantities of interest. In order to use the method for our study of the liquid-solid interface we must first provide an adequate description of the substrate surface.

SURFACE MOLECULAR DYNAMICS (SMD)

In most MD studies periodic boundary conditions (pbc) are used. 3D pbc's are appropriate for the description of bulk properties, a different situation is presented for a system which contains a surface. While for a two-dimensional system the question of pbc's is simple, a semi-infinite system is a much harder problem. Space limitations will allow us to present here only a brief description of our procedure. A schematic description of the system is shown in Fig. 1.

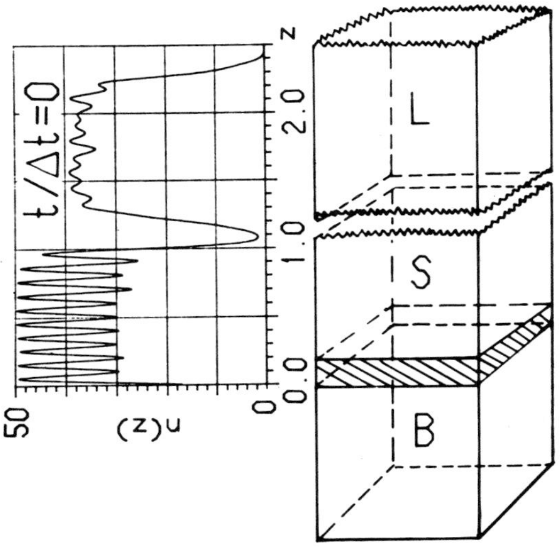

Figure 1: Schematic description of the bulk (B) - surface (S) - liquid (L) system. The bulk block posses 3D pbc's, the S and L posses 2D pbc's. The coupling region is hatched. Also included is the particle number versus z profile of the system at t=0, consisting of an equilibrated surface and supercooled liquid film at $T^{*}=0.4$. Distance in units of 7.94σ.

The "bulk block" (B) and "surface block" (S), consist each of 500 particles, interacting via a 6-12 Lennard-Jones potential. The B-system possesses 3D pbc's and the S-system possesses only 2D pbc's and is free in the z direction. Consequently, while the dynamics of particles in the S-system is influenced by that of particles in the B-system, the reverse statement does not hold, (this is done in the spirit of a bulk being an infinite reservoir whose dynamics and properties should not be influenced by surface effects). To provide for Newton's third-law (or fluctuation-dissipation), the first 3 layers of the S-B interface are used as a "coupling region" on which we impose (time-step by time-step) a scaling of velocities such that the average kinetic energies (kinetic temperature) in these layers is equal to the bulk ones. The integration of the equations of motions is performed using a predictor-corrector method [8] with a time step $\Delta t^*=0.0075$ [9] and the evolution of the S and B systems is synchronized at each time-step. The S-B system has been equilibrated as an fcc crystal (the density of the B system is adjusted to yield a vanishing average equilibrium pressure) exposing the (001) face, at a temperature $T^*=0.4$ (Ar melts at $T^*\simeq 0.7$). The above SMD technique has an advantage over slab configurations or calculations in which a static bulk or a random matrix are used.

SURFACE - LIQUID SYSTEM

Having equilibrated the surface described above we prepare an equilibrated liquid sample. In the following we describe the results for a supercooled L-J liquid film (2D pbc's and free in the third direction) consisting of 500 particles, at $T^*=0.4$ (see block L in Fig. 1. other liquids such as: bulk 3D pbc's sample, and a liquid film at $T^*=0.737$ have also been investigated by us). Once equilibrated the L system is positioned at a distance d_o (chosen such that the smallest distance between a surface and a liquid particle is equal to 1.12σ, i.e. the location of minimum of the L-J potential) and the evolution of the coupled solid surface-liquid system is followed. Sample particle-number versus z profiles at various time steps are shown in Fig. 2. It is observed that while the layers in the solid are well defined, those in the liquid fluctuate, averaging to the density profile of a liquid film. However as time progresses a permanent layering of the liquid occurs. In fact stratification of the liquid in the z (001) direction precedes the achievement of intralayer good crystalline order. Note also that the topmost layer distance of the surface (layer 10) first expands (in fact it is expanded at t=0) and upon solidification of the liquid it contracts to the bulk spacing. In the following, layers in the liquid region are defined as regions in space whose thickness in the (001) direction is that of the next to the top-most layer of the solid surface (layer 9). Samples of the layer kinetic temperatures $T^*=k_BT/\varepsilon$, defined as the mean kinetic energy of particles in layer ℓ, and layer potential energies versus time are shown in Figs. 3 - 6. Also shown are

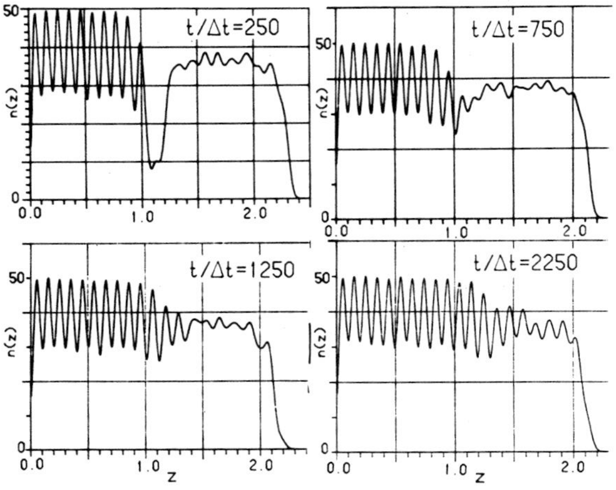

Figure 2: Particle number versus z at 4 different times during the evolution of the system. Note the expansion and subsequent contraction of the surface top-most layer spacing (layer 10), and the layering in the liquid region. The small peak of $z \simeq 1.1$, at the 250'th time step corresponds to the embryonic cluster (see text).

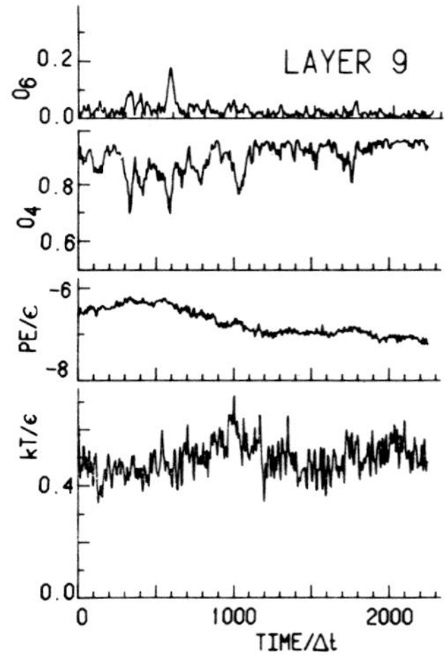

Figure 3: The kinetic temperature T^*, potential energy (PE) and orientational order parameters O_4 and O_6 (Eq. 1) for layer 9. Notice the peaking around $t/\Delta t=1000$ in T^*, (also occuring in Figs. 4-6) corresponding mainly to the expulsion of latent heat of ordering in the (001) direction , the initial increase in PE associated with disordering, and subsequent stabilization.

the orientational order parameters O_4 and O_6 defined as

$$O_n^{\ell}(t)=|N_\ell^{-1} \sum_{I,J\varepsilon\ell} N_{I,nn}^{-1} \exp(in\,\Theta_{IJ}) \times \quad (1)$$

$$\textcircled{H}(R_{nn} - |\vec{r}_I - \vec{r}_J|)|^2 , \quad n = 4,6$$

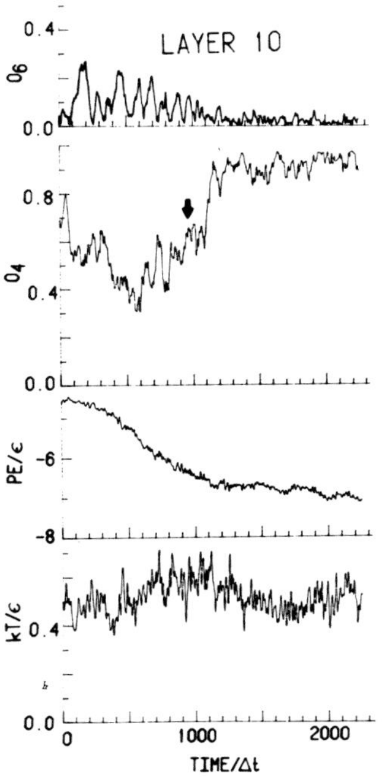

Figure 4: Same as Fig. 3, for layer 10 (top surface). Notice the stabilization of the layer, and the order disorder phenomena indicated by lowering of O_4 and peaks in O_6.

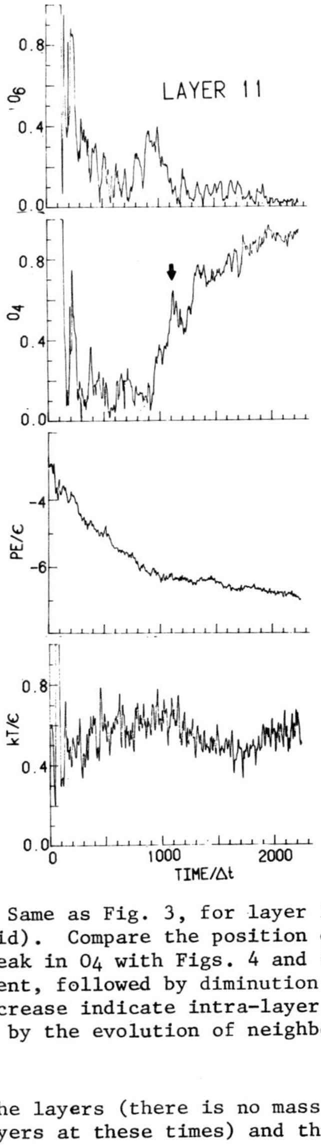

Figure 5: Same as Fig. 3, for layer 11 (crystallizing fluid). Compare the position of the arrowed peak in O_4 with Figs. 4 and 6. It's establishment, followed by diminution and subsequent increase indicate intra-layer ordering, influenced by the evolution of neighboring layers.

where N_ℓ is the number of particles in layer ℓ at time t, Ⓗ is the Heavyside step function, R_{nn} is the neighbor distance, $N_{I,nn}$ is the number of neighbors to particle I within a sphere of redius R_{nn}, Θ_{IJ} is the "bond" angle between I and J with reference to an arbitrary direction. O_4 and O_6 take the values 1 and 0 respectively for a perfect cubic crystal, while in the liquid state O_6 is generally larger then O_4. Observation of the T^* plots reveals that they undergo a maximum around t /Δt =1000 for layers 9-13. This is also the time at which the monotonic decrease in the layers potential energies achieves a plato. Notice that while at t=0 layers 9 and 10 (solid surface) were less stable, less coordinated, then a deeper layer they, along with the crystallized layers of the liquid, achieve enventually a bulk value. The time variation of the orientational order parameters reveals the dynamically coupled nature of the process. While the maximum in T^* indicates the expulsion of latent heat of ordering and is associated mainly with the "layering" of the fluid (and occurs at approximately the same time for a wide region of the liquid), the variations in O_4 and O_6 reflect intralayer ordering. Following the peak in O_4 designated by an arrow from layer to layer it is seen that it shifts to longer times for higher layers and in consequtive layers the peaking of O_4 is associated with a minimum in a neighboring layer. This systematic variations reflect the thermal coupling of the layers (there is no mass transport between layers at these times) and the transient order - disorder transformations which they undergo during the evolution of the system. Similar behaviour is observed in O_6. We observe that layer 10 (and even 9) initially partially disorder (roughened?) and eventually order (smoothed?). We have also observed that in case of a surface which contains a planar defect an annealing occurs. It is of interest to comment upon the embryonic stage of the crystallization process. We have observed that the nucleation of the first epitaxial layer involves a cluster of supercooled liquid atoms rather then single particle adsorption at random sites described by lattice-gas models. This cautions against the direct extrapolation of theories of gas-solid epitaxial crystallization to liquid phase epitaxy.

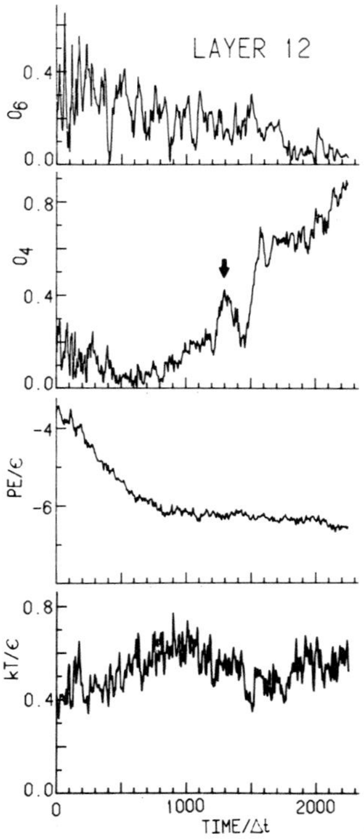

Figure 6: Same as Figs. 3-5, for layer 12.

Finally, the crystallization occuring in our system is interesting by itself, since it involves a meta-stable state of the fluid and no temperature gradient (the velocity of crystallization determined from our results is approximately 100 meters/sec). Our investigation is the first molecular dynamics study of this non-equilibrium system. The method has the advantage that it allows for heat dissipation via a thermal reservoir, thus allowing to follow the approach to equilibrium. A detailed account of this study, results concerning the crystallization of a hot liquid and data pertaining to other quantities such as: structure factors, $S(\vec{k})$, pair distribution functions and Voronoi Polyhydra will be reported elsewhere.

We gratefully acknowledge most valuable help and advice by Aneesur Rahman. The assistance and services of the GIT computer center, Rand Childs and Jerry Segers in particular, proved invaluable to this work.

REFERENCES

*Work supported by U.S. DOE contract No. EG-S-05-5489.

1. H. A. Wilson, Phil. Mag. 17, 283 (1900).
2. J. Frenkel, Physik Z. Sovjet union 1, 498 (1932).
3. D. P. Woodruff, The Liquid State (Cambridge Univ. Press, 1973), chap. 8.
4. J. W. Cahn. Acta Met. 8, 554 (1960).
5. J. C. Brice in Current Topics in Materials Science, Vol. 2, Eds. E. Kaldis and H. J. Scheel (North-Holland, Amsterdam, 1977), pp. 572, and references therein.
6. H. Müller-Krumbhaar, in ibid , pp. 116; G. H. Gilmer and K. A. Jackson, in ibid, pp. 80.
7. In particular see the recent studies of homogeneous nucleation by C. S. Hsu and A. Rahman, J. Chem. Phys. 70, 5234 (1979) and ibid. 71, 4974 (1979).
8. A. Rahman, Phys. Rev. 136, A405 (1964).
9. Reduced units are used throughout the paper. The reduced time is $t^*=\mu\sigma^2/\varepsilon$ where μ is mass, σ and ε are the 6-12 Lennard-Jones parameters. For Argon σ=3.4Å, $\varepsilon/k_B=120^oK$, and $t^*=1.82\times10^{-12}$ sec. Energy is in units of ε. The potential cutoff is 2.5σ. In the figures, length is given in units of b=7.94σ. Layers are numbered increasingly from the surface bulk interface, with layer 10 being the top-most layer of the solid surface. t=o is the time the equilibrated B-S and L systems are put together.

Sinha, ed. Ordering in Two Dimensions

THE MELTING OF TWO DIMENSIONAL SOLIDS

J. TOBOCHNIK and G. V. CHESTER

Laboratory of Atomic and Solid State Physics
Cornell University
Ithaca, New York 14853

ABSTRACT

We have simulated two dimensional systems of 1024 particles interacting through a Lennard-Jones potential using a Monte Carlo constant density ensemble. We present data on the energy, pressure, and elastic constants and in addition discuss the behavior of particle trajectories, orientational correlations and defects such as dislocations and disclinations. We compare our results with the Halperin-Nelson theory of melting as well as a first order interpretation of melting.

We have simulated a system of 1024 particles interacting through the standard Lennard-Jones potential. The particles are contained in a rectangular box with periodic boundary conditions. The densities used are $\rho\sigma^2$ = 0.888, 0.856, and 1.143 where σ is the hard core size parameter of the Lennard-Jones potential. The two lowest densities are sufficiently dense to avoid negative pressures in the solid phase. Our data are based on Monte Carlo runs of between 5,000 and 75,000 passes through the system. We have computed the energy, pressure, structure function, angular correlation function, and various elastic constants. In addition we have drawn pictures showing the trajectories of particles during Monte Carlo runs of various durations, snapshots of the bond angle configuration, and snapshots of the positions of disclinations and dislocations determined by looking at the voronoi polygons.

According to the Halperin-Nelson (HN) theory [1], melting occurs in two steps. At the first step bound pairs of dislocations begin to unbind, resulting in a jump in the elastic constant $K \equiv 4\mu(\mu+\lambda)/(2\mu+\lambda)$ from 16π to 0. Here μ and λ are the usual Lame coefficients divided by k_BT/a^2 where a = lattice spacing. In Figure 1 we show our data for K. At the two lower densities our results are remarkably close to those predicted by HN theory. The curve is a fit to the theoretical form $K^{-1} = (1-ct^{\nu})/16\pi$ where t is the reduced temperature $(T_m-T)/T_m$ and ν is a universal exponent equal to 0.370. Our fit shows that it is within a few percent of the theoretical predictions. At the higher density our results are inconsistent with HN theory.

Theory predicts a rather broad peak in the specific heat. As shown in Figure 2 we find a sharp bending in the energy at T_m which indicates that there is a sharp peak in the specific heat. This result is similar to that found in the planar model [2] which is also believed to follow the Kosterlitz-Thouless theory [3]. Thus, a sharp peak may not preclude the validity of the rest of the theory. This sharp bend in the energy could also be due to the system entering a region of two phase coexistence as in three dimensions [4]. We would then expect a similar bend when the system leaves the two

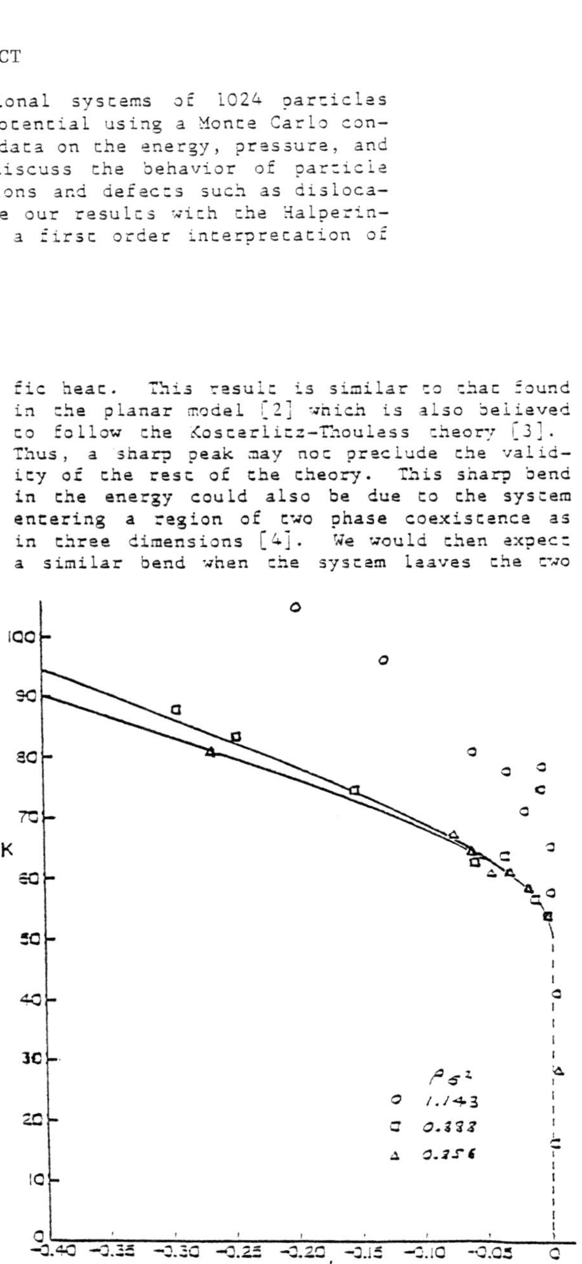

FIGURE 1. Elastic constant, K, vs. $t = (T_m-T)/T_m$. Error bars are of order 10% near $t = 0$.

phase region. The second bend for our two lower densities occurs very close to the first bend. The energy at our highest density has two clearly distinct changes, the second about 12% of T_m above the first. We expect that the width of the 2 phase region divided by T_m would decrease with increasing density as found from free energy calculations of Toxvaerd [5] in 2-D and 3-D work by Hansen [6]. Thus, it is likely that our two bends in energy at our lower densities are part of a single transitional region and not the boundaries of a two phase region. The pressure behaves qualitatively in the same way as the energy.

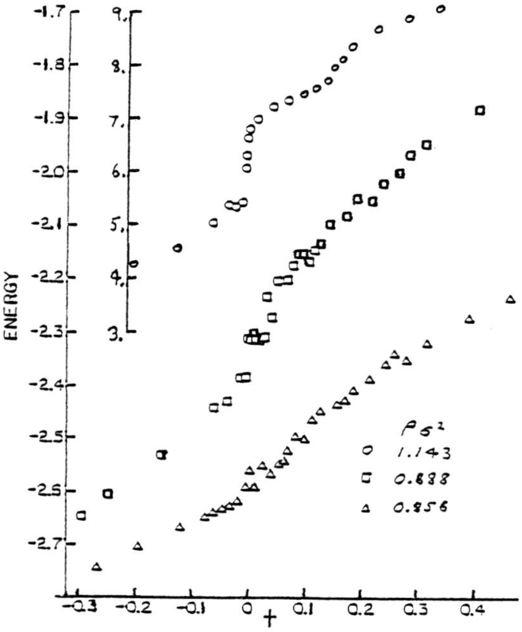

FIGURE 2. Energy vs. $t = (T_m - T)/T_m$

One of the more controversial features of the HN theory is the existence of a hexatic phase where unbound dislocations exist as bound disclination pairs. The nature of the hexatic phase is such that it is difficult to distinguish it from a two phase region. We have computed an orientational correlation function C(r) somewhat different from that of McTague [7] in that we use all the neighbors within a short distance of each particle to determine the bonds and we locate the bonds of each particle at a position midway between the two particles comprising the bond. McTague averages over the six nearest neighbors of a particle and locates the bond at the particle. Our results show that in the solid C(r) goes rapidly to a constant value and in the liquid it decays exponentially. There is however an intermediate region about 20% of T_m wide in which C(r) decays slowly across the system. In this region there are huge fluctuations in the value of C(r) for large r.

Pictures of trajectories of particles in this intermediate region show areas of order and areas of disorder over about five to ten thousand Monte Carlo passes. This could be two phase coexistence or a hexatic phase with critical fluctuations. One must remember that our longest computer simulations correspond to durations of around 10^{-8} seconds. Pictures of the bond angle configurations look similar to pictures of the spin configurations of the low temperature phase of the planar model as they should if there was a hexatic phase.

Our results for the defects are similar to those of McTague. We have a dislocation core energy at the lower densities of about 10 $k_B T_m$, suggesting a dilute concentration of defects where HN theory should apply.

In conclusion, we find that our simulation studies support some but not all of the predictions of HN theory at our lower densities. There is some evidence for a first order transition, however this evidence could arise from the very long time fluctuations which make the melting transition appear first order for the short times available for computer simulations. The transition at the higher density appears to be a more likely candidate for a first order transition.

REFERENCES

1. D.R. Nelson and B. Halperin, Phys. Rev. B19, 2457 (1979).
2. Jan Tobochnik and G.V. Chester, Phys. Rev. B20, 3761 (1979).
3. J.M. Kosterlitz and D.J. Thouless, J. Phys. C6, 1181 (1973).
4. See paper by F. Abraham in these proceedings.
5. S. Toxvaerd, J. Chem. Phys. 69, 4750 (1978).
6. J.P. Hansen, Phys. Rev. A2, 221 (1970).
7. See paper by J. McTague et al. in these proceedings.

SECTION II

C. THEORY OF 2-D PHASE TRANSITIONS

MELTING OF ANISOTROPIC LAYERS

S. OSTLUND and B. I. HALPERIN

Department of Physics, Harvard University

Cambridge, Mass. 02138

ABSTRACT

Using the ideas of Kosterlitz and Thouless to describe dislocation mediated melting of 2-D solids, we consider the melting of anisotropic layers of molecules. Depending on the symmetry of the Burgers vector of the dislocation which is most prone to unbind, we find new types of melting behavior. In the most interesting case, the properties of the melted phase are described by three characteristic lengths, where there are crossovers between regimes of 2-D solid-like, smectic-like, nematic-like and quasi-isotropic behavior. Close to melting there are divergences in the anisotropic properties of the crystal, due to one type of dislocation being free, but the other type being effectively bound.

INTRODUCTION

The theory of dislocation mediated melting has recently been applied in detail to the melting of a regular triangular 2-D solid [1,2,3]. It was predicted that, in addition to a conventional solid and liquid 2-D phase, there should be an intermediate liquid-crystal-like "hexatic" phase, where quasi-long range order of bond orientations persists. Here, we generalize the melting problem to include anisotropic solids, and find behavior which differs significantly from that of the regular triangular solid.

Phase transitions in isolated layers of smectic liquid crystals are an important potential application of two dimensional melting theory. In particular, smectic layers may be formed of rod-like molecules, whose axes are aligned along a vector at an angle away from the normal to the layer. The projection of the orientation vector onto the plane of the layer will have a preferred orientation relative to the bonds in the solid. The oriented molecules will, in turn, cause shear of the lattice [4] although the positions of the molecules might otherwise tend to form a regular triangular lattice within the layer. In the simplest case which we consider in detail, the molecules align either along or halfway between the bond directions, and the solid retains a rectangular symmetry with two perpendicular symmetry axes.

A layer of nonspherical molecules adsorbed on a crystalline surface may form an anisotropic 2-D solid, having a lattice incommensurate with the substrate under appropriate conditions. The theory described below is also applicable to such a solid.

Anisotropic solids have the possibility of very interesting behavior, since all elementary dislocations are not equivalent, as they are in the regular lattice. In a uniaxial solid, two equivalent dislocations, hereafter labeled Type I, have their Burgers vector along a reflection symmetry axis, and four dislocations, (Type II), equivalent with each other, but inequivalent with the first type, lie at angles of $\pm\phi_o$ from the reflection axis. We define the x-axis to be the reflection axis coincident with the elementary lattice vector.

The energy of the solid is given by

$$\frac{H_o}{k_B T} = \frac{1}{2}\int d^2r\, C_{ijkl}\, U_{ij}\, U_{kl} \qquad (1)$$

where C_{ijkl} is the elasticity tensor and U_{ij} is the strain field.

The solid phase has dislocations which are tightly bound in pairs. When the temperature is raised, the pairs unbind, and destroy the crystalline order.

The coefficient of the logarithm in the energy of an isolated dislocation of type α is $K_\alpha \ln(A)$ for the anisotropic solid, where A is the area of the system and K_α is a constant determined by the elasticity tensor. According to the ideas of Kosterlitz and Thouless, either type I or II dislocations would unbind at lower temperature, depending on the relative magnitude of the K_α. A new, highly anisotropic phase, with properties of both a solid and a liquid-crystal occurs if the other type dislocation remains bound. This "almost" happens in the case when K_I is less than K_{II}.

If K_{II} is less than K_I , apriori, type II dislocations unbind at lower temperature than type I. But since a pair of type II dislocations which are oriented at angles of $\pm\phi_0$ from the positive x-axis, is indistinguishable from a single type I dislocation, when observed with a resolution large compared to their separation, it follows that type I dislocations will also be unbound. We call this "type II" melting. If K_I is less than K_{II}, however, a highly anisotropic system occurs close to the melting temperature, where free dislocations of type I , but none of type II occur, over a very large relative length scale. Note that there is no way to add type I dislocations together to form type II. We call this situation "type I melting".

We shall see that in both type I and type II melting, the phase just above T_m is characterized by algebraic decay ("quasi-long range order") of orientational correlations, in the limit of large length scales. The melted phase may be described loosely as a two-dimensional nematic, with $\vec{n}$ playing the role of the director in a nematic. (Unlike the nematic, however, the orientations of $\vec{n}$ and $-\vec{n}$ are distinguishable in the present case.)

If the temperature T is raised sufficiently, one will eventually reach a point where disclinations appear in the molecular orientation, and bond orientation fields. The quasi-longrange orientational order is lost [2]. This second transitions will not be discussed here; rather we shall concentrate on the properties fo the nematic phase, on various intermediate length scales, close to the melting temperature.

TYPE I MELTING

The 2-D nematic may be described by a Hamiltonian

$$\frac{H_N}{k_B T} = \frac{1}{2}\int d^2r\,[K_x(\frac{\partial\phi}{\partial x})^2 + K_y(\frac{\partial\phi}{\partial y})^2] \qquad (2)$$

where ϕ is the angle of orientation of orientation of the projection n of the molecular axis in the x-y plane. (The local orientation of nearest neighbor bonds will be locked to the molecular orientation, so that ϕ may also be interpreted as the bond orientation field.) Note that for molecules aligned side by side, K_x is the Frank constant for splay in the molecular orientation, and K_y is the Frank constant for bend. The situation is reversed when the molecules prefer to lie end to end.

In principle, the nematic phase should become isotropic ($K_x = K_y$) in the limit of very long lengths [5]. In practice the minimum length scale ξ_I for this quasi-isotropic behavior can be astronomically large close to T_m, and one will rather see anisotropic nematic behavior ($K_x = K_y$) in a large range of length scales, $\xi_N(T) << L << \xi_I(T)$. The minimum length ξ_N for nematic behavior also diverges rapidly, as $T\rightarrow T_m$. In the case of type I melting, for lengths in a range $\xi_S(T) < L < \xi_N(T)$, only the type I dislocations are effectively unbound, and the system behaves like a 2-D smectic. On a length scale L smaller that ξ_S , the system retains the properties of the two-dimensional solid.

Just above the melting temperature, the three characteristic lengths obey

$$\xi_S \propto \exp(t^{-\frac{1}{2}}) \qquad (3)$$

$$\xi_N = \xi_S^{p+1} \qquad (4)$$

$$\xi_I = \exp(\xi_S^2) \qquad (5)$$

where p is a nonuniversal number greater than zero, and $t = \text{const}|T_m - T|$. On a length scale in the range $\xi_S(T) < L < \xi_N(T)$, the system may be described by a smectic-like Hamiltionian, with molecules arranged in rows parallel to the x axis [6,7]:

$$\frac{H_s}{k_B T} = \frac{1}{2}\,B\int d^2r\,[(\frac{\partial u}{\partial y})^2 + \lambda^2\,(\frac{\partial^2 u}{\partial x^2})^2] \quad (6)$$

where $u(\vec{r})$ is the displacement of the rows in the y direction, and B and λ are coefficients.

In this length range, dislocations with Burgers vector in the x direction are unbound, while there is a vanishingly small density of bound dislocations with Burgers vector components along the y direction, corresponding to the occurence of incomplete rows. The coefficient λ diverges like ξ_S , while the coefficient B remains finite for $T\rightarrow T_m$.

Close to T_m , for type I melting, the Frank constants should be very anisotropic, with $K_x << K_y$. In particular as the temperature approaches T_m, we find

$$K_x^o \propto \xi_S^2 \qquad (7)$$

$$K_y^o \propto \xi_N^2 \qquad (8)$$

$$K_y^o/K_x^o \propto \xi_S^p \qquad (9)$$

where K_x^o and K_y^o are the "bare" Frank constants measured on the length scale $L > \xi_N$. The Frank constants are renormalized by fluctuations in the orientation field, in such a manner that they tend to become equal at very long length scales, $L > \xi_I$. [5]

If observations are made on a fixed length scale L, large compared to the molecular spacing, one may pass through all of the above regimes with increasing temperature. One would observe solid-like behavior for temperatures slightly above T_m, smectic-like behavior when $\xi_S(T) < L < \xi_N(T)$ and nematic-like behavior when $\xi_N(T) < L$. The quasi-isotropic behavior may be observed if one can reach a regime with $\xi_I(T) < L$. Of course, these different regimes are not separated by sharp phase transitions; they are simply different regimes of the same 2-D liquid-crystal phase, with short rangetranslational order and quasi-long range order for the orientational order parameter $\exp[i\phi(r)]$.

Smectic-like behavior on an intermediate length scale may be revealed in principle by x-ray or neutron scattering, since we expect the Bragg scattering perpendicular to the type I lattice vector to be strong and narrow, reflecting the longer range correlations along this direction in the lattice.

As the temperature approaches the melting temperature T_m from below, weak singularities occur in the compliances S_{1111}, and S_{1212}, with

$$[S_{ijkl}(t) - S_{ijkl}(0)] \propto t^{\frac{1}{2}} \qquad (10)$$

The compliance tensor S_{ijkl} is the inverse to the elasticity tensor , in the sense that

$$\sum_{kl} S_{ijkl}\, C_{klmn} = \frac{1}{2}\,(\delta_{im}\delta_{jn} + \delta_{in}\delta_{jm}) \qquad (11)$$

All other compliances have dominant analytic behavior, with very weak singularities which will not be detectable experimentally. K_I, which is a relatively complicated function of the compliances, approaches the universal constant 4, and reduces to $K_R/4\pi$, defined in previous work [2], as the isotropic limit of our equations are taken . We incorporate a factor of $(k_BT)^{-1}$ in K_I and K_{II}, so that these quantities are dimensionless.

OTHER MELTING BEHAVIOR

In the case $K_I > K_{II}$, type II dislocations are more prone to unbind than type I. But since a type I dislocation can be constructed by adding two type II dislocations, it follows that the type I core energies are roughly twice the type II core energy. Although anisotropy remains, there are no real divergences in anisotropic behavior as the temperature approaches the melting temperature. In particular, there is no smectic-like phase, and the Frank constants behave like

$$K_x \propto K_y \propto \xi_S^2 \qquad (12)$$

as T approaches T_m. The length ξ_S is given by Eq. 3. The ratio of Frank constants does not diverge, but goes to a (nonuniversal) constant.

The nonzero elastic constants all display weak singularities of the type $t^{\frac{1}{2}}$. K_{II}, again a relatively complicated function of the compliances approaches the universal constant 4. The structure factor diverges with a half width which scales like ξ_S^{-1} at all Bragg points.

It is to be noted that the only two-dimensional solid for which the dislocation picture gives $\bar{\nu}$, the power of 1/t in the exponent in Eq. 3, not equal to 1/2 is the regular, triangular solid. General anisotropic lattices, square, and rectangular lattices all have $\bar{\nu} = 1/2$. Only in the isotropic triangular solid do interacting triplets of dislocations affect the behavior at the melting temperature. This gives the anomalous value to $\bar{\nu} = .369...$, found in [2,3].

The results described above can be further generalized. For an anisotropic solid where there is less symmetry than the uniaxial solid, the dislocation melting picture always gives behavior consistent with type I, since two elementary lattice vectors cannot add to form a third.

EFFECTS OF A SUBSTRATE

The effect of a substrate on the melting of an adsorbed layer has been considered in some detail [2] when the adsorbate lattice has hexagonal symmetry, and many of the modifications of that theory can be generalized in the present case. If the substrate periodicity is incommensurate with the adsorbate, or if it is commensurate at a sufficiently high order, then the most important effect of the substrate is to introduce a discrete set of preferred directions for the adsorbate structure. Expanding about one of the preferred orientations, one may represent the effect of the substrate by a term in the Hamiltonian of the form

$$\frac{H^A}{k_B T} = \frac{1}{2} \gamma \int d^2r \; \theta^2(\vec{r}) \qquad (13)$$

where the angle θ describes the deviation of the bond orientations from the favored direction and γ is a new effective elastic constant.

The new elastic term breaks rotational invariance and modifies the interactions between dislocations in the regular solid. This has relatively little effect on the melting transition at the temperature T_m where free dislocations appear, but the coupling to the substrate does affect the nature of the orientational order in the melted phase. Instead of the quasi-long range order predicted for the nematic phase in the absence of a substrate, there should now be true long range order in the orientations. The disclination unbinding transition between hexatic phase and isotropic and liquid phases which occurs at a temperature T_i (higher than T_m) in the absence of a substrate, may be suppressed entirely or will be modified substantially, according to whether there exists one or several distinguishable favored orientations for the adsorbate relative to substrate at temperatures just above T_m. The substrate anisotropy will lead to a set of favored orientations for the tilt orientations, as well as for the bonds in the adsorbate. An important consequence is that the presence of a small density of free dislocations with Burgers vector in the x-direction does not now provide screening sufficient to eliminate the logarithmic interaction between remaining type II dislocations. These dislocations may remain bound above the first melting temperature T_m, until a second transition temperature T_m' is reached where the coefficient of the logarithm falls below the critical value $4k_B T$.

In the temperature range $T_m < T < T_m'$, one then predicts quasi-long range order for the correlation of the order parameter $\exp(2\pi i u/a_0)$, and the system may be properly described as a 2-D smectic. (Here a_0 is the distance between the rows.)

The stabilizing effect of the substrate interaction may be readily understood. The orienting forces lead to a modified smectic Hamiltonian, of the form

$$\frac{H'}{k_B T} = \frac{1}{2} \int d^2r \; [B(\frac{\partial u}{\partial y})^2 + \gamma(\frac{\partial u}{\partial x})^2] \qquad (14)$$

since at sufficiently long wavelengths, the term proportional to λ may be neglected in Eq. 6. The Hamiltonian has the same form as for a two-dimensional planar spin model (X-Y model). By analogy with the X-Y model, one can readily establish the necessary condition for stability of the 2-D smectic phase

$$(\gamma B)^{\frac{1}{2}} > \frac{8\pi}{a_0^2} \qquad (15)$$

where γ and B are the macroscopic (renormalized) values of the elastic constants.

REFERENCES

1. J.M Kosterlitz, and D.J. Thouless, J. Phys. C6, 1181 (1973); J.M Kosterlitz, J. Phys. C7, 1046 (1974)
2. D.R. Nelson and B.I. Halperin, Phys. Rev. B19, 2457 (1979).
3. A.P. Young, Phys. Rev. B19, 1855 (1979)
4. D.R. Nelson, and B.I. Halperin, Phys. Rev. B. (in press)
5. D.R. Nelson and R.A.Pelcovits, Phys. Rev. B16, 2191 (1977)
6. P.S. Pershan, J. Appl. Phys., 45 1590 (1977).
7. J. Toner and D.R.Nelson, preprint, (1980)

Sinha, ed, Ordering in Two Dimensions

RENORMALIZATION OF THE VORTEX DIFFUSION CONSTANT IN SUPERFLUID FILMS

R. G. PETSCHEK and A. ZIPPELIUS

Harvard University, Cambridge, Ma. 02138

ABSTRACT

We study the dynamics of vortices in superfluid helium films near the Kosterlitz Thouless transition. The diffusive and convective motion of vortices can be described by a Langevin equation. In the long wave length, low frequency limit the density response is diffusive. Dynamic screening effects give rise to a renormalization of the diffusion constant, while the convective parameter remains unchanged. As the transition temperature is approached the diffusion constant shows a universal cusp, reaching a finite value at T_c.

INTRODUCTION

The critical properties of thin superfluid He^4 films have been of great interest recently. Kosterlitz and Thouless [1] proposed a low temperature superfluid phase, in which bound vortex-antivortex pairs coexist with smoothly varying phase fluctuations. At a critical temperature T_c bound vortex pairs unbind, thereby destroying superfluidity and driving a transition to a normal phase. The most striking prediction of the theory is a universal jump in the superfluid density at T_c [2]. Subsequently Ambegaokar et al.[3] and Huberman et al.[4] combined the Kosterlitz Thouless theory with a phenomenological picture of vortex motion. They calculated the dissipation near T_c and found good qualitative agreement with the experiments of Bishop and Reppy [5].

In this paper we present a systematic analysis of vortex motion on long length and time scales. Starting from a phenomenological Langevin equation, we eliminate successively vortex pairs with small separations. This gives rise to a renormalization of the equation of motion of the remaining vortices. In the hydrodynamic limit all modifications due to the screening of vortex pairs can be expressed in terms of a static dielectric function and a renormalized diffusion constant. In the limit $T \to T_c$ the diffusion constant shows a cusp-like singularity, reaching a finite value at T_c.

VORTEX MOTION

Consider a thin He^4 film on a substrate, which is driven with a velocity v_n. The thickness of the film is assumed to be small compared to a bulk correlation length and variations in the film thickness as well as variations in the temperature are neglected. Our aim is to calculate the density and current response of quantized vortices in the film. We define a vortex as a point defect, characterized by its position $\vec{r}^{\nu}$ and its vorticity $n^{\nu} = \pm 1$. The circulation around the vortex is given by

$$\oint_c \vec{v}_s(\vec{r}) \cdot d\vec{r} = \frac{h}{m} n^{\nu} \qquad (1)$$

Where c is a contour enclosing a vortex, m is the mass of a helium atom and $\vec{v}_s(\vec{r})$ is the local superfluid velocity.

Following ideas of Hall and Vinen [6] as elaborated by Ambegaokar et al. [3] one can write down an equation of motion for a collection of vortices moving in the local superfluid velocity field. Balancing drag and Magnus forces yields

$$\frac{d\vec{r}^{\nu}}{dt} = - n^{\nu} \frac{Dh\rho_s^{\circ}}{mk_B T} \hat{z} \times \vec{v}_s(\vec{r}^{\nu}) - (C-1)\vec{v}_s(\vec{r}^{\nu}) + \vec{\eta}^{\nu} \qquad (2)$$

We have assumed a time independent driving force v_n so that we can work in the frame in which $v_n = 0$, ρ_s^o is the superfluid density integrated across the film thickness and $\hat{z}$ is the normal to the surface. The diffussion and the drift constants D and C can be expressed in terms of phenomenological drag coefficients, describing interactions with the substrate and with thermal excitations (see ref. 3). The $\vec{\eta}^\nu$ are fluctuating gaussian white noise sources, whose second moment is related to the diffusion constant

$$\langle \eta_i^\nu(t)\eta_j^\mu(t')\rangle = 2D\delta_{ij}\delta_{\nu\mu}\delta(t - t') \qquad (3)$$

At sufficiently low frequencies the local superfluid velocity may be written in the form

$$\vec{v}_s(\vec{r}^\nu) = \frac{h}{m}\,\hat{z} \times \vec{\nabla} \sum_{\mu\neq\nu} n_\mu G(\vec{r}^\nu,\vec{r}^\mu) + \vec{u}_s(\vec{r}_\nu) \qquad (4)$$

The first term gives the flow field due to vortices at $\vec{r}^\mu$, $G(\vec{r},\vec{r}^\mu)$ is the solution of

$$\left(\frac{\partial^2}{\partial x^2} + \frac{\partial^2}{\partial y^2}\right) G(\vec{r},\vec{r}^\mu) = 2\pi\delta(\vec{r}-\vec{r}^\mu) \qquad (5)$$

subject to the boundary condition $G(\vec{r},\vec{r}^\mu) = 0$ for $\vec{r}$ on the edges of the sample. Far from the boundary and the core region

$$G(\vec{r},\vec{r}^\mu) \simeq \ln\frac{|\vec{r} - \vec{r}^\mu|}{a} + \tilde{C} \qquad (6)$$

where we have introduced a cutoff a, chosen to be the vortex core diameter, $\tilde{C}$ is a positive constant, related to the core energy. The test field $\vec{u}_s(\vec{r})$ is assumed to have infinitesimal amplitude and to be slowly varying in space and time so that retardation effects can be ignored, as we have already assumed in 4. In particular we restrict ourselves to the case $\vec{\nabla}\cdot\vec{u}_s(\vec{r})=0$, so that $\hat{z} \times \vec{u}_s$ can be written as the gradient of a scalar potential V

$$\frac{h\rho_s^o}{mk_BT}\, n^\mu\, \hat{z} \times \vec{u}_s(\vec{r}_\mu,t) = \vec{\nabla}_\mu V(\vec{r}_\mu;t) \qquad (7)$$

which we take to vary sinusoidally in space and time

$$V(\vec{r};t) = \frac{V_0}{\varepsilon}\, e^{-i\omega t}\, e^{i\vec{k}\cdot\vec{r}} \qquad (8)$$

Here ε is a dielectric function, not including the effect of vortices.

FOKKER PLANCK EQUATION

We first consider the Langevin equation without convection, i.e. C = 1, and rewrite eqn. 2 as a Fokker Planck equation [7] for the N vortex distribution function $\Gamma_n(\vec{r}^\nu)$ (9)

$$\frac{\partial}{\partial t}\Gamma_N = D \sum_\nu \vec{\nabla}_\nu\cdot \left(\vec{\nabla}_\nu + (n_\nu\vec{\nabla}_\nu V) - \sum_{\mu\neq\nu} n_\nu n_\mu (\nabla_\nu U(\vec{r}_{\nu\mu}))\right)\Gamma_N$$

where the following abbreviations have been introduced,

$$U(\vec{r}_{\nu\mu}) = J\, G(\vec{r}_{\mu\nu}) \quad \text{with } J = \left(\frac{h}{m}\right)^2 \frac{2\pi\rho_s^o}{k_BT} \qquad (10)$$

If only bound vortex pairs are present in the film, Γ_{2m} is of the order of y^{2m}, where $y = \exp(-\mu/k_BT)$ is the vortex fugacity and μ the vortex chemical potential. Truncating the hierachy of dynamic distribution functions, we can generate a fugacity expansion, analogous to the one used in the equilibrium problem [1, 8]. Instead we consider a situation in which a finite density Γ_o of free positive and negative vortices have been excited. We choose $J\Gamma_o/k^2 \ll 1$ and $\Gamma_o D/\omega \ll 1$ so that each vortex will move independently of the fields of the other free vortices and so that the vortices will not recombine on the time scale of the experiment. The distribution function of free vortices in the absence of bound pairs is given as the solution of

$$\frac{\partial}{\partial t}\Gamma_1^{(\pm)}(\vec{r}_1,t) = D(\nabla_1^2 \pm \vec{\nabla}_1\cdot(\vec{\nabla}_1 V))\Gamma_1^{(\pm)}(\vec{r}_1,t) \qquad (11)$$

The vorticity density response is

$$\frac{\delta\Gamma_1}{\delta V}(\vec{k},\omega) = \frac{\delta}{\delta V(k,\omega)}(\Gamma_1^+(\vec{r}_1,t)-\Gamma_1^-(\vec{r}_1,t))e^{i\omega t-i\vec{k}\cdot\vec{r}_1}$$

$$= -2\varepsilon^{-1}\Gamma_o/(1 - \frac{i\omega}{Dk^2}) \qquad (12)$$

which allows us to define a diffusion constant

$$D^{-1} = \lim_{k^2\to 0}\ \lim_{\frac{i\omega}{k^2}\to 0}\ \lim_{\Gamma_o D/\omega\to 0} k^2 \frac{\partial}{\partial(i\omega)} \ln\frac{\delta\Gamma_1}{\delta V}(k,\omega) \qquad (13)$$

The total density of vortices is determined by

$$\frac{\partial}{\partial t}\Gamma^+(\vec{r}_1,t) = D(\nabla_1^2 + \vec{\nabla}_1\cdot(\vec{\nabla}_1 V))\Gamma^+(\vec{r}_1;t)$$

$$+D\vec{\nabla}_1\int d\vec{r}_2(\vec{\nabla}_1 U(\vec{r}_{12}))\ \{\Gamma^{+-}(\vec{r}_1\vec{r}_2;t) - \tfrac{1}{2}\Gamma^{++}(\vec{r}_1,\vec{r}_2;t)\} \qquad (14)$$

We want to consider small deviations from equilibrium only and replace $\Gamma^{+}(\vec{r}_1;t)$ in the second term of eqn. 14 by its equilibrium value Γ^{+}_{eq}. The two vortex distribution function $\Gamma^{+-}(\vec{r}_1,\vec{r}_2)$ can be related to the 3 vortex distribution functions

$$\Gamma^{+-}(\vec{r}_1,\vec{r}_2) = \Gamma_2^{+-}(\vec{r}_1,\vec{r}_2) + \tfrac{1}{2}\int d\vec{r}_3 \Big[\Gamma^{+--}(\vec{r}_1,\vec{r}_2,\vec{r}_3) - \Gamma^{-}(\vec{r}_3)\Gamma_2^{+-}(\vec{r}_1,\vec{r}_2)\Big] + \tfrac{1}{2}\int d\vec{r}_3 \Big[\Gamma^{+-+}(\vec{r}_1,\vec{r}_2,\vec{r}_3) - \Gamma^{+}(\vec{r}_3)\,\Gamma_2^{+-}(\vec{r}_1,\vec{r}_2)\Big] \qquad (15)$$

and a similar equation for $\Gamma^{++}(\vec{r}_1,\vec{r}_2)$. Here $\Gamma_2^{+-}(\vec{r}_1,\vec{r}_2)$ is the density of pairs, neglecting all other vortices. The second and third term give the change in the 2 vortex distribution due to the presence of a third vortex. A factor 1/2 has been included to account for the fact that two vortices of the same sign are involved. Since we want to calculate screening effects at lowest order in $\Gamma_0 y^2$ only, we close the hierarchy at the level of Γ_3 and neglect contributions of order Γ^2 or y^4. If eqn.15 is substituted into eqn.14 the resulting equation for Γ^{+} is rather complicated, since memory effects are important for finite frequencies. However, in the hydrodynamic limit we can expand all distribution functions as a power series in ω. The term independent of ω yields the static density response. In the long wavelength limit all modifications due to screening vortex pairs can be expressed in terms of an effective dielectric constant

$$\varepsilon_{eff}^{-1} = \varepsilon^{-1}\left\{1 - 2J\pi^2 y^2 \int_1^{\infty} d\left(\frac{r}{a}\right)\left(\frac{r}{a}\right)^{3-J}\right\} \qquad (16)$$

This is just the result of the equilibrium theory [2]. To go further we keep terms linear in ω and calculate the dynamic density response function to lowest order in ω. Detailed calculations give a result of the same analytic form as eqn. (12) but with D replaced by an effective diffusion constant

$$D_{eff}^{-1} = D^{-1}\left\{1 + Jy^2\pi^2 \int_1^{\infty} d\left(\frac{r}{a}\right)\left(\frac{r}{a}\right)^{3-J}\right\} \qquad (17)$$

where use has been made of eqn. 13.

The main contributions to ε_{eff} and D_{eff} come from configurations in which one vortex antivortex pair is tightly bound and the other vortex is comparitively far away. If the convective terms are included in the Fokker-Planck equation (eqn. 14) these results remain unchanged. To obtain the renormalization of C, one has to calculate the current response. We find that vortex pairs do not affect C, as expected.

SINGULARITIES NEAR THE TRANSITION

As J approaches 4, perturbation theory breaks down due to infrared divergences. Following refs. 8,9 we break up the integral into two parts

$$\int_1^{\infty} d\left(\frac{r}{a}\right) \rightarrow \int_1^{e^{\ell}} d\left(\frac{r}{a}\right) + \int_{e^{\ell}}^{\infty} d\left(\frac{r}{a}\right) \qquad (18)$$

The small r parts of the integral can be absorbed into a redefinition of ε and D. Rescaling the remaining integrations so that they again reach from 1 to ∞, we find equations of the same form as eqns. 16,17 but with modified coefficients ε, D and y. In the limit $\ell \to 0$ we can construct differential recursion relations

$$\frac{dJ^{-1}}{d\ell} = 2\pi^2 y^2 \qquad (19a)$$

$$\frac{dy}{d\ell} = \left(2 - \frac{J}{2}\right) y \qquad (19b)$$

$$\frac{d\ell nD}{d\ell} = -4\pi^2 y^2 \qquad (19c)$$

The first two equations are identical to those obtained by Nelson and Kosterlitz [2]. Using their results we can solve (19c) to find the renormalized diffusion constant $D_r = \lim_{\ell \to \infty} D(\ell)$ in the vicinity of T_c. At $T = T_c$, D_r has a finite value and shows a cusp like singularity as $T \to T_c$,

$$\frac{D_r(T)}{D_r(T_c)} = (1 + b\, t^{1/2}), \quad t = \left|\frac{T-T_c}{T_c}\right| \qquad (20)$$

where b is a nonuniversal constant. Furthermore there is a universal relation between $D_r(T)$ and $J_r(T)$

$$\lim_{T \to T_c} \left(\frac{D_r(T) - D_r(T_c)}{J_r(T) - J_r(T_c)}\right) = \frac{1}{8} D_r(T_c) \qquad (21)$$

We find that the constant C remains unchanged by the presence of vortex-antivortex pairs, i.e. it will not show a cusp at the phase transition. Further details of this work will be published elswhere[10].

We would like to acknowledge useful conversations with B. I. Halperin. This work has been supported in part by the National Science Foundation under grant DMR 77-10210. One of us (AZ) has been supported by a grant from the Deutsche Forschungsgemeinschaft.

REFERENCES

1. J.M.Kosterlitz and D.J.Thouless, J.Phys. C6, 1181 (1973); J.M.Kosterlitz J.Phys. C7, 1046 (1974).
2. D.R.Nelson and J.M.Kosterlitz, Phys. Rev. Lett. 39, 1201 (1977).
3. V.Ambegaokar, B.I.Halperin, D.R.Nelson, and E.D.Siggia, Phys. Rev. Lett. 40, 783 (1978) and Phys. Rev. B21 (1980 in press).
4. B.A.Huberman, R.J.Myerson and S.Doniach, Phys. Rev. Lett. 40, 780 (1978).
5. B.J.Bishop and J.D.Reppy, Phys. Rev. Lett. 40, 1727 (1978).
6. See W.F.Vinen, Prog. Low Temp. Phys. 3, 1 (1961) and references therein.
7. V.Ambegaokar and S.Teitel, Phys. Rev. B19, 1667 (1979).
8. D.R.Nelson, Phys. Rev. B18, 2318 (1978); D.R.Nelson and B.I.Halperin, Phys. Rev. B19, 2457 (1979).
9. J.José, L.P.Kadanoff, S.Kirkpatrick, and D.R. Nelson, Phys. Rev. B16, 1217 (1977).
10. R.G.Petschek and A.Zippelius, to be published.

CRITICAL AND TRICRITICAL BEHAVIOR IN A TWO-DIMENSIONAL ANTIFERROMAGNET

D. P. LANDAU and J. TOMBRELLO

University of Georgia, Athens, Georgia 30602

R. H. SWENDSEN

IBM Zurich Research Laboratory, 8803 Rüschlikon, Switzerland

A Monte Carlo Renormalization Group method has been used to study the critical and tricritical behavior of an Ising antiferromagnet on a square lattice with competing interactions. For small values of uniform magnetic field we find two relevant eigenvalues which are the same as the zero field thermal and magnetic eigenvalues. At the tricritical point $kT_t/J \approx 1.29$ we find four relevant eigenvalues which determine the (non-classical) tricritical exponents as well as the tricritical-critical crossover. Along the first order phase boundary we find attraction to a discontinuity fixed point.

INTRODUCTION

Substantial interest in tricritical behavior of simple magnetic systems has been generated over the past few years. Tricritical behavior has been studied in a number of three-dimensional models [1] and it is now known that the tricritical exponents are classical with possible modification by logarithmic corrections. Progress in the understanding of two-dimensional tricritical behavior has been slow. Early Monte Carlo results [2] showed that tricritical exponents were probably non-classical; however, we now know that finite size rounding is quite important near the transition and more data are needed to fully interpret the Monte Carlo results. The two-dimensional Blume-Capel model was studied [3] using a Kadanoff lower-bound renormalization transformation, and the square lattice Ising antiferromagnet has been examined using a cell-cluster renormalization group approximation [4] and using an ε-expansion method [5]. Although there are some points of agreement between the different treatments, there are also inconsistencies in either tricritical exponents or the location of T_t. We here report the results of a study of a simple Ising antiferromagnet using a Monte Carlo Renormalization Group Technique and compare the results with those obtained from standard Monte Carlo and real-space renormalization group methods.

MODEL AND METHOD

We consider a spin-½ Ising antiferromagnet on a square lattice with Hamiltonian

$$\mathcal{H} = J\sum_{nn}\sigma_i\sigma_j - \frac{J}{2}\sum_{nnn}\sigma_i\sigma_k + H\sum_i\sigma_i \qquad (1)$$

where $J>0$ and the first and second sums are over nearest-neighbor (nn-) and next-nearest-neighbor (nnn-) pairs respectively. This model was studied using the Monte Carlo Renormalization Group (MCRG) method developed by Swendsen [6]. We consider a simple real space RG transformation which divides the lattice into "blocks" of spins and assigns to each block a new spin whose value is determined by the spin value of the majority of the spins in the block. We have used the five site block (scale factor $b=\sqrt{5}$) first proposed by van Leeuwen [7]; a block is made up of a spin and its four nnn-spins. Hence, all spins within a block are on the same antiferromagnet sublattice. A standard Monte Carlo method is used to generate spin-states and the block-spin transformation is then used to renormalize the lattice n-times. Various spin-correlation functions are calculated for original as well as the n-transformed lattices and are used to estimate the matrix elements of the linearized transformation matrix $T^*_{\alpha\beta}$. The eigenvalues of $T^*_{\alpha\beta}$ yield exponent estimates following each iteration (application of the RG transformation) and we then examine the variation of the exponent estimates with successive iterations. We separate the $T^*_{\alpha\beta}$ for even and odd operators to improve accuracy. For the present problem the odd operators involve staggered interactions e.g. the sublattice magnetization. This method was applied for a range of field and temperature for 50×50 lattices with periodic boundary conditions and the convergence of the eigenvalue estimates was used to locate the phase boundary.

RESULTS AND DISCUSSION

We found a typical Ising antiferromagnetic-paramatnetic phase boundary which was 2nd order at low fields and which became 1st order at high fields and low temperature. Data were taken at low, constant field values to study the phase diagram in the vicinity of the Neél temperature T_N. The behavior showed the same general characteristics: one relevant eigenvalue y_1^e was obtained from the even interactions and one y_1^o from the odd interactions. At $T_c(H)$ these eigenvalues quickly iterated to the 2-dim Ising values $y_1^e=y_T=1.0$, $y_1^o=y_H=1.875$. Above and below $T_c(H)$ the eigenvalues iterated away from the Ising fixed point. As the field increased the behavior of the eigenvalues began to show changes, and studies were made by fixing the temperature and sweeping the magnetic field across the phase boundary. Typical results at $\sim T_c(H)$ for $kT/J=1.5$, $H/J=3.89$ are shown in Table 1. In the

TABLE I

RG Iteration	Number of Interactions	y_1^e	y_2^e	y_1^o	y_2^o
1	1	1.574		1.896	
	2	1.564	.210	1.896	
	3	1.567	.190	1.893	.952
	4	1.566	.362	1.891	.881
	5	1.566	.349	1.892	.881
	6	1.569	.349		
2	1	1.474		1.876	
	2	1.445	.148	1.876	
	3	1.441	.127	1.881	.835
	4	1.431	.334	1.879	.811
	5	1.435	.333	1.881	.809
	6	1.436	.277		
3	1	1.362		1.841	
	2	1.315	-.110	1.841	
	3	1.300	-.116	1.872	.716
	4	1.293	.119	1.870	.662
	5	1.298	.137	1.874	.655
	6	1.281	-.494		

MCRG ANALYSIS CRITICAL EXPONENTS OF THE 2-DIM ISING ANTIFERROMAGNET ON A 50X50 LATTICE FOR kT/J=1.50, H/J=3.89 WITH 1.5X10^4 MCS AFTER DISCARDING 5X10^3 MCS. THE RENORMALIZATION GROUP TRANSFORMATION WAS APPLIED EVERY 5 MCS.

1st iteration a total of four relevant eigenvalues appeared. The largest odd eigenvalue y_1^o quickly iterated to the Ising value and the smaller eigenvalue y_1^o iterated toward zero. The largest even eigenvalue y_1^e iterated slowly towards the Ising critical value $y_T=1.0$ and the second even eigenvalue y_2^e iterated towards zero. As the temperature was further reduced the eigenvalues began to "stick" at the values obtained on the first iteration. Below $kT/J=1.29$, hysteresis was observed and y_1^o showed a tendency towards the discontinuity fixed point value $y_H=2.0$. From our data we estimate that the tricritical point is located at $kT_t/J\approx1.29$, $H_t/J\approx3.95$. This corresponds to $T_t/T_N=0.34$ which is quite close to the estimate obtained from ordinary Monte Carlo investigation [2]. In contrast, the real space RG estimate [4] is $T_t/T_N=0.22$. The exponent values obtained from the first iteration were virtually identical for fields and temperatures near the tricritical point, but differences began to appear in later iterations. An MCRG analysis at the estimated tricritical point is shown in Table II. The eigenvalues for the first two iterations are

TABLE II

RG Iteration	Number of Interactions	y_1^e	y_2^e	y_1^o	y_2^o
1	1	1.707		1.921	
	2	1.706		1.921	
	3	1.708	.533	1.918	1.059
	4	1.708	.531	1.917	.989
	5	1.708	.621	1.917	.988
	6	1.708	.625		
2	1	1.690		1.915	
	2	1.690		1.915	
	3	1.690	.615	1.911	1.043
	4	1.690	.616	1.910	1.025
	5	1.690	.683	1.911	1.025
	6	1.690	.682		
3	1	1.607		1.886	
	2	1.609		1.886	
	3	1.608	.336	1.898	.993
	4	1.608	.336	1.896	.934
	5	1.608	.386	1.896	.933
	6	1.609	.381		

MCRG ANALYSIS OF "TRICRITICAL" EXPONENTS OF THE 2-DIM ISING ANTIFERROMAGNET ON A 50X50 LATTICE FOR kT/J=1.29, H/J=3.950 WITH 1.5X10^4 AFTER DISCARDING 5X10^3 MCS. THE RENORMALIZATION GROUP TRANSFORMATION WAS APPLIED EVERY 5 MCS.

very similar, and we believe that the slight variations in the third iteration is a consequence of a small error in the location of the tricritical point. Our results therefore suggest the appearance of four relevant tricritical eigenvalues: $y_1^e \approx 1.68$, $y_2^e \approx 0.6$ and $y_1^o \approx 1.91$, $y_2^o \approx 1.0$. The uncertainties in these values are still unclear. The small number of RG-iterations and the statistical errors make it difficult to locate the tricritical point in a two-dimensional space of coupling constants. The lack of convergence comparable to that which has been obtained by this method for other models [6] indicates that the "best"eigenvalue estimates as well as the location of the tricritical point are not precise. It is, hence, difficult to estimate the errors from all sources. The corresponding tricritical exponents are clearly non-classical with our best estimates being $\nu_t=0.60$, $\delta_t=21.2$, $\gamma_t=1.08$ with a critical-tricritical crossover exponent $\phi_t=0.36$. For comparison we note that that Nienhuis and Nauenberg [4] obtained $y_1^e=1.85$, $y_2^e=0.65$ from a real space RG analysis. In addition, a recent variational RG calculation [8] for the q-state potts model predicts that the tricritical exponents for q=2 (the Ising model) are $y_1^e=1.81$, $y_1^o=1.93$. We therefore find very good overall agreement for y_1^o with a disagreement for y_1^e which is non-trivial but which is probably within the uncertainties of these calculations. Since the RG cluster approximation results for y_1^e are slightly greater than ours, the estimate $\phi=0.352$ is almost identical to ours.

Clearly we need more data on larger lattices to allow more RG iterations and thus a more precise determination of the tricritical point and tricritical exponents. Further work is in progress.

Acknowledgements

This research was supported in part by the National Science Foundation through Grant DMR76-11329-A01 and through the Undergraduate Research Participation Program.

REFERENCES

1. See, for example, E. K. Riedel and F. J. Wegner, Phys. Rev. Letters 29, 349 (1972); F. Harbus and H. E. Stanley, Phys. Rev. B8, 1156 (1973); D. P. Landau, Phys. Rev. B14, 4056 (1976); D. M. Saul, M. Wortis and D. Stauffer, Phys. Rev. B9, 4964 (1974).
2. D. P. Landau, Phys. Rev. Letters 28, 449 (1972); B. L. Arora and D. P. Landau, AIP Conf. Proc. 10, 870 (1973).
3. T. W. Burkhardt, Phys. Reb. B14, 1196 (1976); T. W. Burkhardt and H. J. F. Knops, Phys. Rev. B15, 1602 (1977).
4. B. Nienhuis and M. Nauenberg, Phys. Rev. B13, 2021 (1976).
5. A. L. Lewis and F. W. Adams, Phys. Rev. B20, 2080 (1979).
6. See e.g. R. H. Swendsen, Phys. Rev. B20, 2080 (1979).
7. J. M. J. van Leeuwen, Phys. Rev. Lett. 34, 1056 (1975).
8. B. Nienhuis, E. K. Riedel, and M. Schick, J. Phys. A13, 631 (1980) and (to be published).

A MICROSCOPIC APPROACH OF THE ANTIFERROELECTRIC TRANSITIONS IN STOICHIOMETRIC β-ALUMINAS : A THREE STATE TWO DIMENSIONAL POTTS MODEL

J.F. GOUYET

Laboratoire de Physique de la Matière Condensée*
Ecole Polytechnique, 91128 Palaiseau, France

It is shown that the Tc=300K transition in silver β-alumina can be explained as an antiferroelectric transition due to the ordering of the electric dipoles created by the off-centered silver cations and oxygen anions in the conducting layers. It was recently suggested that when T→Tc with $(T-Tc)/Tc>10^{-2}$, the correlation length ξ increased in the D=2 layers with a critical behavior $\xi \sim (T-Tc)^{-\nu}$ which was associated with the D=2, s=3 Potts universality class. We show here the microscopic details of this model that gives an upper value for the theoretical Tc around 440K, and that corresponds to a true D=2 three states Potts system.

1. INTRODUCTION

β-alumina is a fast ion conductor. Its structure [2] is composed of spinel blocks separated by conduction planes in which cations as Ag^+, K^+, Na^+, NH_4^+... can be inserted. A unit cell in a conducting plane is represented in Fig.I.

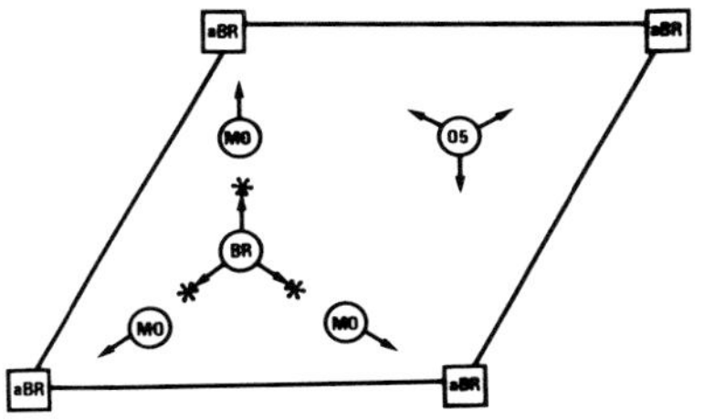

Fig.I. Main available sites for cation in a conducting plane.
(BR)....Beever-Ross site
[aBR] anti-Beever-Ross site
(05)....oxygen bridge between spinel blocks
(MO)....mid-oxygen site
* off-centered-position

Oxygen atoms (05) bridge the different spinel blocks while different sites are available for cations : during the conduction the cations jump from an occupied BR site (Beever-Ross site) to an empty aBR site (anti-Beever-Ross site) then to another empty BR site. In stoichiometric β-alumina, the cations are located in the BR sites. X-ray scattering experiments [4-6] show a 2D ordered structure (Fig.II) ; this structure corresponds to an hexagonal super lattice structure with $a'=a\sqrt{3}$ (a being the high temperature lattice parameter) [3-5]. The low temperature phase is an ordering of Ag^+ and O^{--} displacements interpreted [1,3] as an order-disorder antiferroelectric transition in two dimensions. This interpretation is supported by this experimental result that the Ag^+ and O^{--} displacements correspond to fixed wells induced by the spinel blocks [4-6].

It was suggested recently [1] that this structural ordering observed around 300K in stoichiometric $Ag^+\beta$-alumina was a realization of the D=2, s=3 Potts model which predicts a correlation length with a critical exponent ν near 0.83, as the experiment gives $\nu_{exp}=0.85\pm0.1$. The 2D-3D cross-over appear in such material at a T* very closed to Tc namely (T*-Tc)/Tc of a few 10^{-2}.[6]

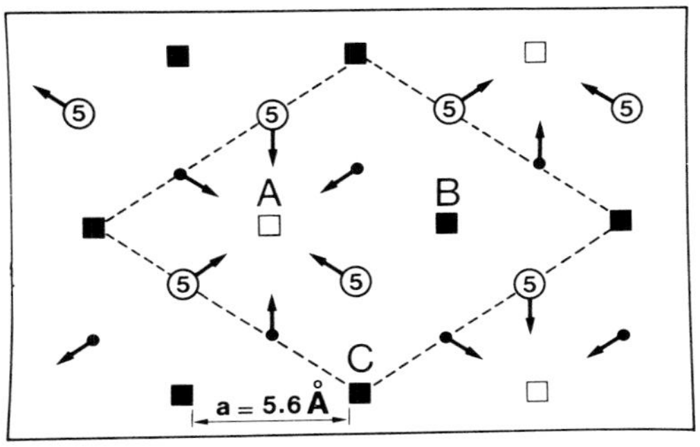

Fig.II. Ordered structure deduced from X-rays experiment [4].
(5) : O^{--} anions ; • Ag^+ cations ;
□ ■ aBR sites.
--- : $\sqrt{3}\times\sqrt{3}$ superstructure unit cell.

2. THE MICROSCOPIC MODEL

The existence of displaced Ag^+ and O^{--} gives rise to interacting electric dipoles that are supposed to be the only relevant interaction [3]. This interaction is written

$$\mathcal{H} = \sum_{\mu\nu} \left\{ \frac{\vec{d}_\mu \cdot \vec{d}_\nu}{r_{\mu\nu}^3} - 3\,\frac{(\vec{d}_\mu \cdot \vec{r}_{\mu\nu})(\vec{d}_\nu \cdot \vec{r}_{\mu\nu})}{r_{\mu\nu}^5} \right\}$$

where the dipoles d_μ can take six different values (3 orientations for O^{--} and 3 orientations for Ag^+). The ground state in a layer is obtained numerically through the following iterative process : we first consider a small cluster (typically an hexagon of alternated Ag^+ and O^{--} around an aBR site as shown on Fig.V for example) and find the lowest energy configurations. We then put two clusters together and determine the lowest energy configurations of this larger cluster from a combination of the lowest states of the initial clusters, and so on iteratively.

The ground state structure is found to depend on the ratio of the electric dipoles associated with the two species Ag^+ and O^{--} revealing a limiting value R=2.583 : the structure represented in Fig.IIIa corresponds to $|d_{Ag^+}/d_{O^{--}}| = D^2 \in [R^{-1},R]$ (Region 1), and has a three fold degeneracy while the structure represented in Fig.IIIb has a $|d_{Ag^+}/d_{O^{--}}| = D^2 \notin [R^{-1},R]$ (Region 2). The ground state of this last structure is infinitely degenerated and Fig.IIIb is only a particular computed realization.

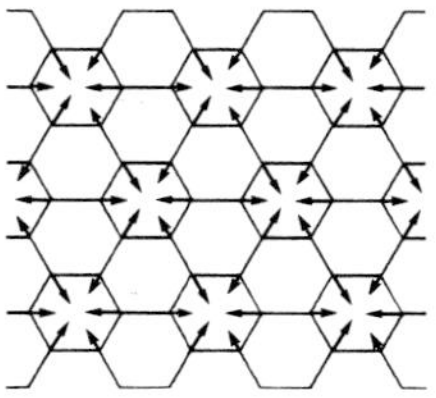
a

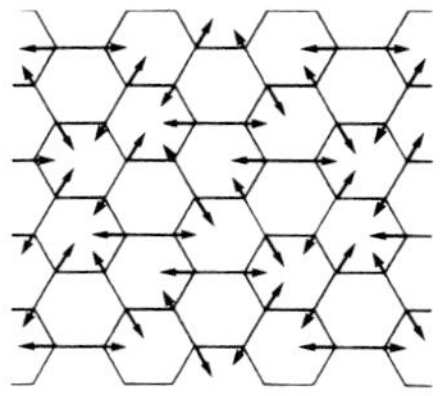
b

Fig.IIIa. Ground state structure of region 1 (three fold degeneracy).

Fig.IIIb. Ground state structure of region 2 (infinite degeneracy when only first neighbours interaction J is taken into account).

From X-ray [4], the observed displacements are 0.8 Å for Ag^+ and 0.3 Å for O^{--} and therefore we can estimate that $D^2 = |d_{Ag^+}/d_{O^{--}}| \simeq 1.33$ (the charges are supposed to be e on Ag^+ and -2e on O^{--}) and we expect from our calculations the structure of Fig.IIIa. This is precisely the structure obtained experimentally [1,4] (Fig.II). The experimental symmetry $P6_3/m$, is smaller than that of Fig.IIIa due to the 3D ordering that is not taken into account in this paper.

We saw that the ground state in region 1 had a three fold degeneracy : if in the hexagonal lattice structure the aBR sites are divided into three classes A, B and C, one state ϕ_1 is obtained by letting the Ag^+ and O^{--} ions move towards the aBR site A (Fig.II) while the two other degenerate states ϕ_2 and ϕ_3 correspond to ions moving towards B and C.

In the following we will show that such a system is isomorphic to a 2D three states Potts models (with a ferro-type interaction if the dipole ratio belongs to region 1). The three states can be associated to the three different displacements of Ag^+ and O^{--} ions as shown on Fig.IV : each set of three hexagons (A B C) is a Potts site i and such a Potts site is precisely equivalent to the $(\sqrt{3}\times\sqrt{3})$ unit cell of Fig.II. To calculate the interaction between two next neighbours Potts sites i and j we will first examine the three ground states degenerate structures (ϕ_1, ϕ_2 and ϕ_3) then the lowest excitations of this system consisting in frontiers between clusters of ϕ_1, ϕ_2 and ϕ_3 types.

3. DETAILED STUDY OF THE GROUND STATE STRUCTURE

Let us first examine the lowest hexagonal dipoles structures : the ground state structure is a pavement of three types of hexagonal tiles. Two tiles are energetically degenerated and correspond to the lowest energy in every case :

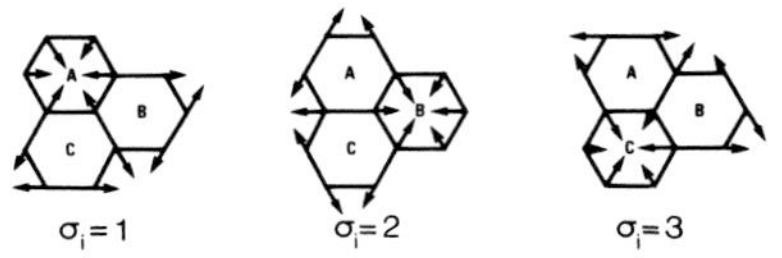

Fig.IV. Three equivalent orderings of the ground state. A set 'i' of A,B,C, aBR sites corresponds to a Potts site and σ_i=1,2,3 distinguishes the three equivalent components. The arrows represent the displacements of Ag^+ (⟶) and O^{--} (→).

They are labelled (+) and (-) recalling the two roundabout of the dipoles (see Fig.V). The third one corresponds to ions moving towards an aBR site and is labelled by (0).

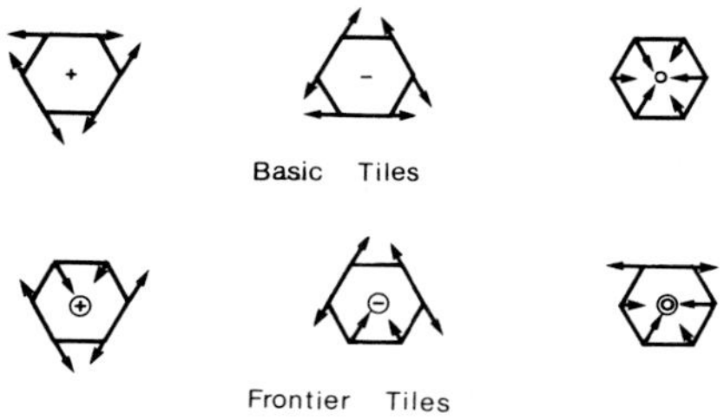

Fig.V. Six kinds of hexagons used to build ground and first excited states. Each basic tile is singly degenerated, while the ⊕,⊖ and ⊚ have a three fold degeneracy. The arrows represent the displacements of Ag^+ (⟶) and O^{--} (→).

Naturally a tile pavement cannot be arbitrarily arranged as each dipole belongs simultaneously to three hexagonal tiles. In fact a state is completely defined by a $(\sqrt{3}\times\sqrt{3})$ pavement. For example the ground state component on Fig.II obtained by letting the Ag^+ and O^{--} ions move towards the aBR is completely defined by a pavement of (o) tiles on A sites, or equivalently by (+) tiles on B sites, or (-) tiles on C sites. The three components labelled ϕ_1, ϕ_2, ϕ_3 are associated with the three different orderings given by the Table I :

		sites		
state	σ_i	A	B	C
ϕ_1	1	0	+	-
ϕ_2	2	-	0	+
ϕ_3	3	+	-	0

Table I

In region 1, the first excited states of this system consist of domains of these three components. So we will now examine in more details the structure of these excited states.

4. THE FIRST EXCITED STATES

It is easier to calculate the ground state energy starting not from the interacting dipoles but from the interacting (0), (+) and (-) hexagons. Then there are, for each component ϕ_1, ϕ_2, ϕ_3, three equivalent manners to calculate their ground state energy depending if the state looked as a $\sqrt{3}\times\sqrt{3}$ pavement of (0), (+) or (-)

tiles. For example for ϕ_1, the energy has the expression :

$$E(\phi_1) = \Sigma_i\, E_o(A_i) + \Sigma_{i<j}\, E_{oo}(A_i,A_j)$$
$$= \Sigma_i\, E_+(B_i) + \Sigma_{i<j}\, E_{++}(B_i,B_j) \qquad (I)$$
$$= \Sigma_i\, E_-(C_i) + \Sigma_{i<j}\, E_{--}(C_i,C_j)$$

where an hexagonal site is completely defined by its coordinates $h=(A_i)\ldots$, that is to say a Potts site i and one of the three "directions" A,B,C on a Potts site. E_o, E_+ or E_- are the energies of (0), (+) or (-) tiles centered on sites A_i, B_i or C_i, while $E_{oo}(A_i,A_j)$ are interaction energies between hexagonal (0) tiles on sites A_i and A_j,$E_{++}(B_i,B_j)$ between (+) hexagonal tiles, etc. For example the energies E_o,E_+,E_- can be expressed in reduced unit (that is to say $a'=a\sqrt{3}=1$ and $(d_{Ag^+})\,|d_{O--}|=1$) as a function of the ratio D^2 :

$$E_{+red.} = E_{-red.} = -\ 203.9 - 43.8(D^2 + D^{-2})$$

$$\text{while} \quad E_{Ored.} = -\ 403.6 + 117.6(D^2 + D^{-2})$$

It is advantageous to consider pavement with (+) or (-) hexagonal tiles for the following reasons: (+) and (-) are the most stable entities and have (as well as (0) tiles) no net dipolar moment. Interaction between tiles varies then roughly as R^{-5} so that the interaction between second neighbouring tiles on the A substructure (or (A) pavement) is $(\sqrt{3})^5 \simeq 16$ times smaller than between first neighbours. This system can be modeled with only first neighbours (A-A, B-B or C-C) interactions.

The list and the shortened notations of the basic (+), (-) and (0) tiles are given on Fig.V. The excited states, corresponding to the formation of frontiers between ϕ_1, ϕ_2 and ϕ_3 domains, generate (⊕), (⊖) and (◎) tiles also shown in Fig.V. Figure VI gives an example of frontiers between ϕ_1, ϕ_2 and ϕ_3 domains.

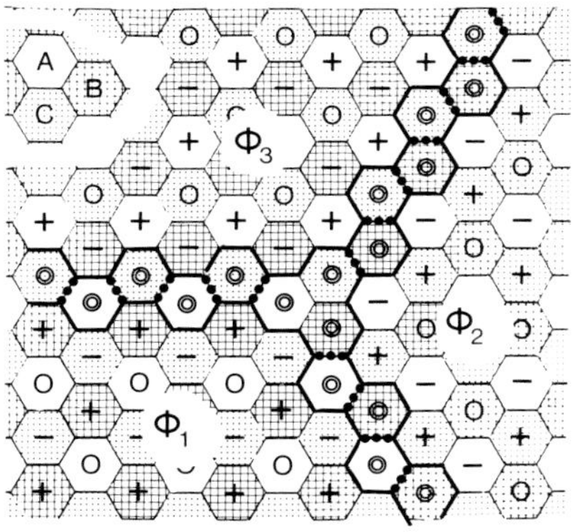

Fig.VI. Representation of the three ground state components and their frontiers using the shortened notation of Fig.V. The aBR site A,B,C are associated with different hexagonal background as shown on the left upper side. The (+)-(-) interactions are represented by sets of three black points.

V. CALCULATION OF THE TOTAL ENERGY OF A DOMAIN STRUCTURE

Let $\mathbf{D}_1$ (resp. $\mathbf{D}_2$, $\mathbf{D}_3$) be the domain occupied by ϕ_1 (resp. ϕ_2, ϕ_3) and let us examine, on the B substructure, the reunion of ϕ_1 and ϕ_3 domains. on $\mathbf{D}_3$, A is occupied by (+), while on $\mathbf{D}_2$, A is occupied by (-). The total energy is a sum of the tiles self-energy,

$$\Sigma_{i\in\mathbf{D}_3}\, E_+(A_i) + \Sigma_{j\in\mathbf{D}_2}\, E_-(A_j)$$

the tile-tile next neighbour interaction,

$$\Sigma_{i,j\in\mathbf{D}_3}\, E_{++}(A_i,A_j) + \Sigma_{i,j\in\mathbf{D}_2}\, E_{--}(A_i,A_j)$$

and the frontier interaction,

$$\Sigma_{i\in\mathbf{D}_3,j\in\mathbf{D}_2}\, E_{+-}(A_i,A_j)$$

Substracting from this sum the ground state total energy, it only remains the frontier energy (i, j are next neighbours Potts sites) :

$$E_{frontier} = \Sigma_{i\in\mathbf{D}_3,j\in\mathbf{D}_2}\, E_{+-}(A_i,A_j) - E_{++}(A_i,A_j)$$

The fundamental parameter in this problem is then the energy difference

$$2J = E_{+-} - E_{++}$$

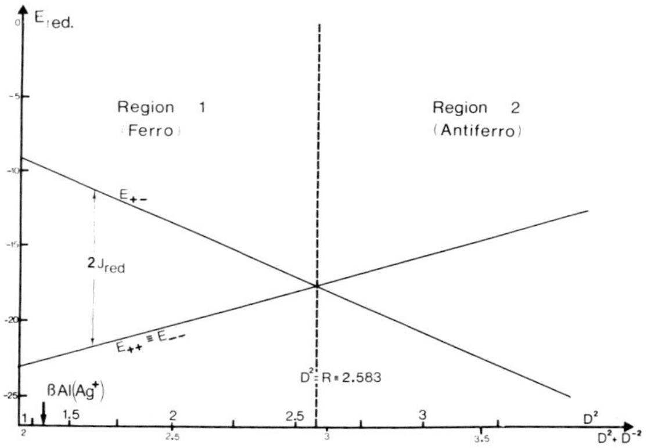

Fig.VII. The energies of the (+)-(+) and (+)-(-) interactions are represented as a function of the dipoles ratio D^2. The value $D^2 = 2.583 \equiv R$ separates the two ferro (region 1) and antiferro (region 2) regions.

Figure VII gives the curves E_{+-} and E_{++} as a function of D^2+D^{-2}. J becomes zero when $D^2=R\equiv 2.583$; in this case there is no need of energy to create a frontier and this value precisely divides the possible structures in two regions with different ground states as shown above. Restricted to B substructure we have an Ising triangular lattice in which the interaction becomes antiferro when D>2.583 (Region 2). The problem is in fact more complex as we have in reality three non-independent Ising systems A, B and C. We will show below that this is a realization of a three states Potts model. All the discussion is limited to the study of region 1.

In region 1, the interaction between (+) or between (-) is attractive so that the disorder is associated with the formation of compact clusters delimited by frontiers. These frontiers correspond to different types of hexagonal tiles, depending on the orientation of the frontiers with respect to the different phases. Figure VIIIa represents a cluster drawn in such a manner that the different orientations are present : frontiers are paved with (⊕), (⊖) or (◎) tiles. The (+)-(-) interaction, equal to 2J, is represented by sets of three black points. The equivalent Potts model is represented by figure VIIIb. Each set of three aBR sites (A,B,C) represent a Potts site 'i'. The interaction between two Potts pseudospin is then (σ_i=1,2,3) :

$$\text{zero if } \sigma_i = \sigma_j$$

$$2J \text{ if } \sigma_i \neq \sigma_j$$

i, j being next neighbours.

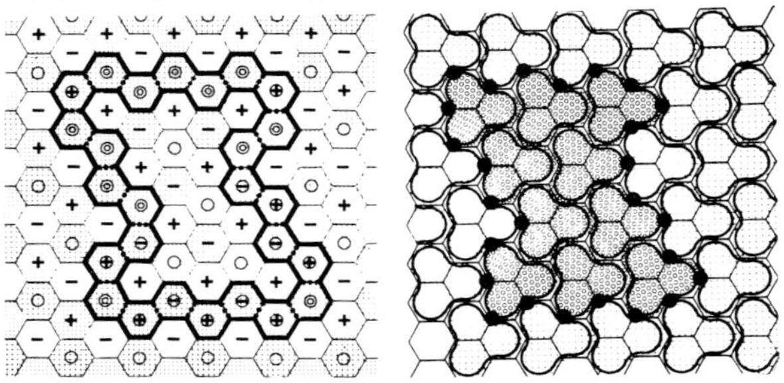

a b

Fig.VIIIa. A cluster of ϕ_3 imbedded into ϕ_2 is represented with all the possible hexagonal excited tiles on the frontiers (see Fig.V). The (+)-(-) interactions are represented by sets of three black points.

Fig.VIIIb. The same cluster as in Fig.VIIIa is now represented using (ABC) Potts sites (the cluster has a σ_i=3 and the outside σ_j=2). The first neighbours interactions are represented by heavy black points.

It can be easily verified that the number of next neighbour interaction is equal to the number of (+)-(-) bonds. The Hamiltonian of our system is therefore a Potts Hamiltonian (σ_i= 1,2,3)

$$H = J\Sigma_{<ij>} (1 - \delta_{\sigma_i\sigma_j}) - \Sigma_{i\alpha} \zeta_\alpha \delta_{\sigma_i\alpha}$$

in which all the ζ_α are zero. This is to our knowledge the first realization of a microscopic Potts model in two dimensions (helium or graphite having a lattice gas Potts structure). It is a triangular ferro Potts system. It was shown [8-9] using a star-triangle transformation that :

$$kT_c = 1.5855 \text{ J}$$

J can be expressed as function of a $J_{red.}$ in reduced unit (a'=1 and $(d_{Ag^+})|d_{O^{--}}|$=1) by

$$J = \frac{(d_{Ag^+})|d_{O^{--}}|}{(a\sqrt{3})^3} J_{red.}$$

where (see Fig.VII) $J_{red.}=7.15[(R-D^2)+(R^{-1}-D^{-2})]$. As, a=5.6 Å, d_{Ag^+}=0.8, $d_{O^{--}}$=-2×0.3, R=2.583 and D^2=1.33, this gives

$$2J = 558.8 \text{ Kelvins}$$

and T_c can be estimated to

$$T_c(\text{theor}) \simeq 443 \text{ K}.$$

This value slightly overestimates the experimental value due to the fact that the effective charge on the O^{--} is certainly less than -2e, that the barycenter of the charges is not necessarily on Ag^+ and O^{--} nuclei and that moreover, the dielectric effect certainly gives a smaller value for the Coulomb interaction [11]. From the general expression for T_c we also see that T_c is the product of a reduced factor depending on the dipole ratio and an interaction proportional to the dipole-dipole interaction. The critical temperature decreases to zero when the dipole ratio increases or when the product $d_{Ag^+}{\times}|d_{O^{--}}|$ becomes smaller. For K^+, $T\ell^+$, H_3O^+ or NH_4^+ cations, either this product is very small, or the ratio D^2 belongs to region 2, accounting for the fact that the corresponding antiferroelectric phase has no transition at finite temperature [10]. Finally the case of Na^+ which presents a transition at 150K will be a good test for this model. Experiments are in progress to determine the displacements of Na^+ and O^{--}.

VI. CONNECTIONS BETWEEN THE TWO DISORDER TRANSITIONS

An important question is to know if the dipoles order-disorder transition can be associated with an increase in the conductivity (i.e. a change in the mean activation energy) or said differently if the BR-aBR order disorder transition corresponds to the same temperature T_c.

Experimentally no other transition was found at temperature from 300K up to 550K.

It is then interesting to examine the effect of domains on the transport phenomena.

Figure IX gives the two possible structures for the lowest excited frontiers. In these two types, dipole moments exist along the dislocations associated with a collective movement of each type of ions Ag^+ and O^{--} along these frontiers.

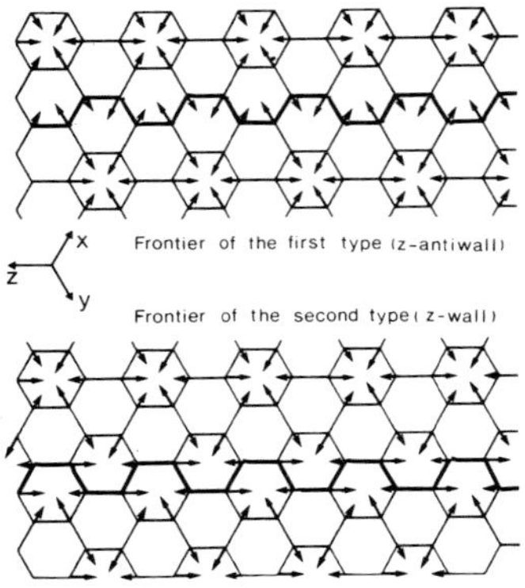

Fig.IX. The two possible kinds of frontiers are drawn with a heavy line. They can have three possible orientations (x, y or z).

When a site BRvac is unoccupied on a frontier, these collective movements naturally favor the jumps BRocc→BRvac, as the barrier is lowered along the BRocc→aBR→BRvac path ; it would be interesting to perform similar calculations to that of Wang, Gaffari and Choi [6] to determine the activation energies on such frontiers.

Acknowledgements.

I want to thank B. Sapoval and H. Arribart for fruitful discussions and criticism of the manuscript.

APPENDIX
ROUGH CALCULATION OF THE ENERGIES

Setting, $d_{Ag^+}=d_O D$ and $d_{O^{--}}=-d_O/D$ we give on Fig.Xa the graph of the energy of the lowest hexagonal tiles as a function of the dipole ratio D^2 or more conveniently of D^2+D^{-2} such that if a tile is symmetric by transposition of d_{Ag^+} and $d_{O^{--}}$, its energy is a linear function of D^2+D^{-2}. In practice to have a more correct idea of the importance of each tile without tedious calculations it is interesting to calculate a mean energy of a tile in a pavement, that is to say to add the internal energy plus half the perimeter energy of an hexagon as this perimeter energy is counted two times in a given (A,B,C) pavement (Fig.Xb).

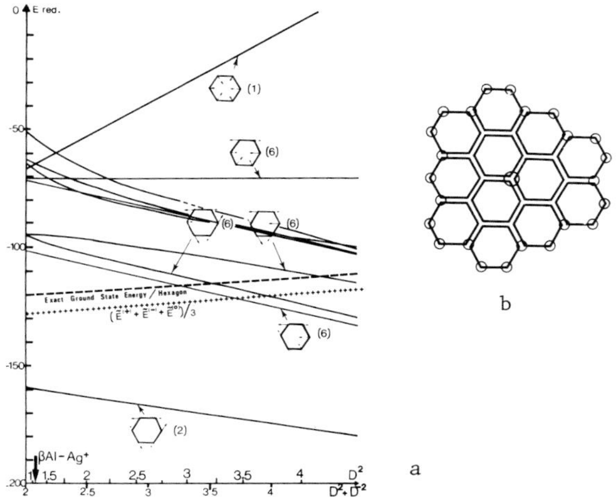

Fig.Xa. Variation of the mean energy of the ground and first excited hexagons as a function of the ratio $D^2=|d_{Ag^+}/d_{O^{--}}|$. Their degeneracy is given in parenthesis.

Fig.Xb. A structure energy can be approximately calculated as a sum of mean energy hexagons (1/2 perimeter + internal energy). By this method only the o• • •◯ interactions are neglected.

The mean total energy, given by the sum of each tile mean energy, only neglects the second neighbours shell of dipole-dipole interaction. However as dipolar interaction is in R^{-3} this gives only qualitative results. We have also represented in Fig.Xa the mean energy per hexagon of the ground state $\tilde{E}=(\tilde{E}_O + \tilde{E}_+ + \tilde{E}_-)/3$ which is found lower by about 10% than the correct ground state energy taken from formula I.

REFERENCES

* Groupe de Recherche 05.0038 du Centre National de la Recherche Scientifique

1. J.F. Gouyet, B. Sapoval and P. Pfeuty, Antiferroelectric Transition in β-Alumina. A realization of the D=2, s=3 Potts model? Journ. Phys. Lett. 41, L115 (1980).
2. For a review of β-Alumina see : Proceedings of the International Conference on Superionic Conductors, Schenectady, edited by W.L. Roth and G.D. Mahan (Plenum, New York, 1976) ; Solid Electrolytes, Topics in Applied Physics Vol. 21, edited by S. Geller (Springer, Berlin 1977) or Solid Electrolytes, edited by P. Hagenmuller and W. Van Gool (Academic Press, New York, 1978).
3. H. Arribart, J.F. Gouyet and B. Sapoval, Fast Ions Transports in Solids, Vashista, Mundy, Shenoy, eds (Elsevier North-Holland, Inc. 1979) p. 569.
4. J.P. Boilot, Ph. Colomban, R. Collongues, G. Collin, R. Comes, Phys. Rev. Lett. 42, 785 (1979).
5. J.P. Boilot, H. Théry, R. Collongues, R. Comes, A. Guinier, Acta Cryst. A32, 250 (1976).
6. J.P. Boilot, Ph. Colomban, G. Collin, R. Comes, X-Ray Diffuse Scattering Study of β-Alumina Sublattice Phase Transition in Stoichiometric β-Alumina, Fast Ion Transport in Solids, Lake Geneva, Wisconsin, USA (1979) p. 243.
7. J.C. Wang, M. Gaffari and S. Choi, J. Chem. Phys. 63, 772 (1975).
8. L. Mittag and M.J. Stephen, J. Phys. A7, L109 (1974) and Phys. Lett. 41A, 357 (1972).
9. D. Kim and R.I. Joseph, J. Phys. C7, L167 (1974).
10. M. Schick, J.S. Walker and M. Wortis, Journ. Phys. 10, C4, 121 (1977). Our present case corresponds to H=0.
11. The 2D-3D cross-over also gives a smaller apparent Tc.

EXAMPLES OF SPIN AND EXTERNAL FIELD DEPENDENT CRITICAL EXPONENTS

VIKTOR URUMOV

Fakultet za Fizika, Univerzitet "Kiril i Metodij",
p.fah 162, 91001 Skopje, Yugoslavia
and
Department of Physics and Department of Mathematics,
University of Florida, Gainesville, Florida 32611, USA

Using the solution of the eight-vertex model which shows a continuous dependence of the critical exponents on the coupling parameters, it is shown that the critical exponents may depend on the spin magnitude S as well as on the magnitude of the external field. The lattice considered is the Union Jack lattice with three-spin interactions. If the central spins are of magnitude S, the specific heat critical exponent α is monotonically increasing or decreasing or constant function depending on the type of anisotropy of the interaction constants. Similarly when staggered magnetic field acting on the central spins is present, $\alpha(H)$ is monotonically decreasing function.

1. INTRODUCTION

One of the central ideas of the modern theory of critical phenomena is that of the so called universality. According to universality the critical exponents are functions of the dimensionality of the space and the order parameter and also depend on the range of the interactions, but are independent of the spin magnitude, the external field magnitude and the details of the Hamiltonian.

There are several exact evaluations of the partition function, mainly for two-dimensional lattices [1]. The well known Baxter solution of the eight-vertex model [2] was put into a spin language form by Wu [3] and Kadanoff and Wegner [4]. The model consists of two interpenetrating square lattices with nearest-neighbor interactions coupled with a four-spin interaction. An interesting feature of the model is that it exhibits a continuous dependence of the critical exponents on the four-spin interaction strength. The model was further generalized by Baxter [5] to include almost all the previously known solutions.

Not so many exact solutions exist for models involving presence of an external field or spins of higher magnitudes. Using different transformation techniques [1] Fisher produced solutions [6] for lattices with alternating arrangements of spins of magnitude S and 1/2 and for a special model of a superexchange antiferromagnet [7] in which an external field acts only on part of the lattice spins.

In the following we generalize the model of Hintermann and Merlini [8] to include general S spins and an external field. The model is mapped onto the Baxter solution [2,3,4] and consequentely gives rise to spin and external field dependent critical exponents.

2. SPIN-DEPENDENT CRITICAL EXPONENTS

Here we consider a model with pure three-spin interactions defined on the Union Jack lattice [8]. The lattice is shown in Fig. 1 and it decomposes in two sublattices of spins of magnitude 1/2 and S, designated by σ and s respectively.

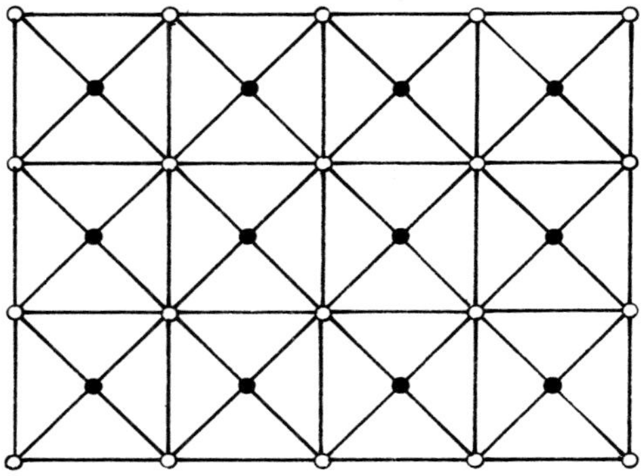

Fig. 1. The Union Jack lattice with 1/2-spins (open circles) and S-spins (black circles).

We consider the general case of anisotropic three-spin interactions involving two 1/2-spins and one S-spin (Fig. 2) defined by the Hamiltonian

$$\mathcal{H} = -\sum_{i,j} (J_1\sigma_{i,j}\sigma_{i+1,j} + J_2\sigma_{i+1,j}\sigma_{i+1,j+1} + J_3\sigma_{i,j+1}\sigma_{i+1,j+1} + J_4\sigma_{i,j}\sigma_{i,j+1})s_{i,j} \quad . \quad (1)$$

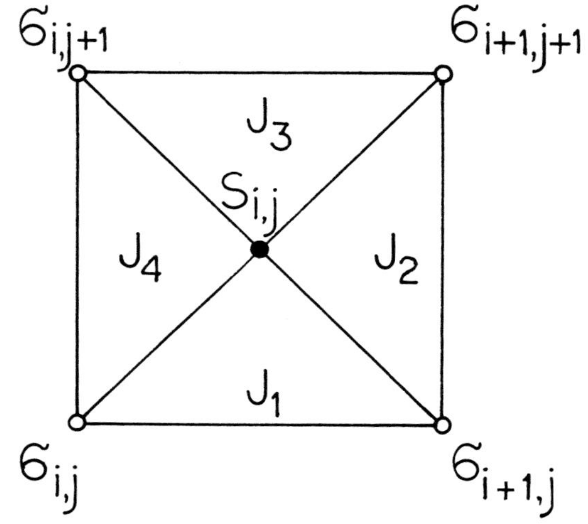

Fig. 2. Unit cell of the Union Jack lattice with anisotropic three-spin couplings.

Performing the partial trace over the S-spins we end up with an effective interaction (K^+ and K^-) between the spins on the diagonals of the unit cell and an effective four-spin interaction (K) among the four spins in the corners of each unit cell. Thus the partition function reduces to the Baxter model partition function

$$Z = A^N \sum_{\{\sigma_i\}} \exp(K^+ \sum_{i,j} \sigma_{i,j}\sigma_{i+1,j+1} + K^- \sum_{i,j} \sigma_{i+1,j}\sigma_{i,j+1} + K \sum_{i,j} \sigma_{i,j}\sigma_{i+1,j}\sigma_{i,j+1}\sigma_{i+1,j+1}) \quad (2)$$

where

$$A = (f_1f_2f_3f_4)^{1/4}, \quad K^+ = \ln(f_1f_2/f_3f_4), \quad K^- = \ln(f_1f_3/f_2f_4), \quad K = 4\ln(f_1f_4/f_2f_3) \quad (3)$$

and

$$f_{1,2} = \frac{\sinh[(2S+1)(K_1 + K_2 \pm K_3 \pm K_4)/8]}{\sinh[(K_1 + K_2 \pm K_3 \pm K_4)/8]}, \quad f_{3,4} = \frac{\sinh[(2S+1)(K_1 - K_2 \mp K_3 \pm K_4)/8]}{\sinh[(K_1 - K_2 \mp K_3 \pm K_4)/8]} \quad (4)$$

and $K_i = J_i/kT$.

The critical temperature equation becomes [3,4]

$$e^{K/8} = |\cosh[(K^+ - K^-)/4]/\sinh[(K^+ + K^-)/4]| \quad (5)$$

and the specific heat critical exponents are determined from

$$\alpha = \alpha' = 2 - \pi/\mu \quad (6)$$

and

$$\mu = \pi/2 + \sin^{-1}[\tanh(K/8)]. \quad (7)$$

We have solved the equation (4) in several possible cases of anisotropy: a) $J_1 = J_2 = J_3 = J_4$; b) $J_1 = J_3 = J$, $J_2 = J_4 = \lambda J$; c) $J_1 = J_2 = J$, $J_3 = J_4 = \lambda J$ and d) $J_1 = J$, $J_2 = J_3 = J_4 = \lambda J$ where λ is a parameter. The results for the critical temperature and the critical exponent α as functions of the spin magnitude S are shown in Table I for the particular case $\lambda = 2$. As one would expect the critical temperature rises as

Table I

THE INVERSE CRITICAL TEMPERATURE $K_c=J/kT_c$ AND THE CRITICAL EXPONENT FOR DIFFERENT S AND $\lambda=2$

S	K_c a)	K_c b,c)	α b)	α c)	K_c d)
1/2	3.525	2.438	0.56803	0.46350	2.054
1	2.063	1.419	0.56295	0.46641	1.199
3/2	1.488	1.022	0.56154	0.46721	0.864
2	1.170	0.803	0.56095	0.46755	0.679
5/2	0.966	0.663	0.56063	0.46773	0.560
3	0.823	0.565	0.56045	0.46783	0.478
7/2	0.718	0.492	0.56033	0.46790	0.416
4	0.636	0.437	0.56025	0.46794	0.369
9/2	0.572	0.392	0.56019	0.46797	0.332

a) $J_1=J_2=J_3=J_4=J$, $\alpha=1/2$.
b) $J_1=J_3=J$, $J_2=J_4=\lambda J$.
c) $J_1=J_2=J$, $J_3=J_4=\lambda J$, the critical temperature is same as for case b.
d) $J_1=J$, $J_2=J_3=J_4=\lambda J$, $\alpha=1/2$.

S is increased. However the behavior of α as a function of S is not easy to predict and as we can see from Table I it depends on the type of anisotropy present in the model. In the case b, α is a monotonically decreasing, while in the case c, it is monotonically increasing function of S. In the other cases α remains constant. The dependence of α on S is very weak and becomes stronger for larger anisotropy. However even for $\lambda=5$, the difference between the extreme values of α barely exceeds 4%. If such a dependence on S exists in real materials, described with different Hamiltonians, it may well be unnoticeable experimentally.

Recently Krinsky and Mukamel [9] argued that the behavior of the Union Jack model with triplet interactions can be attributed to the dimensionality of the order parameter (n=3) which differs from the case solved by Onsager (n=1). Obviously the change of the magnitude of the central spins does not affect the dimensionality of the order parameter. Another model due to Jüngling [10] which is also mapped onto the Baxter model can be similarly generalized and one could expect a similar dependence on S. Approximate treatments of different models have not produced dependence on S of the critical exponents [11].

3. CRITICAL EXPONENTS DEPENDENT ON THE EXTERNAL FIELD

If an external field is added to the Hamiltonian (1) it is not possible to perform the mapping onto the Baxter model. This is not possible even in the case when the external field acts only on the s-spins (for simplicity here we assume that their magnitude is also 1/2, although the subsequent derivation can be appropriately modified to be valid for general S). Taking the partial trace over the s-spins as before, we obtain besides the effective interactions which appear in (2), effective interactions between the nearest-neighbor σ-spins. However to the effective nearest-neighbor interaction of a pair of σ-spins contribute both adjacent unit cells after the partial trace over their respective s-spins is taken. The nearest-neighbor interaction can vanish if the interaction constants J_i in adjacent cells are of opposite sign. The most general case when this can be achieved is when the interaction constants are related as follows

$$J_1 = J_3 = J, \; J_2 = J_4 = \lambda J \tag{8}$$

where λ is arbitrary parameter characterizing the anisotropy. Obviously the alternating sign of the interactions will cause antiferromagneting ordering of the s-spins below the critical temperature. Also, a sufficiently strong field will destroy the antiferromagnetic order.

Thus we assume that an external field H is coupled to the s-spins only, or in other words that the σ-spins have zero magnetic moment. The contributions to the Hamiltonian (1) is $-\mu H \sum_i s_i$. The effective interactions of the equivalent Baxter model are

$$K^+ = K^- = [\ln(c_1 c_3 / c_2 c_4)]/8,$$

$$K = [\ln(c_1 c_2 c_3 c_4 / c_0^4)]/8 \tag{9}$$

where

$$c_{1,2} = 2 \cosh \; (2K_1 \pm 2K_2 + L), \tag{10}$$

$$c_{3,4} = 2 \cosh \; (-2K_1 \mp 2K_2 + L), \quad c_0 = 2 \cosh L$$

and $K_{1,2} = J_{1,2}/kT$, $L = \mu H/kT$. We should note that the effective interaction constants (9) remain unchanged by the transformation $J_i \to -J_i$ (or $H \to -H$). They are equal for all the unit cells of the lattice.

Using the criticality condition for the Baxter model we find the equation

$$\sqrt{\cosh 2qK_c + \cosh 4(1+\lambda)K_c} - \sqrt{\cosh 2qK_c + \cosh 4(1-\lambda)K_c} = 2\sqrt{2} \cosh qK_c \tag{11}$$

which determines the critical temperature $K_c = J/kT_c$ as a function of the magnetic field $q = \mu H/J$. The equation was solved numerically and the transition curves are shown on Fig. 3

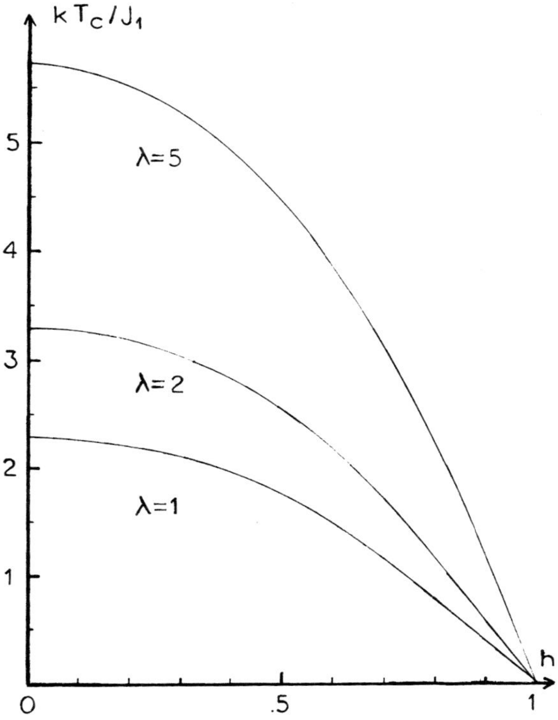

Fig. 3. The reduced critical temperature kT_c/J as a function of the reduced magnetic field $\mu H/2(J_1+J_2)$ for different anisotropies $\lambda = J_2/J_1$.

for several values of the anisotropy parameter λ. The critical temperature vanishes when the external field is sufficiently strong $\mu H = 2J_1 + 2J_2$ to overcome the ordering tendencies of the alternating triplet interactions. It is possible to solve (10) in the limiting cases of weak external field or low critical temperature. In the former case the transition temperature falls from its maximum value following a quadratic dependence on the field, while in the latter case the transition field falls away from its maximum linearly with temperature, similarly to the results obtained by Fisher [7]. Knowing the critical temperature, the critical exponent of the specific heat α can be obtained from (6). Fig. 4 shows the monotonic dependence of α on the field. Using the asymptotic solutions of (10) we find that at low fields α decreases with the square of the magnetic field, while when H approaches its maximum, α exponentially approaches its limiting value of 1/2. The increase of the anisotropy λ generally makes α larger.

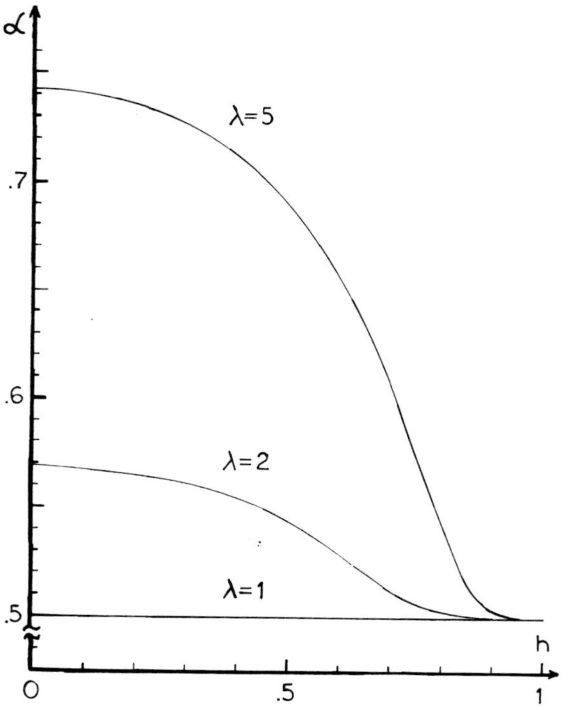

Fig. 4. The specific heat critical exponent α as a function of the reduced external field $\mu H/2(J_1+J_2)$ for different anisotropies $\lambda = J_2/J_1$.

4. CONCLUSION

In this paper we have demonstrated explicitly by exact solution of two-dimensional models with triplet interactions that the critical exponents may depend on the magnitudes of the spin and the external field. The dependence of the critical exponent α on the spin is weak, whereas it depends more strongly on the field, the variation being most pronounced at higher fields, about 1/2 of the critical field which destroys the phase transition. The idea of the existence of such dependences was present in the literature for quite a while [12,13,14] but it was abandoned, largely due to the appeal of the universality hypotehsis and to the inconclusiveness of the estimates of the critical exponents obtained from short series expansions.

REFERENCES

1. I. Syozi in C. Domb and M. S. Green, Phase Transitions and Critical Phenomena, Vol. I, Academic Press, London, 1972.
2. R. J. Baxter, Phys. Rev. Letters 26, 832 (1971).
3. F. Y. Wu, Phys. Rev. B4, 2312 (1971).
4. L. P. Kadanoff and F. J. Wegner, Phys. Rev. B4, 3989 (1971).
5. R. J. Baxter, Ann. Israel Phys. Soc. 2, 37 (1977).
6. M. E. Fisher, Phys. Rev. 113, 969 (1959).
7. M. E. Fisher, Proc. Roy. Soc. A254, 66 (1960).
8. A. Hintermann and D. Merlini, Phys. Lett. 41A, 208 (1972).
9. S. Krinsky and D. Mukamel, Phys. Rev. B16, 2313 (1977).
10. K. Jüngling, Z. Physik B24, 391 (1976).
11. See for example L. Sneddon, J. Phys. C11, 2823 (1978).
12. H. E. Stanley and T. A. Kaplan, J. Appl. Phys. 38, 977 (1967); H. E. Stanley, ibid. 40, 1546 (1969).
13. R. J. Elliott and C. Wood, J. Phys. C4, 2359 (1971).
14. P. Prelovšek and I. Sega, J. Phys. C11, 2103 (1978).

Sinha, ed. Ordering in Two Dimensions

QUANTUM FRUSTRATION ON TRIANGULAR LATTICE : RENORMALIZATION GROUP APPROACH

K. A. PENSON*, R. JULLIEN, S. DONIACH** AND P. PFEUTY

Université Paris-Sud, Laboratoire de Physique des Solides, Bât. 510, Centre d'Orsay, 91405 Orsay, France

The ground state properties of several antiferromagnetic (AF) quantum spin models with external fields on a triangular lattice are investigated using a real space renormalization group method. The quantum analog of frustration in classical systems is studied on Ising and generalized (including Dzialoshinsky-Moriya type) $S = 1/2$ XY models. The evidence is presented that, compared to F case, the quantum frustration is reflected by a strong reduction of critical fields, important increase of ground state energy and the appearance of special scaling properties of correlations. A generalized $S = 1/2$ Heisenberg model (with AF Ising and F XY parts) with a field in the Z-direction is considered. This last model incorporates quantum corrections to classical models of monolayer of He on graphoil. Some experimental implications of frustration effects are discussed.

The study of physical model systems displaying the phenomenon currently referred to as frustration started from a pioneering work of Wannier [1]. He had shown the impossibility of antiferromagnetic ordering on a triangular lattice. By changing the sign of Ising interaction, F → AF, the ordered doubly degenerate ground state goes over into an infinitely degenerate ground state with a finite entropy. The ordering vanishes at $T = 0$ and does not reapear for $T > 0$, and at the same time the ground state energy per spin increases from $-3J$ to $-J$. The appearance of strongly degenerated "frustrated" ground states with supressed ordering and peculiar critical properties is characteristic for a large number of theoretical models describing uniform, impure and disordered systems [2].
A sufficiently general definition of a frustrated state is still lacking [3]. Until now almost all the theoretical investigations of frustration-type phenomena have been limited to classical models.

It has been recognized for some time that also in quantum systems this kind of phenomena may exist [4]. However, the complexity of quantum systems for $D > 1$ renders their systematic investigations difficult and the results accumulated until present are of rather qualitative character [5].

Only recently a new, systematic approach to study the ground state of quantum systems has been developed. This real-space renormalization group method was initially conceived to study the uniform systems [6]. The method was subsequently extended and critically studied for various uniform and disordered systems [7]. In this paper we describe several applications of the above method to various quantum spin systems which may have frustrated-type ground state.

We shall be mainly concerned with the ground state properties of a generalized Heisenberg model

$$H = \sum_{\substack{<ij> \\ \alpha=X,Y,Z}} \frac{1}{2} J^{\alpha\beta}_{ij} S^{\alpha}_{i} S^{\beta}_{j} - h \sum_{i} S^{Z}_{i}, \qquad (1)$$

where S^{α}_{i} is the α-th component of $S = \frac{1}{2}$ spin, i. e. the α-th Pauli matrix on a site i and $<ij>$ represents a nearest neighbors pair on a triangular lattice. The relevant situation here is when at least one of coupling constants $J^{\alpha\alpha} > 0$; in other words if some of interactions are of AF-type. In this case one cannot eliminate the AF sign by any gauge transformation on one sublattice, because (1) is a two-body interaction and the triangular lattice has three sublattices. One is therefore expecting that the presence of some AF interactions in (1) would have an important effect on the ground state as compared to pure F case. The investigation of (1) with the AF interactions on $D > 1$ lattices is hampered by the lack of exact results which might have served as guidelines. The perturbation-theory type approaches cannot be very effective because of expected (high) degeneracy of ground state. A possible alternative is to adapt the real-space RG method for quantum systems [6] to study the "frustrated" situations. We are going to describe now the RG study of several special cases of (1), which range from the simplest of all situation (A) (Ising model), through a generalized XY-like model (B) to (C) : the Heisenberg-Ising model always with at least one $J^{\alpha\alpha} > 0$. This approach to "frustrated" situations is based on several studies of ours [7] of quantum systems with simpler ground states.
The method consists of dividing the lattice into blocks which are solved exactly. For every block a number of low lying states is

retained. They serve as a truncated basis in which new values of coupling constants are recalculated. The recursion relations are iterated towards their fixed points. Some of fixed points may signal a phase transition.

(A) By setting $\frac{1}{2} J_{ij}^{xx} = J > 0$ (and $J_{ij}^{\alpha\beta} = 0$ otherwise) we obtain the simplest quantum spin model : the Ising model with transverse field h. For h = 0 the Wannier state is its ground state. The way how $h \neq 0$ affects this ground state is expected to be different from F ($J < 0$) case, where the ground state is a doublet up to $(h/J)_c \approx 5$ [8]. We applied the RG method using a hexagonal block of 7 sites which itself has lattice symmetry and for h = 0 is frustrated [9]. The analysis of recursion relations indicate that for $|h/J| < |(h/J)_c| \sim 1.41$, the ground state remains highly degenerate without ordering. At $(h/J)_c$ a gap Δ opens with $\Delta \sim \left((h/J) - (h/J)_c\right)^s$, where s = 0.8, accompanied by an exponential decay of X-X correlation functions. The coherence length ξ diverges as $\xi \sim \left((h/J)-(h/J)_c\right)^{-\nu}$ with $\nu = 0.8$. For $|(h/J)|<|(h/J)_c|$ the correlation function $G_x(R)$ have a power-law behaviour $\sim R^{-\eta}$ with $\eta = 0.32$; for h = 0 $\eta = 0.5$ exactly. At $(h/J)_c$ η jumps to $\eta = 0.68$. The exact value of $h = 0$ ground state energy E_o/N was recovered and it was found that $|E_o(F)| > |E_o(AF)|$ for all $|h/J| < (h/J)_c$, a characteristic result of quantum frustration. This 2D quantum model corresponds to a 3D classical system consisting of AF triangular layers connected by n. n. F bonds. This classical model possesses the ground state degeneracy $\sim N^{2/3}$ but without entropy. We predict that this last system will have a transition at $T_c > 0$ to a low-T phase without order but with power-law behaviour of the correlation functions as for the classical XY model in 2D.

(B) A generalized XY model can be written as

$$H=-\frac{J}{2}\sum_{<ij>} \{\cos\theta(S_i^x S_j^x+S_i^y S_j^y)+\sin\theta(S_i^x S_j^y-S_i^y S_j^x)\}-h\sum_i S_i^z \quad (2)$$

where h > 0, J > 0 and θ is a parameter. The expressions for $J_{ij}^{\alpha\beta}$ of (1) are immediately obtained. H of (2) is a quantum extension of the classical interaction $-J\cos(\theta - \phi_{ij})$ (ϕ_{ij} is the angle between neighboring spins). It includes F and AF XY models ($\theta = 0$ and π respectively) as well as Dzialoshinsky-Moriya (DM) model, $\theta = \frac{\pi}{2}$. For $\theta \neq 0$ no exact informations about the ground state (GS) are available. First indications for frustration effects were obtained only very recently for $\theta = \pi$, h = 0 : an important increase of GS energy-compared to the F case- was deduced from finite cell calculations [10]. Another indication is the exact value for the critical field above which the gap opens : $(h/J)_c = 3\cos(\theta - 2n\,/3)$, $(2n-1)\pi/3 < \theta \lessapprox (2n+1)\pi/3$, which we obtained by Fourier transforming (2). Similarly to case (A) a strong reduction of critical fields ($h_c(AF) = 0.5\, h_c(F)$) results. Subsequently both regions h = 0 and $h \approx h_c$ were treated by quantum RG method with two levels kept at each iteration using hexagonal blocks. The frustration was introduced gradually by varying θ. The values of $(h/J)_c$ (θ) were reproduced within ~ 10 % accuracy. In addition an important increase of G. S. energy (for h = 0, $-E_o/N = 1.5$ and 0.88 for F and AF respectively) was observed. The principal result is that the fully frustrated AF case has its own characteristic scaling properties, whereas <u>any</u> other situation like DM etc..., has the properties of F case. It is instructive to compare our results with classical ($S = \infty$) AF XY model on a triangular lattice. Here also the GS energy is raised and the GS is more complicated than in the F case [11].

(C) The correspondence between the spin $\frac{1}{2}$ Ising model and a classical lattice fluid is well known [12]. This correspondence can be extended to the quantum case for which the relation between the $S = \frac{1}{2}$ Heisenberg-Ising model and the quantum liquid has been established [13]. Following the derivation of [13] the Hamiltonian of the quantum lattice fluid reads

$$H= -\sum_{<i,j>} \left(JS_i^z S_j^z + \frac{K}{2}(S_i^x S_j^x+S_i^y S_j^y)\right) - h\sum_i S_i^z \quad , \quad (3)$$

where the coupling constants of (3) on a triangular lattice are $J = -u$, $K=16\hbar^2/3ma^2 > 0$ and $h = \mu - 2\hbar^2/ma^2 - 3u/2$. Here, u is the n. n. potential energy, μ, m and a are chemical potential, atomic mass and lattice spacing respectively and K is a quantum hopping term. In the classical limit $K \to 0$ and (3) becomes the Ising model with magnetic field. For helium monolayers adsorbed on grapholl $u > 0$ [14] and for K = 0 the resulting AF Ising model and generalizations thereof were extensively studied using <u>classical</u> RG techniques [15]. The extreme quantum limit, u = 0, was recently investigated at T = 0 using <u>quantum</u> RG approach [7]. It is therefore of interest to analyse the quantum corrections to classical models of helium, well represented by (3) with J < 0 and K > 0. It is generally believed that the classical theory of Schick et al. is a good description of the high temperature side of the transition, while for the low T and $T \to 0$ properties this theory appears inapplicable. The T = 0 part of phase diagram of (1) (J < 0) can be obtained from energy considerations. For K = h = 0 the infinitely degenerate Wannier state is the ground state (i. e. $n = \frac{1}{2}$ for the lattice gas). For K = 0 and weak magnetic fields this degeneracy is lifted and a threefold degenerate paramagnetic phase appears, with a magnetic moment formed by two (out of three) sublattices having the same $\sum_i S_i^z$ (so called Potts phase, $n = \frac{1}{3}$ ($\frac{2}{3}$)). Finally, for $|h/J| > 6$ an ordered ferromagnetic phase (n = 1, (0)) results. For the moment, no exact results exist concerning the effect of finite quantum tunneling K on the phase diagram. We have made RG calculations using a hexagonal block to investigate this effect. For the time being we have limited ourselves to two important limits $h \approx 0$ and $h \approx 6|J|$, still keeping two levels at each iterations. The main result is that for both $h \approx 0$ and $|h/J| \approx 6$ $K^* = 0$ is an unstable fixed point, i. e. $(K^{(n+1)}/J^{(n+1)}) = C^{(n)}(K^{(n)}/J^{(n)})$ with $C^{(n)} > 1$ for all n. Hence the system tends towards a stable $(K/J)^* = \infty$ fixed point. Physically it means that as long as an infinitesimally small K is present in the system its physical behaviour is that of the $J \equiv 0$ situation ; or, the system is described by the ferromagnetic XY model. Therefore in this limit a superfluid phase [16] will always result in the presence of infinitesimal tunneling. We do not yet know the stability of the Potts phase in the presence of finite tunneling. We are currently investigating this question [17]. However it seems likely that the Potts phase will become unstable for some $K > K_c$ ($K_c \geqslant 0$). If this will turn out to be the case it would mean that the inclusion of quantum corrections to otherwise classical models of helium drastically change their low - T phase diagrams, perhaps indicating the stability of a superfluid phase.

* Work supported by the Deutsche Forschungsgemeinschaft, Bonn
** On leave from Department of Applied Physics, Stanford University, Stanford, Ca 94305, USA.

REFERENCES

1. G. H. Wannier, Phys. Rev. 79, 357 (1950)
2. An exhaustive list of references can be found in B. Derrida, Thesis, 1979 (unpublished)
3. P. W. Anderson, in "Ill-Condensed Matter", R. Balian et al., Eds. North-Holland, 1979
4. P. Fazekas and P. W. Anderson, Phil. Mag. 30, 423 (1974)
5. A. Sütő and P. Fazekas, Phil. Mag. 35, 623 (1974)
6. S. D. Drell, M. Weinstein and S. Yankielowicz, Phys. Rev. D 14, 487 (1976)
7. K. A. Penson, R. Jullien and P. Pfeuty, Phys. Rev. B (1980) and references therein.
8. K. A. Penson, R. Jullien and P. Pfeuty, Phys. Rev. B 19, 4653 (1979)
9. K. A. Penson, R. Jullien and P. Pfeuty, J. Phys. C 12, 3967 (1979)
10. L. G. Marland and D. D. Betts, Phys. Rev. Lett. 43, 1618 (1979)
11. J. Villain, J. Phys. C 10, 4793 (1977)
12. M. E. Fisher, J. Math. Phys. 5, 994 (1964)
13. D. D. Betts, in "Phase Transitions and Critical Phenomena", vol. 3, C. Domb and M. S. Green, eds. (Academic Press, 1974)
14. A. N. Berker, S. Ostlund and F. A. Putnam Phys. Rev. B 17, 3650 (1978)
15. M. Schick, J. S. Walker, M. Wortis, Phys. Rev. B 16, 2205 (1977)
16. In fact, it is a quantum XY model, but it seems likely that at $T \neq 0$ its properties will be analogous to those of a classical XY model.
17. K. A. Penson, R. Jullien, S. Doniach and P. Pfeuty, to be published.

Sinha, ed. Ordering in Two Dimensions

AB PERCOLATION ON A TRIANGULAR LATTICE

T. MAI and J. W. HALLEY

School of Physics and Astronomy, University of Minnesota, Mpls., Minn. 55455

ABSTRACT

We have studied the AB percolation problem by simulation on a triangular lattice. The problem is interesting because 1) AB percolation is known not to occur on a square lattice; 2) the universality class for AB percolation has not been established and 3) applications to chemisorption are possible. Our results are 1) AB percolation occurs on the triangular lattice with $p_c^{\pm} = .5 \pm (.2855 \pm .0005)$. 2) The probability P_{AB} of AB percolation near $p_c^{\pm}$ behaves as $P_{AB} \propto |p - p_c^{\pm}|^{\beta_{AB}}$ with $\beta_{AB} = .121 \pm .004$. This, and the value of $\beta = .14$ for ordinary percolation indicate that AB percolation on a triangular lattice may belong to a different universality class from ordinary percolation. We also present results on γ_{AB} and the scaling function for the percolation probability P_{AB}.

We have previously [1] explored an extension of the percolation problem in which one studies the statistical distribution of clusters linked by nearest neighbor bonds connecting unlike atoms in an ensemble of samples prepared in exactly the same way as in the ordinary percolation problem. We call this new percolation problem the AB percolation problem. Earlier we showed that AB percolation (in the sense that an infinite cluster occurs with finite probability in the thermodynamic limit) does not occur for any concentration p of atoms of one of the two types (called A) in the sample on a square lattice. A numerical study indicated that AB percolation on a simple cubic lattice existed with critical exponents close to those for ordinary (A) percolation. A solution on the Bethe lattice gave AB exponents identical to those for ordinary percolation.

These results raised intriguing questions about the lower critical dimensionality of AB percolation. The proof of the non-existence of AB percolation on the square lattice depended on the fact that two sub-lattices exist in the square lattice. In the present note we report a numerical study of AB percolation on the triangular lattice. We have found that AB percolation exists on the triangular lattice and have determined the exponents β and γ. The results indicate that AB percolation on a triangular lattice may be in a different universality class from ordinary percolation in two dimensions.

We have done standard Monte Carlo simulations of samples of sizes L x L where L = 60, 100, 120, 280. We average data for 20 samples for L = 60, 100, 120 and 3 samples for L = 280. The probability of an infinite cluster approximated as the size of the largest cluster divided by L^2 is shown in Figure 1 for L = 280. To estimate p_c and β from this data,

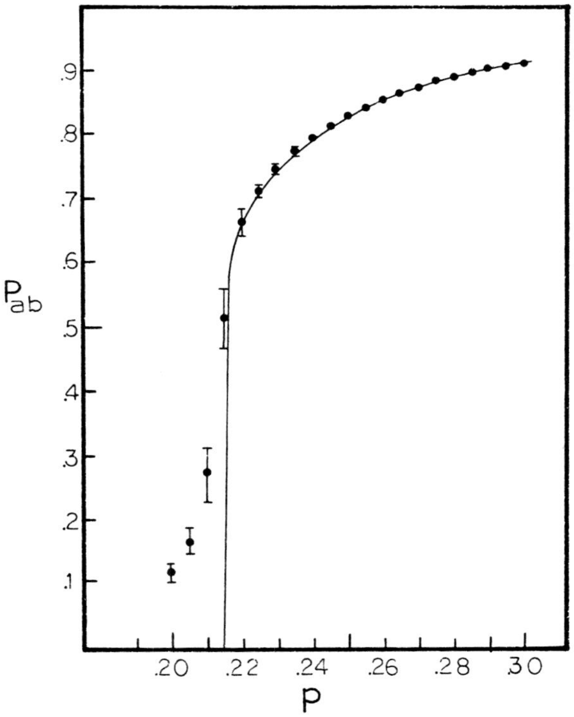

Figure 1. The probability P_{AB} that a site lies in an infinite AB cluster as a function of the concentration p on a triangular lattice. Dots are data for 3 samples of size 280 x 280. The solid line is a fit to $P_{AB} = A\,|p - p_c|^{\beta}$ as explained in the text.

we first fitted the form $P_{AB} = A\,|p-p_c|^{\beta_{AB}}$ in a region $\frac{\Delta p}{p_c} >> L^{-1/\nu}$ (The estimation of the range will be described elsewhere.) with the results shown in Figure 2. We found $p_c = .2145 \pm .0004$, $A = 1.2317 \pm .0740$ for $L = 280$.

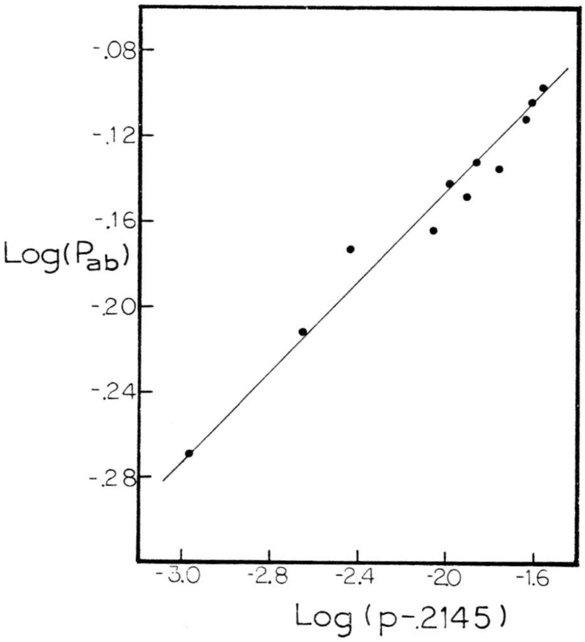

Figure 2. Log (P_{AB}) as a function of log (p-.2145). .2145 is the best fit value of p_c. Data are for 5 realizations of a 280 x 280 lattice.

We also obtained $\beta_{AB} = .121 \pm .004$ which is different from the accepted [2] value of $\beta_A = .14$ for ordinary percolation.

We have made a similar analysis to determine the exponent γ. We computed $\chi = \sum_s n(s)s^2/\sum n(s)$ where n(s) is the number of clusters of size s.

Data for $p < p_c$ are shown in Figure 3. From the 200 x 200 sample (averaging over five samples) we find $\gamma_{AB} = 1.564 \pm .182$, and $p_c = .215 \pm .001$ in good agreement with p_c determined from the P_{AB} data. While the accuracy of our estimates of γ_{AB} is not as good as that of β_{AB}, it clearly deviates from the value $\gamma_A = 2.3$ for ordinary percolation in two dimensions [2].

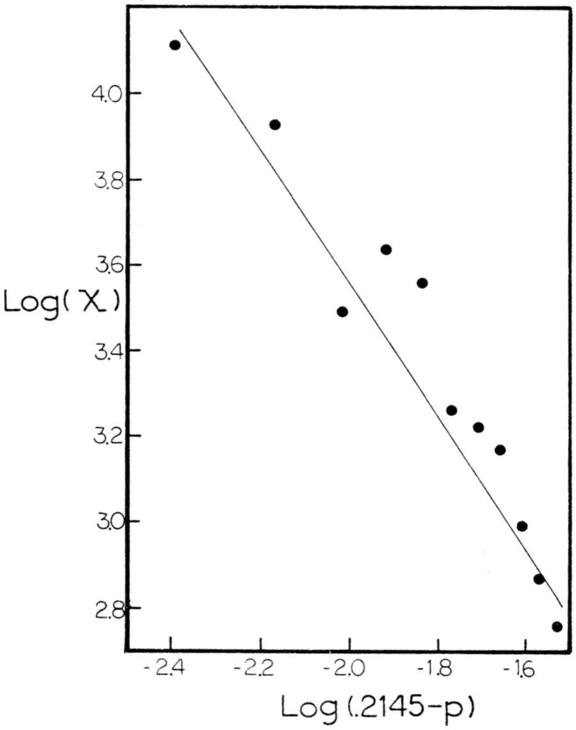

Figure 3. Log ($\sum ns^2/\sum ns$) as a function of log (.2145-p). .2145 is the best fit value of p_c. Data are for 5 realizations of a 200 x 200 lattice.

Finally, we have made a preliminary attempt to determine ν by computing the scaling function $P_{AB}\,L^{\beta/\nu}$ as a function of $\frac{|p-p_c|}{p_c}\,L^{1/\nu}$. We show such plots for $p_c = .2145$ and $\beta = .121$ in Figure 4 for $.18 \leq p \leq .24$ for L = 60, 100, 120, 280 and for $\nu = 0.97$. We have not yet analysed these data quantitatively and can only say that they indicate $0.8 \leq \nu_{AB} \leq 1.2$.

In conclusion, our data on β_{AB} and γ_{AB} indicate that AB percolation on a triangular lattice may be in a different universality class from ordinary percolation. Many interesting questions remain. Under what conditions do we expect AB percolation to differ from ordinary percolation? So far, we know that it is the same on the Bethe

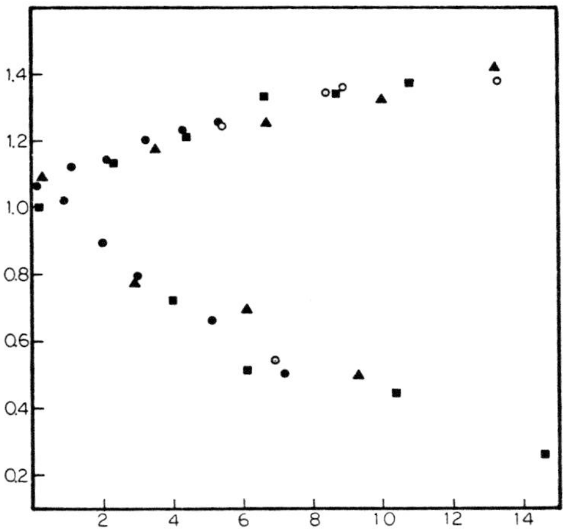

Figure 4. The scaling function $P_{AB}\, L^{\beta/\nu}$ as a function of $L^{1/\nu}\, (\frac{p-p_c}{p_c})$ for L = 60, 100, 120, 280, β = .121 p_c = .2145, ν = .97. The vertical axis is $P_{AB}\, L^{\beta/\nu}$. The horizontal axis shows $\frac{p-p_c}{p_c}\, L^{1/\nu}$.

lattice and apparently different on the triangular lattice. If it is different for all dimensions below 6 then this should be revealed by an ε expansion using the equivalent spin model we previously derived. [1] If it is only different on frustrated lattices then we should find a different result on fcc than on simple cubic lattices.

REFERENCES

1. M. Barma, J. W. Halley and M. Stephen (to be published).

2. A. B. Harris and S. Kirkpatrick, Phys. Rev. B16, 542 (1977).

Sinha, ed. Ordering in Two Dimensions

STRINGS, VORTICES, MELTING AND THE CLOCK MODEL

R. SAVIT

Physics Department, University of Michigan, Ann Arbor, MI 48109

ABSTRACT

We show that the "topological excitations" of the N-state clock model can be thought of as a collection of interacting strings (domain boundaries) and vortices. The two Kosterlitz-Thouless phase transitions of this system (which exist for $N \geq N_c$) can be understood as being caused by a sequential condensation of the two types of excitations. The picture bears a close relationship to two-stage two dimensional melting upon which we comment. We also briefly describe arguments indicating that the phase structure of the clock model is the same on a square or triangular lattice - in particular, N_c is the same. This information should be useful for analyzing data on physical systems.

I would like to discuss today some properties of the unusual phase structure expected for the two-dimensional Z_N-symmetric clock model, (also called the vector Potts model) and other models which should be in the same universality class. By way of background let me first remind you about some of the properties of this model [1].

The clock model Hamiltonian is given by

$$-(kT)^{-1}\mathscr{H} = \frac{\beta}{2} \sum_{\langle\rangle} s_i \cdot s_j^+ + \text{h.c.} \qquad (1)$$

where $s_i = \exp(i2\pi q_i/N)$ is a complex spin on the site i, q_i is an integer $0 \leq q_i < N$, and the sum is over all nearest neighbors. In the partition function one sums independently over all the q_i. For the moment let us suppose that (1) is defined on a square lattice. For N less than a certain critical value, N_c, (which is believed to be 5) the model (1) has a single second order phase transition. However, for $N \geq N_c$ the clock model has two distinct Kosterlitz-Thouless type phase transitions [2] occuring at temperatures T_1 and T_2. For $T<T_1$ the system is in a ferromagnetic phase in which $\Gamma(r) \to$ const. $\neq 0$ as $r \to \infty$, where $\Gamma(r) \equiv \langle s_k s_\ell \rangle$ with $|k-\ell|=r$. For $T_1<T<T_2$ $\Gamma(r) \sim r^{-\eta}$ as $r \to \infty$, as in the low temperature phase of the two-dimensional x-y model. Finally, for $T>T_2$ $\Gamma(r) \sim e^{-\mu r}$ for large r. As N increases, T_1 approaches zero ($\propto 1/N^2$ for large enough N), and T_2 approaches a non-zero constant. Thus in the $N \to \infty$ limit the ferromagnetic phase disappears and we recover the d=2 x-y model.

There are three aspects of this model and its phase structure about which I will speak. First I will present a simple physical picture for the phase transitions in this system which shows that the two transitions can be thought of as being due to sequential condensations of two types of topological excitations which the system possesses: strings and vortices [3]. Second, I want to describe some studies of clock model-like systems on a triangular lattice which indicates that the phase structure of a Z_N-symmetric model on this lattice is the same as that of the corresponding model on a square lattice [4]. In particular, N_c should be independent of the lattice shape. Finally, I want to briefly comment on the similarity between the phase structure of the clock model and the phase structure suggested by recent work for the problem of two dimensional melting.

For $N \geq N_c$, the N-state clock model on a two dimensional square lattice has two Kosterlitz-Thouless-like phase transitions. To understand the physical origin of these transitions we first ask what kinds of topological excitations the theory possesses. Since Z_N is a discrete symmetry, it is clear that we can have closed domain boundaries which separate regions in which the spins are aligned in one direction from regions in which they are pointing in a different direction. Such boundaries obviously satisfy a Z_N algebra. They have a "strength" from one to N/2 (or(N-1)/2 if N is odd increasing as the magnitude of the angle between neighboring spins increases. They also have an arrow whose direction depends on whether the spins rotate in a clockwise or counter-clockwise direction as we cross the boundary.

Now in the limit $N \to \infty$ we recover the x-y model for which the topological excitations are vortex points. For large enough but finite N, we might expect to see ersatz vortices-precursors of the true vortices which emerge in the $N \to \infty$ limit. Suppose we try to produce such a vortex-like structure in the N-state clock model. We want to create a state in which the spins rotate by 2π as we traverse a closed circuit surrounding the center of the vortex. There are obviously many such configurations we can make but, all else being equal, the most important ones should be those of least energy. Now roughly speaking, the minimum energy configuration will be produced when the total spin rotation of 2π is divided equally among nearest neighbors as we move around a circle centered on the position of the vortex. For a Z_N-symmetric theory, if we are within $\mathcal{O}(N)$ lattice spacings of the vortex center, the 2π rotation can be equally divided among the spins, just as if we were producing a vortex in the x-y model. However, if we move more than $\mathcal{O}(N)$ lattice spacings away from the vortex center, the spins can no longer divide equally the 2π rotation among themselves, and the best we can do is to form N wedge shaped domains of spins whose orientations differ from their neighbors' by $2\pi/N$. Thus, a Z_N vortex looks like a bicycle wheel with a hub of $\mathcal{O}(N)$ lattice spacings, radiating N spokes which are domain boundaries whose arrows are all pointing out or all pointing in. We now see that the domain boundaries may be thought of as forming a set of currents, $j_{\mu;i}$, which are integer-valued, mod N, and which are divergenceless, mod N. That is $\Delta_\mu j_{\mu;i}=NQ_i$, where Q_i is an integer and represents the strength of a Z_N-vortex. The i refers to a position in the lattice, μ is a direction on the lattice, and Δ_μ is a difference operator defined by $\Delta_\mu \alpha_i=\alpha_i-\alpha_{i-\hat{\mu}}$. Of course, divergenceless configurations of j_μ are just closed domain boundaries. (Strictly speaking, the $j_{\mu;i}$ and the Q_i are associated with the links and sites of the dual lattice. See ref. 3 for details).

Returning to our bicycle wheel, and remembering that domain boundaries have a finite energy per unit length, it is clear that the energy of such a configuration is enormous, diverging linearly with the (linear) size of the system. However, we can produce vortex-antivortex pairs with finite energy by bending the bicycle spokes of a vortex around and letting them converge to produce an antivortex. We now have a vortex-antivortex pair bound by N strings (the domain boundaries). Because the strings have finite energy per unit length, the energy of our vortex antivortex pair will grow linearly with their separation - i.e., they are bound by a linear potential (at least at low enough temperatures).

The two phase transitions of the clock model for $N \geq N_c$ can be understood as sequential condensations of the two types of excitations, strings and vortices. The low temperature phase which is ferromagnetic has only very small closed string loops and tightly bound vortex-antivortex pairs. For large enough N, the finite energy per unit length of the strings is proportional to $1/N^2$. But the entropy of a string is, modulo logarithms, also proportional to its length. Thus, a simple energy versus entropy argument indicates that it becomes free energetically favorable to have many very long strings for $T>T_1 \propto 1/N^2$. That is, for $T>T_1$ the strings become floppy. This floppy string phase has a massless excitation (i.e., $\Gamma(r) \sim r^{-\eta}$) and is like the low temperature phase of the x-y model [1].

Now if the strings become floppy, one might suppose that the vortex pairs which are held together by strings will also dissociate at $T=T_1$. In general, however, this is not the case. What does happen is that the potential binding the vortex-antivortex pairs is softened from linear to logarithmic in their separation [3]. This can be understood by thinking about the x-y model. The logarithmic potential between vortices in that model is really due to the fact that the U(1) spins can differ infinitesimally in orientation for their nearest neighbors. This causes the integral $\int_0^\Lambda \frac{d^2k}{|\underset{\sim}{k}|^2} e^{i\,\underset{\sim}{k}\cdot\underset{\sim}{r}}$ to diverge logarithmically as $r \to \infty$, providing the logarithmic potential. In the clock model case the spin orientations cannot differ infinitesimally from their neighbors. But if we imagine doing a thermal average over all sets of configurations of domain boundaries, then in an average sense the spins can have infinitesimally different orientations from their neighbors. Thus, nailing down a vortex-antivortex pair and averaging over configurations of floppy strings joining them produces an effective logarithmic potential between the vortices.

This sets the stage for the second transition at $T=T_2$ (to a first approximation T_2 is independent of N) which is triggered by the usual Kosterlitz-Thouless vortex unbinding mechanism[2]. For $T>T_2$, the system is completely disordered and is like the high temperature phase of the x-y model. In understanding this scenario it is very important to realize that the "screening" of the binding potential between the vortices by the strings becoming floppy is essential for the second transition to proceed.

If $N<N_c$, then we expect the clock model to have only one second order transition. We can understand this in our picture by recalling that $T_1 \sim 1/N^2$. For $N<N_c$, $T_1>T_2$, so that by the time the strings become floppy, the entropy times temperature of the vortex-antivortex pairs is already large enough so that they immediately fall apart at $T=T_1$ and complete disordering takes place at a single transition. This sort of picture also lets us qualitatively understand why in the case of a single transition the transition is "harder" than either of the transitions which occur when $N \geq N_c$. In case of a single transition the ground state energy changes more abruptly than it does in either of the two transitions which occur when $N \geq N_c$, since for $N<N_c$ both strings and vortices condense at the same temperature. [3,5]

It would certainly be interesting to find a physical two dimensional system which evidences these two phase transitions. Since a triangular lattice is the structure of choice for so many surfaces, it becomes interesting to ask whether a clock model on a triangular lattice has the same phase structure as one on a square lattice. Of course, a strong belief in universality implies that the phase structure is the same. On the other hand, this three phase structure is rather unusual and it is certainly not a priori obvious that N_c should be independent of the lattice shape. Furthermore, the proof that there must be at least two phase transitions for large enough N required the use of a periodic Gaussian form of the clock model which is self-dual on a square lattice [1]. Since the dual of a triangular lattice is a honeycomb lattice, such models on a triangular lattice will not generally be self-dual and so it is not clear that an analogous proof can be carried through.

Indeed, except for special values of N, the clock model and it's periodic Gaussian version are not self-dual on a triangular lattice. However, it is possible to define a modified periodic Gaussian model (PGM) on a triangular lattice which differs from the original PGM in that there are vortices at the centers of alternating elementary cells of the triangular lattice, rather than in every cell as in the original PGM (and the original clock model) [4]. Thus, in this modified PGM the vortices form a (dual) triangular lattice with the same lattice spacing as the original triangular lattice. This modified PGM is self-dual. When arguments analogous to those of ref. 1 are carried out for this model one arrives at an equation (actually, an inequality) for N_c^Δ, the critical value of N for this modified PGM on a triangular lattice. What is found [4] is that there must be at least two phase transitions for $N \geq N_c^\Delta$, where N_c^Δ is the first integer greater than or equal to $2\pi\beta_\Delta^*\sqrt{3}$. β_Δ^* is the vortex unbinding temperature for the x-y model vortices (on a triangular lattice) which exist in the $N \to \infty$ limit of the modified PGM on a triangular lattice. By comparison $N_c^\square$, the critical value of N for the PGM (and clock model) on a square lattice, is the first integer greater than or equal to $2\pi\beta_\square^*$, where $\beta_\square^*$ is the analogous vortex unbinding temperature for the U(1) PGM on a square lattice. Simple energy versus entropy arguments indicate that the two transition temperatures are related by $\beta_\square^* = \beta_\Delta^* \sqrt{3}$, which implies that $N_c^\Delta = N_c^\square$. Thus, the lattice structure does not seem to affect N_c, and assuming, as is likely, that the modified and original PGM's on the triangular lattice are in the same universality class, then N_c for the clock model on the square and triangular lattices should be the same.

I want to close with a few remarks on the similarity between the phase structure of the clock model and the two stage theory of two dimensional melting [6]. Both these systems have an underlying global Z_N-symmetry, and so it is perhaps not too surprising that their phase properties should be so similar. Each system has (at least for $N \geq N_c$) two Kosterlitz-Thouless phase transitions generated by very similar mechanisms. The dislocation condensation responsible for the lower temperature transition in the melting problem is analogous to the domain boundary condensation of the clock model. As a result of the condensation of the dislocations the interactions between disclinations in the crystal are screened and softened to a logarithmic interaction, just as the domain boundary condensate screens the vortex-antivortex interactions in the clock model. The second phase transition is caused by disclination unbinding in the theory of two-dimensional melting, and by vortex-antivortex pair unbinding in the clock model.

Not only are the phase transitions caused by similar mechanisms, but the three phases of the two theories are closely related. Indeed, the clock model spin-spin correlation function, $\Gamma(r)$, has the same long range behavior in the three phases of the clock model that the rotational correlation function has in the corresponding phases of the melting theory. Unfortunately, there is no simple analogue of the melting theory's translational correlation function in the clock model. This can be understood by counting degrees of freedom: For each lattice site in the clock model there is one (rotational) degree of freedom while for each crystal lattice site in the melting theory there are two degrees of freedom - displacements in the x and y directions. To generate in a simple way the analogue of the translational correlation function one could consider a model similar to the clock model, but in which the magnitude as well as the orientation of the spin is allowed to vary. For example, a judiciously chosen version of complex ϕ^4 field theory with a Z_N-symmetric source term would probably do.

REFERENCES

1. S. Elitzur, R. Pearson and J. Shigemitsu, Phys. Rev. D19, 3698 (1979); A. Ukawa, P. Windey and A. Guth, Princeton University preprint (1979); J. Cardy, UCSB preprint no. TH-30 (1978). A similar phase structure was also found in a slightly different context by J. José, L. Kadanoff, S. Kirkpatrick, and D. Nelson, Phys. Rev. D16, 1217 (1977).
2. J. Kosterlitz and D. Thouless, J. Phys. (Paris) C6, 1181 (1973).
3. M. Einhorn, R. Savit, and E. Rabinovici, University of Michigan preprint no. UM HE 79-25 (1979) to be published in Nuclear Physics B, "Field Theory and Statistical Systems".
4. R. Savit, University of Michigan preprint no. UM HE 80-3 (1980).

5. T. Banks, Tel-Aviv University preprint (1979).

6. B. Halperin and D. Nelson, Phys. Rev. Lett. 42, 121 (1978); Phys. Rev. B19, 2457 (1979).

Published 1980 by Elsevier North Holland, Inc.
Sinha, ed. Ordering in Two Dimensions

GAUGE FIELDS AND ORDER*

M. ÚLEHLA

Solid State Division, Oak Ridge National Laboratory
Oak Ridge, Tennessee 37830

ABSTRACT

In this work we wish to apply the recently proposed geometric description of order in solids to the problem of melting in 2 dimensions. After a brief description of the geometric approach we shall show how the dislocation mediated theory of 2d melting arises in this formulation.

INTRODUCTION

Suggestions of finite temperature phase transitions for classical spins with continuous rotational symmetry in two dimensions,[1-5] work of Jancovici[6] and others on two dimensional harmonic solids[6-8] and pioneering work of Kosterlitz and Thouless[9] gave new impetus to the study of order and melting in two dimensions.[10-12] The basic notion in this approach to the study of melting is the idea that the phase transition is driven by the dissociation of bound pairs of topological singularities (dislocations, vortices or simply charges); the low temperature phase is characterized by the absence of free charges (confinement in the bound states) whereas the high temperature phase has characteristics of a (overall neutral) gas of interacting charges.

It is the purpose of this paper to apply the recently proposed geometric point of view of the description of order in solids[13] to the particular case of two dimensional solids and show how the dislocation mediated theory of melting[10] arises in this formulation. We first summarize the geometric approach to order.

GEOMETRIC DESCRIPTION OF ORDER

In this section we exploit certain geometric concepts pertaining to the <u>local</u> as well as to the <u>connective</u> properties of the dynamical variables describing the solid.[13] These considerations lead to the notion of local symmetries and the ensuing need for introduction of gauge fields.[14,15,16,21] The phases of the solid can be labeled by the choice of the structure group for the gauge fields, i.e., depending on the choice of the structure group G (which of course depends on temperature and dynamics) the phase may be crystalline or fluid. As dynamical variables relevant to the description of order we can take the (vector) field describing the matter distribution, i.e., positions of the particles $\vec{u}(\vec{x})$. We assume that for any $\vec{x}$, the dynamical variable $\vec{u}(\vec{x})$ can range over a certain manifold of values denoted by $T_{\vec{x}}$. The manifold or fiber $T_{\vec{x}}$ erected over a point $\vec{x}$ depends on a physical situation under study (and will be constrained by dynamical considerations, i.e., a choice of a Hamiltonian H); but it could be, for example, a point, a finite collection of points, a loop, an (open) portion of a plane or volume etc.[17] We further assume that this manifold is sufficiently well behaved that a group of transformations G_0 can be associated with it in the sense that the totality of the manifold $T_{\vec{x}}$ can be generated by the action of the group G_0 acting on a reference vector $\vec{u}_0$ based at $\vec{x}$, and that further, $T_{\vec{x}}$ is invariant under a change in the choice of reference vector $\vec{u}_0$.[18]

By this construction one allows for a possibility of a <u>local</u> gauge freedom. The possibility for <u>realization</u> of this local symmetry depends on the form of the Hamiltonian (or Lagrangian) one chooses to describe the physical system. If the Hamiltonian is a purely local function (i.e., a potential) of the dynamical variables, then the symmetry can be realized; if however, as is usually the case, the Hamiltonian contains a kinetic energy part that couples dynamical variables at nearby points $\vec{x}$ and $\vec{x}'$ (derivative or various nearest neighbor interactions) the <u>local</u> invariance will in general not be <u>realized</u> and the connectivity properties of the manifold of fibers $T_{\vec{x}}$ will become important.

The totality of the fibers $T_{\vec{x}}$ over $\vec{x}$ for all $\vec{x}$ is called the fiber bundle, denoted T. As mentioned above, to complete the geometrical description of the physical system (and of T), we need to specify, in addition to the local structure of T, the connectivity properties of T, i.e., we need to specify the effects on the dynamical variables $\vec{u}(\vec{x})$ as we vary $\vec{x}$. This is accomplished by making a particular choice for a group of transformations G, the structure group of T.[13,19] The group G, which in general need not be the same as G_0, defines the connectivity properties of T; its elements are the connections, specifying a "twist" the fiber bundle undergoes as one ranges from point $\vec{x}$ to point $\vec{x}'$.[13] In a familiar way now a gauge field (alluded to in the title) can be introduced via the notion of "gauge-covariant derivative," and identified with a connection of the structure group G.[20,21,13]

We shall define a solid to be in an ordered state, (a prototypical crystal) when the structure group of the fiber bundle is globally the trivial identity group. When the structure group becomes larger than the identity group but a subgroup of G_0, the local set of possible configurations is increased, the solid is being disordered and we shall speak of an amorphous or glassy solid.

Any global configuration is characterized by a particular distribution of the gauge fields associated with the structure group, but two configurations of gauge fields related by a nonsingular gauge transformation yield the same partition function as a result of the explicit (built-in) local symmetry of the Hamiltonian. Thus, the partition function is not a functional of a gauge configuration itself but rather of those characteristics of a configuration that are invariant under a gauge transformation.

CONNECTION WITH 2D THEORY OF MELTING

Consider a two-dimensional solid. We shall formulate the theory on a lattice because of the finite size of the atoms forming the solid, and as dynamical variables we shall take the two-dimensional vector field $\vec{u}_\alpha(\vec{i})$, $\alpha = 1,2$, describing displacements of the atoms from lattice sites. As a local fiber we take R_2, although as discussed in the Introduction, when the atoms form a solid, most of R_2 is in fact not available to the local dynamical variables, as the atoms perform mainly small oscillations around the (equilibrium) lattice sites. It is convenient to write the dynamical variables as

$$\vec{u}(\vec{i}) = e^{\delta(\vec{i})} e^{i\theta(\vec{i})} \quad . \tag{1}$$

For a model Hamiltonian take a standard nearest neighbor coupling,

$$H = K \sum_{\langle\vec{i}\vec{j}\rangle} u_\alpha(\vec{i}) u_\alpha(\vec{j}) = K \sum_{\langle\vec{i}\vec{j}\rangle} e^{\delta_{\vec{i}}+\delta_{\vec{j}}} \cos\,(\theta_{\vec{i}}-\theta_{\vec{j}}) \ . \tag{2}$$

It is convenient to perform an analytic continuation to imaginary values of δ so that Hamiltonian (2) becomes symmetric in δ and θ. Then

$$H = K \sum_{\langle\vec{i}\vec{j}\rangle} \cos(\delta_{\vec{i}}-\delta_{\vec{j}}) \cos(\theta_{\vec{i}}-\theta_{\vec{j}}) \ . \tag{3}$$

Hamiltonian (3) is invariant under two separate global transformations, i.e.

$$\delta H = H'-H = 0 \ , \text{ if } \theta_{\vec{i}} \to \theta'_{\vec{i}} = \theta_{\vec{i}} + \alpha, \text{ for all } \vec{i} \tag{4}$$

and similarly for $\delta_{\vec{i}}$.[23] In order to introduce disorder into the system, i.e., allow for local variation in the fiber bundle, we introduce into (3) a pair of gauge fields $A_{\vec{i}\vec{j}}$ and $B_{\vec{i}\vec{j}}$

$$H = K \sum_{\langle\vec{i}\vec{j}\rangle} \cos(\delta_{\vec{i}} - \delta_{\vec{j}} - B_{\vec{i}\vec{j}}) \cos(\theta_{\vec{i}}-\theta_{\vec{j}}-A_{\vec{i}\vec{j}}) \tag{5}$$

so that (5) now will be locally gauge invariant under either transformation. Namely, if we have $\theta_{\vec{i}} \to \theta'_{\vec{i}} = \theta_{\vec{i}} + \alpha_{\vec{i}}$ and[24]

$$A_{\vec{i}\vec{j}} \to A'_{\vec{i}\vec{j}} = A_{\vec{i}\vec{j}} + \alpha_{\vec{i}} - \alpha_{\vec{j}} \tag{6}$$

then

$$\delta H = 0 \ , \tag{7}$$

and similarly for a gauge transformation on δ_i.

It is convenient to streamline the notation. Instead of denoting a nearest neighbor by position $\vec{j}$, we shall use the site $\vec{i}$ and a direction μ. Then defining the finite difference operator Δ_μ as $\Delta_\mu\theta_{\vec{i}} = \theta_{\vec{i}}-\theta_{\vec{i}-\mu}$, the Hamiltonian (5) can be rewritten as

$$H = K \sum_{(\vec{i},\mu)} \cos(\Delta_\mu\delta_{\vec{i}}-B_\mu(\vec{i})) \cos(\Delta_\mu\theta_{\vec{i}}-A_\mu(\vec{i})) \ . \tag{8}$$

Then we can write the partition function in a fixed configuration of gauge degrees of freedom $\{B_\mu(\vec{i}), A_\mu(\vec{i})\}$ as

$$Z\{A_\mu(\vec{i}),B_\mu(\vec{i})\} = \int D\delta_{\vec{i}} \int D\theta_{\vec{i}} \ \exp\left\{ K \sum_{(\vec{i},\mu)} \cos(\Delta_\mu\delta_{\vec{i}}-B_\mu(\vec{i}))\cos(\Delta_\mu\theta_{\vec{i}} - A_\mu(\vec{i}))\right\} \tag{9}$$

where the integrals denote normalized integrations over unit circles.

To rewrite the partition function (9) in gauge invariant way by characterizing it as a distribution of topological charges we define these as

$$\Phi_A(\vec{i}) = \varepsilon_{\mu\nu}(\Delta_\mu A_\nu(\vec{i}) - \Delta_\nu A_\mu(\vec{i})) \quad (10a)$$

$$\Phi_B(\vec{i}) = \varepsilon_{\mu\nu}(\Delta_\mu B_\nu(\vec{i}) - \Delta_\nu B_\mu(\vec{i})) \quad (10b)$$

where $\varepsilon_{\mu\nu}$ is the completely antisymmetric tensor. The Poisson summation formula and duality transformations discussed by José et al.[25] yield

$$Z\{\Phi_A(\vec{i}),\Phi_B(\vec{i})\} = \int D\delta_{\vec{i}} D\theta_{\vec{i}} \prod_{<\vec{i}\mu>} \sum_{S_{\vec{i}\mu}} \sum_{R_{\vec{i}\mu}} \exp\Big\{ iS_{\vec{i}\mu}(\Delta_\mu \delta_{\vec{i}} - B_\mu(\vec{i})) + iR_{\vec{i}\mu}(\Delta_\mu \theta_{\vec{i}} - A_\mu(\vec{i})) + \tilde{V}(S_{\vec{i}\mu},R_{\vec{i}\mu})\Big\} \quad (11)$$

where $e^{\tilde{V}(S_{\vec{i}\mu},R_{\vec{i}\mu})}$ is a Fourier transform of e^H. Integrating over $\theta_{\vec{i}}$ and $\delta_{\vec{i}}$

$$Z\{\Phi_A(\vec{i}),\Phi_B(\vec{i})\} = \sum_{\{S_{\vec{i}\mu}\}} \sum_{\{R_{\vec{i}\mu}\}} \exp\Big\{ \sum_{\vec{i}\mu} -iS_{\vec{i}\mu}B_\mu(\vec{i}) - iR_{\vec{i}\mu}A_\mu(\vec{i}) + \tilde{V}(S_{\vec{i}\mu},R_{\vec{i}\mu})\Big\}$$

$$\prod_{\vec{i}} \delta(\Delta_\mu S_{\vec{i}\mu})\delta(\Delta_\mu R_{\vec{i}\mu}) \ . \quad (12)$$

The zero divergence conditions in (12) can be satisfied explicitly by setting

$$S_{\vec{i}\mu} = \varepsilon_{\mu\nu}\Delta_\nu n_{\vec{i}} \ , \quad (13a)$$

$$R_{\vec{i}\mu} = \varepsilon_{\mu\nu}\Delta_\nu m_{\vec{i}} \ . \quad (13b)$$

Then

$$Z\{\Phi_A(\vec{i}),\Phi_B(\vec{i})\} = \sum_{\{n_{\vec{i}}\}} \sum_{\{m_{\vec{i}}\}} \exp\Big\{ \sum_{\vec{i}} in_{\vec{i}}\Phi_B(\vec{i}) + im_{\vec{i}}\Phi_A(\vec{i}) + \tilde{V}(\varepsilon_{\mu\nu}\Delta_\nu n_{\vec{i}}, \varepsilon_{\mu\alpha}\Delta_\alpha m_{\vec{i}})\Big\} . \quad (14)$$

To satisfy the invariance of (12) under global shifts $S_{\vec{i}\mu} \to S_{\vec{i}\mu} + S$, $(R_{\vec{i}\mu} \to R_{\vec{i}\mu} + R)$, each charge Φ_α, α = A,B must satisfy a neutrality condition[26]

$$\sum_{\vec{i}} \Phi_\alpha(\vec{i}) = 0 \ , \quad \alpha = A,B \ . \quad (15)$$

From (14) we see why it is appropriate to call the objects Φ_A,Φ_B charges, i.e., they appear in the functional integral as sources of the harmonic (in Villain approximation) degrees of freedom. Provided the Villain approximation[27] to the potential H can be made, the partition function (14) reduces, with the use of $\varepsilon_{\mu\nu}\varepsilon_{\mu\sigma} = 2\delta_{\nu\sigma}$, to

$$Z\{\Phi_A(\vec{i}),\Phi_B(\vec{i})\} = \sum_{\{n_{\vec{i}}\}} \sum_{\{m_{\vec{i}}\}} \exp\Big\{ \sum_{\vec{i}} in_{\vec{i}}\Phi_B(\vec{i}) + im_{\vec{i}}\Phi_A(\vec{i}) - \frac{1}{K}(\Delta_\mu n_i)^2 - \frac{1}{K}(\Delta_\mu m_i)^2 \Big\} \quad (16)$$

In (16) an (infinite) constant has been factored out. The harmonic degrees of freedom in (16) can be integrated and we finally obtain the partition function in terms of the overall neutral vector Coulomb gas, the starting point for the study of melting in 2 dimensions.[10]

$$Z\{\Phi_A(\vec{i}),\Phi_B(\vec{i})\} = Z_{harmonic} \exp\Big\{ K \sum_{\vec{i}\vec{j}} \Phi_\alpha(\vec{i})\Phi_\alpha(\vec{j})\ G(\vec{i}-\vec{j}) \Big\} \quad (17)$$

In Eq. (17) $Z_{harmonic}$ is the contribution to the partition function of the harmonic degrees of freedom, the distribution of (α=A,B), charges is restricted to only overall neutral configurations and the propagator is given for large distances by

$$G(r) \approx \ln r + \text{const., if } r \gg 1 \ . \quad (18)$$

REFERENCES

*Research sponsored by the Division of Materials Sciences, U. S. Department of Energy under contract W-7405-eng-26 with the Union Carbide Corporation.

1. H. E. Stanley and T. A. Kaplan, Phys. Rev. Lett. 17, 913 (1966).
2. H. E. Stanley, Phys. Rev. Lett. 20, 589 (1968).
3. M. A. Moore, Phys. Rev. Lett. 23, 861 (1969).
4. F. J. Wegner, Z. Phys. 206, 465 (1967).
5. V. L. Berezinskii, Zh. Eksp. Teor. Fiz. 59, 907 (1970).
6. B. Jancovici, Phys. Rev. Lett. 19, 20 (1967).
7. H. J. Mikeska and H. Schmidt, J. Low Temp. Phys. 2, 371 (1970).

8. Y. Imry and L. Gunther, Phys. Lett. A 29, 483 (1969).
9. J. M. Kosterlitz and D. J. Thouless, J. Phys. C 6, 1181 (1973).
10. D. R. Nelson, Phys. Rev. B 18, 2318 (1978).
11. A. P. Young, Phys. Rev. B 19, 1855 (1979).
12. D. R. Nelson and B. I. Halperin, Phys. Rev. B 19, 2457 (1979).
13. M. Úlehla, to be submitted to Phys. Rev. B (1980); geometric notions are also discussed for instance in W. Drechsler and M. E. Mayer, Fiber Bundle Techniques in Gauge Theories, Springer Verlag, Berlin, 1977.
14. Similar ideas may occur in the physics of the spin and gauge systems. See for instance, G. Toulouse, Commun. Phys. 2, 115 (1977); X. Dzyaloshinskii and Y. Volovik, J. Phys. (Paris).
15. E. Fradkin, B. A. Huberman, and S. M. Shenker, Phys. Rev. B 18, 4789 (1978).
16. J. B. Kogut, Rev. Mod. Phys. 51, 659 (1979).
17. In general, we will require that this manifold not have a boundary.
18. I wish to thank D. Mermin for helpful discussions regarding this point.
19. Although there is some autonomy in the choice of G, it may depend on temperature (see Section IV).
20. E. S. Abers and B. W. Lee, Physics Reports 9C, 1, (1971).
21. S. Coleman, Secret Symmetry, 1973 Ettore Majorana Lectures.
22. F. Wegner, J. Math. Phys. 12, 2259 (1971).
23. Notice that the global gauge invariance on δ holds for Hamiltonian (2) as well provided one rescales the (so far) arbitrary coupling K.
24. This implies we have made the choice $A_{ij} = - A_{ji}$, and similarly for B_{ij}.
25. J. V. José, L. P. Kadanoff, S. Kirkpatrick and D. R. Nelson, Phys. Rev. B 16, 1217 (1977).
26. Since the sum over lattice points is discrete this implies that in particular all charges $\Phi_\alpha(i)$ of the same kind (α = A,B) must be integral multiples of a common "elementary" charge (perhaps different for A and B).
27. J. Villain, J. Phys. (Paris) 36, 581 (1975).

Sinha, ed. Ordering in Two Dimensions

HONEYCOMB LATTICE GAS: APPLICATION TO ADSORBED SYSTEMS

W. KINZEL and M. SCHICK

Department of Physics, University of Washington, Seattle, WA 98195

A.N. BERKER

Department of Physics, Massachusetts Institute of Technology

Cambridge, Mass. 02139

ABSTRACT

The usual lattice-gas model of adsorbed systems is viewed as the pure, or vacancy-free, limit of a q=2 Potts lattice gas and treated by real-space renormalization group methods. This greatly improves the results over previous calculations for the honeycomb lattice appropriate to a preplated substrate. By comparing the resulting phase diagram with that observed experimentally we are able to deduce the effective Helium-Helium interaction. The nearest-neighbor repulsion is found to be considerably reduced from its bare Lennard-Jones value due, presumably, to quantum mechanical correlations, and further neighbor attractions are enhanced.

I. INTRODUCTION

The lattice gas has proved to be an extremely useful model in the study of order-disorder transitions in adsorbed systems. The geometry of the lattice is, of course, determined by the structure of the substrate. For example, the basal planes of graphite provide a triangular array of adsorption sites. As graphite is a common substrate, the triangular lattice gas has been the object of much attention [1,2]. Other common adsorbates provide a honeycomb lattice of adsorption sites. Examples are the (111) face of f.c.c. and b.c.c. materials and graphite which has been pre-plated with a close-packed layer of noble gas. Nevertheless, the honeycomb lattice has received much less attention than the triangular lattice. The possible continuous phase transitions which can be exhibited by adsorbed systems on such a lattice are known [3] but few phase diagrams for various interactions have been calculated. It is our purpose to provide an accurate calculation of the phase diagrams of those order-disorder transitions which are dominated by nearest-neighbor repulsions. This produces a continuous transition in the universality class of the two-dimensional ferromagnetic Ising model [3]. We also include further neighbor attractions which introduce first-order transitions and a multicritical point. We outline the renormalization group (RG) calculation and present some results concerning the transition in the next section. They are a substantial improvement over previous RG calculations. We compare, in Section III, our phase diagram with that observed in the He/Kr/graphite system [4] in order to obtain information about the effective interactions in that system. Our principle conclusion is that the nearest-neighbor He-He repulsion is greatly reduced from the bare Lennard-Jones value, an effect which is due, presumably, to quantum correlations, and that the further neighbor attractions are substantially enhanced.

II. RENORMALIZATION GROUP CALCULATION

The lattice gas is governed by the Hamiltonian

$$H = \tfrac{1}{2}\sum_{i,j} V_{i,j} n_i n_j - \mu \sum_i n_i$$

where μ is the chemical potential. As is well-known the model is equivalent to an Ising model in a magnetic field and it is convenient to discuss the problem in terms of the magnetic analogy. When the nearest-neighbor interactions of the lattice gas are repulsive, the case which we consider, those of the Ising model are antiferromagnetic. For applied magnetic fields which are not too large, the ground state of the magnet consists of alternating up and down spins. Equivalently the honeycomb lattice can be decomposed into two triangular sublattices. In the ground state one sublattice consists entirely of up spins and the other of down spins.

We treat the problem by position-space RG methods and decompose the sites into cells of four sites as shown in Fig. 1. Note that the cells also form a honeycomb lattice. In the ordered state, the central spin of a cell has the opposite orientation of the three outer spins. Following Berker et al. [2] we shall map disordered states of the cell to an empty state, or vacancy. Unlike Ref. [2], however, in which the vacancies were introduced in a one-time-only prefacing transformation, we follow Nienhuis et al. [5] and treat the original

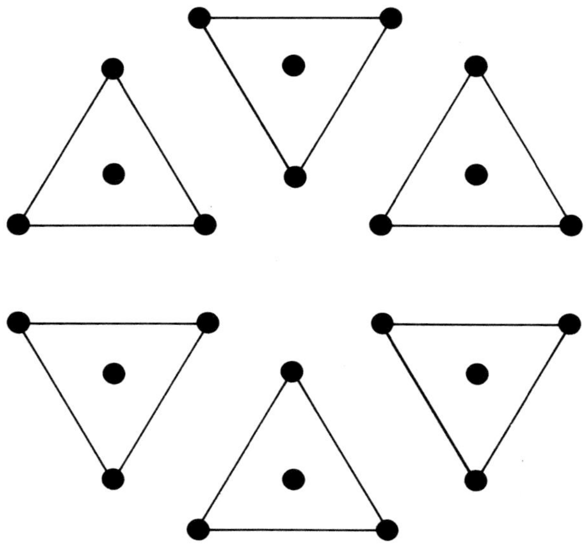

Fig. 1. Decomposition of honeycomb lattice into cells.

Ising lattice as the vacancy-free limit of the two-state Potts lattice gas [2] and apply the same RG transformation from the first step onward.

We consider that at each site there is a variable τ_i which is zero if the site is vacant or unity if the site is occupied. In the latter case there is a second variable S_i which can take the values ± 1. In our approximation we consider two cells only, periodically repeated, so that, in the presence of a magnetic field the most general (reduced) Hamiltonian we need consider is

$$-\beta H = \sum_{\langle ij\rangle} \tau_i\tau_j [K+J\delta_{S_iS_j}+L(\delta_{S_i,+}+\delta_{S_j,+})] - \sum_i \tau_i(\Delta+B\delta_{S_i,+}) \tag{1}$$

where the first sum is over nearest-neighbor pairs. Note that if S_i could take on more than two values there would be an additional term $M\tau_i\tau_j\delta_{S_iS_j}\delta_{S_i,+}$ in the first summation. Our approximation is completely determined upon specification of the projection operator which we take to be a product of projection operators for each cell. Our choice for the cell projection operator is most easily specified by defining

$$u \equiv \sum_{i=1}^{3} \tau_i S_i - \tau_4 S_4$$

where site four is the central site of the cell. This parameter measures the number of right minus wrong spins in the cell where right and wrong are defined relative to one of the order states. The absolute value of u takes its largest value, four, in either ordered state.

Our choice of projection operator is as follows. If $|u|>2$, the renormalized cell variables take the values $\tau'=1$, $S'=\mathrm{SIGN}(u)$. If $|u|<2$, the cell is mapped to a vacancy $\tau'=0$. For the intermediate case $|u|=2$ the cell is mapped to a vacancy $\tau'=0$ with weight $W \equiv 1 + \frac{1}{2}(\sum_{L=1}^{4}\tau_L-4)p$ and to $\tau'=1$, $S'=\mathrm{SIGN}(u)$ with weight 1-W. This choice permits the initial vacancy-free configuration to generate vacancies. The parameter p is at our disposal. We fix it by requiring that the transition temperature in zero field of the nearest-neighbor Ising antiferromagnet which we obtain in our approximation be equal to its exact value [6] $J^{-1}=0.759$. This fixes p at 0.3 and completes the specification of the RG transformation.

From the critical fixed point of our RG transformation we obtain $y_T=0.93$ (exact $y_T=1$) and find that the magnetic field is irrelevant as it should be. We further find the minimum ordering density for the lattice gas with only nearest-neighbor repulsions to be 0.55. The estimate by transfer-matrix methods of this quantity [7] is 0.58. The results of our RG calculation which employs vacancies can be compared to that of Mahan and Claro [8] who used the same choice of cells but employed no vacancies. This procedure yields $y_T=0.55$ and a relevant magnetic field. The maximum ordering temperature obtained is $J^{-1}=0.588$ and the minimum ordering density is 0.58.

Further neighbor attractions are easily accomodated in our approximation. Although they do not alter the form of the renormalized Hamiltonian, Eq.(1), which is considered, they do alter the initial conditions of the RG flow. Upon introducing further neighbor attractions, we find that a line of first-order transitions is introduced which meet the line of second-order transitions at a tricritical point. The critical exponent obtained for this point is $y_T=1.98$ which is to be compared to the value $y_T=1.8$ conjectured by Nienhuis et al. [5].

It appears from the above that our simple approximation which employs vacancies yields rather good results for the honeycomb lattice gas.

III. APPLICATION

Tejwani et al. [4] have obtained the phase diagram of He adsorbed on Kr plated graphite. There exists a line of second-order transitions with a ferromagnetic Ising-like specific heat, a line of first-order transitions and a tricritical point. The most complete data was taken when the Kr was apparently in registry with the graphite substrate. This could be ascertained from the fact that upon increasing the Kr coverage the maximum transition temperature of the He order-disorder transition increased as would be expected.

We have used the approximation outlined above and compared it with the experimental data in order to determine the interaction between the He atoms adsorbed on the Kr/graphite substrate. We consider the interactions out to fifth neighbors and denote them by V_1, V_2, V_3, V_4, and V_5. We assume that V_2/V_3, V_2/V_4, and V_2/V_5 are equal to the values they would have using the bare Lennard-Jones 6-12 potential at the appropriate distances. The values of V_1 and V_2 are then determined by adjusting them until the calculated values of two temperatures agree with the experimental ones. These are the maximum transition temperature 3.1°K and the tricritical temperature 1.9°K. The resulting phase diagram is shown in Fig. 2 where it is compared with the experimental results. The fit is rather good. The values of the interactions which result are, in degrees Kelvin, $V_1 = 3.41$, and $V_2 = -3.75$. The remaining values follow from the assumed ratio and are $V_3 = -1.6$, $V_4 = -0.31$, $V_5 = -0.15$. For comparison, the bare Lennard-Jones values V(r) at the same distances are V(2.46Å) = 13.37, V(4.26Å) = -1.82, V(4.92Å) = -0.79, V(6.51Å) = -0.15, V(7.38Å) = -0.08.

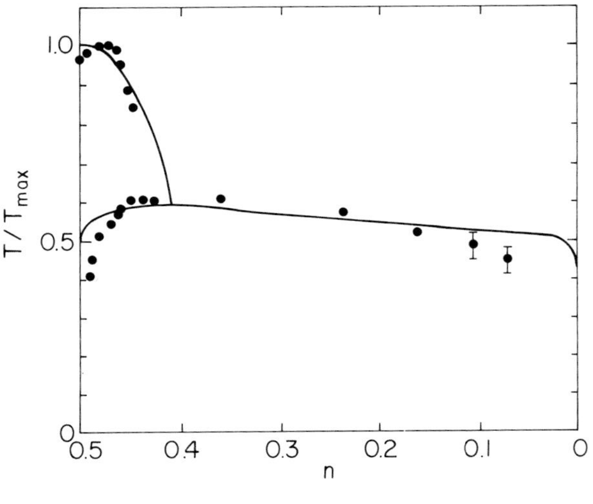

Fig. 2. Comparison of calculated and experimentally measured phase diagrams.

It is at first surprising that the magnitude of the ratio $|V_2/V_1| = 1.1$ is so large. However this might have been anticipated from the fact that the experimental value of the tricritical temperature is about 0.6 of the maximum transition temperature. Previous calculation on the square lattice [9] indicates that this requires a large ratio of $|V_2/V_1|$.

Secondly we note that V_1 is substantially less than the bare Lennard-Jones value. However it must be remembered that the V_1 of the lattice gas approximation is derived from a matrix element of the inter-particle potential taken between states localized on nearest-neighbor sites. Due to correlated motion of the helium, the repulsive interaction can be substantially reduced. Lastly it is seen that the further neighbor attractions are enhanced over their bare values. The source of this enhancement is unclear. We have also investigated the effect of three-particle interactions and have found that their inclusion does not enhance the fit to the experimental data.

In conclusion, we have performed a simple RG transformation on the honeycomb lattice gas which treats it as the vacancy-free limit of a two-state Potts lattice gas. The results appear to be rather good. We have then applied this model to the experimental system of He/Kr/graphite and found that the measured phase diagram can be well fit and have used the fit to extract the effective He-He interactions.

Support from the National Science Foundation under grants DMR79-20785 and 79-26405 is gratefully acknowledged.

REFERENCES

1. M. Schick, J.S. Walker, and M. Wortis, Phys. Rev. B16, 2205 (1977); B. Mihura and D.P. Landau, Phys. Rev. Lett. 38, 977 (1977); S. Ostlund and A.N. Berker, Phys. Rev. Lett. 42, 843 (1979).
2. A.N. Berker, S. Ostlund, and F.A. Putnam, Phys. Rev. B17, 3650 (1978).
3. E. Domany, M. Schick, and J.S. Walker, Phys. Rev. Lett. 38, 1148 (1977); E. Domany and M. Schick, Phys. Rev. B9, 3828 (1979).
4. M.J. Tejwani, O. Ferreira, and O.E. Vilches, Phys. Rev. Lett. 44, 152 (1980).
5. B. Nienhuis, A.N. Berker, E.K. Riedel, and M. Schick, Phys. Rev. Lett. 43, 737 (1979).
6. R.M.F. Houtappel, Physica (Utrecht) 16, 425 (1950).
7. L.K. Runnels, L.L. Combs, and J.P. Salvant, J. Chem. Phys. 47, 4015 (1967).
8. G.D. Mahan and F.H. Claro, Phys. Rev. B16, 1168 (1977).
9. B. Nienhuis and M. Nauenberg, Phys. Rev. B13, 2021 (1976).

CALCULATION OF FINITE-SIZE EFFECTS AND DETERMINATION OF CHARACTERISTIC SUBSTRATE SIZES

D.J.E. CALLAWAY and M. SCHICK

Department of Physics, University of Washington, Seattle, WA 98195

ABSTRACT

We show how real-space renormalization group methods can be used to extract the characteristic size of homogeneous adsorption areas from experimental data. We first calculate the expected specific heat signal from a single system of finite size. Results for the square lattice are in excellent agreement with exact results of Ferdinand and Fisher. Then the substrate is treated as a distribution of systems of differing sizes. Employing a honeycomb lattice and comparing with the data of Tejwani, Ferreira, and Vilches we deduce that 1) the characteristic size of homogeneous adsorbing regions of Kr-plated graphite is about 140 Å and 2) about 50% of the adsorbed helium does not take part in the order-disorder transition.

I. INTRODUCTION

In the experimental investigation of critical phenomena, the effects of finite size must always be considered. In particular, sufficiently close to the transition temperature of the infinite system, the thermodynamic functions of the finite and infinite systems differ significantly. Although those of the former change rapidly with external fields, they do not exhibit the singularities characteristic of the latter. The evolution of the singularities as the size of the system increases is predicted by finite-size scaling theory [1]. Complete information on the thermodynamic functions of the finite system can be obtained in some cases where the infinite system is exactly soluble [2]. However, in the general case, approximations must be employed. It seems to have been generally recognized that renormalization-group (RG) methods are particularly well-suited for this problem, yet only a single such calculation has been carried out [3]. Furthermore, no calculation addresses the experimentally relevant problem of a sample composed of a distribution of sub-systems of various sizes.

In this paper we present a calculation of the thermodynamic properties of a finite-size two-dimensional system by means of an RG calculation which differs significantly from that of Berker and Ostlund [3]. Our approximate results are in good agreement with the exact results of Ferdinand and Fisher [2]. We then show how the calculation can be combined with a distribution of subsystems of varying size and compared with experimental results to obtain characteristics of the substrate such as the largest size of homogeneous regions available for adsorption. We consider as an example the specific heat of the order-disorder transition exhibited by He on Kr-plated graphite studied by Tejwani et al [4]. By fitting the measured specific heat to that due to an assembly of systems of different size, we find that the resulting distribution is skewed, as expected , and characterized by a mean size of about 140 Å and a mean-square-deviation of 106 Å. Furthermore we find that some 50% of the adsorbed atoms do not take part in the transition. Presumably they are adsorbed on sites of strong binding or are otherwise trapped. Both of these results support previous observations.

II. RENORMALIZATION GROUP METHOD

The renormalization group transformation maps a Hamiltonian $H_N(K,\{S\})$ governing N spins S and characterized by the couplings K to a Hamiltonian of the same form $H_{N'}(K',\{S'\})$ governing $N'< N$ spins S' and characterized by a set of couplings K'. The mapping can be written

$$\exp[H_{N'}(K',\{S'\}) + Ng] = \sum_{\{S\}} P(\{S\},\{S'\})\exp H_N(K,\{S\}) \tag{1}$$

where P is a projection operator subject to

$$\sum_{\{S'\}} P(\{S\},\{S'\}) = 1 \,. \tag{2}$$

The renormalized couplings K' and the spin independent function g are both functions of the couplings K

$$K' = K'(K) \tag{3}$$
$$g = g(K) \tag{4}$$

It follows from (1) and the definition of the free energy per spin f that

$$f(K,N) = g(K) + b^{-d}f(K',b^{-d}N) \tag{5}$$

where we have written $N' = b^{-d}N$ with d the dimensionality.

With the aid of Eqs. (3) and (4) this equation can be iterated

$$f(K,N) = \sum_{n=0}^{m} \frac{g(K^{(n)})}{b^{nd}} + \frac{f(K^{(m+1)},b^{-(m+1)d}N)}{b^{(m+1)d}} \tag{6}$$

where $K^{(n)}$ is the n'th iterate and $K=K^{(0)}$. To determine the free energy per spin of the infinite system f(K) which is the limit of f(K,N) as N increases without limit, we take lim $N \to \infty$ followed by $m \to \infty$. For a finite system however f(K,N) can be obtained by iterating to a value of m chosen so that the number of spins $b^{-(m+1)d}N$ is small enough to permit the explicit calculation of $f(K^{(m+1)}, b^{-(m+1)d}N)$ which is then inserted into (6).

The task now is to formulate a simple approximation which, in the limit $N \to \infty$, yields the correct singularities of the thermodynamics functions. Berker and Ostlund [3] treat the q-state Potts lattice gas, an approximation to the lattice gas model of adsorbed systems, by an extremely simple approximation method. They adjust q so that the observed singularities of the infinite system are reproduced. The severity of the approximation, however, is indicated by the fact that in order to obtain the measured thermal exponent of the q = 3 state Potts model, the value of q in the approximation must be set to q = 20.8.

We consider lattice-gas models of adsorption or their Ising model equivalents. Our method is lattice sensitive, and we treat first the square lattice with nearest-neighbor interactions only. The lattice of sites is divided into cells of four sites each in the usual way and the projection operator is taken to be a product of cell-projection operators. The latter can be written

$$P(S',\{S\}) = \tfrac{1}{2} + AS'(S_1S_2S_3+S_1S_2S_4+S_1S_3S_4+S_2S_3S_4) + BS'(S_1+S_2+S_3+S_4). \quad (7)$$

If we denote by V_3 and V_4 the values taken by P when S' equals +1 and three or four respectively of the S_i are +1 then

$$A = (2V_4-4V_3+1)/16, \quad (8)$$

$$B = (2V_4+4V_3-3)/16 \quad (9)$$

The "majority rule" projection operator corresponds to $V_3=V_4=1$. Although this simple projection operator works well at high and low temperatures, it does not yield the exact critical exponents in any approximation employing a few cells. Our method consists in employing a K-dependent projection operator in which the weights V differ from the majority-rule values only in the neighborhood of the fixed point K^*. For the square lattice we employ a two-cell approximation with periodic boundary conditions. The projection operator is of the form Eq. (7)-(9) with

$$V_n(K) = 1 + (\tilde{V}_n-1)\exp[-(K-K^*)^2/2\sigma^2] \quad (10)$$

for n = 3,4. We require that the critical temperature $(K^*)^{-1}$ and the thermal exponent y_T be equal to their exact values 2.269 and 1 respectively. These requirements determine $\tilde{V}_3 = 1.081$, $\tilde{V}_4 = 1.002$, values which indicate that the deviation of the projection operator from majority rule is never very large. The parameter σ is chosen so that the recursion relations smoothly approach those of the majority rule. A value of 0.1 suffices.

Each iteration of our recursion relations reduces the degrees of freedom by a factor of four. At the last step the free energy per spin of a periodically continued cluster of two spins is substituted into Eq. (6). Therefore after n iterations we are evaluating the free energy of a rectangular Ising system of 2^{2n+1} spins with periodically continued boundaries. The specific heat is obtained by iterating the first derivative of the free energy and numerically differentiating once.

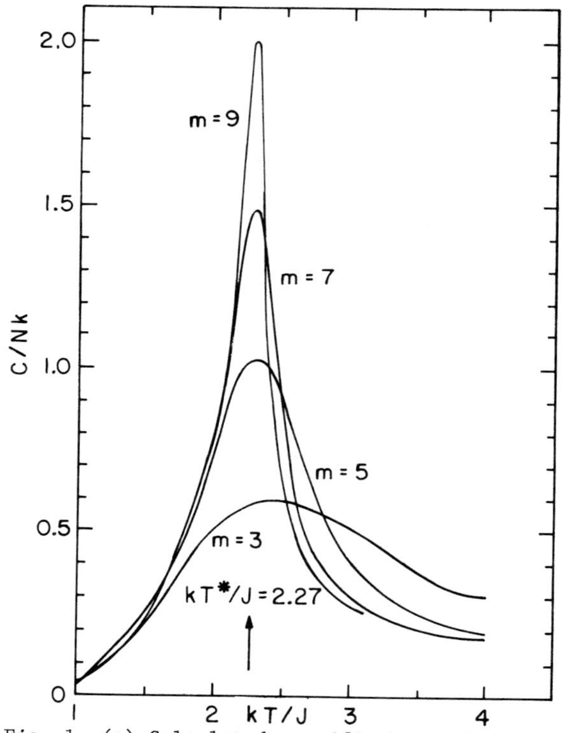

Fig. 1. (a) Calculated specific heat of square lattice of 2^m spins.

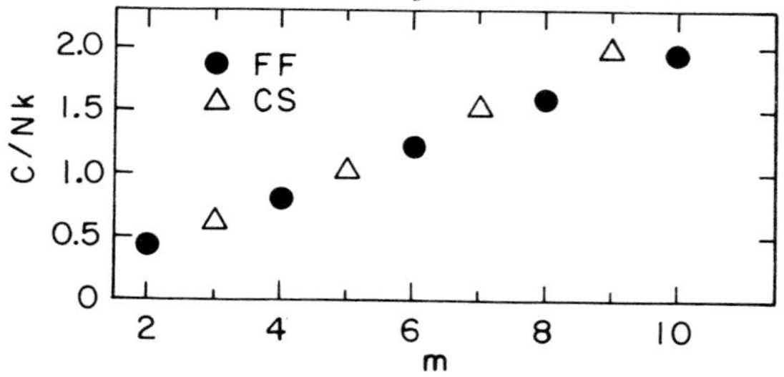

(b) Maximum specific heat of square lattice of 2^m spins. Circles - exact results of Ref. 2; triangles - approximate results of this paper.

Our results for the specific heat per spin of rectangular periodic systems with 2^3, 2^5, 2^7, and 2^9 spins are displayed in Fig. 1a. They are in qualitative agreement with the exact calculation of Ferdinand and Fisher [2]. The peak shapes are quite similar and in both cases the peak temperatures are higher than the critical temperature. Although the systems considered in [2] are of different size and shape so no detailed quantitative comparison is possible, we compare in Fig. 1b the peak heights of systems with 2^m spins as calculated in [2] and by us.

We next consider the honeycomb lattice-gas with repulsive nearest-neighbor interactions or its equivalent Ising model with antiferromagnetic interactions. The sites are grouped into cells of four sites each as shown in Fig. 2. The projection operator is taken to be a product of cell projection operators which are written

$$P(S',\{S\}) = \tfrac{1}{2} + AS'S_1S_2S_3 + BS'(S_1+S_2+S_3)$$

where S_1, S_2, S_3 are the spins on the perimeter of the cell. If we denote by V_2 and V_3 the value of P when S' = +1 and two or three of the spins S_1, S_2, S_3 are equal to +1 then $A = -\tfrac{1}{4}(3V_2-V_3-1)$, $B = \tfrac{1}{4}(V_2+V_3-1)$. The majority rule values are $V_2=V_3=1$ and in this form the projection operator was used by Mahan and coworkers [5]. In analogy with our work on the square lattice we write V_2, V_3 in the form of Eq. (10) for n = 2,3 and again require that $(K^*)^{-1}$ and y_T be given by their exact values 1.52 and 1. This yields $\tilde{V}_2 = 1.044$ and $\tilde{V}_3 = 1.003$. The parameter σ is again chosen to be 0.1. Specific heats of finite systems of 2^{2n+1} spins (or 2^{2n} atoms in the lattice gas) can now be obtained as before. The results are qualitatively similar to those of the square lattice.

III. SURFACE CHARACTERIZATION

Substrates used in adsorption experiments rarely consist of a single crystal face but rather are composed of several facets or grains of varying sizes. Thus an experimental measurement arises from a composite of the responses of systems of different sizes. If the composite system contains $M(N_i)$ sub-systems with N_i atoms each and a specific heat per atom c_i, the specific heat of the composite C_{TOT}/N is

$$C_{TOT}/N = \sum_i M(N_i)N_ic_i / \sum_i M(N_i)N_i = \sum_i \rho(N_i)c_i$$

where $\rho(N_i)$ is the fraction of atoms in systems of size N_i and $N = \sum_i M(N_i)N_i$. As the specific heats c_i have been calculated for systems of size $N_i = 2^{2i}$, a distribution $\rho(N_i)$ can be inserted into the above and a specific heat predicted or, conversely, the experimental specific heat can be fit and the distribution determined. Before carrying out the latter, there are two additions to the above we shall make. First this expression assumes that all adsorbed atoms take part in the order-disorder transition. It makes no allowance for the possibility that some of the atoms are adsorbed on "hot spots" with large heats of adsorption (edges, steps, etc.) or are otherwise lost to the order-disorder process. As this appears to be likely, we include in the above a term denoted by i = 0 characterized by non-zero N_0 but $c_0 = 0$. Second our lattice-gas specific heats contain no contribution from the kinetic energy, a contribution which is temperature dependent. We remedy this by fitting experimental data only in the vicinity of the transition temperature and adding to the above an analytic background B_0+B_1T. Thus we shall fit experimental data to the form

$$C_{TOT}/N = \sum_{i=0}^{\ell} \rho(N_i)c_i + B_0 + B_1T \qquad (11)$$

Fitting parameters are ℓ values of $\rho(N_i)$, B_0 and B_1. Clearly the precise values of $\rho(N_i)$ will be unreliable but we expect that certain characteristics of the resulting distribution will be stable with respect to variations in the value of ℓ for example.

We have employed such a procedure on the experimental data taken by Tejwani et al [4] on the system He/Kr/graphite foam. To test the stability of the calculated distribution we have used Eq.(11) with the largest system $N_\ell = 2^{2\ell}$ having ℓ= 6,8, and 10. The main features of the distribution were quite insensitive to these changes and the fit was excellent in all cases as might be

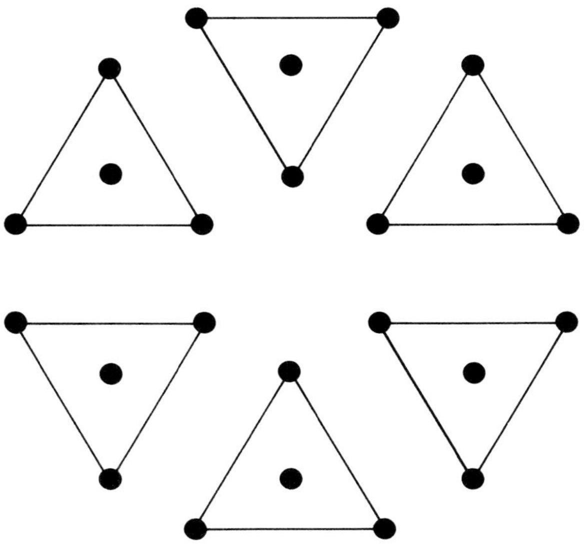

Fig. 2. Division of honeycomb lattice into cells.

TABLE I

Fraction ρ_i of atoms adsorbed on areas of characteristic size L_i. See text for i = 0+1.

i	0+1	2	3	4	5	6	7	8
L_i(Å)		15.9	31.7	63.5	127	254	508	1016
ρ_i(%)	49.97	4.56	7.46	8.01	13.1	16.3	0.5	0.1

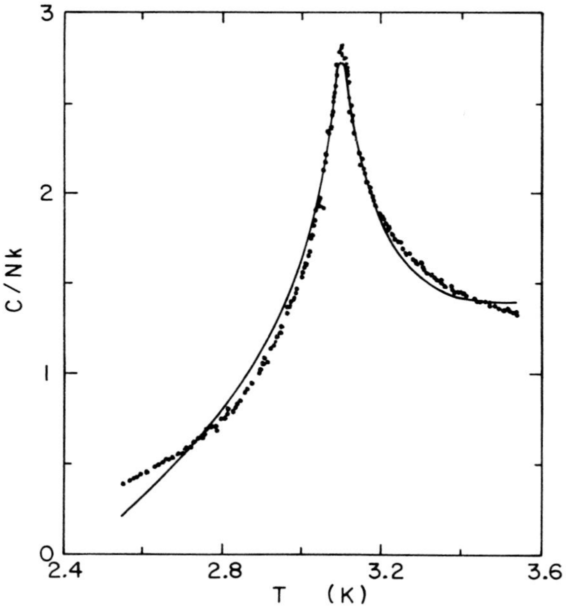

Fig. 3. Comparison of experimental data Ref. 4 and best fit from Eq. 11.

expected (see Fig. 3). These features are as follows. First there is a large peak shared by $\rho(N_0)$ and $\rho(N_1)$, the fraction of atoms which do not contribute at all and the fraction in the smallest cluster of four atoms. As such a cluster can hardly be said to undergo a transition we lump these two together. Their sum is equal to 0.50. Thus of the total number of adsorbed atoms, 50% do not participate in the transition at all. A qualitatively similar result has been noted in other experiments [6]. Second, the distribution of the atoms which do undergo the transition is skewed with the largest fraction adsorbed on small platelets of the order of 100 Å but a small percentage adsorbed on sizes up to 1000 Å. (The size can be obtained from the N_i, the geometry of the honeycomb lattice and the fact that the lattice gas density is ½.) If the 50% is ignored, the mean size of the adsorbing platelets is 140 Å with a standard deviation of 106 Å. The distribution with $\ell = 8$ is shown in Table I.

The characteristic size of adsorbtion areas is typically determined from lineshapes obtained from diffraction experiments together with a substrate model which includes the effects of a single characteristic length and the distribution of orientations of the crystallites [7]. This combination of experiment and modeling yields characteristic sizes of the order of 100 Å for Grafoil [7,8], and 450 Å for ZYX [9]. Thermodynamics measurements have also been combined with substrate models to obtain estimates of characteristic sizes for Grafoil [10] of 100 Å and for graphite foam [11] of 1200 Å. No results of either kind are available for Kr-plated foam.

We have shown that an alternative or perhaps supplementary procedure to determine characteristic sizes is to make use of the measured thermodynamic functions. This is possible because, as indicated, these functions of finite systems can be easily and reliably calculated.

Support from the Department of Energy and National Science Foundation under grant DMR 79-20785 is gratefully acknowledged.

REFERENCES

1. M.E. Fisher and M.N. Barber, Phys. Rev. Lett. 28, 1516 (1972).
2. A.E. Ferdinand and M.E. Fisher, Phys. Rev. 185, 832 (1969).
3. A.N. Berker and S. Ostlund, J. Phys. C 12, 4961 (1979).
4. M.J. Tejwani, O. Ferreira, and O.E. Vilches, Phys. Rev. Lett. 44, 152 (1980), M.J. Tejwani, dissertation, Univ. of Washington (unpublished).
5. K.R. Subbaswamy and G.D. Mahan, Phys. Rev. Lett. 37, 642 (1976); G.D. Mahan and F.H. Claro, Phys. Rev. B16, 1168 (1977).
6. See for example A. Thomy and X. Duval, Journal de Chimie Physique 66, 1966 (1969), E.M. Hammonds, P. Heiney, P.W. Stephens, R.J. Birgeneau, and P. Horn (to be published).
7. J.K. Kjems, L. Passell, H. Taub, J.G. Dash, and A.D. Novaco, Phys. Rev. B13, 1446 (1976).
8. K. Carneiro, Journal de Physique C4-1 (1977).
9. P.M. Horn,,R.J. Birgeneau, P. Heiney, and E. M. Hammonds, Phys. Rev. Lett. 41, 961 (1978).
10. T.T. Chung and J.G. Dash, Surf. Sci. 66, 559 (1977).
11. M. Beinfait, J.G. Dash and J. Stoltenberg, Phys. Rev. B21, 2765 (1980).

SECTION II

D. INTERCALATED MATERIALS

Sinha, ed. Ordering in Two Dimensions

ENTHALPY CHANGES OF THE PHASE TRANSITIONS IN GRAPHITE-HALOGENS*

J. C. WU, J. S. CULIK,† and D. D. L. CHUNG**

Department of Electrical Engineering, Carnegie-Mellon University, Pittsburgh, PA 15213

ABSTRACT

Differential scanning calorimetry (DSC) has been performed on graphite-halogens (Br_2, ICl) to investigate the enthalpy changes that accompany the phase transitions in these materials. For each intercalate species, two or more reversible transitions have been observed. X-ray and electron diffraction results have shown the correspondence of some of these DSC peaks with phase transitions associated with the intralayer intercalate ordering and the interlayer intercalate layer ordering. The concentration dependence of the enthalpy change (in cal/mole intercalate) of a transition shows different trends for intralayer and interlayer transitions.

INTRODUCTION

Graphite intercalation compounds exhibit phase transitions associated with the intralayer and interlayer ordering of the intercalate [1]. These transitions affect the stability [2] and the properties [3,4] of these compounds. Therefore, information on the transitions is essential for the application of graphite intercalation compounds at various temperatures.

This paper focuses on the use of differential scanning calorimetry (DSC) to study the intralayer intercalate ordering and the interlayer intercalate layer ordering in graphite-halogens (Br_2, ICl). The intralayer intercalate ordering refers to the ordering of the intercalate molecules within an intercalate layer; the interlayer intercalate layer ordering refers to the correlation between different intercalate layers, such as the $\alpha\beta\gamma\delta$ stacking sequence in stage 1 graphite-K (C_8K) [5].

EXPERIMENTAL TECHNIQUE

This work was performed by using a Perkin-Elmer DSC-2 differential scanning calorimeter, which is programmed to vary the average temperature of the sample pan and the reference pan (typically empty) at a constant rate, such as 10^oK/min. Differential power is supplied to keep the temperature of the sample pan equal to that of the reference pan. As the average temperature changes, the power supplied to the sample with respect to the reference is constant, except when the sample undergoes a phase transition that involves latent heat. When this occurs, the power supplied to the sample changes with respect to that supplied to the reference and results in a maximum or minimum on the strip-chart recorder trace. The transition temperature is defined as the temperature corresponding to the intersection of the leading edge of the DSC trace with the baseline.

Samples were prepared by vapor phase intercalation of highly-oriented pyrolytic graphite (HOPG), which had typical dimensions of ∿4 x 4 x 0.5 mm. Graphite-ICl (graphite-Br_2) of various intercalate concentrations were prepared by using ICl-CCl_4 (Br_2-CCl_4) solutions. X-ray diffraction was used to obtain information on the stage.

In order to determine the enthalpy change per mole of intercalate due to a phase transition, sample weight was obtained both before and after a DSC run with a Perkin-Elmer AD-2Z microbalance, which had an uncertainty of ±0.002 mg.

During the DSC measurement, the sample was mounted in an unsealed aluminum pan with an aluminum cover; the sample holder was purged with dry argon at 20 cc/min.

RESULTS

Graphite-ICl

Shown in Fig. 1 are DSC thermograms obtained during heating for graphite-ICl of various intercalate concentrations. Two endothermic peaks were observed at 307^oK and 314^oK. On cooling, exothermic peaks were observed at 310^oK and 302^oK. The temperatures of both peaks are independent of the intercalate concentration,

*Research sponsored by the National Science Foundation.
†Present Address: Solarex Corp., 1335 Piccard Dr., Rockville, MD 20850.
**Also in the Department of Metallurgy and Materials Science.

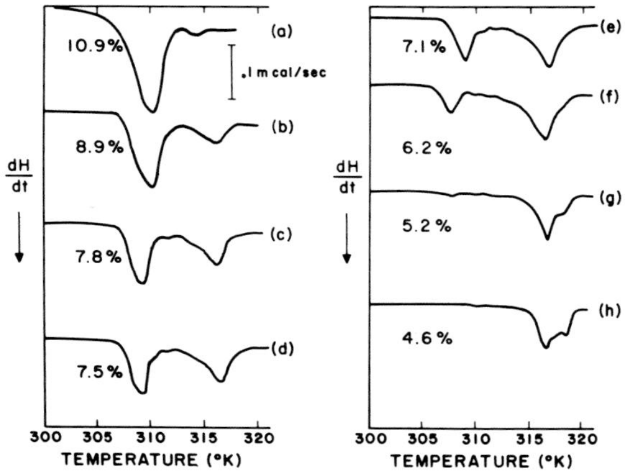

Fig. 1 DSC thermograms of graphite-ICl of various intercalate concentrations (in mole % ICl) during heating.

but the size of each peak is strongly dependent on the intercalate concentration. The different intercalate concentrations (in mole % ICl) were obtained by desorption of a stage 1 compound for various lengths of time. The size of the 307°K peak decreases as the ICl concentration decreases, whereas that of the 314°K peak increases as the ICl concentration decreases. The concentration dependence of the enthalpy change (in cal/mole intercalate) of each peak is graphically shown in Fig. 2. The enthalpy change of the 307°K peak is 500 cal/mole at 12.5 mole % ICl (stage 1) and decreases monotonically to zero at ∿5 mole % ICl (stage 2-3). On the other hand, the enthalpy change of the 314°K peak is zero at 12.5 mole % ICl, increases monotonically to 570 cal/mole ICl at ∿4 mole % ICl, and is constant for concentrations less than ∿4 mole % ICl.

To identify the two DSC peaks with structural transitions, consider results of electron and X-ray diffraction. The X-ray diffraction results of Turnbull and Eeles [6] showed that graphite-ICl at 304°K has a three-dimensional crystal structure for stage = 1, but is nearly two-dimensionally ordered for stage ≥ 2. A two-dimensional ordering means that there is ordering within an intercalate layer but there is no correlation between different intercalate layers. The electron diffraction work of Chung [7] showed that graphite-ICl undergoes in-plane melting of the intercalate at 316 ± 10°K. This observation has recently been confirmed by X-ray diffraction [8]. Combination of the above diffraction results indicates that (i) for stage 1 graphite-ICl, there is a phase transition at a temperature between 304 and 316°K, because the intercalated ICl forms a three-dimensional structure at 304°K and is disordered at temperatures above 316 ± 10°K, (ii) for graphite-ICl of stage ≥ 2, there is a change from nearly two-dimensional order to disorder at 316 ±10°K. Therefore, the 314°K DSC peak is identified as ude to the melting of the intercalate layers. Since the temperature of the phase transition associated with the stacking order of the intercalate layers should be lower than that of the melting of the intercalate layers, we interpret the 307°K DSC peak as due to a phase transition associated with the stacking order of the intercalate layers.

This interpretation of the 307°K DSC peak is consistent with the concentration dependence of the apparent enthalpy change (Fig. 2). The decrease of the apparent enthalpy change of the 307°K peak as the intercalate concentration decreases is expected from the fact that no interlayer phase transition between 304°K and 316°K was observed for stage ≥ 2. As the originally stage 1 sample undergoes desorption, i.e., as the intercalate concentration decreases, the sample becomes a mixture of stages. Since only stage 1 contributes to the 307°K peak's enthalpy change, which is normalized by the total ICl concentration to be in the unit of cal/mole ICl, the enthalpy change decreases and is zero when there is no stage 1 component in the sample. Since a partially desorbed sample of concentration corresponding to that of a pure stage 2 compound is a mixture of stages and thus is partly stage 1, the stage 1 component vanishes at an overall sample concentration between that corresponding to a pure stage 2 compound and that of a pure stage 3 compound. This explains the fact that the extrapolation of the enthalpy change vs. concentration curve for the 307°K peak in Fig. 2 intersects the zero enthalpy change axis at a concentration between stage 2 and stage 3 rather than at the stage 2 concentration.

To give further evidence for the association of the 307°K peak with the stage 1 component of a sample, we prepared stage 2 graphite-ICl by using a solution of ICl in CCl_4. X-ray diffraction confirmed that there was no stage 1 component in the sample. Indeed, no DSC peak was observed in this sample at 307°K.

The interpretation of the 314°K DSC peak as the melting of the intercalate layers is also consistent with the concentration dependence of the enthalpy change (Fig. 2). Since the in-plane bonding is essentially not affected by the change in concentration, the enthalpy change of the intralayer phase transition should be independent of the concentration. Fig. 2 shows that the enthalpy change of the 314°K peak is independent of the concentration for low concentrations and decreases to zero at high concentrations. This observation can be explained by assuming that the three-dimensional structure of the intercalate in a stage 1 compound is completely disordered on heating

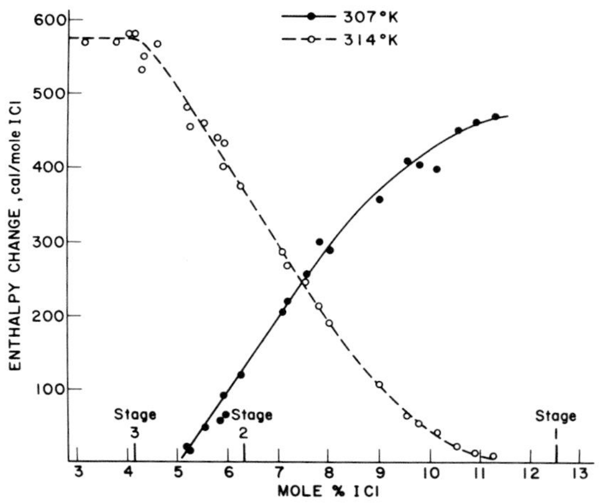

Fig. 2 Dependence of enthalpy change on intercalate concentration for the 307°K and 314°K peaks of graphite-ICl

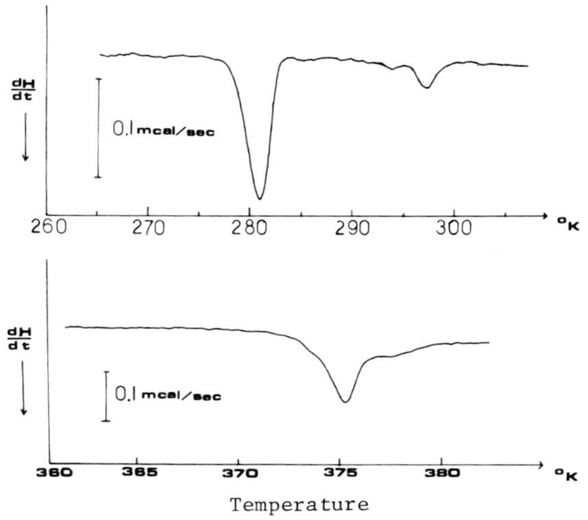

Fig. 3 DSC thermogram of graphite-Br_2 during heating

through 307°K. In other words, both the in-plane order and the stacking order of the intercalate layers are lost at 307°K. This assumptiom is suggested by the diffraction data of Turnbull and Eeles [6], but further X-ray diffraction work is needed for its verification. This assumption indicates that a completely stage 1 compound does not exhibit the 314 °K DSC peak, but exhibits only the 307°K peak. Thus, only stages $\geq$ 2 contribute to the 314°K peak's enthalpy change.

To give further evidence for the association of the 314°K peak with components of stage $\geq$ 2, we prepared stage 2 graphite-ICl, i.e. no stage 1 component. The enthalpy change of the 314°K peak of this sample remained constant at $\sim$570 cal/mole ICl as the sample desorbed. Thus the enthalpy change is indeed independent of the concentration for stage $\geq$ 2.

Graphite-Br_2

Shown in Fig. 3 is the DSC thermogram of graphite-Br_2 obtained during heating. Three endothermic peaks are observed in the temperature range 220-390°K. They occur at 277°K, 297°K, and 373°K. On cooling, exothermic peaks were observed at 373°K and 286°K. The 373°K peak has been identified as due to the melting of the intercalate layers, as indicated by electron [7] and X-ray [8] diffraction results.

The concentration dependence of the apparent enthalpy changes of the 277°K and 373°K peaks is shown in Fig. 4. Because of exfoliation, which occurs at $\sim$ 350°K for high concentration graphite-Br_2 [4], information on the 373°K peak could only be obtained on dilute samples. The data on the 277°K peak were obtained during

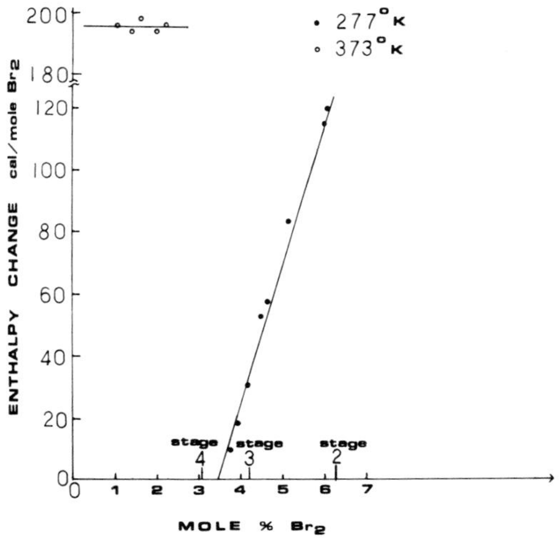

Fig. 4 Dependence of enthalpy change on intercalate concentration for the 277°K and 373°K peaks of graphite-Br_2

desorption of a stage 2 graphite-Br_2 compound. Comparison of Figures 4 and 2 indicates that the 277°K peak of graphite-Br_2 behaves like the 307°K peak of graphite-ICl. The apparent enthalpy change of the 277°K peak reaches zero at a concentration between that of pure stage 3 and that of pure stage 4. As in the study of the 307°K peak on graphite-ICl, we prepared a stage 3 graphite-Br_2 compound by using a Br_2-CCl_4 solution. X-ray diffraction indicated that no stage 2 component existed in the sample. NO DSC peak was observed in the sample at 277°K.

Therefore, only the stage 2 component contributes to the enthalpy change of the 277°K peak of graphite-Br_2. We tentatively interpret this peak as due to a phase transition associated with the stacking order of the intercalate layers and that only stage 2 exhibits such a transition. Due to exfoliation, the behavior of stage 2 at 373°K is not known. The three-dimensional ordering of stage 2 at 77°K is suggested by the X-ray diffraction results of Eeles and Turnbull [9].

DISCUSSION

Diffraction and DSC results strongly suggest that stage 1 graphite-ICl loses both interlayer and intralayer intercalate ordering at 307°K and that the intercalate in graphite-ICl of stage $\geq$ 2 undergoes two-dimensional melting at 314°K.

Consider the interlayer and intralyer bonding between intercalate molecules. The distance between nearest intercalate layers in stage 1 graphite-ICl is ∿7Å. Although the in-plane crystal structure of graphite-ICl is not completely known, a rough estimate of the average in-plane distance between adjacent ICl molecules is ∿ 8.5 Å, which is the value in stage 1 graphite-K[5]. Thus the interlayer and intralayer spacings between nearest intercalate molecules are comparable in stage 1 graphite-ICl. This implies that, in stage 1 graphite-ICl, the intercalate layers are so close together that the bonding between ICl molecules in different layers is comparable to that between ICl molecules in the same layer. Thus, when the interlayer bonding is broken as the temperature is raised, the intralayer bonding is broken too. This is a possible explanation for the fact that stage 1 graphite-ICl loses both intralayer and interlayer intercalate ordering at 307°K.

For stage 2 graphite-ICl, the distance between adjacent intercalate layers is 10.35 Å. This relatively large distance reduces the interaction between intercalate layers so that the intercalate ordering is nearly two-dimensional even at temperatures below 307°K.

REFERENCES

1. J. S. Culik and D. D. L. Chung, Mater. Sci. Eng. 37, 213 (1979).
2. J. S. Culik and D. D. L. Chung, Mater. Sci. Eng. 44, (1980).
3. D. G. Onn, G. M. T. Foley and J. E. Fischer, Mater. Sci. Eng. 31, 271 (1977)
4. A. R. Ubbelohde, Proc. Roy. Soc. A304, 25 (1968).
5. W. Rüdorff and R. Schulze, Z. Anorg. Allg. Chem. 227, 156 (1954).
6. J. A. Turnbull and W. T. Eeles, 2nd Conf. on Industrial Carbon and Graphite, 1965, Soc. Chem. Ind., London, 1966, p.173.
7. D. D. L. Chung, J. Electron Mat. 7, 189 (1978).
8. K. K. Bardhan and D. D. L. Chung, this proceedings.
9. W. T. Eeles and J. A. Turnbull, Proc. Roy. Soc. London, Ser. A, 283, 179 (1965).

X-RAY DIFFRACTION STUDY OF THE PHASE TRANSITIONS IN GRAPHITE-HALOGENS*

K. K. BARDHAN

Department of Physics

and

D. D. L. CHUNG

Department of Metallurgy & Materials Science
And Department of Electrical Engineering

Carnegie-Mellon University
Pittsburgh, PA 15213 USA

ABSTRACT

X-ray diffractometry has been performed on the a-face of dilute graphite-halogens (Br_2, ICl) based on highly-oriented pyrolytic graphite to study the intralayer ordering and its associated phase transitions. In contrast to the low stage samples used in earlier work, the samples used were desorbed compounds. Two structural transitions were observed in graphite-Br_2 at 373.7 °K and 226 °K; one was observed in graphite-ICl at 314 °K. The 373.7 °K transition in graphite-Br_2 and the 314 °K transition in graphite-ICl are associated with the melting of the ordered intercalate layers into liquidlike layers. The 226 °K transition in graphite-Br_2 corresponds to the change of streaks to spots in the (0001) zone electron diffraction pattern of graphite-Br_2 on cooling from room temperature to liquid nitrogen temperature. In addition, in graphite-Br_2, the first observation was made of a phase transition involving the shifting of various x-ray diffraction peaks in various directions at ~340 °K. This transition is tentatively interpreted as a commensurate-incommensurate phase transition.

Although considerable X-ray diffraction work has previously been done to study the crystal structure of graphite intercalation compounds, most of the information reported is concerned with the stage ordering (the stacking order of the carbon and the intercalate layers), as obtained from 00ℓ diffraction lines. Information about the in-plane structure of the intercalate layers, as can be obtained from hk0 lines, is scant. This is because most of the X-ray diffraction work was conveniently performed by reflection of the X-ray beam off the c-face (the cleavage plane) of compounds based on highly-oriented pyrolytic graphite (HOPG). We report here a reflection geometry in which the X-ray beam is reflected off the a-face of compounds based on HOPG. This method is complementay to electron diffraction; the latter remains the most convenient method of observing the (001)reciprocal lattice plane [1], whereas the former provides quantitative intensity measurement.

In order to perform X-ray diffraction by reflection off the a-face, we used samples of which the surface area along the c-direction is much greater than that perpendicular to the c-direction. The samples were obtained from HOPG with a large thickness along the c-direction by cutting with a low speed diamond saw. Because optical flatness is not required for X-ray diffraction work, no subsequent surface preparation procedure [1] was necessary. The orientation of the sample with respect to the X-ray beam is illustrated in Fig. 1. Due to the texture of HOPG, all the diffraction vectors nearly lie on the basal plane, so that hk0 lines are observed. Of importance is that the geometry shown in Fig.

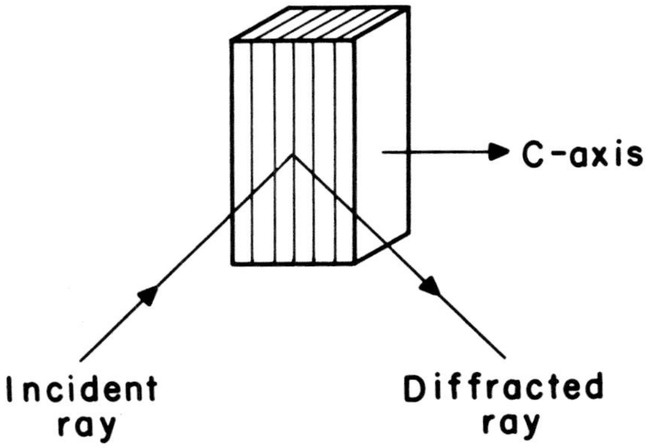

Fig. 1 Geometry of X-ray diffraction experiment

*Research sponsored by the National Science Foundation

1 for an ideal HOPG sample does not permit the observation of any lines other than hk0 lines. This implies that this geometry yields hk0 lines that are more intense than the corresponding ones obtained on powder samples.

Because of intercalate desorption, which increases in rate as the temperature increases and becomes anomalously fast near the in-plane melting temperature [1], X-ray diffraction was performed as a function of temperature on samples which had been desorbed at a temperature above the temperature range of interest to the diffraction experiment. The use of desorbed samples minimized the variation of the intensity due to desorption. This is in contrast to earlier work on graphite-alkali metals [5], which used low stage samples without prior desorption. On the other hand, the use of desorbed samples restricts the investigation to relatively dilute (high stage) samples. However, because of the independence of the intralayer intercalate ordering on the intercalate concentration in graphite-halogens [4], this restriction is not of concern to the study of the in-plane phase transitions in graphite-halogens.

The X-ray diffraction pattern of dilute graphite-Br_2 at 298.8 °K is shown by a full line in Fig. 2. The pattern was obtained with the geometry illustrated in Fig. 1, using CuKα radiation (λ= 1.542Å). Besides the various hk0 intercalate peaks and the 100 graphite peak, there is a relatively intense graphite 002 peak at q = 1.87$Å^{-1}$, where q is defined as

$$q = \frac{4\pi \sin\theta}{\lambda}$$

The 002 peak is observed in this geometry possibly because of scattering from the edge.

Similar measurements have been made on graphite-Br_2 compounds covering the entire range of intercalate concentration. The diffraction pattern was found to be qualitatively independent of the intercalate concentration, except that the relative intensities of the intercalate peaks and graphitic peaks increased with intercalate concentration. The observation that the in-plane diffraction pattern does not change with intercalate concentration is consistent with electron diffraction results [1].

As the sample temperature was raised, the intercalate peaks gradually decreased in intensity and finally vanished at 373.7 °K, leaving diffused peaks characteristic of a disordered phase with a short-range order. Shown by a dotted line in Fig. 2 is the X-ray diffraction pattern of graphite-Br_2 at 376 °K. This transition at 373.7 °K corresponds to the order-disorder transition associated with the melting of the intercalate layers. The transition temperature obtained is in agreement with the results of electron diffraction [1] and calorimetry [2].

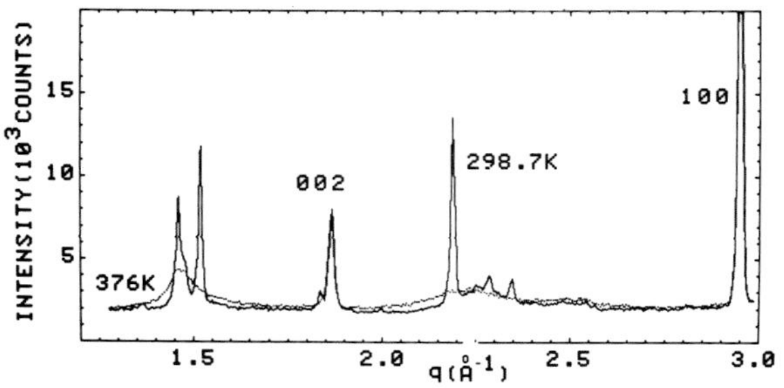

Fig. 2 X-ray diffraction patterns of graphite-Br_2 at 298.8 °K (full line) and 374.4 °K (dotted line)

Three intercalate peaks at different temperatures below 373.7 °K are shown in Fig. 3. Of significance is that the peak positions shift as the temperature is changed. The variation of the position of the peaks at q ∿1.46 $Å^{-1}$ and 2.2 $Å^{-1}$

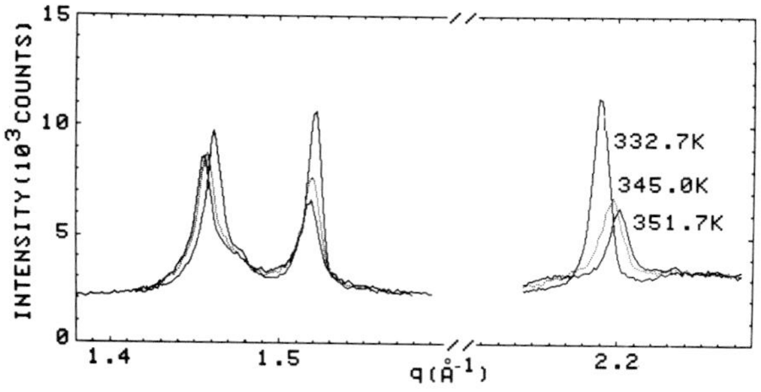

Fig. 3 A portion of the X-ray diffraction pattern of graphite-Br_2 at 332.7 °K, 345.0 °K, and 351.7 °K. All three peaks are due to the intercalate.

with temperature is shown in Fig. 4. Note that the peaks shift in opposite directions and the shift occurs for both peaks at ∿340 °K. This is interpreted as due to a commensurate-incommensurate phase transition at ∿340 °K.

In addition to the 373.7 °K transition, another structural transition was observed in graphite-Br_2 at ∿226 °K. Figure 5 shows the diffraction pattern obtained at ∿80.0 °K. In the range of q from 2.2 to 3.0$Å^{-1}$, numerous small sharp peaks appear on cooling below ∿226 °K. This transition involves only some of the in-plane superlattice peaks and corresponds to the change of streaks to spots in the (0001) zone electron diffraction pattern on cooling graphite-Br_2 from room temperature to liquid nitrogen temperature [1,3].

Figure 6 shows the integrated intensity of the

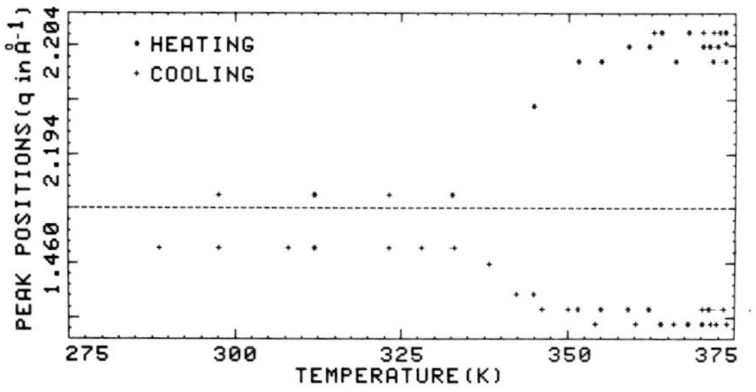

Fig. 4 Variation of peak position with temperature for the peaks at q∿1.46 $Å^{-1}$ and 2.2 $Å^{-1}$.

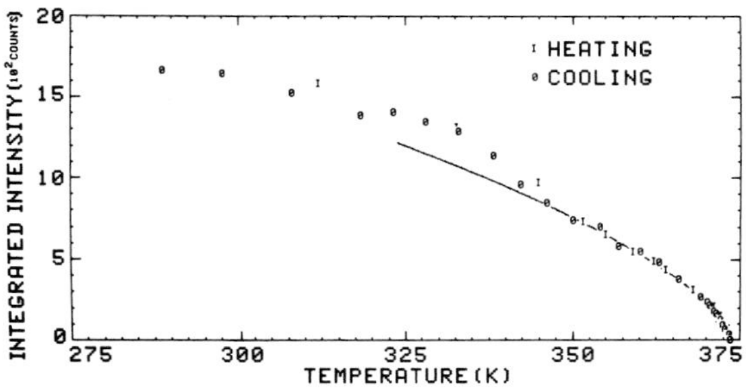

Fig. 6 Variation of the integrated intensity with temperature for the peak at q∿1.52 $Å^{-1}$

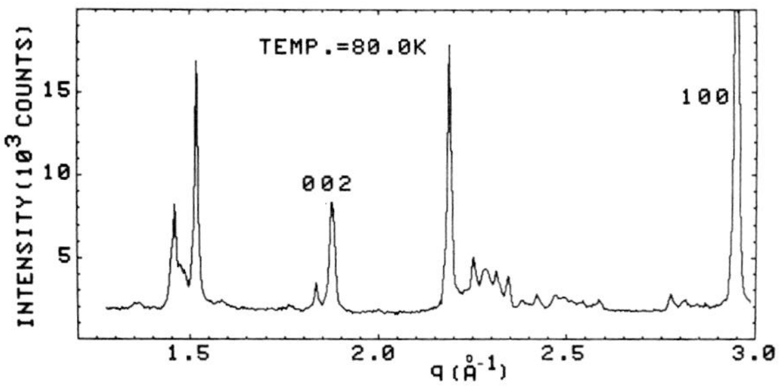

Fig. 5 X-ray diffraction pattern of graphite-Br_2 at 80.0 °K.

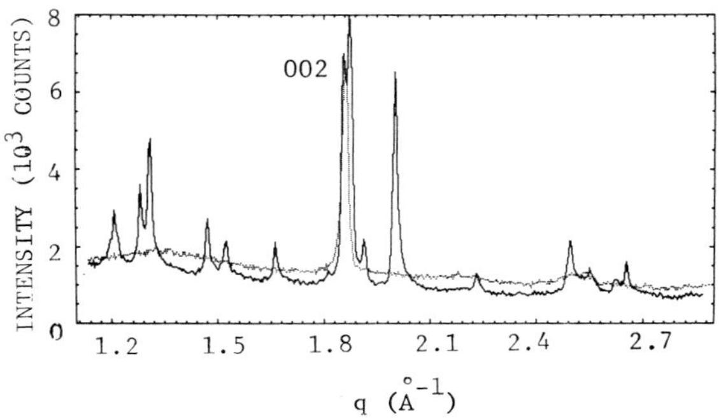

Fig. 7 X-ray diffraction patterns of graphite-ICl at 79.5 °K (full line) and 321 °K (dotted line)

intercalate peak at q ∿1.52$Å^{-1}$ as a function of temperature. The intensity falls as the temperature increases and becomes zero at 373.7 °K. In addition, a decrease in intensity was observed at ∿340 °K. This observation correlates with the appearance of an incommensurate phase at ∿340 °K. Also shown in Fig. 6 is the fit of the intensity vs. temperature curve to a power law, which gives the critical exponent β=0.32 for the 373.7 °K transition. Because of the incompletely known in-plane crystal structure of graphite-Br_2 the interpretation of this value of β is not pursued in this work.

The intensities of the various peaks have been found to vary rather smoothly with temperature before vanishing at 373.7 °K indicating that the 373.7 °K transition is of a higher order type.

Similar investigation was also carried out on dilute graphite-ICl, in which one reversible transition was observed at 314 °K, in agreement with electron diffraction [1] and calorimetry [2] results. Figure 7 shows X-ray diffraction patterns of graphite-ICl at 79.5 °K and 321 °K, respectively. This transition is similar in nature to the 373.7 °K transition of graphite-Br_2, i.e., it involves the disordering of the intercalate layers on heating, as indicated by electron diffraction results [1].

ACKNOWLEDGEMENTS

Stimulating discussions with Professor R. B. Griffiths and Dr. D. M. Butler of Carnegie-Mellon University are gratefully acknowledged.

REFERENCES

1. D. D. L. Chung, J. Electron. Mat. 7, 189 (1978).
2. J. S. Culik and D. D. L. Chung, Mater. Sci, Eng. 37, 213 (1979).
3. W. T. Eeles and J. A. Turnbul, Proc. Roy. Soc. (London) A283, 179 (1965).

Published 1980 by Elsevier North Holland, Inc.
Sinha, ed. Ordering in Two Dimensions

ORDER-DISORDER TRANSITION AND CHARGE-DENSITY WAVES IN GRAPHITE INTERCALATION COMPOUNDS

R. S. MARKIEWICZ

General Electric Research and Development Center
P. O. Box 8, Schenectady, New York 12301

ABSTRACT

It is suggested that the order-disorder transition in acceptor intercalated graphite compounds is driven by the ordering of the ionized fraction of molecules within a single layer. The resulting phase transition is three-dimensional, but depends only weakly on stage. The periodic ionic potential in turn induces a charge density oscillation of the holes in the bounding graphite layer.

CHARGE TRANSFER MECHANISM

Materials intercalated into graphite generally act as electron donors or acceptors, transferring electrons or holes to the graphitic layers. The nature of this charge transfer is not well understood, and many fundamental questions remain unanswered. In most cases, the fraction (f) of charge transferred per intercalant molecule is not well known. Nor is it understood how, if f is less than one, the charge is shared among intercalant molecules.

There is growing evidence [1] that for alkali metal atoms, which act as donors, the intercalant atoms form an s-like metallic conduction band at the Fermi level, but with $f<1$. On the other hand, for some of the strongly acidic acceptor species (AsF_5, SbF_5), a simple ionic interpretation seems valid. As proposed by Bartlett, et al [2], the neutral pentafluorides disproportionate according to

$$3MF_5+2e^-\rightarrow 2MF_6^-+MF_3$$

(M=As, Sb). If this reaction were carried to completion, it would lead to a charge transfer f=2/3, but substantially lower charge transfer is inferred from the electronic properties of the system [3]. Indeed recent experiments [4] have shown that AsF_5 gas can be pumped out of the intercalated graphite in vacuum. More direct evidence comes from structure studies [5] based on the (00ℓ)-x-ray reflection spectra, which show the presence of both AsF_5 and AsF_6^- molecules in the intercalate layers, in a ratio consistent with electronic values of f. While f shows some variation with stage and type of graphite used (natural flake vs. HOPG), it is in the range ∿.25-.50 for AsF_5, .12-.21 for SbF_5[6]. For HNO_3-intercalated graphite, the charge transfer is estimated to be ∿0.2-0.25 [7].

IONIC SUPERLATTICE AND PHASE TRANSITIONS

Pietronero, et al [8] have shown that the free carriers introduced into the graphite are predominantly localized on the carbon layers bounding the intercalant layer, screening out the electric fields produced by the excess charge in that layer, (although the electric fields are still strong enough to produce the characteristic staging of the compounds [9]).

Screening within the intercalant layer is much less effective, since the screening charge is localized on a separate atomic layer (separated by ∿4Å for AsF_5). Indeed, even if the holes were completely localized immediately above and below the ionized MF_6^- molecule ("perfect screening"), there would still be a quadrupolar repulsive force between any two ions. For the observed values of f, this long-range repulsion should cause the ions to order into a superlattice at a fairly high temperature.

A rough estimate of the transition temperature can be given as follows. For rational values of f, the ground state will be a simple ionic lattice in which the ions stay as far apart as possible. Fig. 1 shows a possible superlattice for f=.25. The remaining sites are occupied by neutral MF_5, MF_3, or vacancies. (In the stoichiometric compounds, there is one intercalated molecule for every eight C-atoms in a layer. The large molecular volume suggests that this is approximately a close packed arrangement). The simplest excited state in this system would be a Frenkel defect, in which an MF_6^- and a neutral molecule interchanged positions. If the energy needed to create a defect, E_D, is known, the melting temperature can be estimated by analogy to AgBr [10]. Here the melting is a first order phase transition, brought about by an avalanching of the number of Frenkel defects. The analogous melting temperature T_c for the intercalant layer is approximately $k_B T_c \approx E_D/9$, depending weakly on f.

The energy E_D can only be estimated very approximately even in such well studied ionic systems as NaCl and CaF_2 [11]. The principal contribution is the Madelung energy: the difference in electrostatic energies between the perfect lattice and the lattice with one Frenkel defect. For the intercalant layer, calculation of this energy requires knowledge of the extent of screening by the holes. This calculation was made both in the perfect screening limit defined above and in the unscreened case (uniform hole density). For f>.25, the two estimates differ by less than a factor of two; an estimate of the hole screening (below) suggests that the unscreened value is likely to be more accurate. The resulting energy may be written $E_D^{(1)} = \eta e^2/\varepsilon b$, where b is the ion-ion separation $\varepsilon \sim 3$ is an average in-plane dielectric constant, and η is a numerical factor which depends on

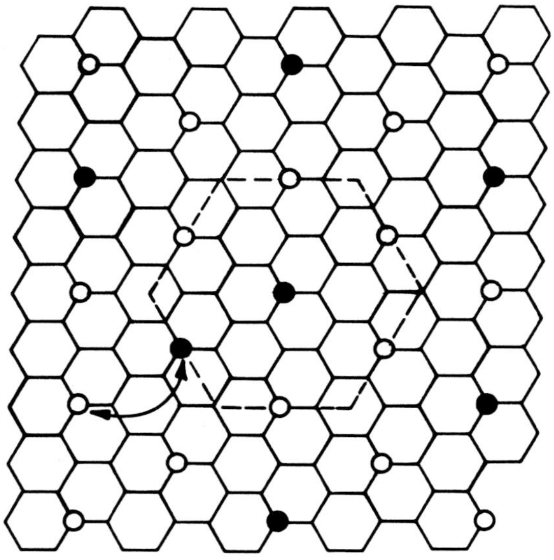

Fig. 1 - Ionic superlattice appropriate to charge transfer f=0.25, superimposed on graphite layer. Open circles = neutral molecules; closed circles = ionized. Arrows show Frenkel defect.

f and on the screening (e.g. for f=.25, η=.76 (unscreened), .38 (perfectly screened)).

The above value will overestimate E_D, since effects of lattice polarization are ignored. The difficulties of such a calculation are discussed in Ref. 11. In the present situation, the correction is estimated by the Jost method, treating the other molecules as a polarizable continuum. The energy is found to be $E_D^{(2)} = -\nu e^2/\varepsilon b_o$, where b_o is the intermolecular distance, $\nu \approx .48\ (1-1/\varepsilon')$, and ε' is the dielectric constant of the molecular medium, estimated to be $\sim$1.8 for AsF_5 from the atomic polarizabilities and the Lorentz-Lorenz relation. This gives $\nu \approx .21$. There may be an additional contribution to the polarizability, since MF_3 has a permanent dipole moment. However there is some question as to how much MF_3 actually goes into the lattice, and it is known that MF_3 de-intercalates rapidly in the absence of an excess pressure of MF_5 gas. Consequently this additional complication is ignored in the present calculation. The calculated values of T_c are plotted in Fig. 2 as a function of f.

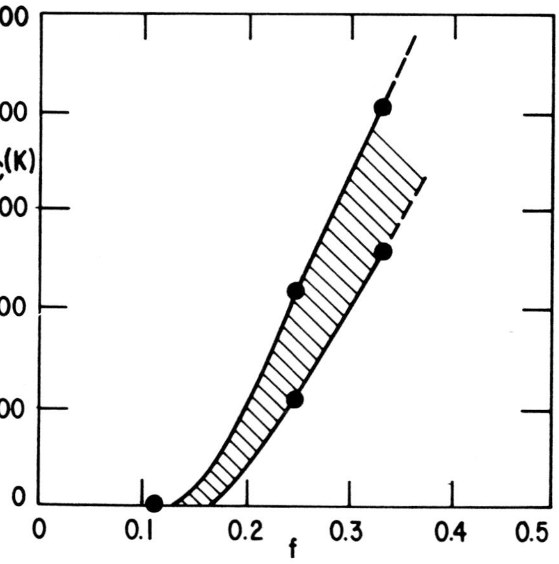

Fig. 2 - Order-disorder temperature as a function of charge transfer. Shaded regions show range of allowed temperatures, assuming varying degrees of hole screening; lower bound = "perfect screening"; upper = uniform hole density. Calculations were made only at the points marked by dots, assuming simple superlattices. The curves are symmetric around f=0.5.

Interactions between molecules in different layers have relatively little effect on the transition temperature. However, once the individual layers have ordered, the net interplane interaction increases in proportion to the number of ions in the lattice, so that the actual transition is three-dimensional. Because of ion-ion repulsion, successive layers will be staggered so that ions do not lie directly above one another. This will generally increase the c-axis repeat distance.

Order-disorder transitions have been observed in graphite intercalated with either AsF_5 or SbF_5, as well as in several other acceptor (and donor) compounds [12-13]. The transition in HNO_3-intercalated graphite is particularly striking [14]. The transition temperature, 250K, is independent of stage, but the intercalant layers are all found to order in a staggered ($\alpha\beta\alpha\beta$) sequence, thus doubling the c-axis periodicity, in stages 1-4. For SbF_5, the critical temperature shows a stronger stage dependence [6]. The experimental values of T_c lie in the range 150-250K, with values of f estimated as lying between .12-.50. While the qualitative agreement with Fig. 1 is gratifying, neither the values of f nor the theoretical polarization corrections are known sufficiently well for any quantitative test of the model.

The present model predicts ordering for the ionized species only, and says nothing about an ordering of the neutral molecules. In both the AsF_5 and SbF_5 systems, there is evidence for residual disorder in the intercalate layer below T_c. For AsF_5, the F-NMR signal, which is motionally narrowed at room temperature, broadens below T_c, but is still much narrower than the linewidth expected for a perfectly ordered system [15]. The line continues to broaden as T is reduced down to at least 136K. In the SbF_5-system, the resistivity shows that an additional disorder-scattering can be quenched in by rapidly cooling the sample below T_c [6].

CHARGE-DENSITY OSCILLATIONS

The periodic ionic potential will substantially perturb the hole distribution on the bounding graphite layer, producing an oscillation in the hole density with the same period. The magnitude of the oscillation can be estimated within the RPA, by treating the ionic potential as composed of three sinusoidal oscillations with wave number $q=2\pi/b$ and amplitude W_q. From the dielectric constant $\varepsilon_{RPA}(q)$, the amplitude of the density oscillation, n_q, is found to be $4\pi n_q e^2/q^2 = W_q(1/\varepsilon_{RPA}-1)$. For the simple superlattice considered here (Fig. 1), $q>2k_F$ and the resulting n_q is small, $\sim$2% of the total hole density. (Values of the Fermi momentum k_F appropriate to $C_{16}AsF_5$ are used [3b]). A Hartree-Fock calculation (minimizing the sum of the hole kinetic plus exchange energy in the presence of the periodic potential) predicts an n_q an order of magnitude larger, showing that hole correlations oppose oscillations with $q>2k_F$. The oscillations are still large enough to have a pronounced effect on electronic properties such as the de Haas-van Alphen oscillations, and may explain the sum and difference frequencies which have been observed [3b].

REFERENCES

1. P. Oelhafen, P. Pfluger, E. Hauser, and H.-J. Güntherodt, Phys. Rev. Lett. 44, 197 (1980); T. Ohno, K. Nakao, and H. Kamimura, J. Phys. Soc. Japan 47, 1125 (1979).
2. N. Bartlett, B. McQuillan, and A. S. Robertson, Mat. Res. Bull. 13, 1259 (1978).
3. (a) B. R. Weinberger, J. Kaufer, A. J. Heeger, J. E. Fischer, M. Moran, and N. A. W. Holzwarth, Phys. Rev. Lett. 41, 1417 (1978); (b) R. S. Markiewicz, H. R. Hart, Jr., L. V. Interrante, and J. S. Kasper, Sol. St. Commun., to be published.
4. E. M. McCarron and N. Bartlett, J. C. S. Chem. Commun., to be published, and M. J. Moran, J. E. Fischer, and W. R. Salaneck, Bull. A. P. S. 25, 334 (1980).
5. J. S. Kasper, R. S. Markiewicz, and L. V. Interrante, to be published.
6. H. A. Resing, F. L. Vogel, and T. C. Wu, Mat. Sci. Eng. 41, 113 (1979).
7. M. J. Bottomley, G. S. Parry, and A. R. Ubbelohde, Proc. Roy. Soc. (London) A279, 219 (1964); P. Touzain, Synth. Met. 1, 3 (1979).
8. L. Pietronero, S. Strässler, H. R. Zeller, and M. J. Rice, Phys. Rev. Lett. 41, 776 (1978).
9. D. R. Hamann and S. A. Safran, Bull. A. P. S. 25, 298 (1980).
10. Z. Matyas, Czech. J. Phys. 4, 14 (1954).
11. C. P. Flynn, Point Defects and Diffusion (Oxford, 1972).
12. J. E. Fischer, Physics and Chemistry of Materials with Layered Structures, Vol. 5, F. Levy, ed., (Riedel, Dordrecht, Holland, 1979).
13. The order-disorder transition in donor compounds (stage $\geq$2) may have a similar interpretation, even though all the atoms are ionized. Compared to stage 1, 1/3 of the intercalant sites are vacant, giving an effective f=2/3.
14. D. E. Nixon, G. S. Parry, and A. R. Ubbelohde, Proc. Roy. Soc. A291, 324 (1966).
15. B. R. Weinberger, J. Kaufer, A. J. Heeger, E. R. Falardeau, and J. E. Fischer, Sol. St. Commun. 27, 163 (1978).

Sinha, ed. Ordering in Two Dimensions

ELECTRICAL AND OPTICAL PROPERTIES OF SELECTED IRON DOPED LAYERED TRANSITION METAL DISELENIDES

C. R. KANNEWURF, J. W. LYDING, M. T. RATAJACK AND J. F. REVELLI*

Department of Electrical Engineering and Computer Science
and Materials Research Center
Northwestern University, Evanston, Illinois 60201, USA

J. F. GARVIN, Jr.+ and R. C. MORRIS

Department of Physics
Florida State University, Tallahassee, Florida 32306, USA

The results of optical and electrical characterization studies for systems of the form Fe_xTSe_2 (T = Zr, Ti, and Nb) are presented. The host materials (TSe_2) range in character from semiconductor (T = Zr) to semimetal (T = Ti) to a superconducting metal (T=Nb); however the iron intercalated compounds in many cases exhibit properties quite different from those of the host. The Fe_xZrSe_2 system exists in two distinct phases, a 1T semiconducting phase for $x < 1/3$ and a 2H metallic phase for $x \approx 1/2$. Low iron concentrations in the Fe_xTiSe_2 and Fe_xNbSe_2 systems result in charge-density wave suppression and magnetic impurity scattering effects, whereas for x = 1/3 both structural and magnetic ordering of the intercalate is observed.

INTRODUCTION

Recent studies of the intercalation properties of materials with layered structures such as the layered transition metal dichalcogenides (LTMD) have resulted in numerous compounds of the form M_xTX_2; where T is a transition metal, X is S, Se or Te, and M is the intercalated specie. Generally, many of the metal intercalation compounds of the group IVb and Vb dichalcogenides are formed with an M from the 3d series of transition metals. A summary of the structure and properties of these intercalation compounds has been reviewed by Subba Rao and Shafer [1]. While much of the early and continuing emphasis with intercalation systems has been directed toward electrochemical applications and the enhancement of the superconducting transition temperature for certain compounds, systematic studies of many physical properties have not been reported. One objective of the present work is to examine transport and related properties as a function of the amount of intercalate concentration x for iron intercalated group IVb and Vb diselenides.

The studies of magnetic as well as transport properties have been of particular interest where the intercalate atom is Fe, Co or Ni [1, 2,3,4]. In this paper we examine specifically the three systems Fe_xTiSe_2, Fe_xZrSe_2 and Fe_xNbSe_2. For low iron concentrations the structure of all these compounds is isotypic to that of the binary host compound, TX_2. However, for higher iron concentrations, structural ordering of the intercalate specie does occur [5]. In the case of the Fe_xTiSe_2 system for $x < 0.2$ the structure remains isotypic to the host [6], but for $x = 1/3$ a unique ordered intercalate structure has been found [5]. Gleizes et al. [7] found that for $x < 0.33$ the 1T polytype of $ZrSe_2$ is retained, but at $x \approx 1/2$ the structure has changed to the 2H hexagonal form similar to many of the group Vb LTMD systems. Voorhoeve and Robbins [8] have studied the Fe_xNbSe_2 system and identified structural ordering similar to that observed in the Fe_xTiSe_2 system. Of particular interest has been the studies of the physical properties as a function of composition and temperature dependence for these systems. In the following sections the results of optical dispersion and electrical transport measurements will be interpreted to identify transitions in conduction behavior as well as the aspects of charge-density wave (CDW) formation and the predominant scattering mechanisms that influence the electronic transport behavior.

OPTICAL PROPERTIES

Each of the systems Fe_xTSe_2 (T = Zr, Ti, and Nb) has been studied as a function of composition by means of optical reflectance measurements. The measurements were performed at room temperature over the frequency range of 33 to 50,000 cm^{-1} using the technique described in [9]. The representative spectra shown in Fig. 1 demonstrate the semiconductor, semimetal nature of the three systems and are characteristic of two basic physical processes; lattice polarizability for T = Zr and Ti and free carrier polarizability for T = Ti and Nb. This results in a dielectric constant for these materials which may be expressed as a superposition of damped Lorentzian oscillators and Drude terms

$$\epsilon^* = \epsilon_\infty + \sum_j \frac{S_j}{\omega_{oj}^2 - \omega^2 + i\omega_{gk}} - \sum_k \frac{\omega_{pk}^2}{\omega(\omega - iG_k)} \quad (1)$$

where ϵ_∞ represents the residual high frequency polarization, the summation over j represents the lattice (phonon) modes, and the summation over k represents the free carrier (plasmon) modes.

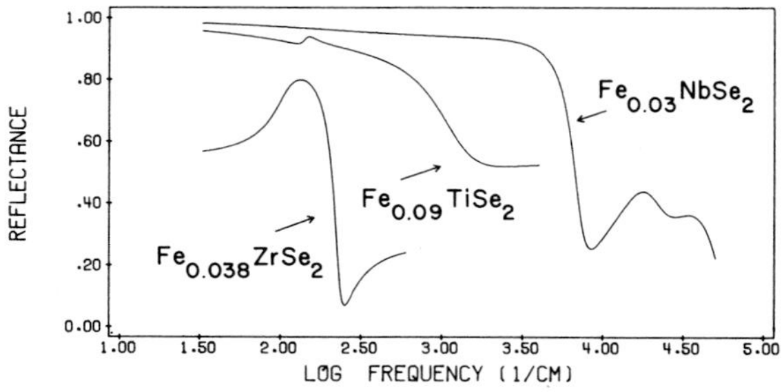

Fig. 1. Reflectance spectra at 300°K for single crystal specimens of Fe_xZrSe_2 (x = 0.038), Fe_xTiSe_2 (x = 0.09) and Fe_xNbSe_2 (x = 0.03).

The reflectance spectrum of Fe_xZrSe_2 is simply characterized by a single phonon mode typical of other group IVb LTMD semiconductors [10]. A detailed examination of this system as a function of iron concentration [11] has shown that throughout the semiconducting phase the iron acts only as to modify the oscillator parameters of the host, consistent with the electrical properties described in the next section. The principal change is an increase in the high frequency polarizability ϵ_∞ and oscillator strength S_1 with iron concentration resulting in invariant TO and LO phonon frequencies with intercalate concentration.

The Fe_xTiSe_2 system, like the Fe_xZrSe_2 system, exhibits a single phonon mode; however, this mode is superimposed on the principal feature of the spectrum which corresponds to plasma-like reflection. The variation of the phonon mode with iron concentration is similar to that of the previously described behavior for Fe_xZrSe_2, characterized by an increase in the high frequency polarizability due to the electronic polarizability of the iron within the van der Waals gap. A previously reported analysis of the infrared reflectance spectrum of the host $TiSe_2$ [12] has suggested that the free carrier behavior is characterized by a two plasmon model. The free carrier behavior of the iron intercalated system is also consistent with this model and an increase in carrier concentration is observed with increasing iron concentration.

A near infrared plasma edge is the dominant feature of the spectrum of Fe_xNbSe_2 with electronic transitions, as determined from a Kramers-Kronig analysis to be at approximately 2.0 and 3.7 eV, producing minor peaks at higher frequencies [13]. An analysis of the spectra obtained as a function of iron concentration using a single plasmon model has shown that unlike Fe_xTiSe_2, no appreciable change in carrier concentration occurs over the range $0 < x < 0.33$. The increase in resistivity with iron concentration at room temperature is thus found to be due to an increase in scattering.

ELECTRICAL TRANSPORT PROPERTIES

Temperature dependent electrical conductivity measurements have been obtained for the Fe_xTSe_2 systems (T = Zr, Ti, and Nb) using standard four-point probe techniques. The host materials range from semiconducting $ZrSe_2$ through semimetallic $TiSe_2$ to metallic $NbSe_2$. This wide range of conduction behavior provides an interesting basis from which a study of the effect of iron intercalation can be made.

As previously described, the Fe_xZrSe_2 system has been shown to exist in two distinct structural phases; a 1T semiconducting phase for $x < 1/3$ and a 2H metallic phase for $x \approx 1/2$. Representative data of the two phases of the Fe_xZrSe_2 system are shown in Fig. 2. For low intercalate concentrations it has been found that the iron acts as a donor-like impurity consistent with a rigid-band scheme in which the iron forms localized donor states within the bandgap of the host. As the iron concentration is increased, the width of these states increases causing a slight reduction in the electrically determined activation energy. For $x \approx 1/2$, the presence of iron results in the promotion of the $ZrSe_2$ sandwiches from a 1T to a 2H polytype. In this phase the zirconium atoms achieve a d^1 electronic configuration similar to the group Vb LTMD systems in which trigonal prismatic coordination of the host transition metal atoms exists. This results in metallic conduction behavior as shown in the insert of Fig. 2.

Unlike $ZrSe_2$, $TiSe_2$ is a semimetal and exhibits a second-order phase transition to a commensurate CDW state at 200°K [14]. This transition has received considerable experimental and

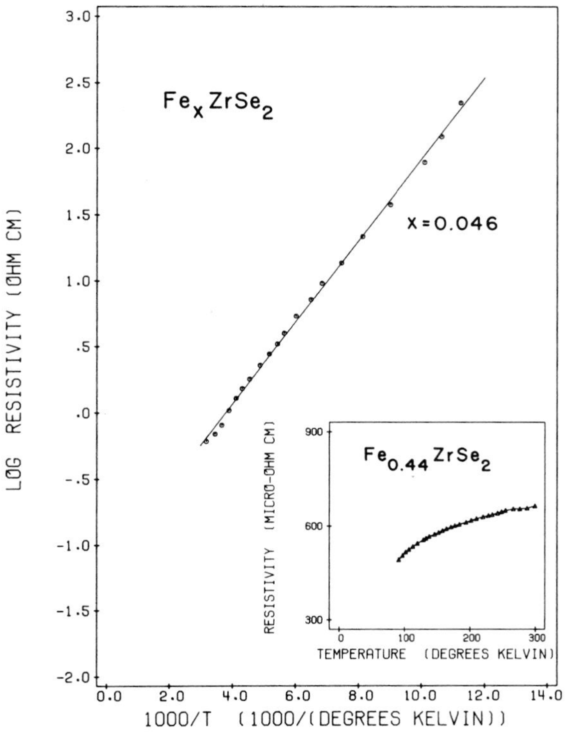

Fig. 2. Temperature dependence of the resistivity for Fe_xZrSe_2 which shows the typical semiconductor behavior for low iron concentration. The insert figure shows the metallic phase for x = 0.44.

theoretical attention in recent years with the phenomenon of electron-hole coupling emerging as the best candidate for the underlying driving force of the transition. The present studies of the Fe_xTiSe_2 system give further evidence to this hypothesis since iron intercalation disrupts the electron-hole carrier balance found in pure $TiSe_2$. A comparison of the electrical resistivity data for the host and one sample with dilute iron concentration is shown in Fig. 3. The anomalous peak in the resistivity of the host composition is due to the formation of the CDW. The insert shows the corresponding data for the x = 1/3 concentration.

As with other first row transition metal compounds, titanium diselenide tends to form non-stoichiometrically with excess Ti atoms occupying octahedral holes within the van der Waals gap. Quantitative chemical analysis has indicated the presence of 1% excess Ti for the host composition used in this study. The suppression of the CDW resulting from carrier imbalances created by small quantities of excess titanium has been elucidated by DiSalvo et al. [14]. Since iron also behaves as a donor, a similar suppression is expected for small amounts of intercalated iron in Fe_xTiSe_2. This is indeed the case where for x = 0.09 the CDW transition is absent as shown in Fig. 3. Although the number of donated electrons is proportional to the iron concentration, a monotonic increase in resistivity is observed with increasing iron concentration for $x < 0.2$. This results from enhanced electron scattering attributable to the localized magnetic moments of iron atoms. For x = 1/3 an ordered intercalate structure occurs [5] in which the magnetic moments interact due to closer proximity. This results in a decrease in the magnetic spin-flip scattering and consequently the lower resistivity. Also, a resistivity transition is observed at lower temperatures. This has been confirmed by magnetic susceptibility measurements to be the onset of antiferromagnetic ordering of the iron below $T_N = 113^oK$, thus further reducing the electron scattering.

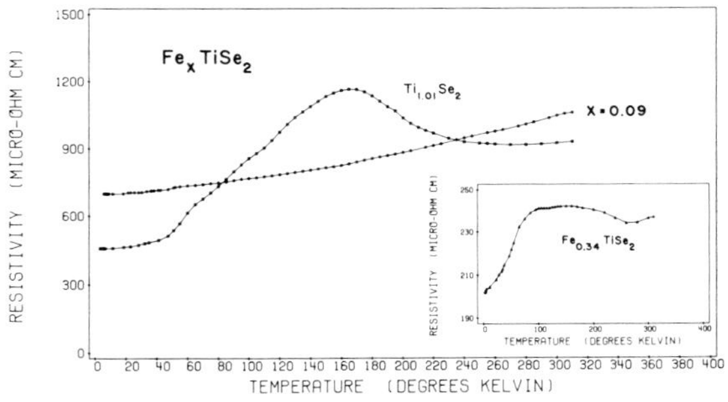

Fig. 3. Comparison of the temperature variation of resistivity for a sample of Fe_xTiSe_2 with x = 0.09 to that of the host $Ti_{1.01}Se_2$. The insert figure shows the resistivity variation for x = 0.34.

The compound 2H - $NbSe_2$ is a superconducting metal with $T_c = 7.25^oK$. Like $TiSe_2$ and other group Vb LTMD systems 2H - $NbSe_2$ exhibits a CDW transition at $T = 40^oK$ [15]; however, unlike $TiSe_2$ the CDW in 2H - $NbSe_2$ shows no commensurate phase. Studies of very low iron concentrations in Fe_xNbSe_2 have shown a rapid suppression of the CDW (-50^oK/at. % Fe) [16]. The present studies have been concerned with the intermediate and high iron concentrations of Fe_xNbSe_2. Resistivity data for three compositions of this system are shown in Fig. 4.

In addition to a rapid quenching of the CDW in 2H - $NbSe_2$, small intercalate iron concentrations result in a localized iron magnetic moment. This gives rise to the low-temperature resistivity behavior characteristic of a Kondo system for x = 0.005 in Fig. 4. For intermediate iron concentrations (x = 0.15 in Fig. 4) the resistivity has been observed to be almost temperature and composition independent. This departure from the low iron concentration behavior has been suggested as a transition to an intermediate iron concentration spin-glass system [17]. For x = 1/3, magnetic ordering similar to that observed in the $Fe_{1/3}TiSe_2$ system is observed as evidenced by the drop in resistivity shown in Fig. 4.

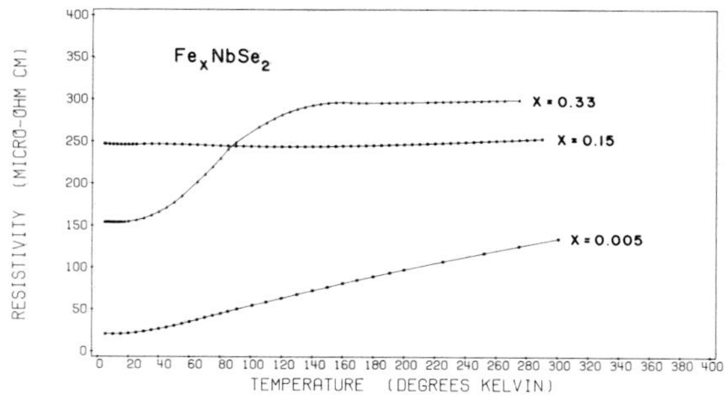

Fig. 4. Temperature dependence of the resistivity for dilute, intermediate and high iron concentrations in Fe_xNbSe_2.

CONCLUSION

This study has been concerned with a comparison of the optical and electrical properties of the iron intercalation systems Fe_xTSe_2 (T = Ti, Zr, Nb). Consistent with the range of behavior observed in the host systems are the effects observed upon iron intercalation. For low iron concentrations, simple donor type behavior is observed in the Fe_xZrSe_2 system whereas magnetic scattering effects and CDW suppression are exhibited by the Fe_xTiSe_2 and Fe_xNbSe_2 systems. The presence of structural intercalate ordering in these systems for higher iron concentrations results in dramatic changes in electrical transport properties. The metallic phase of the Fe_xZrSe_2 system at $x \approx 1/2$ represents the first observation of a 2H polytype in a IVb LTMD system. In addition to structural intercalate ordering in the Fe_xTiSe_2 and Fe_xNbSe_2 systems at x = 1/3, the presence of a low temperature antiferromagnetic state has been identified in these systems.

ACKNOWLEDGMENTS

The research at Northwestern University was supported under the NSF-MRL program through the Materials Research Center of Northwestern University (Grant No. DMR 76-80847). The research at Florida State University was supported by the National Science Foundation (Grant No. 78-61212).

FOOTNOTES

*Present address: Xerox Corporation, Webster Research Center, Webster, New York 14580.

+Present address: Texas Instruments Inc., Dallas, Texas 75222.

REFERENCES

1. G. V. Subba Rao and M. W. Shafer in "Intercalated Layered Materials," (F. A. Levy, Ed.) D. Reidel Pub. Co., Dordrecht (1979) p. 99.
2. A. R. Beal, in "Intercalated Layered Materials," (F. A. Levy, Ed.) D. Reidel Pub. Co., Dordrecht (1979) p. 251.
3. R. H. Friend, A. R. Beal and A. D. Yoffe, Phil. Mag. 35, 1269 (1977).
4. D. A. Whitney, R. M. Fleming and R. V. Coleman, Phys. Rev. B15, 3405 (1977).
5. H. Goodman, J. A. Ibers, J. W. Lyding, M. T. Ratajack and C. R. Kannewurf, Manuscript in preparation.
6. Y. Arnaud, M. Chevreton, A. Ahouandjinou, M. Danot and J. Rouxel, J. Solid State Chem. 18, 9 (1976).
7. A. Gleizes, J. Revelli and J. A. Ibers, J. Solid State Chem. 17, 363 (1976).
8. J. M. Voorhoeve and M. Robbins, J. Solid State Chem. 1, 134 (1970).
9. M. T. Ratajack, K. Kishio, J. O. Brittain and C. R. Kannewurf, Phys. Rev. B21, 2144 (1980).
10. G. Lucovsky, R. M. White, J. A. Benda and J. F. Revelli, Phys. Rev. B7, 3859 (1973).
11. M T. Ratajack, C. R. Kannewurf, J. F. Revelli and J. B. Wagner, Phys. Rev. B17, 4674 (1978).
12. G. Lucovsky, W. Y. Liang, R. M. White and K. R. Pisharody, Solid State Commun. 19, 303 (1976).
13. M. T. Ratajack, C. R. Kannewurf, J. F. Garvin, Jr. and R. C. Morris, Bull. Am. Phys. Soc. 24, 474 (1979).
14. F. J. DiSalvo, D. E. Moncton and J. V. Waszczak, Phys. Rev. B14, 4321 (1976).
15. R. C. Morris, Phys. Rev. Lett. 34, 1164 (1975).
16. K. Noto, S. Morohashi, K. Arikawa and Y. Muto, Physica 99B, 204 (1980).
17. J. F. Garvin, Jr. and R. C. Morris, Phys. Rev. B21, 2905 (1980).

SECTION II

E. 2-D MAGNETISM

MAGNETIC COUPLINGS IN 2-D ^{3}He SOLID LAYERS

M. Héritier, G. Montambaux and P. Lederer

Laboratoire de Physique des Solides+, Université de Paris-Sud

Centre d'Orsay, 91405 Orsay, (France)

We study the magnetic polarisation induced around vacancies in 2-D quantum crystals of fermions in the limit of vanishing exchange interactions. The lowest vacancy state is not the fully polarised ferromagnetic configuration. Various magnetic configurations are investigated. We give indications that the "vortex" configuration, in which each spin makes a $2\pi/3$ angle with its neighbours, might be more stable than a partially polarised one. This analysis is applied to ^{3}He solid layers adsorbed on graphite.

In the highly dense layers formed by liquid ^{3}He in contact with graphite, a new antiferromagnetic coupling arising from virtual exchange of particles between the solid layer and the liquid should predominate.

We have published recently a study of ^{3}He triangular adsorbed layers on grafoil [1]. We argued in that work that the ground state of the 2-D crystal in the presence of one vacancy might have zero total spin and that the magnetic configuration which minimizes the kinetic energy of the vacancy is a "vortex" configuration (hereafter the V-configuration). As a result we stated that the experimentally observed tendency to ferromagnetism [2] for ^{3}He in 2-D confined geometries could not be ascribed to vacancies contrary to the case of bcc ^{3}He [3].

On the other hand, Iordanskii [4] reaches a different conclusion than ours about the same system. He finds that the ground state with one vacancy is ferromagnetic, with a magnetization one third the saturation magnetization. (We note hereafter this ground state the I-configuration). A few remarks are in order. Iordanskii uses a variational procedure. He works within a subspace of wave functions obtained from the wave function with maximum total spin by flipping spins on a variable number of lattice sites. It is thus quite clear that his "ground state" wave function is not an eigenfunction of the total spin operator. The set of wave functions he uses is constrained to be a single vector of the particular basis employed to span the whole space. In particular, it has non-zero overlap with the wave functions which have spin ranging from maximum total spin to 1/3 the maximum total spin. It is the lowest energy state within this particular subspace of wave functions, with the constraint mentioned above.

Our wave function [1] also is not an eigenfunction of the total spin ; it is a complicated linear combination of wave functions with spin ranging from the maximum total spin to zero total spin. It provides a spin configuration which is efficient in suppressing closed loops, as argued in [1].

In order to decide which wave function has the lowest energy, one should use comparable schemes. It would be very useful to have moments of the density of states for each configuration up to a large order and compare the extrapolated band edges, given by

$$\omega_o^2 = \lim_{n\to\infty} \frac{M_{n+2}}{M_n}$$

where M_n is the nth moment of the density of states. However, computation of the moment of the I-configuration for $n > 8$ is hopelessly cumbersome. We are thus left with useful indications from lower order moments, not with undisputable results.

The indications are the following : i) the geometrical center of the band is given by

$M_3/M_2 = a_2$

ii) The total band width is given by

$$W = \frac{4\sqrt{M_2M_4 - M_2^3 - M_3^2}}{M_2}$$

iii) the quantity $R = \frac{16M_2}{W^2}$ is smaller when states are more concentrated near the band center, defined by $M_1=0$

We obtain the following results (taking the intersite tunneling frequency t = 1)

- for the I-configuration : $a_2^I = 0.667$

$W^I = 9.141$

$R^I = 1.149$

- for the V-configuration $a_2^V = 0.5$

$$W^V = 9.165$$
$$R^V = 1.143$$

We conclude that the I-configuration and the V-configuration have nearly the same band width but the geometrical center of the latter is 0.167 lower in energy. Both bands have quite similar compacity.

As a result, the indications are that the V-configuration seems to provide a lower ground state energy than the I-configuration. In that case the conclusions of [1] hold at low enough temperature. However, if $\Delta\varepsilon$ is the energy difference between the I-configuration and the V-configuration, one should expect ferromagnetic like fluctuations to be present in the system down to $k_B T^* \sim \Delta\varepsilon$. From the present moment estimate, $\Delta\varepsilon \sim 0.17\ t \sim 40$ mK. (This figure cannot be taken at face value).
Large effects are expected in the low density solid commensurate with the substrate, where ground state vacancies may exist.

We now turn to a situation which arises when liquid ^{3}He is in contact with graphite. Highly dense adsorbed solid ^{3}He layers are formed because of the strong Van der Waals attraction with the surface. Our main point is that by far the main indirect exchange mechanism between ^{3}He atoms in the solid layer via the liquid is a nearest neighbour antiferromagnetic coupling. The mechanism is as follows : an atom of the solid layer can be emitted into the liquid, leaving a vacant site. In this 2-D quantum crystal, the vacancy can tunnel to a neighbouring site, which can be filled again by an atom of the liquid. These virtual transitions are only possible when the spins on the involved sites are parallel and, as shown below, they increase the energy because intermediate states with bonding vacancy wave function, i. e. positive vacancy kinetic energy, are more easily excited. Thus the coupling is antiferromagnetic.

The solid layer is a quantum crystal with a tunneling frequency t very much larger than the exchange frequency J which we find convenient to describe by a simple Hubbard model with exactly one atom per lattice site (vacancies are neglected)

$$H_o = -t \sum_{i,j} C^+_{i,\sigma} C_{j\sigma} + u \sum n_{i\uparrow} n_{i\downarrow} \quad (t > 0)$$

In the first term the sum is taken over nearest neighbour sites (i) and (j). $C^+_{i\sigma}$ ($C_{i\sigma}$) is the creation (annihilation) operator of a ^{3}He atom Wannier state on site i with spin σ. The second term, in which $n_{i\sigma} = C^+_{i\sigma} C_{i\sigma}$ represents the fermion hard core repulsion.

In this simple model, the nearest neighbour direct (as opposed to the indirect one discussed here) exchange constant is $J = 2\frac{t^2}{u}$. Here $J \sim 1\mu$K whereas t is certainly at least as large as in the bulk solid at the melting pressure : $t \gtrsim 100$ mK (see ref. 6 and references therein).

Furthermore we allow for exchange of particles with the liquid, by introducing an interaction term

$$H_i = \sum_{q,i} t_{s\ell}(\vec{q})\, e^{i\vec{q}.\vec{R}_i}\, a^+_{q\sigma} C_{i\sigma} + \text{h. c.}$$

where R_i is the position of lattice site i, $a^+_{q\sigma}$ is the creation operator of an ^{3}He atom in the liquid with wave vector q and spin σ and h. c. means the hermitic conjugate of the first sum. The strength of the coupling $|t_{s\ell}(\vec{q})|$ is likely to be of the order of t. The chemical potential ϕ of the ^{3}He atoms in the liquid, measured from the ground state of H_o, depends on the strength of the Van der Waals attraction and thus on the given substrate. It seems to range from 5 K to 50 K. Thus the following inequalities are likely :

$$t \sim |t_{s\ell}(q)| << \phi << u$$

Then a straightforward third order perturbation calculation (to first order in t/ϕ and second order in $t_{s\ell}/\phi$) gives a contribution to the energy

$$\Delta E^{(3)} = \sum_{i,j} \tilde{J}_{ij} (S_i . S_j + \frac{1}{4})$$

where the sum is taken over nearest neighbour sites and

$$\tilde{J}_{ij} = 4t \sum_{q} \frac{|t_{s\ell}(q)|^2 \cos \vec{q}(\vec{R}_i - \vec{R}_j)}{(\phi + \varepsilon(q))^2} (1 - f(\varepsilon_q))$$

where $\varepsilon(q)$ is the Hartree Fock energy of a ^{3}He atom with wave vector $\vec{q}$ in the liquid, measured from ϕ and $f(\varepsilon)$ is the Fermi distribution function. (Correlation effects in the liquid are neglected). $\tilde{J}_{ij}$ is found to be positive for nearest neighbour atoms, so that the coupling is antiferromagnetic. Because fermion interactions are much weaker in the liquid than in the solid layer ($\phi << u$), the indirect coupling $\tilde{J}$, although a 3rd order process, is not negligible compared to the 2nd order superexchange $J = 2t^2/u$ for acceptable values of the parameters. For example, taking t = 200 mK, $|t_{s\ell}(q)| \simeq 100$ mK and $\phi \simeq 10$ K, we obtain $\tilde{J} \simeq 40\ \mu$K while $J \simeq 3\mu$K. The ratio $\tilde{J}/J \sim \frac{u}{t} (\frac{t_{s\ell}}{\phi})^2$ is much larger than 1 because of the unusually high value for $u/t \sim 10^5$.
Our indirect coupling is different from the RKKY-like indirect coupling [7]. The latter is long ranged, ferromagnetic at short distance and it

is a fourth order interaction $\sim t_{s\ell}^4/\phi^3$, so that it should be much smaller than ours. Cyclic three-spin exchange [8] within the solid layer is a ferromagnetic mechanism. Since the upper limit for the strength of any exchange mechanism which does not require the presence of the liquid is set experimentally at 3 μK, 3-spin exchange is negligible compared with the process described in this paper [6]. We propose that the latter is responsible for the nearest neighbour antiferromagnetic interaction J = 60 μK observed in an NMR study of liquid ^3He in contact with carbon black [9].

In conclusion, we point out that the strongest exchange mechanisms (by one or two orders of magnitude) within the adsorbed ^3He layers are antiferromagnetic ; furthermore vacancies do not favour ferromagnetism, so that the tendency to ferromagnetism which is observed for liquid ^3He in confined geometries is not to be explained by a ferromagnetic interaction within the adsorbed solid layer, but should probably be ascribed to the high density liquid near the surface [10].

REFERENCES

1) M. Héritier and P. Lederer, Phys. Rev. Lett. 42, 1068, (1979)
2) See references in D. Spanjaard, D. L. Mills and M. T. Béal-Monod, J. Low Temp. Phys. 34 307 (1979)
3) M. Héritier and P. Lederer, J. Phys. Lett. 38, L 209 (1977)
4) S. V. Iordanskii, Pis'ma Zh. Eksp. Teor. Fiz. 24, 608, (1977)
5) Gaspard, Thèse, Orsay (1975)
6) M. Héritier, Journal de Physique Lett. 40, L-451 (1979)
7) J. B. Sokoloff and A. Widom, Proceeding of the Int. Conf. on Quantum Crystals (Fort-Collins, Colorado, 1977)
8) J. M. Delrieu (private communication)
9) A. I. Ahonen, T. A. Alvesalo, T. Haavasoja, M. C. Veuro, Phys. Rev. Lett. 41, 494, (1978)
10) M. T. Béal-Monod and S. Doniach, J. Low Temp. Phys. 28, 175, (1977)

Sinha, ed. Ordering in Two Dimensions

2D MAGNETISM IN ITINERANT NEARLY MAGNETIC FERMI LIQUIDS

M. T. BEAL-MONOD

Physique des Solides,* Universite de Paris-Sud, 91405 Orsay, France

A. THEUMANN

Physics, Polytechnic, 333 Jay St., Brooklyn, N. Y. 11201, U.S.A.

Two types of confined geometry effects are considered within the paramagnon theory of nearly magnetic Fermi liquids at very low temperature. Huge spin fluctuations near a surface or an interface with another medium may, either enhance the overall Pauli susceptibility, the effective mass of the linear temperature specific heat... over the corresponding bulk values, or, if the spin-spin interaction exceeds some critical value smaller than unity, the surface region undergoes a magnetic transition and switches to a 2D (or quasi 2D) itinerant ferromagnetic behavior while the bulk still remains paramagnetic. These two types of effects may explain recent data obtained in normal liquid He^3; they may be relevant too in catalysis and for induced triplet pairing superconductivity.

THE THEORETICAL MODEL

In recent years, a lot of interest has been devoted, both theoretically and experimentally, on properties of nearly magnetic Fermi liquids near a surface (free surface or interface with a wall) and more generally in confined geometry, specially for liquid He^3 [1].

We first recall that, in the bulk, in a sample of large dimensions so that surface effects are negligible, the Stoner magnetic instability of a strongly interacting itinerant fermion system follows from the balance between the kinetic and potential energies of the fermions. The paramagnon picture [2] reduces the Landau Fermi liquid description to a strong, contact repulsion between fermions of opposite spins; in mean field, the instability occurs when the repulsion I reaches the value of the characteristic energy E_F of the free fermions, or $\bar{I}=1$, with $\bar{I}=IN(E_F) \propto I/E_F$, where $N(E_F)$ is the density of states at the Fermi level E_F. Thus one has for the bulk spin susceptibility χ_B, at very low temperature:

$$\bar{I} < 1, \ \chi_B = \chi_{Stoner} = \chi_{Pauli} [1 - \bar{I}]^{-1} \quad (1a)$$

bulk paramagnetic, close to be magnetic when $\bar{I} \to 1$;

$$\bar{I} \geq 1, \ \chi_B = \chi_{Curie\text{-}Weiss} = \frac{\chi_{Curie}}{1 - T_o/T} \quad (1b)$$

bulk itinerant ferromagnetwith T_o of order $T_F = (1-\bar{I}^{-1})^{1/2}$ in the Stoner theory.

In the following, we consider those bulk Fermi liquids which remain paramagnetic down to OK, i.e. $\bar{I} \leq 1$: the transition never occurs except possibly at OK. In that case however, strong spin fluctuations, "paramagnons", enhance all the properties of the Fermi liquid; the characteristic temperature of these fluctuations T_{sf} is much smaller than in absence of interactions: $T_{sf} = (1 - \bar{I})T_F << T_F$ when $\bar{I}$ is close to 1. It has been shown [3] that, in 3 dimensions, and due to quantum effects at OK, the critical exponents are the mean field ones, so that the real enhancement appears to the power 1 as in the Stoner mean field expression (1a).

We now turn to take into account the finite size of such a system, when the surface to bulk spins ratio is no longer negligible, so that surface effects play a role [4-6]. A surface, like any other perturbation (impurity...), in a system which possesses a Fermi surface, induces Friedel type oscillations in all properties of that system. Then, at a distance z from the surface in the direction perpendicular to it, the local susceptibility resembles $(1 - \bar{I}osc.(k_F z))^{-1}$, [4] containing an oscillating function of z and of the Fermi momentum k_F, the oscillations being the strongest close to the perturbation (i.e. close to the surface) and vanishing far away, deep in the bulk (where $osc.(z \to \infty) \to 1$). This explains physically why, near a surface, within some range Δz (model dependent), even if $\bar{I}$ is < 1 (bulk paramagnetic), the average product $<\bar{I}\ osc.(k_F z)>_{\Delta z}$ may, -i- either be < 1 but larger than $\bar{I}$ so that paramagnon effects will be more enhanced than in the bulk, and then the overall Stoner enhancement averaged over the whole sample S_{eff} is larger than the bulk one $S = 1/(1 - \bar{I})$, or, -ii- it may even exceed 1, in which case the local mean field susceptibility blows up. This last situation occurs all the more easily that $\bar{I}$ is closer to 1. It has been shown [4-6] that it occurs when the bulk value $\bar{I}$ exceeds some critical value $\bar{I}_c$, where $\bar{I}_c$ depends on the potential exerted by the surface or the interface on the single particle wave function of the fermion system, in particular on the Van der Walls attraction to the surrounding medium ($\bar{I}_c$ decreases when the attraction to the surface increases). In other words, while the instability criterion is reached in the bulk for $\bar{I}=1$, it is reached, near the surface, for a value $\bar{I}_c$ smaller than 1. To summarize, if one phenomenologically [7] incorporates the spatial Friedel oscillations in a space dependent function $\bar{I}(\vec{r})$ and if $n(\vec{r})$ is the number of spins at the distance $\vec{r}$ from a given origin in the center of the sample one expects the following results:

CASE 1: $\bar{I} < \bar{I}_c < 1$, bulk and surface regions both paramagnetic down to OK with the effective overall constant susceptibility, χ_{eff} given by:

$$\frac{\chi_{eff}}{\chi_B} = \frac{1-\bar{I}}{N}\int_V \frac{n(\vec{r})d^3\vec{r}}{1-\bar{I}(\vec{r})} = \frac{1-\bar{I}}{1-\bar{I}_{eff.}} = \frac{S_{eff.}}{S} > 1 \quad (2)$$

with $\bar{I} < \bar{I}_{eff.} < 1$

where N is the total number of spins in the total volume V of the sample and $\int n(\vec{r})\, d^3\vec{r} = N$ (if $V\to\infty$, $\bar{I}(\vec{r}) = cst = \bar{I}$). The confined geometry sample has a larger susceptibility than the corresponding infinite system. In that case, the specific heat is still like in the bulk, proportional to the temperature at very low temperature, but with a coefficient $\gamma_{eff.}$ larger than the corresponding bulk one γ_B:

$$\frac{C_{eff.}}{C_B} = \frac{\gamma_{eff.}}{\gamma_B} = \frac{(m^*/m)_{eff.}}{(m^*/m)_B} > 1 \quad (3)$$

We recall that the mass enhancement m^*/m entering γ(or $\gamma_{eff.}$) is related to $\bar{I}$ (or $\bar{I}_{eff.}$) in the paramagnon model by:

$$\frac{m^*}{m} = 1 + \frac{9}{2}\,\bar{I}\,\ln\left[1 + \bar{p}_1^{\,2}\,\bar{I}\,\{12(1-\bar{I})\}^{-1}\right] \quad (4)$$

where the momentum cut-off $\bar{p}_1 = p_1/p_F$ is of order 1 (this has been shown to be indeed the case in bulk liquid He^3, in a paramagnon analysis [8] of neutron data).

CASE 2: $\bar{I}_c < \bar{I} < 1$, bulk paramagnetic but the surface region has switched to an itinerant ferromagnet with a Curie-Weiss temperature T_o, finite in mean field (but which does not correspond to a physical transition as will be explained below); in that case one has two distinct contributions entering the overall effective susceptibility:

$$\chi_{eff.} = \chi^{(1)}_{eff.} + \chi^{(2)}_{eff.} \quad (5)$$

$\chi^{(1)}_{eff.}$ is of the Stoner types (1a) and (2), temperature independent at very low T; it corresponds to the susceptibility of the interior of the sample of volume V_1 (and a number of spins N_1), where the oscillations are weak enough so that the corresponding $\bar{I}_{eff.}$ is < 1; $\chi^{(2)}_{eff.}$ in contrast, has the Curie Weiss form (1b), i.e. proportional to $(T-T_o)^{-1}$ above T_o, and thus strongly temperature dependent; it corresponds to the susceptibility due to the N_2 spins in the volume V_2 close to the surface, where the oscillations are strong enough so that $\bar{I} < 1 < \bar{I}_{eff.}$ The condition for $\chi^{(2)}_{eff.}$ to be measurable is of course that $N_2 = \int_{V_2} n_2(\vec{r})d^3\vec{r}$ is not negligible compared to $N_1 = \int_{V_1} n_1(\vec{r})d^3\vec{r}$ (with $N_1 + N_2 = N$). Similarly to (5) one will have two distinct contributions to the specific heat:

$$C_{eff.} = C^{(1)}_{eff.} + C^{(2)}_{eff.} \quad (6)$$

coming respectively from the N_1 spins of the interior V_1 of the sample and the N_2 spins of the surface region V_2.

We concentrate now on the $T\to 0$ behavior of the V_2 contribution. Since it will concern a few layers close to the surface, it will have a two (or quasi-two) dimensional character. It is well known that 2 dim. ferromagnetism, in general, is a very difficult problem; even worse, very little is known on 2 dim. itinerant ferromagnetism [9]; in that case the difficulty arises primarily from the flatness of the noninteracting susceptibility at zero frequency $\chi^o(q,0)$, versus q, for all q between 0 and $2k_F$, so that the Stoner instability $1-I\,\chi^o(q,0)=0$ is, in principle, fulfilled for a continuum of q values between 0 and $2k_F$, while in 3 dim. $\chi^o(q,0)$ is maximum at q=0, and the instability occurs for that unique value. Therefore in 2 dim. the meaning of the Stoner criterion appears puzzling since the resulting magnetic instability is, a priori, as well of the ferromagnetic type as of antiferromagnetic ones. One then tries to renormalize $\chi^o(q,\omega)$ in 2 dim.: one studies the Landau-Ginzburg-Wilson Lagrangian [10] which describes interacting fluctuations (2 dim. "paramagnons"); the aim being to examine whether the renormalized $\tilde{\chi}^o(q,\omega)$ has got a maximum at one well defined value of q. Usually in critical phenomena [10] the coefficients of the quartic term and the following ones in the Lagrangian, are supposed to be well-behaved and assimilated to constants inthe rescaling procedure. In the present problem however, the vertices involved in those coefficients, which identify to the interactions between two, three...fluctuations, are formed, to lowest order, by fermion closed loops. Such multitail diagrams have been previously analyzed, but in 3 dimensions [11] where it was shown that they are all the more singular at $2k_F$ that the number of outgoing tails, i.e. outgoing fluctuations (4,6...), is higher. For 3 dim. paramagnons, such a study is not relevant since the magnetic instability has already been reached for increasing I when $I^{-1} = \chi^o(o,o)$. But in 2 dim. the analogous singular behavior of the corresponding vertices is indeed meaningful. We have thus studied these singularities in 2 dim. We find no singularity below $2k_F$ (moreover the 4,6... tail diagrams identically vanish for zero momentum). But at $2k_F$, we find the main singularity for the 4 tail diagram (in the coefficient of the quartic term of the Lagrangian) to be even stronger in 2 dim. (power law singularity =-3/2) than in 3 dim. (power law singularity =-1). The consequence is the following: one is not allowed to treat the coefficients of the ϕ^4, ϕ^6...terms in the Lagrangian of interacting 2 dim. paramagnons as constants as usually done [10] and as supposed to be so in [9]; moreover it is not obvious that their high singularities will be wiped out in any further integrations over q and ω to calculate the renormalized $\tilde{\chi}^o$ (q,ω) and its q dependence. The algebra becomes rapidly incredibly difficult [12].

At that stage, since this 2 dim. itinerant magnetism problem seems as marginal [13] as the 2 dim. localized one, we invoke universality arguments to propose that our 2 dim. itinerant fermion problem might behave like the 2 dim. localized Heisenberg spin system of ref. 14 . Moreover, since in our problem the role of the 3rd dimension ($\perp$ to the surface) is crucial since the Friedel oscillations (responsible for the surface instability) occur along that direction, we assume that our system in volume V_2, may keep a memory of the 3 dim. case and will be of ferromagnetic type (like any paramagnon problem between 2+ε dim. and 3 dim. [15]). It might even be a real cross over problem between 2 and 3 dim. But as a first step we treat it as a 2 dim. ferromagnetic case. Then, according to the mapping to the results of ref. 14 , we expect the following to happen:

$$\chi^{(2)}_{eff.} \sim \begin{cases} (T/T_o)^3 \exp.(4\pi T_o/T) & T\to 0 \quad (7a) \\ (T - T_o)^{-1} & T > T_o \quad (7b) \end{cases}$$

In other words $\chi^{(2)}_{eff.}$ is finite at T_o and becomes ∞ only at T=0, and the ordered phase is never reached (except possibly if the problem is proved to be a cross over one between 2 and 3 dim.) T_o thus appears as to be just the "high T" extrapolation of $(\chi^{(2)}_{eff.})^{-1}$ and does not correspond to any real instability as noted earlier. Similarly, one can deduce the very low temperature behavior of the specific heat from [14]:

$$C^{(2)}_{eff.} \sim \exp.\ (4\pi T_o/T) \quad , \quad T\to 0 \qquad (8)$$

so that $C^{(2)}_{eff.}$ will become ∞ at T=0. Therefore, at very low temperature we will have:

$$\begin{cases} \chi_{eff.} \sim A(1-\bar{I}_{eff.})^{-1} + B(T/T_o)^3 \exp.(4\pi T_o/T)) & (8a) \\ & T\to 0 \\ C_{eff.} \sim \gamma_{eff.}\ T + D \exp.(4\pi T_o/T) & (8b) \end{cases}$$

APPLICATIONS TO EXPERIMENTS

We have emphasized above the distinction between CASE 1 where the spin fluctuations are strongly enhanced near the surface, but not sufficiently to produce a surface instability, so that the thermodynamic properties are similar to the bulk ones although enhanced compared to them, and the CASE 2 where a surface instability would occur in mean field at T_o, but for which the 2 dim. character makes the actual transition to occur only at T=0, with however different thermodynamic behaviors near the surface and far from it. We now examine the possible applications of our theory:

CASE 1: a) it has been shown [16] that triplet pairing superfluidity ("p-wave" type) of bulk liquid He^3 is favored by spin fluctuations. It was thus proposed [17] that in small grains of nearly magnetic metals (high purity Pd for instance) which are expected to exhibit stronger spin fluctuations than the bulk due to surface enhancement effects, "p-wave" type superconductivity may be induced at very low temperatures. Such type of superconductivity may have been observed in $CeCu_2Si_2$ [18]; reducing the size of those samples is expected to favor the effect; indeed since in confined geometry $\bar{I}_{eff.}$ ought to be greater than $\bar{I}_B$, the superconducting transition may be higher [19];

b) it has been proposed [20] that strong spin fluctuations could favor hydrogenation reactions in catalysis; since spin fluctuations are stronger near the surface, here too, reducing the dimensions of the sample will all the more favor those reactions;

c) it was recently observed [21] that Fermi liquid droplets in liquid-solid solutions of the He isotopes exhibit a low temperature specific heat linear in temperature with an anomalously high coefficient; one of the possible explanations [21] is that if the droplets are pure liquid He^3, they may have an $\bar{I}_{eff.}$ larger than the high pressure bulk value (0.95); for instance $\bar{I}_{eff.}$ of order 0.985 to 0.988, would imply through (4) a mass enhancement $(m^*/m)_{eff.}$ $\sim$ 10 fitting the data.

CASE 2: a) this case has been proposed [22][6][1] to apply to the observed anomalous susceptibility of liquid He^3 confined in grafoil, or between small particles of carbon black or alumina [23][1] in the mK range; in all these experiments, χ was shown to exhibit a Curie-Weiss temperature dependence superimposed to the Stoner constant contribution of the bulk; moreover, at the lowest temperatures, deviations to the Curie-Weiss contribution, were observed, possibly extrapolating to an infinite χ at T=0. Furthermore in presence of a magnetic field, the authors of ref. [14] showed that for the 2 dim. Heisenberg spin system, $\chi H/T$ is a universal function of the product $(HT^2 \exp(4\pi J/T)$ where J is the nearest neighbor interaction; (formula (7a) was extracted from the H$\to$0, T$\to$0 expansion of the above formula). It was shown [22][6] that plots of the various data of [23] fitted well such a universal law.

b) in the region $T > T_o$, simultaneous measurements of $\chi(T)$ and the longitudinal relaxation time T_1 of liquid He^3 in confined geometries

(Godfin et al. in [1]) exhibited, besides the Curie-Weiss behavior for χ, a T_1 increasing linearly with T; this contrasts with the bulk behavior $(T_1)_B \sim T^{-2}$; a theoretical relation between χ and T_1 was derived in [24] based on the surface effects of CASE 2 above, which allowed to qualitatively interpret the observed behavior for T_1.

c) the experimentalists of ref. 1 did not measure the specific heat in the same conditions, so one cannot check if, corresponding to the anomalous $\chi(T)$, one would obtain an anomalous C(T) as we expect here. However a series of recent experiments [25] at zero and finite pressures, on, liquid He^3, exhibited in the normal phase, just above the superfluid temperature, and superimposed to the bulk contribution (proportional to the temperature), an anomalous contribution increasing for decreasing T. If wall effects play a role in that experiment (but this is not at all clear for the time being), this extra contribution might be linked to our above contribution (8b) which (were it not the superfluid phases [26]) would diverge exponentially for $T \to 0$. Obviously further experiments are needed in that case, including measurements of $\chi(T)$ and C(T) on the same samples, at various pressures.

d) experimental evidences seem to have been found [27] for the existence of 2D band ferromagnetism or huge Stoner enhancements in thin Pd films.

REFERENCES

*Laboratoire associe au C.N.R.S.

1. See the Proc. of the 15th Int. Conf. on Low Temp. Phys., Grenoble, France, (1978), Jour de Physique C6, 279 to 293, (1978).
2. S. Doniach and S. Engelsberg, Phys. Rev. Lett., 17, 750 (1966); N.F. Berk and J.R. Schrieffer, Phys. Rev. Lett., 17, 433, (1966).
3. M.T. Beal-Monod and K. Maki, Phys. Rev. Lett., 34, 1461, (1975); J. Hertz, Phys. Rev. B14, 1165, (1976).
4. M.T. Beal-Monod, P. Kumar and H. Suhl, Sol. State Comm., 11, 855, (1972); E. Zaremba and A. Griffin, Sol. State Com. 13, 169, (1973) and Can. Journ. Phys., 53, 891, (1975); E. Zaremba, Thesis, Univ. Toronto (1974) unpublished.
5. J.P. Muscat, M.T. Beal-Monod, D.M. Newns and D. Spanjaard, Phys. Rev., B11, 1437, (1975); S.C. Ying, L.M. Kahn and M.T. Beal-Monod, Sol. State Com., 18, 359, (1976).
6. D. Spanjaard, D.L. Mills and M.T. Beal-Monod, Jour. Low Temp. Phys. 34, 307, (1979) and ref. 1.
7. As noted in [4-6], it is not possible to have an analytical formula for χ near the surface, since it is the solution of an integral Bethe-Salpeter equation with a nonuniform, nor separable, kernel. However the behavior of χ was obtained from computer calculations in the above references, and the possibility of a surface magnetic instability was proven analytically in [5].
8. M.T. Beal-Monod, Jour. Low Temp. Phys., 37, 123, (1979) (a misprint yields in formula (35) the values of p_1 to be read 0.9775 and 1.1134 respectively; see erratum in Jour. Low Temp. Phys. 39, 231, (1980)).
9. See for instance in M. Gabay and M.T. Beal-Monod, Phys. Rev. B18, 5033, (1978) and refs. therein and M. Gabay, 3rd Cycle thesis, Paris (1977) unpublished.
10. See for instance K.G. Wilson and J. Kogut, Phys. Rep. 12 C, 76, (1974).
11. E.G. Brovman and Yu Kagan, J.E.T.P., 36, 1025, (1973).
12. The details will be published elsewhere.
13. This was noted previously in M.T. Beal-Monod Phys. Lett. 59A, 157, (1976).
14. E. Brezin and J. Zinn-Justin, Phys. Rev. B14, 3110 (1976).
15. M.T. Beal-Monod, Phys. Letters, 54A, 197, (1975). See also [9].
16. A. Layzer and D. Fay, Int. Jour. Magn. 1, 135, (1971); P.W. Anderson and W.F. Brinkman, Phys. Rev. Lett., 30, 1108 (1973).
17. M.T. Beal-Monod, Sol. State Com. 32, 357 (1979).
18. F. Steglich et al., Phys. Rev. Lett. 43, 1892, (1979).
19. K. Levin, O.T. Walls, Phys. Rev. B17, 191, (1978).
20. J.R. Schrieffer and R. Gomer, Surf. Science 25, 215 (1971); H. Suhl, J.H. Smith and P. Kumar, Phys. Rev. Lett., 25, 1442 (1970).
21. B. Hebral, A.S. Greenberg, M.T. Beal-Monod, M. Papoular, G. Frossati, H. Godfrin and D. Thoulouze, to be published.
22. M.T. Beal-Monod and S. Doniach, Jour. Low Temp. Phys. 28, 175, (1977).
23. A.I. Ahonen et al. Journ. Phys. C9, 1665, (1976), Phys. Rev. Lett. 41, 494 (1978), and in ref. 1; H.M. Bozler et al., Phys. Rev. Lett. 41, 490, (1978) and in ref. 1; H.I.P. Godfrin et al. in ref. 1.
24. M.T. Beal-Monod and D.L. Mills, Jour. Low Temp. Phys., 30, 289, (1978) and in ref. 1.
25. T.A. Alvesalo et al., Phys. Rev. Lett., 43, 1509, (1979) and 44, 1076 (1980).
26. As noted in [22] and [23] for liquid He^3, in reduced dimensions one has a combination of normal and superfluid phases at ultralow T; but the superfluid phase may disappear, when the dimensions are smaller than the superfluid coherence length.
27. G. Bergman, Phys. Rev. Lett. 43, 1357, (1979); M. B. Brodsky and A. J. Freeman, to be published.

CRITICAL DYNAMICS OF 2D PLANAR MAGNETS*

D. L. HUBER

Department of Physics, University of Wisconsin, Madison, WI 53706

ABSTRACT

We investigate the critical dynamics of the Kosterlitz-Thouless transition in 2D planar magnets. Attention is focused on the critical region above T_c. If we ignore correlations between spin waves and vortices the transverse spin autocorrelation function is a product of spin wave and vortex factors. If correlations between vortices are also neglected the vortex factor decays at a rate proportional to the product of the density of free vortices and the vortex self-diffusion constant. The longitudinal spin autocorrelation function is the sum of spin wave and vortex terms. The latter is proportional to the product of the vortex density and the vortex velocity autocorrelation function. Possible tests of the theory are discussed.

INTRODUCTION

One of the most interesting developments in critical phenomena in recent years has been the recognition of the unique properties of phase transitions in two dimensions for systems having two-component order parameters [1,2]. In such systems the transition, often referred to as the Kosterlitz-Thouless transition, is associated with the unbinding of topological defects. In the case of two dimensional planar (easy-plane) magnets with negligible anisotropy in the basal plane these defects assume the form of spin vortices. Although recent experiments on He films [3,4], along with Monte Carlo studies of planar models [5], support the theory developed in Refs. 1 and 2, there is as yet no firm evidence for the existence of a vortex gas in real magnetic materials.

TRANSVERSE DYNAMICS

Most of the studies of the Kosterlitz-Thouless transition in spin systems have focused on static effects which are probed in measurements of the specific heat and susceptibility. Relatively little work has been done on the dynamic behavior. In an earlier paper [6] we outlined a semi-phenomenological theory for the transverse spin autocorrelation function which is applicable in the region immediately above the critical point. In the theory the angle ϕ characterizing the orientation of the spin in the basal (XY) plane is written as the sum of spin wave and vortex terms, viz. [1,2]

$$\phi = \phi_s + \phi_v, \tag{1}$$

where ϕ_s and ϕ_v are the spin wave and vortex contributions respectively.

The essential approximation of the theory is to neglect correlations between ϕ_s and ϕ_v, an approximation first introduced by Kosterlitz and Thouless [1,2] in their analysis of the equilibrium behavior. When this is done the transverse autocorrelation function C(t) defined by

$$C(t) = (1/2)[\langle\{S_x^n(t),S_x^n(0)\}\rangle+\langle\{S_y^n(t),S_y^n(0)\}\rangle], \tag{2}$$

where {A,B} denotes the symmetrized product, assumes the form

$$C(t) = C_s(t)C_v(t). \tag{3}$$

Here $C_s(t)$ and $C_v(t)$ are the spin wave and vortex factors, respectively. The latter is written

$$C_v(t) = \mathrm{Re}\langle\exp[i(\phi_v^n(t)-\phi_v^n(0))]\rangle, \tag{4}$$

where ϕ_v^n is the vortex contribution at site n and Re denotes real part.

As shown in Ref. 6 when the fluctuations are treated in a Gaussian approximation and correlations between vortices are explicitly neglected Eq. (4) reduces to

$$C_v(t) = \exp[-2\pi n \ln(L/a)\int_0^t d\tau (t-\tau)F(\tau)]. \quad (5)$$

Here n is the density of free vortices, L is a macroscopic length, and a is the lattice constant. The symbol F(t) denotes the vortex velocity autocorrelation function

$$F(t) = (1/2)\langle\vec{V}(t)\cdot\vec{V}(0)\rangle. \quad (6)$$

In the long-time limit $C_v(t)$ reduces to

$$C_v(t) = \exp[-2\pi n \ln(L/a) D_v |t|], \quad (7)$$

where D_v is the vortex self-diffusion constant

$$D_v = \int_0^\infty dt\ F(t). \quad (8)$$

The results shown in Eqs. (2-8) have a simple physical interpretation. The motion of the vortices leads to a time-dependent ϕ_v which in turn gives rise to the factor $C_v(t)$ in the autocorrelation function. Aside from the fact that they move in response to applied (and dipolar) magnetic fields [1,2] and that the vortex gas probably has no propagating excitations [7], little is known about vortex dynamics. The irreversible processes giving rise to vortex diffusion presumably arise from vortex - impurity scattering and from residual vortex-vortex and vortex-spin wave interactions.

It is evident that in the long-time limit the transverse autocorrelation function decays exponentially at a rate which is proportional to the density of free vortices. As the transition is approached from the high temperature side n is expected to vary as ξ_+^{-2} where ξ_+, the correlation length, diverges as

$$\xi_+ \sim a \exp[b(T/T_c-1)^{-\nu}], \quad (9)$$

with $b_+ \sim 0.9$ and $0.5 \lesssim \nu \lesssim 0.7$ [5]. Thus as $T \to T_c^+$ $C_v(t)$ shows critical slowing down, provided D_v does not diverge more rapidly than ξ_+^2. Above T_c on physical grounds we expect $C_s(t)$ to decay on a time scale set by the exchange frequency. This is likely to be much more rapid than the decay of $C_V(t)$ since the latter is associated with the movement of large scale defects. As a consequence the critical slowing down of the vortex factor is expected to have a negligible effect on the time dependence of the transverse autocorrelation function.

LONGITUDINAL DYNAMICS

In a recent note [8] we pointed out that unlike the transverse case the spin waves and vortices contribute additively to the longitudinal autocorrelation function. In the case of a Heisenberg magnet with hard-axis anisotropy AS_z^2, A>0 when spin wave-vortex correlations are neglected the longitudinal autocorrelation function has the form [8]

$$\langle\{S_z^n(t),S_z^n(0)\}\rangle=(2A)^{-2}[f_s(t)+\langle\frac{d\phi_v^n(t)}{dt}\frac{d\phi_v^n(0)}{dt}\rangle], \quad (10)$$

while for the XY model A is replaced by zJ where J is the exchange integral and z is the number of nearest neighbors. From Ref. 6 we have the result

$$\langle\frac{d\phi_v^n(t)}{dt}\ \frac{d\phi_v^n(0)}{dt}\rangle = 2\pi n \ln(L/a)F(t), \quad (11)$$

neglecting correlations between vortices. Here F(t) denotes the vortex velocity autocorrelation function, Eq. (6).

From (10) and (11) we see that the autocorrelation function is the sum of spin wave and vortex terms. The fourier transform of $f_s(t)$ is expected to have appreciable weight out to frequencies on the order of the exchange frequency. In contrast the fourier transform of F(t) is likely to be confined to very low frequencies. Thus in relation to the spin wave part the vortex term in the fourier transform of (10) appears as a central peak with relative weight $2\pi n(2A)^{-2}\ln(L/a)\langle\vec{V}\cdot\vec{V}\rangle\langle S_z^{n2}\rangle^{-1}$, which is probably extremely small.

DISCUSSION

In this paper we have presented results of a semi-phenomenological theory for the spin autocorrelation functions in two dimensional planar magnets which applies to the critical region above T_c. The essential approximation in the theory was the neglect of the correlations between spin waves and vortices. The presence of the free vortices is reflected in a slowly varying factor in the transverse autocorrelation function. In contrast, the spin wave and vortex terms contribute additively to the longitudinal function. Since the time scale associated with vortex motion is much longer than that of the spin waves the vortex contribution to the Fourier transform of $\langle\{S_z^n(t), S_z(0)\}\rangle$ appears as a central peak.

Information about longitudinal spin correlations can be obtained from inelastic neutron scattering measurements. In order to avoid confusing vortex and hydrodynamic (diffusive) central peaks the measurements should be carried out at momentum transfers which are outside the hydrodynamic regime. Unfortunately, because the relative weight of the vortex term is small the central peak may be hard to detect.

A more promising method of detecting vortices involves measurements of the nuclear magnetic resonance linewidth in a weak field perpendicular to the easy plane. In this configuration the homogeneous linewidth contains a term proportional to the integral [9]

$$\int_0^\infty dt\langle\{S_z^n(t),\ S_z^n(0)\}\rangle. \qquad (12)$$

From this we infer that the vortex contribution to the linewidth, $(1/T_2)_v$, is given by

$$(1/T_2)_v = 2\pi n A_{\parallel}^2 (2A)^{-2} D_v \ell n(L/a), \qquad (13)$$

where D_v is the self-diffusion constant, Eq. (8) and $A_{\parallel}$ is the coupling constant in the longitudinal hyperfine interaction $A_{\parallel} I_z S_z$. Since the vortex velocity autocorrelation function decays much more slowly than $f_s(t)$ it is possible that $(1/T_2)_v$ is comparable to the spin wave contribution, $(1/T_2)_s$, which can be written

$$(1/T_2)_s = a^2 (2A)^{-2} \int_0^\infty dt f_s(t). \qquad (14)$$

Finally we note that the explicit form of the vortex contribution to the longitudinal and transverse autocorrelation functions is a consequence of the neglect of correlations between vortices except in so far as they influence the vortex velocity autocorrelation function. Were we to take these correlations into account F(t) would be replaced by a multivortex correlation function. However the slow decay of $C_v(t)$ and the central peak in the fourier transform of the longitudinal autocorrelation function would still be present provided the spin wave and vortex dynamics are characterized by different time scales.

REFERENCES

*Supported by the NSF under the grant DMR-7904154.

1. J. M. Kosterlitz and D. J. Thouless, J. Phys. C 6, 1181 (1973).
2. J. M. Kosterlitz, J. Phys. C 7, 1046 (1974).
3. I. Rudnick, Phys. Rev. Lett. 40, 1454 (1978).
4. D. J. Bishop and J. D. Reppy, Phys. Rev. Lett. 40, 1727 (1978).
5. J. Tobochnik and G. V. Chester, Phys. Rev. B 20, 3761 (1979).
6. D. L. Huber, Phys. Lett. 68A, 125 (1978).
7. S. Trimper, Phys. Lett. 70A, 114 (1979).
8. D. L. Huber, Phys. Lett. 76A, 406 (1980).
9. T. Moriya, Progr. Theor. Phys. 16, 23, 641 (1956).

Note added. Phase transitions in quasi-two dimensional magnets have been discussed in a recent paper by S. Hikami and T. Tsuneto (Progr. Theor. Phys. 63, 387 (1980)). Also, P. C. Hohenberg, B. I. Halperin, and D. R. Nelson (preprint) have discussed the critical dynamics of two dimensional superfluid helium. They postulate a correlation function for the order parameter which is equivalent to that given by Eqs. (3), (7), and (8) with $C_s(t) \sim t^{-1/4}$ (cf. Ref. 6).

Published 1980 by Elsevier North Holland, Inc.
Sinha, ed. Ordering in Two Dimensions

RECONSTRUCTION VS. MAGNETISM AT (111) Si AND Ge SURFACES

C. T. WHITE W. E. Carlos*

Naval Research Laboratory
Washington, DC 20375

The competition between reconstruction and itinerant magnetism at vacuum cleaved (111) Si and Ge surfaces is investigated through a many-body model of the surface dangling bond band. Results are obtained by applying a generalized self-consistent field approximation to the Hamiltonian with the variational space general enough to include the possibility of coupled reconstructed and magnetic solutions. Within our approach, it is found that when reconstructed and magnetic solutions exist they are mutually exclusive over large regions of the parameter space. Furthermore, no region of coupled magnetic and reconstructed solutions is found. Our results support the assumption made in many places in the literature of neglecting the possibility of local magnetic moments in studying the electronic structure of the 2x1 reconstructed surface.

1. INTRODUCTION

Over the last several years many experimental [1-6] and theoretical [7-14] studies have been made of Si and Ge surfaces. Low energy electron diffraction (LEED) studies of such surfaces formed by ultrahigh vacuum cleavage and unannealed, have yielded a crystallographic surface structure appropriate for a 2x1 unit mesh. It is generally agreed [1-14] (in accordance with Haneman's [1] original ideas) that this 2x1 LEED pattern arises from the displacement of alternate rows of surface atoms inward and outward resulting in the formation of a rippled (2x1) surface.

The driving mechanism for this 2x1 reconstruction is thought to be a dehybridization effect which as pointed out e.g. by Harrison [11] can be understood in simple chemical terms. More specifically, first focus on a particular threefold coordinated surface atom generated by the vacuum cleavage and its associated dangling hybrid state. Now within a chemical picture assume that the three hybrids used to form bonds between this surface atom and its three associated back atoms are each constrained to point toward a different back atom. This assumption should be a reasonable approximation. Then as this surface atom is moved toward (away from) the plane of its three back atoms these three back bond hybrids should become more sp^2-like (p-like) and hence the associated orthonormal dangling hybrid state becomes more p-like (s-like) and is raised (lowered) in energy. Now consider the plane formed by the surface atoms if the dangling bonds were constrained to be singly occupied. Then the above described dehybridization effects would be expected to favor a distortion of the surface atoms out of this plane which generates a 2x1 structure. Explicitly, by raising and lowering alternate rows of surface atoms to form ridges and valleys about this plane and transferring electrons from the valleys to the ridges, one would expect to lower the energy of the system.

Of course the dangling bond states interact with one another (both directly and indirectly) which broaden these surface states into a surface dangling bond band (SDBB) which lies near to the semiconductor gap and is half filled. In this instance the essentials of the above ideas still apply and the formation of the 2x1 will produce a gap at E_F in this SDBB if the shifting of the one electron levels due to the distortion is large enough. Indeed such a gap in the SDBB of Si has been observed experimentally [3] and attributed to this mechanism. Additional experimental evidence exists in support of this picture for Si such as Ibach's observation [6] of surface-optical-phonon-slow electron interactions in the 2x1 structure.

The above implies that in any satisfactory treatment of this SDBB neither the local geometry of the surface nor the average number of electrons at a particular site should be assumed but rather determined in a self-consistent fashion through some energy minimization approach. Such studies using cluster approaches of varying degrees of sophistication [11-13] have been carried out leading to results consistent with the above simplified chemical analysis. Furthermore, within a tight-binding framework energy minimization treatments of the infinite system have been made e.g. by White and Ngai [10] to investigate the temperature dependence of the Si-2x1 reconstruction and subsequently by Chadi [14] using a more complex model, to study the magnitudes of the distortions.

A common feature of many of these energy minimization approaches has been their neglect of spin density fluctuations in carrying out the analysis. On the other hand the SDBB is on the order of a few tenths of eVs wide [3] and the Coulomb repulsion between two electrons in the same dangling orbital is expected to be the same order of magnitude [10]. Hence it could be favorable in this surface system to form effective local magnetic moments in order to partially minimize the Coulomb repulsive energy in analogy with results obtained [15] from a Hubbard model.

In the next section of this paper we will investigate the aforementioned competition between the 2x1 reconstruction and magnetism within perhaps the simplest model that contains the essential ingredients of the problem. Our results show that when the 2x1 reconstructed system exists it is by and large not energetically favorable to generate local magnetic moments associated with the SDBB although U can be large. Rather the effects of U can be incorporated simply through a renormalization of the effective electron pairing interaction generated by the reconstruction in accord with assumptions that have been made either implicitly or explicitly in the literature.

Before going on it is important to note that the cleavage prepared 2x1 reconstructions of (111) Si and Ge surfaces are transformed irreversible to 7x7 and 2x8 structures respectively upon annealing at several 100C. One might think that this fact would call into question any energy minimization studies of the 2x1 based on properties of the ideal (111) surfaces. However, the 2x1 reconstruction although e.g. further stabilized by cleavage steps is thought to be intrinsic to pristine (111) Si and Ge surfaces and as such represents a well-defined local minimum in the actual free energy of the system. Thus, energy minimization studies of the 2x1 carried out within models which allow only for this reconstruction should be proper. This is also true for the kind of studies that we report here which essentially involve testing the stability of the spin paired 2x1 against the formation of a reconstructed 2x1 with local magnetic moments.

2. THE MODEL: BASIC APPROACH AND RESULTS

The model we have studied is described by the Holstein-Anderson-like [16,17] Hamiltonian

$$H = \sum_{i\sigma} (\varepsilon_h - \lambda X_i) n_{i\sigma} + \sum_{ij\sigma} V_{ij} a^\dagger_{i\sigma} a_{j\sigma} + U\sum_i n_{i\uparrow} n_{i\downarrow} + (1/2)C\sum_i X_i^2 + \sum_i (P_i^2/2M) + \lambda \sum_i X_i, \quad (1)$$

where $a^\dagger_{i\sigma}$ and $a_{i\sigma}$ are the usual creation and annihilation operators respectively for electrons of spin σ in the Wannier state centered at the site i, $|i\sigma\rangle$, $n_{i\sigma} = a^\dagger_{i\sigma} a_{i\sigma}$ is the number operator and the sites $\{i\}$ form a triangular lattice. The energy ε_h entering the first term on the right hand side (RHS) of Eq. 1 is simply the dangling hybrid energy in the absence of correlations. The quantity λX_i accounts for the shifting of ε_h due to dehybridization effects as the threefold coordinated surface atom is displaced a distance X_i. This displacement is measured perpendicularly from the surface plane that would form if the locally averaged number of electrons per site were contained to be one. The second sum on the RHS of Eq. 1 is a hopping term which accounts for the overlap of the dangling orbitals, while the next Hubbard-like [18] term allows for the Coulomb repulsion between two electrons in the same dangling hybrid. The last three terms entering Eq. 1 simply represent the Hamiltonian for a group of independent harmonic oscillators, one for each threefold coordinated surface atom of mass M. Each of these oscillators is displaced in such a way as to require that if $\sum_\sigma \bar{n}_{i\sigma} = 1$ then $\bar{X}_i = 0$ where $\bar{X}_i$ represents the average displacement of the atom i and $\bar{n}_{i\sigma}$ the average number of electrons of spin σ at the site i. The elastic energies entering the model, $C\sum_i X_i^2/2$, arise principally from the bending and stretching of the backbonds that occur as the surface atoms are distorted.

To treat the Hamiltonian (1) employ a generalized self-consistent field approach. This entails linearizing the manybody terms according to the approximations

$$\lambda X_i n_{i\sigma} \approx \lambda X_i \bar{n}_{i\sigma} + \lambda \bar{X}_i n_{i\sigma} - \lambda \bar{X}_i \bar{n}_{i\sigma} \quad (2a)$$

and

$$U n_{i\uparrow} n_{i\downarrow} \approx U \sum_\sigma \bar{n}_{i-\sigma} n_{i\sigma} - U\bar{n}_{i\uparrow}\bar{n}_{i\downarrow}, \quad (2b)$$

where the terms $\lambda\bar{X}_i\bar{n}_{i\sigma}$, $U\bar{n}_{i\uparrow}\bar{n}_{i\downarrow}$ are present to prevent double counting of the interaction. The replacements (2) generate from the manybody Hamiltonian (1) an effective onebody Hamiltonian $H_{eff} = H_{el} + H_{HO}$ where H_{el} (H_{HO}) is the effective Hamiltonian for the electrons (harmonic oscillators) and $[H_{el}, H_{HO}] = 0$. The quantities $\{\bar{X}_i\}$ can now be easily calculated from H_{HO} and they turn out to be

$$\bar{X}_i = (\lambda/C)\,(\sum_\sigma \bar{n}_{i\sigma} - 1) \quad (3)$$

so that the lattice coordinates $\{\bar{X}_i\}$ can henceforth be eliminated in favor of the electronic coordinates $\{\bar{n}_{i\sigma}\}$. The quantities $\bar{n}_{i\sigma}$ can be obtained from H^σ_{el} which however depends through Eqs. (2,3) on $\{\bar{n}_{j\sigma}\}$ and thus we obtain an infinite set of self-consistent equations.

To make further progress we must restrict this set of equations. This we do by allowing for only two types of sites (A,B) arranged over the surface in such a manner as to produce a 2x1 structure. Since in the SDBB we have only one electron per site we have the constraint that

$\Sigma(n^A_\sigma+n^B_\sigma) = 2$ where e.g. n^A_σ represents the average number of electrons of spin σ at a site labeled A. Furthermore, we require that there be no net z component of the total spin so that $(n^A_\uparrow-n^A_\downarrow) = -(n^B_\uparrow-n^B_\downarrow)$.

We now have only two self consistent equations which can be put into the form

$$n^A_\sigma = -\frac{1}{\pi}\mathrm{Im}\int_{-\infty}^{\infty} dEfg_{i\sigma}(E^+;n^A_\uparrow,n^A_\downarrow), \qquad (4)$$

where $g_{i\sigma} = \langle\sigma i|(E^+-H^\sigma_{el})^{-1}|i\sigma\rangle$, $E^+ \equiv (E+i\varepsilon)$ with $\varepsilon\to o^+$ and f is the fermi function. In order to solve Eqs. (4) we must obtain the Green functions $g_{i\sigma}$ which can be done by going into the Bloch representation. Here however we will obtain these Green functions approximately by replacing the actual triangular lattice by a Cayley tree with (1) the same number of nearest neighbors and (2) the sites A,B arranged over this lattice in a manner to reproduce properly the local 2x1 environment.

At this point rather than solve Eqs. 4 directly it is more convenient to first introduce the effective scattering strengths

$$\varepsilon^A_\sigma = \varepsilon_h+Un^A_{-\sigma}-(\lambda^2/C)(n^A_\uparrow+n^A_\downarrow-1). \qquad (5)$$

After a considerable amount of algebra it can then be shown that the total free energy per site of the system, F, can be expressed as

$$F(x,y) = K+\int_o^x n(z)dz+\int_o^y n(z)dz-(x+y)/2 + (x-y)^2/(4U) + (x+y)^2/4R , \qquad (6)$$

where $x = \varepsilon^A_\uparrow, y = \varepsilon^A_\downarrow, R = (2\lambda^2/C)-U$, K is independent of x,y and the functional form of n(z) is given by Eq. 4 with $H^\sigma_{el} = z\Sigma_{i\varepsilon A}|i\sigma\rangle\langle\sigma i|-z\Sigma_{i\varepsilon B}(i\sigma\rangle\langle\sigma i|)+\Sigma_{ij} V_{ij}|i\sigma\rangle\langle\sigma j|$. Note in obtaining (6) we have taken $\varepsilon_h = -U/2$ so that the electronic chemical potential is zero. This results in no loss of generality within our approximation. It is now straightforward that at points (x_o,y_o) such that $(\partial F/\partial x) = (\partial F/\partial y) = 0$ the self-consistent Eqs. 4 are satisfied.

Consideration of the function n(z) shows that within the present approximation n(z) decreases monotonically from one to zero as z varies from $-\infty$ to ∞. Furthermore, the function $n'(z) = dn(z)/dz$ has only one stationary point which occurs at z=0 where n(z) = 1/2 and $|n'(z)|$ is a maximum. In addition, $n(-z) = -(n(z)-1)$.

One can now show [19] that the above described properties of n(z) are sufficient to produce the phase diagram displayed in Fig. 1. In region 5 of this diagram the only self-

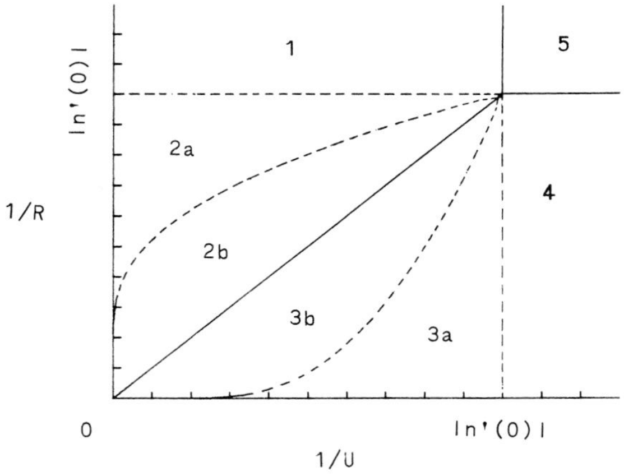

Fig. 1. Illustrates the different magnetic and reconstructed regions described in the text.

consistent solution that exists corresponds to $x_o=y_o=0$ hence one has neither the possibility of reconstruction nor magnetism. (Note, by reconstruction we mean distortion about the plane which is defined when all $\{X_i\}$ are zero.) On the other hand in region 1 of Fig. 1 there are three stationary points of F(x,y). Two of these are stable, degenerate in energy, and correspond to purely magnetic solutions ($x_o=-y_o\neq 0$) while the third occurring at $x_o=y_o=0$ now becomes a saddle point. The transformation from region 5 to region 1 occurs continuously. In region 4 there are also three stationary points with two of these stable and degenerate in energy. However now the stable solutions correspond to the purely reconstructed solutions $x_o=y_o\neq 0$. The transformation from region 4 to region 5 is continuous. In region 2a (3a) only the two degenerate purely magnetic (reconstructed) solutions are stable with the two degenerate purely reconstructed (magnetic) solutions representing saddle points.

In contrast in region 2b (3b) the two degenerate purely reconstructed (magnetic) solutions become locally stable however the two degenerate purely magnetic (reconstructed) solutions are lower in energy. Hence a first order phase transition occurs as one proceeds across the line U=R from region 2b to 3b. This first order transition could be smeared out of small fluctuations that we have neglected here. One would expect though that smearing would become less and less effective as $1/R = 1/U \to 0$.

The above results then show that within the present model it is never favorable to build local magnetic moments on top of the self-consistent obtained 2x1 reconstructed solution. There is a warning here however which involves the fact that in obtaining this result we have employed a Cayley tree which requires particle-hole symmetry in the electronic density of states. This feature however is not present in the actual triangular lattice. On the other hand our Cayley approximation becomes progressively better as the surface atoms distort to form the 2x1 structure with the electrons and holes being more and more confined to ridges and valleys respectively. Indeed, analysis of the experimentally implied density of states [3] indicates that we are well into the region where the particle-hole symmetry is almost restored. Thus our results indicate that if one were to employ the actual triangular lattice to treat the 2x1 one would expect at most only very small effective magnetic moments coupled to the reconstructed solution which can be neglected. This of course is what we set out to show.

In conclusion it is noteworthy that the self-consistent field properties of the Hamiltonian (1) detailed in Fig. 1 follow assuming any single band lattice structure so long as n(z) has properties similar to those we have detailed above. Furthermore, in treating these systems, temperature enters simply by changing $|n'(0)|$. Also note in obtaining Fig. 1 we have not assumed that V_{ij} is restricted to only nearest neighbor interactions. Application of the present formalism along all of these lines will be made elsewhere.

*NRC-NRL Resident Research Associate

REFERENCES

1. D. Haneman, Phys. Rev. 170, 705 (1968).
2. W. Mönch, Surface Sci. 63, 79 (1977); Surface Sci. 86, 672 (1979) and refs. therein.
3. G. Chiarotti, S. Nannarome, R. Pastore, and P. Chiaradia, Phys. Rev. B4, 3398 (1971).
4. M. Erbudak and T. Fischer, Phys. Rev. Lett. 29, 732 (1972).
5. J. Lander, G. Gobeli and J. Morrison, J. Appl. Phys. 34, 2298 (1963).
6. H. Ibach, Phys. Rev. Lett. 27, 253 (1971).
7. J. Appelbaum and D. Hamann, Phys. Rev. B12, 1410 (1975).
8. K. Pandey and J. Phillips, Phys. Rev. Letts. 34, 1450 (1975).
9. M. Schlüter, J. Chelikowsky, S. Louis and M. Cohen, Phys. Rev. Lett. 34, 1385 (1975).
10. C. T. White and K. L. Ngai, J. Vac. Technol. 15, 1237 (1978); Phys. Rev. Lett. 41, 885 (1978).
11. W. Harrison, Surf. Sci. 55, 1 (1976).
12. W. Verwoerd and F. Kok, Surface Sci. 80, 89 (1979).
13. W. Goddard, III, and J. Barton, J. Vac. Sci. Technol. 15, 1273 (1978).
14. D. Chadi, Phys. Rev. Lett. 41, 1062 (1978).
15. See e.g. E. Economou, C. White and R. DeMarco, Phys. Rev. B18, 3984 (1978).
16. T. Holstein, Ann. Phys., N.Y. 8, 325 (1959).
17. P. Anderson, Phys. Rev. Lett. 34, 953 (1975).
18. J. Hubbard, Proc. Roy. Soc. London, Ser A, 281, 401 (1964).
19. C. T. White, in preparation.

SECTION II

F. MOLECULAR MONOLAYERS, BILAYERS, MEMBRANES AND LIQUID CRYSTALS

Sinha, ed. Ordering in Two Dimensions

THE ANISOTROPY OF SELF-DIFFUSION IN THE LAMELLAR PHASE*

P. UKLEJA

Physics Department, Southeastern Massachusetts University, N. Dartmouth, MA 02747

J. W. DOANE

Liquid Crystal Institute and Department of Physics, Kent State University, Kent, OH 44242

ABSTRACT

The method of pulsed field gradient nuclear magnetic resonance [1] was used to measure the anisotropy $D_\perp/D_\parallel$ of the self-diffusion coefficients in the stacked lamellar phases of two lipid systems: potassium palmitate (K-Pal) plus deuterated water and dipalmitoylphosphatidylcholine (DPPC) plus deuterated water. $D_\perp$ and $D_\parallel$ are, respectively, the coefficients for self-diffusion perpendicular and parallel to the planar normals of the lamella. The measurements were performed on aligned samples at temperatures above the gel phase. Values of about 40 were obtained for the anisotropy of the diffusion of the water molecules in both systems as well as for K-Pal molecules. For the DPPC only a lower bound could be determined which was found to be ∿20.

INTRODUCTION

In systems of stacked bilayers such as the potassium-water system in the lamellar phase, the self-diffusion of molecules is normally believed to be limited to two dimensions. Thus the self-diffusion coefficient $D_\perp$ for motion perpendicular to the plane would be expected to be several orders of magnitude larger than $D_\parallel$, the coefficient for motion parallel to the normal. Literature on actual measurements of $D_\perp/D_\parallel$ in layered liquid crystal systems has been meager with direct measurements of self-diffusion reported only for some smectic thermotropic liquid crystals and special lyotropic lamellar systems. For example, measurements on smectic A liquid crystals using a pulsed magnetic field gradient[2] have given values between 1.5 and 10 for the anisotropy $D_\perp/D_\parallel$. Similar measurements reported for an aligned lamellar phase sample of lithium perfluoro-octanoate gave surprisingly low anisotropies of only 1.3 for the water molecules[3]. In this paper we present results of measurements using again a pulsed NMR technique[1] on stacked bilayers of potassium palmitate (K-Pal) plus water and on dipalmitoylphosphatidylcholine (DPPC) plus water where the anisotropies in the self-diffusion coefficients of water and lipids were in all cases found to be much greater than 20. In contrast, when we applied the same technique to a smectic A material, we found little measureable anisotropy, with $D_\perp/D_\parallel$ < 1.

In the basic spin-echo experiment, two radio frequency pulses are applied to the sample with an echo signal appearing at a later time. If the sequence is repeated, but this time pulses of magnetic field gradient $\vec{g}$ are applied, one between the two r.f. pulses and the other between the second r.f. pulse and the echo, the echo height is reduced depending upon the rate of molecular self diffusion. In an isotropic medium, the attenuation of the echo depends on the width δ of the gradient pulses, the spacing Δ between their leading edges, the magnitude g of the field gradient and the coefficient of diffusion D according to the formula:

$$A = \exp[-Dg^2\gamma^2\delta^2(\Delta-\delta/3)] \qquad (1)$$

where γ is the gyromagnetic ratio for the spin observed. In a medium such as the bilayer where the diffusion is anisotropic but still has an axis of cylindrical symmetry parallel to the bilayer normal, D can be expressed as:

$$D = D_\parallel \cos^2\theta + D_\perp \sin^2\theta \qquad (2)$$

where θ is the angle between the bilayer normal and the direction of the field gradient. The NMR technique has the following virtues: 1) No probe molecules are needed. 2) It is sensitive to motion in one direction and that direction can be varied. 3) The time during which diffusion occurs (roughly, Δ) can be varied from about 100 microseconds to several seconds in some cases. For samples in the smectic A and lamellar phases dipole-dipole interactions usually make this experiment difficult to perform. However, in a well-aligned sample these interactions can be drastically reduced [4,5] by directing the normal to the planes, $\vec{N}$, so that it makes an angle of 54.74° (the "magic angle") with the magnetic field.

MATERIALS

Potassium Palmitate Multibilayer Samples. Potassium palmitate was prepared by mixing palmitic acid (Sigma Chemical Co., 2.0g) with an equal molar amount of KOD in 1.1g of D_2O at 60°C and stirring for 5 minutes. Purification was achieved by recrystallization from absolute ethanol to give 1.5g of salt.

Lamellar bilayers were obtained by a mixture of potassium palmitate with 28% D_2O and an estimated 2% of HOD. Aligned multibilayer samples for the diffusion studies were prepared from a stack of 30 thin (120 μm) slides with no spacers between the slides other than the potassium palmitate-water mixture. The glass slides were treated with a silane surfactant, n,n-dimethyl-n-octadecyl-3-aminopropyltrimethoxysilyl chloride for homeotropic alignment by the procedure of Powers and Clark[6]. The stack of slides was then placed in an 8mm diameter NMR sample tube and sealed at a point immediately above the stack.

DPPC Multibilayer Samples: Commercially obtained (Sigma Chemical Co.) anhydrous 1,2-dipalmitoylphosphatidylcholine (DPPC) was mixed with 30 wt. % of D_2O and an estimated 2% of HOD in a nitrogen atmosphere. This mixture was used to make an aligned multibilayer sample using the same method as for the sample above [6]. Additional D_2O was later added by exposure to the saturated vapor at temperatures above the lamellar-gel transition. From the NMR signal the ordered portion of the sample was determined to be about 50% of the total. In the experiment, the sample was originally ordered at 80°C. When the temperature was changed, it was done slowly (about 0.2°C/min) and allowed to remain at the final temperature until the amplitude of a spin-echo was stable.

Smectic A Sample: A one gram sample of 4-n-pentylphenylthio-4'-n-octyloxybenzoate (8S5) was aligned by cooling from the nematic phase to the smectic A phase at 58.4°C in the presence of a strong (13.8 kG) magnetic field.

INSTRUMENTATION

The ^1H NMR signals were taken at 59 MHz using a BRUKER, Model B-KR322S pulsed NMR spectrometer modified with a more sophisticated pulse programmer, a Biomation Model 802 transient recorder and a Nicolet NIC-80 mini-computer. The field gradient coils were wound in a quadrupole configuration [7] and were capable of supplying a field gradient of 540 G/cm with a rise time (10% to 90%) for the current of about 50μs. The gradient pulses were placed so as to end at least 4ms before any r.f. pulse or echo. The gradient strength was calibrated to a value of D in water of 2.3×10^{-5} cm^2/s at 25°C and gave the value of 2.85×10^{-8} cm^2/s at 25°C for dry glycerol in agreement with Tomlinson [8]. The direction of the field gradient was set to coincide with the normal, $\vec{N}$, in one of the two magic angle positions. The angle between the gradient and the magnetic field was later estimated to be 51 ± 1° by measuring the attenuation at several angles for the echo from the small amount of HOD in the DPPC sample. The proton NMR echo from the HOD showed virtually no dipole-dipole interaction so that the echo height with no field gradient was independent of the angle between the magnetic field and $\vec{N}$. With the field gradient pulses on, the angular variation was due to the variation of D as in Eqn. (2).

RESULTS

In all experiments the attenuation of the echo was measured for several values of the gradient pulse width δ and the logarithm plotted versus the quantity $\delta^2(\Delta-\delta/3)$. For a single species of diffusing molecule the result should be a straight line projecting to an attenuation of one as δ approaches zero. The value of Δ was sometimes varied to see if the value obtained for D depended on the time of the diffusion measurement as it would in some cases of restricted diffusion. No evidences of restricted diffusion were found. The diffusion coefficients were measured at two magic angles with θ=3.7° or 105.7° and the values of $D_\perp$ and $D_\parallel$ deduced, using Eq. (1), to be

$$D_\parallel \simeq 1.005\ D(3.7°) - .005\ D(105.7°) \text{ and}$$

$$D_\perp \simeq 1.083\ D(105.7°) - .083\ D(3.7°) \quad (3)$$

K-Pal-Water Multibilayer Sample: The plots of tenuation versus $\delta^2(\Delta-\delta/3)$ showed curvature for small values of δ which came about due to the more rapidly diffusing HOD molecules which contribute about 18% of the echo. This could be subtracted out by repeating the measurements away from the magic angle where the echo was due only to the HOD molecules. The results for the K-Pal molecules using 6-8 points were: $D_\perp$ =24.9, $D_\parallel$ =0.56 x $10^{-7}cm^2/s$ at 100°C and 15.9 and 0.27 x $10^{-7}cm^2/s$ respectively at 75°C, giving anisotropies of 44 and 69 at the two temperatures. The errors are about 5% in $D_\perp$ and 10-15% in $D_\parallel$. Diffusion times of 10-35 ms were used.

The signals from the HOD were quite small but adequate to determine values for $D_\perp$ of 1.8 and 1.1 x $10^{-5}cm^2/s$ at 100°C and 75°C respectively. The anisotropies were estimated to be at least 25 at the two temperatures.

DPPC-Water Multibilayer Samples: In this sample only about one-third of the echo at the magic angle (no field gradients) was due to the lipid as was seen from the echo height away from the magic angle. The subtraction of the background worked to give values for $D_\perp$ but only gave upper bounds for $D_\parallel$ for the lipid, since the apparatus would not measure values $\lesssim 10^{-8} cm^2/s$. The values measured were $D_\perp = 7.7$ and $D_\parallel < 0.6 \times 10^{-7} cm^2/s$ at 80°C and 5.0 and $< 0.2 \times 10^{-7} cm^2/s$ at 65°C, giving anisotropies of at least 13 and 25.

The data for the HOD molecules by themselves showed the existence of more than one diffusing species due to the relatively large amount of unaligned sample. The estimates for the anisotropies were about 50 between 50° and 80°C with values for $D_\perp$ varying from 1.5 to $3.5 \times 10^{-5} cm^2/s$ in that temperature range.

$\bar{8}$S5: For this sample, using times of 10 to 20 ms between r.f. pulses, essentially no echo was seen except when $\vec{N}$ was within a degree of the magic angle. The results of the measurements in the smectic A phase of $\bar{8}$S5 at 58.4°C are $5.70 \pm .13$ and $3.51 \pm .15 \times 10^{-7} cm^2/s$ for $D_\parallel$ and $D_\perp$ and an anisotropy of 0.55 ± 0.04. The diffusion times used were 6-11 ms. Measurements in the isotropic phase at 91°C gave the result for D of $1.65 \pm .05 \times 10^{-6} cm^2/s$ using diffusion times of 10 to 20 ms and various pulse sequences.

DISCUSSION

Our experimental apparatus was limited to values of D larger than $\sim 10^{-8}$, depending somewhat upon the preparation of the sample. Only values larger than this could be measured with reasonable certainty. With modest improvements the technique could be extended to $10^{-9} cm^2/sec$ but it would take imaginative techniques to extend the pulse gradient method to measure slower diffusion rates than this. Our reported values of $D_\parallel$ for K-Pal were well within the measureability of the apparatus and technique and we feel confident of these results. On the other hand, $D_\parallel$ for DPPC was below the capabilities of the experiment and only an upper bound could be estimated for this quantity. It could be that $D_\perp/D_\parallel$ for the bilayers in DPPC is very much larger than that for the bilayers in K-Pal which we measured to be ~ 40.

The values of $D_\perp/D_\parallel$ for water could be more accurately measured by larger concentrations of HOD. Our error in these reported values was only introduced by low intensity signals.

Finally, we comment on the unexpected results for $D_\perp/D_\parallel$ measured in the smectic A phase of $\bar{8}$S5. A value of less than one is characteristic of diffusion in the nematic phase. The measured result of a similar value in the smectic A phase could reflect the near second order character of the nematic-smectic A phase transition in this material.

ACKNOWLEDGEMENTS

The authors acknowledge Philip Bos for his design and construction of the pulse programmer without which this experiment would have been impossible. Acknowledgement is also due Nuno Vaz and Maria Vaz for their assistance in sample preparation.

*This research was supported in part by a grant from the National Institutes of Health GM-22599.

REFERENCES

1. E.D. Stejskal and J.E. Tanner, J. Chem. Phys. 42, 288 (1965).
2. G.J. Krüger, H. Spiesecke and R. Weiss, Phys. Lett. 51A, 295 (1975).
3. Gordon J.T. Tiddy, J. Chem. Soc. Faraday Transactions I, Vol. 73 (1977).
4. J.W. Doane and R.S. Parker, Magnetic Resonance and Related Phenomena, XVIIIth Cong. Ampere, 410 (1973).
5. S.B.W. Roeder and E. Elliot Burnell, J. Chem. Phys. 64, 1848 (1976).
6. L. Powers and P.S. Pershan, Biophysical Journal 20, 137 (1977).
7. I. Zupancic and J. Pirs, J. Phys. E 9, 79 (1976).
8. D.C. Tomlinson, Mol. Phys. 25, 735 (1972).

Sinha, ed, Ordering in Two Dimensions

LASER MEASUREMENTS OF HEAT TRANSFER THROUGH THE ORDERED INTERFACE OF WATER CONTAINING REVERSED MICELLES

J.R. LALANNE, B. POULIGNY and E. SEIN
University of Bordeaux I and C.N.R.S. Research Center P. Pascal
University Domain, 33405 Talence, France

J. BUCHERT and S. KIELICH
Institute of Physics, University A. Mickiewicz
60780, Poznan, Poland

We present here recent results obtained on microemulsions by using a variation of the well-known thermal blooming technique used, these last years, for the measurement of thermal conductivity in fluids. An infrared, microsecond, laser pulse is used to develop the thermal effect. The decay of the thermal lens is analysed by a c.w., non absorbed, auxilliary wave. This dual beam technique is, for the first time, used for studying selective heat transfer in microemulsions. More precisely, we have induced a T-jump in inversed micelles containing water and followed the conductive heat transfer through the ordered interface between the dispersed phase and the surrounding liquid (carbon tetrachloride). Values of thermal diffusitivity and conductivity are reported and discussed.

INTRODUCTION

We report here what we believe to be the first application of the laser induced thermal blooming to Colloïd Science, and ,more precisely, to the study of heat transfer at the liquid-liquid interface of microemulsions. The physical leading idea is very simple. Figure n°1 shows the supposed structure of inversed micelles filled with water in an apolar medium (microemulsion water/oil) {1}.

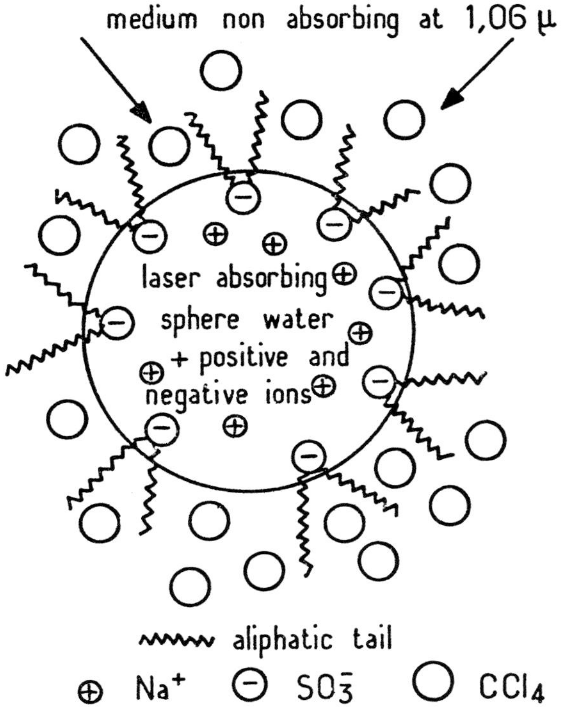

Figure n°1

The spheres are surrounded by an apolar medium (carbon tetrachloride). The "core" and the tails of the surfactant are absorbers for near IR wave (1060 nm delivered by noedymium glass laser) while carbon tetrachloride is totaly transparent for this wavelength. We can then suppose, with a good approximation, that the laser pulse at 1060 nm will be selectively absorbed by the micelle. It would be then very interesting to study the relaxation of heat through the ordered interface,after inducing a laser T-jump in the "cores".

EXPERIMENT

We use the thermal induced defocusing technique {2} for our experiments. It can be considered as an extension of the well-known self-defocusing of laser beams. A short laser pulse, partially absorbed by the sample, generates phase changes which induce defocusing of a non-absorbed laser wave. The thermal lens effect decay leads to the measurement of the thermal diffusivity constant of the sample. The experimental set-up is shown in figure n°2.

Figure n°2

The laser oscillator is already described {3}. It is here used, at 1060 nm, in the free running mode. TEM_{00} is selected by a pinhole of adapted diameter inserted in the optical cavity. Since the analysis depends upon the assumption of a Gaussian transverse distribution of the heating energy, such a selection has been carefully controled. A Pockels cell is used as an optical shutter, selecting one of the free running oscillations of the oscillator. The pulse duration is about 1 µs at half-height. After amplification, the energy of the pulse is about 10 mJ and the highest value of the associated electric field is about 0.8 x 10^6 V.cm^{-1}. Such a long impulsion is specially well adapted for heating. The probe wave is given by a weak power c.w. He-Ne laser (∿ 1 mW). Seven

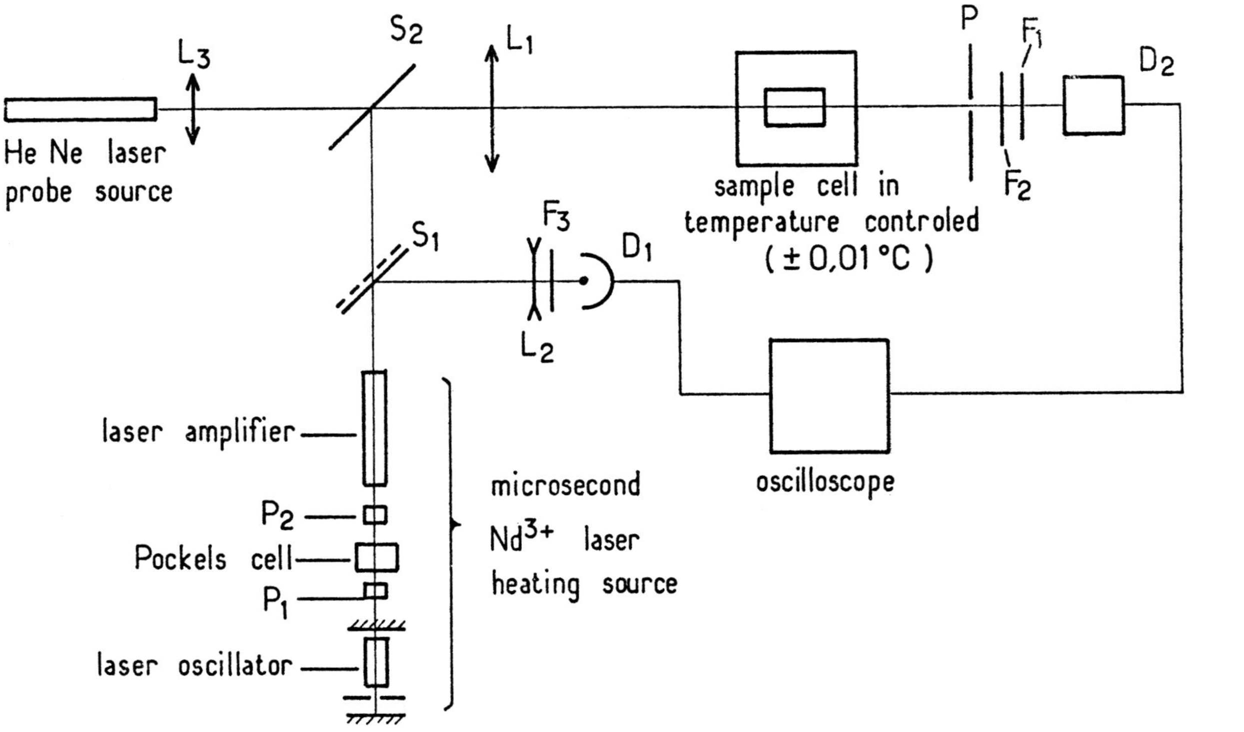

S_1, S_2 : beam splitters
L_2 : divergent lens
P : pinhole ϕ 1 mm
P_1, P_2 : Glan polarizers
L_1 : convergent lens f = 80 cm
F_1, F_2, F_3 : filters
D_1, D_2 : photodiode or photocell detectors

Figure n°2: Experimental set up

solutions have been studied corresponding to a progressive dilution of micelles in carbon tetrachloride.

ANALYSIS

Neglecting convection effects, the temperature distribution is obtained from the heat conduction equation (in cartesian coordinates) :

$$-\frac{1}{D_{th}}\frac{\partial(\Delta T)}{\partial t}+\nabla^2(\Delta T)=-\frac{1}{\Lambda}A\exp\left(-\frac{2r^2}{\omega_0^2}\right)\exp(-bz)\delta(t) \qquad (1)$$

The propagation of optical waves occurs on the Oz axis. D_{th} is the thermal diffusivity constant of the sample ; ΔT the induced temperature change, Λ the thermal conductivity ($D_{th}=\Lambda/\rho C_p$, ρ=density of the liquid, C_p = specific heat) ; A is a constant$\cdot\exp(-\frac{2r^2}{\omega_0^2})$ takes into account the radial distribution of TEM_{00} (ω_0 is the radius of the heating wave to the 1/e height, b is the absorption coefficient in cm^{-1} at 1060 nm and $\delta(t)$ the Dirac function. Spatial Fourier transform and the introduction of boundary conditions lead to a solution of equation (1) which is obtained in the form :

$$\Delta T(r,z,t)=\frac{\tau_L A D_{th} Y(t)\exp\left[-r^2/5\left(D_{th}+\frac{\omega_0^2}{8}\right)\right]}{2\Lambda(D_{th}t/\omega_0^2+1/8)}$$

$$\times\sum_{m=1}^{\infty}\frac{m\sin(m\Pi z/L)\exp\left[-D_{th}(m\Pi/L)^2\right]\left[1-(-1)^m\exp-bL\right]}{(Lb)^2+(m\Pi)^2} \qquad (2)$$

where τ_L is the time constant of the heating pulse and Y(t) the Heaviside function. It can be easily shown that :

$$\frac{\tau_L A D_{th}}{2\Lambda}=\Delta T_o$$ (initial temperature jump) and the constant A is found to be :

$$A=\frac{0.48\ b\ W_0}{\Pi\ \omega_0^2\ \tau_L}$$

where W_0 is the energy of the heating pulse. The induced refractive index distribution in the cell is then calculated within the approximation of weak thermal effect :

$$n(r,z,t) = n_0 + \frac{\partial n}{\partial T}\,\Delta T(r,z,t) \qquad (3)$$

Induced phase shift can be then calculated :

$$\Delta\psi = -\frac{k r^2 L \frac{\partial n}{\partial t} \Delta T_o Y(t)(1+\exp -bL)}{2\Pi\omega_0^2 (D_{th} t/\omega_0^2 + 1/8)^2} \sum_{\substack{m=1 \\ (m=odd)}}^{\infty} \frac{\exp\left[-D_{th} t \left(\frac{m\Pi}{2}\right)^2\right]}{(Lb)^2 + (m\Pi)^2} \qquad (4)$$

where k is the amplitude of the wave vector. From the induced phase shift given by equation (4), it is possible to perform a rapid relative determination of the ratio $D_{th}/D_{th,water}$, where D_{th} refers to an unknown sample. With some approximations, the laws of gaussian optics lead, after some calculations, to :

$$\frac{D_{th}\, t_{1/2}}{\omega_0^2} = cte \qquad (5)$$

and the relative measurements of time at half-height $t_{1/2}$ and $t_{1/2,water}$, lead to the determination of $D_{th}/D_{th,water}$ (for the same beam radius ω_0).

Let us now present the obtained results.

TABLE N°1

MEASURED VALUES OF SPECIFIC HEAT, $\rho\, C_p$; THERMAL DIFFUSIVITY AND THERMAL CONDUCTIVITY AT (25.0 ± 0.1)°C

sample	volume fraction of micelles	C_p $cal.g^{-1}.K^{-1}$ (±0.02)	$\rho\, C_p$ $cal.cm^{-3}.K^{-1}$ (±0.02)	$10^3\, D_{th}$ $cm^2.s^{-1}$ (±0.03)	Λ $cal.cm^{-1}.s^{-1}.K^{-1}$ (±0.02)
M	0.48	0.37	0.50	0.32	0.16
1	0.36	0.32	0.45	0.66	0.30
2	0.24	0.28	0.41	0.68	0.28
3	0.12	0.24	0.37	0.63	0.23
4	0.08	0.23	0.35	0.80	0.28
5	0.06	0.22	0.35	0.85	0.29
6	0.05	0.22	0.34	0.87	0.30
CCl_4	0	0.20[b]	0.33	0.77	0.25[a]

a) from D. Solimini, J. Appl. Phys., 37, 3314 (1966)

b) from Handbook of Chemistry and Physics

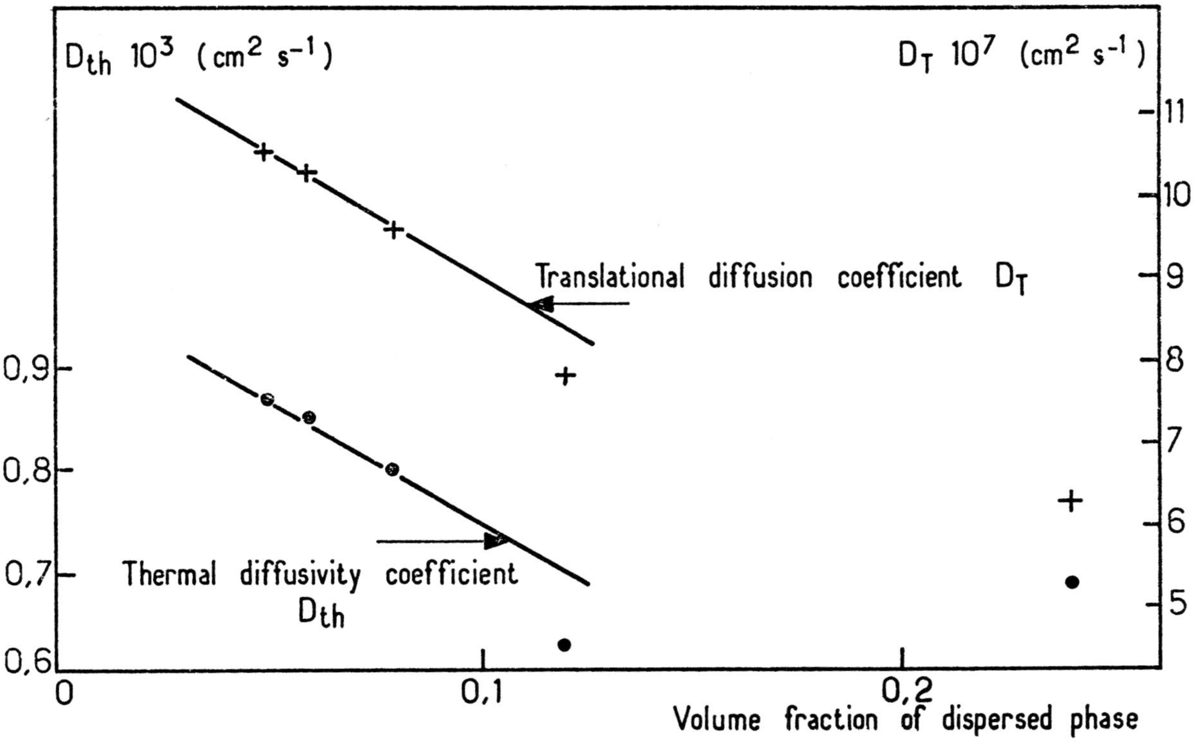

Figure n°3: Comparison of variations of mass translationnal diffusivity D_T and thermal diffusivity D_{th} with volume fraction of micelles (at 25°C).

RESULTS

Table n°1 reports our main experimental results. The micelles filled with water are described under the name of "dispersed phase". Volume fraction ϕ of such micelles is calculated to be in the range 0.05-0.48. The specific heat C_p has been obtained from auxilliary measurements described elsewhere {3}. The thermal diffusion coefficient D_{th} (values in the range $3x10^{-4}$ - $9x10^{-4}$ $cm^2.s^{-1}$) is found to decrease when increasing volume fraction ϕ of micelles, and to vary more strongly than in a recent work {4}, performed on a more complex system by forced Rayleigh scattering. Figure n°3 reports the compared values of D_{th} and D_T, where D_T is the translational mass diffusion coefficient measured, on the same samples, by photon correlation spectroscopy {1}. An important point is that extrapolated values at ϕ=0 of thermal diffusivity ($D°_{th}$ # $9.9x10^{-4}cm^2.s^{-1}$) and thermal conductivity ($\Lambda°$ # $3.3x10^{-4}cal.cm^{-1}.s^{-1}.K^{-1}$) are largely different from the values tabulated for pure carbon tetrachloride (respectively $7.7x10^{-4}$ $cm^2.s^{-1}$ and $2.5x10^{-4}cal.cm^{-1}.s^{-1}$ K^{-1}). Such an experimentaly revealed difference seems to contain information about the heat transfer at the interface. Moreover, the variations with ϕ of the two transport coefficients seem strongly connected and linearly varying with ϕ for ϕ < 0.1. Such a result implies that the collisions between micelles and surrounding carbon tetrachloride molecules play the main role in the propagation of heat in such a medium. It must be compared to a recent one {5}, obtained on oriented liquid crystals phases, which leads to a relatively small value of the mean free path of thermal phonons in the materials.

CONCLUSION

In this paper, we tried to show how the use of long duration laser pulses, which are now in rapid development in many laboratories, can be used for the study of heat transfer in liquids and, more precisely, at the interface of microemulsions. The weak absorbance of such solutions, at the usual laser wavelengths, leads to the "choice" of the induced defocusion technique which seems to give results with acceptable accuracy (within 10 %). Moreover, such a technique does not need the use of absorbing centers (dyes) which are often incorporated in colloïds systems without any proof for not disturbing them. Original results have been reported here ? Such preliminary results, if confirmed, can be used for calculations of heat transfer through liquid-liquid interfaces.

REFERENCES

1 . All structural and chemical informations concerning the studied system can be found in: E. Sein, J.R. Lalanne, J. Buchert, and S. Kielich, J. Colloid. Int. Sci. 72, 363 (1979)

2 . R.C. Leite, R.S. Moore, and J.R. Whinnery, Appl. Phys. Lett. 5, 141 (1964)

3 . E. Sein, Ph. D Thesis n°285, University of Bordeaux 1, may 6 th 1980

4 . C. Allain, A.M. Cazabat, D. Langevin, and A. Pouchelon, C.R. Acad. Sci. Paris 288, 363 (1979)

5 . F. Rondelez, W. Urbach, and H. Hervet, Phys. Rev. Lett. 41, 1058 (1978)

Sinha, ed. Ordering in Two Dimensions

THE CORRELATION LENGTH AND LATERAL COMPRESSIBILITY OF PHOSPHOLIPID BILAYERS IN THE PRESENCE OF THERMODYNAMIC DENSITY FLUCTUATIONS

M.J. ZUCKERMANN D. PINK*

Physics Department, McGill University
Montréal H3A 2T8 Canada
*Physics Department, St. Francis Xavier University
Antigonish, Nova Scotia B0H 1C0 Canada

Pink's ten state statistical model for the gel-fluid (main) phase transitions of phospholipid bilayers is used in conjunction with linear response theory to obtain expressions for the coherence length, ξ, and the lateral compressibility for wet lipid bilayers in the presence of thermodynamic density fluctuations near the temperature T_f of the main phase transition. Let E_B be the free energy barrier experienced by an ion as it penetrates the bilayer. It is shown that the static and the density fluctuation effects contribute almost equally to E_B at T_f when the range ξ_0 of interaction between ions and the polar heads is greater than half a molecule diameter.

1. INTRODUCTION

The mechanism of the permeability of phospholipid bilayers to ions is of considerable importance from the point of view of charge transfer through biological membranes. There is, at present, much interest in the change in permeability when the bilayer undergoes a phase transition, and in this paper we shall attempt to evaluate the relative importance of some mechanisms.

Wet phospholipid bilayers are quasi two-dimensional lipid-water mixtures, which undergo a first order phase transition from a solid or a gel to a liquid crystalline or fluid phase [1]. This phase transition is known as the main phase transition to distinguish it from the pretransition [2] and its transition temperature will be written as T_f. The main phase transition is characterized as follows: the hydrocarbon chains of the lipid molecules go from a mostly rigid configuration in the gel phase to a highly degenerate rotational isomeric state in the liquid crystalline phase which includes many gauche bonds (chain melting). At the same time the two-dimensional approximately triangular lattice representing the spatial arrangement of the hydrocarbon chains in the gel state melts to form a two-dimensional fluid. The heat of transition for the main transition is mostly accounted for by chain melting.

Many dynamic properties of lipid bilayers are greatly enhanced in the critical region close to the main transition temperature, e.g., the ionic permeability,[3,4] which exhibits a cusp at T_f. In particular, the enhancement of the ionic permeability has been described in terms of thermodynamic density fluctuations of the bilayer in the critical region by Doniach [5] and in terms of the static lateral compressibility of the bilayer by Pink [6] following Nagle and Scott's hypothesis [7]. Doniach's theory and data fitting [5] were based on a Landau-Ginsburg description of the fluctuations derived from a qualitative two-state Ising-like model for the main phase transition [8]. Pink's fit [6] to the Na^+ permeability out of dipalmitoyl phosphatidylcholine (DPPC) vesicles was obtained using a ten-state model [6] for the main phase transition in conjunction with the standard expression for the lateral compressibility.

The purpose of this communication is to obtain a general expression for the lateral compressibility which includes both the static contribution by Pink [6] and the thermodynamic density fluctuations of Doniach [5] so that a numerical evaluation of their relative importance can be made.

2. THEORY

In this section we present an expression for the response to an external probe in the random phase approximation (RPA) for a lipid bilayer. This expression was obtained by use of Pink's ten-state model for the main phase transition of lipid bilayers from a gel to a liquid crystal state [6,9].

Pink's Hamiltonian [6] can be written:-

$$H = \sum_{i,n=0}^{9} (E_n + \Pi_e A_n) L_{ni} - \frac{J_0^M}{2} \sum_{\langle ij \rangle} \sum_{m,n=0}^{9} I_m^W I_n^W L_{mj} L_{nj} \quad (1)$$

$E_n(i=0,\ldots 8)$ are the intra-molecular energies of nine lipid hydrocarbon chain configurations which describe the gel phase. Each configuration has a surface area per chain of less than 26 Å^2 and all except configurations of lowest energy (i=0; the all-*trans* state) are degenerate. E_9 is the energy of the high energy "melted" configurational (nine) state which describes the liquid crystal phase of the bilayer. This state is taken to have an area of 34 Å^2 and a degeneracy $D_9 = 6 \times (3)^{M-6}$ where M is the number of carbon atoms in the chain. L_{ni} is the projection operator for a hydrocarbon chain in the nth state at site i of a triangular lattice. Π_e($\simeq$ 30 dyne/cm) is effective internal surface pressure corresponding to the interaction between the lipid polar heads, and A_n is the cross-sectional area per lipid chain in the nth state. J_0^M is the coupling constant of the attractive Van der Waals interaction between nearest-neighbour hydrocarbon chains and I_m^W is an interaction parameter related to the order parameters of the individual carbon-carbon bonds of the chains in their m^{th} states (see reference 6).

The details of the calculation for the linear response of the bilayer to an external probe will be published elsewhere [10]. The results are as follows : let V be a physical quantity pertaining to a single lattice site. Then the operator representing V can be written as follows:-

$$V = \sum_{i,n} g_i V_n L_{ni} \qquad (2)$$

Let the operator representing the external probe X be written

$$H_{ex} = \sum_{i,n=0}^{9} \delta Y_i X_n L_{ni} \qquad (3)$$

It can then be shown that the linear response $\langle\delta V\rangle$ to the external probe X can be written as follows for the Hamiltonian (1) in the RPA

$$\langle\delta V\rangle = - X_{VX} \{N^{-1} \sum_q g_q \delta Y_q + \frac{X_{I_W V} X_{I_W X}}{X_{VX}} N^{-1} \sum_q Z J_0^M f_q g_q (1 - Z J_0^M f_q X_{I_W I_W})^{-1} \delta Y_q\} \qquad (4)$$

where Z(=6) is the co-ordination number of the triangular lattice. g_q is the lattice transform of g_i of (2), δY_q is the lattice transform of δY_i, and X_{FG} is written as follows for two physical quantities F_n and G_n

$$X_{FG} = \frac{1}{k_B T} \{\langle F_n G_n \rangle_{MF} - \langle F_n \rangle_{MF} \langle G_n \rangle_{MF}\} \qquad (5)$$

$\langle \dots \rangle_{MF}$ is the thermodynamic average in the molecular field approximation and T is the absolute temperature. Finally f_q is a structure factor given by

$$f_q = \frac{1}{Z} \sum_{\Delta} e^{iq\cdot\Delta} \qquad (6)$$

Δ is a vector between nearest neighbour sites.

3. RESULTS

Use of the continuum approximation for q in conjunction with equation (4) gives the following expression for the correlation length $\xi(T)$ in terms of the temperature dependent lattice parameter $\xi_{oo}(T)$ of the triangular lattice:-

$$(\xi/\xi_{oo})^2 \doteq Z J_0^M X_{I_W I_W} [4(1 - Z J_0^M X_{I_W I_W})]^{-1} \qquad (7)$$

For calculations related to ionic permeability of lipid bilayers, we choose $X_n = V_n = A_n$ and $\delta Y_q \equiv \delta\Pi_q$, the lattice transform of the perturbation of surface pressure of the bilayer by an ion. Following Doniach [9] , $\delta\Pi_q$ will be assumed to have a Gaussian form with range parameter ξ_o. Finally $\langle\delta V\rangle$ is identified as ΔE_B, the change in the free energy barrier experienced by an ion penetrating the bilayer. Then, from (4), $\Delta E_B = -X_G(\xi_o) |\delta\Pi_0|^2$, where $X_G(\xi_o)$ is given by :-

$$X_G(0) = X_{AA} + \Gamma(T) K_o(\xi_{oo}/\xi) \qquad (8a)$$

$$X_G(\xi_o) = X_{AA}(1-e^{-x})/x + \frac{\Gamma(T)}{2} e^{\mu} E_1(\mu) \qquad (\xi_o > \xi) \qquad (8b)$$

where $x = (8\Pi/\sqrt{3})\mu$ and $\mu = (\xi_o/\xi)^2$. $X_G(0)$ is the lateral compressibility and $\Gamma(T)$ ($\simeq$1) is a calculable quantity related to the functions X_{FG} of (4).

For DPPC bilayers, the maximum value of the correlation length $\xi(T)$ occurs at the main transition temperature T_f = 314K. In this case (7) gives $\xi(T_f) = 0.71\ \xi_{oo} = 4.49$ Å for the liquid crystalline phase. Further from (8a) it can be shown that the maximum contribution of the fluctuations to $X_G(0)$ occurs in the liquid crystalline phase at the main transition temperature T_f= 314K and amounts to only 13 % of the static compressibility, $X_{AA}(T_f)$, for the same phase. Hence the static contribution dominates for $\xi_o = 0$.

When $\xi_o \neq 0$, (8b) shows that the fluctuating term becomes progressively more important and is now larger than the static term. The maximum value of $X_G(\xi_o)$ occurs in the fluid phase at T_f = 314 K and it can be shown that $X_G(\xi_o)$ at T_f decreases as ξ_o/ξ_{oo} increases, since only fluctuations with a range greater than ξ_o contribute, and that the fluctuation contribution and the static contribution to $X_G(\xi_o)$ become comparable in magnitude for values of $\xi_o > \xi_{oo}$. The fluctuating contribution becomes between 1.5 to 2 times larger than the static contribution.

Following Pink [6], the free energy barrier experienced by an ion as it penetrates the bilayer can be written :-

$$E_B = E_1 + E_B \qquad (9)$$

where ΔE_B is the contribution from both the static lateral compressibility and the density fluctuations. From (9) E_B becomes :-

$$E_B = E_1 - E_2(\xi_o)\ X_G\ (\xi_o) \qquad (10)$$

where $E_2 (= |\delta\Pi_o|^2)$ depends on the interaction between the ions and the polar heads. Pink fitted expression (10) to the data of Papahadjopoulos et al. for the permeability of Na^+ ions out of DPPC vesicles for a range of temperature near T_f in the absence of fluctuations and in the case where $\delta\Pi$ is a δ-force $(\chi_G(\xi_o) = \chi_{AA})$. E_1 and E_2 were found to be 7.0×10^{-13} ergs/Å^2 respectively. We have calculated the change in E_2 needed to maintain the barrier height when density fluctuations are included. When $\delta\Pi$ is a δ-force and hence $\xi_o = 0$, $E_2(0)$ is reduced to 0.71×10^{-13} ergs/Å^2 because the fluctuations only increase the value of the lateral compressibility slightly. Much greater values of $E_2(\xi_o)$, however, are needed to maintain the same barrier height in the liquid crystalline state at T_f for $\xi_o/\xi_{oo} > 1$. For example, $E_2(\xi_o) = 2.04 \times 10^{-13}$ ergs /Å^2 when $\xi_o/\xi_{oo} = 2.275$.

No attempt has been made to fit the experimental permeability data using (10) since the range ξ_o of the interaction between the ions and the polar heads of the lipid molecules is not known. It is unlikely that the interaction is a δ-force and quite probable that ξ_o is at least one lattice spacing. The ionic permeability must therefore be described in terms of both the static contribution and the density fluctuation contribution.

4. REFERENCES

1. D. Chapman, Quart. Rev. Biophys. 8,185(1975).
2. D. Chapman, R.M.Williams and B.D. Ladbrooke, Chem. Phys. Lipids 1, 445 (1967).
3. D.Papahadjopoulos, K. Jacobson, S. Nir and T.Isac, Biochim. Biophys. Acta 311, 330 (1973).
4. S.N.W.Wu and H.M. McConnell, Biochim.Biophys. Res. Commun. 55, 484 (1973).
5. S. Doniach, J. Chem. Phys. 68, 4912(1978).
6. A. Caillé, D.A. Pink, F. de Verteuil and M.J. Zuckermann, Can.J. Phys. (1980) in press. This is a review article which gives details of the ten-state model and its various applications. Other theoretical models for the main phase transition are also reviewed in this article.
7. J.F.Nagle and H.L. Scott, Biochim. Biophys. Acta 513, 236 (1978).
8. A. Caillé, A. Rapini, M.J.Zuckermann,A.Cros and S. Doniach, Can. J. Phys. 56, 348(1978).
9. D.A. Pink, T.J. Green and D. Chapman, Biochemistry 19, 345 (1980); D.A.Pink and C.F. Carroll, Phys. Lett. 66A, 157 (1978).
10. D.A. Pink and M.J. Zuckermann, to be published.

Published 1980 by Elsevier North Holland, Inc.
Sinha, ed. Ordering in Two Dimensions

PHASE TRANSITIONS AND CRITICAL BEHAVIOUR OF A PHOSPHOLIPID BILAYER MODEL*

A. GEORGALLAS and D. A. PINK

Theoretical Physics Institute, St. Francis Xavier University
Antigonish, N. S., Canada

and

M. J. ZUCKERMANN

Department of Physics, McGill University, 3600 University Street,
Montreal, Quebec, Canada

ABSTRACT

A model is described which has been used in theoretical studies of a variety of phenomena relating to biological membranes. The Hamiltonian describing this model can be mapped onto an Ising Hamiltonian with a temperature dependent magnetic field, and this field varies linearly with temperature in the critical region. We discuss the effect that this temperature-dependent effective field has upon the critical indices and show explicitly that $\beta = 1/\delta$ for this model.

*Supported in part by NSERC of Canada and UCR, St. Francis Xavier University.

The first-order gas to liquid crystal phase transition in pure phospholipid bilayers has been described by a lattice model using the Hamiltonian [1,2]

$$\mathcal{H} = -\frac{J_o}{2} \sum_{<ij>} \sum_{nm} J(n,m)\, \mathcal{L}_{in} \mathcal{L}_{jm}$$

$$+ \sum_{i} \sum_{n} (\Pi A_n + E_n)\, \mathcal{L}_{in} \qquad (1)$$

Here <ij> refer to nearest neighbor lattice cites at each of which one hydrocarbon chain of each phospholipid molecule is localized. Each chain can be in one of ten states: (a) an all-trans ground state g with internal energy 0, cross-sectional area A_g and degeneracy 1, (b) eight intermediate states 2,...,q with internal energies, areas and degeneracies $\{E_n, A_n$ and $D_n\}$, n = 2,...,q, and (c) a highly excited "melted" state e with energy, area and degeneracy E_e, $A_e = 34\ \text{Å}^2$ and D_e. Because the density change is < 4% [3] through the first-order transition temperature, T_f, we define $A_n = A_g L_g / L_n$ where L_n is the chain length projected onto the chain director $\mathcal{L}_{im}$ is a chain projection operator for site i with a chain in state m. Π is an effective lateral pressive acting on the chains due to the existence of polar interactions involving the phospholipid polar groups which maintain the bilayer in existence.

In order to study (1) we simplify it by two approximations: (i) mean field calculations [1] show that the states g, 2 ("jog"), 5 ("kink") and e account for nearly all the probability of occupation. We shall then restrict the sum over states in (1) to these four states. (ii) Transitions between the states g, 2 and 5 occur on a time scale of $\sim 10^{-11}$ secs, while those between these states and e have a time scale of $\sim 10^{-6}$ secs. In addition we note that the interactions between the states g, 2 and 5 are nearly the same. We shall therefore replace these three low-lying states by one effective ground state, G, which is described by temperature dependent internal energy, area and degeneracy

$$\mathcal{Y}_G = \sum_{n=g}^{S} Y_n D_n e^{-\beta(E_n + \Pi A_n)} / Z_G$$

$$Z_G = \sum_{n=g}^{S} D_n e^{-\beta(E_n + \Pi A_n)} \qquad (2)$$

where $Y_n = A_n, E_n, D_n$ and $\mathcal{Y}_n = \mathcal{A}_n, \mathcal{E}_n, \mathcal{D}_n$. Finally, because the energies A_n are nearly the same, we shall ignore them [4]. The Hamiltonian is now

$$H = \frac{-J_o}{2} \sum_{<ij>} \sum_{nm} J(n,m)\, L_{in}\, L_{jm}$$

$$+\sum_i \sum_n (\Pi A_n + E_n)\, L_{in}\, J(G,G) = 1,\ G(G,e)$$

$$= J(n,e)\ (n{=}g,2,5),\ J(e,e) = J(e,e) \tag{3}$$

where n,m = G or e, with $\mathcal{D}_e = D_e$, $A_e = A_e$, $E_e = E_e$. If we map $L_{iG} = 1/2\ (1 + \sigma_i)$, $L_{ie} = 1/2\ (1 - \sigma_i)$ where $\sigma_i = \pm 1$ then (3) takes the form of an Ising model with a temperature-dependent external field.

$$H = \frac{-J}{2} \sum_{<ij>} \sigma_i\sigma_j - h(T,\Pi) \sum_i \sigma_i,$$

$$J = \frac{J_o}{4}\,[1 - 2J(G,e) + J(e,e)],$$

$$h(T,\Pi) = \frac{J_o}{4}\, q[1 - J(e,e)] + \frac{1}{2}\,\Pi(A_e - A_G)$$

$$+ \frac{1}{2}\,(E_e - E_G) \tag{4}$$

where q is the lattice coordination number. Although the degeneracies of the states G and e are $_G$ and D_e, and are unequal, we can simplify the calculations by noting that we can write $= e^{n}$ so that summation over $_i$ contributes the following to the partition function:

$$\mathcal{D}_G\, e^{\beta h(T,\Pi)+\ldots} + \mathcal{D}_e\, e^{-\beta h(T,\Pi)+\ldots}$$

or

$$e^{\beta[h(T,\Pi)-(\ell n\mathcal{D}_e-\ell n\mathcal{D}_G/2\beta)+\ldots}$$

$$+\ e^{-\beta[h(T,\Pi)-(\ell n\mathcal{D}_e-\ell n\mathcal{D}_G]/2\beta+\ldots} \tag{5}$$

where we have factored a term $\exp[(\ell n\mathcal{D}_G+\ell n\mathcal{D}_e)/2\beta]$. We see that the effective Hamiltonian can then be written as

$$H = \frac{-J}{2} \sum_{<ij>} \sigma_i\sigma_j - H(T,\Pi) \sum_i \sigma_i \tag{6}$$

$H(T,\Pi) = h(T,\Pi) - (\ell n\mathcal{D}_e-\ell n\mathcal{D}_G)/2\beta$, is explicitly temperature dependent through the different degeneracies of the +1 and -1 states. In (6) these are included in H and need not be further considered. It is clear from (6) that a phase transition can occur only when $H(\Pi,T)$ changes sign. As has been shown by Georgallas et al. [4] this effective field term $H(\Pi,T)$ may be expanded about T_c to form

$$H(T,\Pi) = H(T_c,\Pi) - \varepsilon F(T_c)\quad \varepsilon{\to}0 \tag{7}$$

Here $= (T-T_c)/T_c$ is the reduced temperature and $F(T_c)$ is a function of T_c alone. Furthermore by the definition of the critical pressure Π_c, $H(\Pi_c,T_c) = 0$ and becomes

$$H(T,\Pi_c) = -\ \varepsilon F(T_c) \sim -\ \varepsilon \to 0 \tag{8}$$

Thus we have an effective field which varies linearly with temperature in the critical region. The authors believe this to be a unique feature of this model and one which leads to some unusual relationships between the critical indices associated with this model. Firstly, it is known [4] that this model exhibits zero magnetization only at the critical temperature T_c. Though this may lead to some doubt as to the suitability of as a choice for the order parameter in this case, we may, nevertheless make some observations concerning the exponent β. This exponent is defined by

$$\langle\sigma\rangle \sim |\varepsilon|^{\beta}\quad \varepsilon{\to}0 \tag{9}$$

which combining with equation (8) leads to

$$\langle\sigma\rangle \sim |H(T,\Pi_c)|^{\beta},\quad T{\to}T_c \tag{10}$$

However by the usual definition of the critical isotherm exponent δ.

$$\langle\sigma\rangle \sim |H(T_c,\Pi)|^{1/\delta}\quad \text{for all } \Pi \tag{11}$$

Comparison of these two equations leads immediately to the conclusion that the β exponent is Π dependent and converges onto 1/δ as $\Pi \to \Pi_c$, that is

$$\beta = 1/\delta\quad \Pi{\to}\Pi_c \tag{12}$$

Work involving relationships between other critical indices for this model is progressing and will be reported shortly.

REFERENCES

1. D. A. Pink, T. J. Green and D. Chapman, Biochemistry 19, 349 (1980).
2. A. Caillé, D. A. Pink, F. deVerteuil and M. J. Zuikermann, Can. J. Phys. in press.
3. J. F. Nagle, Proc. Natl. Acad. Sci. USA 70, 3443 (1973).
4. A. Georgallas, D. A. Pink and M. J. Zukermann, to be published.

Sinha, ed. Ordering in Two Dimensions

PHASE-TRANSITION OF THE MEMBRANE ASSOCIATED WITH THE BLOOD-BRAIN BARRIER VIA INTERACTION WITH DRUGS---IMPORTANCE FOR THE TRANSPORT OF DRUGS FROM THE BLOOD TO THE BRAIN

R.L.P. Vimal and R.R. Sharma

Department of Physics, University of Illinois
Chicago, Ill. 60680

Making use of the decorated Ising model we have studied the interaction of a drug with the membrane associated with the blood-brain-barrier to cause the phase transition in the membrane. Both the polar and nonpolar drugs have been considered. The influence of the phase transition for the transport of the drugs to the brain through the barrier has been discussed. The brain-up-take index has been inferred to be larger for the polar drugs than for the nonpolar drugs.

I. INTRODUCTION

It is extremely valuable to understand how drugs (and other substances) pass from the blood stream to the brain. The barrier which is to be crossed by the drugs is referred to as the blood-brain barrier. If the details of the permeability properties of the blood-brain barrier [1] are known, it is possible to extract useful information concerning the possible selection of appropriate drugs which could be used for curing various diseases (e.g., meningeal leukemia).

The blood-brain barrier is composed of endothelial cells of the brain capillary, which are joined together to form tight junctions (zonulae occludents) with no vesicles or fenestrae (small opening) thereby having attributes of restricting inter-cellular exchange of substances. The specific barrier sites are cerebral blood vessels, choroid plexus and arachnoid membrane. The blood-brain barrier and its function have not yet been understood owing to the complex nature of the associated membrane and the lack of knowledge of the interactions between the substances and the barrier. It is important to know how the specific barrier sites coordinate their functions to permit easy access to some drugs to the brain and restrict the others and regulate metabolism of central nervous system.

One requires to study not only the details of the structure of the biomembranes and cell junctions but also the basic mechanisms responsible for the transport of the drugs and the permeability of the cell membranes along with the various interactions involved. The interactions of the drugs with the membrane play an important role in modifying the characteristics of the membrane and hence its permeability properties related to drugs. The aim of the present article is to report our findings that, because of the interactions with the drugs, the membrane associated with the blood-brain barrier undergoes the phase-transition from the ordered state to the disordered state. In certain cases it is possible to obtain a sufficiently low value of the transition temperature for the membrane owing to the interactions with the drug, which in turn furnishes an explanation why certain drugs exhibit a relatively high probability characteristics for the intake by the brain.

II. THEORY

In general the membranes possess an ordered structure; their properties have been the subject of discussion by several authors [2-9]. Recently, Jahnig [2] has studied the molecular theory of lipid membrane order incorporating the inter- and intramolecular interactions of the hydrocarbon chains. Chaugeux et al. [6] have investigated the cooperative effects in a membrane triggered by the ligands bound to the membrane. Hill [5] has combined his own treatment of the effects of the electric fields on the membrane with the results of Chaugeux et al. to infer that the membranes exhibit phase transitions as a result of the change in the membrane potential or in a bound ligand concentration or both. Ghosh and Sengupta [9] have made an attempt to connect stimulus-response behavior of membranes with the structural phase transition at various temperatures.

A biomembrane is a phospholipid bilayer which has inserted in it special proteins, steroids (e.g. cholesterols) and enzymes that serve to promote processes such as the transport and diffusion of substances and metabolic syntheses and probably provide pores for the flow of ions. We assume that the drug molecules interact with the phospholipids of the membrane. We wish to investigate the effect of this interaction to determine how the membrane changes its characteristics. This constitutes a complex problem which could, however, be simplified by assuming that a membrane is a two-dimensional array of lipid molecules each possessing a net electric dipole moment. The dipolar picture can be justified by observing that the complex ions such as trimethylamine ion, $-N^{+}\equiv(CH_3)_3$ present in biomembrane phospholipids contain permanent electric dipole moments [10]. The lipid molecules interact with one another, with the proteins (and cholesterols and enzymes) contained in the membrane and with the external drug molecules. The proteins and drug molecules may have permanent dipole moments or induced dipole moments. It is possible to write the Hamiltonian H for the system in "magnetic language" making use of the two-dimensional decorated Ising model [11]. Accordingly,

$$H = -J \sum_{<i,j>} \sigma_i \sigma_j \tag{1}$$

where J denotes the effective interaction between the nearest neighbor phospholipids on the Ising lattice indicated by the symbol $<i,j>$ and σ_i and σ_j assume the values ± 1. The effective interaction includes, besides the direct interaction between the nearest neighbors of the Ising lattice, the indirect interactions arising through the mediated effects of the interaction with the proteins (and cholesterols and enzymes) and also with the external drug molecules. Thus,

$$J = J_o + \frac{1}{2} kT\epsilon_p n_p \ln\left[\cosh\left(2J_p/kT\right)\right] + \frac{1}{2} kT\epsilon_D n_D \ln\left[\cosh\left(2J_D/kT\right)\right] \tag{2}$$

where J_o is the direct coupling constant which arises from the (direct) electrostatic dipolar interaction between the phospholipids; J_p is the indirect coupling constant between the proteins and phospholipids whereas J_D is the indirect coupling constant between the drug and the phospholipids; k is the Boltzmann constant and T is the absolute temperature; n_D and n_p are the numbers which determine the intermediate paths involved in the interactions and, in our case, depend also on the relative concentrations of the interacting species. The factors ϵ_p and ϵ_D assume the values +1 or −1 based on the mechanisms that are responsible for the indirect interactions between the interacting species (phospholipids, proteins and drug molecules).

Onsager has given the exact solution of the two-dimensional Ising model for the phase transition which may be expressed as

$$|J|/kT = 0.4407 \tag{3}$$

We make use of Eq. (3), in conjunction with J given by Eq. (2), to obtain the transition temperature for our present case.

III. RESULTS AND CONCLUSION

The direct coupling constant J_o can be estimated from the knowledge of the dipole moments (25 to 1000 D) of the membrane molecules as deduced by Wei [10]. Identifying the direct dipole-dipole interaction with J_o, taking the dielectric constant of the membrane fluid equal to 80 and the distance between the nearest neighbor phospholipids equal to 5Å, we obtain J_o/k as −450°K for the dipole strength of 25D where we have assumed that the dipoles lie perpendicular to the plane of the membrane.

As for the indirect coupling constant J_p, we follow the procedure given by Ohki and Fukuda [12]. Assuming that the dipoles on the protein molecules are the induced ones due to the dipoles on the phospholipids we find that J_p/k is of the order of 710°K and ϵ_p is +1. The value of the indirect coupling constant J_D depends on the nature of the drug involved, the distance between the drug molecule and the membrane phospholipids and the concentration of the drug in the blood stream. Also, the dipoles on the drug molecules may be a permanent or induced one. To be general, we have considered various reasonable values of the dipoles on the drug molecules to estimate J_D.

If the drug molecules are polar and assumed to lie about 2.7 Å above the membrane, the value of J_D/k comes out to be 269°K with $\epsilon_D = -1$. On the other hand, if the drug molecules are nonpolar, one obtains J_D/k to be about 297°K with $\epsilon_D = +1$.

Figure 1 depicts the variation of $|J|/kT$ vs T for three curves: (i) membrane without the effect of the drug, (ii) membrane interacting with polar drug molecules and (iii) membrane interacting with nonpolar drug molecules. For Fig. 1 we have taken $n_p = 1$ considering that only one interaction path with the proteins is involved and that equal concentrations of proteins and phospholipids occur in a membrane. As for n_D we have taken it to be 0.1 which gives the right arterial concentration (about 3 mM) of the drug in the blood. Perusal of Fig. 1 reveals that in the absence of the drug the transition temperature T_c is 330°K which increases to 350°K for the nonpolar drug and decreases to 310°K for the polar drug. Below the transition temperature the system is in the ordered (ferroelectric) state and above this temperature, it is in the disordered state. Because of the changes in T_c due to the influence of the drug the brain-

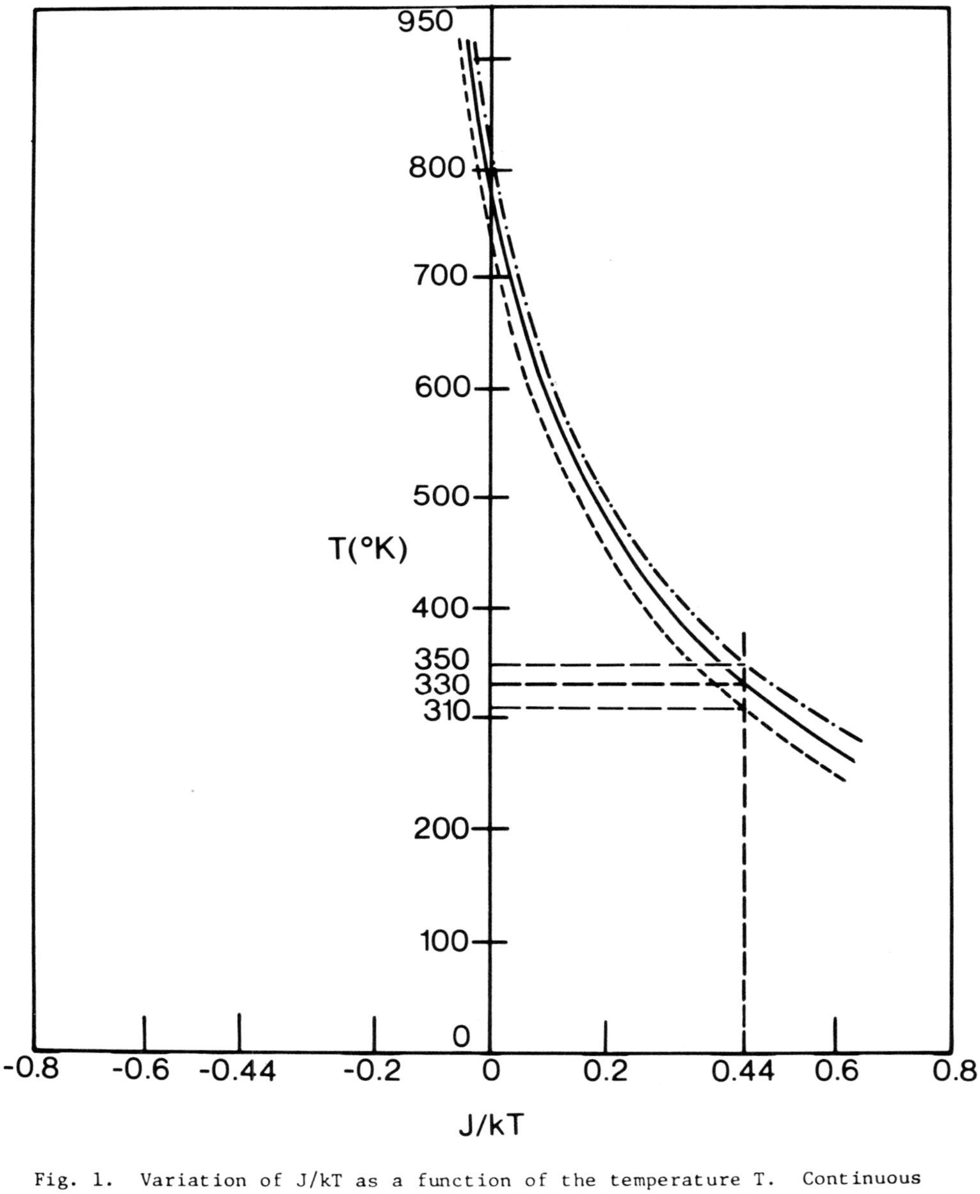

Fig. 1. Variation of J/kT as a function of the temperature T. Continuous curve: membrane without drug; dash-dot curve: membrane interacting with nonpolar drug and dash curve: membrane interacting with polar drug.

uptake-index (BUI) or extraction (E) is expected to increase for the polar drug molecule and decrease for the nonpolar drug molecule. This is a remarkable result and helps in understanding in particular, why D-glucose (which is optically active) has much higher BUI than the L-glucose (which is optically inactive). The variation of T_c as a function of the drug concentration and the estimation of the BUI of glucose will be presented elsewhere [13,14].

ACKNOWLEDGMENT

The authors are thankful to Professor R. Greenberg for valuable discussions.

REFERENCES

1. S.I. Rapport, "Blood-Brain Barrier", Raven Press, New York, 1976.
2. F.Jahnig, J. Chem. Phys. 70, 3279 (1979; see also F. Jahnig, K. Harlos, H. Vogel and H. Eibl, Biochem. 18 1399 (1979); F. Jahnig, Biophys. Chem. 4, 309 (1976.
3. S.P. Almeida, J.D. Bond and T.C. Ward, Biophys. J., 11, 995 (1971.
4. J.F. Nagle, Proc. Natl. Acad. Sci. USA, 70, 3443 (1973); J.F. Nagle, J. Chem. Phys. 58, 252 (1973.
5. T.L. Hill, Proc. Natl. Acad. Sci.USA, 58, 111 (1967).
6. J.-P. Chaugeux, J. Thiery, Y. Tung and C. Kittel, Natl. Acad. Sci., 57, 335 (1967).
7. H. Trauble, M. Teubner, P. Woolley and H. Eibl, Biophys. Chem. 4, 319 (1976).
8. P.A. Forsyth, Jr., S. Marcelja, D.J. Mitchell and B.W. Ninham, Biochim. et Biophys Acta 469, 3351 (1977).
9. P.K. Ghosh and D. Sengupta, J. Theo.Bio. 73, 609 (1978).
10. L.Y. Wei, Bull. Math. Biophys. 33, 187 (1971).
11. E.H. Fradkin and T.P. Eggarter, Phys. Rev. A14, 495 (1976).
12. S. Ohki and N. Fukuda, J. Theo. Bio. 15, 362 (1967).
13. R.L.P. Vimal and R.R. Sharma, Bull. Am. Phys. 25, 23 (1980); H.L-. Anderson, Physiological Review, 59, 305 (1979).
14. R.L.P. Vimal and R.R. Sharma, Brain Res. (to be published).

Published 1980 by Elsevier North Holland, Inc.
Sinha, ed. Ordering in Two Dimensions

RIGIDITY OF MONOLAYERS SPREAD ON WATER

B. M. ABRAHAM and K. MIYANO

Argonne National Laboratory, Argonne, Illinois 60439

and

J. B. KETTERSON

Northwestern University, Evanston, Illinois 60201

ABSTRACT

The shear modulus of a stearic acid monolayer spread on water has been measured as a function of surface pressure at 21 C. Similar measurements were made on a stearyl alcohol film at two temperatures, 20.7 C and 3.0 C. The stearic acid film was in the "liquid condensed" state (L-2) and the stearyl alcohol was in the liquid state at 20.7 C and in both the liquid and the "solid" states at 3.0 C. The stearic acid L-2 film displayed a viscous relaxation accompanied by a residual shear strain which increased with compression. By contrast, the stearyl alcohol film showed no detectable viscous or shear strain. It is concluded that the asymmetry of the head groups contribute to making stearic acid more rigid than stearyl alcohol and that the contribution due to van der Waal's attraction between the hydrocarbon tails is not significant.

INTRODUCTION

The changes and discontinuities in slope displayed in the pressure-area (π-A) diagrams of compressed monolayers led the early workers in this field to postulate phase changes analogous to those observed in three dimensions. However, there were frequently more features than could be accounted for by the conventional gas - liquid - solid phases, which led workers to postulate the existence of intermediate liquid phases. Technical limitations, which may have prevented a detailed investigation of the postulated phases, together with the absence of quantitative theories, contributed to keeping the field dormant for many years. Currently interst in two dimensional systems has been revived, principally by the work of Kosterlitz and Thouless [1] from the physical side; and by the recognition that bilayers play an active role in cellular function from the biological side.

In order to gain greater insight into the physical processes that generate the features observed in the π-A diagram of compressed monolayers spread on water, we have added two additional probes to the conventional Langmuir film balance. These probes are: (1) a capillary wave generator/detector system for measuring dynamic properties such as surface mediated damping and frequency dependent surface tension, and (2) an apparatus to measure the shear modulus of a film. Only measurements of the shear modulus as a function of compression will be discussed in this presentation. The results are preliminary in the sense that meniscus effects and end corrections have yet to be considered. Nonetheless, the results, though qualitative, are unusual and are, we believe, relevant to the physics of two dimensional systems.

EXPERIMENTAL

Details of the balance and trough will be described elsewhere [2]. Suffice it to say that the balance has a resolution of about 1 millidyne/cm of surface pressure and can cover the range up to 80 dynes/cm, at constant sensitivity. A water-glycol mixture could be circulated through the trough in order to stabilize or to change the temperature.
Our approach to the measurement of shear moduli is straight forward. Three parallel vanes project into the water surface. The outer vanes are equidistant from the central vane which is part of the transducer system. The outer vanes are connected by a yoke to a displacing mechanism. The transducer of the shear modulus balance was assembled by modifying a 200 microampere DC meter. An extension was added to the pointer to carry the force receiving vane; the pointer itself carried a small cylindrical lens, a chord cut from a 6 mm dia. Lucite rod, which focussed a light beam on the surface of a differential photo-diode. A force sensed by the vane displaced the light beam which produced an error signal. This signal was amplified and fed back to the meter

coil to supply the restoring force. The magnitude of the current required is a measure of the shearing force. All three vanes were made of platinum foil 0.25 mm thick and 3 cm wide. The vanes were fired to red heat before mounting on the trough.

The de-ionized water used for these experiments had a resistivity greater than 18 megohm-cm. Water doubly distilled from alkaline permanganate gave identical results so the complicating procedure of distillation has been eliminated. The water was considered satisfactory for an experiment if, on compression by a factor of 16, a surface pressure no greater than 250 millidynes/cm developed. The quantity of surfactant added was such that when fully expanded each molecule occupied 40 $Å^2$; when compressed past the "solid"-liquid transition the compressional ratio was only about 2 and the pressure, depending on the substance, varied from 12 - 20 dynes/cm.

Surfactant solutions of stearic acid and stearyl alcohol were prepared by dissolving high purity compounds in spectroscopic grade normal heptane. After spreading an aliquot of the stock solution, 100 microliters, approximately thirty minutes were allowed for system equilibration before commencing the compression. The shear modulus was measured before and after spreading. The film was then compressed at the rate of 0.3 $Å^2$ per min. to about 1 dyne/cm before measuring the shear modulus again.

The compressional rate was then reduced to half and a series of shear measurements were made at several pressures well into the "solid" region. The shear measurements were made by translating the outer vanes by 100 microns, measuring the force on the balance as a function of time, translating the vane an equivalent amount in the opposit direction and again recording the force on the balance. Vane separation, outer to center, was 7.5 mm.

RESULTS

Stearic acid was selected to be the first surfactant to study because the phase diagram is simple and shows no intermediate phases. However, a strain free film is difficult to obtain. The strain is evident by the fact that the central vane always registers a force which does not completely relax when the film is compressed beyond the point where it begins to show viscous relaxation. In the attempt to obtain a strain free film, a total of 200 microliters of n-heptane were dropped on the film in front and in back of the vanes. In all cases the balance pressure dropped immediately and the vane force disappeared. As soon as the solvent layer evaporated, the original condition reappeared. Although these attempts to produce a strain free film failed, they did lead to some unusual observation relating to solvent effects.

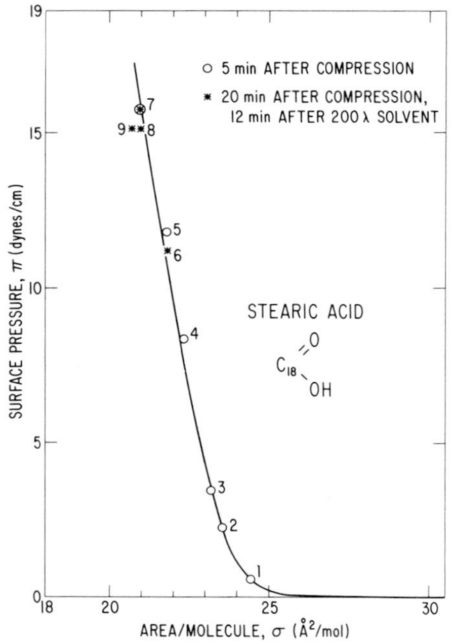

Fig. 1. Pressure-area diagram for stearic acid at 21 C.

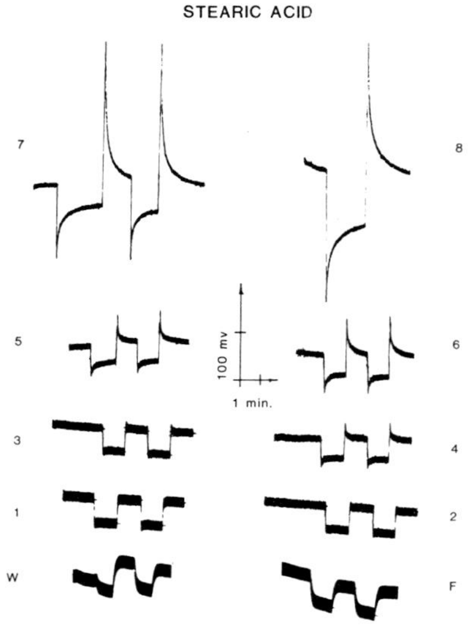

Fig. 2. Shear force of a stearic acid film as a function of time at various pressures.

Figure 1 illustrates the pressure-area phase diagram of stearic acid at 21 C. The numbered points on the diagram correspond to the numbered traces displayed in Figures 2 and 3. At these positions on the phase diagram the compression was arrested to make a shear measurement. The traces shown in Figs. 2 and 3 are the actual strip chart traces. All were taken at the identical sensitivity and identical displacement of the shearing vanes, 0.01 cm. The traces labelled W and F were made on clean water and on the water immediately after spreading the film. The residual displacement seen here results from the asymmetry of the meniscus on the central vane. After a displacement of the outer vanes, the central vane, which is the same width as the outer vanes, will extend slightly beyond the outer vanes on one end but will be between them on the other. The magnitude of the effect was not anticipated and corrective modifications would have been awkward at this stage. The width of the traces arises from hunting about the equilibrium position. As the film becomes more viscous, it can be seen that the oscillations are damped considerably.

There are several interesting features displayed by these traces. As the film is compressed, a viscous relaxation becomes evident, and becomes more pronounced as the compression increases. A residual displacement in excess of that seen with clean water is also seen. These effects show up in the "liquid condensed" phase well below the "liquid-solid" transition. An order of magnitude calculation of the shear modulus at point 9, using the meter sensitivity of 4.3 x 10^{-3} dynes/mv, vane width of 3 cm, vane spacing of 0.75 cm and displacement of 0.01 cm, gives a value of 20 dynes/cm. This is in reasonable agreement with Mouquin and Rideal [3] who obtained a value of 10 dynes/cm. The Mouquin and Rideal measurement appears to be the only one reported in the literature. It was made by a different technique and at an unknown surface pressure. Under the circumstances, the agreement is gratifying.

The behavior of the stearyl alcohol is markedly different. Figure 4 illustrates the pressure-area phase diagram taken at 20.7 C and a portion taken at 3.0 C. Using the Harkins and Copeland terminology [4], the 20.7 C trace shows the transition from liquid to superliquid and the 3.0 C trace shows the transition from liquid to solid. The 20.7 C trace was made by compressing the film in the same manner as stearic acid with the numbers corresponding to compression arrests to make a shear measurement. At point 4 the film was cooled, and after equilibrium had been established the trace at 3.0 C was made, point 5. The temperature-pressure phase diagram published by Harkins and Copeland is shown in Figure 5; the pressure-temperature course followed in this experiment has been added to the diagram. The pressure drop from point 4 to point 5 reflects the temperature coefficient; cooling was carried out at constant area.

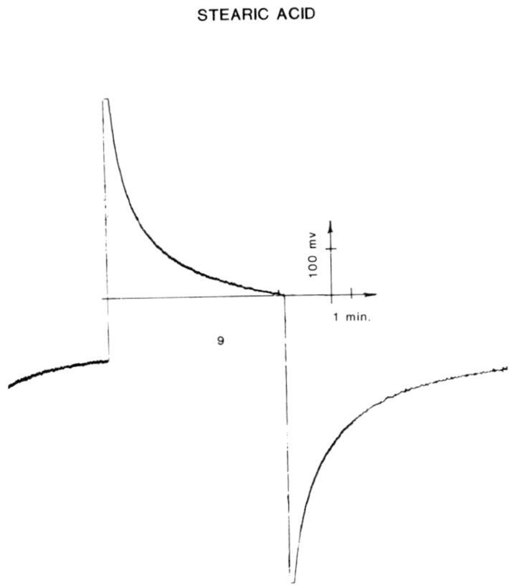

Fig. 3. Shear force of a stearic acid film as a function of time at 15 dynes/cm surface pressure.

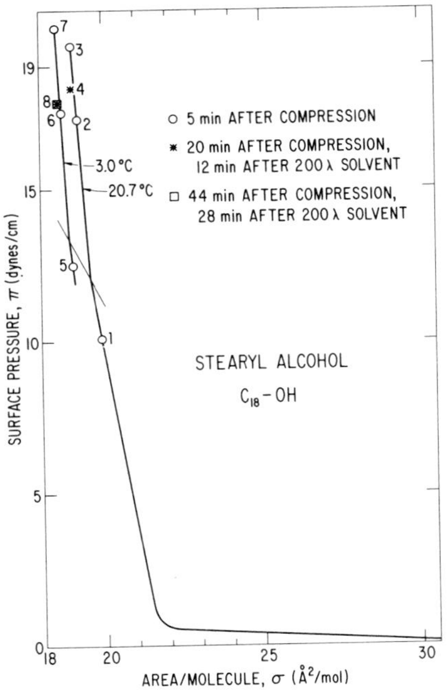

Fig. 4. Pressure-area diagram for stearyl alcohol at 20.7 C and 3.0 C.

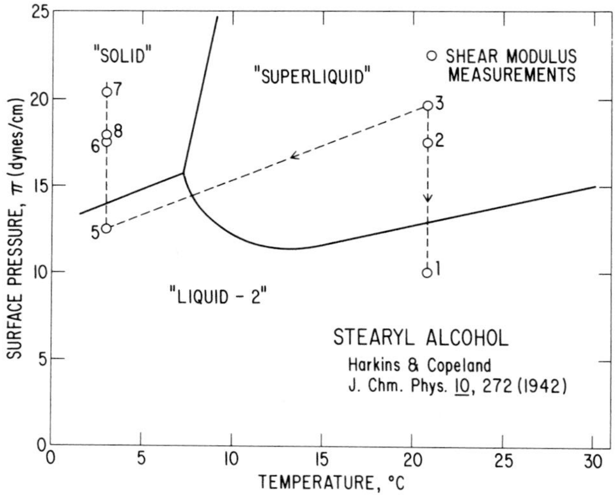

Fig. 5. Pressure-temperature diagram of a compressed stearyl alcohol film (Harkins & Copeland) and course followed by this experiment.

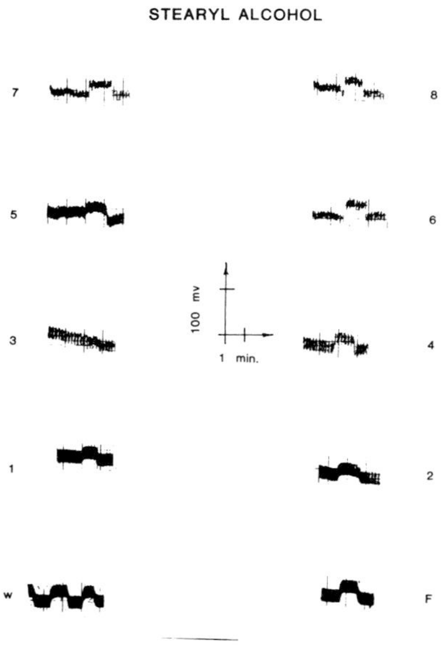

Fig. 6. Shear force of stearyl alcohol film as a function of time at various pressures.

The shear modulus at corresponding points is displayed in Figure 6. Although stearyl alcohol has been presumed to be more viscous than stearic acid [5] there is absolutely no evidence for it from these measurements. Furthermore, the film exhibits no solid behavior. Obviously more work is required to place these measurements on a quantitative basis; nonetheless, the results contradict the accepted dogma about these two materials.

The traces presented in Figs. 2, 3 and 6 measure a rigidity of the film arising from direct interactions between surfactant molecules, stearic effects, as well as interactions between the surfactant via the aqueous substrate. Rigidity can be imparted to the film by short range molecular ordering arising from steric effects such as that proposed by Rabinovitch et.al to explain the relaxation phenomena observed with stearic acid films [6]. This would imply that the asymmetry of the head group is more important than the van der Waal attraction between adjacent hydrocarbon tails.

REFERENCES

1. J. M. Kosterlitz and D. J. Thouless, J. Phys. C. Solid State Phys., 6, 1181 (1973).

2. B. M. Abraham, K. Miyano, K. Buzard and J. B. Ketterson, RSI August 1980, to be published.

3. H. Mouquin and E. K. Rideal, Proc. Roy. Soc. A114, 690 (1927).

4. W. D. Harkins and L. E. Copeland, J. Chem. Phys. 10, 272 (1942).

5. E. Boyd and W. D. Harkins, J. Am. Chem. Soc. 61, 1188 (1939). L. Fourt and W. D. Harkins, J.Phys. Chem. 42, 897 (1938).

6. W. Rabinovitch, R. F. Robertson and S. G. Mason, Can. J. Chem. 38, 1881 (1960).

Published 1980 by Elsevier North Holland, Inc.
Sinha, ed. Ordering in Two Dimensions

SOLVENT EFFECTS ON SURFACTANT MONOLAYERS ADSORBED ON WATER

B. M. ABRAHAM and K. MIYANO
Solid State Science Division
Argonne National Laboratory
Argonne, IL 60439

and

J. B. KETTERSON
Department of Physics and Astronomy
Materials Research Center
Northwestern University
Evanston, IL

We report the effect on the surface pressure which results when droplets (up to 200 microliters) of various solvents are added to the air surface of a monolayer of stearic acid spread on a water substrate. Depending on the solvent, the initial effect may be an increase or decrease in surface pressure; when the solvent evaporates, the final pressure will be less than the initial pressure in the former instance and the original pressure will be restored in the latter.

Introduction

The study of the surface pressure ,π, (difference in surface tension between a clean water surface and one covered by a surfactant film) as a function of the area density of surfactant, A, is a field that has received a great deal of attention. [1] The most commonly used tool has been the Langmuir balance which is designed so that a force transducer separates the uncovered water surface from the film covered surface. Initially the surfaces on both sides of the transducer are uncovered. A known concentration of surfactant is applied to one side and the area density is then varied with a movable barrier that reduces the area available to the film. One may determine in this way $\pi(A,T)$ where T is the absolute temperature.

In order to facilitate dispersion of the surfactant into a monolayer, it is applied to the water dissolved in a volatile solvent which itself will spread spontaneously over the water. When the solvent evaporates, the surfactant remains at an area density lower than that which corresponds to a coherent monolayer. Although almost all solvents ultimately form lenses in equilibrium with a "gaseous" monolayer of solvent, the initial tendency is to spread to a multilayer before retracting to form the lens. The thermodynamics of this behavior was developed by Harkins [2] by 1940.

With the foregoing as background, one can ask the following question: If a surfactant is laid down on the surface of water in the manner described and then compressed to form a coherent monolayer, what will be the effect on the surface pressure and what will be the behavior of a droplet (morphology, evaporation rate, etc.) of any solvent placed on the film?

Our motivation for asking this question was to attempt to form films of stearic acid free of non-uniform strain. We had observed that compressed films displayed very long relaxation times; i.e., a compressed film held at a constant area density showed a long term decrease in surface pressure. Leakage of surfactant past barriers and seals was very quickly eliminated as a possible cause. The rationale for pursuing this question was that the solvent would dissolve the surfactant and spread over the surface. As the liquid evaporated, a strain-free homogeneous film might then remain. The results were quite surprising and suggested a new area of film research.

Experimental Method

The Langmuir balance used in these investigations was a newly designed instrument capable of high differential accuracy (1 millidyne/cm) at a high surface pressure (80 dynes/cm). A complete description of the instrument will appear elsewhere.[3] All solvents were spectroscopic grade and the compounds were highly purified specialty chemicals. The water was de-ionized and had a resistivity greater than 18 megohm-cm. The water was considered satisfactory for use if a surface pressure no greater than 250 millidynes/cm developed when the surface area was reduced by a factor of 16.

Results

The results reported here are limited to a single surfactant, stearic acid, although similar results were obtained with stearyl alcohol. After spreading the film, it was compressed into the so-called liquid-condensed phase to a pressure between 6-15 dynes/cm. A drop of solvent was then placed near the center of the surfactant covered surface. If the solvent was n-hexane or n-heptane, a large flat lens was formed that very slowly evaporated. On an uncovered surface, 100 microliters evaporated in a minute; on the surfactant covered surface it usually took 8 minutes. Figure 1 illustrates the course of the surface pressure with time after successively 25 microliters then twice 75 microliters of n-heptane were applied to a stearic acid film.

The drop in surface pressure implies an increase in area available to the surfactant molecules. This additional area can be calculated from the π-A diagram. The data relevant to the experiment, from which Fig. 1 was obtained, are given in Table I. Note that area increase scales with the size of the droplet surface.

It is concluded from the penultimate and the last columns in the table that both the upper and lower surface of the solvent lens were coated with surfactant molecules. This situation is illustrated pictorially in Fig. 2. As the droplet evaporates the diameter of the lens shrinks and the surfactant on the upper surface is redeposited on the water to finally reconstitute the original film.

The very slow evaporation rates are consistent with this model; it is well known that a film of surfactant molecules greatly reduces the evaporation rate. Here it must be kept in mind that a lens has a smaller area for evaporation than a spread film. Nonetheless, the evaporation rates are still quite smaller than expected for such highly volatile solvents. The behavior we observed with n-hexane and n-heptane is in distinct contrast with our observations with diethyl ether, chloroform, and benzene. For these latter solvents we observed an increase in pressure on application of a droplet. In the case of ether, much surface activity was observed at the application site but no lens was observable; chloroform and benzene both formed lenses. On evaporation of these solvents it was observed that the film pressure drops below the inital value and subsequently does not recover its initial starting pressure. This implies an effective loss of surfactant which is probably due to conversion of some of the material from a monolayer state to a multilayer or bulk state.

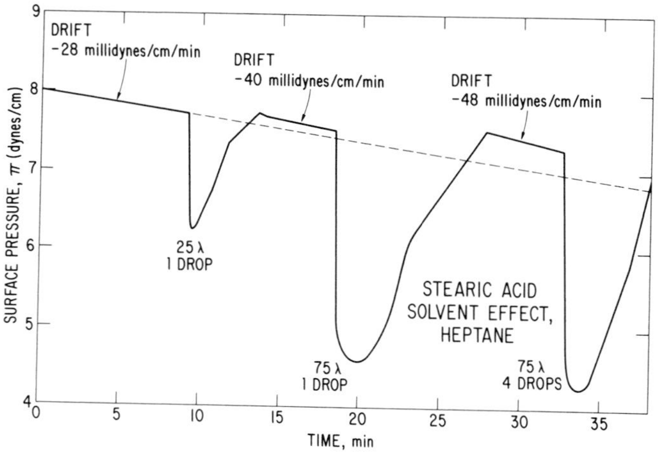

Fig. 1, Trace shows the time dependence of the surface pressure following the application of droplets of n-heptane to a compressed stearic acid film.

Some earlier experiments on the effect of benzene on the surface pressure of stearic acid films have been reported.[4] Washburn et al observed that if a droplet of benezene was deposited on the surface when the stearic acid surface pressure exceeded the initial spreading pressure of benzene (∿ 10 dyne/cm), the overall film pressure droped significantly. This would imply that some material is being removed from the film by the droplet; a slow evaporation rate was also observed. The observation was interpreted as being due to the smaller area for evaporation since the droplet remained as a lens rather than spreading. The possibility of molecules being adsorbed on the upper surface was not discussed. While the smaller evaporating area is relevent, as discussed earlier, the presence of surfactant on the evaporating surface is also important. For stearic acid pressures less than 10 dyne/cm, the total surface pressure increases slightly implying a contribution from a spread benzene film.

We now describe a final experiment which clearly indicates that stearic acid films are in a 2D solid state. A droplet of n-heptane was depostied on a compressed stearic acid film. The film was then rapidly expanded. Initially no change was observed in the heptane lens. However, at a critical point a narrow channel emerged from the lens and rapidly propagated over the surface. We interpret this as the heptane spreading on the clean surface created by a crack appearing in the expanding film.

Finally, we wish to point out that a study of the diameter of a large lens formed when a known amount of material is deposited can be used to obtain information on surface tensions; from the lens diameter and volume of material one can compute the lens thickness, h. If we denote the surface tensions of the air-water, water-solvent, and solvent-air interfaces as σ_{12}, σ_{13}, and σ_{23} respectively, then it is easy to show from surface and gravitational energy equilibrium considerations that

$$\sigma_{12} = \sigma_{13} + \sigma_{23} - \frac{h^2 g}{2} \rho_3 (1 - \frac{\rho_3}{\rho_2})$$

where ρ_2 and ρ_3, are the density of water and solvent respectively and g is the acceleration due to gravity. σ_{12} can be determined with the balance, and all quantities in the last term on the right can be determined. A measurement of either σ_{13} or σ_{23} would determine the remaining one. A measurement of the upper

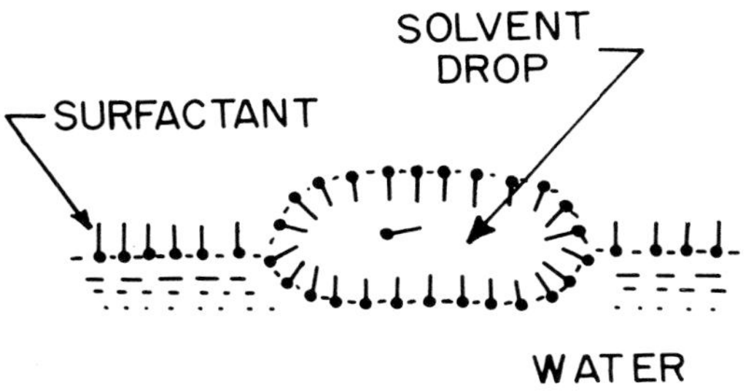

Fig. 2, A pictorial representation of the arrangement of stearic acid surfactant molecules associated with a lens of n-heptane on water.

TABLE I

Effect of n-Heptane on a Stearic Acid Film

Initial surface pressure: 8 dynes/cm

Vol. Added, Microliters	$\Delta\pi$ dynes/cm	Δ Area, $\AA^2$ /molecule	$\frac{\Delta A_{75}}{\Delta A_{25}}$	$\left(\frac{V_{75}}{V_{25}}\right)^{2/3} = \left(\frac{D_{75}}{D_{25}}\right)^2$
25	-1.45	+0.20	-	-
25	-2.88	+0.40	2.0	2.1
75	-3.00	+0.43	2.1	2.1

surface contact angle together with σ_{12} would also permit a complete evaluation of the three surface energies.

In conclusion, we have demonstrated a number of interesting phenomena associated with small droplets applied to water in the pressure of surfactants; which clearly suggest that much further work should be done in this field.

REFERENCES

1. For a review of this field we suggest G. L. Gains "Insoluable Monolayers at a Liquid Gas Interface" John Wiley (Interscience), New York (1966).

2. W. D. Harkins, Chem. Rev. 29, 385 (1941). W. D. Harkins, "The Physical Chemistry of Surface Films", (Reinhold Pub. Corp. New York 1952).

3. B. M. Abraham, K. Miyano, J. B. Ketterson, and K. Buzard, Review of Scientific Instruments, (to be published).

4. E. R. Washburn, L.F. Transue, and T. J. Thompson, J. Am. Chem. Soc. 63, 2742 (1941).

Sinha, ed. Ordering in Two Dimensions

THE POLYMERIZATION AND CHARACTERIZATION OF DIACETYLENE MONOLAYERS FORMED AT THE GAS WATER INTERFACE

D. R. DAY

Dept. of Electrical Engineering, M.I.T., Cambridge, Mass. 02139

J. B. LANDO

Dept. of Macromolecular Sci., Case Western Reserve University

Cleveland, Ohio 44106

H. RINGSDORF

Institute of Chemistry, University of Mainz, Mainz, West Germany

Various surface active diacetylene compounds have been spread and polymerized at the gas-water interface. Due to the extreme anisotropic polarizability and the crystalline nature of the polymerized films, the monolayer morphologies were observable between crossed polars in an optical microscope. Monolayers spread from chloroform under a chloroform saturated atmosphere and then polymerized have resulted in monolayer crystallites over three millimeters in diameter. Bilayers of such films are self supporting and can span macroscopic size holes. Electron diffraction of polymer bilayers has shown over four orders of diffraction. The spacings and intensity data collected from thirteen independent reflections have been used to refine the molecular packing within the monolayer.

INTRODUCTION

The ability of certain surface active diacetylene compounds to undergo polymerization in monolayers at the gas-water interface has been previously demonstrated [1]. The occurance of the monolayer polyreaction (Fig.1) was shown to be dependent on the crystallinity of the monolayer which is in turn a function of tail length and temperature [1,2].

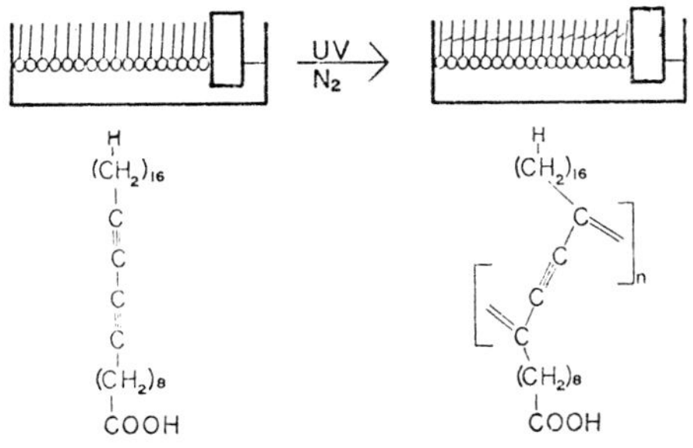

Figure 1

Monolayer Polymerization Reaction

Through the addition of various metal-hydroxide salts to the water substrate, the corresponding metal salts of the diacetylene compounds can be formed and also polymerized. Both the free acid and metal salt polymerized films exhibit a high structural integrity and have shown the ability to cover macroscopic pores (0.5 mm.) in a substrate dipped through the monolayer. In order to attain a better understanding of the nature of these self supporting films, the following structural study was undertaken.

EXPERIMENTAL

Due to the highest structural integrity of the lithium salt monolayers all discussion will be limited to the following compound:

$$H-(CH_2)_{16}-C{\equiv}C-C{\equiv}C-(CH_2)_8-COO^- \ Li^+$$

Monolayers formed from this compound were polymerized at room temperature under a nitrogen atmosphere. The reaction was carried out using U.V. radiation (254 nm.) with an intensity of $5 mW/cm^2$. A Reichert optical microscope was used for observing the birefringence of the polydiacetylene monolayers deposited on glass. The self supporting monolayer films were deposited directly on electron microscope grids for normal observation and diffraction in a Jeol JEM 100 electron microscope.

RESULTS AND DISCUSSION

Optical Microscopy: Polydiacetylene monolayers are invisible to normal light microscopy as a result of their small thickness (35 Å). When placed between crossed polarizers however, a faint birefringence is present enabling the crystalline morphology to be observed. Ordinarily, a monolayer, even if crystalline, would exhibit no birefringence as a result of negligable rotation of the passing light vector. This is offset in polydiacetylene monolayers by the extreme anisotropic polarizability which is a result of the conjugated nature of the polymer backbone (Fig. 1). Unfortunatly the polymer backbones are not present in unreacted monolayers causing them to exhibit no birefringence.

The morphology of the polydiacetylene monolayers was found to resemble the spherulitic nature of other common polymers. In this case however, it is more analogous to a disk. In an effort to determine whether the observed morphology was present in the initial unreacted film or formed after reaction, some of the reaction conditions were varied.

In an attempt to vary reaction rate and final degree of reaction, the U.V. intensity and exposure time were altered respectively. Upon examination of these films, varying degrees of birefringence were observed but the morphology remained the same. It was found that drastic changes in the morphology could be achieved however by altering the conditions under which the initial monomer monolayer was formed. By diluting the diacetylene-chloroform solution (which is dropped on the water surface to form the monolayer) with hexane, a poor solvent, a mosiac block morphology resulted. This and other similar experiments strongly suggest that the morphology is determined when the monomer monolayer is initially formed and is carried through the reaction to the final polymer film. With this knowledge, very large monolayer crystallites were formed by saturating the atmosphere above the water surface with chloroform (a good solvent) before formation of the monolayer. This allows a longer time for the two dimensional crystallization to occur and results in films with two dimensional crystals as large as three millimeters in diameter.

Electron Microscopy: For observation in the electron microscope, two monolayers were deposited on an electron microscope grid and shadowed with gold at 45°. The bilayer structure (two monolayers) acts as a laminate and is much more resistant to heavy radiation than a single layer.

Very flat featureless films are always observed at all magnifications. The exceptions are holes and tears where many filaments stretching across the gap are observed. These filaments range from tenths of millimeters down to 100 Å's in width and demonstrate the high degree of chain orientation in these films.

Electron Diffraction: Bilayers usually exhibit two superimposed electron diffraction patterns arising from each of the monolayers comprising the bilayer. The misregistration in angle between the two patterns is random and depends only on the way in which the two films happened to sandwich together. In some areas a misregistration of 0° occurs and the two patterns merge into one.

The two unit cell dimensions resulting from the orthorhombic pattern are b and c which are 8.1 and 4.9 Å's respectively. X-ray diffraction from multi-bilayer samples has shown a repeat distance (perpendicular to the monolayer plane) of 70 Å's. The maximum theoretical bilayer thickness for a bilayer of this compound is also 70 Å's indicating that the hydrocarbon tails stand at exactly 90° with respect to the monolayer plane. The combination of electron and x-ray diffraction data yields a unit cell where a=70 Å, b=8.1 Å, c=4.9 Å and $\alpha = \beta = \gamma = 90^{\circ}$. The systematic absense of $k+\ell=2n+1$ reflections also indicates a cell centering in the a-b plane.

Structure Refinement: A great deal is known about the monolayer structure before any refinement is done. The polymer backbone which should have a theoretical repeat length of 4.9 Å's can be assumed to lie along the c axis (4.9 Å's). As mentioned above, the hydrocarbon side chains are also known to be exactly perpendicular to the monolayer plane. This causes a superposition of many atoms in the projection along a (perpendicular to the monolayer plane). The major unknown variable is θ, the angle between the hydrocarbon planar zig-zag and the b axis.

A series of mathematical models were constructed using standard bond lengths and angles for a number of θ values. The best agreement between calculated intensities for the models and observed values was found at $\theta = 0^{\circ}$ with a R factor of 10%. An excellent agreement with the model and the Patterson map (a plot of interatomic vectors determined from the observed intensities) also resulted at 0° with two major peaks corresponding to the C-C and C-H bonds.

The solution was non-unique however in that a slight modification also generates the same agreement values. The modification has a glide plane along b and perpindicular to c instead of a cell centering. The cell centered and glide plane structures are indicated in figure 2A and 2B respectively.

Figure 2

A) Cell Centered b-c Projection

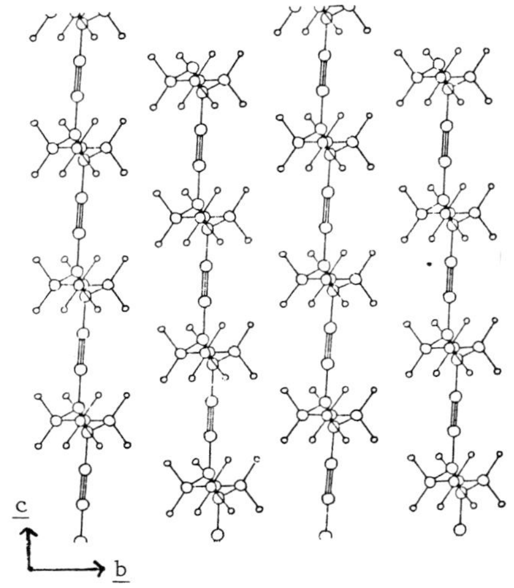

B) Glide Related b-c Projection

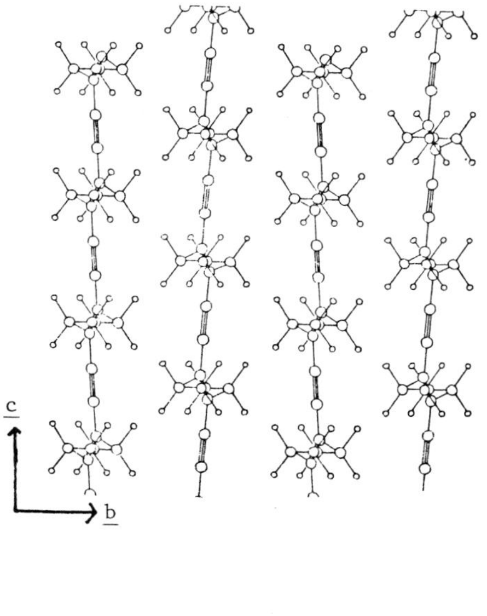

As a result of the similarity of a majority of the atomic positions in both structures, a pseudo cell centering systematic absence would be present for the glide plane structure (Fig. 2B). The correct structure was determined through diffraction of unreacted monolayer films.

Due to the inability of the unreacted monolayers to support themselves, they were deposited on thin carbon supports and then suspended from E.M. grids. Diffraction patterns showed a systematic absense of 0k0=2n+1 reflections indicating a glide plane symmetry. Because a major symmetry change is highly improbable during the solid state reaction, the polymer glide plane structure in figure 2B is the most probable one. The projections along the a and b axes of the final refined model are indicated in figure 3 (Li^+ has been omitted).

Figure 3

Refined Polydiacetylene Monolayer Structure

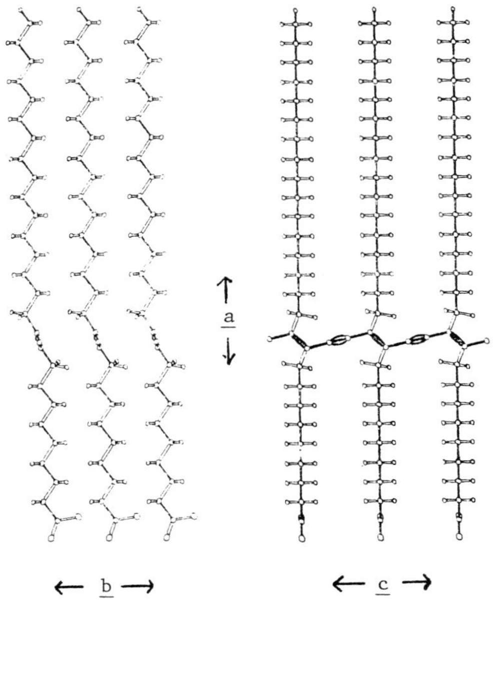

CONCLUSION

Birefringence, electron microscopy and electron diffraction studies have shown polydiacetlyene monolayers to be highly ordered in three dimensions. The polymer backbones lie totally within the monolayer plane and within a given crystalline domain are all parallel to one another. Using these techniques, high integrity ultra thin and flat polymer films can be produced.

REFERENCES

1. D.R. Day and H. Ringsdorf, J.Polym. Sci. (Letters),16,205 (1978)
2. D.R. Day, Doctoral Dissertation, Case Western Reserve University (1980)

Published 1980 by Elsevier North Holland, Inc.
Sinha, ed. Ordering in Two Dimensions

THE PHASE TRANSITION IN THE SOLUTIONS OF SOME QUATERNARY AMMONIUM SALTS

A. LIBACKYJ

84-22, 107 Avenue, Jamaica, N.Y. 11417

ABSTRACT

Aqueous solutions of some quaternary ammonium salts (Hyamine) separate into two liquid phases when a simple electrolyte is added to them. A light scattering study of the homogeneous system prior to the phase transition shows that in addition to the conventional double layer effects, the addition of the electrolyte to these systems decreases the surface potential of the particles. This decrease is determined apparently by the ion-dipole interaction between the group $\overset{1+}{N} - \overset{\delta-}{CH_2} - \overset{\delta+}{\bigcirc}$ of the molecule and the Cl ion in solution.

Aqueous solutions of some long chain quaternary ammonium salts, such as Hyamine, undergo a phase transition: Homogeneous solution → two liquid phases, under the action of a simple electrolyte (NaCl) (Coacervation).

From the light scattering measurements the micellar molecular weights and micellar charges were derived [1].

The micellar molecular weight increases rapidly with increased electrolyte concentration and that increase is essentially exponential (Fig. 1). The slope of the exponential increase of the molecular weight increases with increased amount of I_2 infused into the colloid system. This leads to the fact that the amount of NaCl necessary for the phase transition decreases exponentially with increased I_2 concentration.

The charge properties of dilute solutions of the systems with and without the phase transition were determined as a function of the added electrolyte. The micellar charge density is expressed as p/mx100, where p is the number of the micellar charge sites and m is the micellar aggregation number. In the case of the systems without the phase transition the parameter p/m x 100 shows a linear dependence upon $\sqrt{c}$, c being the concentration of added electrolyte. For systems with the phase transition, however, p/m x 100 shows a linear dependence on $\sqrt{c} \times e^{-bc}$, where b is a constant characteristic of a given system and temperature (Figs. 2, 3).

The observed difference between the charge properties of the systems with and without the phase transition can be explained as follows: The addition of the electrolyte to the system with the phase transition produces a decrease in the surface potential of the micelles. By way of contrast, the addition of the electrolyte to the system without the phase transition does not appreciably alter the potential of the system. In this latter case the observed charge effects may be related to the double layer phenomenon exclusively.

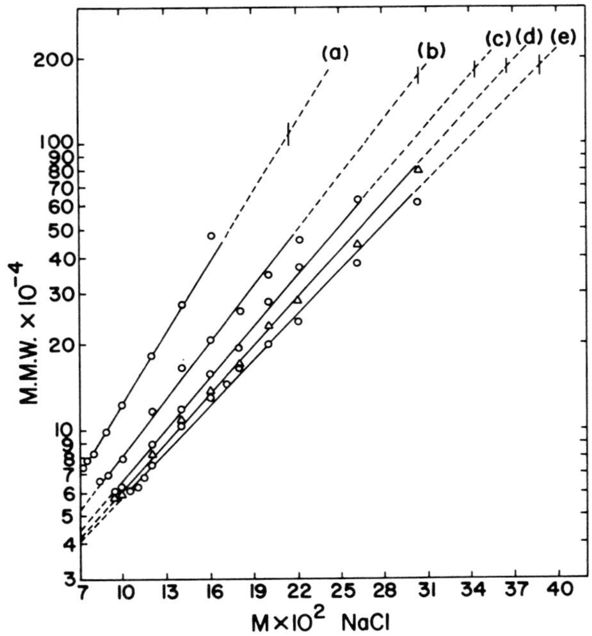

Fig. 1. Micellar Molecular Weights (M.M.W.) of Hyamine 1622-I_2-NaCl-H_2O systems as a function of NaCl concentration at 30° (a) I_2Hy=R=5x10^{-2}; (b) R=2x10^{-2}; (c) R=1x10^{-2}; (d) R=0.5x10^{-2}; (e) R=0

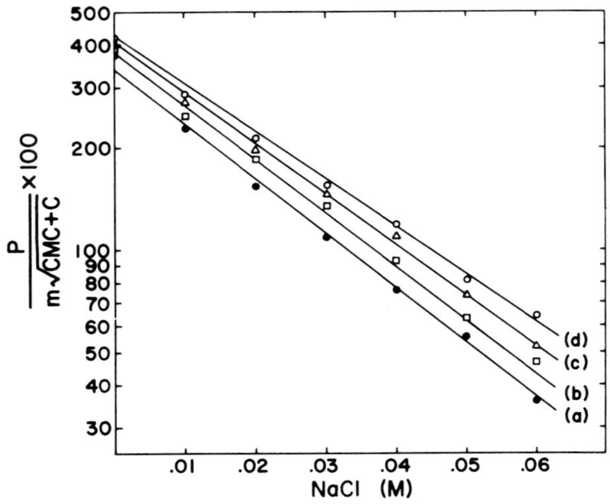

Fig. 2. Log plot of the ratio of charges to the square root of ionic strength as a function of NaCl concentration. (a) Hy-NaCl $R=5.0\times10^{-2}$; (b) Hy-NaCl $R=2.0\times10^{-2}$; (c) Hy-NaCl $R=1.0\times10^{-2}$; (d) Hy-NaCl R=0

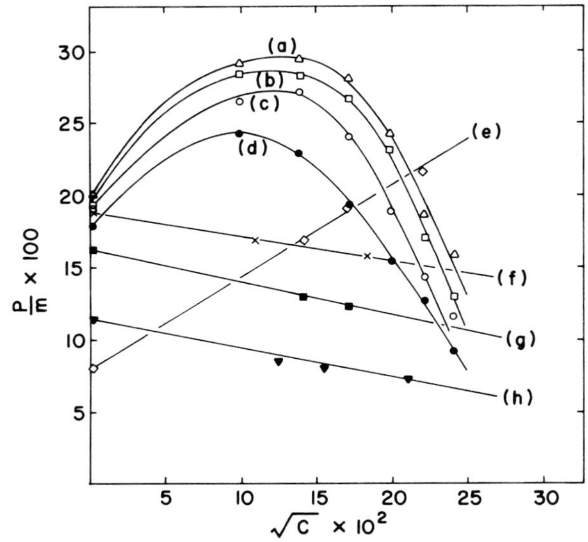

Fig. 3. Charges of non-coacervating systems as a function of electrolyte concentration.

(a) $C_{12}NH_3Cl$-NaCl, 25°C.

(b) $NaC_{12}SO_4$-NaCl, 25°C.

(c) C_{14} trimethyl ammonium bromide-KBr, 25°C.

(d) Hy-Na acetate, 30°C.

Fig. 4 shows the electrolyte dependence of the double layer thickness for the system with the phase transition. Two opposing effects are acting in this case: the effect of $\sqrt{c}$ is to decrease $1/\kappa$ and that of e^{-bc} is to increase $1/\kappa$. A maximum in charge corresponds to a minimum of $1/\kappa$, after which the latter cannot but increase.

The fact of the decrease of the surface potential indicates that, in addition to the conventional double layer effects, which take place at a constant surface potential, some kind of interaction must be present between the solution of the electrolyte and the micellar surface in the case of the systems with the phase transition [2].

The slopes of the variation of the molecular weight and charges with the concentration of the electrolyte can be taken as the measure of this interaction. These slopes increase linearly with increased I_2 concentration and this, in turn, leads to a decrease in cec, i.e., the minimum electrolyte concentration necessary for the phase transition. The observed decrease in cec is also linear with I_2 concentration.

It is possible to explain these observations in terms of the ion-dipole interaction between Cl ions in the solution and the micellar surface. The slopes of the variation of the molecular weights and the charges at constant temperature are proportional to the energy of the interaction. On the other hand, the energy of the ion-dipole interaction is proportional to $1/r^3$, where r is the distance between dipole and ion. $1/r^3$ can be

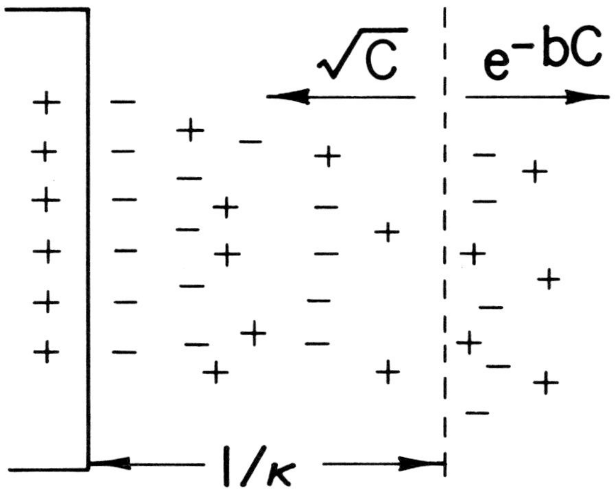

Fig. 4. Electrolyte dependence of the double layer thickness for coacervating systems.

equated to 1/V, where V is the volume in which dipoles are contained and 1/V is, in turn, equal to C, - the concentration of dipoles in a given volume. Thus, the energy of the ion-dipole interaction depends linearly on the dipole concentration, if the concentration of ions remains constant.

The observations described above represent such a case. One can thus conclude, at least tentatively, that in systems undergoing the phase transition a Cl (or other) ion can attach itself to the micellar surface by means of the secondary interaction, namely the ion-dipole interaction. The phase transition is therefore related to the presence in the molecule of a permanent dipole, the magnitude of which is increased by infusion of I_2. In Hyamine and in other compounds which undergo this type of the phase transition the following group is present:

$N - CH_2 -$  . This configuration is absent in
+1 $\delta-$ $\delta+$

the compounds which do not undergo the phase transition. The positive charge on N atom can attract π - electrons from the benzene ring, creating a dipole. The infused I_2 also withdraws π - electrons from the benzene ring, increasing the magnitude of the dipole. In the compounds without the phase transition this configuration is absent, thus no significant permanent dipole is present in the molecule and no interaction with the electrolyte is possible.

By means of the ion-dipole interaction a Cl ion can attach itself to the micellar surface and become a permanent part of it and neutralize some of the positive charge of the surface. This leads to a decrease in the surface potential shown previously. This leads, in turn, to a change in the energy of interaction between micelles and to a redistribution of the micellar sizes as manifested by the increased molecular weight. When the micelles are finally so discharged that they cannot stay apart, the solution separates into two liquid phases.

The effect of temperature, at constant electrolyte concentration, on this phase transition is the following: with decreased temperature the slope of the variation of the molecular weight with electrolyte increases and cec decreases. The slopes were determined from the log plot of the molecular weight. Log of the slope is then plotted versus 1/T, where T is the absolute temperature. From this latter plot the energy of activation of the phase transition is determined, equal approximately to 5 Kcal./mole. This value is in agreement with the postulate that this phase transition is determined largely by a secondary interaction, namely the ion-dipole interaction.

REFERENCES

1. I. Cohen and A. Libackyj, J. Colloid Sci. 19, 560-570, (1964).

2. A. Libackyj, J. Colloid Interface Sci., Vol. 52, No. 1, 199 (1975).

Published 1980 by Elsevier North Holland, Inc.
Sinha, ed. Ordering in Two Dimensions

PARAMETRIC MODEL FOR ISOTROPIC-NEMATIC TRANSITIONS

I. Muscutariu

Physics Department, University of Timisoara, 1900 Timisoara, Romania

I. Motoc

Chemistry Research Centre, Bul. M. Viteazu 24, 1900 Timisoara, Romania

O. Dragomir

Institute Computing Technique, 1900 Timisoara, Romania

ABSTRACT

The transition temperatures of two classes of nematic liquid crystals are parametrized in terms of specific molecular dimensions.

We discuss a parametric model [1] connecting the chemical structure and isotropic-nematic transitemperature [2,3] t_c (°C). Two series of compounds are studied, namely:

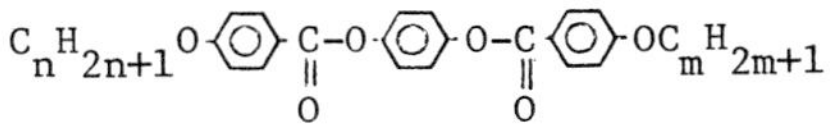

$$C_nH_{2n+1}O\!\!-\!\!\bigcirc\!\!-\!\!\underset{O}{\overset{\|}{C}}\!-\!O\!\!-\!\!\bigcirc\!\!-\!\!O\!\!-\!\!\underset{O}{\overset{\|}{C}}\!\!-\!\!\bigcirc\!\!-\!\!OC_mH_{2m+1}$$

(I)

m = n, n+1, 111 , 8 ; n = 1,2, ... , 8.

$$C_nH_{2n+1}\!-\!O\!-\!\underset{O}{\overset{\|}{C}}\!\!-\!\!\bigcirc\!\!-\!\underset{O}{\overset{\|}{C}}\!-\!O\!\!-\!\!\bigcirc\!\!-\!O\!-\!\underset{O}{\overset{\|}{C}}\!\!-\!\!\bigcirc\!\!-\!\underset{O}{\overset{\|}{C}}\!-\!OC_nH_{2n+1}$$

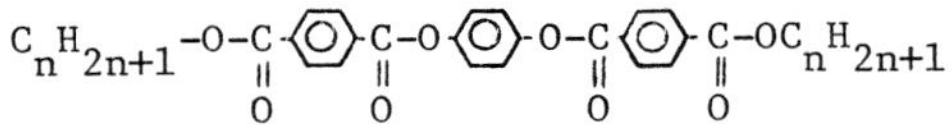

(II)

n = 1,2, ... ,6. The t_c values are taken from Ref. [4].

The dependence of t_c on the dimensions of alkyl groups (i.e., the length, L, and the widths in four perpendicular directions, B_1, B_2, B_3, B_4, in Å) was investigated. The involved L and B parameters were computed by means of the STERIMOL program [5]. The alkyls were considered in the most stable conformations.

Illustratively, Table I lists L and B parameters for some alkyls.

Table I

L AND B PARAMETERS FOR C_nH_{2n+1}

n	L	B_1	B_2	B_3	B_4
1.	3.00	1.52	1.90	1.90	2.04
2.	4.11	1.52	1.90	1.90	2.94
3.	5.05	1.52	1.90	1.90	3.49
4.	6.17	1.52	1.90	1.90	4.42

For the compounds (I) we computed, using the least squares method, the following equations[*]:

$$t_c = 326.57(\pm2.37) \\ - 16.38(\pm0.41)\ B_4(C_mH_{2m+1}) \tag{1}$$

$$(n=35,\ r=0.87,\ s=13.83,\ F=47.46)$$

$$t_c = 291.51(\pm2.87) \\ - 14.73(\pm0.70)\ B_4(C_nH_{2n+1}) \tag{2}$$

$$(n=35,\ r=0.79,\ s=16.70,\ F=27.52)$$

$$t_c = 328.36(\pm2.31) \\ - 11.81(\pm0.28)\ L(C_mH_{2m+1}) \tag{3}$$

$$(n=35,\ r=0.87,\ s=13.49,\ F=50.73)$$

$$t_c = 292.82(\pm2.84) \\ - 10.59(\pm0.48)\ L(C_nH_{2n+1}) \tag{4}$$

$$(n=35,\ r=0.80,\ s=16.53,\ F=28.43)$$

Using the molar refractivity, MR, as a measure of the Van der Waals volume[**] of the alkyl, one gets:

$$t_c = 308.87(\pm1.16) \\ - 1.67(\pm0.02)\left[MR_{C_nH_{2n+1}} + MR_{C_mH_{2m+1}}\right] \tag{5}$$

This equation is significant at >99.99 percentile level (Fisher statistics). The Eq. (5) is in good agreement with results of Maier and Saupe [7] and Kimura [8]. These authors pointed out that:

$$t_c = const_1 + const_2 \cdot F(\text{length and} \\ \text{width of molecule}) \tag{6}$$

In addition, Eq. (5) reveals that the dipole-dipole interactions (via the distance between the segments of the molecules) are the dominant interactions in the class of nematics considered here. This statement is argued by the experimental model II. In this case, the best equation is:

$$t_c = 379.44(\pm 12.45) - 9.38(\pm 0.65)\ MR_{C_nH_{2n+1}} \qquad (7)$$

To conclude, we conisder that the parametric models are very useful to establish relations between molecular features and liquid crystal properties. These equations may be used for "molecular design" of liquid crystals with desired properties.

*n - number of points used in regression; r - correlation coefficient; s - standard deviation; F - Fisher statistics.

**We proved [6] that MR has an electronic and a steric component in proportion of 1:9.

REFERENCES

1. D. J. Newman, Aust. J. Phys., 31, 489 (1978).
2. M. J. Stephen and J. P. Straley, Rev. Mod. Phys., 46, 617 (1974).
3. S. Chandrasekhar and N. V. Madhusudana, Acta Cryst., A27, 303 (1971).
4. D. Demus, H. Demus, and H. Zaschke, "Flüssige Kristalle in Tabellen," Deut. Ver. Grunstoffind., Leipzig, 1974, p. 82-83.
5. A. Verloop, W. Hoogenstraaten, and J. Tipker, in "Drug Design" Vol. 7, ed. E. J. Ariens, Academic, New York 1976, p. 165-207.
6. I. Motoc and F. Kerek, J. Chim. Phys. (in press).
7. W. Maier and A. Saupe, Z. Naturforsch., 14a, 882 (1959).
8. H. Kimura, Phys. Letts., 47A, 173 (1974).

Published 1980 by Elsevier North Holland, Inc.
Sinha, ed. Ordering in Two Dimensions

MECHANICAL MEASUREMENTS ON FREELY-SUSPENDED LIQUID CRYSTAL FILMS

R. PINDAK, D. J. BISHOP, D. D. OSHEROFF AND W. O. SPRENGER

Bell Laboratories
Murray Hill, New Jersey 07974

Freely-suspended films of smectic liquid crystals were prepared between a disk and support ring and two techniques developed to study their mechanical properties. In the first technique the disk was part of a 524 Hz, high Q, torsional oscillator. Changes in resonant frequency and amplitude were related to the shear modulus and damping coefficient of the film. In the second technique a torsional fiber was attached to the disk and a torque applied. The strain relaxation in the film was determined by monitoring the rotation of the disk to its zero-torque position. Measurements were made through the smectic A and B temperature range in films of 40.8 that were from 4 to 24 molecular layers thick. Pronounced features were observed both at and above the bulk B to A transition temperature.

Freely suspended films of smectic liquid crystals with thicknesses from two to hundreds of molecular layers are proving to be important both in characterizing bulk smectic phases and as systems for studying phase transitions in two dimensions.[1-2] We have previously reported the first mechanical measurements of the smectic B-A transition in such films.[3] We used a low frequency torsional oscillator to measure the temperature dependence of the in-plane shear modulus and dissipation of freely suspended films of butyloxybenzylidene octylaniline (40.8) as a function of the number of molecular layers. In this article we review these results and present new strain relaxation measurements that confirm our reported conclusions. Three observations were made: (a) Smectic B films of all thicknesses have a solid-like response. (b) The smectic B-A transition proceeds in at least two separate steps. The lower temperature transition is the bulk melting and is accompanied by a sharply changing shear modulus and a dissipation peak. The higher temperature transition, which we tentatively identify as surface melting, shows a slowly changing shear modulus and a shoulder in the dissipation. (c) Finally, both the shear modulus and dissipation show additional structure at temperatures below the smectic B-A transition for which we have no explanation.

The nature of two dimensional (2D) melting and the behavior of the 2D shear modulus μ was recently studied by Halperin and Nelson.[4] Building on ideas due to Kosterlitz and Thouless[5] they constructed a dislocation mediated theory of melting. As in the Kosterlitz-Thouless theory at the 2D melting temperature T_m they predict a universal jump in the shear modulus μ. In the limit that the other Lamé constant $\lambda >> \mu$, this jump will be of magnitude $\mu_D = \frac{4\pi k_B T_m}{a_o^2}$ where a_o is the lattice spacing. In 40.8 $a_o = 5 \times 10^{-8}$cm which gives a predicted jump $\mu_D = 220$ ergs/cm^2. This value can be at most a factor of 2 higher if the limit $\lambda >> \mu$ is not satisfied. In addition Halperin and Nelson make the observation that if a 2D crystal melts by the appearance of a small density of free dislocations, then it should first melt into the intermediate hexatic phase and then into a 2D liquid. The hexatic phase would have only short-range positional order but quasi-long-range bond orientational order. It would not be expected to support a shear at low frequencies.

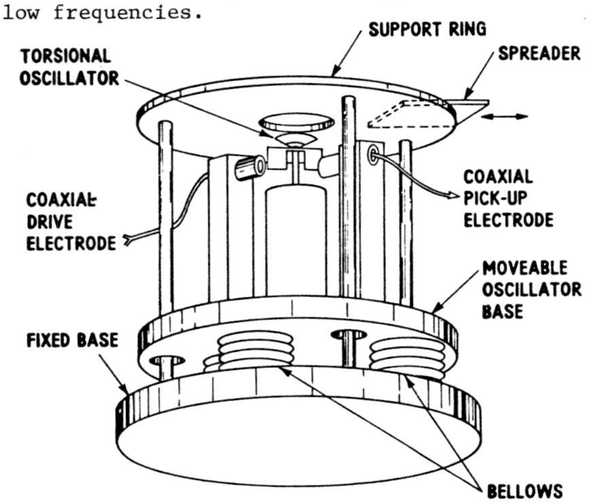

Fig. 1 The experimental cell is shown. The liquid crystal film to be measured is supported between the support ring and the torsional oscillator.

The first technique that we used to measure the shear modulus in liquid crystal films involved constructing the apparatus shown in Fig. 1. The main element is a high-Q BeCu torsional oscillator. It has a disk-shaped top of radius R_1 = 0.59cm supported by a 1mm diameter torsional rod. Attached to the oscillator is a metal plate that forms part of the pick-up and drive capacitors. In operation a quantity of 40.8, in its smectic A phase, was placed around the hole in the support ring and then, using a glass spreader, a freely suspended film of the liquid crystal was drawn across the hole (radius R_2 = 0.70cm). The film's thickness was determined by measuring its optical reflectivity. In this fashion it was possible to produce uniform films from two to hundreds of layers in thickness. After a film was formed the movable oscillator base was raised by pressurizing a set of three bellows until the oscillator, wet with liquid crystal, touched the film. The liquid crystal film then formed an annular region between the torsional oscillator and the support ring. The oscillator was phase locked in the standard way with the high Q mechanical oscillator being the frequency determining element of a feedback loop consisting of an amplifier, phase shifter and a zero-crossing detector. The resonant period of the oscillator could be resolved to 1 part in 10^5 and the amplitude to 1 part in 10^3. The oscillator torsional elastic constant K_o was calibrated to an absolute accuracy of 5% by placing known masses on the oscillator and measuring the shift in period. At T_o = 56°C, the resonant frequency of the oscillator was 524Hz and $K_o = 4.9 \times 10^6$ ergs. The dissipation Q_o^{-1} of the unloaded oscillator was determined at T_o by measuring the 1/e ringdown time with the result that $Q_o(T_o)$ = 770. For the unloaded oscillator the amplitude of oscillation at the outer edge of the disk was 7Å giving a shear angle of 1×10^{-7} radians.

Using the calibrated oscillator, changes in its mechanical properties due to an applied film can be related to the material parameters of the film. In particular, the torsional elastic constant of the film K_F is related to the resonant period P of the oscillator-film system by the equation

$$P = 2\pi(I_o/(K_o + K_F))^{1/2} \qquad (1)$$

where I_o is the moment of inertia of the oscillator. The moment of inertia of the film I_F has been neglected since $I_F/I_o < 10^{-6}$ for the films studied. The 2D shear modulus μ of the film is related to K_F by:

$$\mu = CK_F = \frac{(R_2^2-R_1^2)}{4\pi R_1^2 R_2^2} K_F \qquad (2)$$

where the geometrical constant C = 0.065cm^{-2}.

Combining equations (1) and (2) we can express the shear modulus in terms of the decrease in period as

$$\mu \simeq 2C\, K_o\, (\Delta P/P_o) \qquad (3)$$

where P_o is the period of the oscillator alone and $\Delta P = P_o - P$. The predicted jump μ_D would give a change in period $\Delta P/P_o = \mu D/2C\, K_o = 4 \times 10^{-4}$ which is within the instrument's resolution. It should be stressed that the oscillator is not sensitive to the presence of a liquid-like phase. The added inertia of a liquid-like phase would increase the period of the oscillator but because $I_F << I_o$ its effect would be negligible. For this reason our measurement would not be sensitive to the hexatic phase which is expected to have a liquid-like response at low frequencies.

The temperature dependence of the oscillator's period P_o and dissipation Q_o^{-1} are shown in Fig. 2. Coupling a 24 layer film of 40.8 to the oscillator leads to a new resonant period P and dissipation Q^{-1} for the oscillator-film system.

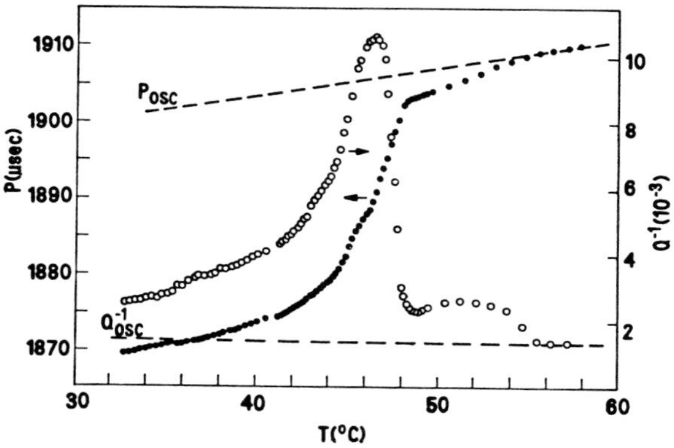

Fig. 2. The period (P) and the dissipation (Q^{-1}) are shown as a function of temperature for a 24 layer film. P_o and Q_o^{-1} are the background period and dissipation of the empty cell.

Above 56°C, the oscillator's period and dissipation are unaffected by the presence of the film. As discussed above, this implies a liquid-like response and is the expected result for the smectic A phase. The surprising result is that this behavior does not continue down to the bulk B-A transition temperature at 49°C. Instead, just below 56°C, a solid-like response (decrease in period) is observed indicating that the film has a finite shear modulus at this temperature. As the temperature is lowered, the shear modulus continues to increase until it reaches a value $\mu \sim 750$ ergs/cm^2. These changes are accompanied by an increase in the dissipation of the system. At 49°C (the bulk B-A transition temperature), there then occurred a sharp increase in the shear modulus accompanied by a peak in the dissipation. Referring to Fig. 3, we have labeled the temperature associated with the onset of the shear modulus by T_1 and the temperature associated with the sharp increase by T_2. For $T \ll T_2$, within the smectic B phase, the shear modulus saturated at a value per layer $\mu_L = 420$ erg/cm^2. This gives a value for the bulk shear modulus $\mu_B = \mu_L/h = 1.5 \times 10^9$ ergs/cm^3 where h is the layer thickness (28.6Å).

To better understand the nature of these transitions at T_1 and T_2, the temperature dependence of P and Q^{-1} was measured as a function of film thickness (Fig. 3). The thinnest film that remained stable enough to be carefully measured was 4 layers and we studied films from 4 to 24 layers. For thin films the curves were repro-

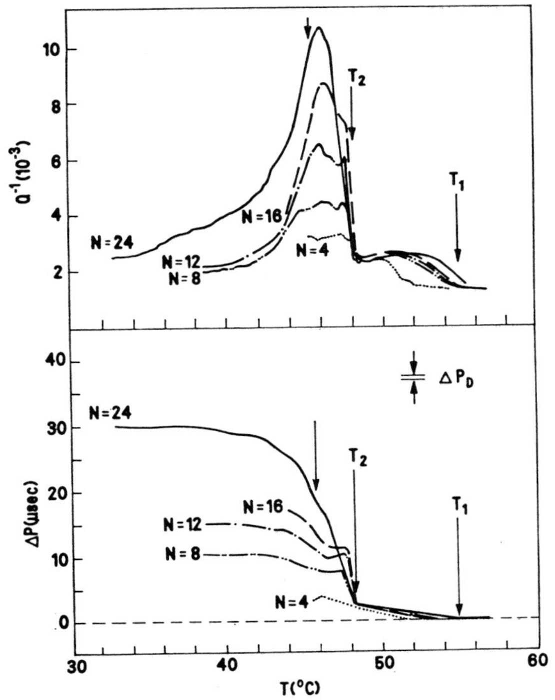

Fig. 3 The dissipation (upper) and period shift (lower) are shown as a function of temperature for different numbers of molecular layers (N). ΔP_D is the expected size of the period shift corresponding to the discontinuous jump in shear modulus μ_D. The curves for N < 24 have been shifted down in temperature such that the feature at T_2 coincides for all the curves.

For the higher temperature transition both the total period change and dissipation increase were found to be <u>independent</u> of film thickness. The fact that it was independent of film thickness, a solid-like response and of magnitude approximately twice the response per layer in the smectic B phase (i.e. $\mu(T_2) \simeq 2\mu_L$ ($T \ll T_2$)) implied that this feature was a surface transition involving the outer layers of the film which remain crystalline to a higher temperature than the bulk. The polar nature of the free surface would tend to preferentially orient the asymmetric molecules with respect to their long axes (random in the bulk).[1] Also, the layer (smectic A) order parameter would be higher at a free surface as evidenced by an increase in the smectic A-nematic transition temperature in thin films. A coupling between either of these order parameters and the in-plane positional order could have resulted in its surface enhancement.

Other features we observed in going from 24 to 4 molecular layers were an increase in T_2 of two degrees and a one degree decrease in T_1. Also, there appeared to be a third feature (indicated by the arrow in Fig. 3) which is not understood. At this temperature both the period change and dissipation leveled off before decreasing at T_2.

The second technique that we used which measured strain relaxation also involved preparing films in an annular geometry.[6] In this case a torsional fiber was attached to the disk and a ducible on heating and cooling, thick films tended to thin on heating so hysteresis effects could not be checked. For the lower temperature transition, it was found that both the total period change and the dissipation maximum scaled with the number of layers. This is what would be expected for the elastic response of a crystal of increasing thickness. Such solid-like response even for thin smectic B films supports recent X-ray measurements on 4O.8 which demonstrated that thick smectic B films are 3D crystals and thin smectic B films are best characterized as 2D crystals.[2]

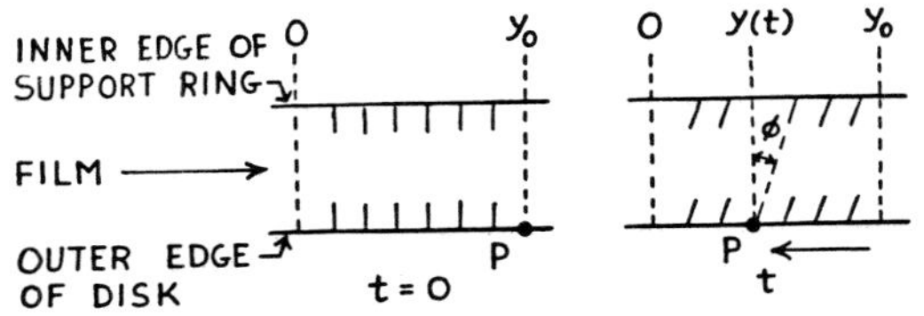

Fig. 4 Strain relaxation experimental geometry. y is the displacement of the disk from its final equilibrium (zero-torque) position and ϕ is the strain angle.

torque applied to the disk which rotated it about an axis normal to the film resulting in a shear stress on the film. As illustrated in Fig. 4, a point P on the disk would rotate from its initial position y_o to its zero-torque position (y = o). This involved elastic response and strain relaxation in the film which could be studied by measuring the disk displacement y as a function of time. The stress, σ, and strain angle, ϕ, are related to the disk displacement by: $\sigma = K'y$ and $\phi = K'y/\mu$ where K' is the torsional elastic constant of the film.

The fractional displacement, y/y_o, 20 msec. after a stress was applied is plotted in Fig. 5 as a function of temperature for an 8 layer film. The data confirm the two step smectic B-A transition observed in the oscillator experiment. Also plotted (Fig. 6) is the fractional displacement for a 20 layer film as a function of the total displacement y_o (proportional to the initial applied stress). For displacements less than ~ 200Å, the strain relaxation is linear. Therefore, the torsional oscillator measurements, with 7Å displacements, were well within the linear regime. This may not be true, however, for films that are only several layers thick.[6]

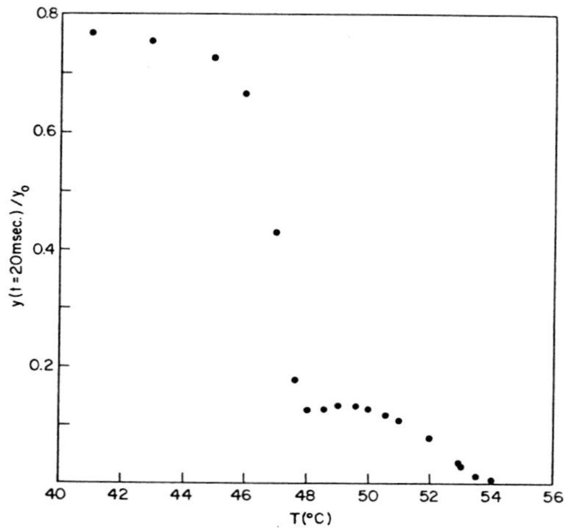

Fig. 5 The relative disk displacement y/y_o, 20 msec after an applied stress, is shown as a function of temperature for an 8 layer film.

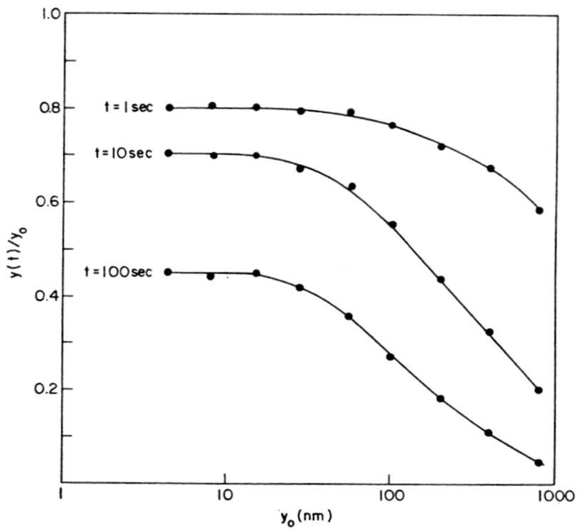

Fig. 6 The relative disk displacement y/y_o is shown as a function of the total displacement y_o for different times after the initial applied stress (proportional to y_o). This data is for a 20 layer film at T = 42°C.

In summary, we have developed apparatus capable of measuring the in-plane shear modulus, dissipation and strain relaxation of thin smectic films. We have observed that smectic B films have a solid-like shear modulus and that smectic A films have a liquid-like response giving further support to the conclusion that the smectic B phase, even in thin films, is a crystalline phase. We have further observed that the smectic B-A transition proceeds in two steps. The lower temperature feature is associated with the bulk melting and the upper feature with an enhancement in the surface melting temperature. Finally we did not find any evidence for a discontinuous jump in μ at either of the transitions. For the 4 layer film (see Fig. 3) this should have been a 10% effect and was not observed. Possible explanations include finite frequency effects on our measurements or impurities in our samples. Further measurements on purified samples and different, more stable compounds are in progress.

REFERENCES

1. C. Rosenblatt, R. Pindak, N. A. Clark and R. B. Meyer, Phys. Rev. Lett. 42, 1220 (1979).
2. D. E. Moncton and R. Pindak, Phys. Rev. Lett. 43, 701 (1979); invited talk by D. E. Moncton at this conference.
3. R. Pindak, D. J. Bishop and W. O. Sprenger, Phys. Rev. Lett. (to be published June 2, 1980).
4. B. I. Halperin and D. R. Nelson, Phys. Rev. Lett. 41, 121 (1978); D. R. Nelson and B. I. Halperin, Phys. Rev. B19, 2457 (1979).
5. J. M. Kosterlitz and D. J. Thouless, J. Phys. C6, 1181 (1973).
6. W. O. Sprenger, D. Osheroff and R. Pindak, to be published.

SECTION II

G. 2-D SUPERCONDUCTIVITY AND TRANSPORT

Published 1980 by Elsevier North Holland, Inc.
Sinha, ed. Ordering in Two Dimensions

MAGNETIC FIELD PROPERTIES OF SUPERCONDUCTING TWO DIMENSIONAL GRANULAR NbN

D.U. GUBSER AND S.A. WOLF

Naval Research Laboratory, Code 6338, Washington, DC 20375

We have investigated the resistive characteristics of the phase incoherent-phase coherent superconducting transition at T_{cj} in two dimensional granular NbN as a function of magnetic field H, from 0 to 10 Tesla. The dynamical scaling function, previously shown to describe the behavior of the voltage-current response near T_{cj}, remains unchanged in a magnetic field provided T_{cj} is replaced by $T_{cj}(H)$. We have also determined the superconducting properties of the phase coherent state below T_{cj}. The behavior of these granular two dimensional superconducting films is consistent with a model involving Josephson coupling between grains, strong magnetic flux pinning and the occurrence of intergranular phase slips rather than fluxoid motion.

1. INTRODUCTION

We have previously reported the resistive R, and current-voltage I-V characteristics of thin (30 Å) granular NbN films as functions of temperature and intergranular coupling strengths [1]. One of the most intriguing observations in these earlier studies was a power law temperature dependence of resistance above a critical temperature T_{cj} and a power law current dependence of voltage at T_{cj}. Indeed, analysis of these data in terms of a dynamical universal scaling function characteristic of critical fluctuation phenomena about a second order phase transition was successful. We have now extended the measurements to include the effects of magnetic field with the result that the critical magnetic field necessary to restore a measurable resistance in the films below T_{cj} was also a power law in $(T_{cj}-T)$ and the same universal scaling function which described the I-V results in 0 field also worked in a field assuming $T_{cj} = T_{cj}(H)$. This behavior is not anticipated from theoretical phase diagrams proposed for granular 2D films [2,3], suggesting that interpretation in terms of flux distribution characterized by individual vortices of a shape and size expected for homogeneous or weak pinning films is not valid.

2. EXPERIMENTAL DETAILS

A planar NbN granular structure as shown in figure 1 was prepared for these studies. Thick NbN pads served as voltage-current contacts while the connecting granular region (1mm x 1mm) was produced by chemical anodization in a manner previously described [1]. Grain size of the NbN particles is estimated to be 80-100 Å nominal diameter and 20-30 Å thick. The room temperature resistance was 2000 ohm/sq, increasing to a value slightly over 3000 ohm/sq at 14K. Since the grain diameter is approximately twice the coherence length of NbN ($\xi_0 \approx 50$ Å), the grains develop a large non zero superconducting order parameter at T_{cg}, the transition temperature of the isolated grain.

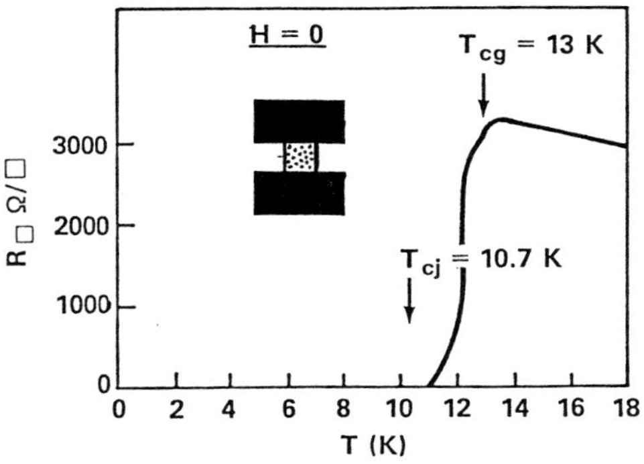

Figure 1 Zero field resistive transition of our granular NbN sample. Insert illustrates sample geometry.

Perpendicular magnetic fields were applied with a superconducting magnet in order to provide a stable, noise free environment. The critical magnetic field H_{c2} was determined as a function of temperature by noting the onset of resistance ($R_{onset} \approx 0.1$ ohm) at low measuring currents ($I_m = 1\mu$amp). Complete $V_T(I,H)$ and $R_H(I,T)$ behavior was mapped for the sample over the field range from 0 to 10 Tesla and temperature range from 2 to 12 Kelvin.

3. EXPERIMENTAL RESULTS

The superconducting transition temperature of the Josephson coupled granular film $T_{cj}(H)$ was determined as shown in figure 2 by monitoring the low current resistance as a function of temperature. The resistance obeyed a power law expression given by

$$R_{I\to 0} = b(H)[d(T-T_{cj}(H))]^{\mu} \qquad (1)$$

over a temperature interval $T-T_{cj} < 1.5K$. The constants and exponents in this equation were determined by fitting to the data. Both d(=14.2) and μ (=3.75) are independent of magnetic field, b(H) (≈ 0.0005) is weakly dependent on field, and $T_{cj}(H)$ is given by

$$T_{cj}(0) - T_{cj}(H) = 1.5H^{0.75} \qquad (2)$$

Equation 2 can be expressed in terms of a critical magnetic field to give the relation (see figure 3)

$$H_{c2} = 0.67\ (T-T_{cj}(0))^{1.33} \qquad (3)$$

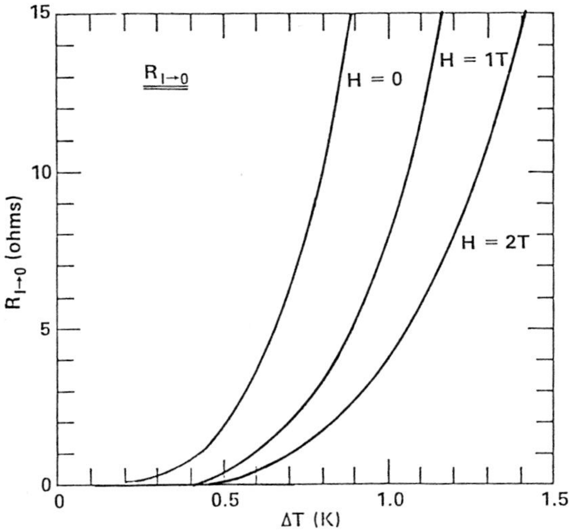

Figure 2 Resistance vs $(T-T_{cj})$ of our sample with 1 amp measuring current at zero, one and two Tesla.

Hence, in the region around T_{cj}, we observe a power law dependence for the critical magnetic field. Power law dependences in $R_{I\to 0}$, I_c, and $V_{T_{cj}}$ (I), which have been reported earlier [1], were all preserved in a field giving the following set of relations:

$$R_{I\to 0} \propto (T-T_{cj}(H))^{\mu} \qquad (4a)$$

$$V_{T=T_{cj}} \propto I^{x} \qquad (4b)$$

$$I_c \propto (T_{cj}(H)-T)^{1/\lambda} \qquad (4c)$$

$$H_{c2} \propto (T-T_{cj}(0))^{p} \qquad (4d)$$

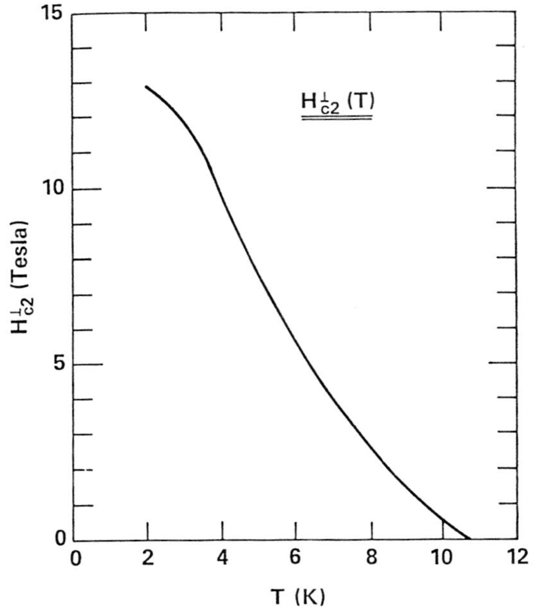

Figure 3 Perpendicular critical magnetic field H_{c2} vs temperature for our granular NbN sample.

where $\mu = 3.75$, $x = 2.8$, $1/\lambda = 2.0$, $p = 1.33$.

These relations can be folded into one universal expression analogous to that previously demonstrated in the zero field case [1].

$$V = b(H)I^{x}\chi\left(\frac{d[T-T_{cj}(H)]}{I^{\lambda}}\right), \qquad (5)$$

where the exponents and the constant d are independent of H. $\chi(0)) = 1$, $\chi(-1) = 0$, and $x - \lambda\mu = 1$. These limiting conditions reproduce equations 4a, 4b, and 4c.

At temperatures below another characteristic temperature T*, the temperature dependence of H_{c2} changes (see figure 4) as does the form of the I-V response curve (see figure 5). For T<T*, there is a discontinuous voltage jump at I_c followed by other current induced discontinuities separating regions of constant differential resistance which increase in value arithmetically after each voltage discontinuity. Increasing the magnetic field rapidly lowers T* until at 1.5 Tesla, a small resistance is measured prior to the discontinuous jump and at 3.0 Tesla the discontinuous features of the curve and the constant voltage regions have disappeared. This change in character of the I-V response with

magnetic field is completely analogous to results obtained by decreasing the coupling interaction between grains [1].

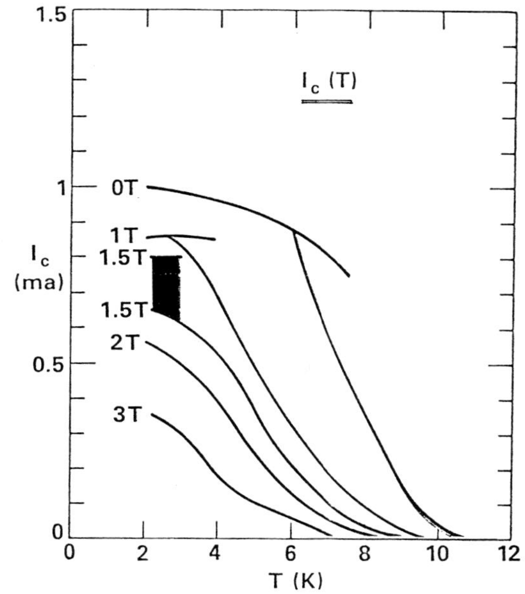

Figure 4 Critical current vs temperature at several magnetic fields.

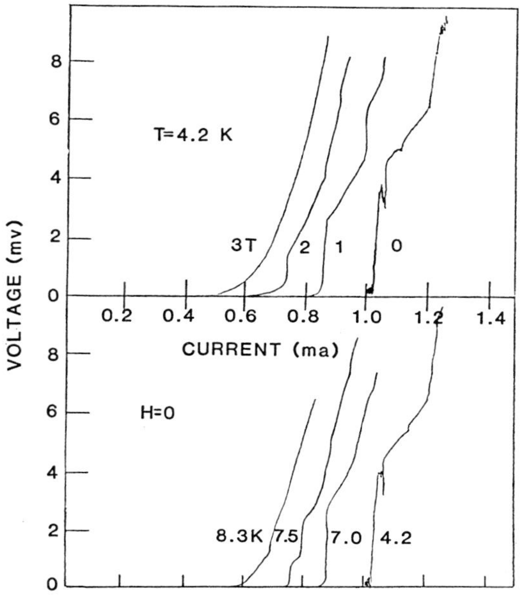

Figure 5 Voltage vs current for our NbN sample. The upper set of curves are for increasing magnetic fields at 4.2K. The lower set of curves are for increasing temperature at zero field.

4. DISCUSSION

A two dimensional topological phase transition in a superconducting film has been predicted due to the long range attractive interaction between magnetic vortices [4]. Unfortunately, neither the temperature dependence of the resistive transition as one approaches T_{cj} from above, nor the current dependence of the voltage near T_{cj} are in agreement with predictions of the model [5]. Likewise, the critical magnetic field data below T_{cj} is inconsistent with theories of 2 dimensional vortex melting [2,3]. No linear vortex flux flow resistance is observed even in high magnetic fields. The only role that magnetic field plays is one of gradually reducing the Josephson coupling strength between grains. This is demonstrated by the fact that the universal scaling function χ depends on H only through the reduction of T_{cj}. Reducing the intergranular coupling strength leaves the scaling function unaltered but reduces T_{cj} [1]. Another feature of reduced intergranular coupling is that the crossover temperature T* is rapidly depressed and eventually goes to zero for finite T_{cj}. This behavior, as earlier noted, is mimicked by increasing the magnetic field strength. We see no evidence of vortex motion or vortex interactions in our films, only a gradual reduction of intergranular coupling strength.

Similiar evidence of the lack of vortex behavior in granular NbN is evidenced in the behavior of wide granular weak link junctions [6]. The experimental high frequency (70 GHz) response of these weak link devices show well formed oscillating steps which are at variance with the vortex flow model but are in remarkedly good agreement with predictions of the resistively shunted junction (RSJ) model, with thermal fluctuations included [6,7]. This behavior would suggest that phase slips rather than well defined vortices are involved in describing the systematics of the bridges even though they have physically large dimensions.

Recently a model for describing the behavior of 2D granular films in terms of a 1D phase slip event instead of 2D vortices has been proposed [8]. It is argued that uniform superfluid flow along a certain direction defines a 1D variation of the superfluid phase. Magnetic flux then enters and leaves the 2D sample not by vortex flow, but rather by the phase slip process whereby the superconducting order parameter is suppressed to zero uniformly in the direction perpendicular to the current flow, permitting the passage of flux. The time scale for such a phase slip event is considerably shorter than that of vortex motion and would account for the behavior of the granular weak link devices in the RSJ framework. The crossover temperature T* and the I-V behavior above and below T* are similar to that observed in 1 dimensional microbridges where the step structure below T* is associated with a localized phase slip center [9].

The temperature dependence of the critical current below T* is characteristic of a Josephson junction which is not expected if the critical current were due to vortex depinning or depairing. Rather, it appears that I_c is an intrinsic property of the granular film. The dynamics of the current induced dissipation in a magnetic field are also dominated by intrinsic properties of the inhomogeneous Josephson coupled array rather than by depinning and flux flow.

We propose that the magnetic flux distribution in the 2D granular NbN films is not that of well defined circular vortices, but rather that the flux is dispersed in an irregular pattern confined to the region between grains. This flux pattern is strongly pinned due to the size and distribution of the superconducting NbN grains. In fact, the flux pinning appears to be so strong that current induced depairing of the superconducting order parameter occurs before flux flow. Present theories of the 2D superconducting state do not treat this strong pinning limit. We believe that phase slip processes occurring in the oxide region between grains must play the dominant role in determining the properties of 2D NbN granular films.

REFERENCES

1. D.U. Gubser, S.A. Wolf, T.L. Francavilla, and J.L. Feldman, Inhomogeneous Superconductors-1979, AIP Conf. Proc. No. 58 Edited by D.U. Gubser, T.L. Francavilla, J.R. Leibowitz, and S.A. Wolf, pg 159 (1980).
2. B.A. Huberman and S. Doniach, Phys. Rev. Lett. 43, 950 (1979).
3. D.S. Fisher, preprint.
4. M. Beasley, J. Mooij, and T. Orlando, Phys. Rev. Lett. 42, 1165 (1979).
5. B.I. Halperin and D.R. Nelson, J. Low Temp. Phys. 36, 599 (1979).
6. J. H. Claassen, E.J. Cukauskas, and M. Nisenoff, [see ref. 1] pg. 169.
7. J.H. Claassen and P.L. Richards, J. Appl. Phys. 49, 4117 (1978).
8. Z. Ovadyahu, preprint.
9. W.J. Skocpol, M. R. Beasley, and M. Tinkham, J. Low Temp. Phys. 16, 145 (1974).

Published 1980 by Elsevier North Holland, Inc.
Sinha, ed. Ordering in Two Dimensions

MICROWAVE CHARACTERIZATION OF GRANULAR AL THIN FILM SUPERCONDUCTORS

K. A. MÜLLER* and M. POMERANTZ

IBM Thomas J. Watson Research Center, Yorktown Heights, N.Y. 10598

We have made simultaneous measurements of microwave properties and the d.c. resistance, R_{dc}, of granular Al superconducting films, by employing wired samples inside a 9.4 GHz microwave cavity. Superconductivity of the sample affects both the losses in the cavity and its frequency. We find that R_{dc} and the microwave loss and frequency have various temperature dependences, proving that these observables are affected differently by properties of the films. The samples span the range expected for three-dimensional to two-dimensional superconductivity. Our results support the idea that a kind of percolation is responsible for the long-range superconductivity in these granular films.

INTRODUCTION

There has been considerable study of the superconducting transition in films thin enough to be considered two dimensional (2d) after it was pointed out [1] that a transition from a bound-vortex to a vortex-plasma [2] might be the commencement of the resistive state. The temperature at which this Kosterlitz-Thouless (K.T.) transition should occur, T_{KT}, compared to the B.C.S., mean field, T_{BCS}, was shown to depend only on the normal state residual resistance per square, $R_{\square}$:

$$T_{KT}/T_{BCS} = [1 + (R_{\square}/23.8K\Omega)]^{-1}. \quad (1)$$

To observe the transition it is desirable to have films whose $R_{\square}$ approach $23.8K\Omega/\square$. A difficulty results here because such films tend to be very thin and hence granular. The question arises whether the K.T. transition, which is conceived as a transition in a uniform superconductor, can occur in materials that are structurally non-uniform. One experimental approach is to try to eliminate inhomogeneity [3]. Another is to acknowledge the presence of inhomogeneity and seek to recognize its effects [4]. Theoretically, one can treat the case of weak and microscopic inhomogeneity. The superconducting correlation lengths become so large that they average out the inhomogeneity [5]. If, conversely, the macroscopic inhomogeneity is large enough a "percolation"-type model may be appropriate [6].

To help characterize granular films we have studied their microwave properties simultaneously with the d.c. resistive transition. The microwave skin depth at our frequency (9.4 GHz) is $\approx 30\mu m$ – greater than the film thickness for these thin and resistive films. (The superconducting penetration depth is also greater than the thickness [7] especially near T_{BCS} where it diverges.) The microwaves therefore completely penetrate the film. Microwaves thus give information about the state of the material even below the temperature at which the d.c. resistance has vanished. In this limit of complete penetration the superconducting electrons present an inductive impedance, L_s, to the microwaves [8], sometimes called the "kinetic inductance" [9], where $L_s^{-1} = n_s e^2/m$. This is a direct measure of the density of Cooper pairs, n_s. In addition, the microwaves are attenuated by normal electrons, either in the metallic grains or in the intervening barriers.

EXPERIMENTS AND DISCUSSION

To observe these effects we placed granular Al films in a microwave cavity [10]. (Along the axis of a 9.4 GHz TE_{01} cylindrical cavity the microwave E was $\perp$ and H was $\|$ to the film surface.) Then, simultaneously, we recorded: the inductance change, i.e. the frequency shift of the cavity, as the superconductivity was quenched by a large external magnetic field; the microwave loss change, which changed the Q and reflection of the loaded cavity; the d.c. resistance as measured by an electrometer connected to the film by pressure contacts.

Samples of thickness about 100 Å were prepared by flash evaporation onto quartz at room temperature in an oxygen atmosphere. Thicker films, 10,000 Å, were e-beam evaporated more slowly [7]. The observations on a 100 Å, hence 2d film, $R_{\square} \approx 280\ \Omega$, are shown in Fig. 1. R_{dc} drops to zero with a temperature dependence that suggests fluctuations at temperatures above the steep part of the transition, as has been reported [11]. The frequency shift measurements verify this. It is

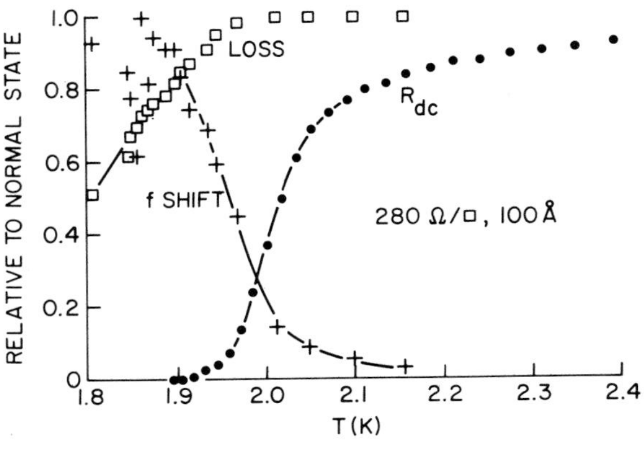

Fig. 1 R_{dc}, microwave loss and cavity frequency shift for a granular Al film of thickness $\approx$ 100 Å, $R_{\square} \approx$ 280 $\Omega/\square$ ($\rho \approx 280\mu\Omega-$cm), as a function of temperature.

proportional to the superconducting pair density n_s, and $n_s \propto <|\psi|^2> = <|\psi_o + \delta\psi|^2>$, where ψ_o is the average Ginzburg-Landau order parameter [12]. Above T_{BCS}, although $\psi_o = 0$, one can nevertheless observe $<|\delta\psi|^2>$, the mean squared fluctuation of the pair density. This gives the tail of the f shift pointing to higher temperatures. As the temperature is lowered the shift shows an approximately linear increase; this conforms to the mean field (B.C.S. or G.-L.) prediction $n_s \propto |\psi_o|^2 \propto (T_{BCS}-T)$. Extrapolating the linear dependence of the shift to zero gives an experimental definition of T_{BCS}. (T_{BCS} = 2.04 K in this case). Note that R_{dc} has fallen about 50% from its normal state value at T_{BCS}, and the microwave loss has not perceptibly changed. Thus, below T_{BCS} there is a temperature range in which there is a finite average pair density, but there are also both d.c. and microwave resistivities. This suggests that the superconductivity is occuring non-uniformly.

This interpretation is supported by comparison with the microwave losses in *bulk* Al where measurements on both clean and dirty samples [13] closely agree with B.C.S. - based theories [14]. For our measuring frequency, $h\nu/kT_{BCS} \approx 0.2$, in bulk Al there would be a drop in microwave resistance within 0.01 K. The observed loss falls in ~ 0.4K, which is grossly different. Loss mechanisms such as a temperature dependent hopping rate would not be switched by applied magnetic fields as the observed ones are. Thermally excited vortices, as postulated for a K.T. ground state, will be unbound and produce loss *above* T_{KT}. For this sample, from Eq. 1 with $R_\square = 280\Omega/\square$, $T_{KT} \approx 2.01$ K. Thus the microwave losses, which persist at least to 1.5 K, cannot be caused by free vortices. A theory for microwave loss by bound vortex pairs and which includes the likely pinning effects is in a preliminary stage [15].

One can check experimentally whether the K.T. state is solely responsible for the observed loss by comparing with a granular film in which the K.T. state should not occur, namely a three-dimensional film. For a film of 10,000 Å with $R_\square \approx 30\ \Omega/\square$, which should be 3d, we observed the results shown in Fig. 2. We found T_{BCS} = 2.15 K and the microwave loss persisted down to at least 1.7 K. This film, then, in which no K.T. phase is expected, but which is granular, shows microwave loss at temperatures well below T_{BCS}. Thus the granularity alone can cause microwave loss. Whether the K.T. bound-vortex state also produces some loss is an open question which could be answered by experiments on non-granular but highly resistive films.

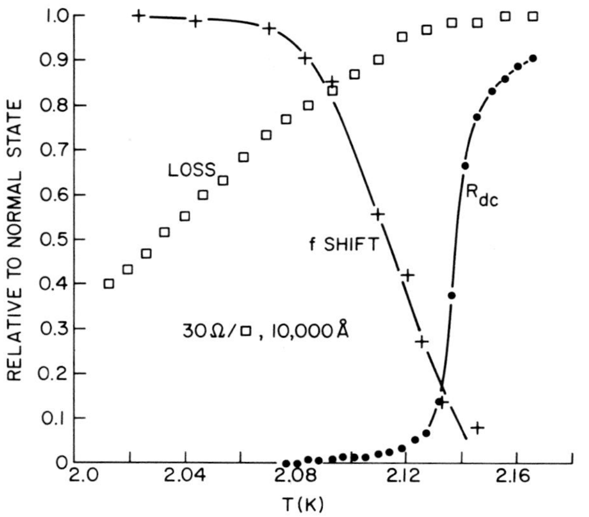

Fig. 2 Same as Fig. 1, except film thickness ≈ 10,000 Å, $R_\square \approx 30\ \Omega/\square$, ($\rho \approx 3000\ \mu\Omega$ cm.).

Our explanation of the large microwave loss in the normal state of these films, and its gradual disappearance in the superconducting state, is that the loss is due to the poorly conducting oxide barriers between the metallic grains. Near T_{BCS} the individual metallic grains become superconducting. As the temperature is further decreased the superconducting phases of adjacent grains lock together. The criterion for this phase-locking is that the Josephson energy be greater than the thermal energy [12] i.e.,

$$(\pi/4)(\hbar/Re^2)\Delta(T)\tanh[\Delta(T)/2kT]>kT, \tag{2}$$

where R is the normal state resistance per unit area of the junction. At low enough T a sufficient number of junctions become superconducting so a continuous path across the sample is formed. This gives the d.c. resistive transition, but according to our microwave absorption there remain non-superconducting (lossy) junctions until much lower temperatures. This "percolation"-type model was advanced by Deutscher et al [6] to explain smeared and lower-T-shifted specific heat, C_p, peaks measured [16] in films like that shown in Fig. 2, but thicker in order to make C_p detectable. Our microwave measurements are consistent with the specific heat data, such that the microwave loss disappears at about the same temperature that the specific heat shows a peak. According to the model this is the temperature at which most of the junctions are superconducting; no loss occurs and there is a C_p anomaly because the phase coherence has become macroscopic. If there are voltages across the junctions there will be quasiparticle tunneling in parallel with the pair tunneling [17]. The quasiparticle tunneling produces loss even if the film is uniformly superconducting. The loss also persists below T_{BCS}. It appears, however, that a uniformly superconducting model [5] does not explain the C_p data in detail.

A consequence of the "percolation" model, which hypothesizes the spreading of superconductivity by the gradual phase locking of grains, is that the d.c. resistive transition should depend on the dimensionality of the structure. It is known, e.g., for a simple cubic (3-d) lattice the conductivity threshold is a fraction = 0.25 of conducting elements and the rest open circuits, whereas for a 2-d square lattice the threshold fraction is 0.5 [18]. We can estimate the relative number of superconducting junctions in our 3-d and 2-d samples by using the data of Figs. 1 and 2, and Eq. (2). We solve Eq. (2) for R(T) which says that any junction which has R < R(T) will be superconducting. The question is: what fraction of junctions have R < R(T) at any given T? This will depend on the distribution of R. From an extension of the data of Figs. 1 and 2, we observed that the microwave loss reached a minimum at T ≈ 1.5 K for the 100 Å film, and at T ≈ 1.7 K for the 10,000 Å film. Assuming these are Ts at which the largest R became locked, and with our measured T_{BCS} for these samples, we used Eq. 2 and B.C.S. theory for $\Delta(T)$ [12] to estimate the *maximum* R, R_m, in each sample. We find $R_m/R_c \approx 1.7$ for the 2-d, and $R_m/R_c = 1.5$ for the 3-d sample where $R_c = \pi\hbar/4e^2$. Thus the maximum junction resistances are rather similar, which is not surprising considering that the T_c's differ by only 5% and electron diffraction also shows the materials to be similar [10]. Because of this similarity we can choose a simple distribution with some confidence that the answer will not depend much on the exact form. Thus we assume a triangular distribution: R varies from zero to R_m and the number of junctions with resistance R increases linearly with R from zero to $R_m/2$, and then decreases linearly to zero at R_m . Then the fraction of junctions with resistance $< R(T)$ is proportional to $(R(T)/R_m)^2$. We can calculate the fraction of superconducting junctions when R_{dc} has decreased to, say, 20% of its normal state value because we know the T from the data and hence R(T) from Eq. 2 at which this occured for both the 2-d and 3-d samples. The result is that the fraction of these junctions is about 10 times larger in the 2-d sample than in the 3-d one. This is in qualitative agreement with the expectation that more super-junctions will be required to produce a given reduction in sample resistance for a 2-d than for a 3-d lattice. We remark that the uniform state model [5] predicts a dependence on the dimensionality, via the number of nearest neighbors, which also is compatible with the above result.

SUMMARY AND ACKNOWLEDGMENTS

By a combination of d.c. resistance and microwave measurements we have observed in a variety of granular Al films (a) measureable density of superconducting electrons above the temperature at which the d.c. resistance vanishes and (b) a microwave loss which persists to temperatures well below the resistive transition. A satisfactory explanation is found in the "percolation" model, in which the phases of ψ in superconducting grains gradually lock together via Josephson tunneling. It is found that the number of junctions needed to reduce the resistance varies with lattice dimensionality, roughly as expected on this model.

We have profited from discussions with Drs. D. S. Fisher, S. Fishman, G. Grinstein, P. Horn, R. Laibowitz, A. Malozemoff and R. Voss. The samples were kindly provided by D. Abraham and Dr. C. M. Knoedler.

REFERENCES

*Permanent Address: IBM Zurich Research Laboratory, 8803 Rüschlikon, Switzerland.

[1] M. R. Beasley, J. E. Mooij and T. P. Orlando, Phys. Rev. Lett. *42*, 1165 (1979).
[2] A.I.P. Conf. Proc. 58, *Inhomogeneous Superconductors-1979* has several pertinent papers. See the review by B. Huberman and S. Doniach, p. 87.
[3] R. E. Glover, III and M. K. Chien, ibid, p. 101.
[4] A. F. Hebard and A. T. Fiory, Phys. Rev. Lett *44*, 291, 620 (E) (1980).
[5] B. R. Patton, W. Lamb and D. Stroud, in ref. 2, p. 13.
[6] G. Deutscher, O. Entin-Wohlman, M. Rappaport and Y. Shapira, ibid, p. 23; and to be published.
[7] D. Abraham, G. Deutscher, R. Rosenbaum and S. Wolf, J. de Phys. Supp. *C6*, 586 (1978).
[8] A. B. Pippard, *Adv. in Electronics and El. Phys. 6*, 1 (1954).
[9] W. A. Little, *Symp. on Physics of Superconducting Devices*, Charlottesville, Va. (1967).
[10] K. A. Müller, M. Pomerantz, C. Knoedler and D. Abraham (to be published) has more details.
[11] A. F. Mayadas and R. B. Laibowitz, Phys. Rev. Lett. *28*, 156 (1972).
[12] M. Tinkham, *Introduction to Superconductivity*, McGraw-Hill, 1975.
[13] J. R. Waldram, Adv. Phys. *13*, 1 (1964) and refs. therein.
[14] D. C. Mattis and J. Bardeen, Phys. Rev. *111*, 412 (1958). P. B. Miller, Phys. Rev. *118*, 928 (1960).
[15] D. A. Browne and S. Doniach, in ref. 2, p. 304.
[16] T. Worthington, P. Lindenfeld and G. Deutscher, Phys. Rev. Lett. *41*, 316 (1978).
[17] D. E. McCumber, J. Appl. Phys. *39*, 2503 (1968).
[18] J. P. Straley, A.I.P. Conf. Proc. No. 40, Ed. Garland and Tanner, p. 118 (1977).

Sinha, ed. Ordering in Two Dimensions

RESISTIVE TRANSITION IN BISMUTH THIN FILMS*

D. VAN VECHTEN† AND V. KORENMAN

University of Maryland, College Park, Maryland 20742

The linear time dependent Ginzburg-Landau equation accurately describes the resistive transition in amorphous thin films of Bi, Ga, and Pb, down to about 80% of the normal state resistance Ro. By including the lowest order effect of the non-linearity we have extended the region of good fit down to about 5% of Ro. The fit parameters vary properly from film to film as a function of Ro. The width of the transition region, as defined by our fits, agrees with the prediction of the Kosterlitz-Thouless theory. Our comparisons are with high precision measurements by Chien and Glover on films of amorphous Bi.

INTRODUCTION

Glover and his coworkers[1,2] have shown that the fluctuation contribution to the conductance of thin films of amorphous Bi, Ga, and Pb is given by the Aslamazov-Larkin[3] expression

$$g_f \equiv \sigma_f d = g_o/\varepsilon \; . \qquad (1)$$

Here g_f is the conductance per square, σ_f the fluctuation conductivity per square, d the film thickness, and ε the reduced temperature ($\varepsilon \equiv (T-Tc)/Tc$). The conductance is related to the measured resistance per square by $g_f = (1/R) - (1/Ro)$ where Ro is the normal state value. Finally g_o is the material independent quantity:

$$g_o = e^2/16\hbar = 1.52 \times 10^{-5}\ \Omega^{-1} \; . \qquad (2)$$

The data fits Eq. (1) typically down to $R/Ro \gtrsim 0.8$, below which non-linear fluctuation effects are expected to dominate. The measured coefficients are within 17% of Eq. (2) for all the 100 or so films studied.

Although originally derived from the microscopic theory, Eq. (1) also follows from the phenomenological time dependent Ginzburg-Landau Equation (TDGLE) in the linear limit (the value of g_o coming from microscopic theory, however). Now the microscopic analysis is incomplete. The correct expression includes the "Maki term" which is in fact required to explain experimental results in granular aluminum and other materials [4]. Nevertheless, Glover's films do follow the prediction of the TDGLE. In this note we show that the range of this agreement is considerably larger than previously believed. Including the first order nonlinear correction to the prediction of the phenomenological theory, we find agreement to values of R/Ro as small as 0.05. As compared with the previous fit to 80% of Ro, this means fitting to a conductance 75 times as great, or to an effective reduced temperature 75 times smaller. We have no new suggestion as to why these particular materials provide such an excellent physical realization of the phenomenological theory. However they are an obvious candidate for comparison with recent theories of two dimensional phase transitions, which are based on Ginzburg-Landau theory.

CONDUCTANCE IN THE TDGLE

The TDGLE can be written

$$\hbar\gamma\partial\psi/\partial t = -\delta F/\delta\psi^* \qquad (3)$$

with ψ the order parameter and F the familiar free energy density

$$F = \hat{a}|\psi|^2 + (\hbar^2/2m)\ |\nabla\psi|^2 + \frac{1}{2}b|\psi|^4 \; . \qquad (4)$$

Here m and b are constant while, with constant a'

$$\hat{a} = a'(T-Tco)/Tco \equiv a'\hat{\varepsilon} \; . \qquad (5)$$

$$\gamma = \pi a'/8kT \; . \qquad (6)$$

Equation (6) comes from the microscopic theory.

There are several ways to find the conductivity. In the linear limit Eqs. (3) and (4) give the order parameter response function Go.

$$Go(k,\omega) = -i\hbar/[\hbar\omega\gamma + i(\hat{a}+\hbar^2k^2/2m)] \; .$$

The classical fluctuation-dissipation theorem can be used to derive the corresponding correlation function, and the conductivity then found by using a Kubo formula, as in Tinkham's book [5]. The result for a film is

$$g_f = 8kT\gamma g_o/\pi\hat{a} \; . \qquad (7)$$

Equation (1) then follows from Eqs. (5), (6), and (7).

With the inclusion of the non-linear terms the inverse response function G is given by

$G^{-1}(k,\omega) = G_0^{-1}(k,\omega) + \Sigma(k,\omega)/\hbar$, where Σ has a diagrammatic expansion of the usual type in powers of b. Here we discuss only the first term, and will describe the inclusion of higher terms elsewhere. Then Σ is the constant 2bn with n the mean squared order parameter. The only effect is to replace $\hat{a}$ everywhere by $a = a'\hat{\varepsilon} + 2bn$. The squared order parameter is

$$n = kT\ \Sigma_q (a+\hbar^2k^2/2m)^{-1}\ .$$

Performing the sum with an isotropic cutoff K, in the thin film limit, we find

$$n = (mkT/2\pi\hbar^2 d)(Kd+\ell n(\hbar K/4mad))\ . \qquad (8)$$

The renormalized conductance expression $g_f = g_o a'/a$ is then combined with the definition of a and with Eq. (8) to give a relation between conductance and temperature

$$(g_o/g_f) + \chi\ \ \ell n(g_o/g_f) = (1+\chi\ \delta)(T/Tco)-1\ . \quad (9)$$

Here δ is a cutoff dependent constant and

$$\chi\ \equiv bmkTco/\pi\hbar^2 a'd\ . \qquad (10)$$

We are ignoring the demonstrably irrelevant temperature dependence of the coefficient of the logarithm. It is essentially Eq. (9) which we have compared with experiment.

One more factor must be considered. Eq. (9) suggests a thickness dependence of the effective Tc, defined as the temperature where the right-hand side of Eq. (9) vanishes. Such a thickness dependence is well known[1,6], but is much stronger than can be accounted for in this way. One solution[6] is to accept the dynamics outlined above, but to modify the boundary condition on the order parameter at the film surface. Fixing the logarithmic derivative of ψ at the surface corresponds to adding a term $(\mu-\nu T)/d$ to the right-hand side of Eq. (9). Accordingly we write Eq. (9) as

$$(g_o/g_f) + \chi\ \ \ell n(g_o/g_f) = \alpha T + \beta \qquad (9')$$

with α and β unconstrained in fitting to the data. There is an implied thickness dependence of α, β, and χ, to which we shall return in the next section.

Finally we remark that this first order analysis is not new. Masker et al.[7] attempted a similar fit to granular aluminum films. Since even the linear result Eq. (1) fails completely for this material, however, no satisfactory fit was found.

DATA AND FIT

Our comparison is with experiments by Chien and Glover[8] who did high precision measurements on 39 amorphous Bi films, with particular emphasis on the low resistance range. They fit all the films to a phenomenological expression whose parameters vary systematically as a function of resistance. Of the 39 films, we recovered the raw data for 12, including films deposited on several different substrates, and with sheet resistances between 8 and 93 Ω. The transition width for these films is only a few millikelvin. An idea of the quality of the measurements is given in Fig. (1), where the symbol width is about equal to the combined R and T error bar in this resistance range.

A typical data set consists of 50 to 60 resistance-temperature pairs, and associated error bars, starting about 2 kelvin above Tc but concentrated in the region of rapidly changing R. Comparison with Eq. (9') is particularly simple. For a given value of Ro, $g_f(T)$ is read off from the data and χ, α, and β determined by a linear least squares fit. Ro is then varied to optimize the fit.

For the films on quartz or roughened glass substrates we fit all but at most one data point to within the stated error bars: for seven of these ten films we fit every point. The lowest resistance data point was generally between 3 and 10% of Ro. The remaining two films, on smooth glass, each had points at smaller R than Eq. (9') could fit, below 5% and 9% of Ro respectively. Figure (1) shows the general quality of fit for one of these two latter films. The actual range of fit extends to the right a distance 100 times the width of the region shown. Figure (2) shows the extreme low resistance range for this same film. Shown with the present fit are the empirical fit of Chien and Glover[8] (middle curve) and the best fit to Eq. (1) (lower curve). The symbol widths in both figures are about equal to the error bars.

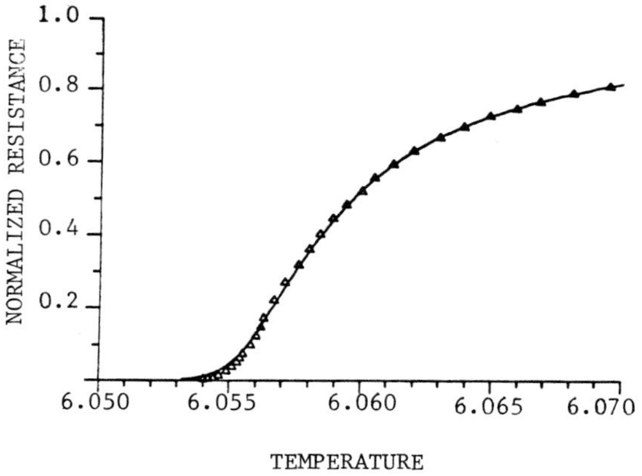

Fig. 1 Experimental points and fit to Eq.(9'). The film is 500 A thick and has resistance per square Ro=30.22Ω.

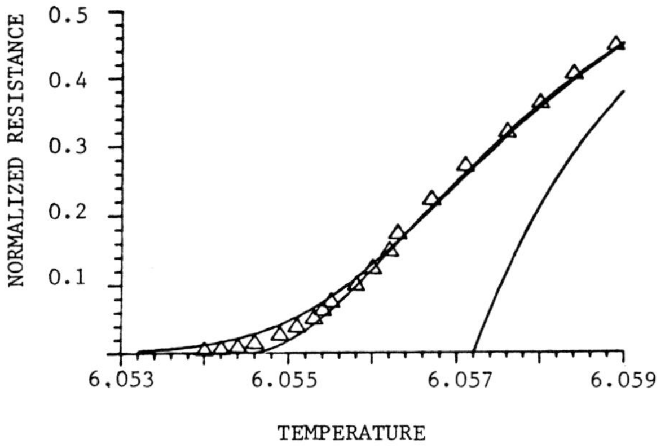

Fig. 2 Low resistance region of the film in Fig. 1. The middle curve is the empirical fit of Chien and Glover. The lower curve is the best fit to Eq.(1).

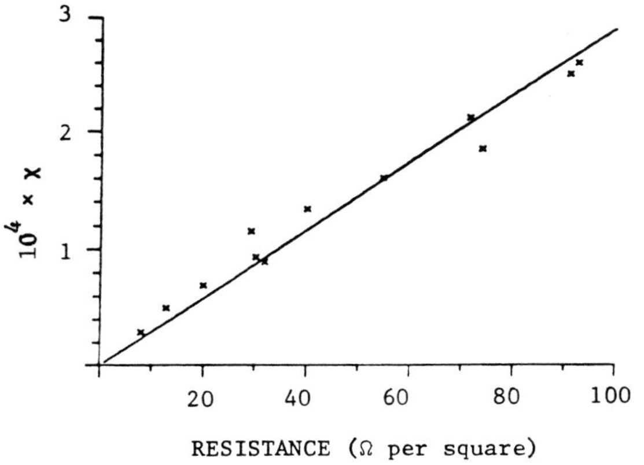

Fig. 3 Fitting parameter χ and least squares fit.

Of the four fitting parameters, Ro is taken as a measure of the film thickness. This assumes a normal state resistivity which does not vary from film to film. This uniformity has been verified to within the accuracy of direct thickness determinations.

Because the resistive transition occurs over such a small temperature region, the fit is not extremely sensitive to the individual values of α and β, and the determination of these is somewhat uncertain. Nevertheless the trend from film to film does bear out the thickness dependence implied in the discussion before Eq. (9'). The length μ characterizing the boundary condition is about 18 Å. The second length ν is consistent with the value $\mu/2$ which is the theoretical prediction. In effect the variation of g_o from film to film follows from the same mechanism which causes Tc to vary.

The ratio $-\beta/\alpha \equiv$ Tc is much more accurately determined, as is the value of χ which sets the scale of the nonlinear term. We return to the values of Tc in the next section.

From Eq. (10) we see that χ should be proportional to 1/d or Ro. Figure (3) shows the values of χ for the 12 films considered. The line is a least squares fit constrained to pass through the origin. It agrees with the data points to within the uncertainty of the latter. It is this dependence of χ on Ro, along with the quality of the individual fits, which convinces us of the essential correctness of the TDGLE for these films.

The line in Fig. (3) has slope $2.9 \times 10^{-6}\ \Omega^{-1}$, corresponding to $\chi = .045/d(\text{Å})$. Now Eq. (10) can be rewritten as $\chi = (16e^2/\hbar^2c^2)\ kT\lambda^2\ \varepsilon/d$ where λ is the temperature dependent penetration depth defined in the usual way in terms of GL free energy parameters. Then our fit predicts a value for λ, $\varepsilon\lambda^2 = (1200\ \text{Å})^2$ which can, in principle, be verified independently.

It is remarkable that the weak coupling theory of superconductivity actually predicts a value for χ/Ro, independent of other material parameters, in the dirty limit. The prediction is that $kT\varepsilon\sigma\lambda^2 = 7\zeta(3)\ \hbar c^3/16\pi^4$. This gives $\chi/\text{Ro} = 7\zeta(3)\ e^2/\pi^4\hbar = 2.1 \times 10^{-5}\Omega^{-1}$, a factor 7 larger than observed. It may be, of course, that the weak coupling theory simply cannot be applied to Bismuth. However, the actual situation seems more complex, as is outlined below.

KOSTERLITZ-THOULESS TRANSITION

Glover and Chien[9] recently compared their fit to the transition in these 39 films to the prediction of Beasley, Mooij and Orlando,[10] based on the Kosterlitz-Thouless theory of phase transitions in two dimensions. In brief, the theory predicts a transition to almost zero resistance at a temperature lower than that of the BCS transition, where a substantial residual resistance should remain. In [10] the prediction is that

$$Tc/T_{KT} = (1 + 2\chi) \qquad (11)$$

<u>with χ the same combination of G-L parameters as appears in Eq. (10).</u>

The theory does not specify how the two transition temperatures are to be extracted from data. Glover and Chien took Tc to be the temperature at which the AL fit to the high temperature data crossed zero resistance, and T_{KT} the temperature at which their phenomenological fit crossed zero. These are the intersections of the two lower curves of Fig. (2) with the temperature axis. The data then showed a dependence of T_c/T_{KT} on Ro consistent with Eq. (11), but with χ a factor of three smaller than the prediction of the weak coupling theory. That is, the transition is narrower than weak coupling theory predicts, although approximately twice as wide as would be predicted from our measured χ. More interesting however is the result of choosing for Tc the value of T where the right-hand side of Eq. (9') vanishes. The value for the film in Figs. (1) and (2) is 6.062 K. The data for all 12 films, with T_{KT} as chosen by Chien and Glover, now agrees with Eq. (11) with the weak coupling value of χ. For eight of the films the ratio $(Tc/T_{KT}-1)/Ro$ is within 10% of this value. For the other four the width is from 20% to 40% too narrow. Note, however, from Fig. (2), that our definition of T_{KT} may systematically understate the actual transition width. Further, the films which agree less well are the thinnest ones, where the accuracy of measurement is worst.

DISCUSSION

We have shown that amorphous Bi thin films have a resistive transition well described by the TDGLE. In the region of onset, treating the nonlinearity to first order suffices, and a meaningful BCS transition temperature can be defined by a fit to the data. The width of the region between this transition temperature and the fall to zero resistance is correctly predicted by the Kosterlitz-Thouless theory. There is, however, a renormalization of the parameter χ which is not understood.

We thank Dr. Chien, and especially Dr. Glover for making their data available to us and for many helpful discussions. Computer facilities for this work were provided by the Computer Science Center of the University of Maryland.

REFERENCES

*Based in part on a thesis submitted by D. Van Vechten to the University of Maryland (May 1979).

†Now at: National Bureau of Standards, Washington, D.C. 20234

1. R. E. Glover, III, Physica 55, 3 (1971).
2. P. J. Silverman, Phys. Rev. B 16, 2066 (1977).
3. L. G. Azlamazov and A. I. Larkin, Phys. Lett. A 26, 238 (1968).
4. R. A. Craven, G. A. Thomas, and R. D. Parks, Phys. Rev. B 7, 157 (1973).
5. M. Tinkham, "Introduction to Superconductivity", (McGraw Hill, New York 1975), p. 250.
6. D. G. Naugle, R. E. Glover III and W. Moorman, Physica 55, 250 (1971).
7. W. E. Masker, S. Marcelja, and R. D. Parks, Phys. Rev. 188, 745 (1969).
8. M. K. Chien, Thesis, University of Maryland, College Park, 1972. M. K. Chien and R. E. Glover III, in "Low Temp. Phys. LT-13" (Plenum Press, New York, 1974) Vol. 3, p. 649.
9. R. E. Glover III and M. K. Chien, presented at the Inhomogeneous Superconductors Conference, Berkeley Springs, West Virginia, November, 1979. AIP Conf. Proc., (to be published).
10. M. R. Beasley, J. E. Mooij, and T. P. Orlando, Phys. Rev. Letters 42, 1165 (1979).

Published 1980 by Elsevier North Holland, Inc.
Sinha, ed. Ordering in Two Dimensions

VORTEX UNBINDING IN SUPERCONDUCTING THIN FILMS?*

P. A. BANCEL[+] and K. E. GRAY

Argonne National Laboratory, Argonne, IL 60439

ABSTRACT

The study of the origin of broad resistive transitions in 2D superconducting films requires extremely homogeneous samples. Granular aluminum evaporated onto freshly cleaved mica substrates are shown to satisfy this criterion. Values of T_c, as a function of resistance per square, are in agreement with a one parameter model of independent grains coupled by the Josephson effect, and not the predictions of the Kosterlitz-Thouless vortex unbinding model.

INTRODUCTION

It is well known that thin films of dirty superconductors exhibit gradual resistive transitions to zero. Several theories have been proposed to explain this[1-5]. Bound vortex pairs have been predicted to undergo a phase transition at the Kosterlitz-Thouless temperature T_{KT}. Thermally activated free vortices are predicted above T_{KT}, which reduce the long range order and provide dissipation when a current is applied.
An implicit assumption of these theoretical ideas and calculations is the uniformity of the "substrate," i.e., the structure of the metallic film which provides the conditions for superconductivity in the conduction electrons. Two parameters which figure prominently in the K-T description[2] of the transition are the normal state sheet resistance $R_{\square}^{N}$, and T_{co}, the BCS transition temperature. An experiment designed to investigate this theory requires a system with a well defined $R_{\square}^{N}$, i.e., a film with a uniform resistance over distances comparable with the fundamental length scales of the theoretical models. For the K-T model this is the correlation length ξ_+, which is typically of order one micron.

We have studied thin(100-200A) films of granular aluminum and NbN made under various conditions including different substrates. The experiments indicate that there can be significant variations of the above parameters on the scale of millimeters when using glass substrates. These variations seem to be attributable to imperfections of the insulating substrates, but are difficult to overcome by conventional treatments. However, freshly cleaved mica substrates do not exhibit these inhomogeneities, and these results will be discussed in relation to the theories.

GRANULAR ALUMINUM ON GLASS SUBSTRATES

Our studies of granular aluminum on glass substrates show severe inhomogeneities in all films investigated. The films were condensed onto clean substrates in oxygen atmospheres of about 0.03 mTorr. By adjusting evaporation parameters, films with $R_{\square}^{N}$ ranging from 10Ω/□ to 10kΩ/□ and higher were made and studied. Various patterns were obtained by photoetching and masking. By comparing the resistive transitions of different sections of such films, each containing many □'s, the effects of non-uniformities became evident. Two such transitions for a single film are shown in Fig. 1. A narrow transition is observed for a large square(1 cm) whose $R_{\square}^{N}$ is 1.97kΩ/□. After photoetching a meander path (80 microns

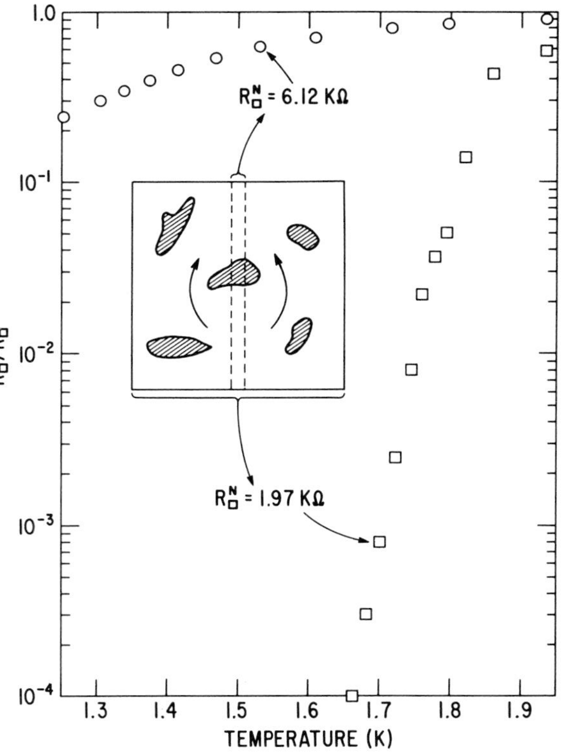

Fig. 1. The resistive superconducting transitions in a large square and a meander path made of the same granular Al sample on a glass substrate.

wide) from this material, $R_\square^N$ increased threefold, and the transition was broadened much more than expected for such an increase. This behavior seems to be due to bad spots which the current avoids in the large square.

A second set of data(Fig. 2) shows results for films of various widths evaporated simultaneously through masks onto the same glass substrate. There are large discrepancies in the resistive transition and $R_\square^N$, including anomalous behavior such as double transitions and very broad transitions. In both cases these were shown to be localized along the length of the strips. However, even the two widest films disagree by a factor of 3 in reduced resistance at the lowest temperatures. Results of lower $R_\square^N$ films($100\Omega/\square$) showed that these cleaner films could be made more uniform, as might be expected. Nevertheless, the resistances for different sections of a meander path varied by up to 10x at low temperatures. Several efforts were made to improve the quality of samples on glass substrates. Neither mechanical nor fire polishing the glass substrates substantially altered these results.

NbN ON SAPPHIRE SUBSTRATES

NbN films(50-200A) were made by reactive sputtering onto sapphire substrates. These films showed improved homogeneity, and different legs of a meander path for one sample matched well throughout its much broader transition. Reduced resistances of about 10^{-5} were within 40%, but the transitions showed fine structure which differed from leg to leg. In addition, in order to obtain transitions in the pumped helium temperature range, $R_\square^N$ must be very large and is therefore significantly activated(3x the room temperature value, compared to the 5-30% for the lower $R_\square^N$ granular Al). The behavior is reminiscent of 2D localization. This effect clouds the concept of $R_\square^N$, making the interpretation difficult.

GRANULAR ALUMINUM ON FRESHLY CLEAVED MICA

One of the best surfaces for cleanliness and uniformity on a atomic scale is freshly cleaved mica. The is borne out by the measured resistive transition shown in Fig. 3. The reduced resistances for 4 sections of a meander path(250 micron wide legs) track

	# $\square$'s	WIDTH (μm)	$R_\square^N$ (kΩ/$\square$)
A	265	10	7.5
B	51	80	1.33
C	18.5	220	2.4
D	3.8	900	2.5
E	1.35	3100	10.4
F	1.0	8200	12.2

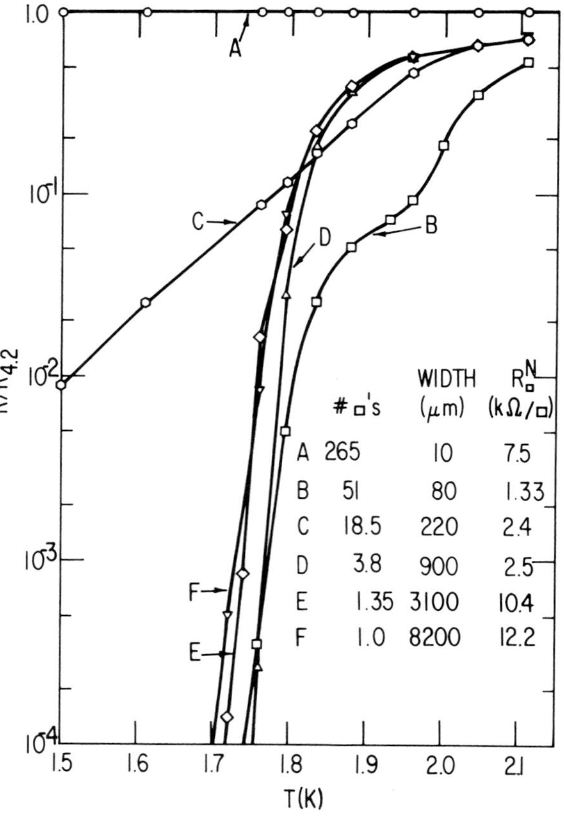

Fig. 2. The resistive transitions for various width strips of granular Al evaporated through a composite mask onto the same glass substrate.

$$\frac{R_\square}{R_\square^N} = C_l b / \sinh^2(b\,\tau_c/\tau)^{1/2}$$

$$\tau_c = 0.17\ R_\square^N/R_\square^c = 0.07$$

$$b = 14$$
$$C_l = 4$$
$$T_{KT} = 1.75\,K$$

$$R_\square^N = 1700\,\Omega$$

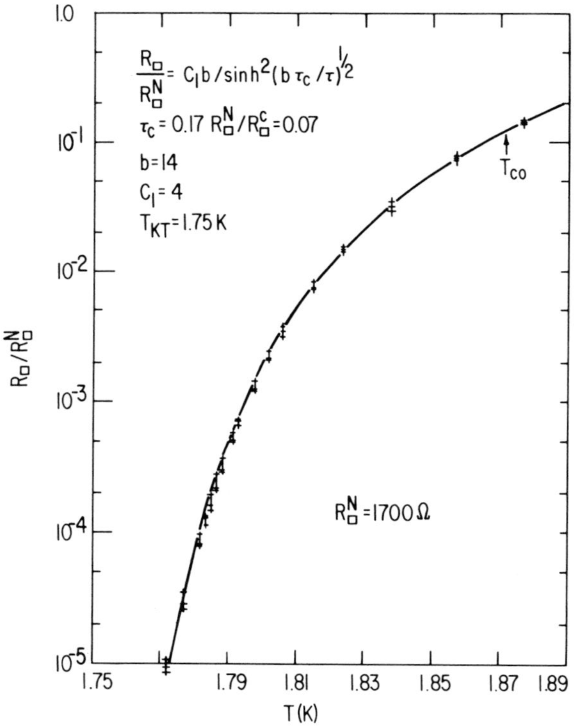

Fig. 3. The resistive transition for 4 legs of a meander path etched out of a granular Al sample deposited onto freshly cleaved mica. The solid line is the theoretical expression[2] with the parameters shown.

within 30% down to values less than 10^{-5}. Considerable care must be taken during evaporation, in addition to using mica, to achieve this degree of uniformity which is evidenced by the results of Fig. 3.

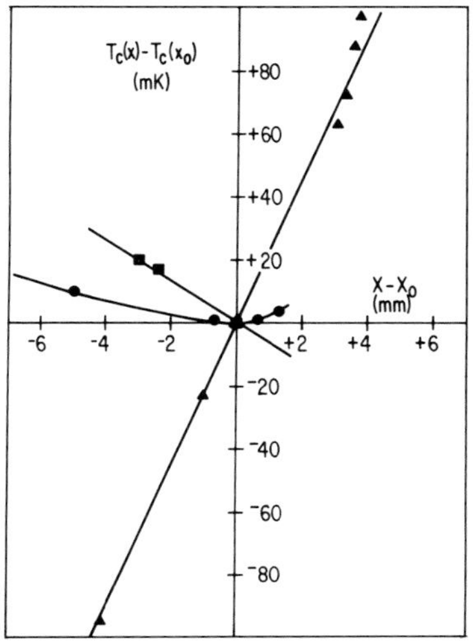

Fig. 4. Variation of T_c with distance from the center of the substrate x_0. Solid circles are for a correctly aligned sample.

Films of various $R_\square^N$ can be made conveniently by changing the source to substrate distance, due to the sensitivity to the relative rates at which Al and oxygen arrive at the substrate. A change of one cm changes $R_\square^N$ by 2x and T_c by 30-40%. However this also emphasizes the importance of careful alignment of the substrate perpendicular to the line of sight from the source. Furthermore, samples must be etched from the exact center of the substrate. These effects are illustrated in Fig. 4, and are crucial since the uniformity of the resistive transition is ultra sensitive to variations of T_c. Note that the variations of $R_\square^N$ resulting from poor alignment are generally too small to measure.

As final evidence for the greater uniformity of granular Al on freshly cleaved mica compared to glass substrates, Fig. 5 shows a compendium of measured T_c versus $R_\square^N$ from many authors. All the results on glass substrates, including our own, form a wide flat band. In contrast the mica results show a much sharper peak and higher T_c values, up to 2.45K. Clearly inhomogeneities will smear out this peak, resulting in a broad band such as is observed for glass substrates.

DISCUSSION

The following discussion will only refer to the very homogeneous samples made on freshly cleaved mica substrates. It is of interest to understand the variation of T_c with $R_\square^N$ shown in Fig. 5,

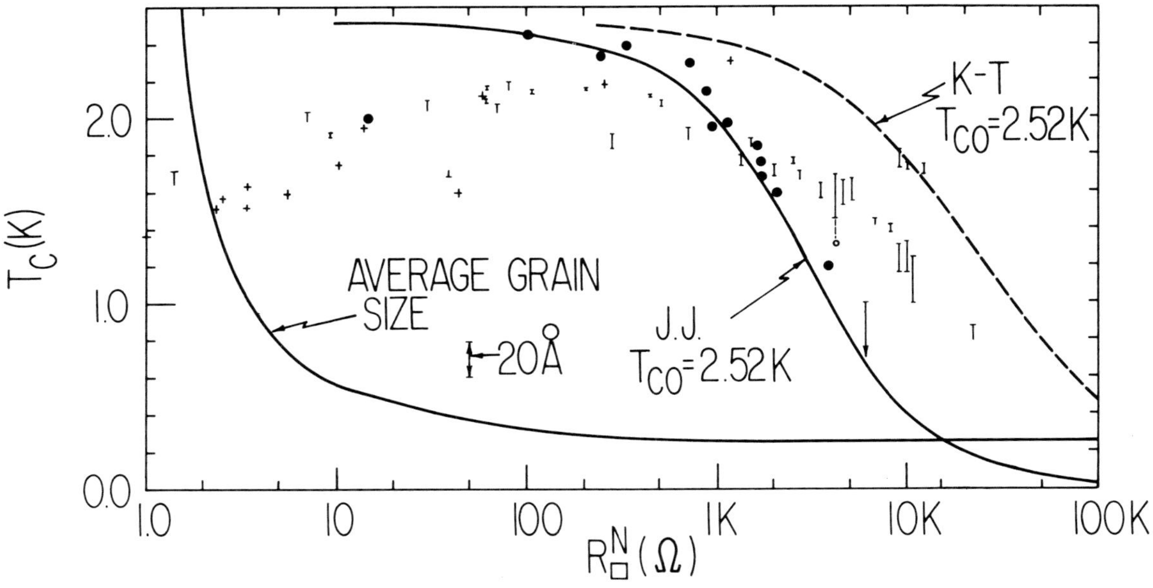

Fig. 5. A compendium of measured values of T_c vs. $R_\square^N$ for granular Al. The error bars refer to reduced resistances of 10^{-2} and 10^{-4}, and the crosses to reported "T_c", both for glass substrates. The solid dots are 10^{-4} points for mica substrates. The curves are K-T:Kosterlitz-Thouless model(Eq. 9 of[3]); J.J.:Josephson junction model; and the average grain size[5].

since it is in very poor agreement with the prediction [3] for Kosterlitz-Thouless vortex unbinding. On the other hand the results can be shown to be in excellent quantitative agreement with a one parameter model of Josephson coupled grains. This model treats each junction independently, ignoring collective behavior and 2D effects. It is well known that the phase coherence across a Josephson junction is lost when kT becomes comparable to the Josephson coupling energy, and Falco, et al., [6] have shown that dc resistance occurs for $kT \gtrsim \hbar I_c/7e$. (For a discussion of higher frequencies, see footnote [7].) The model consists of uniform sized grains in the plane of the film [5]. The average resistance of each junction is then $R_\square^N$, and I_c is given by [8]

$$I_c = [\pi\Delta(T)\tanh\Delta/2kT]/2eR_\square^N \quad (1)$$

In order to determine Δ, the transition temperature of the grains, T_{co} must be known, and this is difficult to access experimentally. However, for the values of $R_\square^N$ of interest, the grain size is shown in Fig. 5 to be independent of $R_\square^N$ and has a narrow distribution [5]. Therefore it is not unreasonable to assume a constant T_{co} which is then the only adjustable parameter of the model. The results of this calculation, using a very reasonable value of $T_{co} = 2.52K$, are shown in Fig. 5 to be in excellent quantitative agreement with the measurements for mica substrates. For lower values of $R_\square^N$, the grain size increases and T_{co} is no longer constant but must approach the bulk transition temperature for Al which is 1.19K. Thus for both glass and mica substrates the transition temperature drops for small $R_\square^N$.

The width of the resistive transition, shown in Fig. 3, must derive from both the width of each individual junctions transition [6] as well as the random variation of junction resistances. The first occurance of a supercurrent is then related to a percolation threshold [5]. Note that in granular Al on glass substrates the occurance of a supercurrent is controlled by bad spots in the film due to substrate imperfections (see Fig. 1).

It should be pointed out that the resistive transition shown in Fig. 3 can be fit extremely well with the theoretical form for the Kosterlitz-Thouless unbinding model[2]. However, since there are three parameters involved, which cannot be independently determined, the meaning of this fit is dubious. It should be noted that our samples on mica substrates are the first ones to exhibit continual curvature in the plot of lnR vs. T, as required for the K-T correlation length.

Additional experiments, beyond the scope of this paper, have been preformed which directly probe the dynamics of vortex unbinding. These have yielded a null result although the experimental sensitivity is almost two orders of magnitude better than the theoretical prediction[2]. These will be discussed in detail in a future publication.

The authors would like to thank R. T. Kampwirth for assistance in making the NbN samples.

REFERENCES

1. W.E. Masker, S. Marcelja and R.D. Parks, Phys. Rev. 188 745 (1969)

2. B.I. Halperin and David R. Nelson, J. Low Temp. Phys. 36 599 (1979)

3. M.R. Beasley, J.E. Mooij and T.P. Orlando, Phys. Rev. Lett. 42 1165 (1979)

4. S. Doniach and B.A. Huberman, Phys. Rev. Lett. 42 1169 (1979)

5. G. Deutscher, O. Entin-Wohlman, S. Fishman and Y. Shapira, Phys. Rev. B June (1980), and G. Deutscher, H. Fenichel, M. Gershenson, E. Grunbaum and Z. Ovadyahu, J. Low Temp. Phys. 10 231 (1973)

6. C.M. Falco, W.H. Parker, S.E. Trullinger and Paul K. Hansma, Phys. Rev. B10 1865 (1974)

7. Note that the restriction on the loss of phase coherence is less severe at higher frequencies. Phase coherence in an r.f. biased SQUID (20-30MHz) has been demonstrated for temperatures at which the weak link of the SQUID has almost its full dc resistance at zero bias [Charles M. Falco, Proc. of 14th Int. Conf. Low Temp. Phys., Matti Krusius and Matti Vuorio, editors (Elsevier, New York, 1974), Vol IV, 242.]. Thus as the frequency of observation increases, so does the transition temperature for phase coherence. This effect can qualitatively explain the frequency dependent conductivity measured in granular Al films on glass substrates [A.F. Hebbard and A.T. Fiory, Phys. Rev. Lett. 44 291, 620(E) (1980)].

8. V. Ambegaokar and A. Baratoff, Phys. Rev. Lett. 10 479 (1963)

---

* Work supported by U.S. Department of Energy

+ Present address: Department of Physics, Northwestern University, Evanston, IL

Published 1980 by Elsevier North Holland, Inc.
Sinha, ed. Ordering in Two Dimensions

NON-METALLIC CONDUCTION IN ELECTRON INVERSION LAYERS AT LOW TEMPERATURES

D. J. BISHOP, D. C. TSUI and R. C. DYNES

Bell Laboratories, Murray Hill, New Jersey 07974

ABSTRACT

We have measured the resistance of electron inversion layers in Si MOSFETS at low temperatures ($\sim$ 50mk) and low electric fields ($\sim$ .1v/meter). At low values of $R_{\square}$ we observe logarithmic dependences of the resistance on both temperature and applied electric field. At $R_{\square} \gtrsim 10\text{k}\Omega$ the resistances cross over to an exponential dependence. Our results will be compared with recent theories of conduction in two dimensions.

There has been much recent experimental and theoretical interest in the subject of electron transport and localization in a quasi-two dimensional (2D) system [1-5]. However such fundamental questions as the nature of metallic conduction, the existence of a mobility edge and the validity of using conductance as a single scaling parameter still remain unanswered. We have measured the electrical conductivity of electron inversion layers in silicon MOSFETS (metal-oxide-semiconductor field-effect transistors) at low temperatures and compare them with recent theoretical ideas and experimental results on metal films [5]. Our measurements show that at sheet resistances $R_{\square} \sim 10\text{k}\Omega/\square$ there is a crossover from an exponential or activated temperature dependence to a logarithmic dependence on temperature and applied electric field. Similar to the previous observations of Dolan and Osheroff [5], we see no evidence for true metallic behavior below $10\text{k}\Omega/\square$. In addition, our results show that these effects scale with $R_{\square}$ of the inversion layer and not with any other single parameter of the system.

Our measurements were performed on N-channel MOSFETS fabricated on (100) and (111) surfaces of P-type silicon. The Si electron inversion layer (shown in Fig. 1) is a 2D electron gas whose density is determined by the applied gate voltage, and can easily be varied. Also the mobility can be varied by applying a substrate bias between the substrate and the inversion layer which moves the electron wavefunction closer to, or further away from the Si-oxide interface.

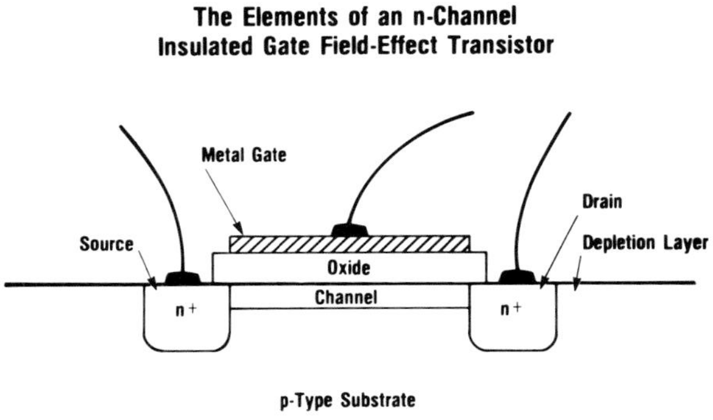

Fig. 1. A schematic picture of a MOSFET (metal-oxide-semiconductor field effect transistor) is shown.

The measurements were performed in a He^3-He^4 dilution refrigerator at temperatures from 50mk to 4k. The four terminal I-V characteristics were measured using a d.c. bridge such that a constant resistance R_S could be subtracted off and small deviations from linearity more clearly observed (see Fig. 2). In Fig. 3 we show representative I-V curves at different temperatures obtained in this fashion with $R_S = 4.9\text{K}\Omega$ from a sample on (111) Si. Note that the curves become non-ohmic at extremely low values of the electric field. The relevant experimental quantities are the zero bias resistances as a function of temperature and the low temperature resistance as a function of applied electric field. The zero bias resistance is shown as a straight line in Fig. 3 for the lowest temperature (0.057K) curve. In Fig. 4 these zero bias resistances are shown as a function of temperature for various electron densities.

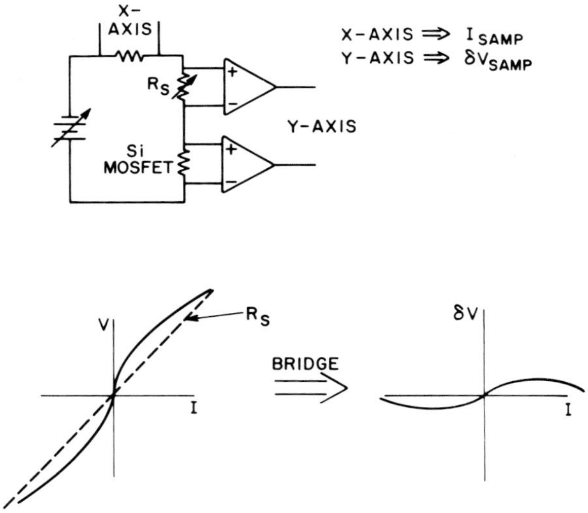

Fig. 2. The d.c. bridge circuit is shown which subtracts off a constant resistance which emphasizes the non-ohmic effects.

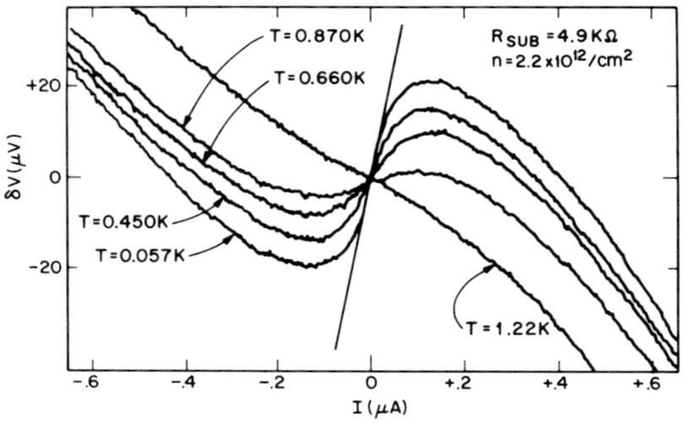

Fig. 3. I-V characteristics are shown for an electron inversion layer at various temperatures. The curves shown are the deviations from linearity after a constant resistance R_S is subtracted off.

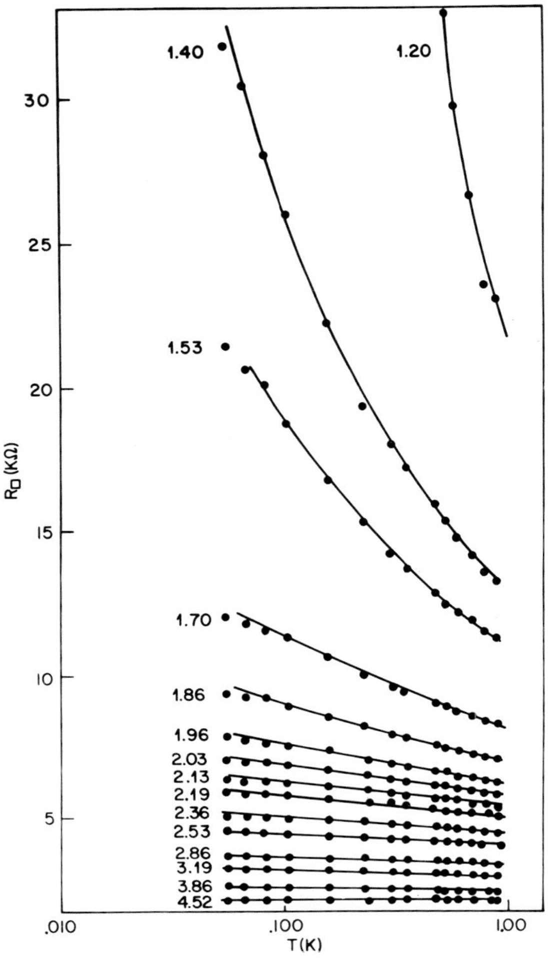

Fig. 4. The zero-bias resistances are shown for various electron densities given in units of $10^{12}/cm^2$.

Measurements of the zero bias resistance as a function of temperature demonstrate two different types of behavior which depend on the $R_\square$ of the channel. In the high resistance regime ($R_\square \gtrsim 10k\Omega$) down to the lowest temperatures studied ($\sim$ 50mK) we find an exponential dependence of zero bias resistance on temperature. This temperature dependence is best fit by the form $R \sim R_o e^{A/T^{1/3}}$, the dependence expected if the dominant conduction mechanism is two-dimensional variable range hopping [6].

In the low resistance regime ($R_\square \lesssim 10k\Omega$), we find that the zero bias resistance decreases logarithmically with increasing temperature for <u>all</u> values of the electron density and $R_\square$. We therefore see no evidence for true metallic conduction at any value of R$_\square$ studied (to date this result holds down to 1000Ω/□ and in metal films to 160Ω/□) [5].

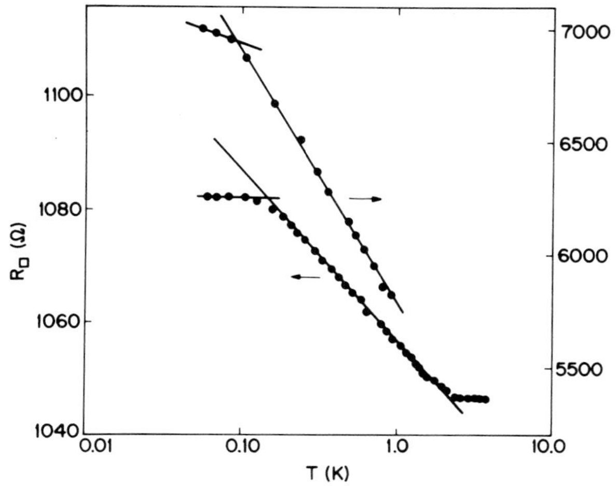

Fig. 5. The zero bias resistance is shown on an expanded scale as a function of temperature for inversion layers with densities of 2.03×10^{12} electrons/cm^2 (right) and 5.64×10^{12} electrons/cm^2 (left).

Representative curves for this logarithmic region are shown in more detail in Fig. 5. At intermediate temperatures we see the logarithmic dependence of the zero bias resistance on temperature. The logarithmic dependence saturates at high temperatures and eventually turns around as seen by other workers [7]. Also at low temperatures this logarithmic dependence either saturates or changes its slope (the data is not adequate to distinguish this). As can be seen in Fig. 5 this "saturation" temperature increases with increasing electron density from T $\lesssim$ 60mK to T ∿ 200mK. The first possible explanation is a simple noise heating effect. This seems unlikely because the noise in our system is low (∿ 1μV) and the "saturation" moves to higher temperatures with decreasing resistance, opposite to conventional noise heating. A possible second explanation is that this is due to the finite size of the sample as at these low temperatures the relevant lengths can be quite long.

From data of the type shown in Fig. 3 we can also extract the electric field dependence of resistance as shown in Fig. 6. For R$_\square$ $\lesssim$ 10KΩ this nonohmic behavior is logarithmic with source-drain voltage. In Fig. 7 the logarithmic slopes of resistance vs. voltage and temperature are plotted as a function of R$_\square$. Both slopes show a linear dependence on R$_\square$ at low enough values of R$_\square$. At R$_\square$ $\gtrsim$ 5KΩ the logarithmic slopes begin to deviate from this linear dependence. At R$_\square$ $\gtrsim$ 10KΩ the resistance vs. temperature and voltage curves are no longer logarithmic and at R$_\square$ $\gtrsim$ 15KΩ they are exponential.

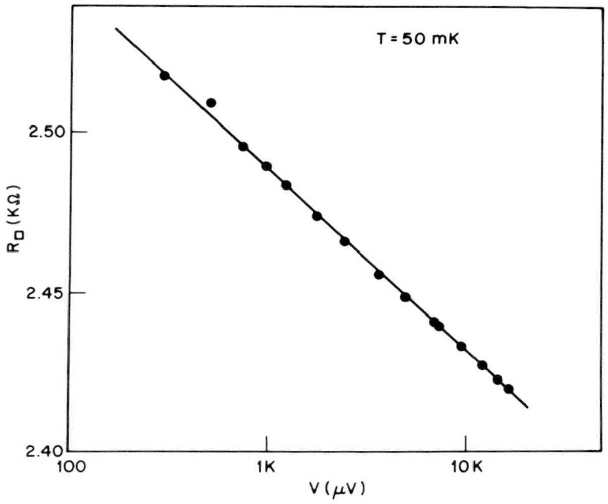

Fig. 6. The logarithmic dependence of resistance on source-drain voltage is shown.

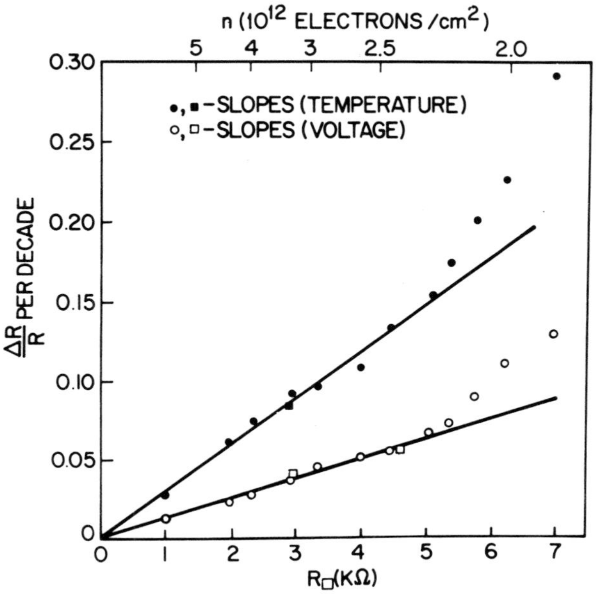

Fig. 7. The logarithmic slopes vs. temperature and voltage are shown as a function of R$_\square$ (T = 1K) (lower scale) and electron density (upper scale).

The logarithmic slopes shown in Fig. 7 have now been measured for MOSFETS with various values of electron mobility, (111) and (100) devices with different electron effective masses, and for metal films. All measurements are in quantitative agreement and form a universal curve when plotted as a function of resistance per square. Therefore these logarithmic effects are independent of density, mobility and effective mass. These measurements provide strong support for the concept of a single scaling parameter in the current theories of conduction in two dimensions. Experimentally this single parameter manifests itself as $R_\square$.

Extending the ideas of Thouless [1], Abrahams, Anderson, Liciardello and Ramakrishnan (AALR) [2], show that the T = 0K conductance of a disordered electronic system depends in a universal manner on its length scale L. They argue that in two dimensions there should exist no true metallic behavior and that for the resistance $R \ll \frac{\pi^2\hbar}{e^2}$ one finds:

$$R_\square(L) = R_\square(L_o)\{1 + \frac{\alpha e^2}{\pi^2\hbar} R_\square(L_o)\ln(L/L_o)\} \quad (1)$$

At finite T, using a diffusion model, Thouless [1] has suggested the geometric mean of the elastic and inelastic mean free paths as the relevant length scale L. It follows then that $L^2 \sim T^{-P}$ with P, the power of T for the appropriate inelastic scattering power, i.e. P = 2 if the appropriate inelastic length scale is determined by electron-electron scattering and P = 2, 3 or 4 for electron-phonon scattering, depending upon phonon dimensionality and dirty or clean limits. Therefore, from (1) the logarithmic slopes vs. temperature should be given by $\frac{\alpha e^2}{\pi^2\hbar}(\frac{P}{2}) R_\square$. From the slope of the solid line in the upper curve in Fig. 3, $(\alpha P/2)$ has the value .52 ± .05. AALR would predict $\alpha P = 3$ for electron-phonon scattering or $\alpha P = 2$ for electron-electron scattering.

Another possibility has been suggested by Altshuler, Aronov and Lee [8]. In two dimensions Coulomb interactions may give logarithmic terms in the conductance which are comparable in size to the effects which have been observed. Unfortunately the present experiments are not sufficient to determine which of the relevant ideas is correct.

Utilizing an electron heating model we can also analyze the voltage slopes shown in Fig. 3. Anderson, Abrahams and Ramakrishnan [4] show that the ratio of the coefficients of $\ln V$ and $\ln T$ for any given $R_\square$ should be a constant given by $2/(2+P')$, where P' is the temperature exponent of the electron-phonon scattering rate. This ratio is independent of the physical mechanism which determines the coefficient of the $\ln T$ dependence. Our data shown in Fig. 3 provide a value for P' of 2.7 ± .5, which suggests an exponent of P' = 3. At these temperatures and elastic scattering rates we are in the dirty limit, $q\,\ell_{elastic} \ll 1 (q = k_B T/hs)$ (S = sound velocity). In this limit P' = 3 is expected as the selection rules for phonon emission should reflect the two-dimensional nature of the Fermi surface.

In conclusion we have measured the conductance of Si MOSFETS at low temperatures and have observed logarithmic dependences on temperature and applied electric field, indicating the lack of true metallic conduction in this 2D electronic system. Our data show that the cross-over from logarithmic to activated behavior is smooth and continuous and that the logarithmic effects are universal and only with $R_\square$ and not with any other parameter of the system. Within current localization ideas we obtain $\alpha P \sim 1$ while one would expect $\alpha P = 3$ for electron-phonon scattering or $\alpha P = 2$ for electron-electron scattering.

We gratefully acknowledge many helpful discussions with G. J. Dolan and P. A. Lee. We have also benefited from discussions with P. W. Anderson and D. D. Osheroff. Finally we are indebted to the superb technical assistance of G. Kaminsky, J. P. Garno and E. J. Benick.

REFERENCES

1. D. J. Thouless, Phys. Rev. Lett. 39, 1167 (1977).
2. E. Abrahams, P. W. Anderson, D. C. Licciardello and T. V. Ramakrishnan, Phys. Rev. Lett. 42, 673 (1979).
3. P. A. Lee, Phys. Rev. Lett. 42, 1492 (1979).
4. P. W. Anderson, E. Abrahams, T. V. Ramakrishnan, Phys. Rev. Lett. 43, 718 (1979).
5. G. J. Dolan and D. D. Osheroff, Phys. Rev. Lett. 43, 721 (1979) and private communications.
6. N. F. Mott, M. Pepper, S. Pollitt, R. H. Wallis and C. J. Adkins, Proc. Roy. Soc. London, A345, 169 (1975).
7. K. M. Cham and R. G. Wheeler, Bull. Am. Phys. Soc. 25, 389 (1980).
8. B. L. Altshuler, A. G. Aronov and P. A. Lee, Phys. Rev. Lett. 44, 1288 (1980).

Published 1980 by Elsevier North Holland, Inc.
Sinha, ed. Ordering in Two Dimensions

CONFERENCE SUMMARY

D. J. THOULESS

Physics Department, Yale University, New Haven, CT 06520

My task in summarizing this Conference is simple. I have to pick out for special mention and emphasis the most significant contributions that have been made in the hundred or so papers presented at the Conference, without causing offense by my omissions or distortions. Also, in my position as an elder statesman of the two-dimensional world, I should point out to the rest of you where the key advances of the next two or three years will be made and what you should be working on when you get back to your laboratories or desks. This task is of course imposed on me as a punishment for having projected myself onto a one-dimensional manifold so that I have not done anything useful in two-dimensions for the past couple of years.

The strongest impression that I am left with by this Conference is that the subject of two-dimensional physics has matured. No longer are we faced with ingenious experiments of doubtful significance, which can be compared with simplistic theories of doubtful validity. We are presented with experimental results which are admirably, and to my mind incredibly, detailed and there is obviously a consensus-questions are raised about fine points, not about fundamentals of interpretation. For adsorbed gases, surfaces of cleaved crystals, for very thin layers of liquid crystals and phospholipid bilayers, for intercalated atoms, and even for electrons on helium, we have been presented with results which would have been regarded as detailed for bulk materials not many years ago. It is a tribute to the enormous increase in sophistication of instrumentation as well as to the physical insight of the experimentalists involved. The weakest showing at this Conference seems to have been in magnetism. So much of what we know about critical phenomena in bulk has come from magnetic systems, and so much of the experimental understanding of dimentionality came from the study of layered magnetic materials, that I find it surprising that we have heard so little about magnetic systems at this Conference. We have also not heard a great deal about superconductivity, which was even earlier than magnetism to direct our attention to lower dimensional systems -- the large correlation length makes reduced dimensionality easier to attain -- but I think that is because a renewed interest in reduced dimensionality is still encountering unresolved problems of interpretation of results and material preparation. A lot more will be known a year from now.

I am not convinced that experimentalists are all meeting continuously varying critical behavior round every corner, as Kadanoff seemed to suggest they should, but many of them have found it to be a very stimulating concept. We have heard of the observation of this in scattering experiments both on adsorbed gases and on liquid crystals. Even if the solid does not melt by dislocation unbinding it comes close to it in many cases, and even if it does not melt to a hexatic phase, there seems to be a liquid crystal phase of the required sort. The Potts model, whose originator long ago gave up physics for traffic flow, has also been pressed into service, so that experimentalists can now produce three-state and four-state Potts models for us. Even the difficult problem of the commensurate-incommensurate phase transitions, obviously of vital importance for understanding adsorbed solid phases, seems to be beginning to make some sense.

For the future we can hope to see experimentalists working with more ideal systems -- for example, substrates whose extent is more closely controlled than grafoil. At the same time theorists have to be better able to cope with impurities and inhomogeneities. I am particularly unimpressed by the relation between the theory of critical phenomena in disordered systems and experimental observations of the same. I caught the phrase about diffuse first order transitions with some interest, although it was immediately denied that it meant what I thought it meant. To get to the theorists' ideal phase transition, with a singularity in observable quantities, quite a variety of extrapolations have to be made from real physical systems to infinite homogeneous systems. Sometimes what looks like a good phase transition turns out not to be -- H_{c2} in a superconductor is probably an example, and the spin glass transition. Does that mean the theory of critical phenomena has nothing to offer?

Finally, a word to our sponsors. Reduced dimensionality used to be the refuge of theorists who could not solve the real physical problems, and I intend no disrespect to Onsager, Lieb, Baxter, etc., who produced masterpieces of theoretical physics. It was never a soft option for experimentalists. Surfaces and interfaces are all around us. Not only is most electronics of the future going to be two or one dimensional, but many chemical reactions of biological or technical importance are catalysed at interfaces.

Author Index

SUBJECT INDEX

8 9898

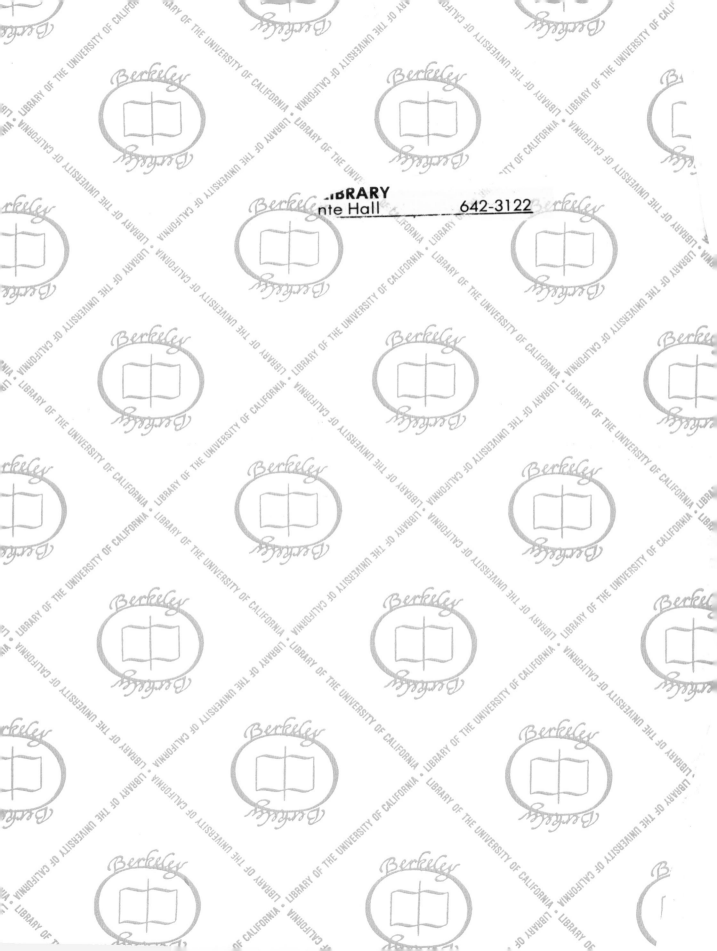